**EMS Textbooks in Mathematics**

*EMS Textbooks in Mathematics* is a series of books aimed at students or professional mathematicians seeking an introduction into a particular field. The individual volumes are intended not only to provide relevant techniques, results, and applications, but also to afford insight into the motivations and ideas behind the theory. Suitably designed exercises help to master the subject and prepare the reader for the study of more advanced and specialized literature.

Jørn Justesen and Tom Høholdt, *A Course In Error-Correcting Codes*
Markus Stroppel, *Locally Compact Groups*
Peter Kunkel and Volker Mehrmann, *Differential-Algebraic Equations*
Dorothee D. Haroske and Hans Triebel, *Distributions, Sobolev Spaces, Elliptic Equations*
Thomas Timmermann, *An Invitation to Quantum Groups and Duality*
Oleg Bogopolski, *Introduction to Group Theory*
Marek Jarnicki and Peter Pflug, *First Steps in Several Complex Variables: Reinhardt Domains*
Tammo tom Dieck, *Algebraic Topology*
Mauro C. Beltrametti et al., *Lectures on Curves, Surfaces and Projective Varieties*
Wolfgang Woess, *Denumerable Markov Chains*
Eduard Zehnder, *Lectures on Dynamical Systems. Hamiltonian Vector Fields and Symplectic Capacities*
Andrzej Skowroński and Kunio Yamagata, *Frobenius Algebras I. Basic Representation Theory*
Piotr W. Nowak and Guoliang Yu, *Large Scale Geometry*

# Joaquim Bruna
# Julià Cufí

# Complex Analysis

Translated from the Catalan
by Ignacio Monreal

Authors:

Joaquim Bruna and Julià Cufí
Department of Mathematics
Universitat Autònoma de Barcelona
Campus de Bellaterra
08193 Cerdanyola del Vallès, Barcelona
Catalonia, Spain

E-mail: bruna@mat.uab.cat
        jcufi@mat.uab.cat

2010 Mathematics Subject Classification: 30-01, 31-01

Key words: Power series, holomorphic function, line integral, differential form, analytic function, zeros and poles, residues, simply connected domain, harmonic function, Dirichlet problem, Poisson equation, conformal mapping, homographic transformation, meromorphic function, infinite product, entire function, interpolation, band-limited function

ISBN 978-3-03719-111-8

The Swiss National Library lists this publication in The Swiss Book, the Swiss national bibliography, and the detailed bibliographic data are available on the Internet at http://www.helveticat.ch.



        Contact address:

        European Mathematical Society Publishing House
        Seminar for Applied Mathematics
        ETH-Zentrum SEW A27
        CH-8092 Zürich
        Switzerland

        Phone: +41 (0)44 632 34 36
        Email: info@ems-ph.org
        Homepage: www.ems-ph.org

Typeset using the authors' TEX files: I. Zimmermann, Freiburg
Printing and binding: Beltz Bad Langensalza GmbH, Bad Langensalza, Germany
∞ Printed on acid free paper
9 8 7 6 5 4 3 2 1

# Preface

Our original purpose in writing this book was to provide a brief manual, perhaps more aptly called a guide book, that would cover the contents of a basic one-semester course in complex analysis as described in most university curricula. The result, however, has been a more extended text that does not fit into a semester course but is rather appropriate for a variety of advanced courses. It also contains some material that is not usually found in the textbook literature of complex variables. For this reason we hope it will prove to be a good complement to many of the references that are commonly used by both students and teachers.

We wrote this book because we wanted to provide something new, not only in presentation but also in content, when compared with the long and still growing list of complex variable textbooks, many of which have become classics. The starting point was to frame complex analysis within the general framework of mathematical analysis. Although it is possible to present – as many texts do – the complex variable as an isolated branch of study in analysis, we have chosen a different option, namely to seek a maximum number of points of contact with other parts of analysis. This has resulted in the inclusion of some sections that are not common in other texts and a new formulation of some classical results. We highlight a few of them below.

In Chapter 3 we give a real version of the theorems of Cauchy and Cauchy–Goursat. The result is a version of Green's formula with very weak regularity assumptions, which serves also for classical theorems of vector calculus. In the same chapter, the presentation of Cauchy's theorem in the context of vector analysis allows us to formulate an approach to the concept of a holomorphic function from a real variable viewpoint, in terms of fields that are simultaneously conservative and solenoidal. The concept of a harmonic function then naturally appears.

Chapter 6 provides a homological version of Green's formula that can be interpreted as a Green's formula with multiplicities. With the help of this formula and a standard process of regularization, a question by Ahlfors is answered affirmatively, about the possibility of modifying the proof of Cauchy's theorem to cover also the case of any locally exact differential form.

Chapter 7 systematically studies harmonic functions and the Laplace operator in the context of real variables in $\mathbb{R}^n$, with emphasis on the special case of dimension 2 and the relation with holomorphic functions. The study includes in detail the properties of the Riesz potential of a measure and its importance in solving Poisson's equation and the Dirichlet and Neumann non-homogeneous problems.

Chapter 9 examines the relationship between Green's function and conformal mapping, which allows one to prove Riemann's theorem using the solution of the Dirichlet problem; we also present Koebe's proof based on the properties of normal families. The existence of solutions to the Dirichlet problem is proved by Perron's method, which is generalizable to any dimension.

In an analogous way to the Poisson equation, which is the inhomogeneous case of the Laplace equation, Chapter 10 deals with the inhomogeneous Cauchy–Riemann equations. The solution in the general case is obtained using the Runge approximation theorem and is applied to study the Dirichlet problem for the $\bar{\partial}$ operator.

Chapter 11 is devoted to the study of zero sets of holomorphic functions, and clearly shows the relationship between this topic and the Poisson equation. This allows us to analyze the distribution of zeros of a holomorphic function in terms of their growth.

Finally, the link between real and complex variables also appears in Chapter 12 with the complex Fourier transform or Laplace transform. We provide a proof of the Shannon–Whittaker theorem, well known in information theory, using methods in Chapter 10 on the decomposition of meromorphic functions in simple elements.

To read this text, it is sufficient to have a good knowledge of the topology of the plane and the differential calculus for functions of several real variables. From there, the book is self-contained and gives rigorous proofs of all statements, including a few issues that tend in many books to be treated somewhat superficially. In this regard we emphasize the study, in Chapter 1, of plane domains with regular boundary, including a treatment of the orientation of the border. This study allows us to formulate a precise version of the classic theorems of complex analysis for domains with regular boundary, which are the most used in applications. However, in Chapter 6, we also give the homological version of the fundamental theorems along the line initiated by Ahlfors, which is more general and relates to topological properties of the domain.

The length and structure of the text allows the reader to pursue a variety of paths through it, and to follow a route at different levels. For example, one can follow a basic course in complex variables with Chapters 1 and 2, Chapter 3, Sections 3.1 to 3.5 and Chapters 4, 5 and 8, without the later Sections 8.8 and 8.9. Another possibility is to use, totally or partially, the contents of Chapters 9, 10, 11 and 12 for an expanding course in complex variables.

Given the initial goal of providing maximum interconnection with other parts of analysis, we have put great emphasis on the role of harmonic functions. We have devoted Chapter 7 to them, which is the longest chapter of the book and can be used as an introduction to potential theory. This chapter can be read independently knowing only the content of Chapter 3; on the other hand, Sections 7.7 to 7.12 may require a level of maturity in mathematics a little higher than the preceding chapters.

In general we have devoted much attention to the details of the proofs. However, in some sections of Chapters 7, 10, 11 and 12 the level of precision is lower than for most of the chapters and this can make reading them a little harder.

Each chapter is divided into sections, each section into subsections. All statements (theorems, propositions, lemmas and corollaries), and also examples, are numbered consecutively within each chapter, only observations are numbered separately. The last section of each chapter contains statements of exercises.

Needless to say, in preparing this book we benefited from the work and experience of previous authors. We express our debt to Ahlfors [1], Burckel [3], Gamelin [7], and Saks–Zygmund [11]. We are grateful to Juan Jesús Donaire for his reading of the original, to Lluís Bruna and Miquel Dalmau who read various parts and to Mark Melnikov who provided us with some exercises. They all have made valuable suggestions. We also thank Ignacio Monreal for the translation into English of the Catalan original text.

Finally, the book would not exist without the excellent typographical work of Raquel Hernández, Maria Julià and Rosa Rodríguez. Our thanks to all of them.

# Contents

# Chapter 1
# Arithmetic and topology in the complex plane

In this chapter the features of the complex plane are studied, attending to its field structure as well as its topological properties. Even though most of them are already known by students of Complex Analysis, a review of the arithmetic of complex numbers and the topology of the complex plane is done, stressing the study of plane domains and regular boundaries. The benefits of complex notation in considering some questions on analytic geometry are also highlighted, such as orthogonal mappings, that will be considered later. Furthermore, the branches of the argument and the notion of index of a plane curve are studied in detail. These questions, even though essentially topological, have great implications for the behavior of holomorphic functions.

## 1.1 Arithmetic of complex numbers

### 1.1.1 Arithmetic operations, modulus and argument

The field of complex numbers $\mathbb{C}$ is the result of joining to the field of real numbers $\mathbb{R}$ an imaginary unit $i$ such that $i^2 = -1$. Then the general expression of a complex number is $z = x + iy$, with $x, y \in \mathbb{R}$; $x$ is called the *real part* of $z$, $x = \operatorname{Re} z$, and $y$ the *imaginary part* of $z$, $y = \operatorname{Im} z$. Associating to $z$ the point of the plane with coordinates $(x, y)$, with respect to a fixed reference system, $\mathbb{C}$ is identified with $\mathbb{R}^2$ and one may speak of the *Argand–Gauss complex plane*. The first coordinate axis is named the *real axis* and the second one the *imaginary axis*.

The number $\bar{z} = x - iy$ is called the *conjugate* of $z$; the mapping $z \to \bar{z}$ is a reflection with respect to the real axis. So one has the relations

$$\operatorname{Re} z = \frac{1}{2}(z + \bar{z}), \quad \operatorname{Im} z = \frac{1}{2i}(z - \bar{z}).$$

The *sum* of the complex numbers $z = a + ib$, $w = c + id$, that is, $z + w = (a + c) + i(b + d)$, may be visualized with the *parallelogram law* of the sum of vectors, identifying $z$ with the vector that begins at the origin of the coordinate system and ends in $z$, and similarly for $w$ (Figure 1.1).

In order to visualize geometrically the *product*

$$zw = (ac - bd) + i(ad + bc)$$

one needs first to talk about the *polar representation* of complex numbers, so the *modulus* and the *argument* of a complex number are now introduced.

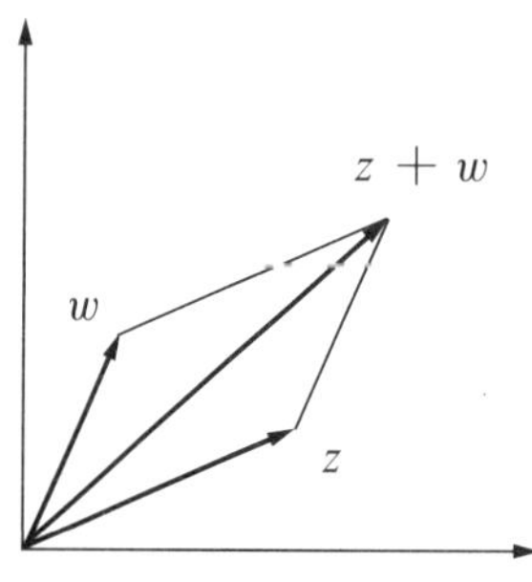


Figure 1.1

The *modulus* of $z = x + iy$ is the distance to the origin of the point $(x, y)$, $|z| = \sqrt{x^2 + y^2} = \sqrt{z \cdot \bar{z}}$. The triangular inequality is easily checked:

$$|z + w| \leq |z| + |w|,$$

$$\left| \sum_{i=1}^{n} z_i \right| \leq \sum_{i=1}^{n} |z_i|,$$

which implies

$$||z| - |w|| \leq |z \pm w|,$$

as well as the equalities

$$|zw| = |z||w|,$$

$$|z_1 z_2 \cdots z_n| = |z_1||z_2| \cdots |z_n|,$$

for $z, w, z_1, \ldots, z_n \in \mathbb{C}$.

With the sum and the product defined above, the complex plane $\mathbb{C}$ is a commutative field in which the inverse of a number $z \neq 0$ is $z^{-1} = \frac{1}{z} = \frac{\bar{z}}{|z|^2}$.

Recall that, by definition, $2\pi$ is the length of the circumference with radius 1 (consequently a circle of radius $r$ has length $2\pi r$). The geometric definition of the trigonometric functions sine and cosine may be stated as follows: if one travels counterclockwise along the circle centered at the origin and with radius 1 a distance $t \geq 0$ starting at the point 1, one ends up in a point denoted by $1_t$; by definition, the real and the imaginary part of this point are, respectively, $\cos t$ and $\sin t$; for $t < 0$ one does the same, but clockwise (Figure 1.2).

Thus, $1_t = \cos t + i \sin t$, and the definition of $\pi$ given above is equivalent to define $\frac{\pi}{2}$ as the first positive zero of the cosine function. The sine and cosine functions are $2\pi$ periodic; moreover, the cosine is an odd function and the sine is even. From previous definitions all the well-known properties of these functions could be proved. In particular, these are functions infinitely differentiable with

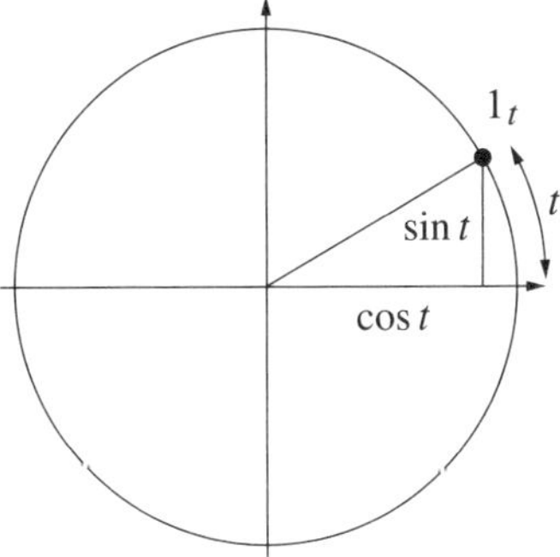


Figure 1.2

respect to the real variable $t$ such that $(\sin t)' = \cos t$, $(\cos t)' = -\sin t$. The addition formulae

$$\cos(t + s) = \cos t \cos s - \sin t \sin s,$$
$$\sin(t + s) = \sin t \cos s + \cos t \sin s \tag{1.1}$$

may also be geometrically justified. Figure 1.3 is a proof of the second formula of (1.1). The equality

$$1_t 1_s = 1_{t+s},$$

which is a consequence of the previous formulae, tells that the mapping

$$t \longrightarrow 1_t$$

is a homomorphism from the additive group of the real numbers onto the multiplicative group $\mathbb{T}$ of complex numbers of modulus 1, with kernel $2\pi\mathbb{Z}$, where $\mathbb{Z}$ represents the ring of integers. In particular de Moivre's formula holds: $(\cos t + i \sin t)^n = \cos(nt) + i \sin(nt)$. The classical double or triple angle formulae for the sine and cosine may be deduced from this one. For example, if $n = 2$, considering separately the real and the imaginary parts in both sides of de Moivre's relation, one has

$$\cos 2t = \cos^2 t - \sin^2 t;$$
$$\sin 2t = 2 \sin t \cos t.$$

Doing the same for $n = 3$, it turns out that

$$\cos 3t = \cos^3 t - 3 \cos t \sin^2 t;$$
$$\sin 3t = -\sin^3 t + 3 \cos^2 t \sin t.$$

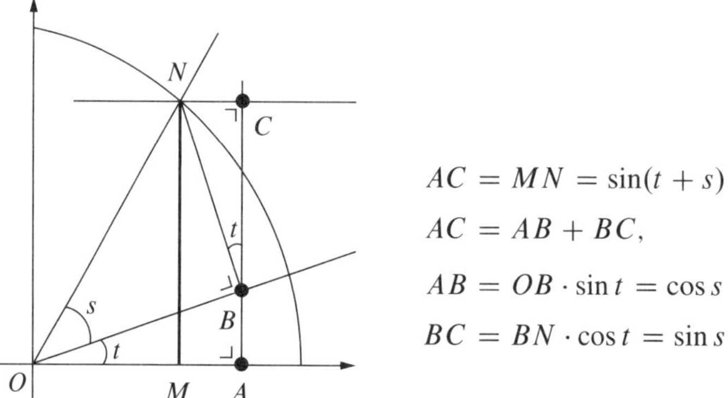

$$AC = MN = \sin(t + s),$$
$$AC = AB + BC,$$
$$AB = OB \cdot \sin t = \cos s \cdot \sin t,$$
$$BC = BN \cdot \cos t = \sin s \cdot \cos t.$$

Figure 1.3

**Example 1.1.** From the previous formulae and using that $\sin \frac{\pi}{2} = 1$, $\sin \pi = -1$, etc. (which is a consequence of the definition of $\pi$) we obtain the well-known trigonometric relations:

$$\sin \frac{\pi}{4} = \cos \frac{\pi}{4} = \frac{1}{\sqrt{2}}; \quad \sin \frac{\pi}{3} = \frac{\sqrt{3}}{2} = \cos \frac{\pi}{6}; \quad \cos \frac{\pi}{3} = \frac{1}{2} = \sin \frac{\pi}{6}. \qquad \square$$

Observe that if $z$ is a complex number, $z \neq 0$, then $z/|z| \in \mathbb{T}$, which implies $z/|z| = 1_\phi$ with $\phi \in \mathbb{R}$.

**Definition 1.2.** The argument of $z \in \mathbb{C}$, $z \neq 0$, denoted by $\arg z$, is any number $\phi \in \mathbb{R}$ such that $\frac{z}{|z|} = 1_\phi$.

Sometimes the notation $\arg z$ is used to refer to the set of all the arguments of $z$. Note that if $\phi$ is an argument of $z$, then $\arg z = \{\phi + 2\pi k, \ k \in \mathbb{Z}\}$. Formally, $\arg z$ is then an equivalence class of $\mathbb{R}/2\pi\mathbb{Z}$; informally, one says that $\arg z$ is determined except for an integer multiple of $2\pi$. This notation and equation (1.1) lead to

$$\arg zw = \arg z + \arg w, \quad \arg \frac{1}{z} = \arg \bar{z} = -\arg z.$$

By definition, *the angle between $z$ and $w$*, where $z, w \neq 0$, or the angle that goes from $z$ to $w$ is $\arg w - \arg z = \arg w\bar{z} = \arg \frac{w}{z}$. These are *oriented angles*; so then, the angle that goes from $w$ to $z$, $\arg z - \arg w$ is the opposite to the one between $z$ and $w$.

**Definition 1.3.** Among all the arguments of $z \in \mathbb{C}$, $z \neq 0$, the only one that belongs to the interval $(-\pi, \pi]$ is called the principal argument of $z$ and is denoted by $\operatorname{Arg} z$.

If $z = x + iy$ with $x > 0$, then $\operatorname{Arg} z = \arctan \frac{y}{x}$ (recall that the inverse tangent function, denoted by $\arctan$, is the inverse of the function $\tan x = \frac{\sin x}{\cos x}$ and it is continuous and bijective from $\mathbb{R}$ to $\left(-\frac{\pi}{2}, \frac{\pi}{2}\right)$). In the other quadrants the principal argument is not $\arctan\frac{y}{x}$. If $z = iy$, $y > 0$, then $\operatorname{Arg} z = \frac{\pi}{2}$; in the second quadrant, $z = x + iy$ with $x < 0$, $y > 0$, $\operatorname{Arg} z = \arctan\frac{y}{x} + \pi$; if $z = x < 0$, then $\operatorname{Arg} z = \pi$; in the third quadrant, $z = x + iy$, $x, y < 0$, $\operatorname{Arg} z = \arctan\frac{y}{x} - \pi$ and finally, when $z = iy$, $y < 0$, it is $\operatorname{Arg} z = -\frac{\pi}{2}$.

The expression $z = |z|1_\phi$, in terms of the modulus and an argument, is called the *polar representation* of $z$; one may also use the notation $z = r_\phi$, where $r = |z|$. A notation based on the exponential function will be introduced later. Multiplication by $z$ may be now visualized easily as a transformation in the complex plane: multiplying by the positive number $|z|$ is a dilation, and multiplying by $1_\phi$ consists of a rotation of an angle $\phi$.

## 1.1.2 Powers and $n$-th roots

An important property of $\mathbb{C}$ is that it is an *algebraically closed* field. This means that every polynomial $P$ of degree $n$ and with complex coefficients has exactly $n$ roots $\alpha_1, \alpha_2, \ldots, \alpha_n$, and $P$ factorizes as

$$P(z) = A(z - \alpha_1) \cdots (z - \alpha_n),$$

where $A$ is a constant. This result is also called the *Fundamental Theorem of Algebra*; a proof of this will be presented later on. In particular, every complex number $z \neq 0$ has $n$ distinct $n$-th roots, that is, there are exactly $n$ solutions of the equation $w^n = z$. Of course this can be proved directly using polar coordinates as follows. Writing $z = |z|1_\phi$ and $w = |w|1_\alpha$, this equation is equivalent to

$$|w|^n 1_{n\alpha} = |z|1_\phi,$$

which imposes $|w|^n = |z|$ and $n\alpha = \phi + 2k\pi$, $k$ integer. Now one checks immediately that the solutions

$$w_k = |z|^{1/n} 1_{\frac{\phi}{n} + \frac{2\pi}{n}k} \quad \text{for } k = 0, 1, 2, \ldots, n-1$$

are all different and that any other solution is one of these. Any of the $n$-th roots of $z$ will be denoted by $z^{\frac{1}{n}}$. Observe that the $n$-th roots of a complex number correspond to points located in the vertices of a regular polygon with $n$ sides.

**Example 1.4.** Calculating the cube roots of $i$, one finds

$$i = 1_{\frac{\pi}{2}}, \quad w_k = 1_{\frac{\pi}{6} + \frac{2\pi}{3}k}, \quad k = 0, 1, 2,$$

which correspond to the points of the Figure 1.4        □

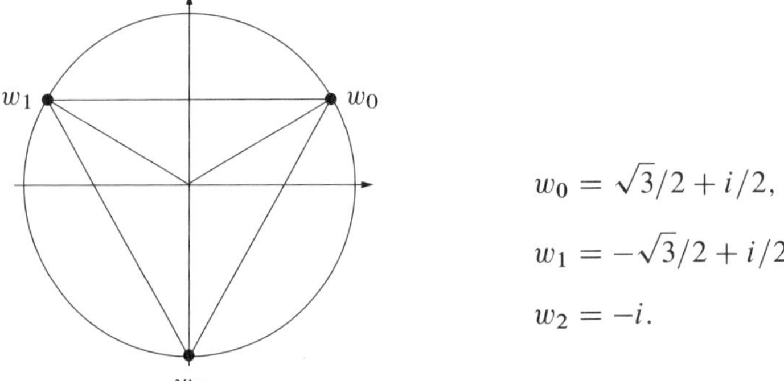

$$w_0 = \sqrt{3}/2 + i/2,$$

$$w_1 = -\sqrt{3}/2 + i/2,$$

$$w_2 = -i.$$

Figure 1.4

The $n$-th roots of 1 are called the *$n$-th roots of unity* and they form a multiplicative group of order $n$. The number $w = 1_{2\pi/n}$ is called the *$n$-th primitive root*, so that all the $n$-th roots of unity are $1 = w^0, w, w^2, \ldots, w^{n-1}$. If $x$ is a positive real number, its positive square root is denoted by $\sqrt{x}$. Let us extend this notation. Considering the two square roots of $z \in \mathbb{C}$, $z \neq 0$, the notation $\sqrt{z}$ will designate the root either with positive real part or in the positive imaginary half-axis; in polar representation,

$$\sqrt{z} = \sqrt{|z|}\, 1_{\frac{\mathrm{Arg}\, z}{2}}.$$

The function $\sqrt{z}$ is called the *principal branch of the square root*. Similarly, the notation $\sqrt[n]{z}$ is reserved for the determination of the $n$-th root that is in the sector $-\frac{\pi}{n} < \mathrm{Arg}\, w \leq \frac{\pi}{n}$.

**Example 1.5.** Consider a polynomial of degree 2, $P(z) = Az^2 + Bz + C$, $A \neq 0$. The equation $P(z) = 0$ may be written, completing squares, as $\left(z + \frac{B}{2A}\right)^2 = \frac{B^2}{4A^2} - \frac{C}{A}$. So the formula

$$z = \frac{-B \pm \sqrt{B^2 - 4AC}}{2A}$$

holds for both solutions of $P(z) = 0$, with $A$, $B$, $C$ complex numbers. $\qquad\square$

Powers with integer exponent have an unambiguous definition,

$$z^n = \underbrace{zz \ldots z}_{n \text{ times}},$$

$$z^{-n} = \underbrace{z^{-1}z^{-1} \ldots z^{-1}}_{n \text{ times}}.$$

If the exponent $q$ is rational and $q = \frac{n}{m}$, then $z^q$ is defined as $z^q = (z^{\frac{1}{m}})^n$. This coincides with $(z^n)^{1/m}$ and does not depend on the expression of $q$. If $z = r1_\phi$ is the polar representation of $z$ and $q = \frac{n}{m}$ is the reduced form of $q$, $z^q$ is the set formed by the $m$ distinct numbers

$$\sqrt[m]{|z|}^n \, 1_{\frac{n}{m}\phi + k\frac{2\pi n}{m}}, \quad k = 0, \ldots, m-1.$$

Briefly, if $\frac{n}{m}$ is the reduced form of $q$ and $z \neq 0$, there are $m$ distinct values of $z^q$. If $w$ is one of these values, then $w^{\frac{m}{n}}$ represents a set of $n$ numbers, one of which is $z$; therefore, one has to be very careful with equations like

$$\left(z^{\frac{n}{m}}\right)^{\frac{m}{n}} = z,$$

which is not correct, because the left-hand side denotes a set of $n$ complex numbers, while the term on the right-hand side is just one of these numbers. The definition of $z^w$ for any complex numbers $z$ and $w$ will be given later on.

### 1.1.3 Field structure

The plane $\mathbb{R}^2$ is identified with $\mathbb{C}$, so it has a field structure. However, it is well known that it is not an ordered field, as $\mathbb{R}$ is.

**Theorem 1.6.** *In the field $\mathbb{C}$ it is not possible to define a total order compatible with the sum and the multiplication operations.*

*Proof.* Let us suppose that there exists an order for the complex numbers such that the following properties hold: 1. Given two distinct points $z$, $w$, either $z < w$ or $w < z$. 2. If $z < w$, then $z + h < w + h$ for all $h$. 3. If $z, w > 0$, then $zw > 0$ (these three properties imply all the usual arithmetic rules). The number $i$ is then positive or negative; let us suppose $i > 0$. Hence $-1 = i^2 > 0$, and summing 1, we obtain $0 > 1$. On the other hand, since $-1 > 0$, we have that $1 = (-1)(-1) > 0$. We conclude then that $0 > 1$ and $1 > 0$ at the same time, which is absurd. We may also arrive at a contradiction starting from $i < 0$. $\square$

On the other hand, it is quite natural to inquire if $\mathbb{R}^n$, for $n > 2$, could have a field structure with $\mathbb{C} = \mathbb{R}^2$ as a subfield. The answer is negative and the situation is described in the following theorem:

**Theorem 1.7.** *No space $\mathbb{R}^n$ with $n > 2$ has a commutative field structure which extends the field structure of $\mathbb{C}$. In the space $\mathbb{R}^4$ one may define a non-commutative field structure, the field of quaternions, which is an extension of $\mathbb{C}$. One may also define in the space $\mathbb{R}^8$ a non-associative field structure, the field of octonions, also an extension of $\mathbb{C}$.*

*Proof.* We will see that one cannot define any commutative field structure extension of $\mathbb{C}$ in $\mathbb{R}^3$, and then we will define the quaternions. Represent the vectors of $\mathbb{R}^3$ as $a + bi + cj$ (that is, the vector $(1, 0, 0)$ is identified with $1$, $i$ is the vector $(0, 1, 0)$ and $j$ is the vector $(0, 0, 1)$), and suppose that there is a multiplication extending the one in $\mathbb{C}$. Observe that any multiplication with these features is completely determined by stating the value of $ij$. Indeed, applying the associativity of the product, it turns out that

$$-j = (i^2)j = i(ij), \quad j^2 = -(ij)^2.$$

Hence, setting $ij = a + bi + cj$ with $a, b, c \in \mathbb{R}$, we obtain

$$-j = ai - b + c(ij) = ai - b + c(a + bi + cj).$$

Now, equating coefficients, one may gather that $c^2 = -1$, contradicting that $c \in \mathbb{R}$.

The idea leading to the quaternions is that, even though it is not possible to define $ij$ as an element of $\mathbb{R}^3$, it could be possible if one had more space, for example, in $\mathbb{R}^4$. This is feasible if one gives up the requirement for the new product to be commutative. Setting $k$ for the vector $(0, 0, 0, 1)$ and

$$ij = -ji = k, \quad ki = -ik = j, \quad jk = -kj = i,$$
$$i^2 = j^2 = k^2 = -1,$$

one may check that $\mathbb{R}^4$ has (non-commutative) field structure. $\qquad\square$

## 1.2 Analytic geometry with complex terminology

### 1.2.1 Lines and circles

It is convenient to state the usual geometric or metric notions of $\mathbb{R}^2$ in complex notation. For this purpose, the complex number $z = x + iy$ and the point $(x, y)$ of the plane, and also the vector from the origin to $z$, are identified.

The Euclidian scalar product between $z = a + ib$ and $w = c + id$ is $ac + bd = \operatorname{Re} z\bar{w}$. Hence $\operatorname{Re} z\bar{w}/|z||w|$ is the cosine of the angle between $z$ and $w$. Since $iw$ is perpendicular to $w$, $(\operatorname{Im} z\bar{w})/|z||w| = -\operatorname{Re} zi\bar{w}/|z||w| = \operatorname{Re} zi\overline{w}/|z||w|$ is the cosine of the angle between $z$ and $iw$ and also the sine of the angle between $z$ and $w$.

Make the change of variables $x = \frac{z + \bar{z}}{2}$, $y = \frac{z - \bar{z}}{2i}$ in the linear equation $Ax + By + C = 0$. It turns out that the general equation of a straight line is of type $\bar{\alpha}z + \alpha\bar{z} + m = 0$, with $\alpha \in \mathbb{C}, m \in \mathbb{R}$. Varying $m$, one obtains all the perpendicular lines to $\alpha$. In parametric form, the equation of the straight line passing through $z_1$ and parallel to the direction $z_2$, $z_2 \neq 0$ is $z(t) = z_1 + tz_2, t \in \mathbb{R}$.

The general equation of a conic is $P(x, y) = 0$, where $P$ is a polynomial of degree 2 with real coefficients: $P(x, y) = \sum_{k+l \leq 2} a_{k,l} x^k y^l, a_{k,l} \in \mathbb{R}$. Replacing $x = (z + \bar{z})/2$, $y = (z - \bar{z})/2i$ leads to an expression of type

$$P(z) = \sum_{k+l \leq 2} \alpha_{k,l} z^k \bar{z}^l,$$

where the quantities $\alpha_{k,l}$ are complex numbers such that $\alpha_{l,k} = \overline{\alpha_{k,l}}$. In general, a polynomial in $z, \bar{z}$ of degree $n$,

$$P(z) = \sum_{k+l \leq n} \alpha_{k,l} z^k \bar{z}^l,$$

takes real values if and only if $\alpha_{l,k} = \overline{\alpha_{k,l}}$. This is obtained imposing the condition $\overline{P}(z) = P(z)$ and equating coefficients. Returning to the case $n = 2$, the general expression of the conics may be rewritten as

$$\bar{\beta} z^2 + \beta \bar{z}^2 + n|z|^2 + \bar{\alpha} z + \alpha \bar{z} + m = 0,$$

with $m, n \in \mathbb{R}$ and $\alpha, \beta \in \mathbb{C}$. For a circle of center $\alpha$ and radius $r$, the equation reads

$$(z - \alpha)(\bar{z} - \bar{\alpha}) - r^2 = |z|^2 - \alpha \bar{z} - \bar{\alpha} z - r^2 + |\alpha|^2 = 0.$$

That is, the circles correspond to $\beta = 0, n \neq 0$ and $\frac{m}{n} - |\alpha|^2 < 0$, in the equation of conics.

## 1.2.2 Conformal and anticonformal linear mappings

The plane $\mathbb{R}^2 = \mathbb{C}$ is an $\mathbb{R}$-vector space. Hence we can consider the mappings $T : \mathbb{C} \to \mathbb{C}$ which are $\mathbb{R}$-linear, that is, the ones such that $T(\lambda_1 z + \lambda_2 w) = \lambda_1 T(z) + \lambda_2 T(w)$ if $\lambda_1, \lambda_2 \in \mathbb{R}, z, w \in \mathbb{C}$. It is common to work with them using the matrix of $T$ with respect to the basis $(1, 0) = 1, (0, 1) = i$,

$$T = \begin{pmatrix} a & b \\ c & d \end{pmatrix}, \quad u, b, c, d \in \mathbb{R},$$

so that $(x, y) \overset{T}{\mapsto} (ax + by, cx + dy)$, $T(1) = (a, c) = a + ci$, $T(i) = (b, d) = b + id$. In terms of $z = x + iy$, the expression is $T(z) = ax + by + i(cx + dy) = \alpha z + \beta \bar{z}$, with $\alpha = \frac{1}{2}(a + d - ib + ic)$, $\beta = \frac{1}{2}(a - d + ic + ib)$. This means that $T(z) = \alpha z + \beta \bar{z}$, where $\alpha, \beta \in \mathbb{C}$ are arbitrary, is the *general expression* of an $\mathbb{R}$-linear mapping. It is immediate to prove that

$$\det T = ad - bc = |\alpha|^2 - |\beta|^2,$$

and hence $T$ is invertible when $|\alpha| \neq |\beta|$. In this case one can express the inverse mapping $T^{-1}$ in complex notation solving the system in $z, \bar{z}$,

$$\alpha z + \beta \bar{z} = w,$$
$$\bar{\beta} z + \bar{\alpha} \bar{z} = \bar{w},$$

which gives $z = T^{-1}(w) = \frac{1}{|\alpha|^2 - |\beta|^2}(\bar{\alpha} w - \beta \bar{w})$.

On the other hand, $\mathbb{C}$ is also a (1-dimensional) $\mathbb{C}$-vector space. So we can consider the mappings $T : \mathbb{C} \to \mathbb{C}$, which are $\mathbb{C}$-linear, that is, the ones such that $T(\lambda_1 z_1 + \lambda_2 z_2) = \lambda_1 T(z_1) + \lambda_2 T(z_2)$, where now $\lambda_1, \lambda_2 \in \mathbb{C}$. Obviously, in this case $T$ is completely determined by $\alpha = T(1)$, because $T(z) = \alpha z$. Of course, $\mathbb{C}$-linear mappings are also $\mathbb{R}$-linear, characterized in this case by the condition $\beta = 0$ if $Tz = \alpha z + \beta \bar{z}$. It is also easy to see that an $\mathbb{R}$-linear mapping $T$ is $\mathbb{C}$-linear if and only if it commutes with the multiplication by $i$ (a $90°$ rotation): $T(iz) = iT(z)$. Analytically, if $T(z) = \alpha z + \beta \bar{z}$, the condition

$$\alpha i z + \beta \overline{(iz)} = T(iz) = iT(z) = i(\alpha z + \beta \bar{z})$$

implies $-\beta i \bar{z} = i \beta \bar{z}$, for all $z$, and so $\beta = 0$.

**Definition 1.8.** An $\mathbb{R}$-linear and invertible mapping $T$ from $\mathbb{C}$ to $\mathbb{C}$ is called conformal if it preserves the size and the orientation of angles, that is, if for any $z, w \in \mathbb{C}$ the angle between $z, w$ is the same as the angle between $Tz, Tw$. Analytically, $T$ is such that
$$\arg Tz - \arg Tw = \arg z - \arg w, \quad z, w \in \mathbb{C}.$$

Clearly this is the same as imposing that $\arg Tz - \arg z = \arg Tw - \arg w$, for all $z, w$. So, the function $\arg Tz - \arg z$ is constant with respect to $z$ when looking at its values in $\mathbb{R}/2\pi\mathbb{Z}$. Since $\arg \frac{Tz}{z} = \arg Tz - \arg z$, this means exactly that all the points $\frac{Tz}{z}$, $z \neq 0$, are in a fixed half-line that contains the origin. However, if $Tz = \alpha z + \beta \bar{z}$, in varying $z \neq 0$ the quantity

$$\frac{Tz}{z} = \alpha + \beta \frac{\bar{z}}{z}$$

covers the circle centered in $\alpha$ and with radius $|\beta|$. Hence it is clear that it only may be included in a ray if $|\beta| = 0$. Thus the following proposition is proved:

**Proposition 1.9.** *Let $T$ be an invertible $\mathbb{R}$-linear mapping from $\mathbb{C}$ to $\mathbb{C}$. Then the following conditions are equivalent:*

a) *$T$ is conformal.*

b) *$T$ may be written as $Tz = \alpha z$, with $\alpha \in \mathbb{C}$, that is, $T$ is $\mathbb{C}$-linear.*

c) *$T$ consists of a rotation of angle $\operatorname{Arg} \alpha$ followed by a dilation by the factor $|\alpha|$.*

In terms of the matrix of $T$, the condition $\beta = 0$ is equivalent to $d - a = i(b+c)$, that is, $a = d$, $b = -c$. This reads either

$$T = \begin{pmatrix} a & -c \\ c & a \end{pmatrix},$$

or $Tz = \alpha z$ with $\alpha = a + ic$. Similarly, the mappings $T(z) = \beta \bar{z}$ are exactly the ones which preserve the size of the angles but change the orientation. These are called either *anticonformal* or $\mathbb{C}$-*antilinear* mappings, and their matrices are of type $T = \left( \begin{smallmatrix} a & c \\ c & -a \end{smallmatrix} \right)$.

If one specifies either a conformal $Tz = \alpha z$ or an anticonformal mapping $Tz = \beta \bar{z}$ to preserve the distances, then $|Tz| = |z|$, that is, either $|\alpha| = 1$ or $|\beta| = 1$. It turns out that $a^2 + c^2 = 1$, which means that the matrix of $T$ is an orthogonal matrix, that is, $T^{-1} = T^t$.

**Definition 1.10.** It is said that the $\mathbb{R}$-linear mapping $T$ from $\mathbb{C}$ onto $\mathbb{C}$ distorts the distances in a constant factor $m > 0$ if $|Tz - Tw| = m|z - w|$, that is, if and only if $|Tz| = m|z|$ for any $z$.

This is equivalent to the fact that $\frac{Tz}{z} = \alpha + \beta \frac{\bar{z}}{z}$ has constant modulus $m$, which may happen only if either $\alpha = 0$ (and then $m = |\beta|$), or if $\beta = 0$ (and then $m = |\alpha|$). So the following has been proved:

**Proposition 1.11.** *An invertible $\mathbb{R}$-linear mapping $T$ from $\mathbb{C}$ onto $\mathbb{C}$ distorts the distances in a constant factor $m$ if and only if it is either $\mathbb{C}$-linear or $\mathbb{C}$-antilinear. In the case $m = 1$ the mapping is a rotation or a rotation followed by a reflection.*

We have shown that the $\mathbb{R}$-linear mappings from $\mathbb{R}^2$ onto $\mathbb{R}^2$ that preserve the size of angles are the $\mathbb{C}$-linear and the $\mathbb{C}$-antilinear ones. How do these transformations behave in the case of the space $\mathbb{R}^n$? The following result answers the question:

**Proposition 1.12.** *An invertible $\mathbb{R}$-linear mapping $T \colon \mathbb{R}^n \to \mathbb{R}^n$ preserves the size of the angles if and only it may be written $T = \lambda O$ with $O$ an orthogonal transformation and $\lambda > 0$.*

Recall that a transformation is orthogonal if it is given in the canonical basis by a matrix $O$ such that $OO^t = O^t O = I$, where $O^t$ is the transpose matrix of $O$ and I is the identity. We will use the notation $|x|$ to denote the norm of the vector $x$ of $\mathbb{R}^n$ and $\langle x, y \rangle$ will be the scalar product of the vector $x, y \in \mathbb{R}^n$.

*Proof.* Preserving the size of the angles amounts to preserving the cosines of these angles, that is,

$$\frac{\langle Tu, Tv \rangle}{|Tu||Tv|} = \frac{\langle u, v \rangle}{|u||v|}, \quad u, v \in \mathbb{R}^n.$$

If $O$ is orthogonal, then it preserves both the scalar product and the norm; thus, if $T = \lambda O$ then

$$\frac{\langle Tu, Tv \rangle}{|Tu||Tv|} = \frac{\langle \lambda^2 Ou, Ov \rangle}{\lambda^2 |Ou||Ov|} = \frac{\langle Ou, Ov \rangle}{|Ou||Ov|} = \frac{\langle u, v \rangle}{|u||v|}.$$

Conversely, let us suppose that $T$ preserves the size of angles and let $e_1, e_2, \ldots, e_n$ be the canonical basis of $\mathbb{R}^n$. Then the vectors $Te_1, Te_2, \ldots, Te_n$ are pairwise orthogonal. Let $O$ be the linear transformation defined by $Oe_i = \frac{Te_i}{|Te_i|}$; since it maps an orthonormal basis into an orthonormal basis, then $O$ is an orthogonal transformation. Besides $O^{-1}T = R$ is another invertible linear transformation which preserves angles and $Re_i = c_i e_i$, $i = 1, \ldots, n$, where $c_i = |Te_i|$. Supposing now that

$$\frac{\langle Ru, Rv \rangle}{|Ru||Rv|} = \frac{\langle u, v \rangle}{|u||v|}$$

with $u = e_i + e_j$ and $v = e_j$, one finds that

$$\frac{(c_i e_i + c_j e_j) \cdot c_j e_j}{(c_i^2 + c_j^2)^{1/2} c_j} = \frac{1}{\sqrt{2}}.$$

Thus $2c_j^2 = c_i^2 + c_j^2$ and therefore $c_i = c_j = \lambda$. This means that $R = \lambda I$, so then $T = \lambda O$. Conformal transformations correspond to $\det O = +1$, and anticonformal ones, to $\det O = -1$. $\square$

Therefore the situation is the same as the 2-dimensional one, in the sense that angle-preserving linear transformations are orthogonal transformations followed by dilations, that is, are *similarities*.

## 1.3 Topological notions. The compactified plane

We will consider the complex plane $\mathbb{C}$ with the usual topology, that is, a basis of neighborhoods of each point $a \in \mathbb{C}$ is the family of open discs, $D(a, r) = \{z \in \mathbb{C} : |z - a| < r\}$, for $r > 0$. We will also use the notation $D_r(a)$ for these discs, and $C(a, r)$ or $C_r(a)$ will denote the corresponding boundary, that is, the circle centered at $a$ with radius $r$. In general, $\partial E$ will denote the topological boundary of the set $E$. The disc $D(0, 1)$ will be called the *unit disc* and will be represented by $\mathbb{D}$. It is convenient, however, to embed the space $\mathbb{C}$ in a wider compact space, obtained by adding the point at infinity.

We will denote by $\mathbb{C}^*$ the compact space obtained by adding to $\mathbb{C}$ the point at infinity, $\infty$. A basis of neighborhoods of $\infty$ consists of the complements of discs centered at the origin, $\{z : |z| > r\} \cup \{\infty\}$. Consider a sequence of complex numbers $(z_n)$ with $\lim_{n \to \infty} z_n = \infty$. This means that for every $r > 0$ one has $|z_n| > r$ for $n$ big enough, that is, $|z_n| \to +\infty$ when $n \to \infty$. Thus, an open set in $\mathbb{C}^*$ which contains $\infty$ has to contain the set $\{z: |z| > r\}$ for a certain $r > 0$. For example, the half plane $\{z: \operatorname{Re} z > a\}$ has $\infty$ as a closure point and the set $\{z: \operatorname{Re} z > 0\} \cup \{\infty\}$ is not an open set of $\mathbb{C}^*$. Any sequence $(z_n)$ of points in $\mathbb{C}$ has a subsequence which converges to a point in $\mathbb{C}^*$; if the sequence is bounded, then it has a subsequence that converges to a point in $\mathbb{C}$, and if it is unbounded, then there is a subsequence tending to $\infty$.

*The stereographic projection* sets a homeomorphism between the sphere $S^2 = \{(x, y, z) \in \mathbb{R}^3 : x^2 + y^2 + z^2 = 1\}$, except for the north pole, and the complex plane. If one considers $\mathbb{C}$ as the equator plane of $S^2$, the point $z$ of the plane is associated in $S^2$ to the intersection point $P$ of $S^2$ and the ray that joins the north pole with $z$. Therefore, the compactified space $\mathbb{C}^*$ is identified with $S^2$, and the north pole with $\infty$ (Figure 1.5). Analytically it may be written as

$$P = \left( \frac{2x}{1 + x^2 + y^2}, \frac{2y}{1 + x^2 + y^2}, \frac{x^2 + y^2 - 1}{1 + x^2 + y^2} \right) \quad \text{if } z = x + iy.$$

This is the reason why we will refer either to the *extended complex plane* $\mathbb{C}^*$ or *Riemann sphere* $S^2$.

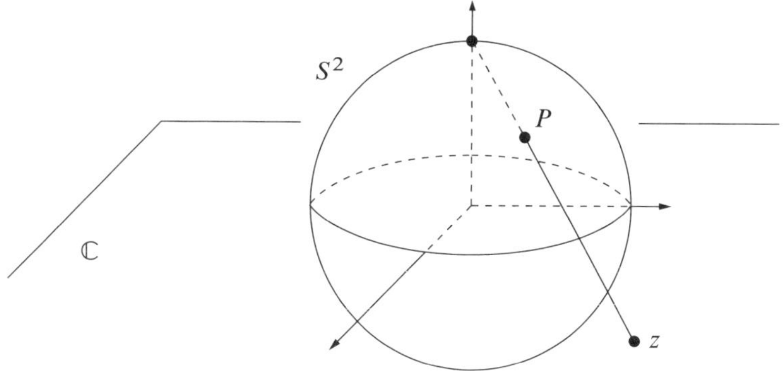


Figure 1.5

We will often consider functions defined on subsets $E$ of $\mathbb{C}$, open in general, which take real or complex values. When speaking about, for example, continuous or differentiable functions, one is considering $E$ as a subset of the plane, with the usual distance, which in complex notation is nothing but $d(z, w) = |z - w|$ for $z, w \in \mathbb{C}$. For example, the fact that a function $f$ defined on $E$ is *continuous* at a point $z \in E$ means that for every $\varepsilon > 0$ there exists $\delta > 0$ such that $w \in E$, $|w - z| < \delta$ implies $|f(w) - f(z)| < \varepsilon$; equivalently, for any sequence $(z_n)$ of points of $E$ with limit $z$, the image sequence $(f(z_n))$ has limit $f(z)$. We will denote by $C(E)$ the set of continuous functions on $E$.

More generally, we may consider continuous functions $f : E \to \mathbb{C}^*$, where $E$ is a subset of $\mathbb{C}^*$ and $f$ also take values in $\mathbb{C}^*$. If $\infty \in E$ the continuity in the point at infinity means that $f(\infty) = l \in \mathbb{C}^*$ and $\lim_{|z| \to \infty} f(z) = l$. In other words: for all $\varepsilon > 0$ there exists $r > 0$ such that either $|f(z) - l| < \varepsilon$ if $|z| > r$, when $l \neq \infty$, or $|f(z)| > 1/\varepsilon$ if $|z| > r$, when $l = \infty$.

Thus, for example, every function defined on $\mathbb{C}^*$ by a polynomial in the complex variable $z$, $P(z) = a_0 + a_1 z + \cdots + a_n z^n$, with $a_i \in \mathbb{C}$, $i = 1, \ldots, n$, $z \in \mathbb{C}$ and $P(\infty) = \infty$, is a continuous function on $\mathbb{C}^*$ if $n \geq 1$. Actually the continuity on $\mathbb{C}$ is evident, and for the point at infinity, the condition $\lim_{|z|=\infty} P(z) = \infty$ is a consequence of the equality $P(z) = z^n \left( \frac{a_0}{z^n} + \frac{a_1}{z^{n-1}} + \cdots + a_n \right)$, $z \in \mathbb{C}$. However, the function defined by a polynomial in two real variables $x$, $y$ is continuous on $\mathbb{C}$ but it may not be on $\mathbb{C}^*$. This fact may be seen taking, for example, $P(x, y) = x + y$.

Throughout this book the notion of *domain* will appear often. A domain of the complex plane (sometimes called a *region*) is an open connected set in $\mathbb{C}$. Besides domains, we will also consider arbitrary open or compact sets of $\mathbb{C}$, so we must review properties of connectivity of these sets, as well as of their complements.

Recall first that the *connected components* of a set $A \subset \mathbb{C}$ are the maximal connected subsets of $A$, and that $A$ is the disjoint union of its components. Any connected component of $A$ is closed in $A$, and therefore also closed in $\mathbb{C}$ if $A$ is closed in $\mathbb{C}$. The components of $A$ are in general not open in $A$; if $A$ is open in $\mathbb{C}$ then the connected components of $A$ are open in $\mathbb{C}$ and, in particular, there are at most a countable quantity of them. When $A$ is not open, there may be an uncountable quantity of connected components. An extreme example is the Cantor set, which is a perfect (all its points are accumulation points), uncountable and totally disconnected set. So the connected components of the Cantor set are each one of its points, which is an uncountable set, and they are not open in $\mathbb{R}$.

Therefore, if $K$ is a compact set of $\mathbb{C}$, the connected components of $\mathbb{C} \setminus K$ are connected open sets of $\mathbb{C}$, that is, a countable quantity of domains of $\mathbb{C}$. Every component $C$ of $\mathbb{C} \setminus K$ satisfies the condition $\partial C \subset K$. Among all components of $\mathbb{C} \setminus K$ there is always one and only one unbounded component: it is the one which contains the points $z \notin K$ with $|z|$ large enough. All the other possible components are bounded and, roughly speaking, they represent the "holes" of $K$. A hole $C$ of $K$ is then a bounded domain of $\mathbb{C} \setminus K$ with $\partial C \subset K$, and there is a countable quantity of them.

If $U$ is an open set in $\mathbb{C}$, then $\mathbb{C} \setminus U$ is closed and its connected components are closed sets in $\mathbb{C}$; their quantity may be not countable. For example this is the case of $U = \mathbb{C} \setminus E$, where $E$ is the Cantor set. If $U$ is an open set, the boundary of every connected component $C$ of $\mathbb{C} \setminus U$ satisfies $\partial C \subset C$, $\partial C \cap U = \emptyset$ and $\partial C \subset \partial U$. When $U$ is open, it is not possible to talk about the "unbounded component" of $\mathbb{C} \setminus U$. It could happen that $\mathbb{C} \setminus U$ either does not have any unbounded connected component (for example, if $U = \mathbb{C} \setminus K$, $K$ compact), or has more than one unbounded connected component (for example, if $U$ is an unlimited band).

Therefore, the intuitive idea of a "hole" of $U$ corresponds to the bounded connected components of $\mathbb{C} \setminus U$, and they may even appear in an uncountable quantity, as seen.

It is immediate to check that a set $A \subset \mathbb{C}$ has no bounded connected components if and only if $A \cup \{\infty\}$ is connected. Then a compact set $K \subset \mathbb{C}$ (respectively an open set $U \subset \mathbb{C}$) has no "holes" (bounded components of the complement in $\mathbb{C}$) if and only if $\mathbb{C}^* \setminus K$ (respectively, $\mathbb{C}^* \setminus U$) is connected. In the case of a compact set $K$, $\mathbb{C}^* \setminus K$ connected is equivalent to $\mathbb{C} \setminus K$ connected. This is because the connected components of $\mathbb{C}^* \setminus K$ coincide with the ones of $\mathbb{C} \setminus K$, adding the point $\infty$ to the unbounded component. However, in the case of an open set $U$, distinct components of $\mathbb{C} \setminus U$ could be connected in $\mathbb{C}^*$ when adding the point $\infty$. For example, if $U$ is a band, $U = \{z \in \mathbb{C} : a < \operatorname{Re} z < b\}$ with $a, b \in \mathbb{R}$, $U$ does not have any hole, $\mathbb{C} \setminus U$ has two components (is not connected), but $\mathbb{C}^* \setminus U$ is connected.

By the above comments, it is more convenient in general to think about the complement of an open or a compact set of $\mathbb{C}$ with respect to the whole extended plane $\mathbb{C}^*$.

**Definition 1.13.** A domain $U$ of the complex plane is called simply connected if its complement with respect to the extended plane, $\mathbb{C}^* \setminus U$, is connected.

Intuitively, simply connected means that the domain $U$ does not have "holes", that is, $\mathbb{C} \setminus U$ does not have bounded components because the only component of $\mathbb{C}^* \setminus U$ contains $\infty$. The term simply connected is also used for a compact set $K$ of the plane such that $\mathbb{C}^* \setminus K$ (or $\mathbb{C} \setminus K$)) is connected.

In the case of a domain with holes, it is appropriate to introduce a name for the domain according to its number of holes.

**Definition 1.14.** A domain $U$ of the complex plane is called $n$-connected or having connection degree $n$ $(n \geq 1)$ if $\mathbb{C}^* \setminus U$ has $n$ connected components. If it has infinite components one says that $U$ has an infinite degree of connection.

The case $n = 1$ corresponds to the simply connected domains of Definition 1.13. In Figure 1.6 there is a bounded domain $U$ with infinite degree of connection, obtained by eliminating from the unit disc the sequence of discs $\overline{D}_n = D(1 - \frac{1}{n}, \frac{1}{n^3})$, $n \geq 2$.

Let us finish this section by showing that any open set of the plane may be covered by something called an *exhaustive sequence of compact sets*.

**Lemma 1.15.** *For any open set $U$ in the plane, there exists a sequence of compact sets $K_n \subset U$, $n = 1, 2, \ldots$ such that:*

a) $K_n \subset \overset{\circ}{K}_{n+1}$, $n = 1, 2, \ldots$, *and* $\bigcup_{n=1}^{\infty} K_n = U$.

b) *For each compact set $K \subset U$ there exists $n \in \mathbb{N}$ such that $K \subset K_n$.*

c) *For each $n \in \mathbb{N}$, every bounded component of $\mathbb{C} \setminus K_n$ contains at least a bounded component of $\mathbb{C} \setminus U$.*

*Proof.* Define

$$K_n = \left\{ z \in \mathbb{C} : d(z, \mathbb{C} \setminus U) \geq \tfrac{1}{n} \right\} \cap \bar{D}(0, n).$$

It is easy to check a) and b). To prove c), let $C$ be a bounded component of $\mathbb{C} \setminus K_n = (\mathbb{C} \setminus \bar{D}(0, n)) \cup (\bigcup_{w \notin U} D(w, 1/n))$. Each set of this union is connected. So if one of them intersects $C$, it will be included in $C$ by the maximality of $C$ (the union of two connected sets with nonempty intersection is connected). Since $C$ is bounded, then $C$ cannot intersect $\mathbb{C} \setminus \bar{D}(0, n)$ and also it is the union of some discs,

$$C = \bigcup_{w_j \notin U} D(w_j, 1/n).$$

In particular, $C$ intersects $\mathbb{C} \setminus U$, so there is a component $F$ of $\mathbb{C} \setminus U$ which intersects $C$. Consequently $F$ is a connected subset of $\mathbb{C} \setminus U \subset \mathbb{C} \setminus K_n$ which intersects $C$, that is, $F \subset C$. $\qquad\square$

The meaning of this lemma is that, although $K_n \subset U$, and therefore $\mathbb{C} \setminus K_n$ is bigger than $\mathbb{C} \setminus U$, $K_n$ does not have other holes than those of $U$ (a component of $\mathbb{C} \setminus K_n$ may likely contain more than one component of $\mathbb{C} \setminus U$). On the other hand, it is clear that any component of $\mathbb{C} \setminus U$ is contained in a component of $\mathbb{C} \setminus K_n$, for all $n$ (Figure 1.6).

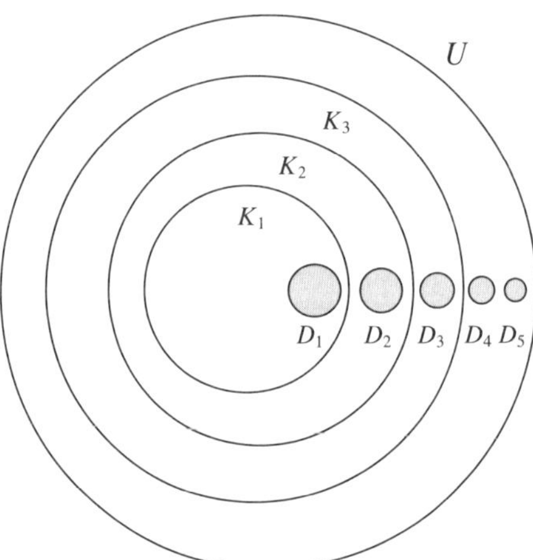


Figure 1.6

# 1.4  Curves, paths, length elements

## 1.4.1  Curves and paths

*Continuous curves* will be used rather often. A continuous curve, or just merely a *curve*, in a domain $U$ will be a continuous mapping $\gamma\colon [a,b] \to U$, where $[a,b]$ is an interval of $\mathbb{R}$. It is also usual to say that $\gamma$ is an *arc of a curve* or, simply, an *arc*. In this context $\gamma^*$ will denote the range of $\gamma$; it is common to visualize a continuous curve as a continuous "thread" in $U$; but, in fact, $\gamma^*$ may be a more general set. For example, it may be a whole square.

**Example 1.16.** The continuous curves that fill up a square are called *Peano curves*. One of the first constructions of these curves was done by Hilbert. It is the following: divide the unit square into four squares numbered consecutively. If one assumes that the mapping one is looking for has to transform the unit interval $[0,1]$ into the whole square, it is natural to impose that by breaking $[0,1]$ into four quarters, the mapping transforms each of these quarters in one of the squares 1, 2, 3, 4. Dividing again each square in 4, and each quarter of $[0,1]$ in 4 equal pieces, one may keep demanding that each $1/16$ of $[0,1]$ is transformed in one of the 16 new squares. One keeps doing this process and enumerating the squares of each generation in such a way that, if one has two consecutive squares of the same generation $m$, the last and the first of the generation $m+1$ of each one are also consecutive. Then each point $t \in [0,1]$ is the limit of the intervals of length $4^{-m}$ containing it, and to each interval corresponds a square of the $m$-th generation, which has side length $2^{-m}$. Associating the point $t$ to the common point of all the squares that correspond to intervals converging to $t$, then it is not difficult to check the continuity of the obtained curve (Figure 1.7).                                                    $\square$

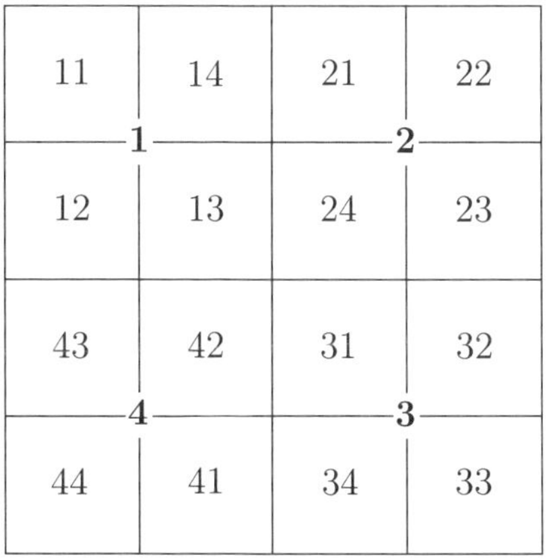

Figure 1.7

The curve $\gamma\colon [a,b] \to U$ is *closed* if $\gamma(b) = \gamma(a)$, and *simple* (non-self-intersecting) if $\gamma(s) \neq \gamma(t)$ for all $s, t \in (a,b)$, $s \neq t$. A simple curve is also called a *Jordan curve* or a *Jordan arc*. The curve is said to be *differentiable* if for any value of the parameter $t_0$ there exists the limit

$$\gamma'(t_0) = \lim_{t \to t_0} \frac{\gamma(t) - \gamma(t_0)}{t - t_0}. \tag{1.2}$$

In terms of the components of $\gamma$, $\gamma(t) = x(t) + iy(t)$, this is equivalent to saying that the real functions $x(t)$, $y(t)$ are differentiable at the point $t_0$, and then $\gamma'(t_0) = x'(t_0) + iy'(t_0)$. If $\gamma'(t_0) \neq 0$, this vector is tangent to the curve at the point $\gamma(t_0)$, that is, the line which contains the point $\gamma(t_0)$ and with direction vector $\gamma'(t_0)$ is tangent to the curve. If $\gamma'(t)$ exists for all $t$ and is continuous on $[a, b]$, one will say that $\gamma$ is of *class* $C^1$. Furthermore, a curve is *regular* if it is of class $C^1$ and has nonzero derivative in each point. If $\gamma'$ exists and is continuous, except at a finite number of points, at any of which $\gamma$ has lateral derivatives, one will say either that $\gamma$ is *piecewise of class* $C^1$, or that $\gamma$ is a *path*. The concept of a piecewise regular curve has a similar meaning, so we will use also the term *regular path*.

The rules for calculating derivatives of sums or products $\gamma(t) + \sigma(t)$, $\gamma(t) \cdot \sigma(t)$ are the usual ones. For example,

$$(\gamma(t) \cdot \sigma(t))' = \gamma'(t) \cdot \sigma(t) + \gamma(t) \cdot \sigma'(t) \tag{1.3}$$

if $\gamma(t)$ and $\sigma(t)$ are differentiable.

Let us study in more detail the example $\gamma(t) = 1_t = \cos t + i \sin t$, $t \in \mathbb{R}$. When $t$ takes values on $\mathbb{R}$, $\gamma(t)$ completes the unit circle $\mathbb{T}$ infinitely many times counterclockwise and

$$\gamma'(t) = -\sin t + i \cos t = i\gamma(t),$$

which tells us that the tangent vector is the radial vector multiplied by $i$, that is, $90°$ rotated. At this point we will leave the notation $1_t$ for the point $\cos t + i \sin t$, introducing the exponential function with a pure imaginary number in the exponent.

Recall that the real Euler number $e$ is characterized as the unique positive real number $a > 0$ that satisfies the property that the exponential function of basis $a$, $t \mapsto a^t$ coincides with its derivative; that is, $(a^t)' = a^t$ if and only if $a = e$. This fact is equivalent to the classical definition of $e$ as $e = \lim_{h \to 0}(1 + h)^{1/h}$. By means of usual differentiation rules, one has that $(e^{u(t)})' = u'(t)e^{u(t)}$ if $u$ is a differentiable function of $t$; in particular, $(e^{\alpha t})' = \alpha e^{\alpha t}$ if $\alpha \in \mathbb{R}$. Since $(1_t)' = i1_t$, as seen above, the notation $1_t$ is usually changed by $e^{it}$, because in this way the rule $(1_t)' = i1_t$ is like the previous one with $\alpha = i$: $(e^{it})' = ie^{it}$. The equation

$$e^{it} = 1_t = \cos t + i \sin t$$

is a *definition* of $e^z$ when $z = it$ is a pure imaginary number. It is called *Euler's identity*. The inverse of $e^{it}$ coincides with its conjugate and is $e^{-it}$. With this notation, the trigonometric functions have the following formulation:

$$\cos t = \operatorname{Re} e^{it} = \frac{1}{2}(e^{it} + e^{-it}),$$

$$\sin t = \operatorname{Im} e^{it} = \frac{1}{2i}(e^{it} - e^{-it}).$$

For the moment we know the definition of $e^z$ for $z = x \in \mathbb{R}$ (from calculus of one real variable functions) and now the definition for $z = it$, a pure imaginary number. The definition of $e^z$ for an arbitrary $z \in \mathbb{C}$ will be given later on.

The curves may be also given in polar coordinates. Considering two continuous functions $R \colon [a, b] \to \mathbb{R}^+$, $\phi \colon [a, b] \to \mathbb{R}$, then

$$\gamma(t) = R(t)e^{i\phi(t)} = R(t)(\cos \phi(t) + i \sin \phi(t))$$

is a continuous curve in $\mathbb{C} \setminus \{0\}$. If $R(t), \phi(t)$ are differentiable, it is immediate to check that $\gamma$ is a differentiable curve and that

$$\gamma'(t) = R'(t)e^{i\phi(t)} + R(t)i\phi'(t)e^{i\phi(t)} = (R'(t) + iR(t)\phi'(t))e^{i\phi(t)}.$$

Later (Theorem 1.23) it will be proved that every continuous curve $\gamma(t), a \leq t \leq b$ which does not contain the origin may be written as $\gamma(t) = R(t)e^{i\phi(t)}$.

**Example 1.17.** If $R(t)$ is an increasing positive function with $\lim_{t \to -\infty} R(t) = 0$, $\lim_{t \to +\infty} R(t) = +\infty$, then

$$\gamma(t) = R(t)e^{it}$$

describes a spiral which expands as it rotates counterclockwise, and we have $\gamma'(t) = R'(t)e^{it} + iR(t)e^{it} = (R'(t) + iR(t))e^{it}$ (Figure 1.8). $\qquad \square$

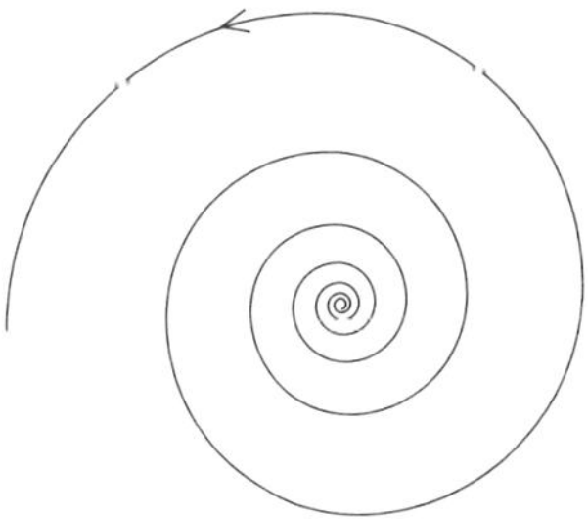

Figure 1.8

It is important not to confuse $\gamma$ with $\gamma^*$, because $\gamma$ includes $\gamma^*$ and the way it is travelled. For example, $\gamma_1(t) = e^{2\pi i t}$, $\gamma_2(t) = e^{2\pi i t^2}$, $0 \le t \le 1$ both describe the unit circle, but with different speeds. This example is a particular case of change of parametrization. In general, $\Gamma(u)$, $u \in [c, d]$ is obtained from $\gamma(t)$, $t \in [a, b]$ by means of a *change of parametrization* if the new parameter $u$ and the original parameter $t$ are linked by a relation $t = \Phi(u)$, where $\Phi$ is a homeomorphism from $[c, d]$ onto $[a, b]$ and $\Gamma(u) = \gamma(\Phi(u))$. Observe that in this case $\Gamma^* = \gamma^*$. Any homeomorphism between intervals is either strictly increasing or strictly decreasing; this fact classifies the parametrizations of $\gamma^*$ into two classes, each one associated intuitively to forward or backward direction. An *orientation* of $\gamma^*$ consists in the selection of one of these two classes. A parametrization $\gamma(t)$, $0 \le t \le 1$ and the parametrization $\tilde{\gamma}(t) = \gamma(1 - t)$, $0 \le t \le 1$, define opposite orientations.

As discussed above, $\gamma^*$ may be quite arbitrary and, in particular, it could be a square. If the curve $\gamma$ is piecewise $C^1$, the set $\gamma^*$ has a more particular structure. Indeed, take a neighborhood of any point $t_0$ for which the derivative exists and does not vanish. Here we can apply the implicit function theorem and the set $\gamma^*$ is, in the neighborhood of $\gamma(t_0)$, of the form $\{(u, v) : v = 0\}$, where $u$, $v$ are certain local coordinates. In this case $\gamma^*$ is a "thread" (Figure 1.9), agreeing with intuition. Consequently, in this case the set $\gamma^*$ has planar measure zero (see Proposition 1.35).

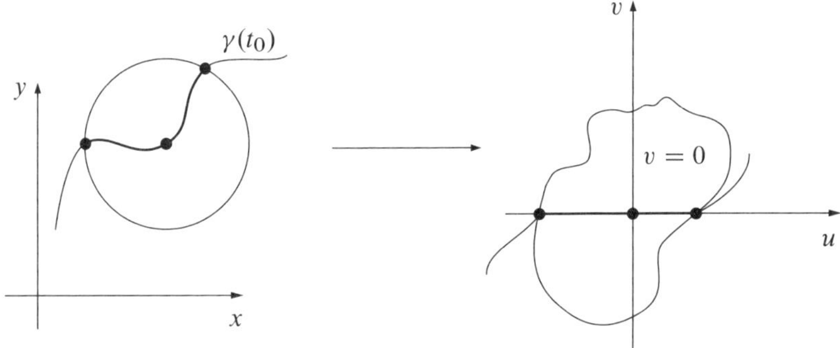

Figure 1.9

## 1.4.2 Length and arc parameter

We now review the notion of length of a curve $\gamma(t)$, $a \le t \le b$. Consider a polygonal curve $P$ inscribed in $\gamma$ and determined by the points $\gamma(t_0), \gamma(t_1), \gamma(t_2), \ldots, \gamma(t_n)$, with $a = t_0 < t_1 < \cdots < t_n = b$. Then its length is

$$L(P) = \sum_{i=0}^{n-1} |\gamma(t_{i+1}) - \gamma(t_i)| = \sum_{i=0}^{n-1} \sqrt{(x(t_{i+1}) - x(t_i))^2 + (y(t_{i+1}) - y(t_i))^2}.$$

$$(1.4)$$

The *length of $\gamma$* is defined by

$$L(\gamma) = \sup\{L(P): \ P \ \text{is an inscribed polygonal curve in } \gamma\}.$$

If $L(\gamma) < +\infty$ one says that $\gamma$ is *rectifiable*. When $\gamma$ is piecewise $C^1$, applying the mean value theorem in every difference of (1.4), one finds that

$$L(P) = \sum_{i=0}^{n-1} \sqrt{(x'(\xi_i))^2 + (y'(\eta_i))^2}(t_{i+1} - t_i), \qquad (1.5)$$

where the points $\xi_i$, $\eta_i$ are intermediate between $t_i$ and $t_{i+1}$. Adding points to the polygonal curve, and taking limits in (1.5), we obtain

$$L(\gamma) = \int_a^b |\gamma'(t)| \, dt.$$

This integral is well defined because by hypothesis $[a, b]$ is decomposed in intervals where $\gamma'$ is continuous. The integrand, $ds = |\gamma'(t)| \, dt$, is the *element of arc length of $\gamma$*. Obviously, the element of arc length is invariant under changes of parametrization, whether it is orientation preserving or not. Indeed, if $t = \Phi(u)$ and $\Gamma(u) = \gamma(\Phi(u))$, one has that

$$|\Gamma'(u)| \, du = |\gamma'(\Phi(u))||\Phi'(u)| \, du = |\gamma'(t)| \, dt = ds.$$

Given a path $\gamma$, denoting by $s(t)$ the length of the arc of the curve between $\gamma(a)$ and $\gamma(t)$,

$$s(t) = \int_a^t |\gamma'(x)| \, dx,$$

one gets $s'(t) = |\gamma'(t)| > 0$; so $t \to s(t)$ is a strictly increasing bijection $\Psi$ from $[a, b]$ onto $[0, L]$, where $L$ is the total length of $\gamma$. Considering the inverse bijection $\Phi = \Psi^{-1}$, one may take $s$ as a new parameter, $\Gamma(s) = \gamma(\Phi(s))$; since the length between $\Gamma(0)$ and $\Gamma(s)$ is $s$, then $|\Gamma'(s)| = 1$. This fact may also be checked observing that $\frac{dt}{ds} = \left(\frac{ds}{dt}\right)^{-1} = |\gamma'(t)|^{-1}$ and that $\Gamma'(s) = \gamma'(t) \cdot \frac{dt}{ds}$. This parametrization is called *arc length parametrization*.

A particular case of a regular curve is the graph of a real-valued function $\phi$ of a real variable $x \in [a, b]$, with continuous derivative, $\gamma(x) = (x, \phi(x))$; in this case $\gamma'(x) = (1, \phi'(x))$, $|\gamma'(x)| = \sqrt{1 + \phi'^2(x)}$ and the formula for the length becomes

$$L(\gamma) = \int_a^b \sqrt{1 + \phi'^2(x)} \, dx.$$

### 1.4.3  Integration with respect to arc length

If $\gamma : [a, b] \to \mathbb{C}$ is a piecewise $C^1$ curve and $h$ is a continuous real-valued function over $\gamma^*$, one may define the *integral of $h$ with respect to arc length $ds$*, $\int_\gamma h\, ds$, as

$$\int_\gamma h\, ds = \int_a^b h(\gamma(t))|\gamma'(t)|\, dt.$$

This integral is well defined; in fact, it is the limit of the Riemann sums

$$\sum_{i=0}^{n-1} h(\gamma(t_i))|\gamma(t_{i+1}) - \gamma(t_i)|,$$

where $t_0 < t_1 < \cdots < t_n$ is a partition of $[a, b]$.

Once again this notion does not depend on a change of parametrization, whether it is orientation preserving or not.

**Example 1.18.** Consider the curve $\gamma(t) = (t, \operatorname{ch} t) = \left(t, \frac{1}{2}(e^t + e^{-t})\right), 0 \le t \le 1$. The tangent vector is $(1, \operatorname{sh} t) = \left(1, \frac{1}{2}(e^t - e^{-t})\right)$ which has modulus $\sqrt{1 + \operatorname{sh}^2 t} = \operatorname{ch} t$ and $ds = \operatorname{ch} t\, dt$. The length is $\int_0^1 \operatorname{ch} t\, dt = \operatorname{sh} 1 = \frac{1}{2}(e - e^{-1})$. If $h(x, y) = xy$, one has that

$$\int_\gamma h\, ds = \int_0^1 h(t, \operatorname{ch} t) \operatorname{ch} t\, dt = \int_0^1 t(\operatorname{ch} t)^2\, dt = \frac{1}{8}(3 + 2\operatorname{sh} 2 - \operatorname{ch} 2). \qquad \square$$

## 1.5  Branches of the argument. Index of a closed curve with respect to a point

### 1.5.1  Branches of the argument

Recall that to any complex number $z \ne 0$ is associated the set $\arg z$ of its arguments.

**Definition 1.19.** If $0 \notin E \subset \mathbb{C}$, a continuous branch (or, briefly, a branch) of the argument in $E$ is a continuous mapping

$$g : E \longrightarrow \mathbb{R}$$

such that $g(z) \in \arg z$ if $z \in E$.

This way $g$ chooses continuously one of the possible arguments of each $z \in E$.

**Example 1.20.** The principal argument, Arg, is not a branch of the argument in $\mathbb{C} \setminus \{0\}$ because at a negative real point $a$ is not continuous, since

$$\lim_{y \to 0,\, y>0} \operatorname{Arg}(a + iy) = \pi = \operatorname{Arg} a,$$

$$\lim_{y \to 0,\, y<0} \operatorname{Arg}(a + iy) = -\pi.$$

However, the function Arg is continuous at all the other points, and so it is a continuous branch of the argument in $\mathbb{C} \setminus \mathbb{R}^-$, which takes values on $(-\pi, \pi)$.    $\square$

Observe that if $g$ and $h$ are two continuous branches of the argument in $E$, then $(g-h)/2\pi$ is a continuous function on $E$ which takes integer values; consequently, if $E$ is connected, it is a constant function. Hence all the existing continuous branches of the argument in a connected set differ one by one in a constant which is an entire multiple of $2\pi$. In $\mathbb{C} \setminus \mathbb{R}^-$, $\operatorname{Arg} z + 2\pi k$, $k \in \mathbb{Z}$ are all the possible branches of the argument.

Sometimes there are no branches of the argument over a set $E$. For example, in $\mathbb{C} \setminus \{0\}$ there are none of them. Indeed, suppose that $g$ is a branch of the argument. Then, as seen above, $g(z) = \operatorname{Arg}(z) + 2\pi k$ for a $k \in \mathbb{Z}$ in $\mathbb{C} \setminus \mathbb{R}^-$, and $g$ could not be continuous on the negative real points.

If $L$ is a ray starting at the origin, say $L = \{z : \operatorname{Arg} z = \alpha\}$, $\alpha \in (-\pi, \pi]$, in $\mathbb{C} \setminus L$ there are branches of the argument given analytically by

$$\operatorname{Arg}(z \cdot e^{i(\pi - \alpha)}) - \pi + \alpha + 2\pi k, \quad k \in \mathbb{Z}.$$

Similarly one talks about *branches of the n-th root* in a set $E$ which does not contain zero. For example, a branch of the square root is a continuous mapping $h \colon E \to \mathbb{C}$ such that $h(z)^2 = z$, $z \in E$. If $h_1, h_2$ are two branches of the square root, then $\left(\frac{h_1}{h_2}\right)^2 = 1$, and so, if $E$ is connected either $h_1 = h_2$ or $h_1 = -h_2$. Observe that if $g$ is a branch of the argument in $E$, then $h(z) = \sqrt{|z|}e^{ig(z)/2}$ is a branch of the square root.

More generally, the branches of the argument or the branches of the roots of nonzero continuous functions $f \colon E \to \mathbb{C}$ are defined as follows: A *branch of the argument of* $f \colon E \to \mathbb{C} \setminus \{0\}$ is a continuous function $g \colon E \to \mathbb{R}$ such that $g(z) \in \arg f(z)$, if $z \in E$; a *branch of the n-th root of* $f \colon E \to \mathbb{C} \setminus \{0\}$ is a continuous function $h$ on $E$ such that $h^n = f$. If there is a branch of the argument of $f$, say $g$, it is clear that there is also a branch of the $n$-th root, just by writing $h = |f|^{1/n}e^{ig/n}$. However, there may exist a branch of a root of a function $f$ but not a branch of the argument of $f$, as shown by the following example.

**Example 1.21.** Consider $w = f(z) = z^2 - 1 = (z+1)(z-1)$ in $E = \mathbb{C} \setminus [-1, +1]$; when $z$ describes the circle centered at 0 with radius 2 ($z = 2e^{i\theta}, 0 \le \theta \le 2\pi$), $z^2 - 1$ describes twice the circle centered at $-1$ with radius 4 so that any branch of the argument will have along this curve a variation of $4\pi$, so it is discontinuous. Instead, the principal branch of $\sqrt{w}$ gives a branch of $\sqrt{f}$. In relation to this example see Exercise 11 of Section 1.7.                                                    □

In the definition of branch of the argument of a function $f$ given above, it is supposed that the domain of $f$ was a part of $\mathbb{C}$, but one may give the same definition when $f$ is defined on any metric or topological connected space $X$. Even though the interest will be specially in the case that $X$ is either an interval of the real line or a domain of the plane, we give a more general definition.

**Definition 1.22.** Let $X$ be a connected space and $f : X \to \mathbb{C} \setminus \{0\}$ a continuous function. A continuous branch (or, briefly, a branch) of the argument of $f$ is a continuous mapping $h : X \to \mathbb{R}$ such that $h(x) \in \arg f(x)$, for any $x \in X$. Equivalently, $f(x) = |f(x)|e^{ih(x)}$, $x \in X$.

Observe that taking $X = E \subseteq \mathbb{C} \setminus \{0\}$, $f(z) = z$, in the previous definition one gets the concept of a branch of the argument given in Definition 1.19.

Two possible branches of the argument of $f$ differ on a function that takes values that are entire multiples of $2\pi$. Since $X$ is connected, it must be a constant multiple of $2\pi$.

If there is a continuous branch $g(w)$ of the argument of the variable $w \in E$ in the image set $E = f(X)$, then $h = g \circ f$ is a branch of the argument of $f$. This condition holds in a neighborhood of each point $x \in X$, taking, for example, the preimage of the disc $D(f(x), |f(x)|)$ by $f$.

Consider the case of a continuous curve $\gamma : [a, b] \to \mathbb{C}$ which does not contain zero, $\gamma(t) \neq 0$. If we want to write $\gamma(t)$ in polar coordinates,

$$\gamma(t) = R(t)e^{i\phi(t)},$$

there is no other option for the term $R(t)$ than $R(t) = |\gamma(t)| = \sqrt{x(t)^2 + y(t)^2}$, if $\gamma(t) = x(t) + iy(t)$ and $R(t)$ is continuous. However, there are infinitely many options for $\phi(t)$ and it is not evident that there exists a continuous selection of $\phi(t)$. Nevertheless it is true: clearly the function $\phi(t)$ is just a branch of the argument of $\gamma$, in the sense of Definition 1.22.

**Theorem 1.23.** *If $\gamma : [a, b] \to \mathbb{C}$ is a continuous curve which does not contain the origin, then there exists a branch of the argument of $\gamma$. Denoting it by $\phi$, then $\phi(t) + 2\pi k$, $k \in \mathbb{Z}$ is the general expression of all the branches of the argument of $\gamma$. If $\gamma$ is differentiable at a point $t \in [a, b]$ then also the function $\phi$ is differentiable, and $\phi'(t) = \operatorname{Im} \frac{\gamma'(t)}{\gamma(t)}$.*

*Proof.* Consider a partition $a = t_0 < t_1 < t_2 < \cdots < t_n = b$ of $[a, b]$ such that $\gamma([t_{i-1}, t_i])$ is inside a disc $D_i$ which does not contain zero; this is possible due to the continuity of $\gamma$ and the compactness of $[a, b]$. If $g_i$ is a branch of $\arg z$ in $D_i$, then $\phi_i = g_i \circ \gamma$ is a branch of $\arg \gamma(t)$ in $[t_{i-1}, t_i]$. At the point $t_i$ one has two arguments of $\gamma(t_i)$, which are $\phi_i(t_i)$ and $\phi_{i+1}(t_i)$, that differ in a multiple of $2\pi$; adding to $\phi_{i+1}$ this entire multiple of $2\pi$, we get by iteration a global branch $\phi$.

Let us check now that if $\gamma$ is differentiable at a point $t$, then also the function $\phi$ is differentiable in $t$ and $\phi'(t) = \operatorname{Im} \frac{\gamma'(t)}{\gamma(t)}$. Suppose without loss of generality that $\gamma(t)$ is in the half plane $\{z : \operatorname{Re} z > 0\}$. Then, in a neighborhood of $t$, $\phi(t)$ differs from the principal branch of $\operatorname{Arg} \gamma(t)$ by a constant and has the same derivative. If $\gamma(t) = (x(t), y(t))$, $x(t) > 0$, then

$$\operatorname{Arg} \gamma(t) = \arctan \frac{y(t)}{x(t)}$$

and by differentiating we get that $\frac{d \operatorname{Arg} \gamma(t)}{dt} = \operatorname{Im} \frac{\gamma'(t)}{\gamma(t)}$. $\qquad\square$

It is important to stress that continuity is required in the parameter $t$. It has been proved that a branch of the argument of $\gamma$ always exists, even though no branch of $\arg z$ may exist on $\gamma^*$.

**Example 1.24.** The curve $\gamma(t) = e^{it}$, $0 \le t \le 2\pi$, has a branch of the argument $h(t) = t$, but on $\gamma^* = \mathbb{T}$ there is no branch of $\arg z$. $\qquad\square$

If $\phi$ is a branch of the argument of $\gamma$, the quantity $\phi(b) - \phi(a)$ does not depend on $\phi$, because two branches differ by a constant. This is called *variation of the argument along* $\gamma$ and it is represented by $\Delta_\gamma \arg$. In the case of piecewise differentiable curves, the fundamental theorem of calculus gives

$$\Delta_\gamma \arg = \operatorname{Im} \int_a^b \frac{\gamma'(t)}{\gamma(t)} \, dt,$$

because $\phi(b) - \phi(a) = \int_a^b \phi'(t) \, dt$, and now Theorem 1.23 may be applied.

### 1.5.2 Index of a closed curve

Let $\gamma : [a, b] \to \mathbb{C}$ be a closed curve that does not contain the origin and $\phi : [a, b] \to \mathbb{R}$ a branch of the argument along $\gamma$. Then $\phi(a)$ and $\phi(b)$ are arguments of the same point, and so $\phi(b) - \phi(a) = \Delta_\gamma \arg$ is an entire multiple of $2\pi$. Hence the following definition makes sense:

**Definition 1.25.** If $\gamma$ is a closed curve that does not pass through the origin, then the integer $\frac{1}{2\pi} \Delta_\gamma \arg$ is called the index of $\gamma$ with respect to the origin, or also the winding number of $\gamma$ around the origin. It is denoted by $\operatorname{Ind}(\gamma, 0)$.

In the case that the closed curve $\gamma$ is piecewise of class $C^1$, the number $\operatorname{Ind}(\gamma, 0)$ may be calculated in the following way: If $\gamma(t) = R(t)e^{i\phi(t)}$, $R(t) = |\gamma(t)|$, then $\gamma'(t) = (R'(t) + iR(t)\phi'(t))e^{i\phi(t)}$ and $\gamma'(t)/\gamma(t) = \left(\frac{R'(t)}{R(t)} + i\phi'\right)$; so, $\mathrm{Re}\,\frac{\gamma'(t)}{\gamma(t)} = \frac{R'(t)}{R(t)}$, which is the derivative of $\log R(t) = \log|\gamma(t)|$, and it turns out that

$$\int_a^b \frac{\gamma'(t)}{\gamma(t)}\,dt = \int_a^b \mathrm{Re}\,\frac{\gamma'(t)}{\gamma(t)}\,dt + i\int_a^b \mathrm{Im}\,\frac{\gamma'(t)}{\gamma(t)}\,dt = \log\frac{|\gamma(b)|}{|\gamma(a)|} + i\,\Delta_\gamma \arg.$$

Thus, for a piecewise $C^1$ closed curve $\gamma : [a,b] \to \mathbb{C} \setminus \{0\}$, one has

$$\mathrm{Ind}(\gamma, 0) = \frac{1}{2\pi i}\int_a^b \frac{\gamma'(t)}{\gamma(t)}\,dt.$$

Roughly speaking, $\mathrm{Ind}(\gamma, 0)$ is the number of rotations one would do by tracking visually from the origin a particle moving over the trajectory $\gamma(t)$. This quantity is invariant under changes of parametrization which preserves orientation and changes the sign with non-orientation preserving changes of parametrizations.

**Definition 1.26.** Let $\gamma$ be an arbitrary closed curve and $z \notin \gamma^*$. The index of $\gamma$ with respect to $z$, denoted by $\mathrm{Ind}(\gamma, z)$, is defined as $\mathrm{Ind}(\gamma - z, 0)$.

In the case of a piecewise $C^1$ closed curve $\gamma$ parameterized by $[a,b]$, one has

$$\mathrm{Ind}(\gamma, z) = \frac{1}{2\pi i}\int_a^b \frac{\gamma'(t)}{\gamma(t) - z}\,dt. \tag{1.6}$$

As a function of $z$, the index is a continuous function in the complement of $\gamma^*$. This fact, in the case of a piecewise $C^1$ curve, which are the ones studied here, may be proved by using the integral expression (1.6). The proof in the case of a continuous closed curve is proposed as an exercise (see Exercise 13 of Section 1.7).

If $\gamma : [a,b] \to \mathbb{C}$ is a closed curve, one knows that the complement of $\gamma^*$ in the plane, $\mathbb{C} \setminus \gamma^*$, has just one unbounded connected component and a countable set of bounded connected components. It is important to note the following property:

*The index $\mathrm{Ind}(\gamma, z)$ as a function of $z$ is constant on each of the connected components of $\mathbb{C} \setminus \gamma^*$ and has value zero in the unbounded component.*

Indeed, $\mathrm{Ind}(\gamma, z)$ is a continuous function on each component and only takes integer values, so it must be constant. The assertion about the unbounded component is clear intuitively and it may be justified also analytically by observing the following: Moving the point $z$ one may suppose that there exists a disc $D$ such that $z \notin D$ and $\gamma^* \subset D$. Then one may take on $D$ a branch of the argument of $w - z$ which, composed with $\gamma$, will give a branch of the argument of $\gamma(t) - z$, which clearly will take the same value for $t = a$ and $t = b$.

**Example 1.27.** a) If one considers a circle centered at the origin with radius $R$ and travelled $m$ times,

$$\gamma(t) = Re^{imt}, \quad 0 \le t \le 2\pi$$

(with $m$ integer) and $z \in \mathbb{C}$, due to the argument above it turns out that:

If $|z| > R$, $\mathrm{Ind}(\gamma, z) = 0$.

If $|z| < R$, $\mathrm{Ind}(\gamma, z) = \mathrm{Ind}(\gamma, 0)$ and

$$\mathrm{Ind}(\gamma, 0) = \frac{1}{2\pi i} \int_0^{2\pi} \frac{mi Re^{imt}}{Re^{imt}} \, dt = m.$$

If $|z| = R$, $\mathrm{Ind}(\gamma, z)$ is not defined.

b) Consider now that $\gamma$ is the edge of a square travelled in a direct sense of orientation (counterclockwise) and $z$ is an interior point of the square. It is intuitively clear that $\mathrm{Ind}(\gamma, z) = 1$. This may be checked also with the method we will now describe for the calculation of an index. An analytic proof of this fact using (1.6) is a bit long and it is proposed as an exercise (see Exercise 12 of Section 1.7).    □

When calculating the index of a closed curve $\gamma$ with respect to a point $z \notin \gamma^*$, aside from the integral formula (1.6) valid for $C^1$ curves, there is a very useful geometric method that is easy to apply. The method is the following:

Consider an arbitrary ray starting at $z$. One may prove that if $\gamma$ is rectifiable, almost all these rays intersect $\gamma(t), t \in [a, b]$ a finite number of times. Here "almost all" is in the sense of the Lebesgue measure, considered on a circle centered at $z$, over which one takes the direction of each ray. Fix one of these rays, say $L$, and let $t_1, t_2, \ldots, t_n \in [a, b]$ be the values of the parameter $t$ for which $\gamma(t) \in L$. Now associate to each $t_j$ for $j = 1, \ldots, n$ a number $\sigma_j$ equal to $\pm 1$ according to the following rule:

$$\sigma_j = +1 \quad \text{if } \gamma \text{ crosses } L \text{ at the point } \gamma(t_j) \text{ in a direct sense,}$$

$$\sigma_j = -1 \quad \text{if } \gamma \text{ crosses } L \text{ at the point } \gamma(t_j) \text{ in an inverse sense.}$$

Then the equation

$$\mathrm{Ind}(\gamma, z) = \sum_{j=1}^{n} \sigma_j \tag{1.7}$$

holds. It is said that $\gamma$ crosses $L$ in a direct sense if it does it in the direction of increasing arguments, and in an inverse sense if it does it in the direction of decreasing arguments, assuming that the origin is placed at the point $z$ (Figure 1.10). The justification of (1.7) is intuitively clear. A rigorous proof is much more subtle and will not be given here.

**Example 1.28.** See Figure 1.11.    □

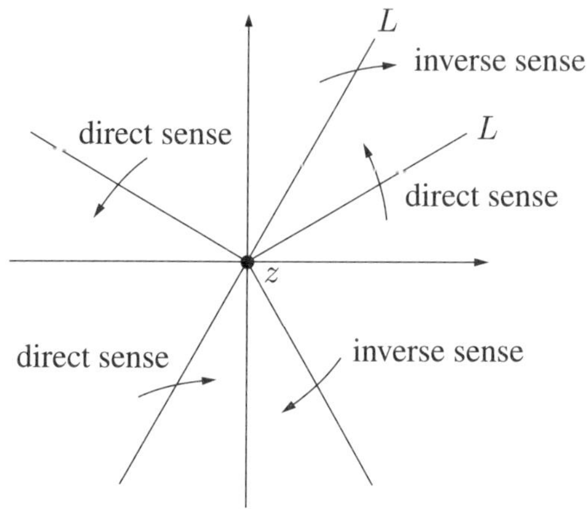


Figure 1.10

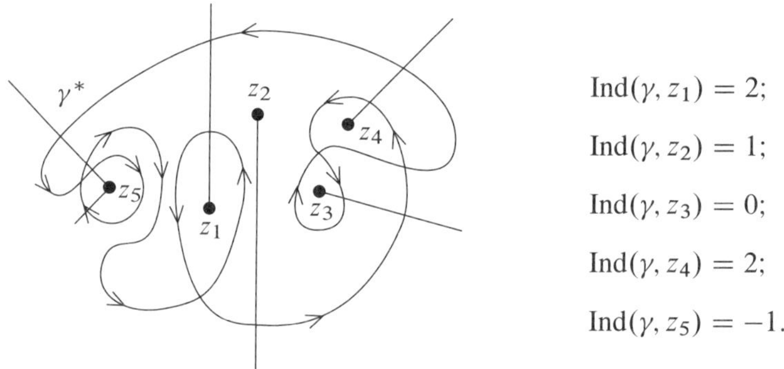


$$\mathrm{Ind}(\gamma, z_1) = 2;$$
$$\mathrm{Ind}(\gamma, z_2) = 1;$$
$$\mathrm{Ind}(\gamma, z_3) = 0;$$
$$\mathrm{Ind}(\gamma, z_4) = 2;$$
$$\mathrm{Ind}(\gamma, z_5) = -1.$$

Figure 1.11

## 1.6 Domains with regular boundary

If $U$ is a domain of the plane, the topological boundary of $U$, $\partial U$, may be unusually complicated. For this reason, in the study of function theory it is often convenient to consider domains with regular boundary. The most natural idea is to assume that the boundary of $U$ is composed by one or several closed curves. In this context the *Jordan curve theorem*, which we recall here, is relevant (see [6], p. 261).

**Theorem 1.29.** *If $\gamma$ is a closed Jordan curve, then the open set $\mathbb{C} \setminus \gamma^*$ has two connected components, a bounded one and an unbounded one. The bounded component is simply connected and $\gamma^*$ is the topological boundary of each one of the components. Moreover, $\mathrm{Ind}(\gamma, z)$ is $0$ if $z$ is in the unbounded component and is either $1$ or $-1$ if $z$ belongs to the bounded component.*

The bounded component of $\mathbb{C} \setminus \gamma^*$ is called the *interior of $\gamma$*, $\mathrm{Int}(\gamma)$, and the unbounded one is the *exterior of $\gamma$*, $\mathrm{Ext}(\gamma)$.

Motivated by this theorem, a *Jordan domain* will be a domain that is the bounded connected component of the complement of a closed Jordan curve. Thus, the boundary of a Jordan domain is a closed Jordan curve. The following statement asserts that this property of the boundary characterizes Jordan domains. Its proof is proposed as an exercise.

**Proposition 1.30.** *If $U$ is a bounded domain of the plane such that $\partial U$ is a closed Jordan curve, then $U$ is a Jordan domain.*

More generally, one can consider domains whose boundary is a 1-dimensional manifold. Recall that a 1-*dimensional manifold* or 1-*manifold* is a separable metric space $M$ which may be covered by a family of open sets of $M$ such that each of them is homeomorphic to the unit interval $(0, 1)$. It is clear that a closed Jordan curve, which is homeomorphic to the unit circle $\mathbb{T}$, is a 1-manifold. If one considers a bounded domain $U$, its boundary $\partial U$ is a compact set which may not be connected. If we assume now that $\partial U$ is a 1-manifold, then any connected component of $\partial U$ is also a 1-manifold which, moreover, is connected and compact. It turns out that each one of these components must be a closed Jordan curve. This is due to the following well-known fact (see [5], p. 127).

**Theorem 1.31.** *Any compact connected 1-manifold is homeomorphic to $\mathbb{T}$ and is therefore a closed Jordan curve.*

So a Jordan domain is exactly a bounded domain of the complex plane whose boundary is a connected 1-manifold.

Another way for requiring a bounded domain $U$ to have a regular boundary is supposing that the compact set $\overline{U}$ is a *submanifold of $\mathbb{R}^2$ with boundary*. Recall that this means that every point $z_0 \in \partial U$ has an open neighborhood $V(z_0)$ of $\mathbb{C}$ so that there is a homeomorphism $\varphi_{z_0} : V(z_0) \to \mathbb{D}$, from $V(z_0)$ onto the unit disc $\mathbb{D}$, such that (see Figure 1.12)

$$
\begin{aligned}
&\text{a)} \quad \varphi_{z_0}(V(z_0) \cap \partial U) = \{z \in \mathbb{D} : \mathrm{Im}\, z = 0\}, \\
&\text{b)} \quad \varphi_{z_0}(V(z_0) \cap U) = \{z \in \mathbb{D} : \mathrm{Im}\, z > 0\}.
\end{aligned}
\tag{1.8}
$$

It is clear that if $\overline{U}$ is a submanifold with boundary, then $\partial U$ is a 1-manifold. This hypothesis restricts the number of components of $\partial U$:

**Proposition 1.32.** *If $U$ is a bounded domain such that $\overline{U}$ is a submanifold of $\mathbb{R}^2$ with boundary, then $\partial U$ has a finite number of connected components.*

The proof is proposed as an exercise. Summarizing, it turns out that if $U$ is a bounded domain and $\overline{U}$ is a submanifold with boundary, then the boundary of $U$

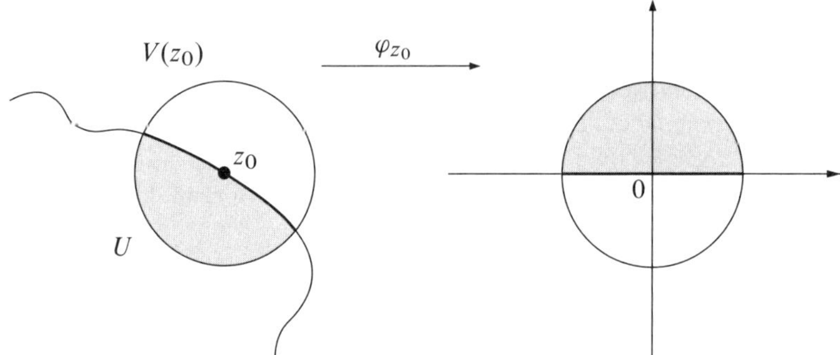


Figure 1.12

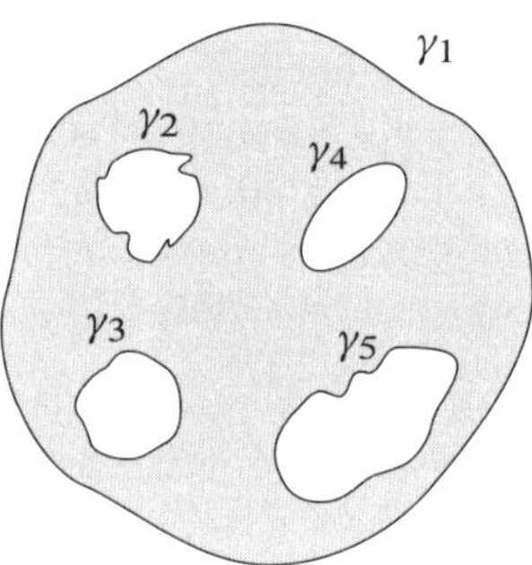


Figure 1.13

is formed by a finite number of closed Jordan curves (Figure 1.13). It is worth mentioning that the opposite of this fact is also true, and one has the following result:

**Proposition 1.33.** *If $U$ is a bounded domain of the plane, then its boundary is the union of a finite number of pairwise disjoint closed Jordan curves if and only if $\overline{U}$ is a submanifold of $\mathbb{R}^2$ with boundary.*

In order to prove it one has to show that if $\gamma$ is a closed Jordan curve, then both the interior of $\gamma$, with the curve $\gamma$ itself, and the exterior of $\gamma$, with the curve itself, are submanifolds with boundary in $\mathbb{R}^2$. This is a consequence of the fact that every homeomorphism between a Jordan curve and $\mathbb{T}$ may be extended to a homeomorphism from $\mathbb{C}$ onto $\mathbb{C}$ (theorem of Schöenflies: see [5], p. 153).

So far all our considerations have been topological, in the sense that no differentiability condition, neither on the curves nor on the homeomorphisms, has been required. But all the equivalences stated remain true if one substitutes Jordan curve by regular Jordan curve, 1-manifold by $C^1$ regular 1-manifold, and submanifold with boundary by submanifold with regular boundary. So one takes as a basic definition the following.

**Definition 1.34.** A bounded domain $U$ of the plane is called a domain with regular boundary (respectively with piecewise regular boundary) if the boundary of $U$ is composed by a finite number of pairwise disjoint regular closed Jordan curves (respectively piecewise regular).

Consider that $U$ is a bounded domain such that $\bar{U}$ is a submanifold of $\mathbb{R}^2$ with regular boundary. Then the homeomorphism $\varphi_{z_0} : V(z_0) \to D$ between a neighborhood of $z_0 \in \partial U$ and the unit disc $\mathbb{D}$ which satisfies (1.8) is a diffeomorphism between $V(z_0)$ and $\mathbb{D}$. A consequence of this fact is that the closed Jordan curves which compose $\partial U$ are regular and so $U$ is a domain with regular boundary in the sense of the Definition 1.34. The converse may be directly proved by using the inverse function theorem.

**Proposition 1.35.** *Let $\gamma(t)$, $0 \le t \le 1$ be a regular closed Jordan curve and $z_0 = \gamma(t_0)$, $0 \le t_0 \le 1$. Then there exists a neighborhood $V(z_0)$ of the point $z_0$, a disc $D_\delta(t_0)$ centered at $(t_0, 0)$ and a diffeomorphism $\varphi$ from $D_\delta(z_0)$ onto $V(z_0)$ such that*

    a) $\varphi(t, 0) = \gamma(t)$ *for* $|t - t_0| < \delta$,

    b) $J_\varphi(t_0, 0) > 0$.

*Now writing $D_\delta^+ = D_\delta(t_0) \cap \{z : \operatorname{Im} z > 0\}$, $D_\delta^- = D_\delta(t_0) \cap \{z : \operatorname{Im} z < 0\}$, and $V^+(z_0) = \varphi(D_\delta^+)$, $V^-(z_0) = \varphi(D_\delta^-)$, then either $V^+(z_0) \subset \operatorname{Int}(\gamma)$, $V^-(z_0) \subset \operatorname{Ext}(\gamma)$ or $V^+(z_0) \subset \operatorname{Ext}(\gamma)$, $V^-(z_0) \subset \operatorname{Int}(\gamma)$ hold (Figure 1.14).*

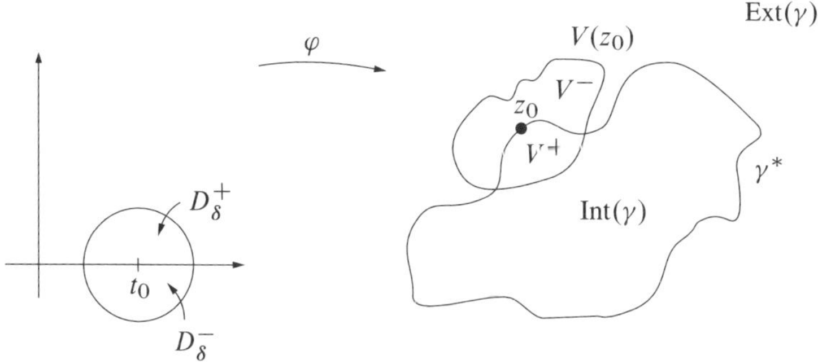


Figure 1.14

Here $J_\varphi$ denotes the Jacobian determinant of the mapping $\varphi$.

*Proof.* Let $x(t)$, $y(t)$ be the components of $\gamma$, that is, $\gamma(t) = (x(t), y(t))$. Since $\gamma$ is regular, $\gamma'(t_0) \neq 0$ and one may assume, for example, that $x'(t_0) \neq 0$. Define on a neighborhood of the point $(t_0, 0)$ the function $\varphi$ by either $\varphi(t, s) = (x(t), y(t)+s)$ if $x'(t_0) > 0$ or $\varphi(t, s) = (x(t), y(t) - s)$ if $x'(t_0) < 0$, so that $J_\varphi(t_0, 0) > 0$.

By the inverse function theorem $\varphi$ is a diffeomorphism of class $C^1$ from a neighborhood of $(t_0, 0)$, say $D_\delta(t_0)$, onto a neighborhood of $\varphi(t_0, 0) = z_0$, say $V(z_0)$, and $\varphi(t, 0) = \gamma(t)$, if $|t - t_0| < \delta$. So the points of $V(z_0) \setminus \gamma^*$ are distributed into two open connected sets denoted by $V^+(z_0)$ and $V^-(z_0)$, which are precisely $\varphi(D_\delta^+)$ and $\varphi(D_\delta^-)$.

Since $V^+$ and $V^-$ must be each one contained in one connected component of $\mathbb{C} \setminus \gamma^*$, the statement is proved. $\qquad\square$

**Corollary 1.36.** *A bounded domain $U$ of the plane is a domain with regular (respectively piecewise regular) boundary if and only if $\bar{U}$ is a submanifold of $\mathbb{R}^2$ with regular (respectively piecewise regular) boundary.*

Let us study in a little more detail the structure of a domain with regular boundary.

**Proposition 1.37.** *Let $U$ be a bounded domain of the plane with regular (or piecewise regular) boundary. Then $\partial U = \gamma_1^* \cup \gamma_2^* \cup \cdots \cup \gamma_N^*$, where $\gamma_i$, $i = 1, 2, \ldots, N$ are pairwise disjoint regular (or piecewise regular) closed Jordan curves such that:*

a) $\gamma_i^* \subset \mathrm{Int}(\gamma_1)$, $\mathrm{Int}(\gamma_i) \subset \mathrm{Int}(\gamma_1)$, $i = 2, 3, \ldots, N$.

b) $\mathrm{Int}(\gamma_2), \mathrm{Int}(\gamma_3), \ldots, \mathrm{Int}(\gamma_N)$ *are pairwise disjoint.*

c) $U = \mathrm{Int}(\gamma_1) \setminus (\overline{\mathrm{Int}(\gamma_2)} \cup \cdots \cup \overline{\mathrm{Int}(\gamma_N)})$.

*Proof.* We already know that $\partial U$ is composed by a finite number of pairwise disjoint regular closed Jordan curves. Suppose now that $\gamma_i$, $\gamma_j$ denote a pair of these curves. Then the proposition is a direct consequence of the following observations:

1) $\gamma_i^* \subset \mathrm{Int}(\gamma_j) \Rightarrow \mathrm{Int}(\gamma_i) \subset \mathrm{Int}(\gamma_j)$.

Indeed, the unbounded component of $\mathbb{C} \setminus \gamma_j^*$ does not meet $\gamma_i^*$, so it must be contained in a component of $\mathbb{C} \setminus \gamma_i^*$ which is nothing but the unbounded component; finally take complements.

2) $\mathrm{Int}(\gamma_i) \cap \mathrm{Int}(\gamma_j) = \emptyset \Rightarrow U \subset \mathrm{Ext}(\gamma_i) \cap \mathrm{Ext}(\gamma_j)$ (Figure 1.15).

Obviously, if $U$ was such that $U \subset \mathrm{Int}(\gamma_i)$, it could not be $\gamma_j^* \subset \partial U$, and analogously for $U \subset \mathrm{Int}(\gamma_j)$.

3) $\gamma_i^* \subset \mathrm{Int}(\gamma_j) \Rightarrow U \subset \mathrm{Int}(\gamma_j) \setminus \overline{\mathrm{Int}(\gamma_i)}$.

Indeed we know that $\mathrm{Int}(\gamma_i) \subset \mathrm{Int}(\gamma_j)$. It is clear that $U \subset \mathrm{Int}(\gamma_j)$ because $\gamma_i^* \subset \partial U$ and it cannot be $U \subset \mathrm{Int}(\gamma_i)$ because $\gamma_j^* \subset \partial U$.

From these considerations and from the fact that $U$ is bounded, we can deduce that there is a Jordan curve and just one of $\partial U$ which contains $U$ in its interior. Denoting this curve by $\gamma_1$, and $\gamma_2, \ldots, \gamma_N$ the remainder, it is easy to see that conditions a), b), c) hold. $\qquad\square$

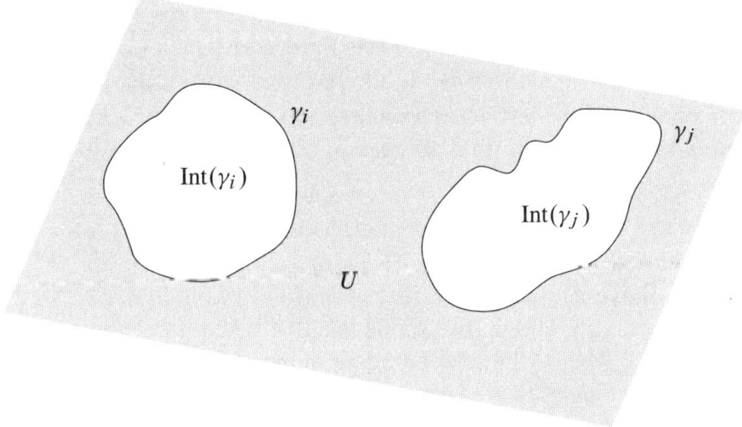



Figure 1.15

Let us finish this section by making some comments about the orientation of the boundary of a domain with regular boundary. Also let us give a precise definition of what is meant by orienting the boundary in the positive sense.

Suppose that $U$ is a bounded domain with regular boundary (or piecewise regular) so that, according to Proposition 1.37, the boundary of $U$ is composed by the closed Jordan curves $\gamma_1^*, \gamma_2^*, \ldots, \gamma_N^*$. Consider each of these curves separately (Figure 1.16). For example, fix $\gamma_1(t)$, $0 \le t \le 1$, and for each value $t_0$ of the parameter consider the point $z_0 = \gamma_1(t_0)$ and a diffeomorphism $\varphi$ like the one in Proposition 1.35. Thus either $V^+(z_0) \subset U$ or $V^-(z_0) \subset U$. In the first case one

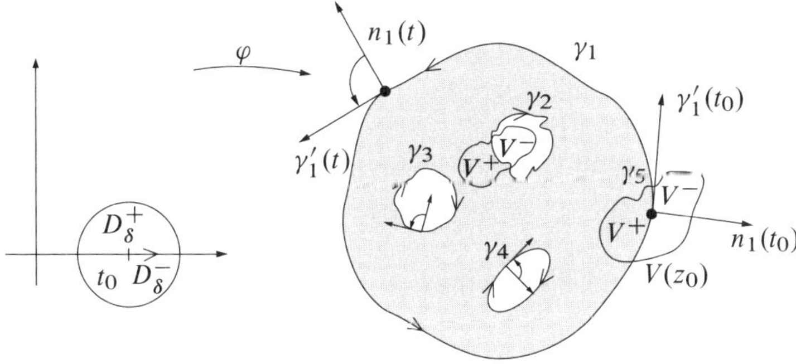



Figure 1.16

chooses the orientation of $\gamma_1^*$ given by the parametrization $\gamma_1(t)$, and in the second one, the opposite orientation given by the parametrization $\gamma_1(1-t)$, $0 \le t \le 1$

(see page 20). Similarly one may give an orientation to the curves $\gamma_2, \ldots, \gamma_N$. The senses of direction when travelling over these curves, $\gamma_1^*, \gamma_2^*, \ldots, \gamma_N^*$, defined this way, give an orientation to $\partial U$ in the *positive sense* (roughly speaking, the domain $U$ remains at the left when travelling on). If $\vec{T}$ is the vector field of tangent vectors to $\partial U$ given by this orientation (that is, the set of vectors $\gamma_i'(t)$ or $-\gamma_i'(1-t)$ for $t \in [0,1]$, $i = 1, 2, \ldots, N$), the vector field $\vec{N} = -i\vec{T}$, which satisfies $\det(\vec{N}, \vec{T}) > 0$, is called *normal exterior field* in $\partial U$.

If $\gamma$ is a closed Jordan curve, by Theorem 1.29 either $\operatorname{Ind}(\gamma, z) = 1$, for all $z \in \operatorname{Int}(\gamma)$ or $\operatorname{Ind}(\gamma, z) = -1$, for all $z \in \operatorname{Int}(\gamma)$. In the first case we say that $\gamma$ is *positively oriented*, and in the second one that it is *negatively oriented*. Now it is natural to ask what is the relation between the orientation of $\partial U$ in the positive sense and the orientation of the Jordan curves $\gamma_1, \gamma_2, \ldots, \gamma_N$, considered in Proposition 1.37, in the spirit of the present discussion. The answer depends on the following lemma:

**Lemma 1.38.** *Let $\gamma$ be a piecewise regular closed Jordan curve, and $U = \operatorname{Int}(\gamma)$ the Jordan domain defined by $\gamma$. Then the positive orientation of $\gamma$ in the sense that*

$$\operatorname{Ind}(\gamma, z) = 1, \quad z \in U,$$

*coincides with the orientation of $\partial U$ in the positive sense.*

*Proof.* Suppose that $\gamma$ is oriented in the positive sense and let us show that there exists a point $z_1 \in U$ with $\operatorname{Ind}(\gamma, z_1) = 1$. Let $V$ be a neighborhood of a certain point of $\gamma$, such as in Proposition 1.35 (with $J_\varphi > 0$), such that $V^+ \subset U$ and $V^- \subset \operatorname{Ext}(\gamma)$. Take $z_1 \in V^+, z_2 \in V^-$; we know that $\operatorname{Ind}(\gamma, z_2) = 0, \operatorname{Ind}(\gamma, z_1) = \pm 1$ and we want to see that $\operatorname{Ind}(\gamma, z_1) = 1$. Break the curve $\gamma$ into two pieces: $\gamma_1^* = \gamma^* \cap V$ and $\gamma_2^* = \gamma^* \setminus \gamma_1^*$. According to the notation of Subsection 1.5.1, denote by $\Delta_\gamma \arg(a)$ the variation of the argument along $\gamma(t) - a$, where $a$ is a fixed point. Then by the definition of index,

$$\Delta_\gamma \arg(z_1) = \Delta_{\gamma_1} \arg(z_1) + \Delta_{\gamma_2} \arg(z_1) = \pm 1,$$

$$\Delta_\gamma \arg(z_2) = \Delta_{\gamma_1} \arg(z_2) + \Delta_{\gamma_2} \arg(z_2) = 0.$$

Hence, due to the fact that $\gamma$ has the orientation induced by the positive one of $U$, it is clear that

$$\Delta_{\gamma_1} \arg(z_1) > 0, \quad \Delta_{\gamma_1} \arg(z_2) < 0$$

and so $\Delta_{\gamma_2} \arg(z_2) > 0$.

Finally, as $\Delta_{\gamma_2} \arg(a)$ is a continuous function of $a$, $|\Delta_{\gamma_2} \arg(z_1) - \Delta_{\gamma_2} \arg(z_2)|$ is as small as wanted if $z_1$ and $z_2$ are very close. Therefore one may achieve that $\Delta_{\gamma_2} \arg(z_1) > 0$, and so it is $\Delta_\gamma \arg(z_1) > 0$, that is, $\operatorname{Ind}(\gamma, z_1) = 1$. $\qquad\square$

Consider now the decomposition $\partial U = \gamma_1^* \cup \gamma_2^* \cup \cdots \cup \gamma_N^*$, under the conditions of Proposition 1.37. Since $U \subset \mathrm{Int}(\gamma_1)$ but $U \subset \mathrm{Ext}(\gamma_i)$, $i = 2, \ldots, N$, it turns out from Lemma 1.38, and from its proof, that if $\partial U$ is positively oriented then $\gamma_1$ is also positively oriented and $\gamma_2, \ldots, \gamma_N$ are negatively oriented, that is,

$$\mathrm{Ind}(\gamma_1, z) = 1 \quad \text{if } z \in \mathrm{Int}(\gamma_1);$$
$$\mathrm{Ind}(\gamma_i, z) = -1 \quad \text{if } z \in \mathrm{Int}(\gamma_i), \quad i = 2, \ldots, N. \tag{1.9}$$

For this reason, from now on we will call *positive orientation* of the boundary of a domain $U$ with regular or piecewise regular boundary the one corresponding to the orientations given by (1.9) for the Jordan curves which form $\partial U$. With the preceding hypothesis, if one writes, by definition,

$$\mathrm{Ind}(\partial U, z) = \sum_{i=1}^{n} \mathrm{Ind}(\gamma_i, z), \quad z \in \mathbb{C} \setminus \partial U,$$

one obtains the following characterization of the points of $U$:

$$U = \{z \in \mathbb{C} \setminus \partial U : \mathrm{Ind}(\partial U, z) = 1\},$$

$$\mathbb{C} \setminus \bar{U} = \{z \in \mathbb{C} \setminus \partial U : \mathrm{Ind}(\partial U, z) = 0\}.$$

## 1.7 Exercises

1. Let $z_1, \ldots, z_N$ be some points of the complex plane with $|z_i| = 1$ for $i = 1, 2, \ldots, N$ and $z_1 + \cdots + z_N = 0$. Show that:

   a) If $N = 3$, the three points must be the vertices of an equilateral triangle.

   b) If $N = 4$, the four points must be located pairwise diametrally opposite.

   c) If $N = 5$, there are infinitely many different solutions from the ones composed by three points forming an equilateral triangle plus two diametrically opposite points.

2. Show that there are some complex numbers $z$ which fulfill the equality

$$|z - a| + |z - b| = |c|,$$

   with $a, b, c \in \mathbb{C}$ given, if and only if $|a - b| \leq |c|$. In this case calculate the maximum and the minimum value of $|z|$ when $z$ belongs to this set of numbers, and also the points where these values are reached.

**3.** Find the value of the sums

$$\sum_{j=0}^{n-1} \sin\left(\frac{jk}{n}\, 2\pi\right) \quad \text{and} \quad \sum_{j=0}^{n-1} \cos\left(\frac{jk}{n}\, 2\pi\right)$$

when $n$ is a natural number and $k$ is an integer not divisible by $n$.

**4.** Prove that the transformation $z \to w$ defined for $z \in \mathbb{C}$ by $w = \alpha\bar{z}$, with $\alpha = |\alpha|e^{i\theta}$, is the symmetry with respect to the line which passes through the origin and has slope $\mathrm{tg}(\theta/2)$, followed by a dilation.

**5.** Find the relation between the coefficients of the equation $z^3 + az^2 + bz + c = 0$ so that the three solutions form an equilateral triangle.

**6.** Given three complex numbers $u, w, z$, consider the numbers $\rho = u + \gamma w + \gamma^2 z$ and $\tau = u + \gamma^2 w + \gamma z$, where $\gamma = e^{2i\pi/3}$.

   a) How do $\rho$, $\tau$ transform if a translation, a rotation centered at the origin or a dilation is applied to $u, z, w$?
   Show that if the numbers $\rho/\tau$ and $\rho'/\tau'$ associated to the points $u, w, z$ and $u', w', z'$, are equal, then the triangles $uwz$ and $u'w'z'$ are similar.

   b) Characterize the triangles $uwz$ for which $\rho = 0$ and $\tau = 0$.

**7.** Let $z_1, z_2, z_3$ be three different points of the complex plane and $R$ the radius of the circle that contains these points. Show the equality

$$\frac{1}{R^2} = \sum_{\sigma} \frac{1}{(z_{\sigma(2)} - z_{\sigma(1)})(\overline{z_{\sigma(3)}} - \overline{z_{\sigma(1)}})},$$

where $\sigma$ ranges over the group of permutations of the set $\{1, 2, 3\}$.

**8.** Find all the continuous homomorphisms from the additive group of $\mathbb{R}$ to the multiplicative group $\mathbb{T}$.

**9.** Show that if the mapping $\Phi\colon \mathbb{C} \to \mathbb{C}$ satisfies $|\Phi(z) - \Phi(w)| = m|z - w|$, with $m > 0$, for any pair of points $z, w \in \mathbb{C}$, then $\Phi(z) = \alpha + T(z)$ with $T$ an $\mathbb{R}$-linear mapping and $\alpha \in \mathbb{C}$. Thus $\Phi$ is either a conformal or an anticonformal mapping followed by a translation.

**10.** If $z, w \in \mathbb{C}^*$, denote by $d(z, w)$ the Euclidian distance in $\mathbb{R}^3$ between the corresponding points to $z, w$ by the stereographic projection onto the sphere $S^2$. Show the following formulae:

$$d(z, w) = \frac{2|z - w|}{\sqrt{(1 + |z|^2)(1 + |w|^2)}} \quad \text{if } z, w \in \mathbb{C},$$

$$d(z, \infty) = \frac{2}{\sqrt{1 + |z|^2}} \quad \text{if } z \in \mathbb{C}.$$

**11.** Show that if there is a branch of $\sqrt{z}$ in an open set $U$ of the plane with $0 \notin U$, there is also a branch of $\arg z$.

**12.** If $\gamma$ is the boundary of a square ranged counterclockwise and $z$ a point inside the square, show analytically that $\mathrm{Ind}(\gamma, z) = 1$.

**13.** Show that if $\gamma$ is a closed curve, then $\mathrm{Ind}(\gamma, z)$ is a continuous function from $z$ on each connected component of $\mathbb{C} \setminus \gamma^*$.

**14.** Let $\gamma_1(t), \gamma_2(t)$ be two closed curves parameterized in $[a, b]$ which do not pass through the origin such that

$$|\gamma_1(t) - \gamma_2(t)| < |\gamma_1(t)|, \quad t \in [a, b].$$

Prove that $\mathrm{Ind}(\gamma_1, 0) = \mathrm{Ind}(\gamma_2, 0)$.

**15.** Let $\gamma_n(t), a \le t \le b, n = 1, 2, \ldots$ be a sequence of closed curves and $\gamma(t)$, $a \le t \le b$, another closed curve such that $\gamma_n(t) \xrightarrow[n \to \infty]{} \gamma(t)$ uniformly on $[a, b]$. Show that if $z$ is a point that does not belong to $\gamma$, then

$$\mathrm{Ind}(\gamma, z) = \mathrm{Ind}(\gamma_n, z)$$

for $n \ge n_0$, $n_0$ big enough.

**16.** Give an example of a closed curve $\gamma$ such that $\mathrm{Ind}(\gamma, z)$ is a function of $z$ which is not bounded when $z \in \mathbb{C} \setminus \gamma^*$.

**17.** Let $\gamma(t), a \le t \le b$ be a closed curve with finite length. Show that there is a sequence of closed polygonal curves $P_n(t), a \le t \le b, n = 1, 2, \ldots$ such that:

    a) $P_n(t) \xrightarrow[n \to \infty]{} \gamma(t)$ uniformly on $[a, b]$.

    b) $L(P_n) \nearrow L(\gamma)$.

    c) $\lim_n \mathrm{Ind}(P_n, z) = \mathrm{Ind}(\gamma, z)$ if $z \in \mathbb{C} \setminus (\gamma^* \cup \bigcup_{n=1}^{\infty} P_n^*)$.

**18.** Let $\gamma(t), 0 \le t \le 2\pi$ be a $C^1$ closed curve such that $\gamma^* \subset \mathbb{T}$, $\mathrm{Ind}(\gamma, 0) = n$ and $L(\gamma) = 2\pi|n|$ for an integer $n$. Prove that if $\varphi(t), 0 \le t \le 2\pi$, is a branch of $\arg \gamma'(t)$, then $\varphi(t)$ is a monotonic function on the interval $[0, 2\pi]$, that is, $\gamma$ describes $|n|$ times $\mathbb{T}$ forward or backward.

**19.** Give an example of a closed Jordan curve $\gamma$, with $L(\gamma) = +\infty$.

**20.** Let $\gamma(t)$, with $a \le t \le b$, be a regular curve, that is, a curve of class $C^1$ with $\gamma'(t) \neq 0, a \le t \le b$. Show that:

a) There exist constants $m, \mu > 0$ and $\delta > 0$ such that

$$m|t - s| \leq |\gamma(t) - \gamma(s)| \leq \mu|t - s| \quad \text{if } t, s \in [a, b] \text{ and } |t - s| < \delta.$$

b) There exists a constant $K > 0$ such that

$$\int_{\{t : |\gamma(t) - z| < r\}} |\gamma'(t)| \, dt \leq Kr$$

for $z \in \mathbb{C}$ and $r > 0$.

**21.** A bounded domain of the plane with regular boundary which is simply connected is a Jordan domain. Is this statement true for every simply connected bounded domain?

**22.** Let $U$ be an open set of the plane such that $C^* \setminus U$ is connected. Show that each connected component of $U$ is a simply connected domain.

**23.** Let $\varphi \colon [a, b] \to \mathbb{R}$ be a function with continuous derivative. Show that the subgraph-type domain

$$U = \{(x, y), \quad a \leq x \leq b, \ 0 \leq y \leq \varphi(x)\}$$

is a simply connected domain with piecewise regular boundary.

# Chapter 2
# Functions of a complex variable

The aim of this chapter is to introduce the concept of function of a complex variable and to describe the most common ones, such as exponential, logarithmic and trigonometric functions. In doing so, we will extend differential calculus to the complex context. The most important way to define new functions of one complex variable is by means of power series. This is the reason why complex power series are studied in detail, stressing the fact that the natural domain of power series, even real ones, is in the complex plane.

The extension of the notion of derivative to functions of a complex variable leads to the concept of a holomorphic function, which is considered both from an analytic and a geometric point of view. Power series define holomorphic functions; but in fact there is a much deeper connection between holomorphic functions and power series because, as one will see in Chapter 4, any holomorphic function is locally the sum of a power series.

The study of functions locally expressed as a sum of a power series, called analytic functions, is then equivalent to the study of holomorphic functions. The last sections of this chapter are devoted to the study of both real and complex analytic functions.

## 2.1 Real variable polynomials, complex variable polynomials, rational functions

Usually, complex-valued functions $f : U \to \mathbb{C}$ defined on a domain $U$ in the complex plane will be considered. They may be represented as $f = u + iv$, where $u = \operatorname{Re} f, v = \operatorname{Im} f$ are real-valued functions defined on $U$.

The simplest functions are the *polynomials of the complex variable* $z$, $P(z) = a_0 + a_1 z + a_2 z^2 + \cdots + a_n z^n$, with $a_j \in \mathbb{C}$, which are functions defined on the whole complex plane $\mathbb{C}$. The number $n$ is called the *degree* of $P$ and is denoted by $\deg(P)$. Recall that due to the fundamental theorem of algebra we can write

$$P(z) = a_n(z - \alpha_1)(z - \alpha_2) \cdots (z - \alpha_n),$$

where the points $\alpha_i$ are the zeros of $P$, counting multiplicities. It is important to distinguish these polynomials from polynomials with two real variables $x$ and $y$. It is clear that if one has a polynomial $P(z)$ and replaces $z$ by $z = x + iy$, then one obtains a polynomial $P(x, y)$ with variables $x, y$ and complex coefficients; one may also write $P(z) = P_1(x, y) + iP_2(x, y)$ with $P_1, P_2$ real-valued polynomials;

for example,

$$z^2 = (x + iy)^2 = x^2 - y^2 + i2xy.$$

But not every polynomial in $x$, $y$ with complex coefficients (or equivalently every polynomial $P_1 + iP_2$ with $P_1$, $P_2$ real-valued) is a polynomial in $z$; for example, $x^2 + y^2$ is not. In fact, one will see below that any polynomial that only takes real values cannot be a polynomial in $z$.

How may one recognize if a polynomial $P(x, y)$ is, in fact, a polynomial $P(z)$ in $z$? This question may be solved algebraically, finding $P(z)$ directly: it is enough to express $x$, $y$ in terms of $z$, $\bar{z}$,

$$x = \frac{z + \bar{z}}{2}, \qquad y = \frac{z - \bar{z}}{2i},$$

substitute these variables in the polynomial $P(x, y)$, and check that there is no term in $\bar{z}$. Another answer for this question is the following:

Let $\gamma(t) = x(t) + iy(t)$ be a complex function of the real variable $t$. In (1.2) we have already defined $\gamma'(t)$, the derivative of this function, as

$$\gamma'(t) = \lim_{h \to 0} \frac{\gamma(t + h) - \gamma(t)}{h}$$

in the points where this limit exists. In this case it reads $\gamma'(t) = x'(t) + iy'(t)$. The formal properties of the derivative $\gamma'(t)$ are the same as in the case of real-valued functions. Hence, for example, if $m \in \mathbb{N}$ then

$$(\gamma(t)^m)' = m\gamma(t)^{m-1}\gamma'(t)$$

which is a consequence of equality (1.3). Then it follows that if $P(z) = \sum_{l=1}^{n} a_l z^l$ is a polynomial in $z$, one has that

$$[P(\gamma(t))]' = \gamma'(t) \sum_l l a_l \gamma(t)^{l-1}.$$

If now, as natural, one denotes by $P'(z)$ the polynomial $\sum_l l a_l z^{l-1}$, one has the following rule:

$$[P \circ \gamma]' = \gamma' P'(\gamma).$$

It turns out that the polynomials in $z$ are the only ones which have this property. Actually, suppose that $P(x, y) = \sum_{k,j} a_{k,j} x^k y^j$ has the property

$$[P \circ \gamma]' = \gamma' Q(\gamma),$$

for some polynomial $Q$ and for all differentiable functions $\gamma(t)$. Taking $\gamma(t)$ parallel to the real axis, $\gamma(t) = t + ib, b \in \mathbb{R}$, one has that $\frac{\partial P}{\partial x} = Q$. Taking $\gamma(t)$ parallel to the imaginary axis, $\gamma(t) = a + it, a \in \mathbb{R}$, one has $\frac{\partial P}{\partial y} = iQ$. Then,

$$\frac{\partial P}{\partial y} = i\frac{\partial P}{\partial x}, \tag{2.1}$$

that is, $ja_{k,j} = i(k+1)a_{k+1,j-1}$. Introducing the numbers $b_{k,j} = k!j!a_{k,j}(-i)^j$, the previous equation means that $b_{k+1,j-1} = b_{k,j}$, that is, by iteration, that $b_{k,j}$ just depend on the value of $k + j = l$. Hence, $b_{k,j} = b_{k+j,0}$. One then has $k!j!a_{k,j} = (k+j)!a_{k+j,0}i^j$, that is,

$$a_{k,j} = \binom{k+j}{k} i^j a_{k+j,0},$$

and so $P(z) = \sum_l a_{l,0}z^l$. Consequently, one has shown analytically that the relation $[P \circ \gamma]' = \gamma' Q(\gamma)$ for a polynomial $Q$ and for any $\gamma(t)$ holds if and only if $P$ is a polynomial in $z$, and in this case $Q = P'$.

The *rational functions* are those of type

$$R(z) = \frac{P(z)}{Q(z)},$$

where $P$, $Q$ are polynomials in $z$. Simplifying the possible reducible common factors, one may suppose that $P$, $Q$ have no common zeros; in this case, the function $R$ is not defined in the set $Z(Q)$ of the zeros of $Q$, which are also called *poles of R*. Hence, $R$ is a continuous function on $\mathbb{C} \setminus Z(Q)$. Every rational function $R$ has a unique decomposition as a sum of a polynomial (which only appears in the case that the degree of $P$ is greater than or equal to the degree of $Q$) and simple fractions, that is, of the kind

$$\frac{a}{(z-\alpha)^k},$$

where $\alpha \in Z(Q)$, $a$ is a constant and $k$ is less than or equal to the multiplicity of $\alpha$ as a zero of $Q$. This result may be proved in a purely algebraic way, but we will also give an analytic proof later on (see Theorem 5.13).

Observe that any rational function, $R = P/Q$, defines a continuous function from $\mathbb{C}^*$ to $\mathbb{C}^*$ taking $R(a) = \infty$ when $a$ is a pole of $R$ and $R(\infty) = \infty$ if $\deg(P) > \deg(Q)$, $R(\infty) = \lim_{|z|\to\infty} R(z)$ otherwise (this will be zero if $\deg(Q) > \deg(P)$ and the ratio of the leading coefficients if they are equal).

## 2.2 Complex exponential functions, logarithms and powers. Trigonometric functions

### 2.2.1 The exponential function

The exponential function of a real variable, $e^x$, is familiar from calculus. The exponential function when the exponent is a pure imaginary number, $e^{iy}$, has been defined in Subsection 1.4.1, as

$$e^{iy} = \cos y + i \sin y. \tag{2.2}$$

The natural extension to any complex exponent is the following:

**Definition 2.1.** For $z = x + iy$, the exponential function of $z$ with basis $e$ is defined as

$$e^z = e^x(\cos y + i \sin y).$$

Using the formulae (1.1) for the sine and cosine of a sum, one may check immediately the following rule:

$$e^{z+w} = e^z \cdot e^w, \quad \text{if } z, w \in \mathbb{C}.$$

This relation is the fundamental property of the exponential function. In particular, one has that $e^{-z} = 1/e^z$ and $e^z \neq 0$, for all $z \in \mathbb{C}$. The equation $e^z = 1$ has infinite solutions: $z = 2\pi ki$, $k \in \mathbb{Z}$. The exponential function is $2\pi i$-periodic: $e^{z+2\pi ki} = e^z$, $k \in \mathbb{Z}$. Furthermore, $|e^z| = e^{\operatorname{Re} z}$, $\operatorname{Im} z \in \arg e^z$ and $e^z = e^w$ if and only if $z - w$ is an entire multiple of $2\pi i$.

The exponential map is bijective from the strip $B = \{z : -\pi < \operatorname{Im} z \leq \pi\}$ to $\mathbb{C} \setminus \{0\}$ and so it is also injective in all horizontal strips of width less than $2\pi$. The horizontal line $y = y_0$ is transformed by the exponential function into the infinite open ray starting at the origin and making an angle $y_0$ with the real axis. The vertical line $x = x_0$ is transformed into the circle centered at the origin and with radius $e^{x_0}$ (Figure 2.1).

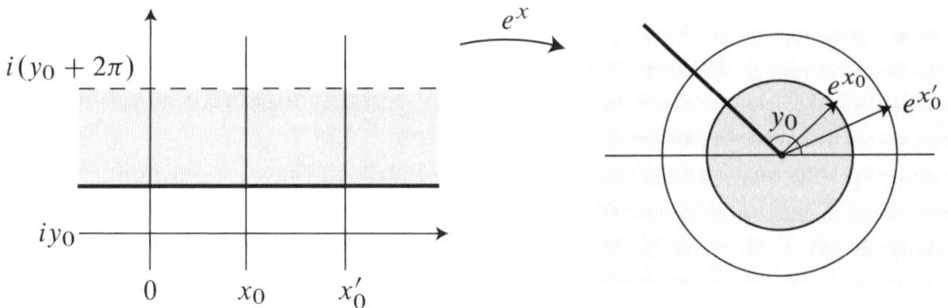


Figure 2.1

A line $y = mx$ with slope $m \neq 0$ is transformed into the curve with parametric equation $x \to e^x \cdot e^{imx}$, which is a spiral (Figure 2.2).

## 2.2.2 Logarithms

As important as the exponential function is its inverse, the logarithm function.

**Definition 2.2.** For $z \in \mathbb{C}$, $z \neq 0$, the logarithm of $z$, $\log z$, is any complex number $w$ such that $e^w = z$.

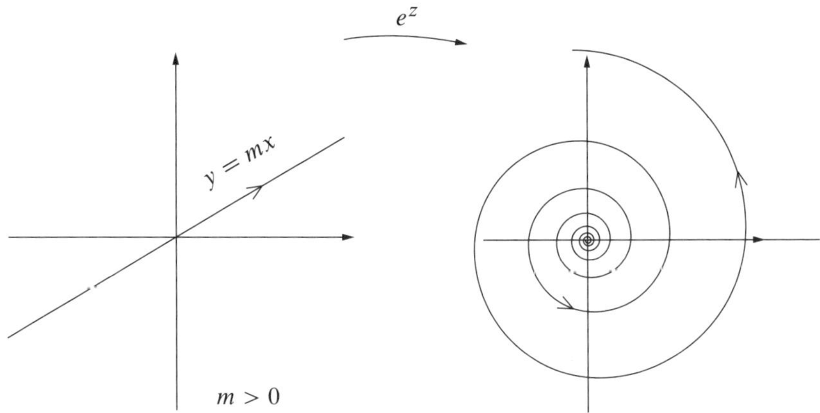


Figure 2.2

The real exponential function is a bijection from $\mathbb{R}$ onto $\mathbb{R}^+$, and so only the positive real numbers have a real logarithm. This defines the (natural) logarithm function $\mathrm{Log}\colon \mathbb{R}^+ \to \mathbb{R}$, inverse of the exponential function. In the complex case, solving the equation $e^w = z$, equivalent to $e^{\mathrm{Re}\, w} e^{i\, \mathrm{Im}\, w} = |z| e^{i\, \arg z}$, it is seen that all the logarithms have the same real part, $\mathrm{Log}\, |z|$, and that its imaginary part is an argument of $z$,

$$\log z = \mathrm{Log}\, |z| + i \arg z.$$

Hence two logarithms of $z$ differ in an entire multiple of $2\pi i$.

**Definition 2.3.** Among all the logarithms of $z$, the one defined by $\mathrm{Arg}\, z$ is called the principal logarithm and it is denoted by $\mathrm{Log}\, z$,

$$\mathrm{Log}\, z = \mathrm{Log}\, |z| + i \, \mathrm{Arg}\, z.$$

When $z$ is a positive real number, this logarithm coincides with the natural logarithm. Hence $\mathrm{Log}$ is a complex function defined in $\mathbb{C} \setminus \{0\}$ and, as in the case of the principal argument function, it is discontinuous on the negative half of the real axis. The rule

$$\log zw = \log z + \log w$$

is valid if interpreted as equality between sets. But the rule $\mathrm{Log}\, zw = \mathrm{Log}\, z + \mathrm{Log}\, w$ is not; for example, for $z = w = -1 - i$ one has $\mathrm{Log}\, z = \mathrm{Log}\, w = \mathrm{Log}\, \sqrt{2} - \frac{3\pi}{4} i$, while for $zw = 2i$ one has $\mathrm{Log}\, zw = \mathrm{Log}\, 2 + \frac{\pi}{2} i$.

As done in the case of the argument, one will speak about *continuous branches of the logarithm*.

**Definition 2.4.** A continuous branch of the logarithm in a connected set $E \subset \mathbb{C}$, not containing zero, is a continuous function $g$ on $E$ such that $e^{g(z)} = z$, if $z \in E$.

More generally, if $f : E \to \mathbb{C}$ is a function defined on any set $E$ and $f(z) \neq 0$ for $z \in E$, then a continuous branch of $\log f$ is a continuous function $g$ defined on $E$ such that $e^{g(z)} = f(z)$, $z \in E$.

The equivalence between continuous branches of the logarithm and continuous branches of the argument is obvious, since the argument is the imaginary part of the logarithm and the real part is completely determined. For example, Log is a continuous branch of the logarithm in the complement of non-positive numbers. In the complement of any infinite ray, and in particular, in any disc not containing zero, there are continuous branches of the logarithm. In Theorem 1.23 the case of curves has been examined, that is, when $E$ is an interval. Also, as a consequence of Theorem 1.23 there is a continuous branch $h(t)$ of the logarithm of $\gamma(t)$ for all continuous curves avoiding the origin. Moreover if $\gamma$ is differentiable, then $h$ is also differentiable and $h'(t) = \gamma'(t)/\gamma(t)$.

### 2.2.3 Complex powers

**Definition 2.5.** Given $z, w \in \mathbb{C}$, $z \neq 0$, we define $z^w$ as the set of complex numbers

$$z^w = e^{w \log z} \quad \text{where } \log z \text{ is any logarithm of } z. \tag{2.3}$$

When $\log z$ is the principal branch, $e^{w \, \mathrm{Log} \, z}$ is called the principal branch of $z^w$. In general, if $f : E \to \mathbb{C}$ is a function defined on a connected set $E$ with $f(z) \neq 0$, $z \in E$, and $w \in \mathbb{C}$, a continuous branch of $f^w$ is a continuous function $g : E \to \mathbb{C}$ such that $g(z) \in [f(z)]^w$, $z \in E$.

Clearly if there is a continuous branch $h$ of $\log f$, then $g(z) = e^{wh(z)}$ is a continuous branch of $f^w$. Since two logarithms of $z$ differ on an entire multiple of $2\pi i$, two possible values of $z^w$ differ on a factor $e^{2k\pi i w}$, $k \in \mathbb{Z}$ and have, then, the same modulus, which is $|z|^w$, if $w$ is a real number. If $w$ is an integer $m$, all these factors are 1. In this case the set of powers is reduced either to the unique value $z^m = z \cdots z$ ($m$ times) if $m > 0$, or $z^m = z^{-1} \cdots z^{-1}$ if $m < 0$. If $w$ is rational with irreducible expression $\frac{n}{m}$, then Definition 2.5 coincides with the one given in Subsection 1.1.2 and $z^{n/m}$ is a finite set of $m$ numbers. If $w$ is irrational, then

$$z^w = e^{w \, \mathrm{Log} \, |z|} \cdot e^{i w \, \mathrm{Arg} \, z} \cdot e^{2k\pi i w}, \quad k \in \mathbb{Z},$$

which is a dense sequence of points in the circle centered at 0 with radius $|z|^w$. If $w = it$ is pure imaginary, one has

$$z^w = e^{it \, \mathrm{Log} \, |z|} \cdot e^{-t \, \mathrm{Arg} \, z} \cdot e^{-t 2k\pi}, \quad k \in \mathbb{Z}.$$

**Example 2.6.** $1^w$ is the set of complex numbers $e^{2\pi i k w}$, $k \in \mathbb{Z}$, and if $w = it$, this set is formed by the real numbers $e^{-2\pi t k}$, $k \in \mathbb{Z}$. $\qquad\square$

Observe that, in general, one has $\log z^w = w \log z + 2\pi \mathbb{Z}i$, understood as an equality of sets. Arithmetic rules such as

$$(z^w)^\eta = z^{w\eta} \tag{2.4}$$

must be properly interpreted. The left-hand side term is $\alpha^\eta$, with $\alpha = z^w$; hence,

$$(z^w)^\eta = e^{\eta \log \alpha} = e^{\eta(w \log z + 2\pi \mathbb{Z}i)} = e^{\eta w \log z + 2\pi \eta \mathbb{Z}i}.$$

From here it is deduced that the set $(z^w)^\eta$ contains the set $z^{w\eta}$, and they are the same only if $\eta$ is an integer. Only if $w, \eta$ are both integers, (2.4) is a valid equality for numbers. Applying it in other cases may lead to absurd conclusions. For example: assume $z \neq 0$; then $z = e^w$ for a certain $w$; let $\eta = w/2\pi i$. Then $z = e^w = e^{2\pi i \eta} = (e^{2\pi i})^\eta = 1^\eta = 1$!!

### 2.2.4 Trigonometric functions

From Euler's identity (2.2) it follows that

$$\cos x = \frac{e^{ix} + e^{-ix}}{2}, \quad \sin x = \frac{e^{ix} - e^{-ix}}{2i}.$$

So the extension of these functions to the complex field is naturally defined by

$$\cos z = \frac{e^{iz} + e^{-iz}}{2}, \quad \sin z = \frac{e^{iz} - e^{-iz}}{2i}, \quad z \in \mathbb{C}. \tag{2.5}$$

It easily follows that $\cos z$, $\sin z$ are $2\pi$-periodic, that is, $\cos(z + 2\pi k) = \cos z$ and $\sin(z + 2\pi k) = \sin z$, $k \in \mathbb{Z}$. It is also immediate to check

$$\cos(z + w) = \cos z \cos w - \sin z \sin w,$$
$$\quad\quad z, w \in \mathbb{C}, \tag{2.6}$$
$$\sin(z + w) = \sin z \cos w + \cos z \sin w,$$

as well as the fundamental relation

$$\cos^2 z + \sin^2 z = 1, \quad z \in \mathbb{C}.$$

Unlike the real case, the functions $\sin z$, $\cos z$ are not bounded functions. For example, $\cos ix = \frac{e^{-x} + e^x}{2} \to \infty$ when $x \to \pm\infty$.

### 2.2.5 Hyperbolic functions

In a similar way we can also consider the *hyperbolic functions of a complex variable $z$*, defined by

$$\operatorname{ch} z = \frac{e^z + e^{-z}}{2}, \quad \operatorname{sh} z = \frac{e^z - e^{-z}}{2}, \quad z \in \mathbb{C}.$$

These functions are linked to trigonometric functions by the relations

$$\operatorname{ch} z = \cos iz, \quad \operatorname{sh} z = -i \sin iz, \quad z \in \mathbb{C}, \tag{2.7}$$

and satisfy the equality $\operatorname{ch}^2 z - \operatorname{sh}^2 z = 1$. The addition formulae

$$\operatorname{ch}(z + w) = \operatorname{ch} z \cdot \operatorname{ch} w + \operatorname{sh} z \cdot \operatorname{sh} w,$$
$$\operatorname{sh}(z + w) = \operatorname{sh} z \cdot \operatorname{ch} w + \operatorname{ch} z \cdot \operatorname{sh} w, \qquad z, w \in \mathbb{C},$$

are also valid. Applying addition formulae (2.6) to $\cos z = \cos(x + iy)$ and keeping in mind the relations (2.7), one has that

$$\operatorname{Re}\cos(x + iy) = \cos x \cdot \operatorname{ch} y, \quad \operatorname{Im}\cos(x + iy) = -\sin x \cdot \operatorname{sh} y,$$

$$\operatorname{Re}\sin(x + iy) = \sin x \cdot \operatorname{ch} y, \quad \operatorname{Im}\sin(x + iy) = \cos x \cdot \operatorname{sh} y.$$

Now one may calculate $|\cos z|$ and $|\sin z|$. Writing always $z = x + iy$ it turns out that $|\cos z|^2 = (\cos x \operatorname{ch} y)^2 + (\sin x \operatorname{sh} y)^2 = \operatorname{ch}^2 y - \sin^2 x(\operatorname{ch}^2 y - \operatorname{sh}^2 y) = \operatorname{ch}^2 y - \sin^2 x$. That is,

$$|\cos z| = \sqrt{\operatorname{ch}^2 y - \sin^2 x} = \sqrt{\cos^2 x + \operatorname{sh}^2 y},$$

and analogously,

$$|\sin z| = \sqrt{\operatorname{sh}^2 y + \sin^2 x} = \sqrt{\operatorname{ch}^2 y - \cos^2 x}.$$

From these equalities we get the estimates

$$|\operatorname{sh} y| \le |\cos z| \le \operatorname{ch} y,$$

$$|\operatorname{sh} y| \le |\sin z| \le \operatorname{ch} y,$$

which show that $|\cos z|, |\sin z| \to \infty$ when $|y| \to \infty$, that is, when one travels away from the real axis. One may also deduce that the growth of $|\cos z|, |\sin z|$ is faster than the one of $|\operatorname{sh} y|$ and slower than the one of $|\operatorname{ch} y|$.

## 2.3  Power series

A very important way of defining new functions is using *power series of the complex variable z*. The natural domain of definition of power series, including the ones of a real variable, is in the complex plane, as will be seen.

### 2.3.1 Series of complex numbers

The series $\sum_{n=0}^{\infty} z_n$ with general term $z_n \in \mathbb{C}$ is said to be *convergent* with sum $S$ if the partial sums $\sum_{n=0}^{N} z_n$ have limit $S$, when $N \to \infty$. If $z_n = x_n + i y_n$ and $S = A + iB$, this is equivalent to the fact that the series of real numbers $\sum x_n$, $\sum_n y_n$ are convergent with sums $A$, $B$, respectively.

The series $\sum z_n$ is said to be *absolutely convergent* if the series of modulus $\sum |z_n|$ converges; once again this is equivalent to the series of real and imaginary parts being absolutely convergent.

**Example 2.7.** Consider the series $\sum_{n=1}^{\infty} \frac{(-1)^n}{n+i}$. Its real part is $\sum \frac{(-1)^n n}{1+n^2}$, an alternating series whose general term has modulus decreasing to zero, whence convergent; the modulus of the general term is comparable to $\frac{1}{n}$, and so the real part is not absolutely convergent. The imaginary part is $\sum \frac{1}{1+n^2}$, with positive general term comparable to $\frac{1}{n^2}$, and so absolutely convergent. Then, the complex series is convergent, but not absolutely convergent. $\qquad\square$

The following result dealing with the different ways of summing a series will be used later on.

**Proposition 2.8.** a) *A series of complex terms is absolutely convergent with sum $S$ if and only if all its rearrangements are convergent with the same sum $S$.*

b) *An absolutely convergent series may be summed in blocks arbitrarily.*

The meaning of this proposition is that if one has a countable family $\{z_\alpha\}$, $\alpha \in A$ such that, in a certain order, their moduli have finite sum, then this family has a well-defined sum $\sum_{\alpha \in A} z_\alpha$, independently of the ordering used (it is said to be a *summable family*). Furthermore, one may calculate the sum using arbitrary blocks; that is, if $A$ is the disjoint union of the sets $A_i$, $i \in I$, then $\sum_{\alpha \in A} z_\alpha = \sum_{i \in I} \sum_{\alpha \in A_i} z_\alpha$.

*Proof.* In part a) it will be only shown that if $\sum z_n$ is absolutely convergent with sum $S$, then any rearrangement is also absolutely convergent with the same sum. The converse for real series may be found in [10], p. 76, and follows for complex series separating real and imaginary parts. Let $\sum w_n$ be a rearrangement of the series $\sum z_n$, supposed absolutely convergent with sum $S$, and write $S_n = \sum_{k=1}^{n} z_k$, $T_n = \sum_{k=1}^{n} w_k$. If $S_m$ contains all the terms of $T_n$, then $\sum_{k=1}^{n} |w_k| \leq \sum_{k=1}^{m} |z_k| \leq \sum_{k=1}^{\infty} |z_n| < +\infty$, so $\sum w_n$ is also absolutely convergent. Now, given $\varepsilon > 0$, choose $n \in \mathbb{N}$ in such a way that

$$|S - S_n| \leq \sum_{k=n+1}^{\infty} |z_k| < \varepsilon.$$

If $T_m$ contains all the terms of $S_n$ and $r \geq m$, one has

$$|T_r - S_n| < \varepsilon$$

and finally,

$$|S - T_r| \leq |S - S_n| + |S_n - T_r| < 2\varepsilon \quad \text{if } r \geq m,$$

that is, $\sum w_n = S$.

For part b), start observing that, since the complete series $\sum_{\alpha \in A} z_\alpha$ is absolutely convergent, it is clear that each block $\sum_{\alpha \in A_i} z_\alpha$ is also absolutely convergent. Write now $S_i = \sum_{\alpha \in A_i} z_\alpha$ if $i \in I$; the series $\sum_{i \in I} S_i$ is also absolutely convergent because if the indexes of $I$ are rearranged in some way, say $i_1, i_2, \ldots, i_n, \ldots$ and one takes a finite sum, then

$$\sum_{k=1}^{n} |S_{i_k}| \leq \sum_{\alpha \in A_{i_1}} |z_\alpha| + \cdots + \sum_{\alpha \in A_{i_n}} |z_\alpha| \leq \sum_{\alpha \in A} |z_\alpha| < +\infty.$$

Order now the indexes of $A$ in the sequence $\{\alpha_1, \alpha_2, \ldots, \alpha_n, \ldots\}$, and given $\varepsilon > 0$, choose $N$ such that

$$\sum_{n > N} |z_{\alpha_n}| < \varepsilon.$$

Now one can find a finite set of indexes $i_1, i_2, \ldots, i_{r_0}$ of $I$ such that $\{\alpha_1, \ldots, \alpha_N\} \subset A_{i_1} \cup \cdots \cup A_{i_{r_0}}$. Then for any $r \geq r_0$ one has

$$\left| S_{i_1} + \cdots + S_{i_r} - \sum_n z_{\alpha_n} \right| \leq \left| S_{i_1} + \cdots + S_{i_r} - \sum_{n \leq N} z_{\alpha_n} \right|$$

$$+ \left| \sum_{n \leq N} z_{\alpha_n} - \sum_n z_{\alpha_n} \right| < 2\varepsilon.$$

This means that $\sum_{\alpha \in A} z_\alpha = \sum_{i \in I} S_i = \sum_{i \in I} \sum_{\alpha \in A_i} z_\alpha.$ $\qquad\square$

To illustrate how to apply these properties consider now the *double series* $\sum_{n,m \in \mathbb{N}} z_{n,m}$. The numbers $\{z_{n,m}\}$ may be thought as the entries of an infinite matrix, with $n$ representing the index for rows and $m$ for columns. One possibility is to sum the terms of the matrix first by rows, $\sum_m z_{n,m}$, and after that, to sum up the results obtained, $\sum_n \sum_m z_{n,}$. Alternatively, one may sum first by columns and then sum up the results. In general both are not equal, that is, the identity

$$\sum_n \sum_m z_{n,m} = \sum_m \sum_n z_{n,m} \tag{2.8}$$

is not true; it could happen that one term makes sense but not the other one, or even that both make sense but have different values. For example, taking $z_{n,m} = 1$ if $n = m + 1$, $z_{n,m} = -1$ if $n = m - 1$ and $z_{mn} = 0$ in other cases, it turns out that $\sum_{n=1}^{\infty} \sum_{m=1}^{\infty} z_{n,m} = -1$ but $\sum_{m=1}^{\infty} \sum_{n=1}^{\infty} z_{mn} = 1$. From Proposition 2.8 it follows that if either $\sum_n \sum_m |z_{n,m}|$ or $\sum_m \sum_n |z_{n,m}|$ is finite, then both members of (2.8) make sense and are equal. This statement is, in fact, Fubini's theorem for series. As an application of this result the multiplication of series of complex terms will be studied now:

**Proposition 2.9.** *Suppose that $\sum_n a_n$ and $\sum_n b_n$ are absolutely convergent with respective sums $A$ and $B$. Then the series $\sum_k c_k$, where $c_k = \sum_{n=0}^{k} a_n b_{k-n}$, is absolutely convergent with sum $AB$.*

*Proof.* Obviously $\sum_k |c_k| \leq \sum_{n,m} |a_n||b_m| = (\sum_n |a_n|)(\sum_m |b_m|) < \infty$, so the series $\sum c_k$ is absolutely convergent. This shows also that the double series $\sum_{n,m} a_n b_m$ is summable (absolutely convergent); summing first in $m$ and then in $n$ we get $AB$, and grouping $a_n b_m$ in blocks defined by $n + m = k$, $k$ fixed, we get $\sum c_k$. $\qquad\square$

The series $\sum_k c_k$ of the previous proposition is called *Cauchy's product* of the series $\sum_n a_n$ and $\sum_n b_n$.

### 2.3.2 Function series

We will need uniform convergence criteria for function series $\sum_n f_n(p)$, where the functions $f_n$ are defined on an arbitrary set $X$. Recall that the concept of uniform convergence is the one that guarantees the continuity of the sum in the case that a topology is defined on $X$ and that each function $f_n$ is continuous on $X$.

**Theorem 2.10** (Weierstrass $M$-test). *If $|f_n(p)| \leq M_n$, for all $p \in X$, $n \geq 1$ and $\sum_n M_n < +\infty$, then the series $\sum_n f_n(p)$ is uniformly convergent on $X$.*

*Proof.* We can check that Cauchy's uniform convergence criterion is satisfied. Indeed, if $m > n$ then for a fixed $\varepsilon > 0$,

$$\left| \sum_n^m f_k(p) \right| \leq \sum_n^m |f_k(p)| \leq \sum_{k=n}^{\infty} M_k < \varepsilon$$

for $n$ big enough, because $\sum M_n < +\infty$. $\qquad\square$

For example, with this criterion one can see that the series $\sum_{n=0}^{\infty} a^n \cos(b^n \pi x)$ with $a, b \in \mathbb{R}$ and $0 < a < 1$ is uniformly convergent on $\mathbb{R}$, and so it defines a continuous function on the real line. Weierstrass used the previous series with $b$ an odd integer and $ab > 1 + 3\pi/2$ to give an example of a continuous function on $\mathbb{R}$, not differentiable at any point.

*Abel's summation by parts formula* is needed to establish a couple of convergence criteria.

**Lemma 2.11.** *Let $(a_n)$, $(b_n)$ be two sequences of complex numbers and write $A_n = a_1 + a_2 + \cdots + a_n$. Then*

$$\sum_{k=1}^{n} a_k b_k = A_n b_{n+1} - \sum_{k=1}^{n} A_k (b_{k+1} - b_k), \quad n \in \mathbb{N}. \tag{2.9}$$

*Proof.* Setting $A_0 = 0$, one has

$$\sum_{k=1}^{n} a_k b_k = \sum_{k=1}^{n} (A_k - A_{k-1}) b_k$$

$$= \sum_{k=1}^{n} A_k b_k - \sum_{k=1}^{n-1} A_k b_{k+1} = \sum_{k=1}^{n} A_k (b_k - b_{k+1}) + A_n b_{n+1}. \qquad \square$$

Abel's summation by parts formula is the discrete version of integration by parts formula. It may be better seen writing it in terms of $A_n$ and $b_n$, as in the previous proof, that is,

$$\sum_{k=1}^{n} (A_k - A_{k-1}) b_k = A_n b_{n+1} - \sum_{k=1}^{n} A_k (b_{k+1} - b_k).$$

This equality corresponds to the formula

$$\int_a^b A'(x) b(x) \, dx = A(x) b(x) \Big|_a^b - \int_a^b A(x) b'(x) \, dx,$$

if we consider the sequences $(A_k - A_{k-1})$ and $(b_{k+1} - b_k)$ as the derivatives of the sequences $(A_k)$ and $(b_k)$, respectively.

**Theorem 2.12** (Dirichlet's test). *Consider a function series $\sum_n f_n(p) g_n(p)$ where the functions $f_n(p)$ are complex-valued and the functions $g_n(p)$ are real-valued for $p \in X$, $n \geq 1$. Denote by $F_n(p) = \sum_{k=1}^{n} f_k(p)$ the $n$-th partial sum of $\sum f_n(p)$ and suppose that there is a constant $M \geq 0$ such that $|F_n(p)| \leq M$, for $n \geq 1, p \in X$. Suppose also that the sequence $(g_n(p))$ is monotonically decreasing, $g_n(p) \geq g_{n+1}(p)$, $p \in X$, $n \geq 1$ and that converges uniformly to zero on $X$. Then the series $\sum_n f_n(p) g_n(p)$ is uniformly convergent on $X$.*

*Proof.* Fix $\varepsilon > 0$. By (2.9), if $n < m$, one has

$$\sum_{k=n}^{m} f_k(p)g_k(p) = \left(\sum_{k=n}^{m} f_k(p)\right)g_{m+1}(p) - \sum_{k=n}^{m}\left(\sum_{\ell=n}^{k} f_\ell(p)\right)(g_{k+1}(p) - g_k(p))$$

(2.10)

and this expression has its modulus dominated by

$$2M g_{m+1}(p) + 2M \sum_{k=n}^{m}(g_k(p) - g_{k+1}(p)) = 2M g_n(p)$$

and then is smaller than $\varepsilon$, for any $p \in X$, if $n$ is big enough. $\qquad\square$

**Example 2.13.** Summing up a geometric sequence and applying the expression of $\sin x$ from (2.5) it turns out immediately that

$$\sum_{k=1}^{n} e^{ikx} = e^{ix}\frac{1 - e^{inx}}{1 - e^{ix}} = \frac{\sin(nx/2)}{\sin(x/2)} \cdot e^{i(n+1)x/2}.$$

Now taking $0 < \delta < \pi$, it follows that

$$\left|\sum_{k=1}^{n} e^{ikx}\right| \le \frac{1}{|\sin(x/2)|} \le \frac{1}{\sin \delta/2} \quad \text{if } \delta \le x \le 2\pi - \delta.$$

Choosing $f_n = e^{inx}$ and $g_n = \frac{1}{n}$ Dirichlet's test guarantees that the series

$$\sum_{n=1}^{\infty} \frac{e^{inx}}{n}$$

converges uniformly on $[\delta, 2\pi - \delta]$, for any $\delta > 0$, $\delta < \pi$. Observe that the Weierstrass $M$-test cannot be applied, since $|e^{inx}| = 1$ and $\sum \frac{1}{n} = +\infty$. $\qquad\square$

**Theorem 2.14** (Abel's test). *Consider a function series $\sum_n f_n(p)g_n(p)$ where the functions $f_n(p)$ and $g_n(p)$ are complex-valued, and suppose that $\sum_n f_n(p)$ is uniformly convergent on $X$ and that there is a number $M \ge 0$ such that, for $p \in X$,*

$$|g_1(p)| + \sum_{n \ge 1} |g_n(p) - g_{n+1}(p)| \le M.$$

*Then the series $\sum_n f_n(p)g_n(p)$ converges uniformly on $X$.*

*Proof.* Given $\varepsilon > 0$, one has $\left|\sum_{k=n}^{m} f_k(p)\right| \le \varepsilon$, if $m > n$ are big enough, for $p \in X$. Now the modulus of (2.10) is bounded by

$$\varepsilon|g_{m+1}(p)| + \varepsilon \sum_{k=n}^{m} |g_k(p) - g_{k+1}(p)| \le 2\varepsilon M,$$

because $|g_\ell(p)| \le M$, for all $\ell \in \mathbb{N}$, $p \in X$, as may be seen by writing $g_\ell(p) = g_1(p) + \sum_{m=1}^{\ell-1}(g_{m+1}(p) - g_m(p))$ and taking absolute value. $\qquad\square$

The hypothesis on the sequence $(g_n)$ in Abel's test is satisfied when it is a monotone uniformly bounded sequence of real-valued functions on $X$.

Clearly, when $X$ is a point, the previous tests become tests for convergence of numerical series. For example, the result according to which an *alternating series* of type $\sum_n (-1)^n a_n$, with $a_n \searrow 0$ is convergent, may be seen as a consequence of Dirichlet's test. More generally, if $(f_n)$ is a decreasing sequence of functions with uniform limit 0 on $X$, then the alternating series $\sum_n (-1)^n f_n(p)$ converges uniformly on $X$.

### 2.3.3 Domain of convergence of a power series

A specially important type of function series are the *power series*.

**Definition 2.15.** A power series centered at a point $a \in \mathbb{C}$ is a function series of the form

$$\sum_{n=0}^{+\infty} c_n (z-a)^n, \quad c_n \in \mathbb{C}.$$

So, a power series is a function series $\sum_n f_n(z)$ with $f_n(z) = c_n(z-a)^n$ a monomial of degree $n$, and the series may be seen as an "infinite degree polynomial". For each value of $z \in \mathbb{C}$ one has a complex number series. Beyond formal aspects, the interesting feature is how to find the *domain of convergence of the power series*,

$$E = \{z \in \mathbb{C} : \text{the series } \sum_n c_n(z-a)^n \text{ is convergent}\},$$

and the properties of the function $f$ defined on $E$ by $f(z) = \sum_n c_n(z-a)^n$. So, $z \in E$ if the sequence of partial sums $\sum_{n=0}^N c_n(z-a)^n$ is convergent in $\mathbb{C}$. Only in particular cases may the partial sums be calculated explicitly and the set $E$ may be obtained from the definition. For example, for the *geometric series* centered at the origin, $\sum_n z^n$, one has

$$\sum_{n=0}^N z^n = \frac{1 - z^{N+1}}{1 - z} \quad \text{if } z \ne 1,$$

and if $z = 1$, the sum is $N + 1$. Hence, the series is convergent if and only if $z^{N+1}$ is convergent, that is, if $|z| < 1$. So, $E = \mathbb{D}$ and the value of the sum is $1/(1-z)$.

Some information about the structure of $E$ is given in the following theorem. Recall that the *upper limit* $\rho$ of a sequence $(x_n)$ of real numbers is defined as

$$\rho = \inf_n \sup_{k \ge n} x_k = \lim_n \sup_{k \ge n} x_k$$

where $\rho = +\infty$ if the sequence is not bounded from above. We will write $\rho = \limsup_n x_n$. When it is finite, $\rho$ is the unique number with the two following properties:

1) For each $\varepsilon > 0$, infinitely many terms $x_n$ of the sequence are greater than $\rho - \varepsilon$.

2) For each $\varepsilon > 0$, all the terms of the sequence after one of them are smaller than $\rho + \varepsilon$.

It turns out that $\rho$ is the maximum of the set of cluster points of the sequence $(x_n)$, that is, in the set of the numbers $m \le +\infty$ for which there is a partial sequence $(x_{k_n})$ convergent to $m$, when $k_n \to \infty$. If the sequence $(x_n)$ is convergent to $m \le +\infty$, then $\rho = m$; if the sequence is the finite union of convergent subsequences, $\rho$ is the biggest of the respective limits.

**Theorem 2.16.** *If $\sum_n c_n(z - a)^n$ is a power series and one writes $R = \frac{1}{\rho}$ where $\rho = \limsup_n |c_n|^{\frac{1}{n}}$ (interpreting that $R = 0$ if $\rho = +\infty$ and $R = +\infty$ if $\rho = 0$), then the series converges uniformly on compact sets of the open disc $D(a, R)$, converges absolutely at each point $z \in D(a, R)$ and diverges out the closed disc $\bar{D}(a, R)$. Hence, $D(a, R) \subset E \subset \bar{D}(a, R)$ and the interior of the domain of convergence $E$ is $D(a, R)$.*

*Proof.* If $|z - a| > R$, then $|z - a|^{-1} < \rho$, and by property 1) of the upper limit, there are infinitely many terms $|c_n|^{\frac{1}{n}}$ which are greater than $|z - a|^{-1}$. That is, there are infinitely many values of $n$ such that $|c_n(z - a)^n| > 1$; so the sequence $(c_n(z - a)^n)$ cannot be convergent to zero and $z \notin E$. Let us show now that the series converges uniformly on every closed disc $\bar{D}(a, r)$, $r < R$: if $|z - a| \le r$, then $|c_n(z - a)^n| \le |c_n| r^n = M_n$, and by property 2) of the upper limit one has $M_n^{\frac{1}{n}} = |c_n|^{\frac{1}{n}} r < r' < 1$ for a number $r'$ and a certain value of $n$. Then, by the root test, the series $\sum_n M_n$ is convergent and the Weierstrass $M$-test assures uniform convergence of the power series on $\bar{D}(a, r)$. $\qquad\square$

The number $R$ is called the *radius of convergence* of the power series, and $D(a, R)$ *is the disc of convergence* (Figure 2.3). The equality $\frac{1}{R} = \limsup_n |c_n|^{\frac{1}{n}}$ is called *Hadamard's formula*.

In the expression of $R$, $c_n$ is by definition the coefficient of $(z - a)^n$; hence, for example, the series $\sum_m z^{2m}$ has $c_n = 1$ if $n$ is even and $c_n = 0$ if $n$ is odd, $\lim_n |c_n|^{\frac{1}{n}}$ does not exist and $\limsup_n |c_n|^{\frac{1}{n}} = 1$, so the radius of convergence is 1. In general, observe that if a power series $\sum_n c_n z^n$ has radius of convergence $R$ and one does the change $z = w^k$, the obtained series, $\sum_n c_n w^{kn}$, has radius of convergence $R^{\frac{1}{k}}$. Remark also that

$$R = \sup\left\{ r > 0 : \sum_n c_n(z - a)^n \text{ converges for } z \in D(a, r) \right\}$$
$$= \sup\left\{ r > 0 : \sum_n |c_n| r^n < \infty \right\}. \tag{2.11}$$

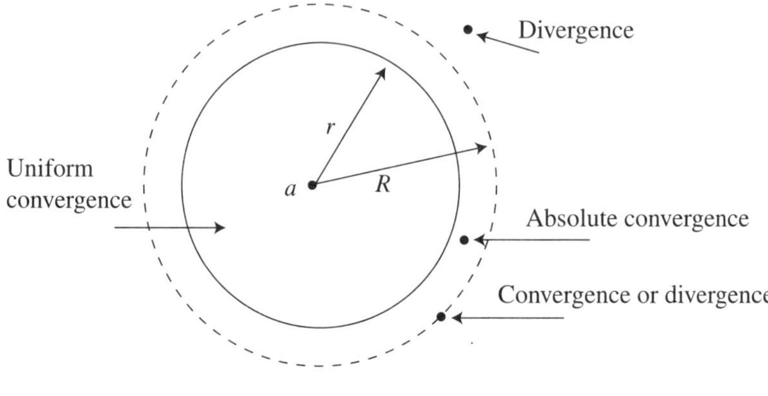


Figure 2.3

The following fact for sequences $(x_n)$, $x_n > 0$, is often useful: if there exists $L = \lim \frac{x_{n+1}}{x_n} \leq +\infty$, then also $\lim x_n^{1/n} = L$. This implies the equality $R = \lim_n \frac{|c_n|}{|c_{n+1}|}$, provided that this limit exists.

It is worth mentioning that, in general, the convergence of the power series is not uniform on the whole disc $D(a, R)$. For example, the geometric series $\sum z^n$ does not converge uniformly on $\mathbb{D}$ because otherwise it would define a bounded function on $\mathbb{D}$, which is not the case. The disc $D(a, R)$ is the biggest open set where the series is convergent. In this disc, the series defines a function $f(z) = \sum c_n(z-a)^n$; obviously $f$ is continuous at every point $z$ of the open disc $D(a, R)$, since the convergence is uniform on every closed disc with radius $r < R$. It is not possible to give a more precise result about the domain of convergence $E$ of a power series.

**Example 2.17.** The geometric series $\sum_n z^n$ has radius of convergence 1 and domain of convergence $E = \mathbb{D}$; the series $\sum_{n \geq 1} \frac{z^n}{n^2}$ also has radius of convergence 1 and converges uniformly on $\bar{\mathbb{D}}$ by Weierstrass' test, because $\left| \frac{z^n}{n^2} \right| \leq \frac{1}{n^2}$ and $\sum_n \frac{1}{n^2} < +\infty$. Hence, in this case $E = \bar{\mathbb{D}}$. $\qquad\square$

These are the two extremal cases; in general, the domain of convergence $E$ consists of the open disc $D(a, R)$ and a subset of the circle $C(a, R)$.

**Example 2.18.** The series $\sum_{n \geq 1} \frac{z^n}{n}$, which has radius of convergence 1, gives the harmonic series at the point $z = 1$, and hence is divergent, while at $z = -1$ it gives an alternating convergent series. We will now prove that at any other point $z \neq 1$ of modulus 1 it is convergent; indeed, this series converges uniformly on compact sets of type $K = \{z : |z| \leq 1, |z - 1| \geq \varepsilon\}$, $\varepsilon > 0$, as it may be seen applying Dirichlet's test with $f_n(z) = z^n$, $g_n(z) = \frac{1}{n}$. Actually, if $z \in K$, then the partial sums $\sum_{n=1}^N z^n = (1 - z^{N+1})/(1 - z)$ are bounded by $\frac{2}{\varepsilon}$, and on the other hand it is clear that $g_n \searrow 0$ uniformly (it converges to zero and does not depend on $z$). In particular, this series defines a continuous function on $\overline{\mathbb{D}} \setminus \{1\}$, which will be determined below. $\qquad\square$

### 2.3.4 Operations with power series

As for polynomials, one may perform some operations with power series: if $\sum_n c_n(z-a)^n$, $\sum_n d_n(z-a)^n$ are two power series, the sum is $\sum_n (c_n+d_n)(z-a)^n$, while the product, as for polynomials, is the series $\sum_n e_n(z-a)^n$, where the coefficients $e_n$ are

$$e_n = \sum_{k=0}^{n} c_k d_{n-k}$$

that is, Cauchy's product computed for each $z \in \mathbb{C}$. For example, the product of the series $\sum_n 2^n z^n$, $\sum_n 3^n z^n$ has the coefficients

$$e_n = \sum_{k=0}^{n} 2^k 3^{n-k} = 3^n \sum_{k=0}^{n} \left(\frac{2}{3}\right)^k = 3^n \frac{1-(2/3)^{n+1}}{1-\frac{2}{3}} = 3^{n+1} - 2^{n+1}.$$

If one has two power series $\sum_n c_n(z-a^n)$, $\sum_n d_n(z-a)^n$ both centered at the point $a$ with respective radius of convergence $R_1$, $R_2$, it is clear using (2.11) that the sum of the two series has radius of convergence, at least, $R = \min(R_1, R_2)$ and, obviously, it defines a function which is the sum of the functions defined by each series. A similar statement is true for the product. If $|z - a| < R$, both series are absolutely convergent at $z$ so Cauchy's product is, by Proposition 2.9. Hence, the product series has convergence radius greater than or equal to $|z - a|$; since $z$ is an arbitrary number with $|z - a| < R$, one obtains that the radius of convergence of the product is also, at least, $R = \min(R_1, R_2)$.

**Example 2.19.** Cauchy's product of the geometric series $\sum_{n\geq 0} z^n$ by itself has $n + 1$ as coefficient of $z^n$, and then

$$\sum_{n}(n + 1)z^n = \left(\frac{1}{1-z}\right)^2. \qquad\square$$

Another important operation with power series is *complex differentiation*, which will be treated below.

### 2.3.5 The exponential function as a complex power series

Recall (by Definition 2.1) that $e^z = e^x(\cos y + i \sin y)$ if $z = x + iy$. Hence, the function $z \mapsto e^z$ extends to the whole plane $\mathbb{C}$ the exponential function $x \mapsto e^x$ of the real variable $x$. Now it will be seen that $e^z$ is also, for any $z \in \mathbb{C}$, the sum of the power series

$$\sum_{n=0}^{\infty} \frac{z^n}{n!}.$$

First of all, it is clear that the radius of convergence of this power series is $+\infty$, since

$$\frac{(n+1)!}{n!} = n + 1 \longrightarrow +\infty.$$

For the time being, let us denote its sum by $E(z) = \sum_n \frac{z^n}{n!}$. Applying Proposition 2.9, it turns out that

$$E(z)E(w) = \sum_n \frac{z^n}{n!} \cdot \sum_m \frac{w^m}{m!}$$

$$= \sum_k \frac{1}{k!} \sum_{n+m=k} \frac{k!}{n!m!} z^n w^m = \sum_k \frac{(z+w)^k}{k!} = E(z+w).$$

In particular, $E(z) = E(x)E(iy) = e^x E(iy)$ if $z = x + iy$, and it remains to check that $E(iy) = \cos y + i \sin y$. Now, in the expression

$$E(iy) = \sum_n \frac{i^n y^n}{n!}$$

the even terms, $n = 2k$, give rise to the real part of $E(iy)$, and the odd terms, $n = 2k + 1$, give rise to the imaginary part. Consequently,

$$\operatorname{Re} E(iy) = \sum_{k=0}^{\infty} (-1)^k \frac{y^{2k}}{(2k)!} = \cos y,$$

$$\operatorname{Im} E(iy) = \sum_{k=0}^{\infty} (-1)^k \frac{y^{2k+1}}{(2k+1)!} = \sin y,$$

and finally

$$E(z) = \sum_0^{\infty} \frac{z^n}{n!} = e^x(\cos y + i \sin y) = e^z, \quad z \in \mathbb{C},$$

as had been stated.

From definitions (2.5) the development in power series of the complex trigonometric functions follows:

$$\cos z = \sum_{n=0}^{\infty} (-1)^n \frac{z^{2n}}{(2n)!}, \quad \sin z = \sum_{n=0}^{\infty} (-1)^n \frac{z^{2n+1}}{(2n+1)!}.$$

## 2.3.6 Convergence at the boundary

Let $\sum_{n\geq 0} c_n(z-a)^n$ be a power series with radius of convergence $R$. Denote by $S$ the set of points $z$ such that $|z-a| = R$ and the power series converges at $z$, that is, $S = E \cap C(a, R)$. If $S$ is non-empty and $m > 1$ is a real number, consider the set

$$S_m = \{z : |z-a| < R,\ d(z, S) \leq m(R - |z-a|)\},$$

where $d(z, S)$ is the distance from $z$ to $S$. If $S$ has only a point $w$, the set $S_m$ has, in a neighborhood of $w$, the shape of an angle with vertex $w$ and measure strictly smaller than $\pi$, depending on $m$, called a *Stolz angle* (Figure 2.4).

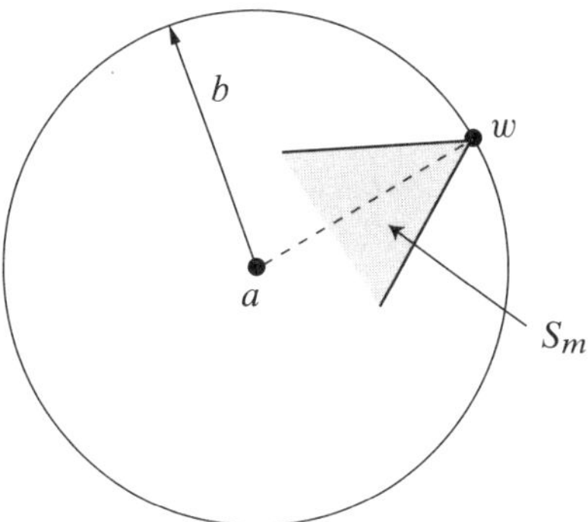


Figure 2.4

For a general $S$, $S_m$ is the union of all Stolz angles with vertex at the points of $S$.

**Theorem 2.20** (Abel). *With the previous notations, suppose that $S$ is non-empty and that the series $\sum_n c_n(z-a)^n$ converges uniformly on $S$. Then the convergence of the series is also uniform on the set $S_m$, for all $m > 1$. In particular, the sum function is continuous on $S_m$ and one has*

$$\lim_{z \to w,\, z \in S_m} \sum_n c_n(z-a)^n = \sum_n c_n(w-a)^n, \quad w \in S.$$

*Proof.* The proof is done only in the case $S$ is a point $w$; without loss of generality, one may suppose $a = 0$, $R = 1$ and $w = 1$. Apply Abel's test (Theorem 2.14) taking $f_n(z) = c_n$, $g_n(z) = z^n$. By hypothesis, the series $\sum_n c_n$ is convergent so that trivially $\sum_n f_n$ converges uniformly (on any set). Let us check now that the condition on $|g_n|$ is satisfied on $S_m = \{z : |z| < 1, |z-1| \leq m(1 - |z|)\}$; actually, one has

$$1 + \sum_{n\geq 0} |z^n - z^{n+1}| = 1 + |1-z| \sum_{n\geq 0} |z|^n = 1 + \frac{|1-z|}{1-|z|} \leq 1 + m. \qquad \square$$

If $f(z) = \sum_n c_n(z - a)^n$ and $w \in S$, the previous theorem says that $\lim_{z \to w} f(z) = f(w)$, if $z$ approaches to $w$ inside $S_m$. In particular, this is true if $z$ tends to $w$ radially, that is, $\lim_{r \to 1} f(rw) = f(w)$. If $w$ is interior to $S$, that is, if there is an arc centered at $w$ contained in $S$, then $f$ is continuous at $w$ (since a whole neighborhood of $w$ contained in $\bar{D}(a, R)$ is inside $S_m$). In particular, if $S$ is the whole circle $C(a, R)$, $f$ is continuous on the whole closed disc.

Abel's Theorem gives a *method for summing* series when applied in the following way. Suppose that $\sum c_n$ is a convergent numerical series and we want to compute its sum; consider the power series $\sum c_n z^n$. Since it converges for $z = 1$, this series has radius of convergence $R \geq 1$ and one has

$$\lim_{r \to 1, r < 1} \sum_n c_n r^n = \sum_n c_n.$$

In the case that $R > 1$ this is true because the sum function is continuous on the open disc of convergence, and if $R = 1$, it is guaranteed by Abel's Theorem. The problem of computing $\sum c_n$ has become the problem of evaluating $\sum c_n r^n$ and then computing a limit. At first glance we have made it more complicated. However, the point is the fact that it could be much easier to compute $\sum c_n z^n$ for $|z| < 1$, using the resources of complex differentiation, which will be shown later, and general operations with power series, instead of computing $\sum c_n$. Observe, however, that there may exist $\lim_{r \to 1} \sum c_n r^n$ even when $\sum c_n$ is not convergent. For example, if $c_n = (-1)^n$ we have

$$\lim_{r \to 1} \sum_{n=0}^{\infty} c_n r^n = \lim_{r \to 1} \frac{1}{1 + r} = \frac{1}{2}.$$

**Example 2.21.** The series $\sum_{n \geq 1} \frac{(-1)^n}{n}$ is convergent because it is an alternating series with the absolute value of its general term decreasing to zero. Consider the power series $\sum \frac{(-1)^n}{n} z^n$; in Example 2.34, using complex differentiation, it will be proved that its sum is $-\mathrm{Log}(1 + z)$ when $|z| < 1$. Now, by Abel's Theorem, one has

$$\sum_{n \geq 1} \frac{(-1)^n}{n} = \lim_{r \to 1} \sum_{n \geq 1} \frac{(-1)^n}{n} r^n = - \lim_{r \to 1} \mathrm{Log}(1 + r) = - \mathrm{Log}\, 2.$$

At a point $z \neq -1$, $|z| = 1$, the series $\sum \frac{(-1)^n}{n} z^n$ is also convergent according to Dirichlet's test, because

$$\left| \sum_{k=1}^{n} (-1)^k z^k \right| = \left| \frac{-z - (-1)^{n+1} z^{n+1}}{1 + z} \right| \leq \frac{2}{|1 + z|}.$$

Hence, Abel's Theorem gives

$$\sum_{n=1}^{\infty} \frac{(-1)^n}{n} z^n = -\operatorname{Log}(1+z), \quad |z| = 1, \; z \neq -1.$$

That is, if $\theta \in [0, 2\pi]$, $\theta \neq \pi$ and $z = e^{i\theta}$, we obtain the formulae

$$\sum_{n=1}^{\infty} \frac{(-1)^n \cos n\theta}{n} = -\operatorname{Log}|1+z| = -\operatorname{Log}\sqrt{2 + 2\cos\theta},$$

$$\sum_{n=1}^{\infty} \frac{(-1)^n \sin n\theta}{n} = -\operatorname{Arg}(1+z) = -\operatorname{Arg}(1 + e^{i\theta}) = -\frac{\theta}{2}.$$

For the particular value $\theta = \frac{\pi}{2}$, $(z = i)$, the second formula becomes the equality

$$\sum_{m=0}^{\infty} \frac{(-1)^m}{2m+1} = \frac{\pi}{4}. \qquad \square$$

## 2.4 Differentiation of functions of a complex variable

For a better knowledge of the properties of functions of a complex variable, one needs to extend differential calculus to the complex field. This is done next.

### 2.4.1 Holomorphic functions

**Definition 2.22.** Let $U$ be an open set in the plane and $f$ a complex function defined on $U$ and $a \in U$. One says that $f$ is $\mathbb{C}$-differentiable at the point $a$ if there exists the limit

$$\lim_{z \to a, \, z \in U} \frac{f(z) - f(a)}{z - a} = \lim_{h \to 0} \frac{f(a+h) - f(a)}{h} = f'(a).$$

The complex number $f'(a)$ is called the (complex) derivative of $f$ at the point $a$.

The definition is, then, analogous to the case of functions of a real variable and $f'(a)$ has the same meaning: it measures the variation of $f$ in a neighborhood of the point $a$. It is worth mentioning, however, that now the increment $h$ takes complex values. In other words, the fact that $f$ is $\mathbb{C}$-differentiable at $a$ means that there is a complex number $\alpha$ such that the increments $f(a+h) - f(a)$ may be approximated to the first order by $\alpha h$, that is,

$$f(a+h) - f(a) = \alpha h + o(h), \quad \lim_{h \to 0} \frac{o(h)}{h} = 0. \tag{2.12}$$

In particular, it turns out that if $f$ is $\mathbb{C}$-differentiable at the point $a$, $f$ is continuous at $a$.

**Definition 2.23.** If the function $f$ is defined on the open set $U$ and it is $\mathbb{C}$-differentiable at all points of $U$, it is said that $f$ is holomorphic on $U$. We will denote by $H(U)$ the set of all holomorphic functions on $U$. The functions which are holomorphic on the whole complex plane are called entire functions.

If $f \in H(U)$, then we define on $U$ the (complex) derivative function of $f$, denoted by $f'$, with value $f'(z)$ at each point $z \in U$. For example, if $n$ is natural, then the function $f(z) = z^n$ is $\mathbb{C}$-differentiable at any point $a$, with derivative $f'(a) = na^{n-1}$ because

$$\frac{z^n - a^n}{z - a} = z^{n-1} + z^{n-2}a + z^{n-3}a^2 + \cdots + za^{n-2} + a^{n-1}$$

has limit $na^{n-1}$ when $z \to a$. The function $f(z) = \bar{z}$ is not $\mathbb{C}$-differentiable at any point because

$$\frac{f(z) - f(a)}{z - a} = \frac{\bar{z} - \bar{a}}{z - a} = \frac{\bar{h}}{h},$$

and if $z$ approaches $a$ horizontally, that is, $z = a + h$, $h \in \mathbb{R}$, then the quotient has value 1, while if $z$ approaches $a$ vertically, that is, $z = a + ih$, $h \in \mathbb{R}$, then the previous quotient takes the value $-1$. In the case of the function $f(z) = |z|^2$, the incremental quotient is

$$\frac{|a + h|^2 - |a|^2}{h} = \bar{a} + a\frac{\bar{h}}{h} + \bar{h}.$$

If $h$ is real-valued, its limit when $h \to 0$ is $\bar{a} + a$, and $\bar{a} - a$ if $h$ is pure imaginary; if $f$ is $\mathbb{C}$-differentiable at the point $a$, these values must coincide and consequently $a = 0$. Since at $a = 0$ the previous limit vanishes when $h \to 0$, one may conclude that the function is $\mathbb{C}$-differentiable only at the point $a = 0$.

This example shows the difference between the real and the complex cases. In the complex case, setting the existence of the limit which defines the derivative is something really restrictive, since there are infinitely many ways of approaching $a$: by horizontal, vertical or sloping lines, along a spiral or any curve which ends at $a$. This will induce holomorphic functions on a domain to behave much better than their real analogues, functions of a real variable which are differentiable on a certain interval.

**Example 2.24.** The exponential function $e^z$ is holomorphic and coincides with its derivative. This is immediate from the definition, since

$$\lim_{h \to 0} \frac{e^{z+h} - e^z}{h} = e^z \lim_{h \to 0} \frac{e^h - 1}{h} = e^z.$$

It is an entire function which extends the real exponential function to the whole plane. Considering the definitions (2.5) and the fact that the usual rules of derivatives are also valid in the complex case, it turns out that the functions $\sin z$, $\cos z$ are also entire functions, and $(\sin z)' = \cos z$, $(\cos z)' = -\sin z$. $\qquad\square$

Compare now the definition of $\mathbb{C}$-differentiable function with the one of differentiable function, considering $f : U \to \mathbb{R}^2$ as a function of two real variables. Recall that $f$ is differentiable at the point $a$ if there is an $\mathbb{R}$-linear mapping from $\mathbb{R}^2$ to $\mathbb{R}^2$ – the differential $df(a)$ – such that

$$f(a + h) - f(a) = df(a)h + o(h), \quad a, h \in \mathbb{R}^2 \simeq \mathbb{C}.$$

Comparing with (2.12), it turns out that $f$ is $\mathbb{C}$-differentiable at $a$ if and only if it is differentiable in the real sense and the differential $df(a)$ is of the form $df(a)(h) = \alpha h$ with $\alpha \in \mathbb{C}$. But, as seen in Proposition 1.9, the $\mathbb{R}$-linear mappings from $\mathbb{R}^2$ to $\mathbb{R}^2$ which are of this type are exactly the $\mathbb{C}$-linear ones. Hence, the first part of the following theorem is proved:

**Theorem 2.25.** *The function $f$, defined on a neighborhood of the point $a \in \mathbb{C}$, is $\mathbb{C}$-differentiable at $a$ if and only if it is differentiable at $a$ and the differential $df(a)$ is $\mathbb{C}$-linear. If $f = u + iv$ with $u$ and $v$ real-valued and $f$ is differentiable at the point $a$, then $f$ is $\mathbb{C}$-differentiable at $a$ if and only if the following equations hold at this point:*

$$
\begin{aligned}
u_x &= v_y, \\
u_y &= -v_x.
\end{aligned}
\tag{2.13}
$$

The equations (2.13), in which the notation $u_x = \frac{\partial u}{\partial x}$, $u_y = \frac{\partial u}{\partial y}$, $v_x = \frac{\partial v}{\partial x}$ and $v_y = \frac{\partial v}{\partial y}$ is used, are called *Cauchy–Riemann equations*.

*Proof.* It has been shown, after Proposition 1.9, that a matrix $A = (a_{ij})_{i,j=1,2}$ is the matrix of a $\mathbb{C}$-linear mapping, that is, a mapping of type $h \to \alpha h$, $\alpha \in \mathbb{C}$, if and only if

$$a_{11} = a_{22},$$

$$a_{12} = -a_{21},$$

and then it is $\alpha = a_{11} + i a_{21}$. In the case of the mapping $df(a)$, with $f = u + iv$, its matrix is the *Jacobian* matrix of $f$, that is, $a_{11} = u_x$, $a_{21} = v_x$, $a_{12} = u_y$, $a_{22} = v_y$. $\qquad\square$

A different way to deduce Cauchy–Riemann equations is the following one: the limit of the incremental quotient must be $f'(a)$ for all directions. If the increment $h$ is real, the limit

$$\lim_{h \to 0} \frac{f(a + h) - f(a)}{h}$$

is $\frac{\partial f}{\partial x}(a) = u_x(a) + i v_x(a)$, while if the increment $h$ is pure imaginary, $h = iy$, it is

$$\lim_{y \to 0} \frac{f(a + iy) - f(a)}{iy} = -i\frac{\partial f}{\partial y}(a) = -i(u_y(a) + i v_y(a)).$$

Equating now real and imaginary parts, Cauchy–Riemann equations may be deduced. Observe that we have also shown that if $f$ is $\mathbb{C}$-differentiable at $a$, then

$$f'(a) = u_x(a) + i v_x(a) = \frac{\partial f}{\partial x}(a) = -i\frac{\partial f}{\partial y}(a).$$

This way, if $J_f$ is the Jacobian matrix of $f$, it turns out that

$$\det(J_f(a)) = u_x(a)^2 + v_x(a)^2 = |f'(a)|^2.$$

Looking again at the example of the function $f(z) = |z|^2 = x^2 + y^2$, one has that $f$ is differentiable at any point, but Cauchy–Riemann equations give $2x = 0$, $2y = 0$, and so they hold only at the origin.

Note that if $f$ is holomorphic on a domain $U$ and $f'(z) = 0$, for all $z \in U$, then $f$ must be constant on $U$ because $f$ is differentiable with $df(z) = 0, z \in U$. This fact is the basis of the following statement:

**Proposition 2.26.** *If a real function $f$ is $\mathbb{C}$-differentiable at a point $a$, then $df(a) = 0$, and so every real holomorphic function on a domain is constant. Every holomorphic function $f = u + iv$ on a domain is completely determined by its real part, except for an additive pure imaginary constant.*

*Proof.* If $v = 0$, the Cauchy–Riemann equations imply $u_x(a) = v_y(a) = 0$, $u_y(a) = -v_x(a) = 0$, and then $df(a) = 0$. $\qquad\qquad\square$

The second assertion of the previous proposition means exactly that if $f = u + iv, g = u + iw$ are holomorphic functions on a domain with the same real part, then $f = g + ic$, with $c \in \mathbb{R}$.

As a corollary of Theorem 2.25, it turns out that if $f = u + iv$ has partial derivatives $u_x, u_y, v_x, v_y$ on a neighborhood of the point $a$ which are continuous at $a$ (so that $f$ is differentiable at $a$) and Cauchy–Riemann equations hold at $a$, then $f$ is $\mathbb{C}$-differentiable at $a$.

### 2.4.2 Holomorphic functions and conformality

We begin this paragraph with a review of the geometric meaning of the differential of a differentiable mapping $f : U \to \mathbb{R}^n$ on an open set $U \subset \mathbb{R}^n$, $n \geq 2$. Recall that the differential of $f$ at the point $a$, $df(a)$, satisfies

$$\lim_{h \to 0} \frac{|f(a + h) - f(a) - df(a)(h)|}{|h|} = 0.$$

If $\gamma : I \to U$, where $I$ is an interval of $\mathbb{R}$, is a differentiable curve (with tangent at all points) which contains $a$, write $a = \gamma(0)$, the composition $f \circ \gamma$ is a differentiable curve which contains $f(\gamma(0)) = f(a)$; its tangent vector at this point is

$$\frac{d}{dt}\bigg|_{t=0} (f \circ \gamma)(t) = \sum_{i=1}^{n} \frac{\partial f}{\partial x_i}(\gamma(t)) \cdot \gamma_i'(t)\bigg|_{t=0} = df(a)(\gamma'(0)).$$

That is, $df(a)$ is a mapping that makes a correspondence between each tangent vector to a curve at the point $a$ and the tangent vector to the image curve at the point $f(a)$.

**Definition 2.27.** A differentiable mapping $f : U \to \mathbb{R}^n$ is conformal at the point $a \in U$ if $df(a)$ is invertible and preserves oriented angles.

This means that the linear mapping $df(a)$ is conformal in the sense of Definition 1.8 (extended to $\mathbb{R}^n$). Bearing in mind the geometric meaning of $df(a)$, the fact that $f$ is conformal at $a$ means that if two curves $\gamma_1$, $\gamma_2$ intersect at $a$ with an angle $\alpha$, then the image curves $f \circ \gamma_1$, $f \circ \gamma_2$ intersect at $f(a)$ also with an angle $\alpha$ (Figure 2.5), understanding that the angle between two curves at a common point is the angle between their tangents at this point.

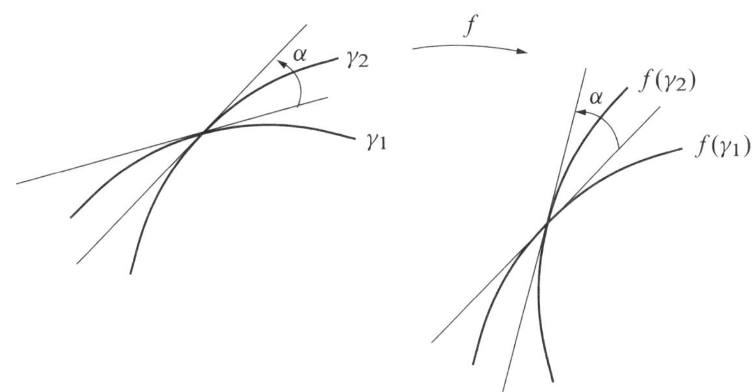


Figure 2.5

In the case $n = 2$, Proposition 1.9 tells us that $f$ is conformal at $a$ if and only if it is $\mathbb{C}$-differentiable at $a$ and $f'(a) \neq 0$. So if the curve $\gamma$ has tangent vector $\gamma'(0) \in \mathbb{C}$ at $a = \gamma(0)$, the tangent vector of $f \circ \gamma$ at $f(a)$ is

$$(f \circ \gamma)'(0) = f'(a) \cdot \gamma'(0),$$

obtained from $\gamma'(0)$ multiplying by $f'(a) = |f'(a)| \cdot e^{i\theta}$, that is, rotating an angle $\theta$ and making a dilation of factor $|f'(a)|$. If all tangent vectors rotate similarly, then oriented angles are clearly preserved. Analytically, for two curves $\gamma_1$, $\gamma_2$, the equality

$$(f \circ \gamma_1)'(0)\overline{(f \circ \gamma_2)'(0)} = |f'(a)|^2 \gamma_1'(0) \cdot \overline{\gamma_2'(0)}$$

represents the invariance of angles.

**Example 2.28.** Consider the transformation $f(z) = e^z$; since $f'(z) = e^z \neq 0$, $f$ is conformal at any point. The horizontal line $y = y_0$ is mapped onto the ray $x \mapsto e^x e^{iy_0}$, the vertical line $x = x_0$ onto the circle centered at 0 and with radius $e^{x_0}$. Horizontal and vertical lines meet at right angles, as do their images. A line with slope $m$, $y = mx$, which meets horizontal lines at an angle $\alpha = \arctan m$, becomes the curve $x \mapsto e^x e^{imx}$, which is a spiral (Figure 2.6). At a point $P = e^x e^{imx} = f((1 + im)x)$, the spiral has tangent $(1 + im)e^x e^{imx}$, which meets the ray starting at the origin and also containing $P$ at an angle $\alpha$.  $\square$

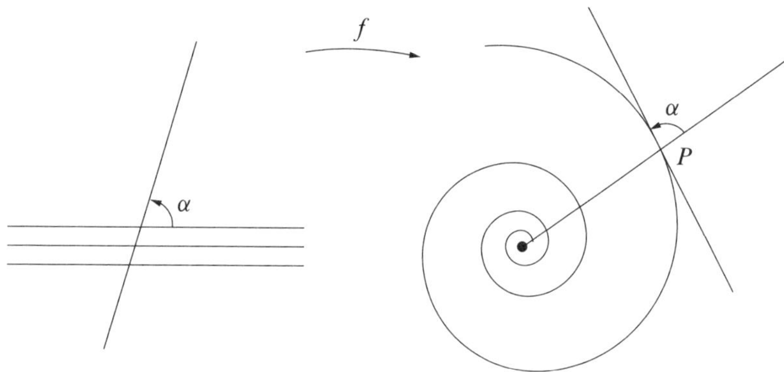


Figure 2.6

The mappings that preserve the size of the angles but change their orientation are called *anticonformal mappings*.

According to Proposition 1.12, one may say now that if $f : U \to \mathbb{R}^n$ is differentiable on the open set $U$ of $\mathbb{R}^n$, then $f$ is conformal or anticonformal on $U$ if and only if for each $x \in U$, one has $df(x) = \lambda(x)O(x)$ with $O(x)$ an orthogonal matrix and $\lambda(x) > 0$. Taking determinants it turns out that

$$\det(J_f(x)) = \det df(x) = \lambda(x)^n \det O(x) = \pm\lambda(x)^n$$

and so, automatically, $\lambda(x) = |\det J_f(x)|^{1/n}$.

Besides linear transformations $T$ of type $\lambda O$, with $O$ an orthogonal transformation, the main examples of conformal and anticonformal mappings on $\mathbb{R}^n$ are the inversions with respect to spheres, which will be discussed in Chapter 8 in the case of the plane. In dimension $n = 2$, it has been seen that any transformation given by a holomorphic function, for example $e^z$, $\cos z$, $\sin z, \ldots$ is conformal. So there are a huge quantity of conformal transformations if $n = 2$, beyond linear transformations, inversions and their compositions. However, if $n > 2$, there are only these ones.

**Theorem 2.29** (Liouville). *If $U$ is a domain in $\mathbb{R}^n$ with $n \geq 3$ and the function $f : U \to \mathbb{R}^n$ is $C^1$ and preserves the size of the angles, then $f$ is the restriction to $U$ of a transformation from $\mathbb{R}^n$ into itself which consists of a finite composition of orthogonal transformations, translations, dilations and inversions.*

This is a theorem difficult to prove. For a function $f$ with more regularity, for example if $f$ has third-order continuous partial derivatives, one may find a proof in [2], p. 105.

### 2.4.3 Complex differentiation and antidifferentiation rules

Using only the definition of derivative, as in the case of a function of a real variable, it is easy to check that the usual rules of differentiation of sums, products and quotients hold. Hence, if $f$, $g$ have complex derivatives at the point $a$, also the functions $f + g$, $fg$ and the quotient $f/g$, when $g(a) \neq 0$, are $\mathbb{C}$-differentiable functions, and one has

$$(f + g)'(a) = f'(a) + g'(a),$$

$$(fg)'(a) = f'(a)g(a) + f(a)g'(a),$$

$$\left(\frac{f}{g}\right)'(a) = \frac{f'(a)g(a) - f(a)g'(a)}{g(a)^2}.$$

The *chain rule* also holds: $(f \circ g)'(a) = f'(g(a))g'(a)$, when $g$ is $\mathbb{C}$-differentiable at $a$ and $f$ is also $\mathbb{C}$-differentiable at $g(a)$. Observe that the chain rule may be also deduced from the general chain rule for differentiable functions in the real sense, because the composition of linear mappings which consist of multiplying by $g'(a)$ and $f'(g(a))$ is the mapping consisting of multiplying by $f'(g(a))g'(a)$.

Hence, every polynomial $P(z) = c_0 + c_1 z + c_2 z^2 + \cdots + c_n z^n$ of the variable $z$ is an entire function and $P'(z) = c_1 + 2c_2 z + \cdots + nc_n z^{n-1}$. Every rational function of the variable $z$, $R(z) = P(z)/Q(z)$ is holomorphic at all the points which are not zeros of $Q$ (supposing the expression of $R$ is irreducible).

Moreover, as in the real case, it may be proved that if $f$ is $\mathbb{C}$-differentiable at the point $a$, bijective on a neighborhood of $a$, with $f^{-1}$ continuous at the point $f(a)$ and $f'(a) \neq 0$, then the inverse function $f^{-1}$ is $\mathbb{C}$-differentiable at $f(a)$ with derivative $\frac{1}{f'(a)}$ (later on it will be proved that if $f$ is bijective on a neighborhood of $a$, automatically $f'(a) \neq 0$). In particular, it turns out that if $g$ is a continuous branch of the logarithm on a neighborhood of the point $b$, then $g$ is $\mathbb{C}$-differentiable at $b$ and its derivative is $g'(b) = \frac{1}{b}$. The proof for this case is the following: if $a = g(b)$, $z$ is close to $b$, and $w = g(z)$, one has that $b = e^a$, $z = e^w$ and the incremental quotient my be written as

$$\frac{g(z) - g(b)}{z - b} = \frac{1}{\frac{e^w - e^a}{w - a}},$$

which has limit $\frac{1}{e^a} = \frac{1}{b}$ when $z \to b$ ($w \to a$).

Locally, every branch of the logarithm of a function $f$ which does not vanish on an open set $U$ (see Definition 2.4) may be obtained by composing $f$ with a branch of the logarithm in the image of $f$. Consequently, one may deduce from the previous rule and the chain rule that if $f$ is holomorphic and non-vanishing on $U$ and there exists a branch $h$ of the logarithm of $f$, then $h$ is also holomorphic on $U$ and has derivative $h' = \frac{f'}{f}$. Similarly, every branch $g$ of a power $f^\alpha$ is also holomorphic and has derivative $g' = \alpha f' g / f$ (that is dangerous to write as $\alpha f' f^{\alpha-1}$).

As in real differential calculus, we can consider also the notion of antiderivative or complex primitive. If $h$ is a continuous function defined on an open set $U$, a *holomorphic antiderivative* or *holomorphic primitive* of $h$ is any holomorphic function $f$ on $U$ such that $f'(z) = h(z)$ if $z \in U$. This relation will be studied in detail later on; in particular, it will be proved that only holomorphic functions may have holomorphic primitives (that is, if $h = f'$ with $f$ holomorphic, then $h$ is holomorphic). Meanwhile we just state some derivation/antiderivation rules between elementary functions. Evidently, the first relations one has to point out are

$$f'(z) = 0 \qquad \text{if and only if} \quad f(z) = c, \qquad\qquad c \text{ constant},$$

$$f'(z) = g'(z) \quad \text{if and only if} \quad f(z) = g(z) + c, \quad c \text{ constant},$$

if the equalities hold for any point $z$ of a domain in the plane. This means that if $h$ has a holomorphic primitive on $U$, $f$, $f(z) + c$ is the general expression of all the holomorphic primitives of $h$ on $U$.

The expression

$$\int h \, dz = f(z) + c$$

will denote the general holomorphic primitive of $h$ without specifying the domain. The reason for the integral notation $\int$ will be explained in the next sections. Hence, one may write

$$\int z^n \, dz = \frac{z^{n+1}}{n + 1} + c \iff (z^{n+1})' = (n + 1)z^n,$$

$$\int e^z \, dz = e^z + c \iff (e^z)' = e^z,$$

$$\int \sin z \, dz = -\cos z + c, \quad \int \cos z \, dz = \sin z + c, \quad \text{etc.}$$

Each differentiation rule may be rewritten as an antidifferentiation rule. The differentiation rule of the product is equivalent to the *antidifferentiation by parts formula*

$$\int f'g \, dz = fg - \int fg' \, dz.$$

For example,

$$\int \text{Log}\, z \, dz = z \,\text{Log}\, z - \int z (\text{Log}\, z)' \, dz = z \,\text{Log}\, z - z + c.$$

*The change of variable formula*

$$\int f(z) \, dz = (z = g(w)) = \int f(g(w))g'(w) \, dw$$

corresponds to the chain rule. For example,

$$\int w e^{w^2} \, dw = (z = w^2) = \int \frac{1}{2} e^z \, dz = \frac{1}{2} e^z + c = \frac{1}{2} e^{w^2} + c.$$

### 2.4.4 The operators $\partial$ and $\bar{\partial}$

In Section 2.1 we asked how one may recognize if a polynomial in the variables $x, y$, with complex coefficients, is a polynomial in $z$. One may use now Cauchy–Riemann equations to answer this question. First let us introduce the differential operators $\partial = \frac{\partial}{\partial z}$ and $\bar{\partial} = \frac{\partial}{\partial \bar{z}}$ defined by

$$\frac{\partial}{\partial z} = \frac{1}{2}\left(\frac{\partial}{\partial x} - i \frac{\partial}{\partial y}\right), \quad \frac{\partial}{\partial \bar{z}} = \frac{1}{2}\left(\frac{\partial}{\partial x} + i \frac{\partial}{\partial y}\right),$$

which may be applied to any function $f$ with partial derivatives of first order. Observe that $\overline{\partial f} = \bar{\partial} \bar{f}$ if $f$ is real. Writing $dx, dy$ in terms of $dz = dx + i \, dy$ and $d\bar{z} = dx - i \, dy$, these operators are the coefficients of $dz$ and $d\bar{z}$ in the differential of a function

$$df = \frac{\partial f}{\partial z} \, dz + \frac{\partial f}{\partial \bar{z}} \, d\bar{z},$$

an equality which may be obtained by substitution of $dx = \frac{1}{2}(dz + d\bar{z})$, $dy = \frac{1}{2i}(dz - d\bar{z})$ in $df = \frac{\partial f}{\partial x} \, dx + \frac{\partial f}{\partial y} \, dy$. It is easy to check that, conversely, the operators of partial differentiation $\frac{\partial}{\partial x}$ and $\frac{\partial}{\partial y}$ are expressed in terms of $\partial$ and $\bar{\partial}$ by means of

$$\frac{\partial}{\partial x} = \frac{\partial}{\partial z} + \frac{\partial}{\partial \bar{z}}, \quad \frac{\partial}{\partial y} = i \frac{\partial}{\partial z} - i \frac{\partial}{\partial \bar{z}}.$$

The main reason why these operators are defined is that Cauchy–Riemann equations may be condensed in a unique equation (with complex coefficients), which is

$$\frac{\partial f}{\partial \bar{z}} = 0. \tag{2.14}$$

This may be checked immediately because writing $f = u + iv$ and supposing $f$ differentiable, the equations (2.13) are equivalent to

$$\frac{\partial f}{\partial y} = i\frac{\partial f}{\partial x} \iff \frac{\partial f}{\partial x} + i\frac{\partial f}{\partial y} = 0 \iff \frac{\partial f}{\partial \bar{z}} = 0.$$

Hence the function $f$, supposed differentiable, is $\mathbb{C}$-differentiable at the point $a$ if and only if it satisfies $\frac{\partial f}{\partial \bar{z}}(a) = 0$, and then $f'(a) = \frac{\partial f}{\partial z}(a)$.

The formalism obtained using these operators is as if $z$, $\bar{z}$ were another coordinate system. Then holomorphic functions are the ones which do not depend on $\bar{z}$ and, in this case, their complex derivatives are the derivatives with respect to $z$. Hence, for example, if one makes a composition of two differentiable functions $f(z), g(w)$ (not necessarily holomorphic) and puts $h = g \circ f$, it turns out that

$$\frac{\partial h}{\partial z} = \frac{\partial g}{\partial w}\frac{\partial f}{\partial z} + \frac{\partial g}{\partial \bar{w}}\frac{\partial \bar{f}}{\partial z}, \quad \frac{\partial h}{\partial \bar{z}} = \frac{\partial g}{\partial w}\frac{\partial f}{\partial \bar{z}} + \frac{\partial g}{\partial \bar{w}}\frac{\partial \bar{f}}{\partial \bar{z}}.$$

When $f$ and $g$ are both holomorphic, these equations become clearly the chain rule $h'(z) = f'(z)g'(w)$. If $\gamma(t)$ is a differentiable curve and taking $h(t) = f(\gamma(t))$ with $f$ differentiable, the chain rule may be expressed as

$$h'(t) = \frac{\partial f}{\partial z}(\gamma(t))\gamma'(t) + \frac{\partial f}{\partial \bar{z}}(\gamma(t))\overline{\gamma'(t)}.$$

In particular, $h'(t) = f'(\gamma(t))\gamma'(t)$ if $f$ is holomorphic. Now one has:

**Theorem 2.30.** *Given a polynomial $P(x, y)$ the following are equivalent:*

a) *$P$ is a polynomial in $z$.*

b) *$P$ is an entire function.*

c) *The equation $\frac{\partial P}{\partial \bar{z}} = 0$ holds on $\mathbb{C}$.*

*Proof.* It is sufficient to comment on the implication c) $\Rightarrow$ a). Consider the auxiliary polynomial $Q(z, w)$ of two formal variables $z$, $w$ defined by $Q(z, w) = P\left(\frac{z+w}{2}, \frac{z-w}{2i}\right)$. Observe that the change $w = \bar{z}$ gives the expression of $P$ in terms of $z, \bar{z}$. By the chain rule, one has $\frac{\partial Q}{\partial w} = \frac{1}{2}\left(\frac{\partial P}{\partial x} + i\frac{\partial P}{\partial y}\right) = \frac{\partial P}{\partial \bar{z}}$. If c) holds, then $Q$ does not depend on $w$, and so $P$ is a polynomial in $z$. $\qquad\square$

Observe that Cauchy–Riemann equations for $P$ have already appeared in (2.1).

## 2.5 Analytic functions of a complex variable

### 2.5.1 Differentiation of power series

Consider a power series $\sum_n c_n(z-a)^n$ centered at the point $a \in \mathbb{C}$ and radius of convergence $R > 0$. The disc $D(a, R)$ is the biggest open set on which the series is convergent, and the function $f(z) = \sum_n c_n(z-a)^n$ is a continuous function on this disc. We will show that $f(z)$ is holomorphic on the disc $D(a, R)$ and that $f'(z)$ is also the sum of a power series, namely of the *derivative power series* of $\sum_n c_n(z-a)^n$, which is obtained by differentiating each of its terms, that is,

$$\sum_n n c_n(z-a)^{n-1}.$$

Observe that this derivative series has also radius of convergence equal to $R$, because

$$\limsup_n (n|c_n|)^{1/n} = \limsup_n |c_n|^{1/n},$$

since $\lim_n n^{1/n} = 1$.

**Theorem 2.31.** *If the power series $\sum_n c_n(z-a)^n$ has radius of convergence $R > 0$, then the function $f(z) = \sum_n c_n(z-a)^n$ is holomorphic on $D(a, R)$ and has derivative $f'(z) = \sum_n n c_n(z-a)^{n-1}$ in this disc.*

*Proof.* Fix $z \in D(a, R)$. Given $\varepsilon > 0$, one has to show that

$$\frac{f(w) - f(z)}{w - z} - \sum_n n c_n(z-a)^{n-1} \tag{2.15}$$

has modulus smaller than $\varepsilon$ for $w$ close enough to $z$. Taking $N$ (to be chosen later), break $f(z) = S_N(z) + R_N(z) = \sum_{n=0}^{N} c_n(z-a)^n + \sum_{n>N} c_n(z-a)^n$ and write the expression (2.15) as the sum of three terms, $\mathrm{I} + \mathrm{II} + \mathrm{III}$,

$$\left(\frac{S_N(w) - S_N(z)}{w - z} - \sum_{n=0}^{N} n c_n(z-a)^{n-1}\right) - \sum_{n>N} n c_n(z-a)^{n-1} + \frac{R_N(w) - R_N(z)}{w - z}.$$

The last term, III, is

$$\sum_{n>N} c_n\left((w-a)^{n-1} + (w-a)^{n-2}(z-a) + \cdots + (w-a)(z-a)^{n-2} + (z-a)^{n-1}\right).$$

If $|z - a| < \rho < R$ and supposing also that $|w - a| < \rho$, one obtains that $|\mathrm{III}| \leq \sum_{n>N} n|c_n|\rho^{n-1}$; since this sum is the tail of a convergent series, one may choose $N$ such that $|\mathrm{III}| < \frac{\varepsilon}{3}$ (whether $w$ is close to $z$ or not; it is enough that

$|w - a| < \rho$). One may assume also that $N$ is big enough so that $|\mathrm{II}| < \frac{\varepsilon}{3}$, because it is also the tail of a convergent series.

Finally, one may get $|\mathrm{I}| < \frac{\varepsilon}{3}$ for $w$ close enough to $z$, because this is simply the derivative of a polynomial. $\qquad\square$

Applying iteratively Theorem 2.31, one gets that in fact $f$ is indefinitely holomorphic, that is, has complex derivatives of any order on the open disc $D(a, R)$. Denoting by $f^{(n)}$ the $n$-th complex derivative of $f$ and taking $z = a$ in the expression of $f^{(n)}(z)$ obtained by differentiating $n$ times, term by term, the series $\sum c_n(z - a)^n$, it turns out that $f^{(n)}(a) = n!c_n$. *A fortiori*, the development of $f$ in power series is

$$f(z) = \sum_{n \geq 0} \frac{f^{(n)}(a)}{n!}(z - a)^n, \quad |z - a| < R,$$

which is, formally, equal to the Taylor series. Hence the following statement is proved.

**Proposition 2.32.** *Let $f : D(a, R) \to \mathbb{C}$. If there exists a power series*

$$\sum_{n=0}^{\infty} c_n(z - a)^n,$$

*convergent on this disc, such that*

$$f(z) = \sum_{n=0}^{\infty} c_n(z - a)^n, \quad |z - a| < R,$$

*then this series is unique. Indeed, $f$ is indefinitely holomorphic and the coefficients $c_n$ are determined by $f$ by means of the relations*

$$c_n = \frac{f^{(n)}(a)}{n!}, \quad n = 0, 1, 2, \ldots .$$

If the hypothesis of Proposition 2.32 holds, it is said that $f$ has a *power series expansion* on the disc $D(a, R)$, with respect to the center $a$. Proposition 2.32 means that given an equality of power series the coefficients can be identified, as in the case of polynomials: if the equality

$$c_0 + c_1(z - a) + c_2(z - a)^2 + \cdots + c_n(z - a)^n + \cdots$$
$$= d_0 + d_1(z - a) + d_2(z - a)^2 + \cdots + d_n(z - a)^n + \cdots$$

holds for $z$ in a neighborhood of the point $a$, then $c_n = d_n$ for $n = 0, 1, 2, \ldots .$

If $f$ has a power series expansion, $f(z) = \sum_n c_n(z-a)^n$, this holds also for $f'$ and $f'(z) = \sum_{n=1}^{\infty} n c_n(z-a)^{n-1}$. In the converse sense, if $f$ is holomorphic on $D(a, R)$ and $f'$ has a power series expansion,

$$f'(z) = \sum_{n=0}^{\infty} d_n(z-a)^n, \quad |z-a| < R,$$

then $f$ may be represented as

$$f(z) = f(a) + \sum_{n=0}^{\infty} \frac{d_n}{n+1}(z-a)^{n+1}, \quad |z-a| < R.$$

Actually, the right-hand term of this equality defines a holomorphic function $g$ on $D(a, R)$ (because the series $\sum \frac{d_n}{n+1}(z-a)^{n+1}$ has the same convergence radius as $\sum d_n(z-a)^n$) which has the same derivative as $f$, $g' = f'$, and that satisfies $g(a) = f(a)$; consequently, $g(z) = f(z)$, if $z \in D(a, R)$.

Differentiating or antidifferentiating power series, one may compute the sum of some power series or even find the power series expansion of certain functions.

**Example 2.33.** The series $\sum_{n\geq 1} n^2 z^n$ has radius of convergence 1 and we want to compute its sum. It may be done by using the sum of the geometric series $\sum z^n$, which is $\frac{1}{1-z}$, and noting that

$$\sum_{n\geq 1} n^2 z^n = \sum_{n\geq 1} (n(n-1) + n)z^n$$

$$= z^2 \sum_{n\geq 2} n(n-1)z^{n-2} + z \sum_{n\geq 1} n z^{n-1}$$

$$= z^2 \left(\frac{1}{1-z}\right)'' + z \left(\frac{1}{1-z}\right)'$$

$$= \frac{2z^2}{(1-z)^3} + \frac{z}{(1-z)^2} = \frac{z(1+z)}{(1-z)^3}.$$

Similarly one could compute the sum of the series $\sum P(n)z^n$ in the case that $P(n)$ is a polynomial in the variable $n$. $\qquad\square$

**Example 2.34.** The function $\mathrm{Log}(1-z)$ that, as already known, is well defined if $1-z$ is not a real negative number, that is, if $z$ is not real and greater than or equal to 1, has a power series expansion in a neighborhood of the origin. Its derivative is $-1/(1-z)$, which has series expansion $-\sum_{n\geq 0} z^n$ for $|z| < 1$ and, then, it turns out that

$$\mathrm{Log}(1-z) = -\sum_{n\geq 0} \frac{z^{n+1}}{n+1} = -\sum_{n\geq 1} \frac{z^n}{n}, \quad |z| < 1.$$

(Figure 2.7). $\qquad\square$

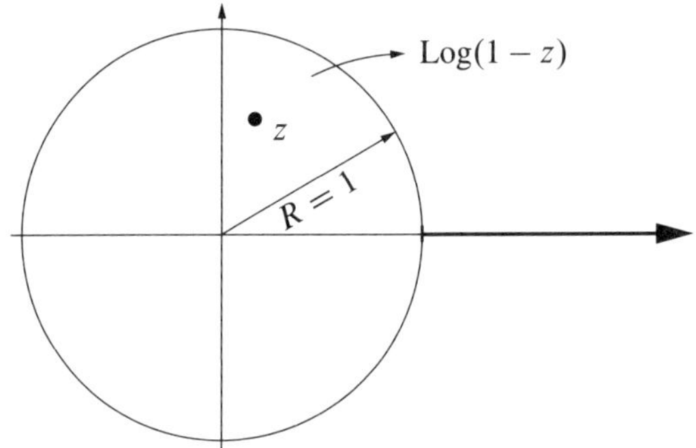


Figure 2.7

**Example 2.35.** Consider now the power series $\sum_{n=1}^{\infty} \frac{z^{n+1}}{n(n+1)}$, which converges on the closed unit disc $\overline{\mathbb{D}}$. If $f(z)$ is the sum of this series, one has that

$$f'(z) = \sum_{n=1}^{\infty} \frac{z^n}{n}, \quad |z| < 1.$$

Bearing in mind Example 2.34 we have $f'(z) = -\text{Log}(1 - z)$. Calculate now the antiderivatives of $-\text{Log}(1 - z)$, integrating by parts,

$$-\int \text{Log}(1 - z)\, dz = \int \text{Log}\, w\, dw \quad (w = 1 - z)$$

$$= w\, \text{Log}\, w - w + c$$

$$= (1 - z)\, \text{Log}(1 - z) + z + c.$$

Then, $f(z) = (1 - z)\, \text{Log}(1 - z) + z + c$. Taking $z = 0$ we obtain $c = 0$ and so

$$\sum_{n=1}^{\infty} \frac{z^{n+1}}{n(n + 1)} = (1 - z)\, \text{Log}(1 - z) + z, \quad |z| < 1. \qquad (2.16)$$

By Abel's Theorem (Theorem 2.20), since the series is convergent at every point $z$ such that $|z| = 1$, writing $z = e^{i\theta}$, one has that

$$\sum_{n=1}^{\infty} \frac{e^{i(n+1)\theta}}{n(n + 1)} = \lim_{r \to 1} [(1 - re^{i\theta})\, \text{Log}(1 - re^{i\theta}) + re^{i\theta}].$$

If $z \neq 1$, this limit is $(1 - z)\, \text{Log}(1 - z) + z$. If $z = 1$, $(1 - r)\, \text{Log}(1 - r) \to 0$ when $r \to 1$ and the limit is 1. Hence, equality (2.16) holds for $|z| \leq 1$, $z \neq 1$. If $z = 1$ it gives $\sum_{n=1}^{\infty} \frac{1}{n(n+1)} = 1$, which is evident, since $\frac{1}{n(n+1)} = \frac{1}{n} - \frac{1}{n+1}$. $\quad\square$

## 2.5.2 Analytic functions

The idea of power series expansion is the basis for the notion of an *analytic function*.

**Definition 2.36.** A function $f : U \to \mathbb{C}$, defined on an open set $U$ of the plane, is said to be analytic on $U$ if, locally, it has a power series expansion, that is, for each point $a \in U$ there is a disc $D(a, \delta) \subset U$, $\delta > 0$, and a power series $\sum_{n=0}^{\infty} c_n z^n$ such that $f(z) = \sum_{n=0}^{\infty} c_n (z - a)^n$ if $z \in D(a, \delta)$.

Observe that both the radius $\delta$ and the series $\sum_n c_n z^n$ may depend on the point $a$. By Theorem 2.31 an analytic function on $U$ is indefinitely holomorphic and each of its derivatives $f^{(n)}$, $n = 1, 2, \dots$, is also analytic on $U$. Later on (Theorem 4.9) it will be shown that, conversely, any holomorphic function on an open set $U$ is analytic on $U$ and, then, indefinitely holomorphic. In Definition 2.36, the disc $D(a, \delta)$ may be arbitrarily small, but it will be proved that the power series expansion holds in the biggest disc centered at $a$ and inside $U$, which has radius $\delta(a) = \operatorname{dist}(a, \mathbb{C} \setminus U)$.

**Example 2.37.** A rational function $R = P/Q$ is an analytic function on $U = \mathbb{C} \setminus Z(Q)$, where $Z(Q)$ is the set of the poles of $R$ (the zeros of $Q$). Actually, $R$ is the finite sum of a polynomial in $z$ and simple fractions of type $\alpha \cdot (z - b)^{-k}$, where $b$ is one of the poles of $R$, $k \in \mathbb{N}$ and $\alpha$ is a constant. A proof of this fact may be found in Theorem 5.13. Hence, it is enough to expand $(z - b)^{-k}$ around a point $a \notin Z(Q)$. Put $r = |a - b| > 0$. If $|z - a| < r$, using the geometric series we can write

$$\frac{1}{z - b} = \frac{1}{z - a - (b - a)} = \frac{1}{\left(\frac{z-a}{b-a} - 1\right)(b - a)} = -\sum_{n=0}^{\infty} \frac{(z - a)^n}{(b - a)^{n+1}}.$$

Differentiating successively this expansion and using Theorem 2.31, we obtain, for $k > 1$, that

$$\frac{1}{(z - b)^k} = (-1)^k \sum_{n=0}^{\infty} \binom{n}{k - 1} \frac{(z - a)^{n-k}}{(b - a)^{n+1}}$$

$$= (-1)^k \sum_{n=0}^{\infty} \binom{n + k}{k - 1} \frac{(z - a)^n}{(b - a)^{n+k+1}}.$$

If $b_1, \dots, b_M$ are the poles of $R$, the expansion of $R$ around $a$ will hold in the disc with radius $\delta(a) = \min\{|a - b_j|, \ j = 1, \dots, M\}$, the distance from $a$ to $\mathbb{C} \setminus U = Z(Q)$.    $\square$

Sometimes it is advisable, in order to expand in power series a certain function, to compute first the power series expansion of its derivative.

**Example 2.38.** The function $f(z) = \mathrm{Log}(1-z)$ is analytic on $U = \mathbb{C} \setminus [1, +\infty)$. To see this, first consider $f'(z) = \frac{1}{z-1}$, the power series expansion around 0 of which is known. Around any point $a$, $a \notin [1, +\infty)$, one has

$$\frac{1}{z-1} = \frac{1}{z-a-(1-a)} = -\sum_{n=0}^{\infty} \frac{(z-a)^n}{(1-a)^{n+1}}, \quad |z-a| < |1-a|. \quad (2.17)$$

In the disc $D(a, |1-a|)$ there exist as many branches of $\log(1-z)$ as branches of $\log w$ in $D(1-a, |1-a|)$; any of these branches, $g$, will have an expansion that will be determined by integrating (2.17):

$$g(z) = g(a) - \sum_{n=0}^{\infty} \frac{1}{n+1} \frac{(z-a)^{n+1}}{(1-a)^{n+1}} = g(a) - \sum_{n=1}^{\infty} \frac{(z-a)^n}{n(1-a)^n}.$$

Consider the branch $g$ which has at $a$ the value $\mathrm{Log}(1-a)$. Then

$$g(z) = \mathrm{Log}(1-a) - \sum_{n=1}^{\infty} \frac{(z-a)^n}{n(1-a)^n}, \quad |z-a| < |1-a|.$$

Write now $\delta(a) = \mathrm{dist}(a, [1, +\infty])$; observe that $\delta(a) = |1-a|$, if $\mathrm{Re}\, a \leq 1$ and $\delta(a) = |\mathrm{Im}\, a| \leq |1-a|$ if $\mathrm{Re}\, a > 1$. Then $\mathrm{Log}(1-z)$ and $g$ are branches of $\log(1-z)$ a $D(a, \delta(a))$ which coincide at $a$ and so they are equal and

$$\mathrm{Log}(1-z) = \mathrm{Log}(1-a) - \sum_{n=1}^{\infty} \frac{(z-a)^n}{n(1-a)^n}, \quad |z-a| < \delta(a),$$

that is, $\mathrm{Log}(1-z)$ may be expanded as a power series around $a$. Observe that, when $\mathrm{Re}\, a > 1$, the series which gives the expansion of $\mathrm{Log}(1-z)$ in $|z-a| < \delta(a)$ converges on a bigger disc, the one centered at $a$ with radius $|1-a|$. If $|z-a| < |1-a|$, the series $-\sum_{n=1}^{\infty} \frac{(z-a)^n}{n(1-a)^n}$ converges to the sum $\mathrm{Log}\left(1 - \frac{z-a}{1-a}\right) = \mathrm{Log}\frac{1-z}{1-a}$ (Example 2.34 changing $z$ by $\frac{z-a}{1-a}$). But $\mathrm{Log}(1-a) + \mathrm{Log}\frac{1-z}{1-a}$ is not always equal to $\mathrm{Log}(1-z)$ if $\mathrm{Re}\, a > 1$; for example, if $a = 2+i$, $z = 2 + \varepsilon i$, one has $\frac{1-z}{1-a} = \frac{-1-\varepsilon i}{-1-i} = \frac{1+\varepsilon i}{1+i} = \frac{(1+\varepsilon)+i(\varepsilon-1)}{2}$, which has principal argument close to $-\frac{\pi}{4}$. Since $\mathrm{Arg}(1-a) = -\frac{3\pi}{4}$, the imaginary part of $\mathrm{Log}(1-a) + \mathrm{Log}\frac{1-z}{1-a}$ is close to $-\pi$. However, the imaginary part of $\mathrm{Log}(1-z)$, $\mathrm{Arg}(1-z) = \mathrm{Arg}(-1-\varepsilon i)$, is near $-\pi$ only if $\varepsilon > 0$, because for $\varepsilon < 0$ it is close to $\pi$ (Figure 2.8).    □

For a polynomial in $z$, $P(z) = \sum_0^N c_n(z-a)^n$, it is a trivial algebraic fact that $P$ is also a polynomial in $(z-b)$, for $b \in \mathbb{C}$; simply write $(z-a)^n = (z-b+b-a)^n$ and use Newton's binomial theorem. There are no matters on convergence because the sum is finite. However, if one has a power series $f(z) = \sum_{n=0}^{\infty} c_n(z-a)^n$ with radius of convergence $R > 0$ and $b \in D(a, R)$, it is not immediate to prove

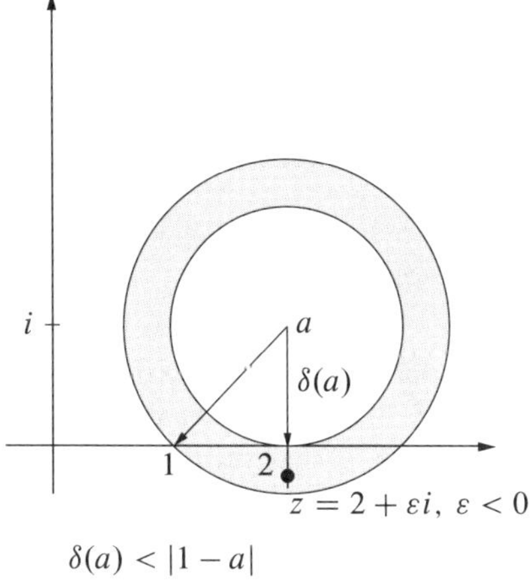


Figure 2.8

that $f$ has a power series expansion in a neighborhood of $b$, that is, that $f$ may be expressed as $f(z) = \sum_n d_n(z - b)^n$ around the point $b$. In other words it is not evident that the sum of the power series $f(z) = \sum_n c_n(z - a)^n$, $z \in D(a, R)$, is an analytic function on $D(a, R)$, the disc of convergence of the series. This result holds and is the content of the following statement.

**Theorem 2.39.** *If* $f(z) = \sum_{n=0}^{\infty} c_n(z - a)^n$ *on* $D(a, R)$ *and* $b \in D(a, R)$, *the series* $\sum_{n=0}^{\infty} \frac{f^{(n)}(b)}{n!}(z - b)^n$ *has radius of convergence greater than or equal to* $R - |a - b|$ *and*

$$f(z) = \sum_{n=0}^{\infty} \frac{f^{(n)}(b)}{n!}(z - b)^n \quad \text{if } |z - b| < R - |a - b|.$$

*Proof.* First note that

$$f^{(n)}(b) = \sum_{m=0}^{\infty} \frac{(n + m)!}{m!} c_{n+m}(b - a)^m, \tag{2.18}$$

$$|f^{(n)}(b)| \leq \sum_{m=0}^{\infty} \frac{(n + m)!}{m!} |c_{n+m}||b - a|^m \tag{2.19}$$

(the power series of $f$ is absolutely convergent on $D(a, R)$). Choose $r$ such that $|b - a| \leq r < R$. Then, using (2.19) and summing up by blocks (Proposition 2.8), it turns out that

$$\sum_{n=0}^{\infty}\left|\frac{f^{(n)}(b)}{n!}\right|(r-|b-a|)^n \leq \sum_{n,m=0}^{\infty}\frac{(n+m)!}{m!n!}|c_{n+m}||b-a|^m(r-|b-a|)^n$$

$$=\sum_{k=0}^{\infty}|c_k|\sum_{m+n=k}\frac{(n+m)!}{m!n!}|b-a|^m(r-|b-a|)^n=\sum_{k=0}^{\infty}|c_k|r^k<+\infty.$$

This shows that the Taylor series of $f$ centered at $b$ has radius of convergence greater than or equal to $r-|b-a|$; since $r$ is arbitrarily close to $R$, this radius is greater than or equal to $R-|b-a|$. Take now a point $z$ such that $|z-b|\leq r-|b-a|$ with $r<R$. The same computation just done shows that the double series

$$\sum_{n,m}^{\infty}c_{n+m}\frac{(n+m)!}{m!n!}(b-a)^m(z-b)^n$$

converges absolutely and, by Proposition 2.8, the terms of the series may be grouped arbitrarily to compute the sum. Making groups for $m+n=k$, as before, we get $\sum_{k=0}^{\infty}c_k(z-a)^k=f(z)$, while summing up first in $m$ and using (2.18) we get the series $\sum_{n=0}^{\infty}\frac{f^{(n)}(b)}{n!}(z-b)^n$. $\qquad\square$

This theorem is also a consequence of the result which says that every holomorphic function on an open set $U$ is analytic on $U$ (Theorem 4.9).

## 2.6 Real analytic functions and their complex extension

The concept of power series and the associated notions (domain of convergence, operations with power series, etc.) may be set out in the context of one real variable motivated, in this case, by approximation by Taylor polynomials. Recall first that a function of one or several real variables is called a $C^r$ *function*, $1\leq r\leq+\infty$, on a domain if it has there continuous $r$-th order partial derivatives.

If $f$ is a $C^N$ function on an open interval $I\subset\mathbb{R}$ and $a\in I$, Taylor's formula reads

$$f(x)=\sum_{k=0}^{N}\frac{f^{(k)}(a)}{k!}(x-a)^k+R_N(x)$$

where $R_N(x)=o(|x-a|^N)$, for $x\to a$, is the remainder term of order $N$. When $f$ is $C^{N+1}$, there are explicit expressions of $R_N(x)$, as Lagrange's formula,

$$R_N(x)=\frac{f^{(N+1)}(\xi)}{(N+1)!}(x-a)^{N+1},$$

with $\xi$ intermediate between $a$ and $x$. The crux of the matter is that $f(x)$ may be approximated by the Taylor polynomial $\sum_{n=0}^{N} \frac{f^{(n)}(a)}{n!}(x - a)^n$ and that the remainder, $R_N(x)$, becomes small as $N$ increases. Naturally, for $f$ a $C^\infty$ function one may ask if $R_N(x) \to 0$ when $N \to \infty$, that is, if the following equality holds:

$$f(x) = \lim_{N \to 0} \sum_{k=0}^{N} \frac{f^{(k)}(a)}{k!}(x - a)^k = \sum_{k=0}^{\infty} \frac{f^{(k)}(a)}{k!}(x - a)^k. \tag{2.20}$$

This leads to consider power series in one real variable $x$, $\sum_{n=0}^{\infty} c_n (x - a)^n$. For these series, definitions and statements of Subsections 2.3.3 and 2.3.4 hold, and also the corresponding version of Theorem 2.31. This means that for the series $\sum_n c_n (x - a)^n$ a radius of convergence $R$ is defined so that the series converges on the interval $\{x : |x - a| < R\}$ and there it defines a function $f(x) = \sum_n c_n (x - a)^n$, which has infinite derivatives with respect to the real variable $x$. Moreover one must have $c_n = \frac{f^{(n)}(a)}{n!}$, and then, the only way of expressing $f(x)$ as a sum of a power series is by means of its Taylor series (2.20).

If the function $f$ is defined on an open interval $I \subset \mathbb{R}$, it is said that $f$ is a *real analytic function* on $I$ if the version of Definition 2.36 for a real variable holds. This means that for each point $a \in I$ there exists a $\delta(a) > 0$ such that (2.20) holds in the interval $\{x : |x - a| < \delta(a)\}$. In particular, if $f$ is analytic on $I$, $f$ is $C^\infty$ on $I$.

In calculus, using Lagrange's formula for the remainder $R_N(x)$, the following equalities are proved:

$$e^x = \sum_{n=0}^{\infty} \frac{x^n}{n!}, \quad \sin x = \sum_{n=0}^{\infty}(-1)^n \frac{x^{2n+1}}{(2n + 1)!}, \quad \cos x = \sum_{n=0}^{\infty}(-1)^n \frac{x^{2n}}{(2n)!}.$$

These expansions hold on the whole real line $\mathbb{R}$. But the Taylor series of a function $f$ is not always convergent for every real number, even when $f$ is $C^\infty$ on $\mathbb{R}$. For example, the equality

$$\frac{1}{1 + x^2} = \sum_{n=0}^{\infty}(-1)^n x^{2n}$$

holds only for $|x| < 1$, even though the function $\frac{1}{1+x^2}$ is $C^\infty$ on $\mathbb{R}$.

All these considerations may be, as said, confined to the context of real variables. But, associating an *interval of convergence* $(a - R, a + R)$ to a real power series $\sum_{n=0}^{\infty} c_n (x - a)^n$, with $a \in \mathbb{R}$, $c_n \in \mathbb{R}$, is an unnecessary limitation. Actually, what has been established by now ensures that if $\sum_{n=0}^{\infty} c_n (x - a)^n$ converges for $|x - a| < R$, then the complex series $\sum_{n=0}^{\infty} c_n (z - a)^n$ converges for $z \in D(a, R)$. This means that the *natural domain* of definition of power series are the discs, not the intervals. The same way, the natural domain of definition of real analytic functions are not the intervals $I \subset \mathbb{R}$, but the symmetric domains $U \subset \mathbb{C}$ with respect to the real axis. A precise statement of this fact will be given in Theorem 4.28. This natural extension of real analytic functions to a complex domain has already been treated in some examples. Indeed, the complex extension of the real exponential function $e^x = \sum_0^\infty \frac{x^n}{n!}$ has been defined as

$$e^z = e^x(\cos y + i \sin y) = \sum_0^\infty \frac{z^n}{n!}$$

if $z = x + iy$ (Subsection 2.3.5). Similarly, for the trigonometric functions, one has

$$\cos z = \frac{e^{iz} + e^{-iz}}{2} = \sum_0^\infty \frac{(-1)^n z^{2n}}{(2n)!};$$

$$\sin z = \frac{e^{iz} - e^{-iz}}{2i} = \sum_0^\infty (-1)^n \frac{z^{2n+1}}{(2n+1)!},$$

which are analytic functions on $\mathbb{C}$, by Theorem 2.39.

Let us start now, for example, from a rational function $R(x) = \frac{P(x)}{Q(x)}$ where $P$ and $Q$ are polynomials in $x$ so that $R(x)$ is analytic on $\mathbb{R} \setminus Z_0(Q)$, where $Z_0(Q)$ are the real zeros of $Q(x)$. Then its complex extension is the rational function $R(z) = \frac{P(z)}{Q(z)}$ which is analytic on $\mathbb{C} \setminus Z(Q)$ where $Z(Q)$ is the set of all (complex) zeros of $Q$ in which $P(z)$ has a zero of smaller order than the one of $Q$ (or does not have any zero).

Furthermore, the complex point of view is useful for a better understanding of some features which may not be completely explained from the real analysis point of view. For example, consider the two expansions

$$\frac{1}{1-x} = \sum_{n=0}^\infty x^n \quad \text{and} \quad \frac{1}{1+x^2} = \sum_{n=0}^\infty (-1)^n x^{2n},$$

which hold if $|x| < 1$. In the first case the fact that the domain of convergence is $(-1, 1)$ may not be surprising here because the function, which has a singularity at $x = 1$, is only defined on the open set $U = \mathbb{R} \setminus \{1\}$, and the radius of convergence is the distance from zero to $U^c = \{1\}$. Instead, in the second case, the function is defined on the whole line $\mathbb{R}$ but its expansion holds only on $(-1, 1)$. Now, thinking about it in terms of complex extension $\frac{1}{1+z^2}$, it is better understood: there are singularities at the points $z = \pm i$, the distance of which to 0 is 1. This explains that here the radius of convergence is 1. With this complex point of view other questions are also clarified. For example, in which intervals does the power series expansion of $\frac{1}{1+x^2}$ around a point $a \in \mathbb{R}$ converge? The answer is the interval centered at

$a$ and with radius $\sqrt{1 + a^2}$, which is the distance from $a$ to the poles $\pm i$ of the function $\frac{1}{1+z^2}$.

The last example consists of the function

$$f(x) = \frac{1}{(1 + x^2)(4 - x^2)},$$

which is real analytic on the interval $I = (-2, 2)$, but has a power series expansion around 0 convergent only on $(-1, 1)$, and not on the whole interval $I$. This last one is the biggest interval centered at 0 contained in the domain $I$ where the function is analytic. The reduction is due, once again, to the complex singularities of the factor $\frac{1}{1+x^2}$.

The sense of all these comments is that analyticity is genuinely a concept of complex variable theory. In the setting of real variables, the notions of $\mathbb{R}$-differentiable function, of indefinitely $\mathbb{R}$-differentiable function and real analytic function are different. There are $\mathbb{R}$-differentiable functions which are not indefinitely $\mathbb{R}$-differentiable, for example $|x|x$, and there are $C^\infty$ functions which are not real analytic, for example

$$f(x) = \begin{cases} e^{-1/x^2} & \text{if } x \neq 0, \\ 0 & \text{if } x = 0. \end{cases}$$

Instead, in the case of a complex variable, it will be seen in Chapter 4 that the three correspondent concepts, $\mathbb{C}$-differentiable (holomorphic) function, indefinitely holomorphic function and complex analytic function, are exactly the same. Due to an abuse of language, when talking about *functions of a complex variable*, one often refers to analytic functions, and similarly when talking about the complex extensions of a function $f(x)$, defined for real values of the variable $x$, one refers to the extensions $F(z)$ (that is, $F(z) = f(x)$ if $z = x \in \mathbb{R}$) which are analytic. Extending a function $f(x)$ to a function $F(z)$ without this requirement of analyticity is trivial, just by taking

$$F(x + iy) = f(x).$$

Moreover, there are countless different ways of extending $f$; for example, one may also use $F(x + iy) = f(x) + iG(y)$ with $G(0) = 0$. However, if $f(x)$ is real analytic the extension of $f$ to a complex analytic function is unique.

There is still another notion of analytic function, which is the one of analytic function of two real variables. A complex- or real-valued function $f$ defined on an open set $U \subset \mathbb{R}^2 \cong \mathbb{C}$ is called an *analytic function of two real variables* on $U$ if for any point $z_0 = (x_0, y_0) \in U$ there exists a disc $D(z_0, r) \subset U$ and a double power series $\sum_{n,m} c_{n,m} x^n y^m$ such that

$$f(x, y) = \sum_{n,m=0}^{\infty} c_{n,m}(x - x_0)^n (y - y_0)^m \quad \text{if } \sqrt{(x - x_0)^2 + (y - y_0)^2} < r.$$

Once more, in this case $f$ is $C^\infty$ and the expansion, which is unique, must be its Taylor series for functions in two variables, that is,

$$c_{n,m} = \frac{1}{n!\,m!} \frac{\partial^{n+m} f}{\partial x^n \partial y^m}(x_0, y_0).$$

Clearly, if $f$ is analytic in the variable $z = x + iy$, then it is also analytic in the variables $x$, $y$. However, in the same way that not every polynomial in $x$, $y$ is a polynomial in $z$, the converse is not true. For example, $\frac{1}{1+x^2+y^2}$ is an analytic function of $x$, $y$ on the whole plane, and is not analytic in $z$ on any open set in $\mathbb{C}$ (since being a non-constant real-valued function, it cannot be holomorphic on any domain, according to Proposition 2.26).

## 2.7 Exercises

1. Prove that the equality $\lim_{n\to\infty}(1+\frac{z}{n})^n = e^z$ holds for any complex number $z$, using the identity:

$$\left(1 + \frac{z}{n}\right)^n = 1 + z + \sum_{p=2}^{n}\left(1 - \frac{1}{n}\right)\cdots\left(1 - \frac{p-1}{n}\right)\frac{z^p}{p!}.$$

2. Show that the function $\cos z$ maps the strip $B = \{z : 0 < \operatorname{Re} z < \pi\}$ onto the domain $U = \mathbb{C} \setminus \{x \in \mathbb{R} : |x| \geq 1\}$ conformally and injectively. Express its inverse function in terms of the logarithmic function.

3. Show that the complex numbers $z \neq 0$ such that $z^z$ takes only real values are all contained in a countable set of lines which are parallel to the imaginary axis, and that there are infinitely many numbers with this property on each line.

4. If $\gamma(t) = re^{it}$, $0 \leq t \leq 2\pi$, $r > 0$, and $h_w(z) = \operatorname{Log}|z - w|$, show the equality

$$\int_\gamma h_w\,ds = 2\pi r \operatorname{Log}(\max(r, |w|)).$$

5. Let $f$ be the holomorphic function defined on the unit disc $\mathbb{D}$ by $f(z) = \sum_{n=1}^{\infty} \frac{z^{2n-1}}{2n-1}$. Show that $f$ is injective on $\mathbb{D}$ and $f(\mathbb{D}) = \{z \in \mathbb{C} : -\pi/4 < \operatorname{Im} z < \pi/4\}$.

6. Compute, for $t \in [-\pi, \pi]$, the sum of the following series:

     a) $\displaystyle\sum_{n=1}^{\infty} \frac{\sin nt}{n}$;

b) $\displaystyle\sum_{n=2}^{\infty}(-1)^n \frac{\cos nt}{n^2-1}$.

7. Given $a, b, c \in \mathbb{R}$ consider $P(z) = ax^2 + 2bxy + cy^2$ where $z = x + iy$. Determine the conditions that $a, b, c$ must fulfill so that there exists an entire function $f$ satisfying $P(z) = \operatorname{Re} f(z)$, $z \in \mathbb{C}$. Find all the entire functions which satisfy it.

8. Let $f(z) = \sum_n c_n z^n$ be the sum of a power series with radius of convergence $R > 0$ and $f'(0) = c_1 \neq 0$. Prove the equality

$$|f(z) - f(w)| = |z - w| \left| \sum_{n=1}^{\infty} c_n (z^{n-1} + z^{n-2}w + \cdots + zw^{n-2} + w^{n-1}) \right|$$

for all $z, w \in D(0, R)$ and, consequently, that $f$ is injective on the disc $D(0, r)$ if $0 < r < R$, and that the following inequality holds:

$$\sum_{n=2}^{\infty} n|c_n|r^{n-1} < |c_1|.$$

9. Let $f(z) = \sum_n c_n z^n$ be the sum of a power series with radius of convergence $R > 0$ and $0 < r < R$. Show:

a)
$$\frac{1}{2\pi} \int_0^{2\pi} |f(re^{it})|^2 \, dt = \sum_n |c_n|^2 r^{2n}.$$

Use this equality to deduce that a bounded holomorphic function on the whole plane must be constant (Liouville's Theorem).

b)
$$\frac{1}{\pi} \iint_{D(0,r)} |f'(z)|^2 \, dx \, dy = \sum_n n|c_n|^2 r^{2n},$$

where $z = x + iy$. If in addition $f$ is injective, then the previous expression is the area of $f(D(0, r))$.

10. Find the expression of Cauchy–Riemann equations in polar coordinates.

11. If $f(z) = \sum_n c_n z^n$, $|z| < R$ and $0 < r < R$, let $L(r)$ be the length of the image curve by $f$ of the circle $\{re^{it} : 0 \le t \le 2\pi\}$. Show the inequality

$$L(r) \ge 2\pi r |c_1|.$$

**12.** Let $f(z) = \sum_n c_n z^n$ with radius of convergence $R > 1$ and $z = x + iy$. Prove:

a)
$$\iint_{\mathbb{D}} |f(z)|^2 \, dx \, dy = \pi \sum_n \frac{|c_n|^2}{n+1}.$$

b)
$$\iint_{\mathbb{D}} |f(z)|^4 \, dx \, dy = \pi \left( \sum_n |c_n|^2 \right)^2.$$

**13.**   a) Show that the function $\arctan z = \tan^{-1} z$ with $\tan z = \frac{\sin z}{\cos z}$ has a holomorphic branch on the unit disc and find all its holomorphic branches.

b) Find the power series expansion around the origin of the function

$$f(z) = z \arctan z - \operatorname{Log} \sqrt{1 + z^2}$$

and its radius of convergence; here arctan indicates a branch of $\tan^{-1}$, and also the principal branch of the square root is taken.

c) Study the behavior of the series in b) at the boundary of its disc of convergence, and explain the relation between the sum of this series and the function $f$. Compute $\sum_{n=1}^{\infty} \frac{(-1)^{n+1}}{2n(2n-1)}$.

**14.** For each $a \in \mathbb{D}$ define $f_a(z) = \frac{z-a}{1-\bar{a}z}$. Show that $f_a$ is one-to-one on $\bar{\mathbb{D}}$, holomorphic on a neighborhood of $\bar{\mathbb{D}}$ and satisfies $f_a(\mathbb{D}) = \mathbb{D}$, $f_a(\bar{\mathbb{D}}) = \bar{\mathbb{D}}$ and $f_a^{-1} = f_{-a}$, on $\mathbb{D}$.

**15.** If $f(z) = \sum_n c_n z^n$ for $|z| < R$, show that, for $0 < r < R$,

$$|c_n| \le \frac{M(r)}{r^n}, \quad n = 0, 1, 2, \ldots$$

where $M(r) = \sup\{|f(z)| : |z| = r\}$ (Cauchy's inequalities).

Deduce from these inequalities that in the case $R = +\infty$ one has that $\sum_n |c_n| r^n \le 2M(2r)$ for all $r > 0$.

**16.** Show that the function $f(z) = \sqrt{|xy|}$, $z = x + iy$, satisfies the Cauchy–Riemann equations at the origin but is not $\mathbb{C}$-differentiable at this point.

**17.** Show that if $P$ is a polynomial, then the zeros of the derivative $P'$ are all in the convex hull of the set of zeros of $P$. As a hint prove that if all the zeros of $P$ are in a half plane, then the zeros of $P'$ are also in the same half plane.

**18.** For any $\alpha \in \mathbb{C}$ the function $f(z) = (1 + z)^\alpha$ has a holomorphic branch on $\mathbb{D}$ such that $f(0) = 1$. Show that, for this branch, the equality

$$f(z) = \sum_{n=0}^{\infty} \binom{\alpha}{n} z^n, \quad |z| < 1$$

holds and study the behavior of the series when $|z| = 1$.

**19.** Let $f(z) = \sum_{0}^{\infty} c_n z^n$ be the sum of a power series with radius of convergence $R > 0$ and suppose that there is a $k \subset \mathbb{N}$ and a function $g$ defined on $D(0, R^k)$ such that $f(z) = g(z^k)$ if $|z| < R$. Prove that the function $g$ is the sum of a power series with radius of convergence $R^k$ and that $c_n = 0$ if $n$ is not a multiple of $k$.

**20.** Consider the polynomial $P(z) = 1 - \sum_{n=1}^{N} \alpha_n z^n$ with $\alpha_1, \ldots, \alpha_N \in \mathbb{R}$ and suppose that the function $f(z) = \frac{1}{P(z)}$ has as power series expansion $f(z) = \sum_{n=0}^{\infty} c_n z^n$ around the origin. Show that the coefficients $c_n$ satisfy the relations $c_0 = 1$, $c_n = \sum_{j=1}^{N} \alpha_j c_{n-j}$ if $n \geq 1$, understanding that $c_k = 0$ if $k < 0$.

Show now that, defining the sequence $(c_n)$ recursively by means of the previous relations starting from $\alpha_1, \ldots, \alpha_N \in \mathbb{R}$, there exists a constant $M > 0$ such that $|c_n| \leq M^n$, for $n \geq 0$. Consequently the power series $\sum_n c_n z^n$ has a radius of convergence $R > 0$ and defines a function $f(z)$ on the disc $D(0, R)$. Check now that $f(z) \cdot P(z) = 1$ and that $R = \min(|\beta_1|, \ldots, |\beta_N|)$, where $\beta_1, \ldots, \beta_N$ are the zeros of the polynomial $P$.

**21.** Show that the power series $\sum_n \frac{z^n}{n^2}$ has radius of convergence equal to 1 and that it converges uniformly on the closed disc $\overline{\mathbb{D}}$. Show now that the sum $f(z)$ of this series satisfies the equation

$$f'(z) = -\frac{1}{z} \operatorname{Log}(1 - z), \quad |z| < 1.$$

Deduce from here that on the domain $U = \{z : |z| < 1, |z - 1| < 1\}$ one has

$$f(z) + f(1 - z) = c - \operatorname{Log} z \cdot \operatorname{Log}(1 - z)$$

with $c = \sum_{n=1}^{\infty} \frac{1}{n^2}$. Use the previous equation to compute the value of $\sum_{n=1}^{\infty} \frac{1}{n^2 2^n}$.

**22.** Let $f, g \in H(\mathbb{D})$ have the power series expansions $f(z) = \sum_{n=1}^{\infty} c_n z^n$ and $g(z) = \sum_{n=1}^{\infty} d_n z^n$ for $|z| < 1$. Show that the series $\sum_{n=1}^{\infty} c_n g(z^n)$ and $\sum_{n=1}^{\infty} d_n f(z^n)$ are uniformly convergent on each compact disc $\overline{D}(0, r)$, $0 < r < 1$, and they define the same holomorphic function on $\mathbb{D}$.

Put now $F(z) = \sum_{n=1}^{\infty} c_n g(z^n) = \sum_{n=1}^{\infty} d_n f(z^n)$ and let $\sum_{n=1}^{\infty} h_n z^n$ be a third convergent power series on $\mathbb{D}$. Prove the equality

$$\sum_{n=1}^{\infty} h_n F(z^n) = \sum_{n,m,k=1}^{\infty} c_n d_m h_k z^{nmk}, \quad |z| < 1.$$

# Chapter 3

# Holomorphic functions and differential forms

In calculus of one real variable, a basic result is the fundamental theorem of calculus, which sets a link between the concepts of differentiation and integration: any continuous function has an antiderivative, and this is given by means of the indefinite integral, up to additive constants. In this chapter this kind of questions for one complex variable functions will be studied. To this end, it is convenient to use the language of vector fields and differential 1-forms. The role of indefinite integrals now will be played by line integrals of fields or forms along paths of the complex plane.

A fundamental result for developing the theory of holomorphic functions is Cauchy's Theorem on cancellation of line integrals of holomorphic functions along closed paths. This result will be presented as a particular case of a version of Green's formula with very weak hypotheses on regularity, which are also valid for the classical theorems of vector calculus.

The second part of the chapter is stated in a real variables context to see that the notion of a holomorphic function corresponds to the notion of a locally conservative solenoidal vector field. Here appears naturally the Laplace operator and the concept of a harmonic function: locally conservative solenoidal fields are locally gradients of harmonic functions. One reaches this way the notion that, from a real variable point of view, generalizes the concept of a holomorphic function to functions defined on open sets of $\mathbb{R}^n$.

## 3.1  Complex line integrals

The definition of a complex line integral entails complex integration over intervals of $\mathbb{R}$ of functions which take *complex values*. If $f(t) = u(t) + iv(t)$ is a continuous function for $t \in [a, b]$, with $u$, $v$ real-valued, one writes

$$\int_a^b f(t)\, dt = \int_a^b u(t)\, dt + i \int_a^b v(t)\, dt.$$

With this definition, the linearity property holds:

$$\int_a^b (\alpha f(t) + \beta g(t))\, dt = \alpha \int_a^b f(t)\, dt + \beta \int_a^b g(t)\, dt,$$

where $f$, $g$ are continuous on $[a, b]$ and $\alpha$, $\beta$ are complex constants. It also holds that

$$\left| \int_a^b f(t)\, dt \right| \le \int_a^b |f(t)|\, dt. \tag{3.1}$$

To prove this inequality, let $\lambda$ be a number with $|\lambda| = 1$ such that $\lambda \int_a^b f(t)\, dt \ge 0$; that is, if $\int_a^b f(t)\, dt = re^{i\theta}$, $r \ne 0$, we put $\lambda = e^{-i\theta}$ (if this integral is zero, there is nothing to prove). Then,

$$\left| \int_a^b f(t)\, dt \right| = \lambda \int_a^b f(t)\, dt = \mathrm{Re}\left[ \lambda \int_a^b f(t)\, dt \right]$$

$$= \int_a^b \mathrm{Re}\,\lambda f(t)\, dt \le \int_a^b |f(t)|\, dt.$$

**Definition 3.1.** If $\gamma(t)$, $a \le t \le b$, is a $C^1$ curve on the complex plane and $f$ is a complex-valued continuous function on $\gamma^*$, the complex line integral $\int_\gamma f(z)\, dz$ of $f$ along $\gamma$ is

$$\int_\gamma f(z)\, dz = \int_a^b f(\gamma(t))\gamma'(t)\, dt.$$

Formally $z$ is replaced by $\gamma(t)$ and $dz$ by $\gamma'(t)\, dt$. Observe that the function to be integrated over $[a, b]$ is a complex-valued function, product of $f(\gamma(t))$ and $\gamma'(t)$, and the result will be a complex number. The integral is well defined here because by hypothesis the integrand is continuous on $[a, b]$. Identically to the case of integrals of real functions or integrals with respect to the arc length $ds$, $\int_\gamma f(z)\, dz$ is the limit of the *complex Riemann sums*

$$\sum_i f(z_i)(z_{i+1} - z_i),$$

where $z_i = \gamma(t_i)$ are the vertices of a polygonal line inscribed on $\gamma^*$. Replacing $(z_{i+1} - z_i)$ by $|z_{i+1} - z_i|$, the corresponding limit is $\int_\gamma f(z)\, ds$, as explained in Subsection 1.4.3. For this reason, it is usually written $|dz|$ instead of $ds$; hence

$$\int_\gamma f(z)\, |dz| = \int_a^b f(\gamma(t))|\gamma'(t)|\, dt.$$

With the given definition of complex line integral, one obviously has

$$\int_\gamma dz = \gamma(b) - \gamma(a).$$

**Example 3.2.** If $\gamma(t) = e^{it}$, $0 \le t \le 2\pi$, and $f(z) = \frac{1}{z}$ we have $\int_\gamma f(z)\,dz = \int_0^{2\pi} \frac{ie^{it}}{e^{it}}\,dt = 2\pi i$. However, $\int_\gamma f(z)|dz| = \int_0^{2\pi} \frac{dt}{e^{it}} = 0$. If $\gamma(t) = (t, t^2) = t + it^2$, $0 \le t \le 1$, and $f(z) = \bar{z}^2$, then $\int_\gamma f(z)\,dz = \int_0^1 (t - it^2)^2(1 + 2it)\,dt = \int_0^1 \left((3t^4 + t^2) - 2it^5\right) dt = \frac{14}{15} - \frac{i}{3}$. $\qquad\square$

Look now at the behavior of the complex line integral with a change of parametrization: $t = \Phi(u)$, $c \le u \le d$, $\Gamma(u) = \gamma(\Phi(u))$. Applying the definitions it turns out that

$$\int_\Gamma f(z)\,dz = \int_c^d f(\Gamma(u))\Gamma'(u)\,du, \qquad (3.2)$$

and since $\Gamma'(u) = \gamma'(\Phi(u))\Phi'(u)$, the integral of (3.2) is

$$\int_c^d f(\gamma(\Phi(u)))\gamma'(\Phi(u))\Phi'(u)\,du.$$

Applying now the change of variable $t = \Phi(u)$, when the reparametrization is orientation preserving, that is, when $\Phi$ is increasing and $\Phi(c) = a$, $\Phi(d) = b$, we get

$$\int_\Gamma f(z)\,dz = \int_a^b f(\gamma(t))\gamma'(t)\,dt = \int_\gamma f(z)\,dz.$$

If the reparametrization changes the orientation, we obtain for $\int_\Gamma f(z)\,dz$ the opposite value of $\int_\gamma f(z)\,dz$.

If $\gamma$ is a piecewise $C^1$ curve and $\gamma_1, \gamma_2, \ldots, \gamma_n$ are the pieces with continuous derivative so that the ending point of $\gamma_i$ is the beginning of $\gamma_{i+1}$, a situation that will be denoted by $\gamma = \gamma_1 + \gamma_2 + \cdots + \gamma_n$, then we will write

$$\int_\gamma f(z)\,dz = \sum_{i=1}^n \int_{\gamma_i} f(z)\,dz.$$

Since the integrals do not depend on parametrizations if they are orientation preserving, there is no need for the parameters of the pieces $\gamma_i$ to vary on disjoint intervals, so it is usual to work with each of them separately.

**Example 3.3.** The integral of the function $z^2$ along the path that consists of $\gamma_1$, which is the arc of the parabola $y = x^2$ that goes from the origin to the point $(1, 1) = 1 + i$, followed by the vertical segment $\gamma_2$, which goes from $1 + i$ to $1$, has two parts: in order to compute $\int_{\gamma_1}$ we use the parametrization $z = t + it^2$, $0 \le t \le 1$, $dz = (1 + 2it)\,dt$ and so we have $\int_0^1 (t + it^2)^2(1 + 2it)\,dt = \int_0^1 (t^2 - 5t^4 + 2it^3(2 - t^2))\,dt = \frac{2}{3}(i - 1)$; to compute $\int_{\gamma_2}$ we can use the parametrization $z = 1 + (1 - t)i$, $dz = -i\,dt$, $0 \le t \le 1$; this is the same as using $z = 1 + ti$, $dz = i\,dt$, where $t$ goes from $1$ to $0$: it turns out that $\int_1^0 (1 - t^2 + 2it)(i\,dt) = 1 - \frac{2}{3}i$. Hence, $\int_{\gamma_1 + \gamma_2} z^2\,dz = \frac{1}{3}$. $\qquad\square$

In the estimations of complex line integrals the following inequality will be often used:

$$\left| \int_\gamma f(z)\,dz \right| \le \int_\gamma |f|\,ds,$$

which is an immediate consequence of the definitions and of (3.1). In particular, if $f$ satisfies the bound $|f(z)| \le M$ on $\gamma^*$, one has

$$\left| \int_\gamma f(z)\,dz \right| \le ML(\gamma). \tag{3.3}$$

To finish this section, observe that the index of a path $\gamma$ with respect to a point $z \notin \gamma^*$, given by (1.6) of Subsection 1.5.2, may be written as a complex line integral

$$\operatorname{Ind}(\gamma, z) = \frac{1}{2\pi i} \int_a^b \frac{\gamma'(t)}{\gamma(t) - z}\,dt = \frac{1}{2\pi i} \int_\gamma \frac{dw}{w - z}. \tag{3.4}$$

This expression will be very useful from now on. For example, using the inequality (3.3), one finds

$$|\operatorname{Ind}(\gamma, z)| \le \frac{L(\gamma)}{2\pi\, d(z, \gamma^*)},$$

where $d$ denotes the Euclidian distance, and then it is clear that the index converges to zero when $z$ tends to infinity. This fact implies that it is necessarily zero in the unbounded component of $\mathbb{C} \setminus \gamma^*$.

If a series $\sum_n f_n(t)$ of real- or complex-valued continuous functions on an interval $[a, b]$ converges uniformly on $[a, b]$, then its sum is a continuous function, the numerical series $\sum_n \int_a^b f_n(t)\,dt$ is convergent and

$$\int_a^b \left( \sum_n f_n(t) \right) dt = \sum_n \int_a^b f_n(t)\,dt.$$

Combining this fact with Definition 3.1 we get the following property:

**Proposition 3.4.** *If $\gamma$ is a path on the complex plane and $(f_n)$ is a sequence of continuous functions on $\gamma^*$ such that $\sum_n f_n(z)$ converges uniformly on $\gamma^*$, then $\sum_n \int_\gamma f_n(z)\,dz$ is convergent and*

$$\int_\gamma \left( \sum_n f_n(z) \right) dz = \sum_n \int_\gamma f_n(z)\,dz.$$

## 3.2  Line integrals, vector fields and differential 1-forms

Complex line integrals are a particular case of line integrals or circulations of vector fields along curves or, equivalently, of the integrals of differential 1-forms (when complex values are allowed). A review of these concepts in their natural context, the domains of $\mathbb{R}^n$, will be done in this section.

A continuous *vector field* $\vec{X}$ on a domain $U$ of $\mathbb{R}^n$ is simply a continuous mapping $\vec{X} \colon U \to \mathbb{R}^n$ that makes a correspondence between a point $x \in U$ and a vector $\vec{X}(x)$, placed with the origin at the point $x$. Velocity fields and force fields are typical examples of vector fields. More generally we can consider vector fields $\vec{X}(x, t)$ which depend also on time, examples of which will be seen in Chapter 7. However, in the following paragraphs only vector fields that are independent of time will be dealt with, known also as *stationary vector fields*.

In the complex plane, with the usual identification, a vector field on a domain $U$ is the same as a complex function defined on $U$. Sometimes it is convenient to look at complex functions as vector fields.

**Example 3.5.** The function $f(z) = z$ defines a radial field. The function $e^z$, which is $2\pi i$ periodic, defines the vector field in Figure 3.1. $\qquad\square$

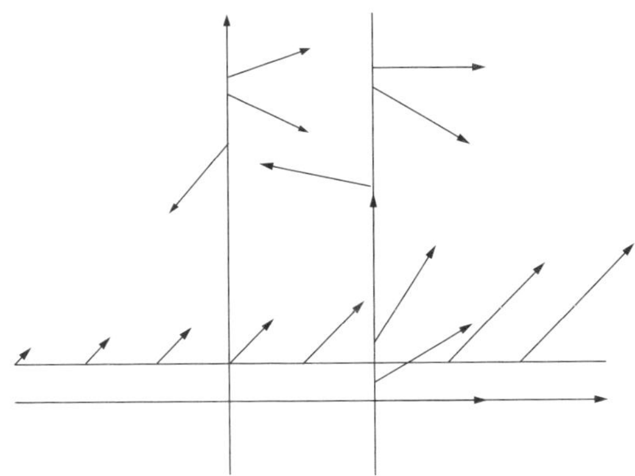

Figure 3.1

An important concept associated to a vector field $\vec{X}$ is the one of *integral curves* or *orbits* of the field. They are $C^1$ curves $\gamma$ which at each point $\gamma(t)$ have tangent vector equal to $\vec{X}(\gamma(t))$. Analytically, this is expressed as

$$\gamma'(t) = \vec{X}(\gamma(t)),$$

or in coordinates, writing $\vec{X} = (X_1, X_2, \ldots, X_n)$ and $\gamma(t) = (x_1(t), x_2(t), \ldots, x_n(t))$, by the differential equations system

$$\frac{dx_i}{dt} = X_i(x_1(t), x_2(t), \ldots, x_n(t)), \quad i = 1, 2, \ldots, n.$$

**Example 3.6.** The orbits of the field $f(z) = \bar{z}$ are the solutions of $x' = x$, $y' = -y$, that is $x(t) = c_1 e^t$, $y(t) = c_2 e^{-t}$. They are hyperbolas. The orbits of $f(z) = iz$ are the circles $x(t) = r\cos t$, $y(t) = r\sin t$. The orbits of the field $f(z) = \frac{1}{\bar{z}} = \frac{x+iy}{x^2+y^2}$ are the solutions of the system

$$x' = \frac{x}{x^2 + y^2},$$

$$y' = \frac{y}{x^2 + y^2}.$$

These equations imply that $\frac{d}{dt}\arctan\left(\frac{y}{x}\right) = 0$, that is, $\arctan\frac{y}{x}$ must be constant on each orbit. The orbits are located then over rays starting at the origin of coordinates. In order to find the parametrization one would better use polar coordinates $x(t) = r(t)\cos\theta(t)$, $y(t) = r(t)\sin\theta(t)$. Then it turns out that $r(t) = \sqrt{2t}$, $\theta(t) = \text{constant}$ (Figure 3.2). $\qquad\square$

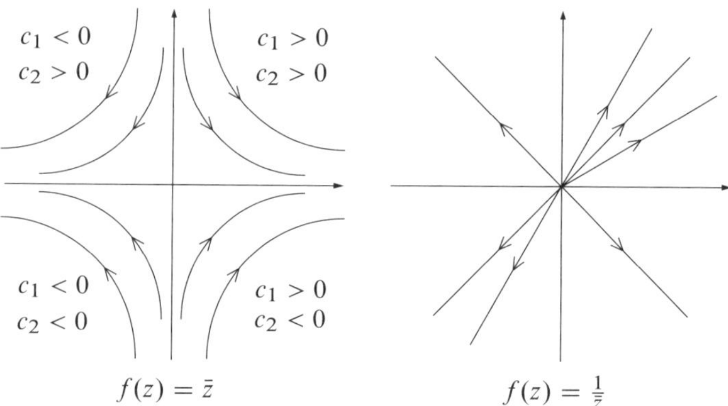


Figure 3.2

A continuous vector field $\vec{X}$ on a domain $U$ of $\mathbb{R}^n$ gives rise to two important notions: *circulations* and *flows*. The concept of flow will be introduced in the next section. The *circulation of $\vec{X}$ along a regular curve* $\gamma$ is, by definition, the sum of the tangential components of the vector field $\vec{X}$ over $\gamma$, that is, if $\vec{T}$ is the unitary tangent vector to $\gamma$, the circulation is

$$\int_\gamma \langle \vec{X}, \vec{T} \rangle \, ds,$$

where $\langle\,,\,\rangle$ is the scalar product in $\mathbb{R}^n$. Analytically, in terms of a parametrization $\gamma(t), a \le t \le b$, since $\vec{T}(\gamma(t)) = \frac{\gamma'(t)}{|\gamma'(t)|}$ and $ds = |\gamma'(t)|dt$ (Subsection 1.4.3), the circulation turns out to be

$$\int_a^b \langle \vec{X}(\gamma(t)), \gamma'(t) \rangle \, dt.$$

This integral makes sense more generally for a piecewise $C^1$-curve $\gamma$. When $\vec{X}$ is a force field, the circulation is the work done by the field to move the unit mass along the orbit. In terms of the components one has

$$\int_\gamma \langle \vec{X}, \vec{T} \rangle \, ds = \int_a^b \left( X_1(\gamma(t))x_1'(t) + X_2(\gamma(t))x_2'(t) + \cdots + X_n(\gamma(t))x_n'(t) \right) dt,$$

for what reason the following notation is also used:

$$\int_\gamma \langle \vec{X}, \vec{T} \rangle \, ds = \int_\gamma X_1 dx_1 + X_2 dx_2 + \cdots + X_n dx_n.$$

**Remark 3.1.** The notions of regular curve, path, tangent vector to a curve or integration with respect to the arc length have been given in Section 1.4 in the case of curves in a domain $U$ of the plane. It is clear that the same notions have identical sense for curves in a domain of $\mathbb{R}^n$.

The expression $\omega = \omega_{\vec{X}} = X_1 dx_1 + X_2 dx_2 + \cdots + X_n dx_n$, that is, what one integrates over $\gamma$ to get the circulation, is called a *differential 1-form*. Vector fields and differential 1-forms are, then, equivalent concepts. The circulation of the field $\vec{X}$ along $\gamma$ is, by definition, the *line integral* of the form $\omega_{\vec{X}}$ along $\gamma$ and it is represented by $\int_\gamma \omega \; (= \int_\gamma \omega_{\vec{X}} = \int_\gamma \langle \vec{X}, \vec{T} \rangle ds)$.

Consider now a complex line integral $\int_\gamma f(z) \, dz$. If $f = u + iv$ and $\gamma(t) = x(t) + iy(t)$, one has

$$\int_\gamma f(z)dz = \int_a^b f(\gamma(t))\gamma'(t)dt$$

$$= \int_a^b (u(x(t), y(t)) + iv(x(t), y(t)))(x'(t) + iy'(t))dt \qquad (3.5)$$

$$= \int_a^b (ux' - vy')dt + i \int_a^b (uy' + vx')dt,$$

which may be written as

$$\int_\gamma f(z) \, dz = \int_\gamma (udx - vdy) + i \int_\gamma (udy + vdx).$$

In terms of the vector field $\bar{f} = (u, -v)$ this last equality means that $\int_\gamma f(z) \, dz$ has the circulation of $\bar{f}$ along $\gamma$ as its real part, and the circulation of $i\bar{f}$ as its imaginary part.

One way of handling simultaneously complex line integrals and vector field circulations is considering differential 1-forms

$$\omega = P(x, y)\, dx + Q(x, y)\, dy,$$

where $P$, $Q$ are *complex-valued* functions. Vector fields that correspond to these forms will have complex components. So, the integrand of a complex line integral may be written $\omega = f(z)\, dz = f(z)\, dx + if(z)\, dy$, that is, $\omega = P\,dx + Q\,dy$, with $P = f, Q = if$. This will be the point of view from now on.

The following theorem may be seen as an $n$-dimensional version of the fundamental theorem of calculus.

**Theorem 3.7.** *Let $\gamma(t)$, $a \le t \le b$, be a path in the open set $U \subset \mathbb{R}^n$ with origin $A = \gamma(a)$ and ending point $B = \gamma(b)$, $h$ a $C^1$ function on $U$ (real- or complex-valued) and let $\vec{X} = \vec{\nabla} h$ be the gradient vector field of $h$. Then*

$$\int_\gamma \langle \vec{X}, \vec{T} \rangle\, ds = h(B) - h(A),$$

*where $\vec{T}$ is the unitary tangent vector to $\gamma$.*

*Proof.* If $x_i(t)$ are the components of $\gamma(t)$, the circulation is

$$\int_a^b \sum_{i=1}^n \frac{\partial h}{\partial x_i}(\gamma(t)) x_i'(t)\, dt = \int_a^b \frac{d}{dt}(h \circ \gamma)\, dt = h(\gamma(b)) - h(\gamma(a)). \qquad \square$$

So the gradient vector field $\vec{X} = \vec{\nabla} h$, with $h$ of class $C^1$ on $U$, briefly $h \in C^1(U)$, has the property that the circulation along a curve depends only on the origin and the ending point of the curve.

**Definition 3.8.** A continuous vector field $\vec{X}$ on a domain $U$ of $\mathbb{R}^n$ is called conservative if the circulation along any curve $\gamma$ in $U$ depends only on the origin and the ending point of $\gamma$. More explicitly, if $\gamma_1$, $\gamma_2$ are two paths in $U$ with the same origin and the same ending point, it holds that

$$\int_{\gamma_1} \langle \vec{X}, \vec{T} \rangle\, ds = \int_{\gamma_2} \langle \vec{X}, \vec{T} \rangle\, ds.$$

One may also say that the circulation of the vector field $\vec{X}$ does not depend on the path in $U$ between two given points.

If $\gamma_1$ and $\gamma_2$ are two paths with the same origin and ending point and $\gamma$ is the path which consists of travelling on $\gamma_1$ followed by the opposite path to $\gamma_2$ (that is, travelled in the opposite direction), then the circulation of a conservative vector field along $\gamma$ is zero. Therefore, the integrals of a field are independent of the path in $U$ if and only if its circulation is zero along every closed path $\gamma$ in $U$. Observe that it is a global concept.

So Theorem 3.7 says that a gradient vector field is a conservative vector field. The opposite is also true.

**Theorem 3.9.** *A continuous vector field $\vec{X}$ on a domain $U$ of $\mathbb{R}^n$ is conservative on $U$ if and only if it is a gradient vector field.*

*Proof.* One has only to show that if $\vec{X}$ is conservative, then there is a function $h \in C^1(U)$ such that $\vec{X} = \vec{\nabla}h$. Fix a point $x_0 \in U$; given $x \in U$ consider any path $\gamma_x$ in $U$ which starts at $x_0$ and ends at $x$ (a path like these exists because $U$ is arc connected) and define

$$h(x) = \int_{\gamma_x} \langle \vec{X}, \vec{T} \rangle \, ds.$$

By hypothesis, the integral does not depend on the chosen path, and so $h$ is well defined. Let us check now that $\frac{\partial h}{\partial x_i} = X_i$ if $\vec{X} = (X_1, X_2, \ldots, X_n)$. This means that

$$\lim_{t \to 0} \frac{h(x + te_i) - h(x)}{t} = X_i(x),$$

where $\langle e_1, e_2, \ldots, e_n \rangle$ is the canonical basis of $\mathbb{R}^n$. Among the paths going from $x_0$ to $x + te_i$ consider $\gamma_x$, which goes from $x_0$ to $x$, followed by the segment $\sigma$ which joins $x$ with $x + te_i$. The incremental quotient is then

$$\frac{1}{t} \int_{\sigma} \langle \vec{X}, \vec{T} \rangle \, ds = \int_0^1 X_i(x + ste_i) \, ds,$$

and it is clear that the limit when $t \to 0$ is $X_i(x)$.     $\square$

Observe that the previous theorem, in the case $n = 1$, is no more than the fundamental theorem of calculus, and it establishes that any continuous function on an interval has an antiderivative which is its indefinite integral. The function $h$ such that $\vec{X} = \vec{\nabla}h$ is called a *potential function* of the vector field $\vec{X}$ and it is completely determined up to constants.

All these notions may be expressed in terms of 1-forms associated to vector fields. If $\omega = \sum_i X_i(x)\,dx_i$, where $\vec{X} = (X_1, X_2, \ldots, X_n)$ is a conservative vector field on $U$ with potential function $h$, $\vec{X} = \vec{\nabla}h$, then the 1-form $\omega$ is said to be *exact on $U$* and it is written $\omega = dh$. Theorem 3.9 may be reformulated saying that a continuous 1-form on $U$ is exact if and only if it has line integral zero along all closed curves in $U$.

A domain $U$ is a *star-like domain* with respect to one of its points $p$ if, given $x \in U$, the segment which joins $p$ with $x$ is contained in $U$. Obviously, a convex domain is star-like with respect to any of its points. When $U$ is star-like, the proof of Theorem 3.9 may be adapted to obtain the following result:

**Theorem 3.10.** *In a star-like domain $U$ of $\mathbb{R}^n$ a continuous 1-form $\omega$ is exact if and only if $\int_{\partial\triangle} \omega = 0$ for any triangle $\triangle \subset U$, and a continuous vector field is conservative if and only if it has circulation zero along the boundary of any triangle contained in $U$.*

It is understood that $\partial\triangle$ is the positively oriented piecewise regular closed Jordan curve which bounds the triangle $\triangle$.

The proof is the same as in Theorem 3.9, but now the chosen path $\gamma_x$ for defining the potential function $h$ at $x$, which goes from $x_0$ to $x$, is the segment joining both points.

**Definition 3.11.** A 1-form $\omega$ is locally exact (equivalently, the associated vector field $\vec{X}$ has locally a potential function) on a domain $U$ of $\mathbb{R}^n$ if for every $a \in U$ there is a ball $B(a, r) \subset U$, $r > 0$ and a function $h \in C^1(B(a, r))$ such that $\omega = dh$ (or $\vec{X} = \vec{\nabla}h$) in $B(a, r)$.

As usual we write $B(a, r) = B_r(a) = \{x \in \mathbb{R}^n : |x - a| < r\}$, with $|\cdot|$ for the norm in $\mathbb{R}^n$.

Considering that balls are star-like domains and using Theorem 3.10 and a decomposition of a triangle into enough smaller triangles, we get the following result.

**Corollary 3.12.** *In a domain $U \subset \mathbb{R}^n$ a continuous 1-form $\omega$ is locally exact if and only if $\int_{\partial\triangle} \omega = 0$ holds for any triangle $\triangle \subset U$. A continuous vector field has locally a potential function if and only if its circulation is zero along the boundary of any triangle inside $U$.*

## 3.3 The fundamental theorem of complex calculus

Now these considerations will be particularized to dimension two and the case of $\omega = f(z)\,dz$. It is worth recalling the notations

$$\frac{\partial f}{\partial \bar{z}} = \frac{1}{2}\left(\frac{\partial f}{\partial x} + i\frac{\partial f}{\partial y}\right), \qquad \frac{\partial f}{\partial z} = \frac{1}{2}\left(\frac{\partial f}{\partial x} - i\frac{\partial f}{\partial y}\right),$$

$$dz = dx + i\,dy, \qquad d\bar{z} = dx - i\,dy.$$

The differential $dh$ of a differentiable function $h$ (real- or complex-valued) may be expressed as

$$dh = \frac{\partial h}{\partial z}\,dz + \frac{\partial h}{\partial \bar{z}}\,d\bar{z}.$$

A 1-form $\omega = P\,dx + Q\,dy$ may be also written $\omega = A\,dz + B\,d\bar{z}$, with $A = \frac{1}{2}(P - iQ)$, $B = \frac{1}{2}(P + iQ)$. That is, $z$ and $\bar{z}$, $\frac{\partial}{\partial z}$ and $\frac{\partial}{\partial \bar{z}}$, $dz$ and $d\bar{z}$ work as if they formed a system of coordinates.

Let us now see when a form of type $\omega = f(z)\,dz$ is exact on a domain $U$ supposing only that $f$ is continuous on $U$. There must be a function $h \in C^1(U)$ such that $dh = \omega$, that is,

$$\frac{\partial h}{\partial z}\,dz + \frac{\partial h}{\partial \bar{z}}\,d\bar{z} = f(z)\,dz.$$

This means that $\frac{\partial h}{\partial \bar{z}} = 0$, that is, $h$ is holomorphic and $h' = f$ on $U$. This way one sees that $f(z)\,dz$ is exact on $U$ if and only if $f$ has a holomorphic antiderivative $h$, in the sense that $h$ has complex derivative $h'(z) = f(z)$, $z \in U$.

Theorem 3.9 gives, as a particular case, the following result. However it is worth repeating the proof for this special case.

**Theorem 3.13** (Fundamental theorem of complex calculus). *Let $f$ be a continuous function on a domain $U$ of the complex plane. Then the complex line integral $\int_\gamma f(z)\,dz$ does not depend on the path in $U$ if and only if $f$ has a holomorphic antiderivative $F$, $F' = f$, on $U$. In this case one has*

$$\int_\gamma f(z)\,dz = \int_\gamma F'(z)\,dz = F(B) - F(A),$$

*where $A$, $B$ are, respectively, the beginning and the end of the path $\gamma$.*

*Proof.* Fix $z_0 \in U$ and for each $z \in U$ choose any path $\gamma_z$ which goes from $z_0$ to $z$ and put

$$F(z) = \int_{\gamma_z} f(w)\,dw.$$

Observe that $F(z)$ is well defined, supposing that the line integral of $f$ does not depend on the path. Therefore, $F$ is, by definition, the indefinite integral of $f$. To prove that $F'(z) = f(z)$, remark that the incremental quotient $\frac{F(z+h)-F(z)}{h}$ at a fixed point $z$ is exactly $\frac{1}{h}\int_\sigma f(w)\,dw$, where $\sigma$ is the segment from $z$ to $z + h$. So it must be seen that this quantity has limit $f(z)$ when $h \to 0$, or equivalently, using that $\int_\sigma dw$ is $h$,

$$\lim_{h \to 0} \frac{1}{h}\int_\sigma (f(w) - f(z))\,dw = 0.$$

However, by (3.3), the previous integral is bounded in modulus by

$$\frac{1}{|h|}\sup\{|f(w) - f(z)| : w \in \sigma^*\}L(\sigma) \le \sup\{|f(w) - f(z)| : w \in \sigma^*\},$$

a quantity which converges to zero with $h$ due to the continuity of $f$.

Conversely, if $f = F'$ and $\gamma$ is given by $\gamma(t)$, $a \le t \le b$, then $f(\gamma(t))\gamma'(t) = F'(\gamma(t))\gamma'(t)$ coincides with $(F \circ \gamma)'(t)$ and so one has

$$\int_\gamma f(z)\, dz = \int_a^b f(\gamma(t))\gamma'(t)\, dt = \int_a^b (F \circ \gamma)'(t)\, dt$$
$$= (F \circ \gamma)(b) - (F \circ \gamma)(a). \qquad \square$$

Theorem 3.10 corresponds to the following one.

**Theorem 3.14.** *Let $f$ be a continuous function on a star-like domain $U$ of the complex plane. Then $f$ has a holomorphic antiderivative $F$, $F' = f$, on $U$ if and only if*

$$\int_{\partial\Delta} f(z)\, dz = 0$$

*for every triangle $\Delta$ inside $U$.*

Thus the parallelism with the real version of the fundamental theorem of calculus is clear. The difference is that, on the real line, to go from $x_0$ to $x$ *there is just one path*, up to reparametrizations which conserve the value of the integral, while in the complex plane there are a lot of them. Therefore, while in a real variable every continuous function has an antiderivative (the indefinite integral), in one complex variable not every continuous function has a holomorphic antiderivative. Only functions with a well-defined indefinite integral have it.

**Example 3.15.** For fixed $a \in \mathbb{C}$ and $m \in \mathbb{N}$, the function $F(z) = \frac{1}{m}(z - a)^m$ is holomorphic and has derivative $F'(z) = (z - a)^{m-1}$ in the whole complex plane $\mathbb{C}$. Hence, $\int_\gamma (z - a)^k\, dz = 0$ if $k$ is a non-negative integer ($k = m - 1$) for any closed path $\gamma$, that is, the form $(z - a)^k\, dz$ has integrals which do not depend on the path.

Considering $F(z) = \frac{1}{m}(z - a)^m$ but now with $m$ a negative integer, $F$ is holomorphic on $U = \mathbb{C} \setminus \{a\}$ and has derivative $F'(z) = (z - a)^{m-1}$. Hence, it is also $\int_\gamma (z - a)^k\, dz = 0$ if $k$ is a negative integer $k \ne -1$ ($k = m - 1$) and $a \notin \gamma^*$. However, if $k = -1$, it is already known that

$$\int_\gamma \frac{dz}{z - a} = 2\pi i \, \mathrm{Ind}\,(\gamma, a),$$

which may be different from 0. Therefore the function $(z - a)^{-1}$ does not have a holomorphic antiderivative on $\mathbb{C} \setminus \{a\}$ $\qquad \square$

**Example 3.16.** The computation of the integral of $z^2$ along the arc of the parabola from 0 to $1 + i$ followed by the segment from $1 + i$ to 1 is done in Example 3.3, and the result is $\frac{1}{3}$, which coincides with the integral along the segment from 0 to 1, easier to compute. $\qquad \square$

The complex fundamental theorem of calculus, exactly as in the real case, allows us to reduce the computation of certain line integrals to the computation of antiderivatives. In one real variable, in order to evaluate for example $\int_0^1 e^{-x^2}\,dx$, one has to use the definition as a limit of Riemann sums and one may only obtain an approximated value. But the fundamental theorem of calculus may be used for functions which have antiderivatives. It is the same in the complex case; according to Theorem 3.13 one has

$$\int_\gamma F'(z)\,dz = F(B) - F(A) \tag{3.6}$$

if $\gamma$ is a path from the points $A$ to $B$, inside an open set $U$ on which $F$ is a holomorphic function.

**Example 3.17.** In order to compute $\int_\gamma \sin z\,dz$ where $\gamma$ is the arc of the curve $y = x^3$, which goes from $(0,0)$ to $(1,1) = 1 + i$, the definition is not applied; just observe that $\sin z$ is the derivative of $-\cos z$, and then the previous integral has value $-\cos(1+i) + \cos 0 = 1 - \cos(1+i)$. $\qquad\square$

Recall that to compute complex antiderivatives we may use analogue rules to the ones for the real case, as explained in Subsection 2.4.3. For example, the *integration by parts formula* sets the equality

$$\int_\gamma f'(z)g(z)\,dz = f(B)g(B) - f(A)g(A) - \int_\gamma f(z)g'(z)\,dz$$

if $\gamma$ goes from $A$ to $B$ inside $U$ and $f, g \in H(U)$. It is no more than the rule of the derivative of a product, using (3.6) for $F = f \cdot g$.

**Example 3.18.** In order to compute $\int_\gamma z \sin z\,dz$, when $\gamma$ is a path joining the points $A$ and $B$, take $u = z$, $dv = \sin z\,dz$; then $du = dz$, $v = -\cos z$, and the indefinite integral $\int u\,dv$ is $uv - \int v\,du = -z\cos z + \int \cos z\,dz = -z\cos z + \sin z$. The defined integral is, therefore, $\sin B - \sin A - B\cos B + A\cos A$. $\qquad\square$

Similarly, the *change of variable formula* establishes that if in the integral $\int_\gamma f(g(z))g'(z)\,dz$ one makes the substitution $w = g(z)$ and $F(w)$ is an antiderivative of $f(w)$, then

$$\int_\gamma f(g(z))g'(z)\,dz = \int_{g(\gamma)} f(w)\,dw = F(g(B)) - F(g(A)),$$

where $\gamma$ goes from $A$ to $B$ and $g(\gamma)$ is the path defined by the mapping $g \circ \gamma$.

**Example 3.19.** Compute $\int_\gamma \frac{z}{1+z^2}\,dz$, where $\gamma$ is the line segment from $0$ to $\frac{1}{\sqrt{2}}(1+i)$. Making the change of variable $w = z^2$ we obtain $\frac{1}{2}\int_\Gamma \frac{1}{1+w}\,dw$, where

$\Gamma$ is the image of this segment by the mapping $z \to z^2$; it is the segment starting at $0$ and ending at $i$; in a neighborhood of this segment $\frac{1}{1+w}$ has antiderivative $F(w) = \mathrm{Log}(1 + w)$ (in fact, the branch of $F(w)$ is defined outside the ray $(-\infty, -1]$); the proposed integral equals $\frac{1}{2}(\mathrm{Log}(1 + i) - \mathrm{Log}\, 1) = \frac{1}{4}\mathrm{Log}\, 2 + i\frac{\pi}{8}$. $\qquad\square$

The existence of holomorphic antiderivatives and of branches of the logarithm are two deeply related matters:

**Proposition 3.20.** *If $f$ is a holomorphic function without zeros on a domain $U$, then there is a branch of the logarithm of $f$ in $U$ if and only if the function $f'/f$ has a holomorphic antiderivative on $U$.*

*Proof.* It has been observed in Subsection 2.4.3 that if $h$ is a branch of $\log f$, then $h$ is holomorphic and $h' = f'/f$. Conversely, suppose that $h$ is holomorphic on $U$ with $h' = f'/f$ and consider the function $F = e^{-h} f$, the derivative of which is

$$F' = -e^{-h} h' f + e^{-h} f' = 0.$$

Therefore, $F$ is a constant $c \neq 0$, and if $c = e^{\alpha}, \alpha \in \mathbb{C}$, then $f = e^{h+\alpha}$ and $h + \alpha$ is a continuous branch of $\log f$ in $U$. $\qquad\square$

**Proposition 3.21.** *Let $K$ be a compact set of $\mathbb{C}$. The following assertions hold:*

a) *If $\alpha \in V_\infty$, the unbounded component of $\mathbb{C} \setminus K$, then there exists a branch of $\log(z - \alpha)$ on a neighborhood of $K$.*

b) *If $\alpha, \beta$ belong to the same bounded component of $\mathbb{C} \setminus K$, then there exists a branch of $\log \frac{z-\alpha}{z-\beta}$ on a neighborhood of $K$.*

*Proof.* Suppose that $V$ is a bounded component of $\mathbb{C} \setminus K$ and that $\alpha, \beta \in V$. Consider a relatively compact neighborhood $U$ of $K$ such that $\alpha, \beta$ are both in the same component of $\mathbb{C} \setminus \bar{U}$ (for example, join $\alpha, \beta$ by means of a polygonal curve $\Gamma \subset V$ and if $\varepsilon = d(K, \Gamma)$, put $U = \{z \colon d(z, K) < \varepsilon\}$). The function $f(z) = (z - \alpha)/(z - \beta)$ will have a branch of the logarithm in $U$ if $f'/f$ has an antiderivative, that is, if

$$0 = \int_\gamma \frac{f'(z)}{f(z)}\, dz = \int_\gamma \frac{dz}{z - \alpha} - \int_\gamma \frac{dz}{z - \beta}$$

for any closed path $\gamma$ inside $U$. Now,

$$\int_\gamma \frac{dz}{z - \alpha} - \int_\gamma \frac{dz}{z - \beta} = 2\pi i \left[\mathrm{Ind}(\gamma, \alpha) - \mathrm{Ind}(\gamma, \beta)\right].$$

Since $\gamma^* \subset U$, it turns out that $\alpha, \beta$ must be in the same component of $\mathbb{C} \setminus \gamma^*$ and, as a consequence, $\mathrm{Ind}\,(\gamma, \alpha) = \mathrm{Ind}\,(\gamma, \beta)$.

In the case that $\alpha \in V_\infty$, one has $\int_\gamma \frac{dz}{z-\alpha} = 2\pi i\, \mathrm{Ind}\,(\gamma, \alpha) = 0$, following a similar argument. $\qquad\square$

## 3.4 Green's formula

The following theorem has a main role in the development of Cauchy's local theory of holomorphic functions.

**Theorem 3.22** (Green's formula). *Let $U$ be a bounded domain of the complex plane with positively oriented piecewise regular boundary and let $\vec{T}$ be the unitary tangent vector to $\partial U$. Let $\vec{X} = (P, Q)$ be a vector field with components $P$, $Q$ differentiable functions on a neighborhood of $\bar{U}$ such that the function $\frac{\partial Q}{\partial x} - \frac{\partial P}{\partial y}$ is continuous on $\bar{U}$. Then the following identity holds:*

$$\int_{\partial U} \langle \vec{X}, \vec{T} \rangle \, ds = \int_{\partial U} P \, dx + Q \, dy = \iint_U \left( \frac{\partial Q}{\partial x} - \frac{\partial P}{\partial y} \right) dx \, dy.$$

Observe that it is not required that the functions $\frac{\partial Q}{\partial x}$, $\frac{\partial P}{\partial y}$ be continuous.

It is convenient to make precise the meaning of the integral of the form $P \, dx + Q \, dy$ along $\partial U$. Recall (Section 1.6) that $\partial U$ is composed of a finite number of Jordan curves $\gamma_1, \gamma_2, \ldots, \gamma_N$. Then, by definition, it is

$$\int_{\partial U} P \, dx + Q \, dy = \sum_{i=1}^{N} \int_{\gamma_j} P \, dx + Q \, dy,$$

understanding that each curve $\gamma_j$ is described by a parametrization which orientates $\gamma_1$ positively and $\gamma_2, \ldots, \gamma_N$ negatively, if $\gamma_1, \gamma_2, \ldots, \gamma_N$ are as in Proposition 1.37. Intuitively, when travelling each $\gamma_j$ the set $U$ must be on the left. Likewise, the unitary tangent vector at the point $\gamma_i(t) \in \partial U$, $i = 1, \ldots, N$, is $\vec{T} = \frac{\gamma_i'(t)}{|\gamma_i'(t)|}$ every time that $\gamma_i'(t) \neq 0$. Concerning the double integral over $U$ of the function $Q_x - P_y$, note that it is taken with respect to Lebesgue measure on the plane, represented either by $dx dy$ or by $dm(z)$, if $z = x + iy$.

The proof of Theorem 3.22 will be done in three steps. The first step corresponds to the particular case in which the domain $U$ is a rectangle. It is worth noting that the proof of Green's formula for a rectangle, let $\bar{U} = [a, b] \times [c, d]$, is very easy if $P$, $Q$ are $C^1$ in a neighborhood of $\bar{U}$, because then the functions $\frac{\partial P}{\partial y}$, $\frac{\partial Q}{\partial x}$ are continuous and an application of the fundamental theorem of calculus and Fubini's

Theorem gives

$$\int_{\partial U} (P\,dx + Q\,dy)$$

$$= \int_a^b P(x,c)dx + \int_c^d Q(b,y)dy - \int_a^b P(x,d)dx - \int_c^d Q(a,y)dy$$

$$= \int_a^b (P(x,c) - P(x,d))\,dx + \int_c^d (Q(b,y) - Q(a,y))\,dy$$

$$= -\int_a^b \int_c^d \frac{\partial P}{\partial y}(x,y)\,dx\,dy + \int_c^d \int_a^b \frac{\partial Q}{\partial x}(x,y)\,dy\,dx$$

$$= \int_{[a,b]\times[c,d]} \left( \frac{\partial Q}{\partial x} - \frac{\partial P}{\partial y} \right) dx\,dy.$$

In the general case, where nothing about each function $\frac{\partial P}{\partial y}$, $\frac{\partial Q}{\partial x}$ is assumed, it is necessary to find a different kind of proof. The one given here is inspired by a proof of the basic fact, in calculus of one real variable, stating that if $f$ is a differentiable function on an open interval $I$ and $f'(x) = 0$, for all $x \in I$, then $f$ is constant on $I$. It is a direct proof, which only uses the definition of the derivative:

One may suppose $[a,b] \subset I$ and prove that $f(a) = f(b)$. Put $I_0 = [a,b]$. If $c$ is the middle point of $[a,b]$, one has

$$\Delta_{I_0}(f) \stackrel{\text{def}}{=} |f(b) - f(a)| \le |f(b) - f(c)| + |f(c) - f(a)|$$

$$\stackrel{\text{def}}{=} \Delta_{[c,b]}(f) + \Delta_{[a,c]}(f).$$

From the two intervals $[a,c]$, $[c,b]$, let $I_1$ be the one on which the absolute value of the variation of $f$ is bigger so that

$$|f(b) - f(a)| \le 2\Delta_{I_1}(f).$$

Repeat the process indefinitely and find a sequence of intervals $I_0 \supset I_1 \supset \cdots \supset I_n \supset \cdots$, each of them a half of the previous one, with length $|I_n| = 2^{-n}(b-a)$ and

$$|f(b) - f(a)| \le 2^n \Delta_{I_n}(f).$$

The intersection of all the intervals $I_n$ is a point $x_0 \in [a,b]$. Given $\varepsilon > 0$, there exists a $\delta > 0$ such that

$$f(x) = f(x_0) + f'(x_0)(x - x_0) + R(x)(x - x_0) = f(x_0) + R(x)(x - x_0)$$

with $|R(x)| \le \varepsilon$, if $|x - x_0| \le \delta$. If $n$ is big enough, any interval $I_n$ is inside the interval $[x_0 - \delta, x_0 + \delta]$ and then

$$\Delta_{I_n}(f) \le \varepsilon |I_n| = \varepsilon 2^{-n}(b-a).$$

With this $|f(b) - f(a)| \le \varepsilon(b-a)$, for any $\varepsilon > 0$ and $f(b) = f(a)$.

The same proof, working with $f(b) - f(a) - \int_a^b f'(x)\,dx$ as a functional of the interval $[a, b]$, proves the fundamental theorem of calculus, that is,

$$f(b) - f(a) = \int_a^b f'(t)\,dt, \quad \text{if } f \in C^1([a, b]).$$

This is, essentially, the proof done next in dimension 2.

*Proof of Theorem* 3.22.

*First step.* The domain $U$ is a rectangle with sides parallel to the axis. Writing $\frac{\partial P}{\partial x} = P_x$, $\frac{\partial P}{\partial y} = P_y$ and similarly for $Q$, put

$$I = I(U) = \int_{\partial U} (P\,dx + Q\,dy) - \iint_U (Q_x - P_y)\,dx\,dy.$$

This notation will be used for any rectangle. Divide the rectangle $U$ into four equal rectangles (Figure 3.3) $U^i$, $i = 1, 2, 3, 4$, such that one has $I(U) = \sum_{i=1}^n I(U^i)$, and let $U_1$ be the one of the four rectangles $U^i$ for which $|I(U^i)|$ is maximum. Therefore,

$$|I| \le 4|I(U_1)|.$$

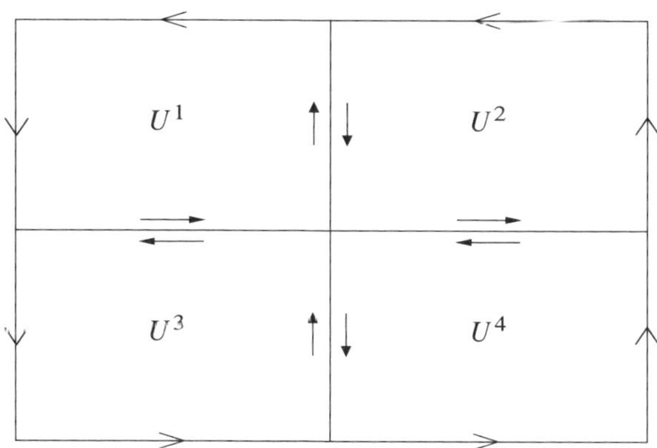


Figure 3.3

With $I_1 = I(U_1)$ one repeats now the process and so on, so one finds rectangles $\bar{U} \supset \bar{U}_1 \supset \bar{U}_2 \supset \cdots \supset \bar{U}_n, \ldots$, each of which is a quarter of the previous one, and

$$|I| \le 4^n |I(U_n)|. \tag{3.7}$$

The intersection of these rectangles $\bar{U}_n$ is a point $z_0 \in \bar{U}$. Given $\varepsilon > 0$ we can write

$$P(x, y) = P(z_0) + P_x(z_0)(x - x_0) + P_y(z_0)(y - y_0) + R_1(z),$$

$$Q(x, y) = Q(z_0) + Q_x(z_0)(x - x_0) + Q_y(z_0)(y - y_0) + R_2(z),$$

where $z_0 = x_0 + iy_0$, $z = x + iy$, with $|R_1(z)|, |R_2(z)| \leq \varepsilon|z - z_0|$, if $|z - z_0| \leq \delta = \delta(\varepsilon)$. By the continuity of $Q_x - P_y$ one may suppose that

$$(Q_x - P_y)(z) = (Q_x - P_y)(z_0) + R_3(z)$$

with $|R_3(z)| \leq \varepsilon$, if $|z - z_0| \leq \delta$. For $n$ big enough, it is $\bar{U}_n \subset D(z_0, \delta)$, and if one assumes that $\bar{U}_n = [a_n, b_n] \times [c_n, d_n]$, one has

$$\int_{\partial U_n} (P \, dx + Q \, dy)$$

$$= \int_{a_n}^{b_n} P(x, c_n) \, dx + \int_{c_n}^{d_n} Q(b_n, y) \, dy$$

$$\quad - \int_{a_n}^{b_n} P(x, d_n) \, dx - \int_{c_n}^{d_n} Q(a_n, y) \, dy$$

$$= P_y(z_0)(c_n - y_0)(b_n - a_n) + Q_x(z_0)(b_n - x_0)(d_n - c_n)$$

$$\quad - P_y(z_0)(d_n - y_0)(b_n - a_n) - Q_x(z_0)(a_n - x_0)(d_n - c_n) + R_n$$

$$= (b_n - a_n)(d_n - c_n)(Q_x(z_0) - P_y(z_0)) + R_n,$$

where

$$R_n = \int_{a_n}^{b_n} [R_1(x, c_n) - R_1(x, d_n)]dx + \int_{c_n}^{d_n} [R_2(b_n, y) - R_2(a_n, y)]dy.$$

Observe that $R_1, R_2$ are continuous functions and, therefore, integrable.

Now if $z \in \partial U_n$, it is $|z - z_0| \leq L_n$, where $L_n$ is the length of the diagonal of $U_n$. Hence, if $P_n$ is the perimeter of $U_n$, it turns out that, for $n$ big enough,

$$|R_n| \leq \varepsilon L_n P_n.$$

But if $L, P$ are the diagonal and the perimeter of $U$, respectively, it is $L_n = 2^{-n}L$, $P_n = 2^{-n}P$, so that

$$|R_n| \leq \varepsilon 4^{-n} LP.$$

One has as well

$$\iint_{U_n} (Q_x - P_y) \, dx \, dy = (Q_x - P_y)(z_0)(b_n - a_n)(d_n - c_n) + \tilde{R}_n,$$

for a certain function $\widetilde{R}_n$ which satisfies $|\widetilde{R}_n| \leq \varepsilon 4^{-n} \cdot A$, where $A$ is the area of $U$ and $n$ is big enough. Hence one has that the two previous integrals has the same *principal part*, so there is a cancellation and one finds

$$|I(U_n)| \leq \varepsilon 4^{-n}(LP + A).$$

Finally, by (3.7) it turns out that $|I| \leq \varepsilon(LP + A)$, for all $\varepsilon > 0$ and consequently, $I = 0$. This finishes the proof of the first step. $\qquad\square$

**Remark 3.2.** a) The same procedure may be used in the case of a triangle $\triangle$ breaking it into the three triangles determined by the vertices and the barycenter of $\triangle$.

b) The arguments used above also prove that the result holds if $\overline{U}$ is a finite union of rectangles with pairwise disjoint interiors.

c) Also a more general version than the one just proved is valid, supposing only that the functions $P$ and $Q$ are continuous on $\overline{U}$, differentiable on $U$ and that the function $Q_x - P_y$ is integrable on $U$. (See: P. J. Cohen, *On Green's theorem*, Proc. Amer. Math. Soc. **10** (1959) 109–112.)

*Second step.* The domain $U$ is of *subgraph* type, that is,

$$U = \{(x, y) \colon a \leq x \leq b,\, 0 \leq y \leq \varphi(x)\}$$

is a domain limited by three segments and by the graph of the function $\varphi\colon$ $[a, b] \to \mathbb{R}$ which is supposed to be $C^1$ (Figure 3.4).

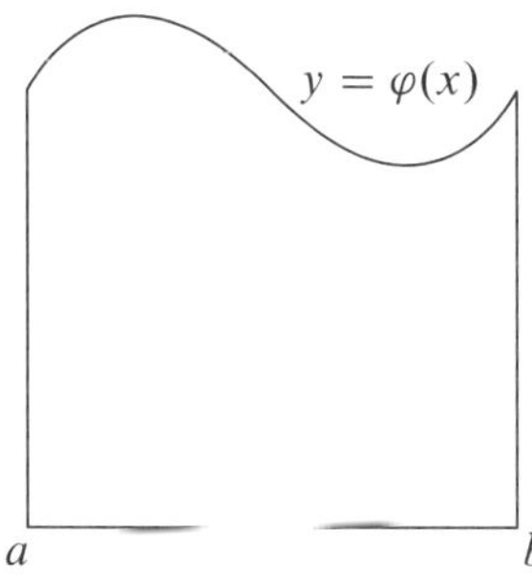


Figure 3.4

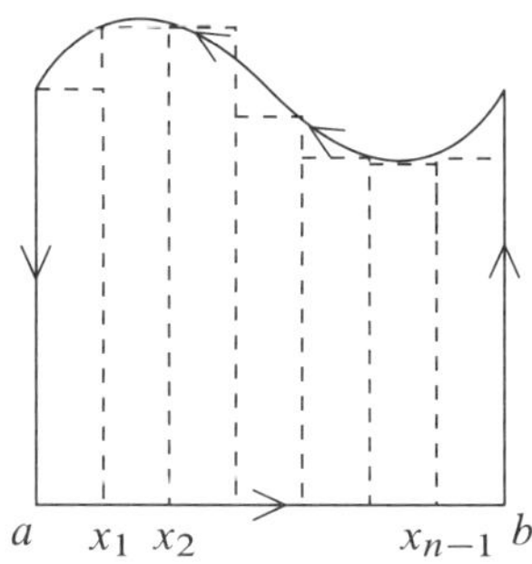


Figure 3.5

The idea in this case is, simply, approximating $U$ by a union of rectangles, as it is done in the definition of the integral with Riemann sums. Given a partition $a = x_0 < x_1 < x_2 < \cdots < x_n = b$ of $[a, b]$, formed by $n + 1$ equidistant points, let $m_i = \inf\{\varphi(x) \colon x_{i-1} \leq x \leq x_i\} = \varphi(\tau_i)$ with $\tau_i \in [x_{i-1}, x_i]$ for $i = 1, 2, \ldots, n$, $\tau_0 = x_0$, $\tau_{n+1} = x_n$. Now consider the domain $R_n$ union of the rectangles $\{(x, y) \colon x_{i-1} \leq x \leq x_i,\, 0 \leq y \leq m_i\}, i = 1, \ldots, n$ (Figure 3.5).

Clearly $R_n \subset \bar{U}$ and

$$\iint_{R_n} (Q_x - P_y)\, dx\, dy \xrightarrow[n\to\infty]{} \iint_U (Q_x - P_y)\, dx\, dy$$

because the function $Q_x - P_y$ is continuous, and therefore integrable on $\bar{U}$. Furthermore since $R_n$ is the union of adjacent rectangles it turns out, according to the first step, that

$$\iint_{R_n} (Q_x - P_y)\, dx\, dy = \int_{\partial R_n} P\, dx + Q\, dy.$$

The difference between $\int_{\partial U} P\, dx + Q\, dy$ and $\int_{\partial R_n} P\, dx + Q\, dy$ is, up to the sign,

$$\int_a^b (P(x, \varphi(x)) + Q(x, \varphi(x))\varphi'(x))\, dx$$

$$-\sum_{i=1}^n \int_{x_{i-1}}^{x_i} P(x, \varphi(\tau_i))\, dx - \sum_{i=0}^n \int_{\varphi(\tau_i)}^{\varphi(\tau_{i+1})} Q(x_i, y)\, dy$$

$$= \left[ \int_a^b P(x, \varphi(x))\, dx - \sum_{i=1}^n \int_{x_{i-1}}^{x_i} P(x, \varphi(\tau_i))\, dx \right]$$

$$+ \left[ \int_a^b Q(x, \varphi(x))\varphi'(x)\, dx - \sum_{i=0}^n \int_{\varphi(\tau_i)}^{\varphi(\tau_{i+1})} Q(x_i, y)\, dy \right] \stackrel{\text{def}}{=} \mathrm{I} + \mathrm{II}.$$

It is clear that I converges to 0 when $n \to \infty$, by the uniform continuity of $P(x, y)$ on $\bar{U}$. The change of variable $y = \varphi(x)$ makes each of the integrals of the second term of II become

$$\int_{\tau_i}^{\tau_{i+1}} Q(x_i, \varphi(x))\varphi'(x)\, dx$$

and then $\mathrm{II} \to 0$ with $n \to \infty$, by the uniform continuity of $Q(x, \varphi(x))\varphi'(x)$.
With this,

$$\int_{\partial U} (P\, dx + Q\, dy) = \lim_n \int_{\partial R_n} (P\, dx + Q\, dy)$$

$$= \lim_n \iint_{R_n} (Q_x - P_y)\, dx\, dy = \iint_U (Q_x - P_y)\, dx\, dy,$$

which finishes the proof of the second step.

Naturally, the same result holds permuting the roles of $x$, $y$, that is, if $U$ is a subgraph domain of a function $x = \varphi(y)$. $\qquad\square$

*Third step.* Let now $U$ be a bounded domain with piecewise regular positively oriented boundary, formed by Jordan curves $\gamma_1, \ldots, \gamma_N$, where each $\gamma_i$ is piecewise regular.

First fix a point $z_0 \in \partial U$ and suppose that $z_0 \in \gamma_1^*$ with $\gamma_1(t) = (x(t), y(t))$, $\gamma_1(t_0) = z_0$. If $z_0$ is a regular point, it is $\gamma_1'(t_0) \neq 0$ and one may assume, for example, $x'(t_0) \neq 0$. Then in a neighborhood of $x_0 = x(t_0)$ the variable $t$ may be written as a function of $x$, $t = t(x)$, and therefore, $y = y(t) = y(t(x)) = \varphi(x)$, that is, $\gamma_1^*$ is the graph of the function $\varphi$ on a neighborhood of $(x_0, y_0)$. Taking now a rectangle $R$ small enough surrounding this point, the graph of $\varphi$ breaks $R$ into two regions

$$R^+ = \{(x, y) \in R, \, y > \varphi(x)\} \quad \text{and} \quad R^- = \{(x, y) \in R, \, y < \varphi(x)\},$$

and necessarily either $R \cap U = R^+$ or $R \cap U = R^-$ (Figure 3.6).

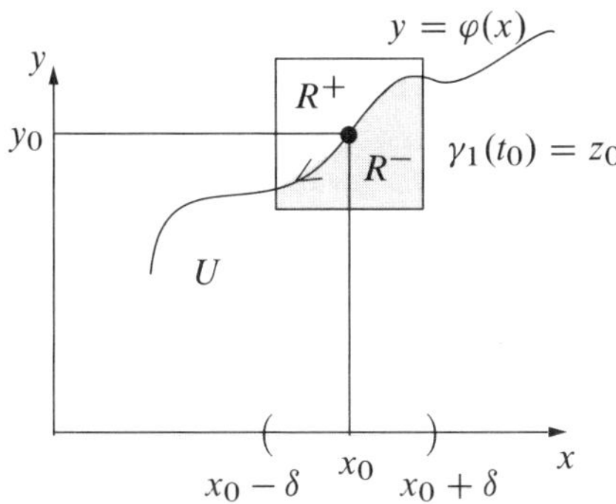


Figure 3.6

Hence there exists a neighborhood $R$ of $z_0$ such that $R \cap \bar{U}$ is either the subgraph type region

$$R \cap \bar{U} = \{x_0 - \delta \leq x \leq x_0 + \delta, \, a \leq y \leq \varphi(x)\} \quad \text{with } \delta > 0, \, a \in \mathbb{R},$$

which has positively oriented boundary, or the result of rotating by $180°$ a region of this type. If the proof started from $y'(t_0) \neq 0$, one would find a neighborhood $R$ of $z_0$ such that $R \cap \bar{U}$ is a subgraph type region, with $x = \varphi(y)$. Observe that for any of these regions Green's formula holds, according to the second step. If the point $z_0 \in \partial U$ is not regular, then $z_0$ has a neighborhood $V$ such that $V \cap \bar{U}$ is the union of two adjoint subgraph regions because $\gamma_1(t)$ is regular for $t < t_0$ and for $t > t_0$, and then Green's formula also holds on $V \cap \bar{U}$ (Figure 3.7).

Now, due to the compactness of the boundary of $U$, it may be covered by a finite number of open rectangles $R_1, \ldots, R_k$ so that every region $R_i \cap \bar{U}$ is of subgraph type (as in the second step). The part of $\bar{U}$ which is not covered

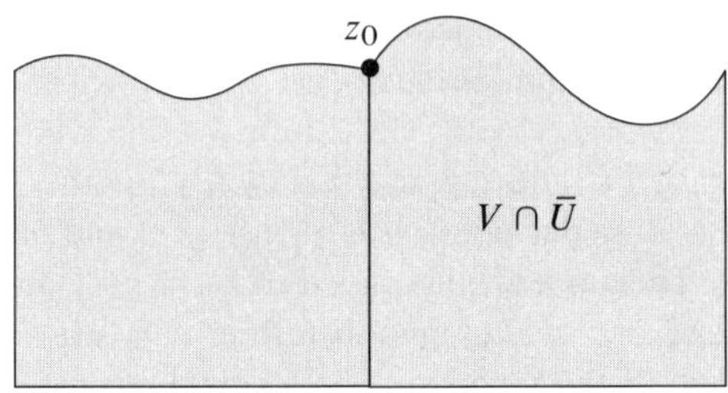


Figure 3.7

by the rectangles $R_i$, that is, $\bar{U} \setminus (R_1 \cup \cdots \cup R_k)$, is a compact set contained in $U$, which may be covered by a union of rectangles with sides parallel to the axis, contained in $U$. We may assume that these rectangles have pairwise disjoint interiors, because a finite union of rectangles may always be chosen such that only their boundaries intersect. Let $R_0$ be the union of these rectangles so that Green's formula clearly holds for $R_0$, according to the first step. Finally consider a $C^1$ partition of unity $\{\rho_0, \rho_1, \rho_2, \ldots, \rho_k\}$ subordinated to the covering $\{R_0, R_1, \ldots, R_k\}$ of $\bar{U}$. To simplify we write $\omega = P\,dx + Q\,dy$, $\omega_i = \rho_i \omega$, $h_i = (\rho_i Q)_x - (\rho_i P)_y$, $i = 0, 1, \ldots, k$. It turns out that

$$\omega = \omega_0 + \sum_{i=1}^{k} \omega_i,$$

where each form $\omega_i$ is differentiable and vanishes outside $R_i$ and $h_i$ is continuous. Therefore

$$\int_{\partial U} \omega_0 = 0, \quad \iint_U h_0 = \iint_{R_0} h_0 = \int_{\partial R_0} \omega_0 = 0$$

and, for the forms $\omega_i$, $i = 1, \ldots, k$,

$$\int_{\partial U} \omega_i = \int_{\partial U \cap R_i} \omega_i = \int_{\partial (R_i \cap \bar{U})} \omega_i = \iint_{R_i \cap \bar{U}} h_i = \iint_U h_i.$$

With this the proof of Green's formula is complete.     $\square$

Note that Theorem 3.22 holds if the functions $P$, $Q$ either are real or complex.

A vector field $\vec{X} = (P, Q)$ and its corresponding 1-form $\omega = P\,dx + Q\,dy$ will be called *differentiable* on a domain $U$ if $P$ and $Q$ are differentiable functions on $U$.

**Corollary 3.23.** *A differentiable 1-form $\omega = P\,dx + Q\,dy$ on a domain $U$ of the plane is locally exact if and only if $P_y = Q_x$ on $U$. A differentiable vector field $\vec{X} = (P, Q)$ has a locally potential function if and only if $P_y = Q_x$. In a star-like domain the same statements hold globally on $U$.*

*Proof.* It is enough to prove the assertion in the case of a star-like domain. If $\omega = dh$ then $h$ is twice differentiable and Schwarz's rule says that $h_{xy} = h_{yx}$, that is, $P_y = Q_x$. Conversely, this condition implies, according to Green's formula

$$\int_{\partial \Delta} \omega = \int_{\Delta} (Q_x - P_y)\, dx\, dy = 0,$$

for each triangle $\Delta \subset U$, and by Theorem 3.10, $\omega$ is exact.    $\square$

With the additional hypothesis that $P, Q$ are $C^1$ the previous corollary may be proved directly without using Green's formula (see Remark 3.5). The differentiable 1-forms on a domain $U$, $\omega = P\,dx + Q\,dy$, which satisfy the condition $P_y = Q_x$, are called *closed forms* on $U$.

## 3.5 Cauchy's Theorem and applications

In the special case $\omega = P\,dx + Q\,dy$ is a closed differentiable 1-form on a neighborhood of $\bar{U}$, the function $Q_x - P_y = 0$ is automatically continuous and Green's formula gives

$$\int_{\partial U} \omega = 0.$$

If the function $f$ is differentiable, imposing that the 1-form $\omega = f(z)\,dz = f(z)\,dx + if(z)\,dy$ is closed means that it must satisfy

$$\frac{\partial f}{\partial y} = \frac{\partial i f}{\partial x} = i\frac{\partial f}{\partial x},$$

an equality that translates Cauchy–Riemann equations. Therefore, $\omega = f(z)\,dz$ is a closed form if and only if $f$ satisfies the Cauchy–Riemann equations $\frac{\partial f}{\partial \bar{z}} = 0$, that is, if $f$ is holomorphic.

In particular one obtains a basic result of complex analysis:

**Theorem 3.24** (Cauchy). *If $U$ is a bounded domain of the plane with positively oriented piecewise regular boundary and $f$ is a holomorphic function on a neighborhood of $\bar{U}$, then*

$$\int_{\partial U} f(z)\,dz = 0.$$

**Remark 3.3.** a) The particular case of Theorem 3.24 when the domain $U$ is a rectangle with sides parallel to the axis is known as *Cauchy–Goursat's Theorem*. Note this result only depends on the first step in the proof of Theorem 3.22.

b) It is easy to see that Theorem 3.24 holds just assuming that the function $f$ is continuous on $\bar{U}$ and holomorphic on $U$ (see Exercise 22 of Section 3.8).

A situation in which Cauchy's Theorem is often used is the following: assume that $U$ is the interior of a positively oriented Jordan curve and that this curve is formed by regular curves $\gamma_1, \ldots, \gamma_N$ such that the ending point of $\gamma_i$ is the starting point of $\gamma_{i+1}, i = 1, 2, \ldots, N - 1$, and the ending point of $\gamma_N$ is the origin of $\gamma_1$. Then, if $f$ is holomorphic on a neighborhood of $\overline{U}$, one has

$$\int_{\gamma_1} f(z)dz = -\sum_{i=2}^{N} \int_{\gamma_i} f(z)dz$$

and it is possible that the integrals $\int_{\gamma_i} f(z)dz$ turn out to be more easily computed than the integral $\int_{\gamma_1} f(z)dz$. This is illustrated in the following example.

**Example 3.25.**  In order to compute the integrals $\int_0^{+\infty} \cos(t^2)dt$, $\int_0^{+\infty} \sin(t^2)dt$ we will show that the following limits exist:

$$\lim_{R \to +\infty} \int_0^R \cos(t^2)\, dt, \qquad \lim_{R \to +\infty} \int_0^R \sin(t^2)\, dt,$$

that is, the integrals exist as improper Riemann integrals (however, they are not absolutely convergent). The entire function $f(z) = e^{iz^2}$ and the paths $\Gamma_1$, $\Gamma_2$ and $\Gamma_3$ of Figure 3.8 will be used. By Cauchy's Theorem one has

$$\int_0^R \cos t^2\, dt + i \int_0^R \sin t^2\, dt = \int_{\Gamma_1} f(z)\, dz = \int_{\Gamma_2} f(z)\, dz - \int_{\Gamma_3} f(z)dz$$

$$= \int_0^R e^{i(te^{i\frac{\pi}{4}})^2}\, d(te^{i\frac{\pi}{4}}) - \int_0^{\frac{\pi}{4}} e^{i(Re^{it})^2}\, d(Re^{it})$$

$$= \int_0^R e^{-t^2} \frac{1}{\sqrt{2}}(1 + i)\, dt - \int_0^{\frac{\pi}{4}} e^{iR^2 e^{2it}}\, Rie^{it}\, dt.$$

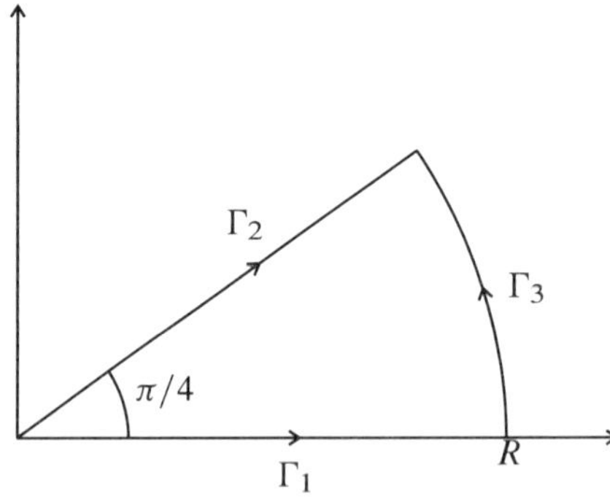


Figure 3.8

Now take limits when $R \to \infty$. The first of these last integrals converges to

$$\frac{1}{\sqrt{2}}(1+i) \int_0^\infty e^{-t^2}\, dt = \frac{\sqrt{\pi}}{2\sqrt{2}}(1+i).$$

On the other hand,

$$\left| \int_0^{\frac{\pi}{4}} e^{iR^2 e^{2it}} R i e^{it}\, dt \right| \leq R \int_0^{\frac{\pi}{4}} \left| e^{iR^2 e^{2it}} \right| dt = R \int_0^{\frac{\pi}{4}} e^{-R^2 \sin 2t}\, dt$$

$$\leq R \int_0^{\frac{\pi}{4}} e^{-R^2 t}\, dt = \frac{1}{R}(1 - e^{-\frac{\pi}{4} R^2}) < \frac{1}{R} \xrightarrow[R \to \infty]{} 0.$$

Therefore, one finally obtains

$$\int_0^\infty \cos(t^2)dt = \int_0^\infty \sin(t^2)dt = \frac{\sqrt{\pi}}{2\sqrt{2}}. \qquad \square$$

**Example 3.26.** Theorem 3.24 is used now to prove that the integral

$$I(\alpha) = \int_{-\infty}^{+\infty} e^{-(x+i\alpha)^2}\, dx$$

does not depend on $\alpha \in \mathbb{R}$. Supposing that this fact holds, and since $I(0) = \int_{-\infty}^{+\infty} e^{-x^2}\, dx = \sqrt{\pi}$ and $I(\alpha) = \int_{-\infty}^{+\infty} e^{-(x^2-\alpha^2+2x\alpha i)}\, dx$, we obtain

$$\int_{-\infty}^{+\infty} e^{-x^2} e^{-2x\alpha i}\, dx = e^{-\alpha^2} \sqrt{\pi}$$

and, taking real parts,

$$\int_{-\infty}^{+\infty} e^{-x^2} \cos 2x\alpha\, dx = e^{-\alpha^2} \sqrt{\pi}.$$

In order to check that $I(\alpha)$ is independent from $\alpha$ consider the function $f(z) = e^{-z^2}$ and the rectangle with vertices $-R, +R, R + \alpha i, -R + \alpha i$. Cauchy–Goursat's Theorem gives

$$\int_{-R}^{+R} e^{-x^2} - \int_{-R}^{+R} e^{-(x+\alpha i)^2}\, dx = \left( \int_{S_1} + \int_{S_2} \right) e^{-z^2}\, dz,$$

where $S_1, S_2$ indicate the vertical sides of the rectangle. Now, if $z = \pm R + it$, then one has

$$|e^{-z^2}| = e^{-\operatorname{Re}(z^2)} = e^{-(R^2-t^2)} = e^{-R^2} e^{t^2} \leq e^{-R^2} e^{\alpha^2}.$$

Hence, both integrals over the vertical sides are bounded by $|\alpha| e^{\alpha^2} e^{-R^2}$, a quantity which converges to zero when $R \nearrow +\infty$, and it turns out that $I(\alpha) = I(0)$ for all $\alpha \in \mathbb{R}$. $\qquad \square$

**Example 3.27.** Now the integral $I = \int_{-\infty}^{+\infty} \frac{\sin t}{t}\, dt$ will be computed; this integral is not absolutely convergent, but it is convergent, and therefore

$$I = \lim_{R \to +\infty} \int_{-R}^{+R} \frac{\sin t}{t}\, dt$$

(see Section 5.8 for a review of types of convergence of the improper integrals). Consider

$$f(z) = \frac{e^{iz} - 1}{z} = \sum_{n \geq 1} \frac{i^n z^{n-1}}{n!},$$

which is an entire function. Observe that, bearing in mind that the real part of the following integral is odd, one has

$$\int_{-R}^{+R} f(x)\, dx = \int_{-R}^{+R} \frac{\cos x - 1 + i \sin x}{x} = i \int_{-R}^{+R} \frac{\sin x}{x}\, dx.$$

Consider the upper semicircle $C_R$ with center at 0 and radius $R$, that is, $C_R = \{z : |z| = R,\ \mathrm{Im}\, z \geq 0\}$. By Cauchy's Theorem, one has

$$\int_{-R}^{+R} f(x)\, dx + \int_{C_R} f(z)\, dz = 0.$$

Then,

$$I = -\frac{1}{i} \lim_{R \to \infty} \int_{C_R} \frac{e^{iz} - 1}{z}\, dz = -\frac{1}{i} \lim_{R \to \infty} \int_0^{\pi} \frac{e^{iRe^{i\theta}} - 1}{Re^{i\theta}}\, iRe^{i\theta}\, d\theta$$

$$= -\lim_{R \to \infty} \int_0^{\pi} \left( e^{iRe^{i\theta}} - 1 \right) d\theta = \pi - \lim_{R \to \infty} \int_0^{R} e^{iRe^{i\theta}}\, d\theta.$$

Now, $|e^{iRe^{i\theta}}| = e^{-R\sin\theta}$ and $\sin\theta \geq \frac{2}{\pi}\theta$ for $0 \leq \theta \leq \pi$; so

$$\left| \int_0^{\pi} e^{iRe^{i\theta}}\, d\theta \right| \leq \int_0^{\pi} e^{-\frac{2}{\pi}R\theta}\, d\theta \to 0, \quad R \to \infty$$

from which $I = \pi$.                                                              □

An important consequence of Cauchy's Theorem is related with the existence of holomorphic antiderivatives and branches of the logarithm of functions. The following corollaries are consequences of Corollary 3.23, Proposition 3.20 and the fact that the form $f(z)\, dz$ is closed exactly when $f$ is holomorphic. In Chapter 6 it will be seen that the statements relative to star-like domains also hold for simply connected domains.

**Corollary 3.28.** *If $f$ is a holomorphic function on a domain $U$ of the plane, then $f$ has locally a holomorphic antiderivative on $U$. If $U$ is star-like, $f$ has a holomorphic antiderivative globally.*

**Corollary 3.29.** *If $f$ is a holomorphic function without zeros on a domain $U$ of the plane, then $f$ has a branch of the logarithm locally on $U$. If $U$ is star-like, $f$ has a branch of the logarithm globally.*

**Example 3.30.** The form $(z - a)^{-1}\, dz$ is closed on $\mathbb{C} \setminus \{a\}$ because $(z - a)^{-1}$ is holomorphic, but is not exact. A function $h$ with $dh = (z - a)^{-1}\, dz$ would be a holomorphic branch of the logarithm of $z - a$ on $\mathbb{C} \setminus \{a\}$, which does not exist. This happens on $\mathbb{C} \setminus \{a\}$; now, in any smaller open set in which there exists a branch $F$ of the logarithm of $z - a$ (for example, the complement of a ray or an arc that starts at $a$ and goes to $\infty$) it is known that $F' = (z - a)^{-1}$. $\qquad\square$

The form $\frac{dz}{z}$ is an example of a non-exact closed form on $\mathbb{C} \setminus \{0\}$. Separating real and imaginary parts it turns out that

$$\frac{dz}{z} = \frac{d(x + iy)}{x + iy} = \frac{dx + i\, dy}{x + iy}$$

$$= \frac{(x - iy)(dx + i\, dy)}{x^2 + y^2} = \frac{x\, dx + y\, dy}{x^2 + y^2} + i\,\frac{x\, dy - y\, dx}{x^2 + y^2}.$$

Observe that the real part does have a potential function:

$$d\left(\frac{1}{2}\operatorname{Log}(x^2 + y^2)\right) = \frac{x\, dx + y\, dy}{x^2 + y^2}.$$

Therefore, if $\frac{dz}{z}$ is not exact, it must be due to the imaginary part, which is the form of Example 3.40.

## 3.6 Classical theorems

The aim of this section is to introduce the necessary concepts to present an approximation to holomorphicity from the real variable point of view, in an arbitrary dimension.

### 3.6.1 Orientable regular submanifolds

The context is now the Euclidian space of dimension $n$, $\mathbb{R}^n$. One starts recalling the concept of *regular submanifold of dimension $k$ with boundary* or *$k$-submanifold with boundary*, which generalizes the concept of regular curve ($k = 1$).

A closed and connected set $M$ in $\mathbb{R}^n$ is called a *regular submanifold of dimension $k$ with boundary* if for any point $p \in M$ there is a neighborhood $U$ of $p$ in $\mathbb{R}^n$, a ball $B = B(0, r) \subset \mathbb{R}^k$ and a mapping $\sigma \colon B \to \mathbb{R}^n$ of $C^1$ class (that is, the components of $\sigma$ are $C^1$), with $\sigma(0) = p$ and differential $d\sigma(t)$ of rank $k$ at every point $t = (t_1, t_2, \ldots, t_k) \in B$ so that $\sigma$ is a homeomorphism either from $B(0, r)$ onto $U \cap M$ or from $B^-(0, r) = \{t = (t_1, t_2, \ldots, t_k) \in B(0, r) \colon t_k \leq 0\}$ onto $U \cap M$. In the first case it is said that $p$ is an *interior* point of $M$, and in the second one $p$ is called a *boundary* point of $M$. The boundary will be denoted by $\partial M$. If $\partial M$ is empty, it is said that $M$ is a *regular submanifold of dimension $k$ without boundary*. The couple $(V, \sigma)$ where $V = B(0, r)$ is called a *parametrization* or *local chart* of $M$ and $(t_1, t_2, \ldots, t_k)$ are the parameters or coordinates of the point $\sigma(t)$; $\sigma(t)$ is analogous to the mapping $\gamma(t)$ which parameterizes a curve ($k = 1$). In general a collection of charts $(V_i, \sigma_i)_{i \in I}$ will be necessary to cover the whole set $M$, $M = \bigcup_i \sigma_i(V_i)$, a collection called an *atlas* of M. In this situation, *the tangent space to $M$ at the point $p$ is the $k$-linear manifold $T_p(M)$ of $\mathbb{R}^n$ which contains all the tangent vectors to the regular curves that pass through $p$ and are contained in $M$*. Analytically and using the previous notations, $T_p(M)$ is the image of $\mathbb{R}^k$ by the linear mapping $d\sigma(0)$ and it is, therefore, generated by the vectors $\frac{\partial \sigma}{\partial t_j}(0)$, $j = 1, 2, \ldots, k$ (Figure 3.9).

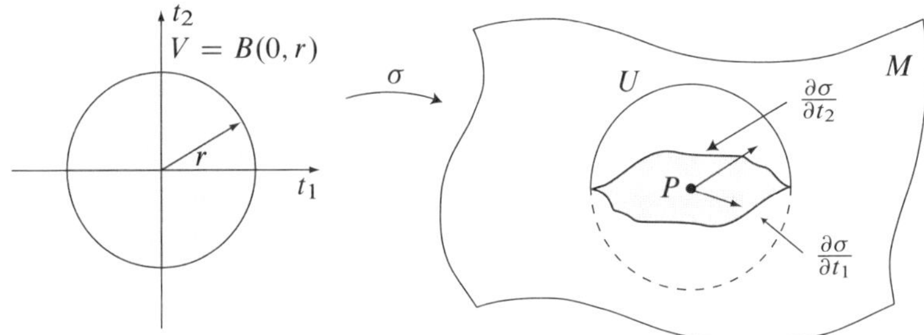


Figure 3.9

The definition of regular submanifold with boundary implies that $\partial M$ is the topological boundary of $M \setminus \partial M$ and it is by itself a regular manifold of dimension $k - 1$: if $p \in \partial M$ and $\sigma$ is as above, then the restriction $\sigma^+$ of $\sigma$ to $\{t \in V \colon t_k = 0\}$ is a local chart of $\partial M$. The $(k-1)$-manifold $\partial M$ has no boundary and symbolically we write $\partial^2 M = \emptyset$.

Similarly to Theorem 1.31, every regular 1-submanifold with boundary $\Gamma$ may be parameterized globally on a closed interval, $\gamma \colon [a, b] \to \Gamma$, and then $\partial \Gamma$ consists of the points $\gamma(a), \gamma(b)$. If $\Gamma$ has no boundary it is parametrized on the torus $\mathbb{T}$, $\gamma \colon \mathbb{T} \to \Gamma$. In the first case one has a simple regular curve in the sense of Subsection 1.4.1, and in the second case one has a simple closed regular curve with empty boundary. In other words, simple regular curves correspond to regular

1-manifolds with boundary (eventually empty). The other extreme case is for $k = n$; then $U = M \setminus \partial M$ is an open set of $\mathbb{R}^n$, and $M$ is its closure. In this case we say that $U$ is an open set with regular boundary.

On every regular $k$-manifold with boundary $M$ there is a $k$-dimensional Lebesgue measure, $dm_k$, which makes it possible to integrate continuous functions on $M$ and gives the $k$-dimensional volume of measurable sets of $M$. With classical notation in $\mathbb{R}^3$, one usually writes $dm_1 = ds$, $dm_2 = dA$ and $dm_3 = dV$. Here, being located in the space $\mathbb{R}^n$, we will use also the notation $dm_n = dV$, $dm_{n-1} = dA$ and $V(X)$, $A(X)$ for the measure of a set $X$, with respect to $dV$, $dA$.

If $(V, \sigma)$ is a local chart of $M$ and $f$ is a continuous function on $M$ vanishing outside $\sigma(V)$, then one has

$$\int_M f\, dm_k = \int_V f(\sigma(t)) \, |\det G_\sigma(t)|^{\frac{1}{2}} \, dt_1 \, dt_2 \ldots dt_k,$$

where $\det G_\sigma(t)$ is the determinant of the Gram matrix of the column vectors of the matrix of $d\sigma(t)$, that is, the one with entries $\langle \frac{\partial \sigma}{\partial t_i}, \frac{\partial \sigma}{\partial t_j} \rangle$, $i, j = 1, \ldots, k$. The quantity $|\det G_\sigma(t)|^{1/2}$ is the $k$-dimensional volume of the parallelepiped generated by the vectors $\frac{\partial \sigma}{\partial t_i}$, column vectors of $d\sigma(t)$.

For $k = n - 1$, the manifolds are called *regular hypersurfaces with boundary*, and for $k = 2$, *surfaces*. If $M$ is a regular hypersurface with boundary and $p \in M$, there are two opposite unitary vectors orthogonal to $T_p(M)$. When it is possible to choose in each point one of these unitary normal vectors in a continuous way on $M$, one says that $M$ is *orientable*. Every regular curve is orientable, but not every hypersurface is; for example, the Möbius strip is not orientable. When $M$ is orientable, then it is so in two different ways, as a curve may be travelled in two different directions. Choosing a unitary tangent vector ($k = 1$) or a unitary normal vector ($k = n - 1$) may be understood as a way of orienting the tangent space $T_p(M)$.

In order to formulate the concept of orientation in general it is necessary to define what orientation means in a linear subspace $F$ of $\mathbb{R}^n$, of dimension $k$. This may be done by separating the basis of $F$ into two classes, grouping in the same class all the basis of $F$ for which the change of basis matrix has positive determinant. When $k = n - 1$, that is, when $F$ is a hyperplane, an orientation of $F$ corresponds to choosing $\vec{N}$, one of the two unitary normal orthogonal vectors to $F$, declaring that a basis $\vec{u}_1, \ldots, \vec{u}_{n-1}$ of $F$ is positive if $\vec{u}_1, \ldots, \vec{u}_{n-1}, \vec{N}$ is a positive basis of $\mathbb{R}^n$ when this space is oriented with the canonical basis. Then a *regular $k$-manifold with boundary $M$* is said to be *orientable* if it is possible to orientate the tangent spaces $T_p(M)$ in a continuous way. In terms of local charts it means that there exists an atlas $(V_i, \sigma_i)$ which covers $M$, $M = \bigcup \sigma_i(V_i)$, so that the mappings $\sigma_i \sigma_j^{-1}$ of changes of coordinates have all positive Jacobian determinants. In this case, the mappings $\sigma_i^+(\sigma_j^+)^{-1}$ also have positive Jacobian determinants; that is, if $M$ is orientable, so is $\partial M$.

An orientation of $M$ and an orientation of $\partial M$ are said to be *compatible* if the basis $\frac{\partial \sigma}{\partial t_j}(0)$, $j = 1, \ldots, k$ of $T_p(M)$ corresponds to the orientation of $M$, and the basis $\frac{\partial \sigma^+}{\partial t_j}(0)$, $j = 1, \ldots, k - 1$, to the one of $\partial M$. There are two particular cases which may be easily understood. When $n = 3$ and $k = 2$, $M$ is a surface of $\mathbb{R}^3$ limited by $\partial M$, which is a simple closed curve. A sense of travelling on $\partial M$ and a selection of normal vector $\vec{N}$ on $M$ are compatible if they fulfill the *right-hand rule*: place your right hand on $M$ near a point of $\partial M$ so that your fingers point in the direction given by the orientation of $\partial M$. Then your thumb points in the direction of $\vec{N}$. The other case is when $M$ is an $n$-manifold with boundary in $\mathbb{R}^n$; then $M \setminus \partial M$ is an open set, and $\partial M$, its boundary, a hypersurface. As said above, an orientation of $\partial M$ corresponds to the selection of a unitary vector field normal to $\partial M$. When $M$ has an orientation given by the canonical basis of $\mathbb{R}^n$, the compatible orientation of $\partial M$ just defined corresponds to a unitary normal vector field $\vec{N}$ called a *normal exterior* vector field on $\partial M$. Then a basis $\vec{u}_1, \ldots, \vec{u}_{n-1}$ of $T_p(\partial M)$ is positive if $\vec{u}_1, \ldots, \vec{u}_{n-1}, \vec{N}$ is positive on $\mathbb{R}^n$.

In this case it is said that $U = M \setminus \partial M$ is an *open set with regular boundary* oriented by the unitary normal exterior vector field $\vec{N}$.

### 3.6.2  Flow of a vector field through a hypersurface

Let $M$ be a regular hypersurface oriented by the continuous selection of a unitary normal vector $\vec{N}$ at each point of $M$. If $\vec{X}$ is a continuous vector field on $M$, the *flow of $\vec{X}$ through $M$* is the sum of the normal components of $\vec{X}$, that is,

$$\int_M \langle \vec{X}, \vec{N} \rangle dA,$$

recalling that $dA = dm_{n-1}$ is the $(n - 1)$-dimensional volume element of $M$.

When $n = 2$, both circulations and flows are computed over oriented $C^1$ curves, and in fact these two concepts are equivalent due to the following observation. Denote by $J$ the linear mapping from $\mathbb{R}^2$ to $\mathbb{R}^2$ given by $Jv = (-y, x)$ if $v = (x, y)$, which in terms of $z = x + iy$ is written as $Jz = iz$; $J$ is, then, the multiplication by $i$. Obviously, $Jv$ is orthogonal to $v$ and for two any vectors $v_1, v_2$ one has $\langle Jv_1, Jv_2 \rangle = \langle v_1, v_2 \rangle$. Then the flow of a vector field $\vec{X}$ through the curve $\gamma$ is written as

$$\int_\gamma \langle \vec{X}, \vec{N} \rangle ds = \int_\gamma \langle J\vec{X}, J\vec{N} \rangle ds$$

and, since $J\vec{N}$ is a vector $\vec{T}$ tangent to $\gamma$, one gets that the flow of $\vec{X}$ is the same as the circulation of $J\vec{X}$ (provided that the selection of the normal vector $\vec{N}$ and the tangent $\vec{T}$ satisfies $\vec{T} = J\vec{N}$).

This way, Green's formula applied to the vector field $J\vec{X}$ gives the following result.

**Theorem 3.31.** *Let $U$ be a bounded domain in the plane with piecewise regular positively oriented boundary and let $\vec{N}$ be the unitary exterior normal vector on $\partial U$. Let $\vec{X} = (P, Q)$ be a differentiable vector field on a neighborhood of $\bar{U}$ such that the function $P_x + Q_y$ is continuous on $\bar{U}$. Then*

$$\int_{\partial U} \langle \vec{X}, \vec{N} \rangle \, ds = \iint_U (P_x + Q_y) \, dx \, dy.$$

In dimension $n > 2$ flows are expressed easily in terms of a parametrization of the hypersurface, as for circulations. Consider first a surface $S$ of $\mathbb{R}^3$. In this case, a parametrization of $S$ is a mapping $\sigma : V \to \mathbb{R}^3$ of class $C^1$, where $V$ is an open set of $\mathbb{R}^2$, which is a homeomorphism between $V$ and $\sigma(V)$, so that the differential $d\sigma(s, t)$ at any point $(s, t) \in V$ has rank 2. The image of $d\sigma(s, t)$, as a linear mapping from $\mathbb{R}^2$ to $\mathbb{R}^3$, is then the tangent plane to $S$ at the point $\sigma(s, t)$. Typically $\sigma(V)$ is the whole $S$ up to some sets of area zero, and one may think that it gives a global parametrization. In this case the vectors $\vec{\sigma}_s = \frac{\partial \sigma}{\partial s}, \vec{\sigma}_t = \frac{\partial \sigma}{\partial t}$ are tangent vectors to $S$, its cross product $\vec{\sigma}_s \wedge \vec{\sigma}_t$ is then normal to $S$, and $\vec{N}(\sigma(s, t)) = \vec{\sigma}_s \wedge \vec{\sigma}_t / |\vec{\sigma}_s \wedge \vec{\sigma}_t|$ is a unitary normal vector. On the other hand, given two vectors of $\mathbb{R}^3$, the area of the parallelogram they define is also the length of their cross product, and so the area element $dA$ of $S$ is expressed, in terms of the parametrization, as $dA = |\vec{\sigma}_s \wedge \vec{\sigma}_t| \, ds \, dt$. All this means that if the orientation is the one given by $\vec{N}$, the flow of a continuous vector field $\vec{X}$ through $S$ is

$$\int_S \langle \vec{X}(\sigma(s, t)), \vec{\sigma}_s \wedge \vec{\sigma}_t \rangle \, ds \, dt.$$

If $\sigma(s, t) = (x(s, t), y(s, t), z(s, t))$, then $\vec{\sigma}_s = (x_s, y_s, z_s)$, $\vec{\sigma}_t = (x_t, y_t, z_t)$, so that $\vec{\sigma}_s \wedge \vec{\sigma}_t$ has components $(y_s z_t - z_s y_t, z_s x_t - z_t x_s, x_s y_t - x_t y_s)$ and the flow is written, if $\vec{X} = (X_1, X_2, X_3)$, as

$$\int_S \left( X_1 \cdot (y_s z_t - z_s y_t) + X_2 \cdot (z_s x_t - z_t x_s) + X_3 \cdot (x_s y_t - x_t y_s) \right) ds \, dt,$$

where the functions $X_i$ are evaluated at $\sigma(s, t)$. Observe that the integrand is formally the result of computing the determinant of the matrix with $\vec{X}$ in the first row, $\vec{\sigma}_s$ in the second one and $\vec{\sigma}_t$ in the third.

For the previous flow one uses also the notation

$$\int_S (X_1 dy \wedge dz + X_2 dz \wedge dx + X_3 dx \wedge dy),$$

understanding that, for example, $dy \wedge dz$ is integrated over $S$ making the substitution $dy = y_s ds + y_t dt$, $dz = z_s ds + z_t dt$ and that $ds \wedge ds = dt \wedge dt = 0$, $ds \wedge dt = -dt \wedge ds = dsdt$.

The expression $\eta_{\vec{X}} = X_1 dy \wedge dz + X_2 dz \wedge dx + X_3 dx \wedge dy$ related to the vector field $\vec{X}$ – whose integral over $S$ gives the flow – is called a *differential 2-form*.

The same computation in dimension $n$ for the flow of the vector field $\vec{X} = (X_1, X_2, \ldots, X_n)$ through the hypersurface $M$ gives

$$\int_M \sum_{i=1}^{n} X_i dx_1 \wedge \cdots \wedge (dx_i)^* \wedge \cdots \wedge dx_n,$$

where $(dx_i)^*$ means that this term is not there. The expression inside the integral is called *differential* $(n-1)$-*form*.

The notion of differential $k$-form, which will not be needed, appears when generalizing these concepts to $k$-submanifolds of $\mathbb{R}^n$.

### 3.6.3  The divergence theorem, the curl theorem and Stokes' theorem for 1-forms

In this subsection the versions of Theorems 3.22 and 3.31 in dimension $n > 2$ will be considered. It will be highlighted how the same procedure to prove Green's formula leads to the *curl theorem* and to the *divergence theorem* with weaker hypotheses than in traditional versions. These theorems will allow to interpret holomorphic functions from the real variable point of view and define a generalization of it to $\mathbb{R}^n$.

In dimension $n > 2$ there is no correspondence between the notions of circulation and flow of a vector field and it is necessary to state a different result for each of these magnitudes.

Let us begin with a study of the flows of a vector field $\vec{X}$. In this case one starts from a bounded domain $U$ in $\mathbb{R}^n$ such that its topological boundary $\partial U$ is a regular hypersurface. One must evaluate the flow of $\vec{X}$ through $\partial U$. The concept of divergence of a differentiable vector field is needed.

Suppose that $\vec{X} = (X_1, X_2, \ldots, X_n)$ is a vector field with differentiable components $X_i$ (*differentiable vector field*) on a neighborhood of a point $p \in \mathbb{R}^n$. Denote by $B_\varepsilon$ the ball $B(p, \varepsilon)$ centered at $p$ with radius $\varepsilon > 0$ and compute approximately the flow of $\vec{X}$ through $\partial B_\varepsilon$ when oriented with the unitary exterior normal $\vec{N}(x) = \frac{1}{\varepsilon}(x - p)$. By Taylor's formula we can write, for $x \in \partial B_\varepsilon$,

$$\vec{X}(x) = \vec{X}(p) + \sum_i \frac{\partial \vec{X}}{\partial x_i}(p)(x_i - p_i) + o(\varepsilon).$$

Next one must consider the normal component of $\vec{X}$ and integrate it over $\partial B_\varepsilon$. The contribution of the constant vector field $\vec{X}(p)$ is zero, because this normal component is an odd function over $\partial B_\varepsilon$; in other words, it is clear that the flow of any constant vector field is zero. The contribution of the vector field $\frac{\partial \vec{X}}{\partial x_i}(p)x_i$ is

$$\int_{\partial B_\varepsilon} \left\langle \frac{\partial \vec{X}}{\partial x_i}(p)(x_i - p_i), \frac{x - p}{\varepsilon} \right\rangle dA.$$

Now, since the function $(x_i - p_i)(x_j - p_j)$ has integral zero if $i \neq j$ and $\frac{c_n}{n}\varepsilon^{n+1}$ if $i = j$ (where $c_n$ is the $(n-1)$-dimensional measure of the unit sphere), we get that the previous integral is exactly $\frac{c_n}{n}\varepsilon^n \frac{\partial X_i}{\partial x_i}(p)$; that is,

$$\int_{\partial B_\varepsilon} \langle \vec{X}, \vec{N} \rangle \, dA = \frac{c_n}{n}\varepsilon^n \sum_i \frac{\partial X_i}{\partial x_i}(p) + o(\varepsilon^n).$$

Bearing in mind that $\frac{c_n}{n}\varepsilon^n$ is the $n$-dimensional volume of $B_\varepsilon$, one finds that at any point $x$ on a neighborhood of which the vector field $\vec{X}$ is differentiable, one has

$$\sum_i \frac{\partial X_i}{\partial x_i}(x) = \lim_{\varepsilon \to 0} \frac{1}{V(B(x,\varepsilon))} \int_{\partial B(x,\varepsilon)} \langle \vec{X}, \vec{N} \rangle \, dA.$$

The function $\sum_{i=1}^n \frac{\partial X_i}{\partial x_i}(x)$ is, consequently, a *density flow* per unit of closed volume and it is called *divergence* of the vector field $\vec{X}$, denoted by $\mathrm{div}(\vec{X})$ or by $\langle \vec{\nabla}, \vec{X} \rangle$. One arrives at the same expression using, for example, cubes instead of balls which contract to $x$. This definition of $\mathrm{div}(\vec{X})$ is *equivalent* to the following theorem:

**Theorem 3.32** (Divergence theorem). *Let $U$ be a bounded domain in $\mathbb{R}^n$ with regular boundary oriented with the unitary exterior normal vector field $\vec{N}$. Let $\vec{X}$ be a differentiable vector field on a neighborhood of $\bar{U}$ with $\mathrm{div}\,\vec{X}$ continuous on $\bar{U}$. Then the flow of $\vec{X}$ through $\partial U$ equals the integral of the divergence on the domain $U$, that is,*

$$\int_{\partial U} \langle \vec{X}, \vec{N} \rangle \, dA = \int_U (\mathrm{div}\,\vec{X}) \, dV.$$

This theorem is proved with the same method as Green's formula and also holds if $\partial U$ is a *piecewise regular hypersurface*, that is, a finite union of regular hypersurfaces with boundary, joined along their boundaries (a cube, for example). For the proof observe first that, as seen above, if $Q$ is a cube of size $\varepsilon$ centered at $p$, then

$$\int_{\partial Q} \langle \vec{X}, \vec{N} \rangle \, dA - V(Q)\,\mathrm{div}(\vec{X})(p) = o(V(Q)).$$

Therefore, if $\mathrm{div}(\vec{X})$ is a continuous function on $\bar{U}$, it is uniformly continuous and one has

$$\int_{\partial Q} \langle \vec{X}, \vec{N}\rangle \, dA - \int_Q \mathrm{div}(\vec{X}) dV = o(V(Q)),$$

uniformly for all cubes $Q \subset U$. Then, by consecutive subdivisions (the essence of infinitesimal calculus!), the divergence theorem is stated for any parallelepiped, due to the cancellation of the flows through consecutive sides of contiguous cubes. This would give the equivalent result to the first step of the proof of Theorem 3.22. The other two steps are proved in a similar way.

**Example 3.33.** Let $U$ be the interior of the ellipsoid of $\mathbb{R}^3$ with equation $x^2 + \frac{y^2}{4} + \frac{z^2}{4} = 1$ and consider the vector field $\vec{X} = (x^2, y, z + y)$. In order to compute the flow of $\vec{X}$ through the boundary of $U$, it is easier to evaluate the volume integral of the divergence, with $\mathrm{div}\,\vec{X} = 2x + 1 + 1 = 2x + 2$. The function $2x$ has integral $0$ on $U$ because it is odd and by the divergence theorem, the flow will be twice the volume of $U$, that is, $\frac{32}{3}\pi$. If we now want to compute the flow through the part $\Sigma$ of the boundary corresponding to $z > 0$, consider the open set $V = \{(x, y, z) \in U : z > 0\}$. Applying the divergence theorem on $V$, we see that the flow through $\Sigma$ plus the flow through the lower lid of $V$, which is the interior of the ellipse $E$ in the plane $xy$ with equation $x^2 + \frac{y^2}{4} = 1$, equals the volume integral of the divergence, which is $\frac{16}{3}\pi$. On the lower lid the unitary exterior normal vector to $\partial U$ is $(0, 0, -1)$, the normal component of $\vec{X}$ is $y$, and the flow through $E$ is zero. Therefore, the flow through $\Sigma$ is $\frac{16}{3}\pi$. $\qquad\square$

The procedure to deal with circulations of a vector field $\vec{X}$ will be similar, introducing the concept of *circulation density*, first in dimension $n = 3$.

Let $\vec{v}_1, \vec{v}_2, \vec{v}_3$ be an orthonormal basis of $\mathbb{R}^3$ with $\vec{v}_3 = \vec{v}_1 \wedge \vec{v}_2$. Let $D_\varepsilon$ be the disc centered at $p \in \mathbb{R}^3$ with radius $\varepsilon$ located on the perpendicular plane to $\vec{v}_3$ which contains $p$, and orient its boundary with the trajectory $\gamma(t) = p + \varepsilon(\cos t)\vec{v}_1 + \varepsilon(\sin t)\vec{v}_2$, $0 \le t \le 2\pi$. This orientation of $\partial D_\varepsilon$ is the one illustrated again with the *right-hand rule*: if the fingers of your right-hand point in the direction in which $\partial D_\varepsilon$ is travelled, then your thumb points in the direction of $\vec{v}_3$ (Figure 3.10). Let $\vec{X} = (X_1, X_2, X_3)$ be a differentiable vector field on a neighborhood of the point $p$ and compute the circulation of $\vec{X}$ along $\partial D_\varepsilon$ using once again Taylor's formula. If $\vec{v}_k = (v_k^1, v_k^2, v_k^3), k = 1, 2$ and $x \in \partial D_\varepsilon$, one has

$$\vec{X}(x) = \vec{X}(p) + \sum_{i=1}^{3} \frac{\partial \vec{X}}{\partial x_i}(p)\varepsilon(\cos t\, v_1^i + \sin t\, v_2^i) + o(\varepsilon).$$

Consider the scalar product of $\vec{X}$ with the tangent vector $\vec{T} = \varepsilon(-(\sin t)\vec{v}_1 + (\cos t)\vec{v}_2)$ to $\partial D_\varepsilon$ and integrate over $\partial D_\varepsilon$. The contribution of the constant $\vec{X}(p)$ is

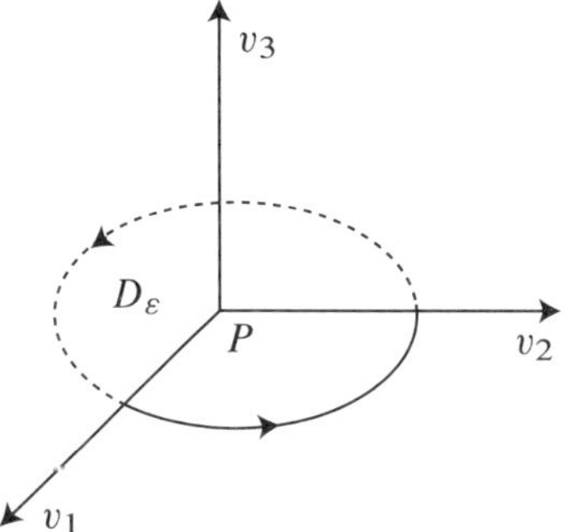


Figure 3.10

zero, the one corresponding to the linear term is

$$\varepsilon^2 \int_0^{2\pi} \left( \sum_{j=1}^{3} ((\cos t)v_2^j - (\sin t)v_1^j) \sum_{i=1}^{3} \frac{\partial X_j}{\partial x_i}(p)((\cos t)v_1^i + (\sin t)v_2^i) \right) dt,$$

and the one of the residual term is $o(\varepsilon^2)$. Each integral

$$\int_0^{2\pi} ((\cos t)v_2^j - (\sin t)v_1^j)((\cos t)v_1^i + (\sin t)v_2^i)) \, dt$$

has value $\pi(v_2^j v_1^i - v_1^j v_2^i)$, in particular 0 if $i = j$. Hence one has

$$\int_{\partial D_\varepsilon} \langle X, \vec{T} \rangle \, ds = \pi \varepsilon^2 \sum_{i,j} \frac{\partial X_j}{\partial x_i}(p)(v_2^j v_1^i - v_1^j v_2^i) + o(\varepsilon^2)$$

$$= \pi \varepsilon^2 \sum_{i<j} \left( \frac{\partial X_j}{\partial x_i}(p) - \frac{\partial X_i}{\partial x_j}(p) \right)(v_2^j v_1^i - v_1^j v_2^i) + o(\varepsilon^2).$$

Now notice that the components of the vector $\vec{v}_3 = \vec{v}_1 \wedge \vec{v}_2$ are precisely the quantities $v_2^j v_1^i - v_1^j v_2^i$ so that in the previous expression appears the scalar product of $\vec{v}_3$ with the vector

$$\mathrm{rot}\,\vec{X} = \left( \frac{\partial X_3}{\partial y} - \frac{\partial X_2}{\partial z}, \frac{\partial X_1}{\partial z} - \frac{\partial X_3}{\partial x}, \frac{\partial X_2}{\partial x} - \frac{\partial X_2}{\partial y} \right),$$

also denoted by $\vec{\nabla} \times \vec{X}$ and called *curl* of $\vec{X}$. So it has been shown that if $D_\varepsilon$ is a small disc centered at $p$, located in the perpendicular plane to a direction $\vec{v}$ and oriented according to the right-hand rule, one has

$$\lim_{\varepsilon \to 0} \frac{1}{\pi \varepsilon^2} \int_{\partial D_\varepsilon} \langle X, \vec{T} \rangle \, ds = \langle \mathrm{rot}\,\vec{X}(p), \vec{v} \rangle.$$

This means that the quantity $\langle \operatorname{rot} \vec{X}(p), \vec{v} \rangle$ is a circulation density per unit of closed area. Consequently the vector $\operatorname{rot} \vec{X}(p)$ gives the axis around which the orbits of the vector field $\vec{X}$ have the maximal tendency to rotate. The last equality is *equivalent* to the curl theorem stated below which may be proved with the same method indicated in the case of the divergence theorem, based on the proof of Theorem 3.22.

**Theorem 3.34** (Curl theorem). *Let $S$ be a regular surface in $\mathbb{R}^3$ with boundary and let $\gamma$ be a regular curve which parameterizes $\partial S$ in such a way that $S$ and $\gamma$ have compatible orientations. Then, if $\vec{X}$ is a differentiable vector field on a neighborhood of $\overline{S}$ with continuous curl on $\overline{S}$, the circulation of $\vec{X}$ along $\gamma$ coincides with the flow of the curl of $\vec{X}$ through $S$, that is,*

$$\int_\gamma \langle \vec{X}, \vec{T} \rangle \, ds = \int_S \langle \operatorname{rot} \vec{X}, \vec{N} \rangle \, dA.$$

**Example 3.35.** Let $\gamma$ be the intersection curve of the unit ball in $\mathbb{R}^3$ with the plane with equation $x + y + z = 0$, oriented according to the right-hand rule when the plane is oriented by the normal vector $(1, 1, 1)$. Compute the circulation of the vector field $\vec{X} = (z, x, xy)$ along $\gamma$. The vector field $\vec{X}$ has a curl given by $(x, 1 - y, 1)$, the normal component of which at the points of the plane is $\frac{1}{\sqrt{3}}(2 + x - y)$. Note that $x$, $y$ are odd on the disc limited by $\gamma$, and conclude that the flow of $\vec{X}$ through this disc and therefore the circulation along $\gamma$ is $\frac{2\pi}{\sqrt{3}}$. $\qquad\square$

The consideration noted before Theorem 3.34 tells how to proceed in dimension $n > 3$; in this case, for a vector field $\vec{X} = (X_1, X_2, \ldots, X_n)$, the computation is exactly the same until getting the expression

$$\sum_{i<j} \left( \frac{\partial X_j}{\partial x_i} - \frac{\partial X_i}{\partial x_j} \right) (v_2^j v_1^i - v_1^j v_2^i)$$

as a density that should be integrated over a surface $S$ with boundary $\gamma = \partial S$ in order to obtain the circulation along $\gamma$. Here $\vec{v}_1, \vec{v}_2$ are vectors which generate the tangent space to $S$. In terms of a parametrization of $S$ given by $\sigma(s, t)$ with domain of parameters $V$, it will be $\vec{v}_1 = \frac{\partial \sigma}{\partial s}, \vec{v}_2 = \frac{\partial \sigma}{\partial t}$, and integrating over $V$ the previous density, one obtains

$$\iint_V \sum_{i<j} \left( \frac{\partial X_j}{\partial x_i} - \frac{\partial X_i}{\partial x_j} \right) (\sigma(s, t)) \left( \frac{\partial \sigma_j}{\partial t} \frac{\partial \sigma_i}{\partial s} - \frac{\partial \sigma_j}{\partial s} \frac{\partial \sigma_i}{\partial t} \right) \, ds \, dt$$

for $\sigma = (\sigma_1, \sigma_2, \ldots, \sigma_n)$.

In an intrinsic way this quantity is written as

$$\int_S \sum_{i<j} \left( \frac{\partial X_j}{\partial x_i} - \frac{\partial X_i}{\partial x_j} \right) dx_i \wedge dx_j, \tag{3.8}$$

where the integrand is a differential 2-form in $n$-variables. Now $dx_i \wedge dx_j$ is integrated over $S$ making the substitution $dx_i = \frac{\partial \sigma_i}{\partial s} ds + \frac{\partial \sigma_i}{\partial t} dt$ and using the fact that $ds \wedge ds = dt \wedge dt = 0$, $ds \wedge dt = -dt \wedge ds = ds\, dt$.

Recall that the vector field $\vec{X}$ with components $X_1, X_2, \ldots, X_n$ is equivalent to the differential 1-form $\omega = \sum_{i=1}^{N} X_i dx_i$. The 2-form raised up in (3.8) is called the *exterior differential* of $\omega$ and represented by $d\omega$. Hence,

$$d\left( \sum_i X_i dx_i \right) = \sum_{i<j} \left( \frac{\partial X_j}{\partial x_i} - \frac{\partial X_i}{\partial x_j} \right) dx_i \wedge dx_j.$$

With these notations the version of the curl theorem for $n > 3$ is the following one:

**Theorem 3.36** (Stokes' theorem for 1-forms). *Let $S$ be a regular surface in $\mathbb{R}^n$ with boundary and let $\gamma$ be a regular curve that parameterizes $\partial S$ in such a way that $S$ and $\gamma$ have compatible orientations. Let $\vec{X} = (X_1, X_2, \ldots, X_n)$ be a differentiable vector field on a neighborhood of $\bar{S}$ with associated 1-form $\omega = \sum_i X_i\, dx_i$ such that the coefficients of the 2-form $d\omega$ are continuous functions on $\bar{S}$. Then*

$$\int_\gamma \langle \vec{X}, \vec{T} \rangle\, ds = \int_S d\,\omega.$$

**Remark 3.4.** Theorem 3.31 is a particular case of the divergence theorem and Green's formula (Theorem 3.22) may be seen as a particular case of the curl theorem (Theorem 3.34). Understanding that the domain $U$ of the plane is a surface in $\mathbb{R}^3$ oriented by the vector $\vec{N} = (0, 0, 1)$ and that the vector field $\vec{X} = (P, Q)$ is a vector field in $\mathbb{R}^3$ with the third component zero, one has rot $\vec{X} = (0, 0, Q_x - P_y)$ and $\langle \text{rot}\, \vec{X}, \vec{N} \rangle = Q_x - P_y$.

### 3.6.4  Closed 1-forms and exact 1-forms

Theorem 3.9 states that a vector field has a potential function if and only if it is conservative. Both conditions are of global type. This subsection is devoted to the study of a local concept weaker than the fact a vector field be conservative or a form be exact.

Let $\vec{X} = (X_1, X_2, \ldots, X_n)$ be a differentiable vector field on a domain $U$ in $\mathbb{R}^n$ with associated 1-form $\omega = \sum_{i=1}^{n} X_i dx_i$. If $\omega$ is locally exact with $\omega = dh$ in a certain region, then the function $h$ is twice differentiable, and by Schwarz's rule one has $\frac{\partial^2 h}{\partial x_i \partial x_j} = \frac{\partial^2 h}{\partial x_j \partial x_i}$. Consequently, $\frac{\partial X_i}{\partial x_j} = \frac{\partial X_j}{\partial x_i}$ $i,j = 1,\ldots,n$. In terms of the exterior differential $d\omega$ it turns out that

$$d\omega = \sum_{i<j} \left( \frac{\partial X_i}{\partial x_j} - \frac{\partial X_j}{\partial x_i} \right) dx_i \wedge dx_j = 0.$$

The differential 1-forms $\omega$ on $U$, with differentiable coefficients, which satisfy the condition $d\omega = 0$ are called *closed* on $U$. This is a local concept. Hence, each locally exact 1-form with differentiable coefficients is closed and the components of a differentiable and locally conservative vector field $\vec{X}$ must satisfy the equations $\frac{\partial X_i}{\partial x_j} = \frac{\partial X_j}{\partial x_i}$. Here the converse will be shown; first a global type result is given.

**Theorem 3.37.** *In a star-like domain of $\mathbb{R}^n$, a differential 1-form with differentiable coefficients is exact if and only if it is closed. A differentiable vector field $\vec{X} = (X_1, X_2, \ldots, X_n)$ is conservative if and only if it satisfies $\frac{\partial X_j}{\partial x_i} = \frac{\partial X_i}{\partial x_j}$, $i,j = 1,2,\ldots,n$ (equations which for $n = 3$ mean* rot $\vec{X} = 0$).

*Proof.* The proof is done in terms of 1-forms. It is enough to see that every closed form $\omega$ is exact; by Theorem 3.10 this is equivalent to

$$\int_{\partial \Delta} \omega = 0$$

for every triangle $\Delta$ included in the domain of $\omega$, a fact which is a consequence of Stokes' theorem (Theorem 3.36). $\qquad\square$

In Chapter 6 it will be proved that this result also holds for simply connected domains of the plane.

**Remark 3.5.** If the domain $U$ of Theorem 3.37 is star-like with respect to the origin, the general expression of the potential function of the vector field $\vec{X}$ is

$$h(x) = \int_\gamma \langle \vec{X}, \vec{T} \rangle \, ds + c,$$

where $\gamma$ is the segment going from $0$ to $x$ and $c$ is a constant. With the additional hypothesis that $\vec{X}$ is $C^1$ (that is, the components of $\vec{X}$ are $C^1$) one may directly prove that $h$ is a potential function: parameterizing the segment $\gamma$ leads to

$$h(x) = \int_0^1 \sum_i (X_i(tx)x_i) \, dt.$$

Hence,

$$\frac{\partial h}{\partial x_j} = \int_0^1 \left( X_j(tx) + t \sum_i x_i \frac{\partial X_i}{\partial x_j}(tx) \right) dt,$$

and using now the hypothesis one finds

$$\frac{\partial h}{\partial x_j} = \int_0^1 \left( X_j(tx) + t \sum_i x_i \frac{\partial X_j}{\partial x_i}(tx) \right) dt.$$

But, the expression inside the integral is $f'(t)$ with $f(t) = X_j(tx)t$, and due to the fundamental theorem of calculus the integral has value $f(1) - f(0) = X_j(x)$.

The same remark is relevant when using other paths to define the potential function. For example, if $\vec{X} = (P, Q)$ is a vector field in the plane satisfying $Q_x = P_y$ and one wants to find the potential function $h(x, y)$, choose $(0, 0)$ as the starting point and let the path be the segment starting at $(0, 0)$ and finishing at $(x, 0)$ followed by the one starting at $(x, 0)$ and ending at $(x, y)$; then

$$h(x, y) = \int_0^x P(t, 0)\, dt + \int_0^y Q(x, t)\, dt.$$

Clearly $h_y = Q$ by the fundamental theorem of calculus, and furthermore,

$$h_x(x, y) = P(x, 0) + \int_0^y Q_x(x, t)\, dt = P(x, 0) + \int_0^y P_y(x, t)\, dt$$

$$= P(x, 0) + P(x, y) - P(x, 0) = P(x, y).$$

In practice, however, it is usual to proceed in the following equivalent way: one wants to solve the equations $\frac{\partial h}{\partial x_i} = X_i, i = 1, \ldots, n$; integrate the first equation in $x_1$, and obtain a constant of integration $C(x_2, \ldots, x_n)$ which depends on $x_2, \ldots, x_n$; later one uses the second equation to determine $C(x_2, \ldots, x_n)$ and so on.

**Example 3.38.** The form $\omega = (\sin xy + xy \cos xy)\, dx + (x^2 \cos xy + y)\, dy$ is closed on the whole plane. Integrating $h_x = \sin xy + xy \cos xy$ with respect to $x$ it turns out that $h(x, y) = x \sin xy + C(y)$, and using $h_y = x^2 \cos xy + y$ yields $C'(y) = y$ and $h(x, y) = x \sin xy + \frac{y^2}{2} + C$ which is the general expression of a potential function of $\omega$. $\square$

It is clear that Theorem 3.37 implies:

**Theorem 3.39.** *A differential 1-form with differentiable coefficients on a domain of* $\mathbb{R}^n$ *is locally exact if and only if it is closed. For a differentiable vector field* $\vec{X} = (X_1, X_2, \ldots, X_n)$ *on a domain* $U \subset \mathbb{R}^n$, *the following properties are equivalent:*

a) $\frac{\partial X_i}{\partial x_j} = \frac{\partial X_j}{\partial x_i}$, $i, j = 1, 2, \ldots, n$ *(rot $\vec{X} = 0$ if $n = 3$) on $U$.*

b) $\vec{X}$ *is locally conservative on $U$.*

c) $\vec{X}$ *has, locally, a potential function on $U$.*

In general, however, a closed form is not exact.

**Example 3.40.** A typical example is the form

$$\omega = -\frac{y}{x^2 + y^2}\, dx + \frac{x}{x^2 + y^2}\, dy$$

on $U = \mathbb{R}^2 \setminus (0,0)$. A computation shows that $d\omega = 0$, but if $\gamma$ is the unit circle travelled in the positive sense – a closed path in $U$ – the line integral $\int_\gamma \omega$ has value $2\pi$. Hence, $\omega$ is not exact on $U$. On the smaller open set $V = \mathbb{R}^2 \setminus \{(x,0),\ x \le 0\}$ the form $\omega$ is exact because the function $\theta(x, y)$ which measures the angle between the ray starting at zero and passing by $(x, y)$ with the positive half axis is a potential function. Analytically, $\theta(x, y) = \arctan \frac{y}{x}$ if $x > 0$, $\theta(x, y) = \arctan \frac{y}{x} + \pi$ if $x < 0$, $y > 0$, $\theta(x, y) = \arctan \frac{y}{x} - \pi$ if $x < 0$, $y < 0$.   $\square$

In this subsection forms on domains of the complex plane ($n = 2$ and $\omega = f(z)\, dz$) have been treated as a particular case of forms on domains of $\mathbb{R}^n$ (arbitrary dimension $n$ and complex differential forms). However, later on, it will be seen that there is a very important difference between the real and the complex case. In the real case there are functions $h$ of $C^1$ class exactly, that is, they are $C^1$ but do not have second-order derivatives. The antiderivative $\phi$ of a continuous non-differentiable function is an example in one variable and $h(x, y) = \phi(xy)$ is an example in two variables. Then the form $\omega = dh$ has continuous coefficients and, by definition, has potential function, so it is exact. Therefore, it has integrals not depending on the path; since it is not differentiable, it makes no sense to ask if it is closed. Hence, in the real case there are locally exact continuous forms which are non-differentiable. In the complex case this is not possible because it will be proved that if the function $f$ has a holomorphic antiderivative, then $f$ is also holomorphic. Therefore, all continuous forms of type $f(z)\, dz$ which are locally exact are automatically differentiable and closed.

## 3.7  Holomorphic functions as vector fields and harmonic functions

### 3.7.1  Solenoidal vector fields

This subsection starts with similar considerations to the ones in Section 3.2, replacing circulations by flows.

**Definition 3.41.** A continuous vector field on a domain $U$ of $\mathbb{R}^n$ is called solenoidal if its flow vanishes on each compact hypersurface without boundary contained in $U$.

Equivalently, a vector field $\vec{X}$ is solenoidal on $U$ if whenever $M_1$, $M_2$ are hypersurfaces on $U$ with the same boundary the flow of $\vec{X}$ through $M_1$ and through

$M_2$ is the same. In the same way as conservative vector fields coincide with gradient vector fields, in dimension $n = 3$ solenoidal vector fields coincide with curl vector fields:

**Theorem 3.42.** *A continuous vector field $\vec{X}$ on a domain $U$ of $\mathbb{R}^3$ is solenoidal if and only if there is a vector field $\vec{Y}$ of class $C^1$ on $U$ such that $\vec{X} = \operatorname{rot} \vec{Y}$.*

*Proof.* The fact that any curl vector field is solenoidal is an immediate consequence of the curl theorem, since the flow of $\operatorname{rot} \vec{Y}$ through a hypersurface equals the circulation of $\vec{Y}$ along the boundary. The proof of the converse is quite more complicated and will not be given here. $\qquad\square$

The vector field $\vec{Y}$ such that $\vec{X} = \operatorname{rot} \vec{Y}$ is called a *potential vector* of $\vec{X}$ and is determined up to vector fields with null curl.

The analogue of Theorem 3.42 in dimension $n > 3$ is better stated in terms of differential forms: a 2-form $\omega = \sum_{i<j} A_{ij} dx_i \wedge dx_j$ continuous on a domain $U$ of $\mathbb{R}^n$ is exact on $U$, that is, there is a $C^1$-class 1-form $\eta$ on $U$ such that $d\eta = \omega$, if and only if $\omega$ has integral zero on any compact surface without boundary contained in $U$. This result holds for arbitrary differential $k$-forms and it is a consequence of a deep theorem, de Rham's theorem (see [13], p. 154).

The concept of solenoidal vector field is a global concept, but we can consider also the local version: a vector field $\vec{X}$ is locally solenoidal on a domain $U$ if every point $p \in U$ has a neighborhood $V \subset U$ such that $\vec{X}$ has zero flow through any regular hypersurface contained in $V$ which is compact and without boundary. The result corresponding to Theorem 3.39 reads as follows.

**Theorem 3.43.** *For a vector field $\vec{X} = (X_1, \ldots, X_n)$ which is differentiable on a domain $U \subset \mathbb{R}^n$, the following properties are equivalent:*

a) $\operatorname{div} \vec{X} = \sum_{i=1}^{n} \frac{\partial X_i}{\partial x_i} = 0$ *on $U$.*

b) *$\vec{X}$ is locally solenoidal on $U$.*

c) *If $n = 3$, a) and b) are equivalent to the fact that $\vec{X}$ has locally a potential vector on $U$.*

*Proof.* The fact that b) implies a) is just a consequence of the definition of divergence of a vector field (Subsection 3.6.3). Conversely, if $\vec{X}$ has zero divergence and $S$ is a compact regular hypersurface without boundary inside a ball contained in $U$, then $S$ is the boundary of an open set $G \subset U$, and the divergence theorem implies

$$\int_S \langle \vec{X}, \vec{N} \rangle \, dA = \int_G \operatorname{div} \vec{X} \, dV = 0.$$

Finally, the equivalence with c) is a consequence of Theorem 3.42. $\qquad\square$

A domain $U \subset \mathbb{R}^n$ is called $k$-*simply connected* if every $k$-submanifold without boundary contained in $U$ is the boundary of a $(k+1)$-submanifold in $U$. It is a topological concept, equivalent to the cancellation of a certain homology group. When $n = 2$, $k = 1$, it will be seen in Chapter 6 that this is the case of simply connected domains introduced in Chapter 1; when $k = n - 1$, it is said that $U$ *has no holes*. In Theorem 3.43 we have used the intuitive fact that balls have no holes. On a domain without holes a differentiable vector field $\vec{X}$ is solenoidal if and only if div $\vec{X} \equiv 0$, due to the divergence theorem. The typical example of a domain with holes is $U = \mathbb{R}^3 \setminus \{(0,0,0)\}$. The newtonian vector field $\vec{X} = \frac{\vec{r}}{r^3}$, with $\vec{r} = (x, y, z)$ and $r = |\vec{r}|$, has div $\vec{X} = 0$ on this domain $U$, but clearly it has flow $4\pi$ through every sphere centered at the origin.

In dimension $n = 3$, one may prove the implication a) $\Rightarrow$ c) of Theorem 3.43 directly, if the additional hypothesis that $\vec{X}$ is $C^1$-class is satisfied. The computation of the potential vector $\vec{Y}$ of a differentiable vector field $\vec{X}$ with zero divergence on a ball can be done by solving the equations

$$X_1 = (Y_3)_y - (Y_2)_z, \quad X_2 = (Y_1)_z - (Y_3)_x, \quad X_3 = (Y_2)_x - (Y_1)_y.$$

A solution satisfying $Y_3 = 0$, which will be called $\vec{Y}_0$, will be found. From the first equation it turns out that

$$Y_2(x, y, z) = -\int_0^z X_1(x, y, t)\, dt + A(x, y),$$

and from the second one,

$$Y_1(x, y, z) = \int_0^z X_2(x, y, t)\, dt + B(x, y).$$

Then the aim is to find the functions $A$, $B$ such that

$$(Y_2)_x - (Y_1)_y = -\int_0^z (X_1)_x(x, y, t)\, dt + A_x - \int_0^z (X_2)_y(x, y, t)\, dt - B_y$$
$$= X_3(x, y, z).$$

Since div $\vec{X} = 0$, the sum of both integrals above is $\int_0^z (X_3)_z(x, y, t)\, dt = X_3(x, y, z) - X_3(x, y, 0)$, so one wants to find $A$, $B$ such that $A_x(x, y) - B_y(x, y) = X_3(x, y, 0)$. One possibility is $A = 0$ and $B(x, y) = -\int X_3(x, t, 0)\, dt$. Once the solution $\vec{Y}_0$ is founded, the general solution is $\vec{Y} = \vec{Y}_0 + \vec{Z}$, where $\vec{Z}$ has zero curl, so it is a gradient by Theorem 3.37.

### 3.7.2 Holomorphic vector fields. Harmonic functions

In Section 3.2 it was seen how the idea of complex line integrals naturally leads to associate the vector field $\bar{f} = (u, -v)$ to the function $f(z) = u(z) + iv(z)$. Then if $\gamma$ is a path in the plane, the real part and the imaginary part of $\int_\gamma f(z)\,dz$ are, respectively, the circulation of the vector field $\bar{f}$ and the circulation of the vector field $i\bar{f}$ along $\gamma$; that is, the circulation of $\bar{f}$ and the flow of $\bar{f}$ over $\gamma$.

In a natural way there is a relationship between holomorphic functions and conservative or solenoidal vector fields. Indeed, suppose that $f = u + iv$ is differentiable as a function of two real variables. According to Theorem 3.39, the vector field $f = (u, v)$ is locally conservative if and only if

$$u_y = v_x$$

and, by Theorem 3.43, is locally solenoidal if and only if

$$u_x = -v_y,$$

which is equivalent to $J\,f = (-v, u)$ being locally conservative. Comparing with Cauchy–Riemann equations, one reaches the following statement:

**Theorem 3.44.** *A differentiable function $f = u + iv$ on a domain $U \subset \mathbb{C}$ is holomorphic if and only if the vector field $\bar{f} = (u, -v)$ is at the same time locally conservative and locally solenoidal.*

This theorem suggests the following definition of holomorphicity from a real variable point of view.

**Definition 3.45.** A differentiable vector field $\vec{X} = (X_1, X_2, \ldots, X_n)$, on a domain $U \subset \mathbb{R}^n$, is called holomorphic if it is locally conservative and locally solenoidal, that is, if it satisfies

$$\frac{\partial X_i}{\partial x_j} = \frac{\partial X_j}{\partial x_i}, \quad i, j = 1, \ldots, n; \quad \operatorname{div} X = \sum_{i=1}^{n} \frac{\partial X_i}{\partial x_i} = 0 \text{ on } U.$$

If $\vec{X}$ is a holomorphic vector field, then $\vec{X}$ is locally the gradient of a function $\phi$. Since $\vec{X}$ is differentiable, $\phi$ is twice differentiable and it follows that

$$\operatorname{div} \vec{X} = \operatorname{div}(\vec{\nabla}\phi) = \frac{\partial^2 \phi}{\partial x_1^2} + \cdots + \frac{\partial^2 \phi}{\partial x_n^2} = 0.$$

**Definition 3.46.** A function $\phi$, twice differentiable on an open set $U$ of $\mathbb{R}^n$, is called harmonic if $\Delta\phi = \frac{\partial^2 \phi}{\partial x_1^2} + \cdots + \frac{\partial^2 \phi}{\partial x_n^2} = 0$ on $U$.

Hence, one gets the following:

**Theorem 3.47.** *Holomorphic vector fields are exactly the vector fields which are locally the gradient of a harmonic function. Holomorphic functions are exactly the functions $f$ which locally satisfy $\bar{f} = \phi_x + i\phi_y$ with $\phi$ harmonic.*

The operator $\Delta$ in Definition 3.46, called *Laplacian or Laplace operator*, is the most important one in mathematical physics. Its properties will be studied in Chapter 7, as well as the properties of harmonic functions.

### 3.7.3 Conjugate harmonic functions

Back to the context of the complex plane, let us analyze in detail the relationship between holomorphic functions and harmonic functions.

Let $f$ be a holomorphic function on a domain $U$. By Theorem 3.47, locally there is a real harmonic function $\phi$ such that $f = \overrightarrow{\nabla}\phi = \frac{\partial \phi}{\partial x} - i\frac{\partial \phi}{\partial y}$, and this is the local general expression of a holomorphic function.

On the other hand, by Corollary 3.28, $f$ has locally a holomorphic antiderivative $F$, and if $U$ is star-like, it is global on $U$. Which relation is there between $F$ and $\phi$? Write the form $f(z)\,dz$ in terms of $\phi$:

$$f(z)\,dz = \left(\frac{\partial \phi}{\partial x} - i\frac{\partial \phi}{\partial y}\right)(dx + i\,dy) = \frac{\partial \phi}{\partial x}\,dx + \frac{\partial \phi}{\partial y}\,dy + i\left(-\frac{\partial \phi}{\partial y}\,dx + \frac{\partial \phi}{\partial x}\,dy\right).$$

The real part of this form is $d\phi$; the imaginary part,

$$d^*\phi = -\frac{\partial \phi}{\partial y}\,dx + \frac{\partial \phi}{\partial x}\,dy,$$

is called the *conjugated form* of $d\phi$. The holomorphic antiderivatives $F$ of $f$ are defined by the equality $dF = f(z)\,dz$. If $F = u + iv$, this reads as

$$du + i\,dv = dF = f(z)\,dz = d\phi + i\,d^*\phi.$$

Therefore,

$$du = d\phi, \quad dv = d^*\phi. \tag{3.9}$$

Conversely, if $F = u + iv$ is differentiable and $u$, $v$ satisfy (3.9), one may suppose $u = \phi$ (up to a constant); now, $F$ has real part $u = \phi$ and the imaginary part $v$ must satisfy $dv = d^*u$, that is,

$$\frac{\partial v}{\partial x} = -\frac{\partial u}{\partial y},$$

$$\frac{\partial v}{\partial y} = \frac{\partial u}{\partial x},$$

which are Cauchy–Riemann equations. Hence, $F$ is holomorphic and $F' = f$. Observe that $v$ is also harmonic.

**Definition 3.48.** Given a harmonic real function $u$ on a domain $U$ of the plane, it is said that a function $v$, differentiable on $U$, is a harmonic conjugate of $u$ on $U$ if $dv = d^*u$, that is, if the function $u + iv$ is holomorphic on $U$.

The previous considerations show:

**Theorem 3.49.** *If $u$ is real harmonic on the domain $U \subset \mathbb{C}$ and $f = \overline{\vec{\nabla}u}$, then the function $u$ has a harmonic conjugate on $U$, $v$, if and only if $f$ has on $U$ a holomorphic antiderivative $F$ and in this case it is $F = u + iv$.*

In the following chapter it will be proved that every holomorphic function is of class $C^\infty$. If $f = u + iv$ is holomorphic, one has $u, v \in C^\infty$, and differentiating Cauchy–Riemann equations leads to $\Delta u = \Delta v = 0$, that is, $\operatorname{Re} f$ and $\operatorname{Im} f$ are harmonic, and, therefore, $f$ is harmonic. The fact that a holomorphic function $f$ is harmonic, with the hypothesis of $f$ being of class $C^2$, is also a consequence of the equality

$$\Delta f = 4 \frac{\partial^2 f}{\partial z \, \partial \bar{z}} \tag{3.10}$$

between the operator $\Delta$ and the operators $\frac{\partial}{\partial z}$, $\frac{\partial}{\partial \bar{z}}$ and the fact that $\frac{\partial f}{\partial \bar{z}} = 0$ when $f$ is holomorphic. Since every harmonic function has locally a conjugate harmonic function (because every holomorphic function has locally an antiderivative), then every harmonic function (in dimension $n = 2$) is of class $C^\infty$. In Chapter 7 it will be shown that this holds in any dimension. In particular, a holomorphic vector field is $C^\infty$.

Observe that a conjugate harmonic function $v$ of $u$ is determined up to a pure imaginary constant. Note also that the condition $dv = d^*u$ implies $\langle \vec{\nabla}u, \vec{\nabla}v \rangle = 0$, that is, the gradients of $u$ and of $v$, and so their level curves, are perpendicular (Figure 3.11).

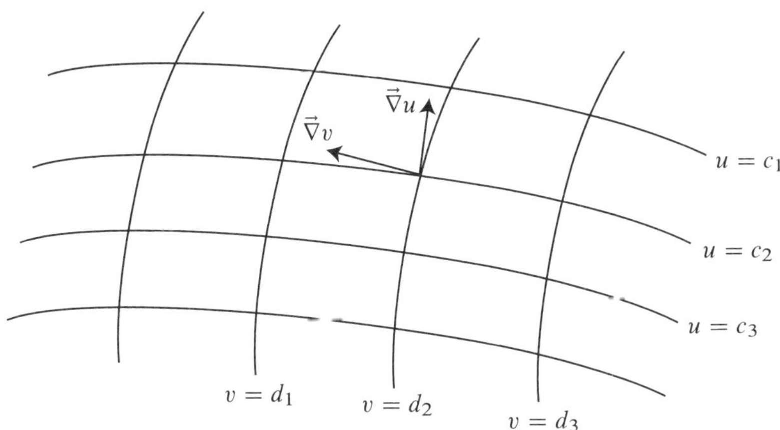


Figure 3.11

All the previous considerations prove:

**Proposition 3.50.** *A real harmonic function $u$ on a domain $U$ has a conjugate harmonic function if and only if the closed form $d^*u$ is exact on $U$, that is, if*

$\int_\gamma d^*u = 0$ *for any closed path $\gamma$ inside $U$, a condition which always locally holds. If $U$ is star-like, every real harmonic function on $U$ has a conjugate harmonic function on $U$.*

As said, the statement of Proposition 3.50 is equivalent to the fact that the holomorphic function $f = u_x - iu_y$ has a holomorphic antiderivative $F$ on $U$ and, in this case, $u = \operatorname{Re} F$.

In general, there does not exist a conjugate harmonic function of $u$, that is, not every real harmonic function on $U$ is the real part of a holomorphic function on $U$. For example, if $u(z) = \operatorname{Log}|z| = \operatorname{Log}\sqrt{x^2 + y^2}$ and $U = \mathbb{C} \setminus \{0\}$, one has $f = u_x - iu_y = \frac{x-iy}{x^2+y^2} = \frac{1}{z}$, which does not have an antiderivative on $U$; therefore, $u$ has no conjugate harmonic function.

On the other hand, the integral $\int_\gamma d^*u$ admits an interesting interpretation: if $\vec{N}$ is the unitary exterior normal vector to $\gamma(t) = (x(t), y(t))$, $\vec{N} = \frac{1}{|\gamma'|}(y', -x')$, then

$$\int_\gamma d^*u = \int_\gamma \frac{\partial u}{\partial x}\, dy - \frac{\partial u}{\partial y}\, dx = \int_\gamma \langle \vec{\nabla} u, \vec{N} \rangle\, ds = \int_\gamma \frac{\partial u}{\partial \vec{N}}\, ds.$$

So the condition $\int_\gamma d^*u = 0$ is equivalent, to $\int_\gamma \frac{\partial u}{\partial \vec{N}}\, ds = 0$ for any closed path $\gamma \subset U$. That is the integral along $\gamma$ of the derivative of $u$ with respect to the normal on $\gamma$ is zero if $\gamma$ is closed. In the previous example $(u(z) = \operatorname{Log}|z|)$ taking polar coordinates one has $u = \operatorname{Log} r$ and for $\gamma$ a circle centered at the origin with radius $R$, $\frac{\partial u}{\partial \vec{N}} = \frac{1}{R}$, it turns out that $\int_\gamma \frac{\partial u}{\partial \vec{N}}\, ds = 2\pi \neq 0$.

The computation of the conjugate harmonic function $v$ of a given harmonic function $u$ is equivalent to the computation of the potential function of the form $d^*u$. As seen in Subsection 3.6.4, on a neighborhood of $z_0 = (x_0, y_0)$ one may either take

$$v(x, y) = \int_{y_0}^y \frac{\partial u}{\partial x}(x, t)\, dt - \int_{x_0}^x \frac{\partial u}{\partial y}(t, y_0)\, dt,$$

or integrate the system $v_x = -u_y$, $v_y = u_x$ in steps. The function $v$ may be also determined computing the holomorphic antiderivative of $f = u_x - iu_y$.

**Example 3.51.** The function $u = x^2 - y^2$ is harmonic. The system which must be solved to find its conjugate harmonic function is $v_x = 2y$, $v_y = 2x$. From the first equation one has $v = 2xy + C(y)$ and substituting in the second one, $C'(y) = 0$; then, $v = 2xy + C$ is the general solution. The function $u + iv = x^2 - y^2 + (2xy + C)i = z^2 + Ci$ is, indeed, holomorphic. In fact: $f = u_x - iu_y = 2x + 2iy = 2z$, the antiderivative of which is $z^2$. $\qquad\square$

To finish, let us identify polynomials in the variables $x$, $y$ that are harmonic functions. A polynomial $P(x, y)$ is also a polynomial in $z$, $\bar{z}$,

$$P(x, y) = \sum_{n,m} a_{n,m} z^n \bar{z}^m,$$

and the coefficients $a_{n,m}$ are univocally given by $a_{n,m} = \frac{1}{n!m!} \frac{\partial^{n+m} P(0,0)}{\partial z^n \partial \bar{z}^m}$. The polynomial takes real values if and only if $P(x,y) = \bar{P}(x,y)$ which means, identifying coefficients, $a_{n,m} = \overline{a_{m,n}}$. By (3.10) this polynomial will be harmonic if and only if $\frac{\partial^2 P}{\partial z \partial \bar{z}}$ is identically zero, which is the same as $nma_{n,m} = 0$, that is, $a_{n,m} = 0$ for $n, m \neq 0$. In conclusion, if $P$ is real and harmonic, $P$ is of the form

$$P(x,y) = a_0 + \sum_n (a_n z^n + \bar{a}_n \bar{z}^n) = a_0 + 2 \operatorname{Re} \sum_n a_n z^n,$$

which gives $P$ as the real part of the holomorphic function $F(z) = a_0 + 2 \sum_n a_n z^n$. Now in the formal expression

$$P(x,y) = a_0 + \sum_n a_n (x+iy)^n + \bar{a}_n (x-iy)^n,$$

one can substitute $x$ by $\frac{z}{2}$ and $y$ by $\frac{z}{2i}$ to find $P\left(\frac{z}{2}, \frac{z}{2i}\right) = a_0 + \sum_n a_n z^n$, and then

$$F(z) = 2P\left(\frac{z}{2}, \frac{z}{2i}\right) - a_0.$$

This way, the holomorphic polynomial $F$ having $P$ as its real part may be found algebraically as well as the conjugate polynomial of $P$, which will be the imaginary part of $F$.

**Example 3.52.** Consider $P(x,y) = x^3 - 3xy^2$, which is clearly harmonic. Then $F(z) = 2\left(\left(\frac{z}{2}\right)^3 - 3\frac{z}{2}\left(\frac{z}{2i}\right)^2\right) = z^3$; the conjugate polynomial is $\operatorname{Im} F = 3x^2 y - y^3$.
□

As a final comment to this section, observe that there are three properties of a domain $U$, each of which implies the next one: a) any closed form on $U$ is exact, b) any holomorphic function on $U$ has a holomorphic antiderivative on $U$, c) any harmonic function on $U$ has a conjugate harmonic function. It has been shown that starlike domains have these three properties. It will be proved later that, in fact, these three properties are equivalent and characterize simply connected domains in the plane.

## 3.8 Exercises

1.  a) Let $f \in C^2(U)$, where $U$ is a domain of the plane. Prove that $f$ is holomorphic on $U$ if the functions $f(z)$ and $zf(z)$ are harmonic on $U$.

    b) Suppose now that $u$ is harmonic on $U$. Show that if the function

    $$u(x+iy) \cdot x$$

    is harmonic on $U$ then $u$ is of the form $u(x+iy) = ay + b$, with $a, b \in \mathbb{R}$.

2. Let $f$ be a differentiable function on a neighborhood of the point $z_0 \in \mathbb{C}$. Show that $f$ is $\mathbb{C}$-differentiable on $z_0$ if and only if

$$\lim_{r \to 0} \frac{1}{r^2} \int_{C_r} f(z)\,dz = 0,$$

where $C_r \equiv C_r(t) = z_0 + re^{it}, 0 \le t \le 2\pi$.

3. Let $f$ be a holomorphic function on a neighborhood of a closed rectangle $R$ except for a finite number of points $z_1, z_2, \ldots, z_N \in \overset{\circ}{R}$ that satisfy $\lim_{z \to z_j}(z - z_j)f(z) = 0$ for $j = 1, 2, \ldots, N$. Prove that $\int_{\partial R} f(z)\,dz = 0$.

4. Compute the integrals

a) $\displaystyle\int_0^{2\pi} e^{\cos\theta} \sin(n\theta - \sin\theta)\,d\theta$;

b) $\displaystyle\int_0^{2\pi} e^{\cos\theta} \cos(n\theta - \sin\theta)\,d\theta$,   with $n \in \mathbb{N}$.

5. Prove that if $f$ is a continuous function on an open convex set $U$ and holomorphic on $U \setminus \{z_0\}$, where $z_0 \in U$, then $\int_\gamma f(z)\,dz = 0$ for any closed path $\gamma$ with $\gamma^* \subset U$.

6. Let $U$ be a star-like domain of the plane and $f \in H(U)$. Show that there exists a function $F \in H(U)$, unique up to an additive constant, such that for every disc $\overline{D}(a, r) \subset U$ one has

$$f(z) = \frac{1}{2\pi i} \int_{C(a,r)} \frac{F(w)}{(w - z)^2}\,dw - \frac{1}{2\pi i} \int_{C(a,r)} \frac{f^2(w)}{(w - z)^2}\,dw,$$

where $z \in D(a, r)$.

7. Prove that if $u_1, u_2$ are harmonic functions on a domain $U$ that are equal on an open subset of $U$, then $u_1$ and $u_2$ are identical on $U$.

8. Write the Laplace operator in the plane in polar coordinates. As an application, prove that if $u$ is harmonic and radial on $\mathbb{C}$, then $u(z) = a\,\mathrm{Log}\,|z| + b$, with $a, b$ constants. Prove also that if $u$ is harmonic and $u(re^{i\theta}) = \varphi(r)\,\psi(\theta)$, with $\varphi, \psi$ functions of one real variable, then either $u(re^{i\theta}) = a\,\mathrm{Log}\,r + b$, or $u(re^{i\theta}) = r^n(a\cos n\theta + b\sin n\theta)$ with $a, b$ constants and $n \in \mathbb{N}$.

9. Let $U$ be an open convex set of the plane, $u_1, u_2$ two harmonic functions on $U$ and $\gamma$ a closed regular path in $U$ with exterior normal vector $\vec{N}$. Show that

$$\int_\gamma \left( u_1 \frac{\partial u_2}{\partial \vec{N}} - u_2 \frac{\partial u_1}{\partial \vec{N}} \right) ds = 0.$$

**10.** If $\gamma = C(0,1)$ and $a \in \mathbb{C}$ with $|a| < 1$, compute

$$\int_{\gamma} \left( \frac{2}{z-a} - \frac{1}{z} \right) dz$$

and prove the equality

$$\int_0^{2\pi} \frac{(1-r^2)dt}{1+r^2 - 2r \cos(\theta - t)} = 2\pi, \quad \text{for } 0 \le r < 1,\ \theta \in \mathbb{R}.$$

**11.** Prove the following inequality:

$$\left| \int_{-r}^{r} \frac{\sin x}{x}\, dx - \pi \right| \le \frac{\pi}{r}, \quad r > 0$$

integrating the function $\frac{e^{iz}-1}{z}$ along the path consisting of $[-r, r]$ followed by $t \to re^{it}, 0 \le t \le \pi$.

**12.** Let $f$ be a $C^2$ function on the disc $D(z_0, R)$ and $0 < r < R$. Show that the equality

$$\int_0^{2\pi} \frac{\partial f}{\partial r}(z_0 + re^{i\theta}) r\, d\theta = \iint_{D(z_0, r)} \Delta f(z)\, dx\, dy$$

holds, with $z = x + iy$. Deduce the mean value property for harmonic functions: if $u$ is harmonic on $D(z_0, R)$ and $0 < r < R$, then

$$u(z_0) = \frac{1}{2\pi} \int_0^{2\pi} u(z_0 + re^{i\theta})\, d\theta.$$

**13.** A continuous function $u$ on an open set $U \subset \mathbb{C}$ is said to be *subharmonic* on $U$ if it satisfies

$$u(z_0) \le \frac{1}{2\pi} \int_0^{2\pi} u(z_0 + re^{i\theta})\, d\theta$$

for $z_0 \in U$ and $r > 0$ such that $\bar{D}(z_0, r) \subset U$. Show that a function $u$ of class $C^2$ on $U$ is subharmonic on $U$ if and only if $\Delta u(z) \ge 0, z \in U$.

**14.** Let $f$ be a holomorphic function on an open set $U$ satisfying $|f(z) - 1| < 1$, $z \in U$. Prove that

$$\int_{\gamma} \frac{f'(z)}{f(z)}\, dz = 0$$

holds for any closed path $\gamma$ contained in $U$.

**15.** If $C_r$ is the circle centered at 0 with radius $r > 0$ travelled once in the positive direction, compute

$$\text{a)} \int_{C_r} \frac{dz}{|z-a|^2}, \quad a \in \mathbb{C}; \qquad \text{b)} \int_{C_r} \left(z + \frac{1}{z}\right)^{2n} \frac{dz}{z}, \quad n \in \mathbb{N}.$$

In particular we can obtain the equality

$$\frac{1}{2\pi} \int_0^{2\pi} \cos^{2n} \theta \, d\theta = \frac{(2n)!}{2^{2n}(n!)^2}.$$

**16.** Prove the equality

$$\frac{1}{1 - 2r\cos\theta + r^2} = \frac{1}{1-r^2}\left(1 + 2\sum_{n=1}^{\infty} r^n \cos n\theta\right) \qquad \text{for } 0 \le r < 1, \ \theta \in \mathbb{R}$$

and use it to compute

$$\int_{C_r} \frac{|z|}{|1-z|^2} \, dz, \quad 0 \le r < 1,$$

where $C_r$ is the circle centered at 0 with radius $r > 0$ positively oriented.

**17.** Consider in $\mathbb{R}^3$ the vector field $\vec{X} = (x^2 - xe^{xz}, xy\cos xyz - 2xy, ze^{xz} - xz\cos xyz + 3x^2)$. Show that $\vec{X}$ is solenoidal and that it has a potential vector of the form $\vec{Y} = (\sin yz, B, C)$, with $B, C$ functions of $x, y, z$. Find the flow of $\vec{X}$ through the surface $S = \{(x, y, z): x^2 + y^2 + 3z^2 = 1\}$ when $S$ is oriented with the exterior normal.

**18.** Consider in $\mathbb{R}^3$ the vector field $\vec{X} = (x^2 \sin^2 \frac{\pi}{2}z, y^2 \sin \frac{\pi}{2} z, z+x)$. Calculate the flow of $\vec{X}$ through the surface $S = \{(x, y, z): z = x^2 + y^2, \ 0 \le z \le 1\}$, oriented with the exterior normal. How does this result change taking an arbitrary function of $z$ instead of $\sin \frac{\pi}{2} z$?

**19.** Find all the harmonic functions of the form $u(x, y) = g(y/x)$ where $g$ has two continuous derivatives. Determine also the holomorphic functions $f$ such that $u = \text{Re } f$.

**20.** If $\gamma$ is a closed path of the plane, which does not pass through the point $z_0$, show directly (without using Subsection 1.5.2) that the value of the integral

$$\frac{1}{2\pi i} \int_\gamma \frac{dz}{z - z_0}$$

is an integer.

**21.** Find an entire function $f$ such that $\operatorname{Re} f = u$ when

a) $u(x, y) = e^{x^2 y^2} \sin 2xy$;     b) $u(x, y) = \dfrac{\sin x \cos x}{\cos^2 x + \operatorname{sh}^2 y}$.

**22.** Prove that if $U$ is a bounded domain of the plane with positively oriented piecewise regular boundary and $f$ is a continuous function on $\bar{U}$ and holomorphic on $U$, then $\int_{\partial U} f(z)\, dz = 0$.

# Chapter 4

# Local properties of holomorphic functions

The aim of this chapter is to establish the basic properties of holomorphic functions that are local, that is, that do not depend on the topological properties of the domain of definition. Among these properties, the one that stands out is the fact that every holomorphic function can be locally expressed in terms of a power series, which leads to the fundamental identity between holomorphic functions and analytic functions of a complex variable. In the local theory, a reproducing formula plays a major role, namely the Cauchy integral formula, deduced directly from a general version of the Cauchy–Green formula.

Once it is known that holomorphic functions are analytic, the structure of their zero sets may be studied and the principle of analytic continuation may be stated. The rest of the chapter is devoted to other properties of holomorphic functions which are consequences either of the Cauchy integral formula or of the fact that every non constant holomorphic function is an open mapping.

## 4.1  Cauchy integral formula

First one needs to recall some basic facts related to integrals, in the sense of Lebesgue, of functions defined on sets of the plane. We will denote by $dm(z)$, or by $dx\,dy$ if $z = x + iy$, the Lebesgue measure on the plane and will consider Lebesgue integrals of type $\int_A h(z)\,dm(z)$ with $A \subset \mathbb{C}$ Lebesgue measurable and $h$ defined on $A$.

Recall that this integral is defined if:

- $h$ is defined at almost every point of $A$ and is a Lebesgue measurable function.

- $\int_A |h(z)|\,dm(z) < +\infty.$

Always, the functions one deals with are defined and continuous on $A$ except for a finite or countable quantity of points of $A$, so that the first condition will automatically hold. The second one implies that if $A$ is broken into shrinking measure sets, $A = \bigcup_j A_j$, choosing $z_j \in A_j$, the Lebesgue sums

$$\sum h(z_j)m(A_j)$$

converge to $\int_A h(z)\,dm(z)$.

An often useful result is the fact that the function $gh$ is integrable when $h$ is integrable and $g$ is measurable and bounded.

The following result will be frequently used.

**Lemma 4.1.** *For $a \in \mathbb{C}$ fixed, the function $\frac{1}{z-a}$ is integrable on every subset of $\mathbb{C}$ of finite measure.*

*Proof.* To prove this claim we have to show that $\int_A \frac{dm(z)}{|z-a|} < +\infty$ if $m(A) < +\infty$. If $B = \{z \in A : |z - a| \geq 1\}$ one has

$$\int_B \frac{dm(z)}{|z-a|} \leq \int_B dm(z) = m(B) \leq m(A) < +\infty.$$

On the other hand, changing to polar coordinates centered at the point $a$, $z = a + re^{i\theta}$, it is $dm(z) = r \, dr \, d\theta$ and it follows that

$$\int_{A \setminus B} \frac{dm(z)}{|z-a|} \leq \int_{|z-a| \leq 1} \frac{dm(z)}{|z-a|} = \int_0^1 \int_0^{2\pi} \frac{r \, dr \, d\theta}{r} = 2\pi. \qquad \square$$

Recall that if $f$ is a differentiable function (in the real sense), the operator $\bar{\partial} = \frac{\partial}{\partial \bar{z}}$ acts on $f$ and gives the function $\bar{\partial} f = \frac{\partial f}{\partial \bar{z}}$ and $f$ is holomorphic when $\bar{\partial} f = 0$.

**Theorem 4.2** (Cauchy–Green formula). *Let $U$ be a bounded domain of the plane with piecewise regular boundary, positively oriented, and let $f$ be a differentiable function on a neighborhood of $\bar{U}$ with $\bar{\partial} f$ continuous on $\bar{U}$. Then, for all $z \in U$, one has*

$$f(z) = \frac{1}{2\pi i} \int_{\partial U} \frac{f(w)}{w - z} \, dw - \frac{1}{\pi} \int_U \frac{\bar{\partial} f(w)}{w - z} \, dm(w). \tag{4.1}$$

This result, which will be shown next, has two evident consequences:

**Corollary 4.3** (Cauchy integral formula). *If $U$ is a bounded domain of the plane with piecewise regular boundary, positively oriented, and $f$ is holomorphic on a neighborhood of $\bar{U}$, then*

$$f(z) = \frac{1}{2\pi i} \int_{\partial U} \frac{f(w)}{w - z} \, dw, \quad z \in U. \tag{4.2}$$

Recall that if $f$ is a function defined on $\mathbb{R}^n$, the *support* of $f$ is the closed set $\mathrm{spt}(f) = \overline{\{x \in \mathbb{R}^n : f(x) \neq 0\}}$.

**Corollary 4.4.** *If $f$ is a differentiable function on $\mathbb{C}$ with compact support and $\bar{\partial} f$ is continuous on $\mathbb{C}$, then*

$$f(z) = -\frac{1}{\pi} \int_{\mathbb{C}} \frac{\bar{\partial} f(w)}{w - z} \, dm(w), \quad z \in \mathbb{C}.$$

Note that the dependence on $z$ is through the kernel $\frac{1}{w-z}$. The kernel $C(w,z) = \frac{1}{2\pi i} \frac{1}{w-z}$ is called the *Cauchy kernel*.

*Proof of Theorem* 4.2. Fix $z \in U$, consider $\varepsilon > 0$ such that $\bar{D}(z,\varepsilon) \subset U$ and let $U_\varepsilon$ be the domain $U_\varepsilon = U \setminus \bar{D}(z,\varepsilon)$. The boundary of $U_\varepsilon$ is composed by the boundary of $U$ and the circle $C(z,\varepsilon)$, travelled in the negative sense (Figure 4.1).

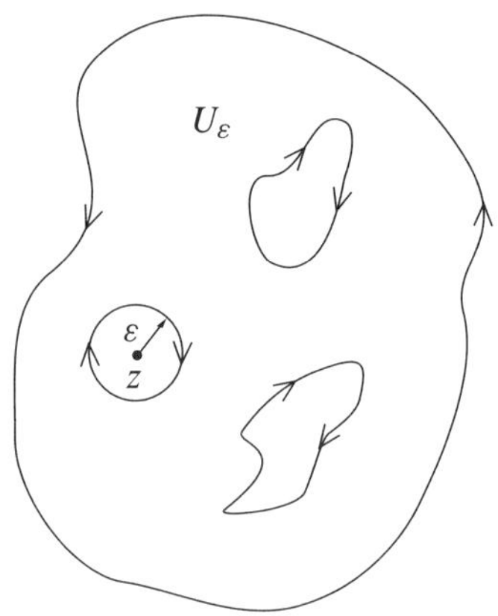

Figure 4.1

So, $U_\varepsilon$ is a bounded domain with piecewise regular boundary, positively oriented. Now apply Green's formula (Theorem 3.22) to the 1-form $\rho = \frac{f(w)}{w-z}\,dw$ on the domain $U_\varepsilon$. If $w = \sigma + i\tau$, one has $\rho = \frac{f(w)}{\omega-z}\,d\sigma + i\frac{f(w)}{w-z}\,d\tau$ and

$$\frac{\partial}{\partial\sigma}\left(i\,\frac{f(w)}{w-z}\right) - \frac{\partial}{\partial\tau}\left(\frac{f(w)}{w-z}\right) = 2i\,\frac{\partial}{\partial\bar{w}}\frac{f(w)}{w-z} = 2i\,\frac{\partial f}{\partial\bar{w}}\frac{1}{w-z},$$

which is a continuous function on $\bar{U}_\varepsilon$. Therefore Green's formula gives

$$\int_{\partial U_\varepsilon} \rho = 2i \int_{U_\varepsilon} \frac{\partial f}{\partial\bar{w}}\frac{1}{w-z}\,dm(w). \tag{4.3}$$

Now,

$$\int_{\partial U_\varepsilon} \rho = \int_{\partial U} \rho + \int_{C(z,\varepsilon)} \rho = \int_{\partial U} \frac{f(w)}{w-z}\,dw - \int_{C(z,\varepsilon)} \frac{f(w)}{w-z}\,dw$$

$$= \int_{\partial U} \frac{f(w)}{w-z}\,dw - i \int_0^{2\pi} f(z + \varepsilon e^{i\theta})\,d\theta$$

which converges to $\int_{\partial U} \frac{f(w)}{w-z}\,dw - 2\pi i f(z)$, as $\varepsilon \to 0$.

Finally, according to Lemma 4.1, the function $w \mapsto \frac{\partial f}{\partial \bar{w}} \frac{1}{w-z}$ is integrable on $U$, so that the integral of the right-hand side of (4.3) converges, when $\varepsilon \to 0$, to the same integral on the domain $U$, and then (4.1) follows. $\qquad\square$

**Remark 4.1.** Corollary 4.4 is based, apparently, on the general Green's formula (Theorem 3.22), which leads to the proof of Theorem 4.2. But, in fact, it follows from the version for rectangles of Green's formula (first step of the proof of Theorem 3.22). Just follow the proof of Theorem 4.2 taking as $U$ a rectangle, $U \supset \mathrm{spt}(f)$, and for $z \in U$ delete a small rectangle $R_z$, centered at $z$. Then $U \setminus R_z$ is the union of eight rectangles and Green's formula holds for this domain.

It is important to pay attention to the structure of formula (4.1): it decomposes a differentiable function $f$ into the sum of two functions, a holomorphic one only depending on the values of $f$ on $\partial U$, and another one which only depends on $\bar{\partial} f$. The formula (4.2),

$$f(z) = \frac{1}{2\pi i} \int_{\partial U} \frac{f(w)}{w-z}\, dw, \quad z \in U,$$

valid when $f$ is holomorphic, is called a *reproducing formula* because it gives the values of $f$ in $U$ just in terms of the values of $f$ on $\partial U$. This is radically new in the sense that there is nothing similar in the case of real differentiable functions. For example, a consequence of (4.2) is that if two holomorphic functions on a neighborhood of $\bar{U}$ coincide at the points of $\partial U$, then they are identical on the whole domain $U$.

**Corollary 4.5.** *If $f$ is an entire function with compact support, then $f$ is identically zero.*

*Proof.* If $D(0, R)$ is a disc big enough, one will have that $f \equiv 0$ on $\partial D(0, R)$; therefore, $f \equiv 0$ on $D(0, R)$. Now let $R \to +\infty$. $\qquad\square$

In contrast, there are differentiable functions, even $C^\infty$ functions, with compact support and not identically zero. For example if $K$ is a compact set of $\mathbb{R}^n$ and $U$ an open neighborhood of $K$, one may always construct a $C^\infty$ function $\varphi, 0 \le \varphi \le 1$, with compact support in $U$ and such that $\varphi \equiv 1$ on $K$.

**Example 4.6.** Let $f$ be a holomorphic function on the unit disc $\mathbb{D}$ such that $\int_{\mathbb{D}} |f(z)|\, dm(z) < +\infty$. Fix $0 \le r < 1$ and let $\varphi$ be a $C^\infty$ function, with compact support contained in $D(0, r')$, $r < r' < 1$, and equal to 1 on a neighborhood of $\bar{D}(0, r)$. The formula (4.1) applied to the function $\varphi \cdot f$ on the disc $D(0, r')$ gives

$$f(z) = -\frac{1}{\pi} \int_D \bar{\partial}\varphi(w) \cdot f(w) \frac{1}{w-z}\, dm(w), \quad |z| \le r.$$

Now letting $M_r = \sup \left\{ |\bar\partial\varphi(w)| \frac{1}{|w-z|} : |z| \leq r, \ w \in \mathrm{spt}\,(\bar\partial\varphi) \right\}$, it turns out that $M_r < +\infty$ (because $\mathrm{spt}(\bar\partial\varphi)$ does not intersect $\bar D(0,r)$), and the inequality

$$\sup_{|z|\leq r} |f(z)| \leq M_r \frac{1}{\pi} \int_{\mathbb D} |f(z)|\, dm(z)$$

is obtained. For example, if $(f_n)$ is a sequence of holomorphic functions on $\mathbb D$ which satisfies $\lim_n \int_{\mathbb D} |f_n(z)|\, dm(z) = 0$, then $(f_n)$ converges to zero uniformly on each compact disc $\bar D(0,r) \subset \mathbb D$. $\qquad\qquad\square$

It is also interesting to stress the particular case of (4.2) in which $U = D(a,r)$ and $z = a$. The parametrization of $\partial U = C(a,r)$ is $w = a + re^{i\theta}, 0 \leq \theta \leq 2\pi$, so that $dw = ire^{i\theta}\, d\theta = i(w-a)\, d\theta$ and one obtains

$$f(a) = \frac{1}{2\pi} \int_0^{2\pi} f(a + re^{i\theta})\, d\theta,$$

which expresses the fact that $f(a)$ is the mean value of $f$ on the circle $C(a,r)$, when $f$ is holomorphic on a neighborhood of $\bar D(a,r)$.

**Definition 4.7.** A continuous function $\varphi$ on a domain $U$ (with real or complex values) is said to have the mean value property on $U$ if, for any closed disc $\bar D(a,r) \subset U$, it holds that

$$\varphi(a) = \frac{1}{2\pi} \int_0^{2\pi} \varphi(a + re^{i\theta})\, d\theta. \tag{4.4}$$

If $\varphi$ is continuous on $U$, $\bar D(a,R) \subseteq U$ and for $0 \leq r \leq R$ one multiplies (4.4) by $r$ and integrates from 0 to $R$, it follows that

$$\int_0^R \varphi(a)r\, dr = \frac{1}{2\pi} \int_0^R \int_0^{2\pi} \varphi(a + re^{i\theta})r\, d\theta dr = \frac{1}{2\pi} \int_{D(a,R)} \varphi(z)dm(z).$$

That is,

$$\varphi(a) = \frac{1}{\pi R^2} \int_{D(a,R)} \varphi(z)dm(z),$$

which is the bidimensional version of the mean value property.

As just seen, holomorphic functions on $U$ and so their real and imaginary parts, which are harmonic functions, have the mean value property on $U$. Later on it will be proved that all harmonic functions have this property and, in fact, the mean value property characterizes them (Theorem 7.7).

## 4.2 Analytic functions and holomorphic functions

In the second chapter (Definition 2.36) the notion of analytic function of the complex variable $z$ has been introduced as a function which, locally, is the sum of a power series of $z$. It has been seen that these functions are *infinitely holomorphic*, that is, they are holomorphic as well as all their derivatives. Next the converse of this result will be proved and the equivalence between holomorphic function and analytic functions will be obtained.

**Theorem 4.8.** *If the function $f$ is holomorphic on the disc $D(a, R)$, then there is a power series $\sum_n c_n(z - a)^n$, centered at $a$ with radius of convergence equal to or greater than $R$, such that $f(z) = \sum_n c_n(z - a)^n$ for $z \in D(a, R)$*

*Proof.* For $r < R$, apply Cauchy integral formula (4.2) to the disc $D(a, r)$ to obtain

$$f(z) = \frac{1}{2\pi i} \int_{C(a,r)} \frac{f(w)}{w - z}\, dw, \quad |z - a| < r.$$

Now the idea is to expand the Cauchy kernel in a power series.

Fix $w$ with $|w - a| = r$; one asks if $\frac{1}{w-z}$ is the sum of a power series in $z$, centered at $a$, on the disc $D(a, r)$. This is easy to see, writing

$$\frac{1}{w - z} = \frac{1}{w - a - (z - a)} = \frac{1}{(w - a)(1 - \frac{z-a}{w-a})}$$

and using the geometric series with common ratio $\xi = \frac{z-a}{w-a}$. Since $|\xi| = \frac{|z-a|}{|w-a|} = \frac{|z-a|}{r} < 1$, one has $(1 - \xi)^{-1} = \sum_{n=0}^{\infty} \xi^n$ and

$$\frac{1}{w - z} = \sum_{n=0}^{\infty} \frac{(z - a)^n}{(w - a)^{n+1}}.$$

Now fix $z$ with $|z - a| < r$; according to the Weierstrass $M$-test (Theorem 2.10) the previous series is uniformly convergent for $w \in C(a, r)$ and one may commute the complex integration with the infinite sum (Proposition 3.4) to obtain

$$f(z) = \sum_n c_n(r)(z - a)^n, \quad |z - a| < r,$$

with $c_n(r) = \frac{1}{2\pi i} \int_{C(a,r)} f(w)(w - a)^{-n-1}\, dw$. For each $r$, $0 < r < R$, one has a power series with coefficients $c_n(r)$ which depend on $r$ and with radius of convergence equal to or greater than $r$ (because it is convergent with sum $f(z)$). But the uniqueness of the series expansion of $f$ (Proposition 2.32) gives that $c_n(r)$ is, in fact, independent from $r$; then we can write

$$c_n = \frac{1}{2\pi i} \int_{C(a,r)} f(w)(w - a)^{-n-1}\, dw, \quad 0 < r < R.$$

Now, for every $z \in D(a, R)$, one just has to consider a number $r > 0$ such that $|z - a| < r < R$, to conclude that $f(z) = \sum_n c_n(z - a)^n$. $\qquad\square$

From this result one has at once:

**Theorem 4.9** (Identification between holomorphic functions and analytic functions).  *If $U$ is an open set of the complex plane, then a function is holomorphic on $U$ if and only if it is analytic on $U$. More precisely, every holomorphic function $f$ on $U$ is indefinitely holomorphic on $U$, and for any point $a \in U$ the Taylor expansion*

$$f(z) = \sum_{n=0}^{\infty} \frac{f^{(n)}(a)}{n!}(z - a)^n \tag{4.5}$$

*is valid on the biggest disc centered at $a$ and included in $U$, that is, on $D(a, \delta(a))$, where $\delta(a)$ is the distance from the point $a$ to the closed set $\mathbb{C} \setminus U$.*

Observe that as established above, this result, together with Theorem 2.31, gives another proof of Theorem 2.39.  On the other hand, Example 2.38 shows that the series (4.5) may have a radius of convergence greater than $\delta(a)$. As well, now it is known that if $f$ is holomorphic on $U$, then $f$ is of class $C^{\infty}$ with respect to real variables on $U$, and so $f$ is harmonic on $U$ (see Section 3.7.3).

Once again one can stress Theorem 4.9 in contrast to the real variable case.  In the frame of real variables, not every differentiable function is infinitely differentiable; for example, $x|x|$ is differentiable at each point and its derivative $|x|$ is not, at $x = 0$. For a more extreme example, consider the antiderivative, $F$, of a continuous function $f$ which is not differentiable at any point. Then $F$ is a differentiable function such that $F' = f$ is not differentiable at any point. Neither is it true that an infinitely differentiable function is real analytic; actually, the analytic functions fulfill the principle of analytic continuation (see Subsection 4.4.2) and, in particular, there is no analytic function with compact support, not identically zero. Therefore, any $C^{\infty}$ function with compact support, not identically zero, is not analytic. Another example, now with non-compact support, is the function

$$f(x) = e^{-\frac{1}{x^2}}, \quad f(0) = 0.$$

This one is $C^{\infty}$, analytic on $\mathbb{R} \setminus \{0\}$, but is not analytic on $\mathbb{R}$, because $f^{(n)}(0) = 0$, $n = 0, 1, 2, \ldots$ (since $f(x) = O(x^n)$, for all $n$); hence $f$ cannot be equal to the sum of its Taylor series at the point 0. The function $f$ is often used to construct $C^{\infty}$ functions with compact support.

The following result is the case $U = \mathbb{C}$ of Theorem 4.9.

**Theorem 4.10.** *The mapping $f \mapsto \left(\frac{f^{(n)}(0)}{n!}\right)_{n=0}^{\infty}$ is a bijection between the space of entire functions and the space of sequences $(c_n)_{n=0}^{\infty}$ such that the series $\sum c_n z^n$ has radius of convergence infinite, that is, $\lim_{n \to \infty} |c_n|^{\frac{1}{n}} = 0$.*

The result stated below may be considered as a characterization of holomorphic functions among the class of continuous functions.

**Theorem 4.11** (Morera's Theorem). *Let $U$ be an open set of $\mathbb{C}$. Then a function $f \in C(U)$ is holomorphic on $U$ if and only if $\int_{\partial\Delta} f(z)\,dz = 0$ for any triangle $\Delta \subset U$.*

*Proof.* If $f \in H(U)$, it is known that $\int_{\partial\Delta} f(z)\,dz = 0$ when $\Delta$ is a triangle contained in $U$, by Cauchy's theorem (Theorem 3.24). Conversely, if $\int_{\partial\Delta} f(z)\,dz = 0$, for any $\Delta \subset U$, Theorem 3.14 asserts that $f$ has locally (for example on each disc contained in $U$) a holomorphic antiderivative. Now by Theorem 4.9 the derivative of a holomorphic function is also holomorphic, so then $f$ is holomorphic on any disc in $U$, that is, $f \in H(U)$. $\qquad\square$

**Example 4.12.** Supposing that $(f_n)$ is a sequence of holomorphic functions on a domain $U$ and $\lim_n f_n(z) = f(z)$, uniformly on any compact subset of $U$, then we can deduce that $f$ is also holomorphic on $U$ (see Theorem 9.3). Actually, $f$ will be continuous (uniform limit of continuous functions on any disc $\overline{D}(a,r) \subset U$) and if $\Delta$ is a triangle in $U$, one will have $\int_{\partial\Delta} f(z)dz = \int_{\partial\Delta} \lim_n f_n(z)dz = \lim_n \int_{\partial\Delta} f_n(z)dz = 0$, by uniform convergence over $\partial\Delta$ and $f_n$ being holomorphic. $\qquad\square$

As a consequence of Morera's theorem we can state the following result about removable singularities.

**Theorem 4.13.** *If $f$ is a continuous function on an open set $U$ and holomorphic on $U \setminus E$, where $E$ is a finite union of points and lines, then $f$ is holomorphic on $U$.*

*Proof.* One must show that $\int_{\partial\Delta} f(z)\,dz = 0$ for any triangle $\Delta \subset U$. If $\Delta$ does not contain any point or line of $E$, Cauchy's theorem says so. Suppose now that $\Delta$ contains a point $a$ of $E$. Then, given $\varepsilon > 0$, $\Delta$ may be decomposed into a union of triangles $\Delta = \Delta_a \cup \Delta_1 \cup \cdots \cup \Delta_m$ in such a way that $a \in \Delta_a$ and $L(\partial\Delta_a) < \varepsilon$ (Figure 4.2) and it follows that

$$\int_{\partial\Delta} f(z) = \int_{\partial\Delta_a} f(z)\,dz + \sum_{j=1}^{m} \int_{\partial\Delta_j} f(z)\,dz.$$

Now $\int_{\partial\Delta_j} f(z)\,dz = 0$ since $a \notin \Delta_j$ and, on the other hand, $f$ being bounded on $\Delta$ because it is continuous on $U$, we get

$$\left| \int_{\partial\Delta_a} f(z)\,dz \right| \leq (\sup_{z\in\Delta_a} |f(z)|)\, L(\partial\Delta_a) \xrightarrow[\varepsilon\to 0]{} 0.$$

If a line $l$ of $E$ intersects $\Delta$, it is easy to see that the integral over $\partial\Delta$ may be decomposed into integrals along triangles with a side on $l$ and each of these is approximated by triangles which do not intersect $l$ (Figure 4.2). $\qquad\square$

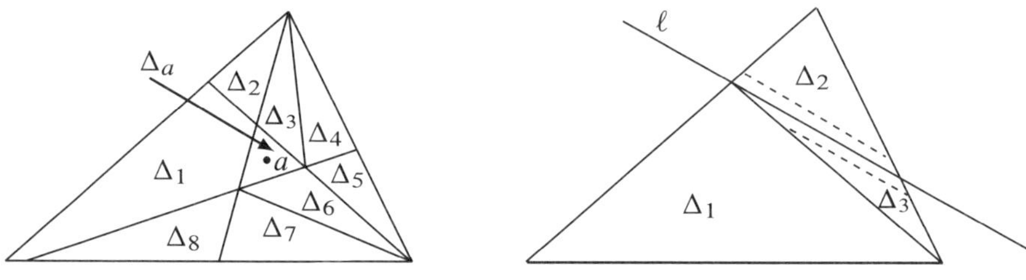

Figure 4.2

In the proof of Theorem 4.8 it has been seen that if $f \in H(D(a, R))$, then

$$\frac{f^{(n)}(a)}{n!} = \frac{1}{2\pi i} \int_{C(a,r)} \frac{f(w)}{(w-a)^{n+1}} \, dw, \quad 0 < r < R. \tag{4.6}$$

This formula is a particular case of the following proposition:

**Proposition 4.14.** *If $U$ is a bounded domain of the plane with piecewise regular boundary, positively oriented, and $f$ is holomorphic on a neighborhood of $\overline{U}$, then*

$$f^{(n)}(z) = \frac{n!}{2\pi i} \int_{\partial U} \frac{f(w)}{(w-z)^{n+1}} \, dw, \quad z \in U, \, n = 1, 2, 3, \ldots. \tag{4.7}$$

*Proof.* It is enough to differentiate (4.2), $n$ times with respect to $z$. The reader may justify that the differentiation under the integral is correct. $\quad\square$

**Example 4.15.** In order to compute the integral $I = \int_{|z|=2} \frac{z^2 e^z}{(z-1)^3} \, dz$, apply (4.6), for $n = 2$, to the function $f(z) = z^2 e^z$. It turns out that $I = \pi i f''(1) = 7\pi e i$. $\quad\square$

**Example 4.16.** Let us compute now $I = \int_{|z|=2} \frac{\sin z}{z^2(1+z^2)} \, dz$. Decomposing in partial fractions $\frac{1}{z^2(1+z^2)} = \frac{1}{z^2} - \frac{1}{2i}\frac{1}{z-i} + \frac{1}{2i}\frac{1}{z+i}$ and applying (4.7) one has

$$I = \int_{|z|=2} \frac{\sin z}{z^2} - \frac{1}{2i} \int_{|z|=2} \frac{\sin z}{z-i} + \frac{1}{2i} \int_{|z|=2} \frac{\sin z}{z+i}$$

$$= 2\pi i \cos(0) - \pi \sin(i) + \pi \sin(-i) = \pi i (2 - e + e^{-1}). \quad\square$$

Observe that, applying (4.2) to the $n$-th derivative $f^{(n)}$ of $f$, it follows that

$$f^{(n)}(z) = \frac{1}{2\pi i} \int_{\partial U} \frac{f^{(n)}(w)}{w-z} \, dw, \quad z \in U, \, n = 1, 2, \ldots. \tag{4.8}$$

The equality between (4.7) and (4.8) comes also from integration by parts, applied iteratively.

## 4.3 Analyticity of harmonic functions. Fourier series

Suppose $u$ is a real harmonic function on a domain $U \subset \mathbb{C}$ and for $a \in U$ fixed, let $D(a, \delta(a))$ be the biggest disc centered at $a$ inside $U$. By Proposition 3.50, $u$ is, on this disc, the real part of a holomorphic function $f$. Consider the expansion of $f$ in power series given by Theorem 4.8,

$$f(z) = \sum_n c_n (z-a)^n, \quad |z-a| < \delta(a), \quad c_n = \frac{f^{(n)}(a)}{n!}.$$

Suppose, without loss of generality, that $a = 0$. Taking real parts in the expression of $f(z)$ as a sum of powers of $z^n$, with $z = x + iy$, one obtains

$$u(x, y) = \operatorname{Re} f(z) = \sum_{n=0}^{\infty} \operatorname{Re}(c_n z^n)$$

$$= \sum_{n=0}^{\infty} \operatorname{Re}[c_n (x + iy)^n] = \sum_{n=0}^{\infty} (\operatorname{Re} c_n)(\operatorname{Re} z^n) - (\operatorname{Im} c_n)(\operatorname{Im} z^n).$$

Computing $\operatorname{Re} z^n$, $\operatorname{Im} z^n$ in terms of $x$, $y$, one gets formally an expression of the type $u(x, y) = \sum_{m,l=0}^{\infty} b_{m,l} x^m y^l$, where $b_{m,l} = (-1)^k \binom{m+l}{m} \operatorname{Re} c_{m+l}$ if $l = 2k$ and $b_{m,l} = (-1)^{k+1} \binom{m+l}{m} \operatorname{Im} c_{m+l}$ if $l = 2k + 1$. However, in order to justify the computation one has to check that the resulting double series is absolutely convergent, so one may sum it by blocks. Taking $\varepsilon = \frac{\delta}{2}$ with $\delta = \delta(0)$ and assuming $|x|, |y| < \varepsilon$, one has $|x| + |y| < \delta$ and

$$\sum_{m,l=0}^{\infty} |b_{m,l}| |x|^m |y|^l \leq \sum_{m,l=0}^{\infty} \binom{m+l}{m} |c_{m+l}| |x|^m |y|^l$$

$$- \sum_{k=0}^{\infty} |c_k| \sum_{m+l=k} \binom{m+l}{m} |x|^m |y|^l$$

$$= \sum_{k=0}^{\infty} |c_k| (|x| + |y|)^k < +\infty.$$

The sum by blocks is then justified (Proposition 2.8) and we obtain

$$u(x, y) = \sum_{m,l=0}^{\infty} b_{m,l} x^m y^l, \quad |x|, |y| < \varepsilon. \tag{4.9}$$

So then the following statement is proved:

**Theorem 4.17.** *Any harmonic function on a domain of the plane is an analytic function of two real variables.*

It has been pointed out in Subsection 3.7.3 that if $P(x, y)$ is a real harmonic polynomial, the polynomial in $z$ which has $P$ as real part and vanishing imaginary part at the origin is $2P\left(\frac{z}{2}, \frac{z}{2i}\right) - P(0,0)$. The previous theorem, saying that each harmonic function is a polynomial on $x$, $y$ with infinite degree, allows us to formulate something similar:

If $u$ is given by (4.9) with $|x|, |y| < \varepsilon = \frac{\delta}{2}$, the double series

$$2 \sum_{m,l=0}^{\infty} b_{m,l} \left(\frac{z}{2}\right)^m \left(\frac{z}{2i}\right)^l - b_{0,0}$$

converges absolutely for $|z| < \delta$ and defines a holomorphic function $f$ which has the harmonic function $u$ as its real part and its imaginary part vanishes at the origin. Formally it may be written

$$f(z) = 2u\left(\frac{z}{2}, \frac{z}{2i}\right) - u(0,0).$$

**Example 4.18.** The function $u(x, y) = \sin x \,\mathrm{ch}\, y$ is harmonic. In this case, taking $f(z) = 2 \sin \frac{z}{2} \,\mathrm{ch}\, \frac{z}{2i} = 2 \sin \frac{z}{2} \cos \frac{z}{2} = \sin z$, it is $u = \mathrm{Re}\, f$ and $f(0) = 0$.    □

From the previous considerations it follows that if $u$ is a real harmonic function on the disc $D(0, R)$, the expansion (4.9) of $u$ in power series of $x$, $y$ around the origin holds for $|x|, |y| < \frac{R}{2}$. However, there is another more interesting expansion of $u$, which holds in the whole disc $D(0, R)$, related with *Fourier series*. One simply takes polar coordinates, $z = x + iy = re^{i\theta}$, and writes $u = \mathrm{Re}\, f$, with $f$ holomorphic on $D(0, R)$, $f(z) = \sum_n c_n z^n$, so that

$$u(z) = \frac{1}{2}(f(z) + \overline{f(z)}) = \frac{1}{2}\left(\sum_{n=0}^{\infty} c_n z^n + \sum_{n=0}^{\infty} \bar{c}_n \bar{z}^n\right)$$

$$= \mathrm{Re}\, c_0 + \frac{1}{2} \sum_{n=1}^{\infty} c_n r^n e^{in\theta} + \frac{1}{2} \sum_{n=1}^{\infty} \bar{c}_n r^n e^{-in\theta}.$$

We get the equality

$$u(z) = \sum_{n=-\infty}^{\infty} d_n r^{|n|} e^{in\theta}, \tag{4.10}$$

which holds for $|z| < R$, with $\bar{d}_n = d_{-n}$, $d_0 \in \mathbb{R}$. If $u$ is harmonic with complex values, we write $u = u_1 + iu_2$, where $u_1$, $u_2$ are real harmonic; then expanding in the previous way each $u_i$, $i = 1, 2$ one will obtain an equality such as the one in (4.10) for $u$, without the restriction $\bar{d}_n = d_{-n}$. Hence, the following result has been proved:

**Theorem 4.19.** *Any harmonic function $u$ on a disc $D(0, R)$ has an expansion, which holds on this disc, of the kind*

$$u(z) = \sum_{n=-\infty}^{\infty} d_n r^{|n|} e^{in\theta} = d_0 + \sum_{n=1}^{\infty} d_n z^n + \sum_{n=1}^{\infty} d_{-n} \bar{z}^n, \quad z = re^{i\theta}, \ 0 \le r < R.$$

The series $\sum_{n=0}^{\infty} d_n z^n$ is called the *holomorphic part* of $u$, and $\sum_{n=1}^{\infty} d_{-n} \bar{z}^n$ is its *antiholomorphic part*. Both series converge absolutely at each point $z \in D(0, R)$ and uniformly on each compact set contained in $D(0, R)$. Observe that

$$d_n = \frac{1}{n!} \frac{\partial^n u}{\partial z^n}(0), \quad d_{-n} = \frac{1}{n!} \frac{\partial^n u}{\partial \bar{z}^n}(0).$$

A holomorphic function on $D(0, R)$ is harmonic; in this case, one just has the terms in $z^n$ with $n \ge 0$ in the expansion given by Theorem 4.19, and there is no term in $\bar{z}^n, n \ge 1$.

Suppose, once again, that $u$ is harmonic on the disc $D(0, R)$ and fix $0 < r < R$. The function $u_r(\theta) = u(re^{i\theta})$ is continuous in $\theta$ and $2\pi$-periodic. The equality (4.10) gives the expansion of $u_r$,

$$u_r(\theta) = \sum_{n=-\infty}^{+\infty} d_n r^{|n|} e^{in\theta}, \tag{4.11}$$

and $C(0, r)$ being compact, it is uniformly convergent with respect to $\theta$. The series (4.11) is the *Fourier series* of the $2\pi$-periodic function $u_r(\theta)$; indeed, multiplying the equality (4.11) by $e^{-im\theta}$ and integrating it yields

$$d_m r^{|m|} = \frac{1}{2\pi} \int_0^{2\pi} u(re^{i\theta}) e^{-im\theta}\, d\theta = \hat{u}_r(m), \quad m \in \mathbb{Z}.$$

These equalities, in the case that $u$ is real and $u = \operatorname{Re} f$, relate the Fourier coefficients of $u_r$, $\hat{u}_r(m)$, with the Taylor coefficients of $f$, $c_m$, since $d_m = \frac{1}{2} c_m$ if $m \ge 1$, $d_m = \frac{1}{2} \overline{c_{-m}}$ if $m \le -1$ and $d_0 = \operatorname{Re}(c_0)$.

If $f$ is holomorphic and $f(z) = \sum_{n=0}^{\infty} c_n z^n$, the Fourier series of the function $f_r(\theta) = f(re^{i\theta})$ is

$$f_r(\theta) = \sum_{0}^{\infty} c_n r^n e^{in\theta}$$

and the equalities between coefficients may be written

$$\frac{f^{(n)}(0)}{n!} = c_n = \frac{1}{2\pi r^n} \int_0^{2\pi} f(re^{i\theta}) e^{-in\theta}\, d\theta = \frac{1}{r^n} \hat{f}_r(n), \quad n \ge 0,$$

which are also the ones obtained from (4.6) parameterizing $C(0, r)$ by $w = re^{i\theta}$.

**Example 4.20.** If $f(z) = \sum_0^\infty c_n z^n$ is holomorphic on the unit disc $\mathbb{D}$ and continuous on $\overline{\mathbb{D}}$, $f$ taken as a function on $\mathbb{T} = \partial\mathbb{D}$ has a Fourier series expansion: $f(\theta) = \sum_{-\infty}^{+\infty} \hat{f}(n)e^{in\theta}$. The equalities above applied to the disc $D(0,r)$, $0 \le r < 1$ and letting $r \to 1$, give $\hat{f}(n) = c_n$ if $n \ge 0$, $\hat{f}(n) = 0$ if $n < 0$. Hence no continuous function on $\mathbb{T}$ may be continuously extended to a holomorphic function on $\mathbb{D}$ if its Fourier coefficients do not satisfy $\hat{f}(n) = 0, n < 0$. For example, the function $\bar{z} = e^{-i\theta}$ on $\mathbb{T}$ has no holomorphic and continuous extension to $\overline{\mathbb{D}}$. $\qquad\square$

## 4.4  Zeros of analytic functions. Principle of analytic continuation

### 4.4.1  Structure of the zero sets of holomorphic functions

It has been seen in Theorem 4.9 that the class of holomorphic functions on a domain $U$ of the plane is exactly the class of analytic functions in the variable $z$ on $U$. Everything we will formulate in this section for analytic functions in $z$ may be applied also to analytic functions of a real variable on an interval $I \subset \mathbb{R}$. Statements will be given, however, for the complex case.

**Lemma 4.21.** *Let $U$ be a domain of $\mathbb{C}$, $f \in H(U)$ and $a \in U$. Then the following conditions are equivalent:*

  a) $f^{(n)}(a) = 0$ *for $n = 0, 1, 2, \ldots$.*

  b) $f(z) = 0$ *for $z$ on a neighborhood of the point $a$.*

  c) $f$ *vanishes on $U$.*

*Proof.* Obviously c) implies a) and b). The fact that a) implies b) is clear, since $f$ being analytic one has $f(z) = \sum_n \frac{f^{(n)}(a)}{n!}(z-a)^n$ on a neighborhood of $a$. One just has to prove that b) implies c). Write

$$A = \{z \in U : f \text{ is zero on a neighborhood of } z\}.$$

If b) holds, $A \neq \emptyset$, and clearly $A$ is open. If we check that $A$ is closed in $U$, then $U = A$, since $U$ is connected, and c) will be proved. Suppose that $z_0 \in U$ is a closure point of $A$ and choose points $z_m \in A$ with $\lim_m (z_m) = z_0$. Since $f$ is zero on a neighborhood of each point $z_m$, it is $f^{(n)}(z_m) = 0$, for all $n, m$, and then $f^{(n)}(z_0) = \lim_m f^{(n)}(z_m) = 0$. Hence $f^{(n)}(z_0) = 0$, for all $n$, and it is already known that this fact implies $f$ vanishes on a neighborhood of $z_0$, that is $z_0 \in A$. $\qquad\square$

If $f$ is holomorphic on the domain $U$, denote by $\mathbb{Z}(f)$ the zero set of $f$ on $U$,

$$\mathbb{Z}(f) = \{a \in U : f(a) = 0\}.$$

By Lemma 4.21, if $a \in \mathbb{Z}(f)$ and $f \not\equiv 0$ on $U$, it cannot be $f^{(n)}(a) = 0$, for each $n \in \mathbb{N}$. Consequently we can consider the natural number $m > 0$ defined by

$$m = m(f, a) = \min\{n \in \mathbb{N} : f^{(n)}(a) \neq 0\}.$$

Hence, $f^{(m)}(a) \neq 0$ while $f^{(i)}(a) = 0$ for $i = 0, 1, \dots, m - 1$. The number $m(f, a)$ is called the *multiplicity or order of the point $a$ as a zero of $f$*. At the point $a$ the expansion of $f$ starts with the term in $(z - a)^m$:

$$f(z) = \frac{f^{(m)}(a)}{m!}(z - a)^m + \cdots = \sum_{n=m}^{\infty} c_n(z - a)^n, \quad c_n = \frac{f^{(n)}(a)}{n!}, \ |z - a| < \delta.$$

Taking out $(z - a)^m$ as a common factor in the previous series, we may write

$$f(z) = (z - a)^m g(z), \tag{4.12}$$

where $g(z) = \sum_{n=m}^{\infty} c_n(z - a)^{n-m} = \sum_{n=0}^{\infty} c_{n+m}(z - a)^n$; it is evident that this last series has the same radius of convergence as $\sum_{n=0}^{\infty} c_n(z - a)^n$ and defines $g$ as an analytic function on $D(a, \delta)$ with $g(a) = c_m = \frac{f^{(m)}(a)}{m!} \neq 0$ and $g(z) = f(z)(z - a)^{-m}$ if $z \neq a$. Since $g(a) \neq 0$, one has $g(z) \neq 0$ on some disc $D(a, \varepsilon) \subset U, \varepsilon > 0$, by continuity. On this disc, the only zero of $f$ is $a$, due to the equation (4.12). Hence, it has been proved that all the points of $\mathbb{Z}(f)$, if $f \not\equiv 0$, are isolated points, that is, $\mathbb{Z}(f)$ is a discrete set. On the other hand it is clear that $\mathbb{Z}(f)$ is a closed set in $U$, because $f$ is continuous. Hence the following result may be stated:

**Theorem 4.22.** *If $f$ is a holomorphic function on a domain $U$, $f \not\equiv 0$, then $\mathbb{Z}(f)$ is a discrete and closed subset of $U$, that is, without accumulation points in $U$. In particular, $\mathbb{Z}(f)$ is a finite or countable set and on each compact subset of $U$ there are a finite number of zeros of $f$.*

**Example 4.23.** A polynomial in $z$ has a finite number of zeros on $\mathbb{C}$. The function $f(z) = e^z$ has no zeros. The function $f(z) = e^z - 1$ has as zeros on $\mathbb{C}$ the points $2\pi k i, k \in \mathbb{Z}$. The function $f(z) = \sin z = \frac{1}{2i}(e^{iz} - e^{-iz})$ has the same zeros as $e^{2iz} - 1$, that is, $k\pi, k \in \mathbb{Z}$. The function $\text{Log } z$ on $U = \mathbb{C} \setminus \mathbb{R}^-$ has a unique zero at the point 1. $\qquad\square$

**Example 4.24.** Consider the function $f(z) = e^{\frac{1+z}{1-z}} - 1$ defined on $U = \mathbb{C} \setminus \{1\}$. The zeros $(z_k)$ of $f$ are given by $\frac{1+z_k}{1-z_k} = 2\pi k i$, that is, $z_k = \frac{2\pi k i - 1}{2\pi k i + 1} = 1 - \frac{2}{2\pi k i + 1}$, which is a set that has $1 \notin U$ as an accumulation point. $\qquad\square$

If $a$ is a zero of an analytic function $f$, $f \not\equiv 0$, on a domain $U$ it has been just seen that the point $a$ has finite multiplicity $m = m(f, a)$. In particular $f(z)$ behaves as $\frac{f^{(m)}(a)}{m!}(z - a)^m$ when $z \to a$,

$$\lim_{z \to a} \frac{f(z)}{\frac{f^{(m)}(a)}{m!}(z - a)^m} = 1, \quad f(z) \sim \frac{f^{(m)}(a)}{m!}(z - a)^m.$$

This means that it is not possible for an analytic function to have an order of cancellation around a zero $a$ which is not an entire power of $z - a$. For example, there cannot exist any holomorphic function $f$ defined on a neighborhood of $0$ such that $|f(z)| \sim |z| |\operatorname{Log} |z||$, because $|z| |\operatorname{Log} |z||$ is an infinitesimal when $|z| \to 0$, which is not equivalent to $|z|^m$ for any value of $m$. The function $f(z) = z \operatorname{Log} z$, which satisfies $|f(z)| \sim |z| |\operatorname{Log} |z|| \, (|z| \to 0)$, is not analytic on any neighborhood of the origin.

If $f \in H(U)$, where $U$ is a domain of $\mathbb{C}$ and $f$ is not identically zero, each point $z_k \in \mathcal{Z}(f)$ has associated its multiplicity $m_k = m(f, z_k)$, as a zero of $f$. Instead of considering separately the sequence $(z_k)$ and its corresponding sequence of multiplicities $(m_k)$, usually one includes in the list of zeros each of them as many times as its multiplicity. So in the resulting sequence

$$(z_1, z_2, z_3, \ldots, z_k, \ldots)$$

there may appear repetitions. Hence one obtains the so-called list of zeros of $f$ *counting multiplicities*.

Evidently, everything we have said about the zeros of $f$, which are the roots of the equation $f(z) = 0$, may be applied to the roots of the equation $f(z) = b$, where $b$ is a fixed complex value. Hence, $f^{-1}\{b\} = \mathcal{Z}(f - b)$ is a discrete and closed set in $U$ if $f$ is holomorphic and not constant on $U$.

If $\mathcal{Z}(f)$ is not finite, it has accumulation points in $\mathbb{C}^*$. Since these points cannot be in $U$, they must belong to $\partial U$; if $U$ is not bounded, $\infty$ is on the boundary of $U$ in $\mathbb{C}^*$ and $\infty$ may be an accumulation point of $\mathcal{Z}(f)$. For example, the zeros $2\pi k i$, $k \in \mathbb{Z}$ of $e^z - 1$ in $\mathbb{C}$ accumulate at $\infty$; the boundary of the domain $U = \mathbb{C} \setminus \{1\}$ is $\partial U = \{1, \infty\}$ in $\mathbb{C}^*$ and the zeros of $f(z) = e^{\frac{1+z}{1-z}} - 1$ in $U$ accumulate at $1$.

If $z_1, \ldots, z_N$ are the zeros of $f$ in a compact set $K \subset U$ (counting multiplicities), one may repeat the factorization (4.12) to obtain

$$f(z) = (z - z_1)(z - z_2) \cdots (z - z_N) g(z),$$

where the function $g$ is holomorphic in $U$ and does not have any zero inside $K$. Actually, if $z_i$ has multiplicity $m_i$, that is it appears $m_i$ times in the list $z_1, z_2, \ldots, z_N$, then $g(z_i) = \frac{f^{(m_i)}(z_i)}{m_i!} \neq 0$; at the other points it is $g(z) \neq 0$ because $f(z) \neq 0$. This process is called *removing the zeros of $f$ in the compact set $K$*.

### 4.4.2 The principle of analytic continuation

The following result is a direct consequence of Lemma 4.21 and Theorem 4.22 applied to the difference of two holomorphic functions. As already mentioned when talking about zeros of holomorphic functions, we again point out that the principle of analytic continuation also holds for real analytic functions on an interval of the real line.

**Theorem 4.25** (Principle of analytic continuation). *Let $f$, $g$ be holomorphic functions on a domain $U$. Then $f(z) = g(z)$ for all $z \in U$ if and only if one of the following equivalent conditions holds:*

  a) *There is a point $a \in U$ such that $f^{(n)}(a) = g^{(n)}(a)$, for $n \geq 0$, that is, $|f(z) - g(z)| = o(|z - a|^n)$, $z \to a$, $n = 0, 1, 2, \ldots$.*

  b) *There is a set $A \subset U$ with some accumulation point in $U$ and $f(z) = g(z)$ for all $z \in A$.*

  c) *There is an open set $V \subset U$ such that $f(z) = g(z)$ for all $z \in V$.*

As an example of a set $A \subset U$ with an accumulation point in $U$ we can consider the set $A = \{a_n : n \in \mathbb{N}\}$ formed by the different terms of a sequence $(a_n)$ which converges to a point of $U$.

**Example 4.26.** If a function $f$, holomorphic on the unit disc $\mathbb{D}$, satisfies $nf\left(\frac{1}{n}\right) = 1, n = 2, 3, \ldots$, then $f(z) = z$, for all $z \in \mathbb{D}$. Indeed, $f(z)$ and $g(z) = z$ coincide on $A = \left\{\frac{1}{n}, n \geq 2\right\}$. However, the function $f(z) = z\left(1 - \sin\frac{\pi}{z}\right)$ also satisfies $nf\left(\frac{1}{n}\right) = 1$ and is not identically $z$. In this case, $f$ is holomorphic on $U = \mathbb{C} \setminus \{0\}$ and $A$ has no accumulation point in $U$. $\qquad\square$

Next an application of the principle of analytic continuation is given. It is said that a domain $U$ of the plane is *symmetric* if for each point $z \in U$ one has that $\bar{z} \in U$. Since $U$ is connected it is evident that, if it is symmetric, then $U \cap \mathbb{R}$ is a non-empty open set of $\mathbb{R}$ and so it is a countable union of open intervals of $\mathbb{R}$.

**Theorem 4.27** (Schwarz's reflection principle). *Suppose that $U$ is a symmetric domain of the plane, $f \in H(U)$ and $f(x)$ is real for $x \in U \cap \mathbb{R}$. Then $f(\bar{z}) = \overline{f(z)}$ for all $z \in U$.*

*Proof.* Consider $g(z) = \overline{f(\bar{z})}$ which is a well-defined function on $U$. If on the disc $D(\bar{a}, r) \subset U$, $f(z)$ is the sum of the series $\sum_n c_n(z - \bar{a})^n$ then $g(z)$ is, on $D(a, r)$, the sum of $\sum_n \bar{c}_n(z - a)^n$. Therefore, $g$ is also analytic on $U$. It may be also seen, alternately, that $g$ satisfies the Cauchy–Riemann equations. Now, if $x \in U \cap \mathbb{R}$ one has $g(x) = \overline{f(x)} = f(x)$. Since $U \cap \mathbb{R} \neq \emptyset$ is an open set (and therefore it has accumulation points inside $U$) we conclude that $g(z) = f(z), z \in U$. $\qquad\square$

**Remark 4.2.** Another way of interpreting the reflection principle is the following. Denote by $U^+ = U \cap \{z : \operatorname{Im} z > 0\}$ the part of the symmetric domain $U$ located in the upper half plane. Suppose that $g \in H(U^+)$ and $g$ has a real limit at the points $x \in U \cap \mathbb{R}$, also denoted by $g$: $g(x) = \lim_{z \to x, z \in U^+} g(z)$. Then the function $f$ defined on $U$ by means of "reflection",

$$
f(z) = \begin{cases} g(z) & \text{if } z \in U^+, \\ g(x) & \text{if } x \in U \cap \mathbb{R}, \\ \overline{g(\bar{z})} & \text{if } z \in U^-, \end{cases}
$$

is holomorphic on $U$ (it is the only possible continuation of $g$ to $U$ which is holomorphic). Indeed, it is clear that $f \in C(U)$ and that it is holomorphic at all the points in $U \setminus \mathbb{R}$; then, by Theorem 4.13, one has $f \in H(U)$.

The principle of analytic continuation allows us to understand correctly a lot of identities of real analysis. Suppose, for example, that $f_1, \ldots, f_N$ are entire functions which satisfy an identity of type $P(f_1, \ldots, f_N)(z) = 0$, only for real values of the variable, where $P$ is a polynomial in $N$ variables. The principle of analytic continuation asserts then that $P(f_1, \ldots, f_N)(z) = 0$ holds identically in $z$. Hence, the fact that $\sin^2 x + \cos^2 x = 1$ for $x \in \mathbb{R}$ is read automatically $\sin^2 z + \cos^2 z = 1$ for $z \in \mathbb{C}$ and, specifying at $z = ix$ with $x \in \mathbb{R}$, it turns out that $\operatorname{ch}^2 x - \operatorname{sh}^2 x = 1$.

As remarked above, all that has been said works for real analytic functions. Indeed, with the help of the principle of analytic continuation one may better understand the relationship between real analytic functions and complex analytic functions, that is, holomorphic functions.

**Theorem 4.28.** *Every real analytic function $f : \mathbb{R} \to \mathbb{C}$ is the restriction to $\mathbb{R}$ of a holomorphic function $F : U \to \mathbb{C}$ defined on a symmetric domain $U$ of the complex plane, that is $\mathbb{R} \subset U$ and $f = F_{|\mathbb{R}}$.*

*Proof.* For each point $a \in \mathbb{R}$ it is known that there exists a radius $R(a) > 0$ such that

$$
f(x) = \sum_{n=0}^{\infty} c_n (x - a)^n, \quad c_n = \frac{f^{(n)}(a)}{n!}, \quad |x - a| < R(a).
$$

The power series $\sum_{n=0}^{\infty} c_n (z - a)^n$ has radius of convergence $\rho$, which must satisfy $\rho \geq R(a)$ and defines a holomorphic function, $F_a$, on $D(a, \rho)$. The restriction of $F_a$ to $(a - \rho, a + \rho)$ is an analytic function on this interval which coincides with $f$ on $(a - R(a), a + R(a))$ and thus, by the principle of analytic continuation, $F_a = f$ on the whole interval $(a - \rho, a + \rho)$. This means that $\rho = R(a)$. Suppose now that $D(a_1, R(a_1)) \cap D(a_2, R(a_2)) \neq \emptyset$, so that both $F_{a_1}$ and $F_{a_2}$ are defined on this intersection; on the part of the intersection that is on $\mathbb{R}$ both functions coincide with

$f$ and once again by the principle of analytic continuation, one obtains $F_{a_1} = F_{a_2}$ on $D(a_1, R(a_1)) \cap D(a_2, R(a_2))$. This means that on the symmetric domain

$$U = \bigcup_{a \in \mathbb{R}} D(a, R(a)),$$

the functions $F_a$ allow us to define a function $F : U \to \mathbb{C}$, equal to $F_a$ on each disc $D(a, R(a))$. By construction $F$ is holomorphic on $U$ and $F\big|_{\mathbb{R}} = f$. $\qquad\qquad \square$

The principle of analytic continuation for analytic functions of two real variables is stated without giving any proof. This property is applied, in particular, to harmonic functions.

**Theorem 4.29.** *Let $f$, $g$ be analytic functions of two real variables on a domain $U \subset \mathbb{R}^2$. Then $f(x, y) = g(x, y)$, for all $(x, y) \in U$, if and only if one of the following equivalent conditions holds:*

a) *There is a point $(x_0, y_0) \in U$ such that $\frac{\partial^{n+m} f}{\partial x^n \partial y^m}(x_0, y_0) = \frac{\partial^{n+m} g}{\partial x^n \partial y^m}(x_0, y_0)$, for $n, m \geq 0$, that is, $|f(x, y) - g(x, y)| = o(\sqrt{(x - x_0)^2 + (y - y_0)^2})^n$, when $(x, y) \to (x_0, y_0)$, $n = 0, 1, 2, \dots$.*

b) *There is an open set $V \subset U$ such that $f(x, y) = g(x, y)$ for all $(x, y) \in V$.*

In this case, it might hold that $f(x, y) = g(x, y)$ for $(x, y) \in A$, where $A$ has accumulation points in $U$, but $f \not\equiv g$. This is due to the fact that the zero sets of an analytic function of the variables $x$, $y$ are not necessarily discrete. For example, the harmonic function $x$ vanishes on the imaginary axis, and it is not identically zero.

## 4.5 Local behavior of a holomorphic function. The open mapping theorem

Consider a holomorphic function $f$ on a domain $U \subset \mathbb{C}$. Suppose that $f$ has no zeros in $U$, and ask whether there is a continuous branch of the logarithm of $f$ on $U$. It has been proved in Proposition 3.20 that this is the case if and only if the function $f'/f$ has a holomorphic antiderivative on $U$. But now it is known that $f'$ is a holomorphic function and so is $f'/f$, so that by Corollary 3.28 it follows:

**Proposition 4.30.** *On a convex open set, every holomorphic function without zeros admits a branch of its logarithm and of its $n$-th root, for all $n \in \mathbb{N}$. Every holomorphic function without zeros on a domain admits, locally, a branch of its logarithm and of its $n$-th root, for all $n \in \mathbb{N}$.*

**Corollary 4.31.** *If $f \in H(U)$ and $f(z) \neq 0$ for all $z \in U$, then the function $\mathrm{Log}\,|f(z)|$ is harmonic on $U$.*

*Proof.* One may check directly that $\Delta \operatorname{Log} |f| = 0$ using the Cauchy–Riemann equations. It is also a consequence of Proposition 4.30 because on each disc $D \subset U$ there is a function $h \in H(D)$ such that $e^h = f$ and then $\operatorname{Log} |f| = \operatorname{Re} h$ is harmonic on $D$. $\qquad\square$

Let $U$ be a domain of the plane, $a \in U$, $f \in H(U)$ and $b = f(a)$. If $f(z) - b$ has a zero with multiplicity $m$ at the point $a$, that is, $f'(a) = 0, \ldots, f^{(m-1)}(a) = 0$ and $f^{(m)}(a) \neq 0$, it is said that $f$ *takes $m$ times the value $b$ at the point $a$.* Let us explain the motivation of this definition. We have

$$f(z) - b = (z - a)^m g(z),$$

where $g \in H(U)$ and $g(a) \neq 0$. Fix a disc $D(a, R)$ on which $g$ has no zeros; by Proposition 4.30, there is a function $h \in H(D(a, R))$ such that $g = h^m$. This means that on $D(a, R)$ we have

$$f(z) - b = [(z - a)h(z)]^m \stackrel{\text{def}}{=} f_1(z)^m. \tag{4.13}$$

Observe that $f_1(z) = (z - a)h(z)$ satisfies $f_1(a) = 0$, $f_1'(a) = h(a) \neq 0$, that is, $f_1$ vanishes once at $a$. It is clear, according to (4.13), that the local behavior of $f$ will be understood if the one of $f_1$ is.

The function $f_1$ is holomorphic on $D(a, R)$, $f_1(a) = 0$ and $f_1'(a) = \lambda \neq 0$. One may assert now that $f_1$ is a homeomorphism of a neighborhood of the point $a$ onto a neighborhood of 0. This fact is a consequence of the inverse function theorem, but one can provide a direct proof as follows. Assume, without lost of generality, that $a = 0$ and $\lambda = 1$, that is, $f_1(z) = z + f_2(z)$ with $f_2'(0) = 0$. Fix a number $0 < \varepsilon < 1$ and find $r < R$ such that $|f_2'(z)| < \varepsilon$ if $z \in D(0, r)$. By the fundamental theorem of calculus (for example, integrating $f_2'$ along the segment that goes from $z$ to $w$) one will have

$$|f_2(z) - f_2(w)| \leq \varepsilon |z - w|, \quad z, w \in D(0, r),$$

and this inequality implies

$$(1 - \varepsilon)|z - w| \leq |f_1(z) - f_1(w)| \leq (1 + \varepsilon)|z - w|, \quad z, w \in D(0, r). \tag{4.14}$$

Let us now prove that if $|\xi|$ is small enough, say $|\xi| < s$, the equation $f_1(z) = \xi$ has a solution on $D(0, R)$ which by (4.14), will be unique and will depend continuously on $\xi$. Notice that the fact that $f_1(z) = \xi$ means $z$ is a fix point of the function $\xi - f_2(z)$. Define, as usual, a recurrent sequence $(z_n)$, with $z_0 = 0$ and $z_{n+1} = \xi - f_2(z_n)$. Since $|z_{n+1} - z_n| = |f_2(z_n) - f_2(z_{n-1})| \leq \varepsilon |z_n - z_{n-1}|$, one will get

$$|z_{n+1} - z_n| \leq \varepsilon^n |z_1 - z_0| = \varepsilon^n |\xi|.$$

Therefore, $|z_n| \le \sum_{i=1}^{n} |z_i - z_{i-1}| \le |\xi| \frac{1}{1-\varepsilon}$; if $|\xi| \le s = r(1-\varepsilon)$, all the points $z_n$ are indeed in $D(0,r)$ and converge to a solution of $f_1(z) = \xi$.

Hence $f_1$ is a local homeomorphism. In this situation it is known (Subsection 2.4.3) that $f_1^{-1}$ is also holomorphic and $(f_1^{-1})'(w) = (f_1'(z))^{-1}$ if $w = f_1(z)$.

Let now $V$ be a neighborhood of the point $a$ such that $f_1$ is a homeomorphism of $V$ onto a disc $D(0,s)$. By (4.13), the image of $V$ by $f$ will be exactly the disc $D(b, s^m)$. Then the following has been proved:

**Theorem 4.32.** *Let $f$ be a holomorphic function on a neighborhood of a point $a \in \mathbb{C}$ that takes the value $b = f(a)$, $m$ times at $a$. Then there is a neighborhood $V$ of $a$ and a disc $D(b, \delta)$ such that $f(V) = D(b, \delta)$ and each value $\xi \in D(b, \delta)$, $\xi \neq b$, has exactly $m$-preimages by $f$ in $V$ (while for $\xi = b$ the $m$-preimages are equal to $a$).*

One can even make more precise the situation described by this theorem by taking the proof up again: given $\xi \in D(b, s^m)$ consider the $m$-roots of $\xi - b$ in $D(0, s)$

$$\xi = b + w_i^m, \quad i = 1, \ldots, m, \ w_i \in D(0, s).$$

For each of the points $w_i$ there is a unique point $z_i \in V$ such that $f_1(z_i) = w_i$, and these values $z_1, \ldots, z_m$ are the $m$ different roots of $f(z) = \xi$ in $V$. Since $f_1^{-1}$ is holomorphic on $D(0, s)$, $f_1^{-1}(0) = a$ and $(f_1^{-1})'(0) = \lambda^{-1}$, we can write

$$z_i = a + \lambda^{-1} w_i + o(|w_i|).$$

The points $w_1, \ldots, w_m$ are uniformly distributed on the angles $\frac{2\pi}{m} k$, $k = 0, \ldots,$ $m - 1$, and so are the points $z_1, \ldots, z_m$, except for a small error, around $a$ and rotated according to $\arg(\lambda^{-1})$ (Figure 4.3). When $\xi \to b$, obviously

$$\max_{i=1,\ldots,m} |z_i - a| \longrightarrow 0,$$

that is, the points $z_1, \ldots, z_m$ collapse to $a$.

**Theorem 4.33** (Open mapping). *If $U$ is a domain of the plane and $f$ is holomorphic on $U$ and non-constant, then $f(U)$ is an open set of $\mathbb{C}$. If $V \subset U$ is open, $f(V)$ is also open.*

*Proof.* If $f$ is non-constant, the hypothesis of Theorem 4.32 holds at each point $a \in U$, by the principle of analytic continuation. Then, every point of $f(V)$ has a neighborhood formed by points which have preimages in $V$. $\qquad\square$

Remark that Theorem 4.32 has as a consequence that $f$ is locally injective at the point $a$ if and only if $f'(a) \neq 0$. In the case of a real variable function just one implication holds, due to the inverse function theorem: if $df(a)$ is invertible, $f$ is locally injective at $a$, but the converse is false. On the other hand, when $f'(z) \neq 0$,

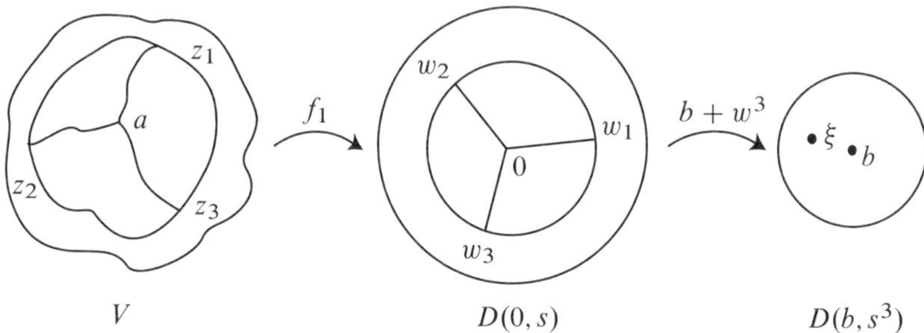


Figure 4.3

for all $z \in U$, $f$ is locally injective on $U$, but it is not necessarily injective on $U$. The clearest example is $f(z) = e^z$.

Observe also that, as a consequence of the open mapping theorem, there cannot be non-constant holomorphic functions on a domain $U$ which take values in a set without interior points (lines, regular curves, ... ). For example, no holomorphic function $f$, different from a constant, may satisfy an equation of type

$$u(\operatorname{Re} f(z), \operatorname{Im} f(z)) = 0 \quad \text{or} \quad v(f(z), \overline{f(z)}) = 0$$

with $u$, $v$ analytic functions in two variables (not identically zero).

## 4.6 Maximum principle. Cauchy's inequalities. Liouville's theorem

In this section some inequalities related to the absolute value of a holomorphic function and its derivatives will be obtained.

**Theorem 4.34** (Maximum principle). *If $f$ is holomorphic and non-constant on a domain $U$ of the plane, then the function $|f|$ cannot have any local maximum in $U$.*

*Proof.* Suppose that $|f|$ has a local maximum at $a \in U$, that is, there is a disc $D(a, R) \subset U$ such that $|f(z)| \leq |f(a)|$ if $z \in D(a, R)$. This prevents $f(D(a, R))$ from being an open set that contains $f(a)$ and contradicts Theorem 4.33. $\square$

**Corollary 4.35.** *Let $U$ be a bounded domain of $\mathbb{C}$ and $f$ a holomorphic function on a neighborhood of $\overline{U}$ or, more generally, $f \in C(\overline{U}) \cap H(U)$. Let $M$ be the maximum of $|f|$ in $\partial U$. Then one has*

$$|f(z)| \leq M \quad \text{for each } z \in U.$$

*In other words, $\max_{\overline{U}} |f| = \max_{\partial U} |f|$.*

*Proof.* Since $\bar{U}$ is a compact set, $|f|$ has a global maximum on $\bar{U}$ which is attained at some point $a \in \bar{U}$. If $a \in U$, then it is a local maximum and by the previous theorem $f$ is constant and the statement is obvious. If $a \in \partial U$, then $|f(a)|$ is also the global maximum on $\partial U$, $|f(a)| = M$ and consequently $|f(z)| \leq M$, $z \in U$. $\square$

**Example 4.36.** The quantity $\sup_{z \in D(1,1)} |e^z|$ coincides with $\max_{z \in \bar{D}(1,1)} |e^z|$, which, by Corollary 4.35, is $\max\{|e^z| : |z - 1| = 1\}$. If $z = 1 + e^{it}$, $0 \leq t < 2\pi$, it turns out that $|e^z| = e^{\mathrm{Re}\, z} = e^{1 + \cos t}$ and the maximum value is $e^2$. $\square$

Corollary 4.35 tells us that an inequality $|f(z)| \leq M$ for $z \in \partial U$ may be extended to $|f(z)| \leq M$ for $z \in U$ when $f \in C(\bar{U}) \cap H(U)$. As an application of this fact one gives the following example:

**Example 4.37.** There exists no sequence of polynomials $P_n(z)$ which converges uniformly to $1/z$ on the unit circle $\mathbb{T}$. Indeed, otherwise one would have

$$\left| P_n(z) - \frac{1}{z} \right| < \varepsilon, \quad n \geq n_0, \ |z| = 1 \ (\varepsilon < 1)$$

and therefore $|z P_n(z) - 1| < \varepsilon$, $n \geq n_0$, if $|z| = 1$. By Corollary 4.35 it would be $|z P_n(z) - 1| \leq \varepsilon$, $n \geq n_0$ for $|z| \leq 1$ and, taking $z = 0$, one arrives at a contradiction. $\square$

Of course, if $f$ is holomorphic and does not have zeros in $U$, taking the function $\frac{1}{f}$ one obtains that $|f|$ cannot have a local minimum in $U$. For $f \in C(\bar{U}) \cap H(U)$ without zeros in $\bar{U}$, one will have

$$\min_{\bar{U}} |f| = \min_{\partial U} |f|.$$

In other words, if $m \leq |f(z)| \leq M$ for $z \in \partial U$ and $f$ does not have zeros in $\bar{U}$, then one also has $m \leq |f(z)| \leq M$ for $z \in U$.

Consider the particular case of Proposition 4.14 corresponding to $U = D(a, R)$. If $f$ is holomorphic on a neighborhood of $\bar{D}(a, R)$, the equation

$$f^{(n)}(z) = \frac{n!}{2\pi i} \int_{\partial D(a,R)} \frac{f(w)}{(w - z)^{n+1}} \, dw, \quad z \in D(a, R) \qquad (4.15)$$

holds. The values of $f$ on $\partial D(a, R)$ determine $f^{(n)}(z)$ on the whole disc by means of this formula. Cauchy's inequalities estimate $|f^{(n)}|$ in terms of $|f|$.

**Theorem 4.38** (Cauchy's inequalities). *If $f$ is holomorphic on a neighborhood of the disc $\bar{D}(a, R)$ and $|f(z)| \leq M$ when $z \in C(a, R)$, then the following inequalities hold:*

$$|f^{(n)}(a)| \leq M \frac{n!}{R^n} \quad \textit{for } n = 0, 1, 2, \ldots . \qquad (4.16)$$

*Proof.* The formula (4.15) for $z = a$ may be written

$$f^{(n)}(a) = \frac{n!}{2\pi R^n} \int_0^{2\pi} f(a + Re^{i\theta})e^{-in\theta}\, d\theta,$$

which implies (4.16) immediately.     □

**Corollary 4.39.** *If $f$ is holomorphic and bounded on the domain $U$, $|f(z)| \leq M$, $z \in U$, then*

$$|f^{(n)}(z)| \leq M \frac{n!}{d(z, U^c)^n}, \quad z \in U, \; n = 0, 1, 2, \dots.$$

*In particular, these inequalities hold if $U$ is bounded, $f \in C(\bar{U}) \cap H(U)$ and $|f(z)| \leq M$ for $z \in \partial U$.*

*Proof.* Apply Theorem 4.38 to the disc $D(z, r) \subset U$, $r < d(z, U^c)$ and let $r$ tend to $d(z, U^c)$.     □

Cauchy's inequalities cannot be improved. For example for $\mathbb{D}$, $f(z) = z^n$ and $a = 0$, the two terms of (4.16) are equal to $n!$.

**Example 4.40.** Let $f$ be a holomorphic function on the unit disc $\mathbb{D}$ satisfying the inequality $|f(z)| \leq \frac{1}{1-|z|}$, $|z| < 1$. Taking $0 \leq r < 1$ and applying (4.16) to the disc $\bar{D}(a, r)$ with $a = 0$ and $M = \sup\{|f(z)| : |z| \leq r\} \leq \frac{1}{1-r}$, we obtain $|f^{(n)}(0)| \leq n! \frac{1}{(1-r)r^n}$, $0 \leq r < 1$. Now one may compute the minimum value of the function $\frac{1}{(1-r)r^n}$ for $0 \leq r < 1$ and it turns out that this minimum is attained for $r = \frac{n}{n+1}$ and has value $\frac{(n+1)^{n+1}}{n^n}$. We have then found the inequality $|f^{(n)}(0)| \leq n! \frac{(n+1)^{n+1}}{n^n}$, $n = 1, 2, \dots$.

**Theorem 4.41** (Liouville's theorem). *If $f$ is an entire and bounded function, then $f$ is constant. Also, a harmonic and bounded function on the whole plane is constant.*

*Proof.* When $f$ is entire, the result is a direct consequence of Corollary 4.39, for $n = 1$, because $d(z, U^c) = +\infty$ and therefore $f' \equiv 0$. If $u$ is harmonic on $\mathbb{C}$, according to Theorem 4.19, and writing $z = re^{i\theta}$ one has

$$u(z) = \sum_{-\infty}^{+\infty} d_n r^{|n|} e^{in\theta}$$

where

$$d_n = \frac{1}{2\pi r^{|n|}} \int_0^{2\pi} u(re^{i\theta})e^{-in\theta}\, d\theta, \quad n \in \mathbb{Z}.$$

If $|u| \leq M$ on $\mathbb{C}$ it turns out that $|d_n| \leq M r^{-|n|}$, for all $r$, and letting $r \to \infty$ we get that $d_n = 0$ if $n \neq 0$, that is $u(z) = d_0$ for $z \in \mathbb{C}$.     □

Hence, an entire function, if it is not constant, cannot be bounded on $\mathbb{C}$. However, its restriction to $\mathbb{R}$ can be bounded as for $f(z) = \sin z$; it can even converge to zero when $x \to \pm\infty$, as for example $f(z) = e^{-z^2}$.

Liouville's theorem provides a proof of the fundamental theorem of algebra.

**Theorem 4.42** (Fundamental theorem of algebra). *Let $P(z) = a_0 + a_1 z + \cdots + a_n z^n$ be a polynomial of degree $n$, with $a_i \in \mathbb{C}$, $i = 0, 1, \ldots, n$, $n \geq 1$. Then $P$ has exactly $n$ roots $\alpha_1, \ldots, \alpha_n \in \mathbb{C}$ (some of which may be counted with its multiplicity) and $P(z) = a_n \prod_{i=1}^{n}(z - \alpha_i)$.*

*Proof.* By iteration it is enough to prove that $P$ has at least one root. Otherwise, the function $f(z) = 1/P(z)$ would be an entire function. Now writing

$$P(z) = z^n \left( a_n + \frac{a_{n-1}}{z} + \cdots + \frac{a_0}{z^n} \right)$$

we see that $a_n z^n$ is the dominating term when $|z| \to +\infty$, and therefore $|P(z)| \to +\infty$ when $|z| \to +\infty$. Hence, $\lim_{|z| \to +\infty} |f(z)| = 0$ and, in particular, $f$ is bounded. By Liouville's theorem, $f$ must be constant and then $P(z)$ would be also constant, but it has been supposed that $\deg(P) \geq 1$. $\square$

It is clear that the corresponding version of Liouville's theorem in one real variable is false; there are analytic functions on the whole real line $\mathbb{R}$ which are bounded and non-constant: $\sin x$, $\cos x$, $\frac{1}{1+x^2}$, etc.

The proof of Theorem 4.42 is applied to any entire function $f$ such that $|f(z)| \to +\infty$ when $|z| \to +\infty$ and one gets the conclusion that $f$ must have a zero. However, later on it will be proved that, in this case, $f$ is a polynomial (see Subsection 5.5.1 and Exercise 20 of Section 4.7).

## 4.7 Exercises

1. Let $U$ be a bounded domain of the plane with positively oriented piecewise regular boundary and let $\varphi(z, w)$ be a continuous function with respect to the two variables for $z \in U$ and $w \in \partial U$. Suppose, in addition, that $\varphi$ is holomorphic as a function of $z$ for each $w \in \partial U$. Show that the function

$$f(z) = \int_{\partial U} \varphi(z, w)\, dw$$

   is holomorphic on $U$ and

$$f^{(n)}(z) = \int_{\partial U} \frac{\partial^n \varphi(z, w)}{\partial z^n}\, dw, \quad n = 1, 2, \ldots, \quad z \in U.$$

**2.** Compute

$$\int_{-\delta}^{\delta} \mathrm{Log}\,|x - z|\,dx$$

for all $z \in \mathbb{C}$ and $\delta > 0$.

**3.** Let $f$ be a holomorphic function on the disc $D(0, R)$, $0 < r < R$, and $h_n \in \mathbb{C}$, $n = 1, 2, \ldots$ with $|h_n| < R - r$ and $(h_n) \xrightarrow[n \to \infty]{} 0$. Defining the functions $\varphi_n$ by

$$\varphi_n(z) = \frac{f(z + h_n) - f(z)}{h_n}, \quad n = 1, 2, \ldots,$$

show that $\lim_{n \to \infty} \varphi_n(z) = f'(z)$ uniformly on the disc $\bar{D}(0, r)$.

**4.** Let $f$, $g$ be two holomorphic functions on a neighborhood of the disc $\bar{D}(a, R)$. Prove the equality

$$\frac{1}{2\pi i} \int_{C(a,R)} \overline{f(z)} g'(z)\,dz = \frac{1}{\pi} \int_{D(a,R)} \overline{f'(z)} g'(z)\,dm(z).$$

Deduce from it that if, in addition, $f$ is injective and $K = f(\bar{D}(a, R))$, then the isoperimetric inequality

$$4m(K) \leq L(\partial K)\,d(K)$$

holds, where $L(\partial K)$ is the length of the curve $\gamma(\theta) = f(a + Re^{i\theta})$, $0 \leq \theta \leq 2\pi$, and $d(K)$ is the diameter of $K$.

**5.** Let $f$ be a holomorphic function on a neighborhood of the disc $\bar{D}(0, R)$. Prove that the integral formula

$$f(z_0) = \frac{1}{2\pi i} \int_{C(0,R)} \frac{R^2 - |z_0|^2}{(z - z_0)(R^2 - \overline{z_0}z)} f(z)\,dz, \quad |z_0| < R$$

holds and consequently

$$u(re^{i\theta}) = \frac{1}{2\pi} \int_0^{2\pi} \frac{R^2 - r^2}{R^2 - 2rR\cos(\theta - \varphi) + r^2} u(Re^{i\varphi})\,d\varphi,$$

$$0 \leq r < R,\ 0 \leq \theta \leq 2\pi$$

if $u$ is harmonic on a neighborhood of $\bar{D}(0, R)$.

**6.** Let $f$ be a holomorphic function on a domain $U$, satisfying

$$f^{(n+1)}(z) = \sum_{j=0}^{n} a_j f^{(j)}(z), \quad z \in U,$$

for $n \in \mathbb{N}$ and $a_0, a_1, \ldots, a_n \in \mathbb{C}$ fixed. Show that $f$ is the restriction to $U$ of an entire function which is a finite linear combination of exponential monomials of type $z^k e^{\lambda z}$ with $k \in \mathbb{N}$, $\lambda \in \mathbb{C}$. Show also that if in addition there exists a point $a \in U$ such that $f(a) = f'(a) = \cdots = f^{(n-1)}(a) = 0$, then $f$ is identically zero on $U$.

**7.** Let $f$ be a holomorphic function on $D(0, R)$ and consider for $1 \le p \le \infty$ the means

$$M_p(f, r) = \left( \frac{1}{2\pi} \int_0^{2\pi} |f(re^{i\theta}|^p d\theta \right)^{1/p}, \quad 0 < r < R, \ 1 \le p < \infty,$$

$$M_\infty(f, r) = \sup_{|z|=r} |f(z)|, \quad 0 < r < R.$$

Show that $M_p(f, r)$ is an increasing function of $r$ for $1 \le p \le \infty$.

*Hint:* Apply Exercise 12 of Section 3.8 to the function $(|f(re^{i\theta})|^2 + \varepsilon^2)^{p/2}$.

Show also that the means

$$\left( \frac{1}{\pi r^2} \int_{D(0,r)} |f(z)|^p dm(z) \right)^{1/p}$$

are increasing functions of $r$ for $1 \le p < \infty$.

**8.** Let $f$ be a holomorphic function on a domain $U$ and suppose there is a disc $\bar{D}(a, r) \subset U$ and a number $p \ge 1$ such that

$$|f(a)|^p = \frac{1}{2\pi} \int_0^{2\pi} |f(re^{i\theta})|^p d\theta.$$

Show that $f$ is constant on $U$. Reach the same conclusion assuming now that

$$|f(a)|^p = \frac{1}{\pi r^2} \int_{D(a,r)} |f(z)|^p dm(z).$$

*Hint:* Reduce to the case $p = 2$ and use Exercise 12 of Section 2.7.

**9.** Let $f_1, f_2, \ldots, f_N$ be holomorphic functions on a domain $U$ and $p \in \mathbb{R}$, $p \ge 1$. Suppose that the function $|f_1|^p + |f_2|^p + \cdots + |f_N|^p$ has a local maximum at a point of $U$. Prove that $f_1, f_2, \ldots, f_N$ are constant on $U$.

*Hint:* Apply Exercise 8 of this section to each $f_i$.

**10.** If $f$ is entire and the function $f(z)e^{-c|z|^\alpha}$, with $c > 0$, $\alpha > 0$, is bounded in modulus by a constant $M > 0$, find the best bound for the function $|f'(z)|e^{-c|z|^\alpha}$.

**11.** Let $f$ be an entire function which satisfies the estimate $|f(z)| = O(|z|^N)$, $|z| \to \infty$, for some $N \in \mathbb{N}$. Show that $f$ is a polynomial of degree smaller than or equal to $N$.

**12.** Let $f$ be a holomorphic function on $D(0, R)$ and for $0 < r < R$ put

$$A(r) = \sup_{|z|=r} \operatorname{Re} f(z).$$

Show that $A(r)$ is an increasing function of $r$ and if in addition $f$ satisfies $f(0) = 0$, then one has

$$\sup_{|z|=r} |f(z)| \le \frac{2r}{R - r} A(r), \quad 0 < r < R.$$

**13.** Let $f$ be a bounded holomorphic function on the strip $U = \{z = x + iy : a < x < b\}, a, b \in \mathbb{R}$ and suppose that $f \in C(\bar{U})$ and $|f(z)| \le 1$ if $z \in \partial U$. Show that $|f(z)| \le 1$ for all $z \in U$.

**14.** Let $f$ be a holomorphic function on the unit disc $\mathbb{D}$, continuous on the closed disc $\bar{\mathbb{D}}$, which satisfies $\operatorname{Re} f(z) \cdot \operatorname{Im} f(z) = 0$ for $z \in \mathbb{T}$. Show that $f$ vanishes on $\mathbb{D}$. Reach the same conclusion assuming now that $f$ satisfies $\operatorname{Re} f(z)^2 - \operatorname{Im} f(z) = 0$, if $z \in \mathbb{T}$. Show that, on the other hand, there exists a function $f \in C(\bar{\mathbb{D}}) \cap H(\mathbb{D})$, satisfying $\operatorname{Re} f(z)^2 + \operatorname{Im} f(z)^2 = 1, z \in \mathbb{T}$.

**15.** Characterize the polynomials in two variables $P(x, y)$ which have the following property: every function $f \in C(\bar{\mathbb{D}}) \cap H(\mathbb{D})$ which satisfies $P(\operatorname{Re} f(z), \operatorname{Im} f(z)) = 0$ for $z \in \mathbb{T}$, must vanish on $\mathbb{D}$.

*Hint:* Consider the connected components of the complement in $\mathbb{C}$ of the set $\{(x, y): P(x, y) = 0\}$.

**16.** Let $f$ be a holomorphic function on the disc $D(0, R)$ and continuous on $\bar{D}(0, R)$. If $f$ is not identically zero, prove that

$$\int_0^{2\pi} \operatorname{Log} |f(Re^{i\theta})| \, d\theta > -\infty.$$

**17.** Show that if $f$ is holomorphic on the domain $U$ then the function $|f|^\alpha$, for $\alpha > 0$, is subharmonic on $U$ (see Exercise 13 of Section 3.8). Show also that $|f|^\alpha$ is harmonic on $U$ if and only if it is constant on $U$.

**18.** Let $U$ be a domain of the plane and $u$ a continuous subharmonic function on $U$. Show that if the maximum value of $u$ is attained at some point of $U$, then $u$ is constant on $U$.

**19.** Let $U$ be a bounded domain of the plane, $u$ a continuous subharmonic function on $U$ and suppose that there exists a constant $M \geq 0$ such that $\limsup_{z \to z_0} u(z) \leq M$ for every point $z_0 \in \partial U$. Show that $u(z) \leq M$ for all $z \in U$.

In particular, if $u \in C(\bar{U})$ is subharmonic on $U$, then $\max_{z \in \bar{U}} u(z) = \max_{z \in \partial U} u(z)$.

**20.** Show that an entire function $f$ which satisfies $|f(z)| \to \infty$ when $|z| \to \infty$ is a polynomial.

**21.** The kernel $K(z, w) = \frac{1}{\pi} \frac{1}{(1 - z \bar{w})^2}$, $z, w \in \mathbb{D}$ is called the *Bergman kernel* of the unit disc. If $f$ is an integrable function on $\mathbb{D}$ with respect to the Lebesgue measure, $f \in L^1(\mathbb{D}, dm)$, then the Bergman transform of $f$ is well defined:

$$Kf(z) = \int_{\mathbb{D}} K(z, w) f(w) \, dm(w) = \frac{1}{\pi} \int_{\mathbb{D}} \frac{f(w)}{(1 - z\bar{w})^2} \, dm(z).$$

Prove the following statements:

a) If $f \in L^1(\mathbb{D}, dm) \cap H(\mathbb{D})$, then $Kf(z) = f(z)$ for $z \in \mathbb{D}$.

b) $L^2(\mathbb{D}, dm) \cap H(\mathbb{D})$ is a closed subspace of the Hilbert space $L^2(\mathbb{D}, dm)$ and if $f \in L^2(\mathbb{D}, dm)$, then $Kf$ is the orthogonal projection of $f$ over this closed subspace.

**22.** Show that there exists a constant $c > 0$ such that

$$\|f\|_2 \leq c \, \|u\|_2,$$

for $f \in H(\mathbb{D})$ with $f(0) = 0$ and $u = \operatorname{Re} f$, where $\| \ \|_2$ is the norm in the space $L^2(\mathbb{D}, dm)$.

# Chapter 5
# Isolated singularities of holomorphic functions. The residue theorem. Applications

This chapter is devoted to the study of functions that are holomorphic on a domain except at isolated points in which the function has no complex derivative or, maybe, is not even continuous. The fact that the function is holomorphic around each of these singular points allows us to determine accurately the nature of the singularities; namely the line integral of a holomorphic function around a singularity may be expressed in terms of a unique number associated to the singularity, that is, the residue of the function. The precise statement of this fact is the Residue theorem, which is an extension of Cauchy's theorem.

Some applications of the residue theorem are given, among which is the outstanding argument principle that allows one to count zeros and poles of meromorphic functions, or Rouché's theorem, that compares the quantity of zeros of two holomorphic functions. Also noteworthy is the use of the calculus of residues for evaluating real integrals, without computing antiderivatives.

## 5.1 Isolated singular points

Recall that $D(a, \varepsilon)$ or $D_\varepsilon(a)$ denotes the open disc with center $a$ and radius $\varepsilon > 0$. From now on, $D'(a, \varepsilon)$ or $D'_\varepsilon(a)$ will represent the *punctured disc* $D(a, \varepsilon) \setminus \{a\}$.

**Definition 5.1.** A function $f$ has an isolated singularity at the point $a$ of $\mathbb{C}$ if $f$ is holomorphic on $D'_\varepsilon(a)$ for some $\varepsilon > 0$. In the case that $f$ may extend to a holomorphic function on the whole disc $D_\varepsilon(a)$, the singularity is said to be removable.

Removable singularities are "false" singularities, that is, actually they are not true singularities. It is an abuse of language, but convenient for introducing the concept of singularity.

Suppose that $f_1$, $f_2$ are holomorphic functions on a disc $D_R(a)$, $R > 0$ and that $f_2$ is not identically zero. If $f_2(a) \neq 0$, then $f = f_1/f_2$ is holomorphic on a disc $D_\varepsilon(a)$, $\varepsilon > 0$. If $f_2(a) = 0$, it is known by Theorem 4.22 that $a$ is an isolated zero of $f_2$; consequently, there is $\varepsilon > 0$, $\varepsilon < R$ such that $a$ is the only zero of $f_2$ in $D_\varepsilon(a)$, and so $f = f_1/f_2$ is defined and holomorphic on $D'_\varepsilon(a)$. It is the situation of Definition 5.1: $a$ is an isolated singularity of $f$, which may be removable or not. Consider the multiplicity, $n$, of $a$ as a zero of $f_2$ : $f_2(z) = (z - a)^n g_2(z)$, where

$g_2(z) \in H(D_\varepsilon(a))$ and $g_2(z) \neq 0$ for all $z \in D_\varepsilon(a)$. If $\alpha = g_2(a)$, this means that $f_2(z) \sim \alpha(z-a)^n$ when $z \to a$. The two following cases are possible.

a) If $f_1(a) \neq 0$, then $f(z) \sim f_1(a)\alpha^{-1}(z-a)^{-n}$ and one sees that $a$ is a non-removable singularity because $\lim_{z \to a} f(z)$ does not exist and $f$ may not extend to a holomorphic function, even not to a continuous function at the point $a$.

b) If $f_1(a) = 0$, then $a$ has multiplicity $m \geq 1$ as a zero of $f_1$ and one may write $f_1(z) = (z-a)^m g_1(z)$ with $g_1(z) \in H(D_\varepsilon(a))$, $\beta = g_1(a) \neq 0$, if $\varepsilon > 0$ has been chosen small enough. Then $f(z) = (z-a)^{m-n} g_1(z)/g_2(z)$ on $D'_\varepsilon(a)$. Since $g_2(z) \neq 0$, for all $z \in D_\varepsilon(a)$, it turns out that for $m \geq n$, $a$ is a removable singularity; the expression $(z-a)^{m-n} g_1(z)/g_2(z)$ makes sense and defines a holomorphic function on the whole disc $D_\varepsilon(a)$. If $m < n$, the same expression shows that $f(z) \sim \beta\alpha^{-1}(z-a)^{m-n}$ does not have finite limit at $a$; in this case $a$ is a true singularity once again.

**Definition 5.2.** If $f$ is holomorphic on a punctured disc $D'_\varepsilon(a)$ and there exist $\alpha \in \mathbb{C}$ and $k \geq 1$ integer such that $f(z) \sim \alpha(z-a)^{-k}$ when $z \to a$, it is said that $f$ has a pole of order $k$ at the point $a$. The number $k$ is called the *multiplicity of the pole* (or *order of the pole*).

Obviously, poles are non-removable singularities and, as a matter of fact, one has $\lim_{z \to a} f(z) = \infty$ if $f$ has a pole at the point $a$. The previous argument proves that if $f_1$, $f_2$ are holomorphic functions on a domain of $\mathbb{C}$ and $f_2$ is not identically zero, then $f = f_1/f_2$ has removable singularities at the zeros of $f_2$ that are also zeros of $f_1$, with multiplicity greater than or equal to the ones corresponding to $f_2$. Moreover, $f$ has poles at the zeros of $f_2$ which are not zeros of $f_1$ or that, being zeros of $f_1$, have smaller multiplicity with respect to $f_1$ than with respect to $f_2$.

**Definition 5.3.** A function $f$ is meromorphic on a domain $U$ if there exists a set $A \subset U$, discrete and closed in $U$, such that $f$ is defined and holomorphic on $U \setminus A$ and has a pole at each point of $A$.

Hence, it turns out that if $f_1$, $f_2$ are holomorphic functions on the domain $U$ and $f_2$ is not identically zero, then $f = f_1/f_2$ is meromorphic on $U$, taking as $A$ the set

$$A = \{a \in U : \text{ multiplicity of } a \text{ as a zero of } f_1$$
$$< \text{ multiplicity of } a \text{ as a zero of } f_2\} \subset Z(f_2).$$

For example, the function $\tan z = \frac{\sin z}{\cos z}$ is meromorphic on the whole plane; their poles are the zeros of the function $\cos z$, that is, the points $z_k = \frac{\pi}{2} + k\pi$, $k \in \mathbb{Z}$ (none of these is a zero of $\sin z$) and all of them have multiplicity $1$.

The rational functions $R = \frac{P}{Q}$ with $P$, $Q$ polynomials, are meromorphic functions on the whole complex plane.

**Example 5.4.** Consider the function $f(z) = e^{\frac{1}{z}}$. This function has a singularity at the origin. It is non-removable because $\lim_{z\to 0} f(z)$ does not exist; for example $\lim_{x\to 0+} f(x) = \infty$, $\lim_{x\to 0-} f(x) = 0$, $f\left(\frac{1}{ik\pi}\right) = e^{ik\pi} = (-1)^k$. This helps to see that $f$ does not have a pole at the origin. $\qquad\square$

This example shows that there are isolated singularities which are neither re-movable singularities nor poles. They are called *essential singularities*. So one has the "false" singularities, the removable ones, and the true singularities, which may be either *essential* or *poles*. In the following theorem the type of singularity of a function $f$ at the point $a \in \mathbb{C}$ is characterized in terms of the behavior of $f(z)$ when $z \to a$.

**Theorem 5.5.** *Suppose that the function $f$ is holomorphic on $D'_\varepsilon(a)$, $\varepsilon > 0$. Then*

a) *The point $a$ is a removable singularity of $f$ if and only if $f$ is bounded on $D'_\delta(a)$ for some $\delta > 0$.*

b) *The point $a$ is a pole of $f$ if and only if $\lim_{z\to a} f(z) = \infty$, that is, for each $M > 0$ there is a $\delta > 0$ such that $|f(z)| > M$ if $0 < |z - a| < \delta$ (equivalently, $\lim_{z\to a} |f(z)| = +\infty$).*

c) *The point $a$ is an essential singularity of $f$ in all the other cases, that is, if and only if $\lim_{z\to a} f(z)$ does not exist in $\mathbb{C}^*$.*

*Proof.* First a) will be proved: if $f$ is holomorphic on $D_\varepsilon(a)$, then $f$ is continuous on $\overline{D_\delta}(a)$ for all $\delta < \varepsilon$, and therefore, bounded on $D_\delta(a)$. Conversely, suppose $|f(z)| \le M$ if $0 < |z - a| < \delta < \varepsilon$. Consider the function $g(z) = (z - a)^2 f(z)$; $g$ is also holomorphic on $D'_\delta(a)$ and $\lim_{z\to a} g(z) = 0$; hence, defining $g(a) = 0$, one has that $g$ is continuous on $D_\delta(a)$. In addition,

$$\lim_{z\to a} \frac{g(z) - g(a)}{z - a} = \lim_{z\to a} \frac{g(z)}{z - a} = \lim_{z\to a} (z - a) f(z) = 0.$$

So $g \in H(D_\delta(a))$, with $g(a) = g'(a) = 0$. Therefore, by (4.12), $g(z) = (z - a)^m g_1(z)$ with $m \ge 2$, $g_1 \in H(D_\delta(a))$; hence, on $0 < |z - a| < \delta$, $f(z)$ coincides with $(z - a)^{m-2} g_1(z)$, which is holomorphic on the whole disc $D_\delta(a)$ (because $m \ge 2$) and $a$ is removable.

Next, assertion b) will be proved. It has been noticed that if $f$ has a pole at the point $a$, then $\lim_{z\to a} f(z) = \infty$. Conversely, assuming that $\lim_{z\to a} f(z) = \infty$, then $g = 1/f$ is holomorphic on some disc centered at $a$ and has limit zero at $a$; by item a) $g$ has a removable singularity at $a$ with $g(a) = 0$. If $m \ge 1$ is the multiplicity of $a$ as a zero of $g$, then $g(z) = (z - a)^m g_1(z)$ with $\alpha = g_1(a) \ne 0$; hence, $f(z) \sim \alpha^{-1}(z - a)^{-m}$ and $f$ has a pole of order $m$ at the point $a$. $\qquad\square$

By definition, $f$ has a pole of order $k$ at the point $a$ when $f(z)(z - a)^k$ has non-zero finite limit at $a$. By the previous theorem, this is equivalent to the fact that

$f(z)(z-a)^k$ has a removable singularity at $a$, that is, $g(z) = f(z)(z-a)^k$ is a function in $H(D_\varepsilon(a))$, with $g(a) \neq 0$. Hence, $f$ has a pole of order $k$ at the point $a$ if and only if it may be written as

$$f(z) = \frac{g(z)}{(z-a)^k}$$

with $g$ holomorphic on a neighborhood of $a$ and $g(a) \neq 0$.

At an essential singular point $a \in \mathbb{C}$, when $z \to a$, $f(z)$ does not approach any value of $\mathbb{C}^*$, finite or not. The following theorem is more specific, saying that the behavior of $f(z)$ when $z \to a$ is chaotic.

**Theorem 5.6** (Casorati–Weierstrass). *If $f \in H(D'_\varepsilon(a))$, $\varepsilon > 0$, and $f$ has an essential singularity at the point $a$, then the set $f(D'_\delta(a))$ is dense in $\mathbb{C}$ for all $\delta$, $0 < \delta < \varepsilon$.*

*Proof.* Assume that for some $\delta < \varepsilon$ the open set $f(D'_\delta(a))$ is not dense; then there are $w \in \mathbb{C}$ and $r > 0$ such that $|f(z) - w| \geq r$ if $0 < |z - a| < \delta$. This means that the function $g(z) = \frac{1}{f(z)-w}$ is bounded and, by item a) of Theorem 5.5 one would have $g \in H(D_\delta(a))$; thus, $f(z) = w + \frac{1}{g(z)}$ would have either a pole or a removable singularity at $a$ and not an essential singularity. $\square$

**Example 5.7.** Analyze $f(D'_\delta(0))$ when $f(z) = e^{\frac{1}{z}}$. If $w = \frac{1}{z}$ and $0 < |z| < \delta$, then $|w| > \delta^{-1}$. The image by $w \mapsto e^w$ of $|w| > \delta^{-1}$ is, in fact, $\mathbb{C} \setminus \{0\}$ (the image of any horizontal strip of width $2\pi$ is $\mathbb{C} \setminus \{0\}$). $\square$

The behavior of a function around a pole is very different from the behavior around an essential singularity, which is described in Theorem 5.6. Indeed, it is known that if the point $a \in \mathbb{C}$ is a pole of $f$, then $\lim_{z \to a} f(z) = \infty$, but one may assert that if $f \in H(D'_\varepsilon(a))$ and $a$ is a pole of $f$, then for all $0 < \delta < \varepsilon$, the image $f(D'_\delta(a))$ is a neighborhood of the point at infinity. Actually, since the point $a$ cannot be an accumulation point of zeros of $f$, one may suppose that $f(z) \neq 0$ if $z \in D'_\delta(a)$. Now the function $g(z) = \frac{1}{f(z)}$ is holomorphic on $D_\delta(a)$ with $g(a) = 0$, and by the open mapping theorem, $g(D_\delta(a))$ contains a neighborhood of $0$, and then $f(D_\delta(a))$ contains a neighborhood of $\infty$.

Furthermore, if $a$ is a pole of $f$ of order $m \geq 1$, the point $a$ is a zero of order $m$ of the function $g(z) = \frac{1}{f(z)}$, because one has $f(z) = \frac{h(z)}{(z-a)^m}$ with $h(z) \neq 0$ if $|z-a|$ is small and, then, $g(z) = (z-a)^m \frac{1}{h(z)}$, with $\frac{1}{h}$ holomorphic and non-vanishing. Applying now Theorem 4.32, it is known that $g$ is $m$ to 1 between a neighborhood of $a$ and a neighborhood of $0$, that is, there exists a neighborhood $V$ of $0$ and $\eta > 0$ such that for all $w \in V$, $w \neq 0$, there are $m$ different points $z_1, \ldots, z_m \in D'_\eta(a)$ with $g(z_i) = w$. Expressing this in terms of $f$, it means that if $a$ is a pole of order $m$ of $f$, there exists a neighborhood $V$ of $\infty$ and $\eta > 0$ such that any point $w \in V$, $w \neq \infty$, has $m$ different preimages in $D'_\eta(a)$.

It is interesting again to compare the behavior of a holomorphic function around an isolated singularity with the analogous situation for functions of one real variable. If $I_\varepsilon(a)$ denotes the interval $(a-\varepsilon, a+\varepsilon)$ and $I'_\varepsilon(a) = I_\varepsilon(a)\setminus\{a\}$, there are countless different asymptotic behaviors of $f(x)$ when $x \to a$ among the functions $f$ which are differentiable or $C^\infty$ on $I'_\varepsilon(a)$. For example,

$$f(x) = |x|^\alpha |\operatorname{Log} |x||^\beta |\operatorname{Log} |\operatorname{Log} |x|||^\gamma$$

with arbitrary $\alpha, \beta, \gamma \in \mathbb{R}$ are examples of $C^\infty$ functions on a punctured neighborhood of $0$, $I'_\varepsilon(0)$, and the behavior is different for each choice of parameters $\alpha, \beta, \gamma$. In complex variables, instead, there are only two different kinds of behavior of $f$, holomorphic around a non-removable singularity $a$: either $|f(z)| \sim \rho|z-a|^{-k}$, for $\rho > 0$ and $k \geq 1$ integer, or $\lim_{z \to a} |f(z)|$ does not exist in $[0, +\infty]$. For example, no function $f$, holomorphic on $D'_\varepsilon(0)$, can be found satisfying

$$|f(z)| \sim |z|^\alpha |\operatorname{Log} |z||^\beta \quad \text{when } |z| \longrightarrow 0,$$

if $\beta \neq 0$. The function $f(z) = \frac{\operatorname{Log} z}{z}$ satisfies

$$|f(z)| = \frac{|\operatorname{Log} z|}{|z|} = \frac{|\operatorname{Log} |z| + i \operatorname{Arg} z|}{|z|}$$

$$= \frac{\sqrt{(\operatorname{Log} |z|)^2 + (\operatorname{Arg} z)^2}}{|z|} \sim \frac{|\operatorname{Log} |z||}{|z|}$$

when $z \to 0$, but it is not defined on any punctured disc centered at the origin.

## 5.2 Laurent series expansion

Theorem 4.8 tells us that every holomorphic function on a disc has a power series expansion. Now we will show that every holomorphic function on a punctured disc $D'(a, R)$ has a similar expansion, accepting powers of $z - a$ with negative exponent.

**Definition 5.8.** A Laurent series centered at $a \in \mathbb{C}$ is a formal series of the type

$$\sum_{n=-\infty}^{+\infty} c_n(z - a)^n, \quad c_n \in \mathbb{C}.$$

Giving a Laurent series is just giving a power series in $z - a$, $\sum_{n=0}^{\infty} c_n(z-a)^n$, and another series, $\sum_{n=-\infty}^{-1} c_n(z - a)^n$, which may be understood as a power series $\sum_{n=1}^{\infty} d_n(w - a)^n$ in $w - a = (z - a)^{-1}$ if $d_n = c_{-n}$. If $\rho_1$ is the radius of convergence of the first series,

$$\rho_1 = (\limsup_{n \to +\infty} |c_n|^{\frac{1}{n}})^{-1},$$

and $\rho_2$ the radius of the second one,

$$\rho_2 = \left( \limsup_{n \to +\infty} |d_n|^{\frac{1}{n}} \right)^{-1} = \left( \limsup_{n \to -\infty} |c_n|^{-\frac{1}{n}} \right)^{-1},$$

then $\sum_{n=0}^{\infty} c_n (z - a)^n$ defines a holomorphic function $f_1(z)$ on $\{z : |z - a| < \rho_1\}$ and $\sum_{n=-\infty}^{-1} c_n (z - a)^n$ defines a holomorphic function $f_2(z)$ on $\{z : |z - a|^{-1} = |w - a| < \rho_2\} = \{z : |z - a| > \rho_2^{-1}\}$.

From now on, we will write $R_1 = \rho_1$, $R_2 = \rho_2^{-1}$ and suppose that $R_2 < R_1$. Then the Laurent series $\sum_{n=-\infty}^{+\infty} c_n (z-a)^n$ defines a function $f(z) = f_1(z) + f_2(z)$ holomorphic on the *annulus* $C(a, R_2, R_1) = \{z \in \mathbb{C} : R_2 < |z - a| < R_1\}$. It is clear that the convergence of $\sum_{n=-\infty}^{+\infty} c_n (z - a)^n$ is absolute at any point of this annulus and uniform on each compact subannulus $\overline{C}(a, r_2, r_1)$ with $R_2 < r_2 < r_1 < R_1$. Moreover, applying Theorem 2.31, it turns out that

$$f'(z) = \sum_{n=-\infty}^{+\infty} n c_n (z - a)^{n-1}, \quad R_2 < |z - a| < R_1.$$

**Definition 5.9.** It will be said that the function $f$, holomorphic on the annulus $C(a, R_2, R_1)$, can be expanded in a Laurent series if it equals the sum of a Laurent series

$$f(z) = \sum_{n=-\infty}^{+\infty} c_n (z - a)^n, \tag{5.1}$$

convergent on the annulus $C(a, R_2, R_1)$.

In this case, the Laurent series expansion is unique because the coefficients $c_n$ are determined by $f$. To see this, take $r$ between $R_2$ and $R_1$ and let $\gamma$ be the circle with center $a$ and radius $r$ positively oriented, $\gamma(t) = a + re^{it}, 0 \le t \le 2\pi$. Using the uniform convergence of the Laurent series on $\gamma^* = C(a, r)$, by (5.1) one finds that

$$\frac{1}{2\pi i} \int_\gamma \frac{f(z)}{(z - a)^{n+1}} \, dz = \sum_{k=-\infty}^{+\infty} \frac{1}{2\pi i} c_k \int_\gamma (z - a)^{k-n-1} \, dz = c_n, \quad n \in \mathbb{Z} \tag{5.2}$$

because $(z - a)^{k-n-1}$ has a holomorphic primitive when $k - n - 1 \neq -1$.

In terms of the parametrization of $\gamma$, the equality (5.2) is

$$c_n = \frac{1}{2\pi} \int_0^{2\pi} f(a + re^{it}) r^{-n} e^{-int} \, dt, \quad n \in \mathbb{Z}, \ R_2 < r < R_1.$$

This expression of the coefficients $c_n$ may be also understood in the following way: $c_n r^n$ is the $n$-th Fourier coefficient, the one that corresponds to $e^{int}$, in the Fourier series expansion of the $2\pi$-periodic function $t \mapsto f(a + re^{it})$, which by (5.1) is

$$f(a + re^{it}) = \sum_{n=-\infty}^{+\infty} c_n r^n e^{int}.$$

The following theorem is the one corresponding to Theorem 4.8.

**Theorem 5.10.** *Every holomorphic function on an annulus may be expanded in a Laurent series.*

*Proof.* Given $f \in H(C(a, R_2, R_1))$ and having fixed a point $z \in C(a, R_2, R_1)$, choose $r_2, r_1$ such that $R_2 < r_2 < |z - a| < r_1 < R_1$ and take $\gamma_1(t) = a + r_1 e^{it}$, $\gamma_2(t) = a + r_2 e^{it}$, $0 \le t \le 2\pi$. Consider the domain $\{z : r_2 < |z - a| < r_1\}$ with its boundary positively oriented. Recall (Section 1.6) that this corresponds to travelling the outer circle $\gamma_1$ in a way that it has index 1 with respect to all points on its interior and that the circle $\gamma_2$ must be travelled with index $-1$ with respect to its inner points. Hence, the Cauchy integral formula gives

$$f(z) = \frac{1}{2\pi i} \int_{\gamma_1} \frac{f(w)}{w - z} \, dw - \frac{1}{2\pi i} \int_{\gamma_2} \frac{f(w)}{w - z} \, dw.$$

In the first integral, write $w - z = (w - a) - (z - a)$; since $|z - a| < r_1 = |w - a|$, it follows that

$$\frac{1}{w - z} = \frac{1}{(w - a) - (z - a)} = \frac{1}{(w - a)\left(1 - \frac{z-a}{w-a}\right)} = \sum_{n=0}^{\infty} \frac{(z - a)^n}{(w - a)^{n+1}}.$$

The expansion is uniformly convergent on $\gamma_1^*$ and it is also uniformly convergent when multiplied by $f(w)$. In the second integral, however, one has $|z - a| > r_2 = |w - a|$, so we can write

$$\frac{1}{w - z} = \frac{1}{(w - a) - (z - a)} = \frac{1}{(z - a)\left(\frac{w-a}{z-a} - 1\right)}$$

$$= -\sum_{n=0}^{\infty} \frac{(w - a)^n}{(z - a)^{n+1}} = -\sum_{n=-\infty}^{-1} \frac{(z - a)^n}{(w - a)^{n+1}}.$$

Integrating these expansions term by term one finds

$$f(z) = \sum_{n=-\infty}^{+\infty} c_n (z - a)^n,$$

where the coefficients $c_n$ are given by

$$c_n = \frac{1}{2\pi i} \int_{|w-a|=r} f(w)(w - a)^{-n-1} \, dw$$

with $r = r_1$ if $n \geq 0$ and $r = r_2$ if $n < 0$. Even though, apparently, these coefficients $c_n$ depend on $r_1$ and $r_2$, the uniqueness of the Laurent series expansion of $f$ makes clear that, indeed, they just depend on $f$.  $\square$

**Example 5.11.** Suppose that $g$ is holomorphic on $D(a, R)$ and $f(z) = \frac{g(z)}{(z-a)^k}$. Then $f$ is holomorphic on the punctured disc $D'(a, R)$. A punctured disc is a particular case of an annulus (when the small radius is zero) and the Laurent expansion of $f$ in $D'(a, R)$ may be obtained by dividing by $(z-a)^k$ the power series expansion of $g$,

$$f(z) = \sum_{n=0}^{\infty} \frac{g^{(n)}(a)}{n!}(z-a)^{n-k} = \sum_{n=-k}^{+\infty} \frac{g^{(n+k)}(a)}{(n+k)!}(z-a)^n.$$

For example, the expansion of $f(z) = \frac{e^z}{(z-a)^k}$ in $\mathbb{C} \setminus \{a\}$ is

$$f(z) = e^a \frac{e^{z-a}}{(z-a)^k} = e^a \sum_{n=0}^{\infty} \frac{(z-a)^{n-k}}{n!}.$$  $\square$

**Example 5.12.** Look for the Laurent series expansion of $f(z) = \frac{1}{z(z-1)}$ on all relevant annuli: $D'(0, 1)$, $C(0, 1, \infty)$, $D'(1, 1)$ and $C(1, 1, \infty)$. For the case of $D'(0, 1)$ divide by $z$ the Taylor expansion of $(z-1)^{-1}$ around the point 0:

$$\frac{1}{z(z-1)} = -\frac{1}{z} \sum_0^{\infty} z^n = -\sum_0^{\infty} z^{n-1}$$

and, similarly, on $D'(1, 1)$ divide by $(z-1)$ the Taylor expansion of $z^{-1}$ around the point 1:

$$\frac{1}{z(z-1)} = \frac{1}{z-1} \frac{1}{1-(1-z)} = \frac{1}{z-1} \sum_0^{\infty}(1-z)^n = -\sum_{-1}^{\infty}(1-z)^n.$$

On $C(0, 1, \infty)$:

$$\frac{1}{z(z-1)} = \frac{1}{z^2 \left(1 - \frac{1}{z}\right)} = \frac{1}{z^2} \sum_0^{\infty} \frac{1}{z^n} = \sum_{-\infty}^{-2} z^n.$$

On $C(1, 1, \infty)$:

$$\frac{1}{z(z-1)} = \frac{1}{(z-1)^2(1 + \frac{1}{z-1})} = \sum_{n=0}^{\infty}(-1)^n(z-1)^{-n-2} = \sum_{n=-\infty}^{-2}(-1)^n(z-1)^n.$$  $\square$

Notice that in Theorem 5.10 it has been shown also that if $f$ is holomorphic on an annulus $C(a, R_2, R_1)$, then $f$ is the sum of a function $f_1$ holomorphic on the disc $D(a, R_1)$ ($f_1(z) = \sum_0^\infty c_n(z-a)^n$) and of a function $f_2$ holomorphic outside the disc $D(a, R_2)$ ($f_2(z) = \sum_{-\infty}^{-1} c_n(z-a)^n$).

All these considerations may be applied in particular if $f$ has an isolated singularity at the point $a$, that is, $f \in H(D_\varepsilon'(a))$ ($D_\varepsilon'(a)$ is the annulus $C(a, 0, \varepsilon)$). In this case, $f$ has a unique Laurent series expansion,

$$f(z) = \sum_{-\infty}^{+\infty} c_n(z-a)^n,$$

where

$$c_n = \frac{1}{2\pi i} \int_{C(a,r)} \frac{f(z)}{(z-a)^{n+1}}\, dz$$

is independent of $r$, $0 < r < \varepsilon$.

Clearly, the singularity is removable when $c_n = 0$, for all $n < 0$. On the other hand, we have seen after Theorem 5.5 that $f$ has a pole of order $k$ at $a$ if and only if it may be written as $f(z) = \frac{g(z)}{(z-a)^k}$ with $g$ holomorphic and $g(a) \neq 0$. This means exactly that the Laurent expansion has a finite number of terms with negative powers and starts with $g(a)(z-a)^{-k}$:

$$f(z) = g(a)(z-a)^{-k} + g'(a)(z-a)^{1-k} + \cdots.$$

Consequently, the singularity is essential if and only if there are infinitely many coefficients $c_n \neq 0$ with $n < 0$.

The part $f_1(z) = \sum_0^\infty c_n(z-a)^n$ is called the *regular part* of $f$ at $a$. The part $f_2(z) = \sum_{-\infty}^{-1} c_n(z-a)^n$ is called the *principal part* of $f$ at $a$. When $a$ is a pole with multiplicity $k$, the principal part is a polynomial in $(z-a)^{-1}$ of degree $k$.

As an application of the Laurent series expansion one may obtain the decomposition of a rational function into a sum of simple fractions and a polynomial. A *simple fraction* is a rational function of type

$$\frac{c}{(z-a)^k}, \quad a, c \in \mathbb{C} \text{ and } k \in \mathbb{N}.$$

**Theorem 5.13.** *Let $R = \frac{P}{Q}$ be a rational function, where $P$ and $Q$ are polynomials without common zeros. Then $R$ has a unique expression as a sum of simple fractions plus a polynomial.*

*Proof.* First, if the degree of the polynomial $P$ is greater than the degree of $Q$, divide $P$ by $Q$, $P = QC + P_1$, to get

$$R = \frac{P}{Q} = C + \frac{P_1}{Q},$$

with $C$ a polynomial and $\deg P_1 < \deg Q$; now we need only decompose $R_1 = \frac{P_1}{Q}$ into a sum of simple fractions.

If $a_1, \ldots, a_k$ are the poles of $R$, which are the same as the ones of $R_1$ and coincide with the zeros of $Q$, let $P_1\left(\frac{1}{z-a_1}\right), \ldots, P_k\left(\frac{1}{z-a_k}\right)$ be the principal parts of the Laurent expansions of $R_1$ around these points, where $P_1, \ldots, P_k$ are polynomials. The function

$$f(z) = R_1(z) - \sum_{j=1}^{k} P_j\left(\frac{1}{z-a_j}\right)$$

is an entire function, that is, holomorphic on $\mathbb{C}$: if $z$ is not one of the points $a_1, \ldots, a_k$, $f$ is $\mathbb{C}$-differentiable at $z$ because so are $R_1(z)$ and all the terms $P_j\left(\frac{1}{z-a_j}\right)$; if $z$ is one of the points $a_j$, $f$ is also $\mathbb{C}$-differentiable at $z$ because so are the functions $P_i\left(\frac{1}{z-a_i}\right)$ with $i \neq j$ and

$$R_1(z) - P_j\left(\frac{1}{z-a_j}\right)$$

is the regular part of $R_1$ at $a_j$ and so is holomorphic. Now, it is clear that $\lim_{|z|\to\infty} f(z) = 0$ and this implies $f \equiv 0$, by Liouville's Theorem. The conclusion is then,

$$R(z) = C(z) + \sum_{j=0}^{k} P_j\left(\frac{1}{z-a_j}\right). \qquad \square$$

**Example 5.14.** Look for the decomposition of $R(z) = \frac{z^4}{(z-i)^2(z+i)}$ into simple fractions. In the proof of Theorem 5.13 it was clarified that $R$ and $R_1$ have the same principal parts at their poles; therefore, one may work directly with $R$. At the pole $i$, in order to find the principal part, one must compute the first two terms of the Taylor expansion of $h(z) = \frac{z^4}{z+i}$, which are

$$h(i) + h'(i)(z - i) = \frac{1}{2i} + \left(-\frac{7}{4}\right)(z - i).$$

Hence, the principal part is $\frac{1}{2i}(z - i)^{-2} - \frac{7}{4}(z - i)^{-1}$. At the pole $-i$, of order 1, one must just evaluate $\frac{z^4}{(z-i)^2}$ at $-i$ and this gives the principal part $-\frac{1}{4(z+i)}$. So

$$R(z) = C(z) + \frac{1}{2i}\frac{1}{(z - i)^2} - \frac{7}{4}\frac{1}{z - i} - \frac{1}{4(z + i)}.$$

The polynomial $C$ has degree smaller than or equal to 1 since $R = \frac{P}{Q}$ implies $\deg(C) \le \deg(P) - \deg(Q)$. In order to compute $C$, write $C(z) = az + b$, divide by $z$ and let $|z| \to +\infty$: we obtain

$$a = \lim_{|z|\to+\infty} \frac{R(z)}{z} = 1.$$

In order to determine $b$, take $z = 0$ and obtain $b = i$. $\qquad\qquad\qquad\qquad\square$

## 5.3  Residue of a function at an isolated singularity

### 5.3.1  The residue theorem

Suppose that the function $f$ has an isolated singularity at a point $a \in \mathbb{C}$, that is, $f \in H(D'_\varepsilon(a))$, $\varepsilon > 0$. According to Theorem 5.10, one may write

$$f(z) = \sum_{-\infty}^{+\infty} c_n (z-a)^n, \quad 0 < |z-a| < \varepsilon$$

with

$$c_n = \frac{1}{2\pi i} \int_{C(a,r)} \frac{f(z)}{(z-a)^{n+1}}\, dz, \quad 0 < r < \varepsilon.$$

It is interesting to highlight the coefficient $c_n$ with $n = -1$,

$$c_{-1} = \frac{1}{2\pi i} \int_{C(a,r)} f(z)\, dz \qquad\qquad (5.3)$$

due to the following fact: $f$ has a holomorphic primitive $F$ on $D'_\varepsilon(a)$ if and only if $c_{-1} = 0$. Actually, if $f = F'$, it is known that the line integral of $f$ along any closed path is zero. Conversely, if $c_{-1} = 0$, then there is no term in $(z-a)^{-1}$ in the expansion of $f$ and all the other terms have an antiderivative, which for the $n$-th term is $\frac{c_n}{n+1}(z-a)^{n+1}$, and the function

$$F(z) = \sum_{\substack{n=-\infty \\ n \neq -1}}^{+\infty} \frac{c_n}{n+1}(z-a)^{n+1}$$

is an antiderivative of $f$ (it is immediate to check that this Laurent series has the same domain of convergence as the one corresponding to $f$).

**Definition 5.15.** If $f \in H(D'_\varepsilon(a))$ the coefficient $c_{-1}$ of the Laurent expansion of $f$ around the point $a$, given by (5.3), is called the residue of $f$ at $a$ and it is denoted by $\mathrm{Res}(f,a)$.

The residue represents, then, the obstacle by which $f$ has no holomorphic primitive on a punctured disc $D'_\varepsilon(a)$, since this exists if and only if $c_{-1} = 0$.

**Example 5.16.** Let $f$ be holomorphic at a punctured disc $D'_\varepsilon(a)$ and let $\gamma$ be a closed path in $D'_\varepsilon(a)$ which does not pass through the point $a$. Then one has

$$\frac{1}{2\pi i} \int_\gamma f(z)\, dz = \mathrm{Ind}(\gamma, a) \cdot \mathrm{Res}(f, a).$$

Actually, one must just integrate along $\gamma$ the Laurent expansion of $f$ around $a$ and apply the definition of residue and (3.4). $\qquad\square$

The following result is a generalization of Cauchy's theorem for functions with isolated singularities.

**Theorem 5.17** (Residue theorem). *Let $U$ be a bounded domain of the plane with positively oriented piecewise regular boundary. Let $V$ be an open set with $\bar{U} \subset V$ and $A \subset V$ a closed discrete set in $V$ such that $A \cap \partial U = \emptyset$, and let $f$ be a holomorphic function on the open set $V \setminus A$. Then*

$$\frac{1}{2\pi i} \int_{\partial U} f(z)\, dz = \sum_{a \in A \cap U} \operatorname{Res}(f, a). \tag{5.4}$$

**Remark 5.1.** The hypotheses mean that each point of $A$ is an isolated singularity of $f$.

The two terms of (5.4) make sense. The left-hand side because $f$ is continuous on $\partial U$, and the right-hand side because in the compact set $\bar{U}$ there are only a finite number of points of $A$ and there are none of them in $\partial U$.

*Proof.* First recall that the boundary of $U$ is formed by closed Jordan curves $\gamma_1, \gamma_2, \ldots, \gamma_N$ with $\gamma_1$ positively oriented and $\gamma_2, \ldots, \gamma_N$ negatively oriented. Let $a_1, \ldots, a_k$ be the points of $A \cap U$ and around each $a_i$ consider a small circle $C_i = C(a_i, \varepsilon)$ which does not intersect $\partial U$, and delete from $U$ the discs $\bar{D}_i = \overline{D(a_i, \varepsilon)}$ (Figure 5.1). The boundary of the domain $\tilde{U} = U \setminus \bigcup_{i=1}^{k} \bar{D}_i$,

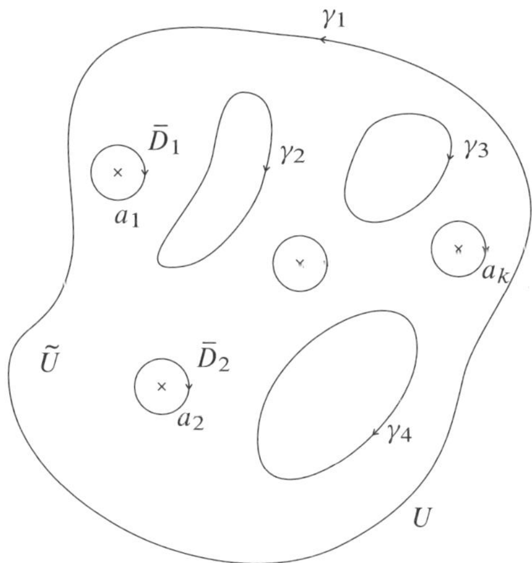


Figure 5.1

piecewise regular and positively oriented, is formed by $\gamma_1, \gamma_2, \ldots, \gamma_N$ and the circles $C_i$, $i = 1, \ldots, k$, negatively oriented, because $C_1, \ldots, C_k$ are in the interior of $\gamma_1$ (see Section 1.6). Now the function $f$ is holomorphic on a neighborhood of $\widetilde{U}$ and Cauchy's theorem (Theorem 3.24) gives

$$\int_{\partial \widetilde{U}} f(z)\, dz = 0,$$

that is,

$$\int_{\partial U} f(z)\, dz + \sum_{i=1}^{k} \int_{C_i} f(z)\, dz = 0. \tag{5.5}$$

Bearing in mind (5.3) and the fact that $C_i$ is travelled in the inverse sense, it turns out that

$$\frac{1}{2\pi i} \int_{C_i} f(z)\, dz = - \operatorname{Res}(f, a_i),$$

which, together with (5.5), gives (5.4). $\qquad\qquad\square$

Cauchy's theorem is the case $A = \emptyset$ of the residue theorem, and Cauchy's integral formula is also a particular case. Actually, it is enough to take $A = \{z\}$, if $z \in U$, and $g(w) = \frac{f(w)}{w-z}$ if $f$ is holomorphic on a neighborhood of $\bar{U}$. Then $\operatorname{Res}(g, z) = f(z)$ and it turns out that

$$\frac{1}{2\pi i} \int_{\partial U} \frac{f(w)}{w - z}\, dw = f(z).$$

### 5.3.2  Calculus of residues

The residue theorem has a lot of applications, but it is clear that these will depend on the possibility of computing the residue of a function at a singular point. This subsection is devoted to explaining how to proceed in some particular cases.

Suppose first that $f$ has a simple pole at the point $a$ so that

$$f(z) = \frac{c_{-1}}{z - a} + c_0 + c_1(z - a) + \cdots, \qquad 0 < |z - a| < \varepsilon.$$

Then it is obvious that

$$\operatorname{Res}(f, a) = c_{-1} = \lim_{z \to a} (z - a) f(z).$$

Many times one has $f = \frac{g}{h}$, with $g, h$ holomorphic on a neighborhood of $a$ and $g(a) \neq 0$, $h(a) = 0$, $h'(a) \neq 0$. Then

$$\operatorname{Res}(f, a) = \lim_{z \to a} (z - a) \frac{g(z)}{h(z)} = \lim_{z \to a} \frac{g(z)}{\frac{h(z)}{z-a}} = \frac{g(a)}{h'(a)}.$$

For example, $f(z) = \frac{e^{iz}}{z^2+1}$ has as residue, at the point $a = i$,

$$\mathrm{Res}(f, i) = \frac{g(i)}{h'(i)} = \frac{e^{-1}}{2i} = \frac{1}{2ie}.$$

Suppose now that $a$ is a pole with multiplicity $k > 1$ of the function $f$, that is,

$$f(z) = \frac{c_{-k}}{(z-a)^k} + \cdots + \frac{c_{-1}}{(z-a)} + c_0 + \cdots, \qquad 0 < |z-a| < \varepsilon.$$

Then, $\mathrm{Res}(f, a) = c_{-1}$ is the coefficient of $(z-a)^{k-1}$ in the expansion of $f_1(z) = (z-a)^k f(z)$, and hence

$$c_{-1} = \frac{f_1^{(k-1)}(a)}{(k-1)!}.$$

**Example 5.18.** The function $f(z) = \frac{e^{iz}}{z(z^2+1)^2}$ has a double pole at $a = i$. Here

$$f_1(z) = (z-i)^2 f(z) = \frac{e^{iz}}{z(z+i)^2}$$

and we get

$$\mathrm{Res}(f, i) = f_1'(i) = -\frac{3}{4e}. \qquad \square$$

**Example 5.19** (Residue of the logarithmic derivative). Let $f$ be a meromorphic function on a neighborhood of some point $a \in \mathbb{C}$. The function $f'/f$, which is also meromorphic, is called the *logarithmic derivative* of $f$; let us compute its residue at the point $a$. For a certain integer $m$, one has

$$f(z) = (z-a)^m g(z), \qquad 0 < |z-a| < \varepsilon$$

with $g$ holomorphic and $g(a) \neq 0$. If $m > 0$, $f$ is holomorphic on $D'(a, \varepsilon)$ and has a zero of order $m$ at the point $a$. If $m < 0$, $f$ has a pole of order $m$ at $a$. Calculating, it turns out that

$$\frac{f'(z)}{f(z)} = \frac{m}{z-a} + \frac{g'(z)}{g(z)},$$

so that, if $m \neq 0$, $f'/f$ has a *simple* pole at the point $a$ and $\mathrm{Res}(f'/f, a) = m$. $\quad \square$

When the singularity is essential, there is no simple and universal rule to compute the residue and each case must be treated particularly. For example, if $f$ is of the form $f(z) = g\left(\frac{1}{z-a}\right)$ where $g$ is an entire function with series expansion

$$g(w) = d_0 + d_1 w + \cdots + d_n w^n + \cdots = \sum_{0}^{\infty} d_n w^n$$

and $g$ is not a polynomial, it is clear that $f$ has an essential singularity at $a$ and that its residue is

$$\operatorname{Res}(f, a) = d_1 = g'(0).$$

For instance $\operatorname{Res}(e^{1/z}, 0) = 1$, $\operatorname{Res}(\cos\frac{1}{z}, 0) = 0$.

One common situation is the following: if $a$ is a pole of order 1 of $f$ and $g$ is holomorphic on a neighborhood of $a$ and $g(a) \neq 0$, then $fg$ has a pole of order 1 at $a$ and $\operatorname{Res}(fg, a) = g(a)\operatorname{Res}(f, a)$. Actually, it is $f(z) = \frac{h(z)}{z-a}$ with $h$ holomorphic on a neighborhood of the point $a$ and $h(a) = \operatorname{Res}(f, a) \neq 0$; therefore $f(z)g(z) = \frac{h(z)g(z)}{z-a}$ has a simple pole at the point $a$ and according to the calculus of the residue at a simple pole, it turns out that $\operatorname{Res}(fg, a) = h(a)g(a) = \operatorname{Res}(f, a)g(a)$.


## 5.4  Harmonic functions on an annulus

In Section 5.2 it was shown that if $f$ is a holomorphic function on the annulus $C(0, R_2, R_1)$, then $f$ is the sum of a Laurent series

$$f(z) = \sum_{-\infty}^{\infty} c_n z^n, \quad R_2 < |z| < R_1.$$

What may be said in the case of a function $u$ which is harmonic on this annulus? If the function $u$ (which is taken real-valued) was the real part of a holomorphic function, that is, had a conjugated harmonic function on $C(0, R_2, R_1)$, then clearly it would be

$$u(z) = \operatorname{Re} \sum_{-\infty}^{\infty} c_n z^n, \quad R_2 < |z| < R_1,$$

for some coefficients $c_n$. But, as it is known, this is not the general situation. In order to see how $u$ may be expanded, one may look for the obstruction to the existence of the conjugated harmonic function of $u$. Considering the function $f$ defined by

$$f(z) = 2\frac{\partial u}{\partial z} = \frac{\partial u}{\partial x} - i\frac{\partial u}{\partial y},$$

which is holomorphic on $C(0, R_2, R_1)$, the existence of a conjugated harmonic function of $u$ is the same as the existence of an antiderivative of $f$. Now it is clear that the obstruction to this problem is the fact that $c_{-1} = \frac{1}{2\pi i}\int_{C(0,r)} f(z)dz \neq 0$ in the Laurent expansion of $f$. Write then

$$a = \frac{1}{2\pi i}\int_{C(0,r)} f(z)\,dz, \quad R_2 < r < R_1,$$

and consider the function $g(z) = f(z) - \frac{a}{z}$, which is holomorphic on the annulus and satisfies

$$\frac{1}{2\pi i} \int_{C(0,r)} g(z)dz = \frac{1}{2\pi i} \int_{C(0,r)} f(z)dz - \frac{1}{2\pi i} \int_{C(0,r)} \frac{a}{z} dz = 0,$$

and therefore has a holomorphic primitive on $C(0, R_2, R_1)$. It is natural to think how the harmonic function $u$ must be modified in order to get $g$ instead of $f$ when taking $\frac{\partial}{\partial z}$. It is enough, considering the harmonic function

$$v(z) = u(z) - a \operatorname{Log} |z|, \quad R_2 < |z| < R_1,$$

to get

$$2\frac{\partial v}{\partial z} = \frac{\partial v}{\partial x} - i\frac{\partial v}{\partial y} = f(z) - \frac{a}{z} = g(z).$$

Observe that the constant $a = \frac{1}{\pi i} \int_{C(0,r)} \frac{\partial u}{\partial z} dz$ is real, since $\int_{C(0,r)} \nabla u ds = 0$. Now $v$ has a conjugated harmonic function on the annulus; that is, there is a Laurent series $\sum_{-\infty}^{\infty} d_n z^n$ convergent on $C(0, R_2, R_1)$ such that

$$v(z) = \operatorname{Re} \sum_{-\infty}^{\infty} d_n z^n.$$

Summarizing, the following statement has been proved:

**Theorem 5.20.** *If $u$ is a real harmonic function on the annulus $C(0, R_2, R_1)$, then there is a Laurent series $\sum_{-\infty}^{\infty} d_n z^n$, convergent on the annulus, and a real constant $a$ such that*

$$u(z) = \operatorname{Re}\left(\sum_{-\infty}^{\infty} d_n z^n\right) + a \operatorname{Log} |z|, \quad R_2 < |z| < R_1, \tag{5.6}$$

*with*

$$a = \frac{1}{\pi i} \int_{C(0,r)} \frac{\partial u}{\partial z} dz \quad \text{for all } r \text{ with } R_2 < r < R_1.$$

Consequently, there is an analog of the statement a) of Theorem 5.5 for harmonic functions.

**Corollary 5.21.** *Let $u$ be a harmonic function on the punctured disc $D'(0, \varepsilon)$.*

a) *If $\lim_{z \to 0} z u(z) = 0$, then there exists a function $u_1$ harmonic on the disc $D(0, \varepsilon)$ and a constant $a$ such that*

$$u(z) = u_1(z) + a \operatorname{Log} |z|, \quad 0 < |z| < \varepsilon. \tag{5.7}$$

b) *If $u$ is bounded on $D'(0, \varepsilon)$, then the origin is a removable singularity of $u$, in the sense that $u$ may be extended to a harmonic function on the disc $D(0, \varepsilon)$.*

*Proof.*  a) The equality (5.6) and the hypothesis of a) imply

$$u(z) = \sum_{-\infty}^{\infty} c_n r^{|n|} e^{in\theta} + a \operatorname{Log} |z| \quad \text{if } z = re^{i\theta}, \ 0 < r < \varepsilon,$$

and it is enough to take $u_1(z) = \sum_{-\infty}^{\infty} c_n r^{|n|} e^{in\theta}$.

b) If $u$ is bounded, the hypothesis of a) holds.  So the equality (5.7) is true and this forces $a = 0$.  $\qquad\qquad\square$

## 5.5  Holomorphic functions and singular functions at infinity

### 5.5.1  Behavior at the point at infinity

In the context of the residue theorem it is convenient to consider the behavior of a function $f(z)$ at the point at infinity, that is, when $|z| \to +\infty$. Similarly to the case of a point of the ordinary plane, this behavior can be regular or singular and, in the singular case, there may be a pole or an essential singularity.

Recall that $S^2$ or $\mathbb{C}^*$ denotes the Riemann sphere, obtained by adding to $\mathbb{C}$ the point at infinity.  The neighborhoods of the finite points $a \in \mathbb{C}$ are the sets that contain an open disc centered at $a$; the "discs" centered at $\infty$ are, by definition, $D(\infty, \varepsilon) = \{\infty\} \cup D'(\infty, \varepsilon)$, where

$$D'(\infty, \varepsilon) = \left\{ z : |z| > \frac{1}{\varepsilon} \right\} = \mathbb{C} \setminus \bar{D}\left(0, \frac{1}{\varepsilon}\right).$$

Hence, one may think that the distance function extends to $\mathbb{C}^*$ with $d(z, \infty) = \frac{1}{|z|}$ and then $D(\infty, \varepsilon) = \{z \in \mathbb{C}^* : d(z, \infty) < \varepsilon\}$. With this topology, $\mathbb{C}^*$ is a compact space.  By definition, the mapping $z \mapsto \frac{1}{z}$ makes a correspondence between the disc $D(0, \varepsilon)$ and the disc $D(\infty, \varepsilon)$.

Suppose that the function $f$ is defined on $D'(\infty, \varepsilon)$ and assume further that $\lim_{z \to \infty} f(z)$ exists; this means that letting $f(\infty) = \lim_{z \to \infty} f(z)$, $f$ is defined on $D(\infty, \varepsilon)$ and it is continuous at the point $\infty$. It is said that $f$ is a *holomorphic function at infinity* if the function $g(w) = f(1/w)$ is holomorphic at the origin. If $f$ is defined on $D(\infty, \varepsilon)$, then $g$ is defined on $D(0, \varepsilon)$ and

$$g(w) = g(0) + \alpha w + O(|w|), \quad \text{for } |w| < \varepsilon, \text{ with } \alpha = g'(0),$$

holds. This means that

$$f(z) = f(\infty) + \alpha \frac{1}{z} + O\left(\frac{1}{|z|}\right) \quad \text{if } |z| > 1/\varepsilon.$$

The value $\alpha = g'(0)$ is also denoted by $f'(\infty)$, and then one has

$$f'(\infty) = \lim_{|z|\to\infty} z(f(z) - f(\infty)).$$

For example, $f(z) = z^k$ is holomorphic at the point $\infty$ if $k$ is a negative integer, with $f(\infty) = \lim_{k\to\infty} z^k = 0$ and

$$f'(\infty) = \lim_{z\to\infty} z z^k = \begin{cases} 1 & \text{if } k = -1, \\ 0 & \text{if } k < -1. \end{cases}$$

If $k = 0$ then $f(\infty) = 1$ and $f'(\infty) = 0$.

Suppose that $f$, defined on $D'(\infty, \varepsilon)$, is holomorphic on this punctured disc. Then $f$ has a Laurent series expansion

$$f(z) = \sum_{-\infty}^{+\infty} c_n z^n, \quad |z| > 1/\varepsilon$$

because the function $g(w) = f(1/w)$ is holomorphic on $D'(0, \varepsilon)$ with expansion

$$g(w) = \sum_{-\infty}^{+\infty} c_n w^{-n} = \sum_{-\infty}^{+\infty} c_{-n} w^n, \quad 0 < |w| < \varepsilon.$$

The principal part of $f$ around the point $\infty$ is $\sum_0^\infty c_n z^n$, which is the part that makes $\lim_{z\to\infty} f(z)$ not exist. It is said that $f$ has a *pole of order $n$ at infinity* if this principal part is a polynomial of degree $n$ or, equivalently,

$$f(z) \sim c_n z^n \quad \text{when } z \longrightarrow \infty$$

(this is equivalent to the fact that $g(w) \sim c_n w^{-n}$ when $w \to 0$). If the principal part is infinite $f$ is said to have an *essential singularity at infinity*; in this case, according to Theorem 5.6, for all $\varepsilon > 0$, $f(D'(\infty, \varepsilon)) = g(D'(0, \varepsilon))$ is dense in $\mathbb{C}$, that is, $f(z)$ has a chaotic behavior when $z \to \infty$.

**Example 5.22.** Every polynomial of degree $n$ has a pole of order $n$ at the point $\infty$. A rational function $R = \frac{P}{Q}$ has a pole at the point $\infty$ if $\deg(P) > \deg(Q)$, and is a holomorphic function at the point $\infty$ if $\deg(P) \leq \deg(Q)$. An entire function which is not a polynomial has an essential singularity at infinity. $\square$

After the proof of the fundamental theorem of algebra (Theorem 4.42), it had been pointed out that, $f$ being entire and $f(z) \to \infty$ when $z \to \infty$, then $f$ has at least a zero on $\mathbb{C}$. Now one has that, under these conditions, $f$ is a polynomial, and we are in fact assuming the hypothesis of Theorem 4.42 (see also Exercise 20 in Section 4.7).

### 5.5.2  Residue of a function at the point at infinity

The residue theorem may be extended to the case that $f$ has also a singularity at infinity. Here we will consider only the equality

$$\frac{1}{2\pi i}\int_\gamma f(z)\,dz = \mathrm{Ind}(\gamma, a)\cdot \mathrm{Res}(f, a), \tag{5.8}$$

where $\gamma$ is a closed path not containing the point $a$. When $a \in \mathbb{C}$, (5.8) has been proved in Example 5.16. When $a = \infty$ the residue of a function at infinity must be properly defined. Suppose that $f$ is holomorphic on a punctured disc, $D'(\infty, \varepsilon) = \{z : |z| > 1/\varepsilon\}$. Since the differential form $f(z)\,dz$ must be integrated, one may look at how the change of variable $w = 1/z$ affects this form. One has

$$f(z)\,dz = f(1/w)\,d(1/w) = -\frac{1}{w^2}f\left(\frac{1}{w}\right)dw.$$

Therefore, it is natural to define $\mathrm{Res}(f, \infty)$ as the residue of $-\frac{1}{w^2}f\left(\frac{1}{w}\right)$ at zero,

$$\mathrm{Res}(f, \infty) = \mathrm{Res}\left(-\frac{1}{w^2}f\left(\frac{1}{w}\right), 0\right).$$

In terms of the Laurent expansion, if

$$f(z) = \sum_{-\infty}^{+\infty} c_n z^n, \quad |z| > 1/\varepsilon,$$

one has

$$-\frac{1}{w^2}f\left(\frac{1}{w}\right) = -\sum_{-\infty}^{+\infty} c_n w^{-n-2}.$$

The term $w^{-1}$ corresponds to $n = -1$, and it turns out that

$$\mathrm{Res}(f, \infty) = \mathrm{Res}\left(-\frac{1}{w^2}f\left(\frac{1}{w}\right), 0\right) = -c_{-1}.$$

Note that infinity may be a removable singularity of $f$ ($f$ holomorphic at the point $\infty$) while $\mathrm{Res}(f, \infty) \neq 0$: for example $f(z) = 1/z$ with $\mathrm{Res}(1/z, \infty) = -1$

If $\gamma$ is a closed path contained in the annulus $\{z : |z| > 1/\varepsilon\}$, since the Laurent expansion is uniformly convergent on $\gamma^*$, it yields

$$\frac{1}{2\pi i}\int_\gamma f(z)\,dz = \frac{1}{2\pi i}\sum_{-\infty}^{+\infty} c_n \int_\gamma z^n\,dz = c_{-1}\,\mathrm{Ind}(\gamma, 0) = -\,\mathrm{Res}(f, \infty)\,\mathrm{Ind}(\gamma, 0).$$

In particular, if $C$ is a circle centered at the origin with radius greater than $1/\varepsilon$, we get

$$\frac{1}{2\pi i}\int_C f(z)\,dz = -\,\mathrm{Res}(f, \infty).$$

Similarly to the case of finite isolated singularities, the residue is the obstruction for $f$ having a holomorphic antiderivative on $D'(\infty, \varepsilon)$.

**Example 5.23.** The function $f(z) = e^{1/z}$ is holomorphic at the point $\infty$ and $\mathrm{Res}(f, \infty) = -1$. The function $g(z) = \frac{1}{z^2} + z$ has a simple pole at the point $\infty$ and $\mathrm{Res}(g, \infty) = 0$. Therefore, if $C$ is a circle around the origin travelled in a direct sense, one has $\int_C e^{1/z}\, dz = -1$, $\int_C (1/z^2 + z)\, dz = 0$. $\qquad\square$

If $U$ is an open set of $\mathbb{C}^*$, a *meromorphic function on $U$* is defined similarly to the case of open sets of $\mathbb{C}$: $f$ is meromorphic on $U$ if there is a set $A \subset U$, discrete and closed in $U$, such that $f$ is holomorphic on $U \setminus A$ and $f$ has a pole at each point of $A$.

**Proposition 5.24.** *Suppose that $f$ is a meromorphic function on the whole Riemann sphere. Then $f$ is a rational function and if $A$ is the set formed by the poles of $f$ and the point at infinity, $A$ is finite and*

$$\sum_{a \in A} \mathrm{Res}(f, a) = 0.$$

*Proof.* The set $A$ is finite because it has no accumulation points on the compact set $\mathbb{C}^*$. If $a_1, \ldots, a_m$ are the points of $A$ and $P_1, \ldots, P_m$ are the corresponding principal parts of $f$ ($P_i$ is a polynomial in $\frac{1}{z - a_i}$ if $a_i \in \mathbb{C}$ and a polynomial in $z$ if $a_i = \infty$), then the function

$$f - \sum_{i=1}^{m} P_i$$

is holomorphic on $\mathbb{C}^*$ and therefore, entire and bounded and so must be constant. Hence, $f$ is a rational function. Let now $C$ be a circle containing inside all the finite points of $A$; by the residue theorem, one has

$$\frac{1}{2\pi i} \int_C f(z)\, dz = \sum_{\substack{a \in A \\ a \text{ finite}}} \mathrm{Res}(f, a).$$

But the left term is also $-\mathrm{Res}(f, \infty)$ and the proof is completed. $\qquad\square$

Observe that the point at infinity must be considered in the set $A$ whether it is a pole of $f$ or not.

## 5.6 The argument principle

In this section the residue theorem will be applied to count the zeros and the poles of meromorphic functions. A general version of the so-called argument principle will be given and the most interesting particular cases will be deduced.

**Theorem 5.25.** *Let $U$ be a bounded domain of the plane with piecewise regular positively oriented boundary. Let $V$ be an open set with $\bar{U} \subset V$, $f$ a meromorphic function on $V$ and $h$ a holomorphic function on $V$. The zeros of $f$ in $V$ are denoted by $\{a_j\}$ and $n_j$ denotes the multiplicity of $a_j$. The set $\{b_j\}$ denotes the poles of $f$ in $V$ and $m_j$ the multiplicity of $b_j$. Suppose that there are no zeros and no poles of $f$ on $\partial U$. Then*

$$\frac{1}{2\pi i} \int_{\partial U} h(z) \frac{f'(z)}{f(z)}\, dz = \sum_{a_j \in U} h(a_j) n_j - \sum_{b_j \in U} h(b_j) m_j. \qquad (5.9)$$

*Proof.* Recall from Example 5.19 that the function $f'/f$, called the logarithmic derivative of $f$, is a meromorphic function on $V$ having a simple pole in each zero and each pole of $f$. Specifically, if

$$f(z) = (z-a)^p g(z), \quad g(a) \neq 0, \quad 0 < |z - a| < \varepsilon,$$

then

$$\frac{f'(z)}{f(z)} = \frac{p}{z-a} + \frac{g'(z)}{g(z)}.$$

Now multiplying this equality by the holomorphic function $h(z)$ yields

$$h(z) \frac{f'(z)}{f(z)} = \frac{h(z)p}{z-a} + h_1(z), \quad \text{with } h_1 \text{ holomorphic on } D'_\varepsilon(a).$$

Therefore,

$$\operatorname{Res}\left(h \frac{f'}{f}, a_j\right) = h(a_j) n_j, \quad \operatorname{Res}\left(h \frac{f'}{f}, b_j\right) = -h(b_j) m_j$$

and equality (5.9) is a consequence of the residue theorem. $\qquad \square$

Taking as function $h$ a constant equal to 1, it follows:

**Corollary 5.26.** *Let $U$ be a bounded domain of the plane with piecewise regular positively oriented boundary, and let $f$ be a meromorphic function on a neighborhood of $\bar{U}$ without zeros or poles on $\partial U$. The number $N$ will denote the total number of zeros of $f$ in $U$ and $P$ the total number of poles in $U$ (counted according to multiplicities) and let $\Gamma = f(\partial U)$. Then*

$$\operatorname{Ind}(\Gamma, 0) = \frac{1}{2\pi i} \int_{\partial U} \frac{f'(z)}{f(z)}\, dz = N - P.$$

*Proof.* The equality between the integral of the logarithmic derivative of $f$ and the integer $N - P$ is (5.9) for $h \equiv 1$. Recall now that $\partial U$ is formed by the union of a finite number of closed Jordan curves: $\gamma_1, \gamma_2, \ldots, \gamma_N$. Therefore, $\Gamma = f(\partial U)$ is the union of $f \circ \gamma_1, f \circ \gamma_2, \ldots, f \circ \gamma_N$ and

$$\mathrm{Ind}(\Gamma, 0) = \frac{1}{2\pi i} \int_\Gamma \frac{d\zeta}{\zeta} = \sum_{j=1}^N \frac{1}{2\pi i} \int_{f \circ \gamma_j} \frac{d\zeta}{\zeta}.$$

Applying now to each integral the change of variable $\zeta = f(z)$, it turns out that

$$\mathrm{Ind}(\Gamma, 0) - \sum_{j=1}^N \frac{1}{2\pi i} \int_{\gamma_j} \frac{f'(z)}{f(z)} \, dz = \frac{1}{2\pi i} \int_{\partial U} \frac{f'(z)}{f(z)} \, dz. \qquad \Box$$

In particular, if $\Gamma = f(\partial U)$ does not wind around the origin, $f$ has the same number of zeros as poles inside $U$, counted according to multiplicity.

**Theorem 5.27** (Argument principle). *Let $U$ be a bounded domain of the plane with piecewise regular boundary, positively oriented, and $f$ a holomorphic function on a neighborhood of $\bar{U}$. Then, if $w \notin \Gamma = f(\partial U)$, one has*

$$\mathrm{Ind}(\Gamma, w) = N(w),$$

*where $N(w)$ denotes the number of roots of the equation $f(z) = w$ inside $U$, counted according to multiplicity.*

*Proof.* Apply Corollary 5.26 to $f(z) - w$ and find

$$\frac{1}{2\pi i} \int_{\partial U} \frac{f'(z)}{f(z) - w} \, dz = N(w).$$

The decomposition of $\partial U$ into Jordan curves and the change of variable $\zeta = f(z)$, done in the proof of Corollary 5.26, give now

$$\frac{1}{2\pi i} \int_{\partial U} \frac{f'(z)}{f(z) - w} \, dz = \mathrm{Ind}(\Gamma, w). \qquad \Box$$

This theorem has several interesting corollaries. The first one says that the image by $f$ of $\partial U$ determines completely the image of $\bar{U}$.

**Corollary 5.28.** *With the same hypotheses as in Theorem 5.27, $f(\bar{U})$ may be obtained from $\Gamma = f(\partial U)$ adding the bounded connected components of $\mathbb{C} \setminus \Gamma$ with non-zero index with respect to $\Gamma$.*

*Proof.* By the argument principle, the points $w \notin \Gamma$ which belong to $f(U)$ are exactly the ones having index greater than or equal to 1 with respect to $\Gamma$, that is,

$$(\mathbb{C} \setminus \Gamma) \cap f(U) = \bigcup_i C_i,$$

where $C_i$ are the bounded components of $\mathbb{C} \setminus \Gamma$ with non-zero index with respect to $\Gamma$. This leads to $f(\bar{U}) = \Gamma \cup \bigcup_i C_i$. Notice that it cannot be assured that $f(U) = \bigcup_i C_i$ because it could happen that $f(U)$ intersects $\Gamma$, that is, a point of $U$ and a point of $\partial U$ have the same image. $\qquad\square$

**Theorem 5.29.** *With the same hypotheses as in Theorem 5.27, let $V$ be a simply connected domain such that $V \supset \Gamma = f(\partial U)$. Then also $V \supset f(\bar{U})$.*

*Proof.* It suffices to show that $\mathrm{Ind}(\Gamma, w) = 0$ for $w \notin V$. If $\partial U = \gamma_1^* \cup \cdots \cup \gamma_N^*$ with $\gamma_j$ a closed Jordan curve, take $\widetilde{\gamma}_j = f \circ \gamma_j$; $\widetilde{\gamma}_j$ is a closed curve of $V$ and we just need to see that $\mathrm{Ind}(\widetilde{\gamma}_j, w) = 0$, $j = 1, \ldots, N$. Actually, $\mathbb{C}^* \setminus V$ is a connected set contained in the unbounded connected component of $\mathbb{C}^* \setminus \widetilde{\gamma}_j$, and so $\mathrm{Ind}(\widetilde{\gamma}_j, w) = 0$ for $w \notin V$. $\qquad\square$

Observe that this theorem implies the maximum principle, taking as $V$ a disc centered at the origin which contains $f(\partial U)$.

**Example 5.30.** If $f$ is holomorphic on a neighborhood of $\bar{U}$ and $f(\partial U)$ is inside an ellipse, $\frac{(\mathrm{Re}\, f(z))^2}{a^2} + \frac{(\mathrm{Im}\, f(z))^2}{b^2} \le 1$, $z \in \partial U$, then $f(U)$ is inside the same ellipse. $\qquad\square$

The following result is a consequence of the argument principle, and helps to compare the number of zeros of some functions.

**Theorem 5.31** (Rouché). *Let $U$ be a bounded domain with piecewise regular boundary and let $f$, $g$ be two holomorphic functions on a neighborhood of $\bar{U}$ with $|f(z) - g(z)| < |f(z)|$ if $z \in \partial U$. Then $f$ and $g$ have the same number of zeros inside $U$ (counted according to multiplicities).*

*Proof.* Observe that the hypotheses imply $f(z) \neq 0$ and $g(z) \neq 0$ if $z \in \partial U$. The function $F(z) = \frac{g(z)}{f(z)}$ is meromorphic on a neighborhood of $\bar{U}$. Consider the boundary of $U$ positively oriented and let $\Gamma = F(\partial U)$.

If $z \in \partial U$, one has $\left|1 - \frac{g(z)}{f(z)}\right| = |1 - F(z)| < 1$, and therefore $\Gamma \subset D(1, 1)$. The argument used in the proof of Theorem 5.29 shows that $\mathrm{Ind}(\Gamma, 0) = 0$. Applying now Corollary 5.26 to the function $F$, it turns out that $N = P$, that is, $F$ has in $U$ the same number of zeros as poles; but the zeros of $F$ are the zeros of $g$ and the poles of $F$ are the zeros of $f$, so the statement is proved. $\qquad\square$

A version of Rouché's theorem for the case that $f$ and $g$ are holomorphic on a neighborhood of an arbitrary compact may be also stated (see Exercise 13 in Section 6.8).

**Examples 5.32.** a) Take $f(z) = 2z^5 + 8z - 1$ and $g(z) = 2z^5$ (which is the dominating term of $f(z)$ for $|z|$ big). If $U = D(0, 2)$, one has for $|z| = 2$,

$$|f(z) - g(z)| = |8z - 1| \le 8|z| + 1 \le 17 < 2^6 = |g(z)|.$$

Hence, $f$ has the same number of zeros as $g$ in $D(0, 2)$, that is, five. Taking now $g(z) = 8z - 1$ and $U = D(0, 1)$ one has for $|z| = 1$,

$$|f(z) - g(z)| = |2z^5| = 2 < 7 \leq 8|z| - 1 \leq |8z - 1| = |g(z)|.$$

Hence, $f$ has the same number of zeros as $g$ in $D(0, 1)$, that is, one. Moreover, this zero is real and positive according to Bolzano's theorem, since $f(0) = -1$ and $f(1) = 9$.

b) Consider the polynomial $P(z) = z^3 - 2z^2 + 4$ and count how many zeros it has in the first quadrant. If $R$ is big enough these zeros (three at most) are inside the domain $U = \{z = re^{it} : 0 \leq r \leq R, 0 \leq t \leq \frac{\pi}{2}\}$. In order to apply the argument principle one must follow the path $\Gamma = P \circ \gamma$, with $\gamma = \partial U$.

When $z = x$ goes from 0 to $R$, $P(x)$ is a positive real number. When $z = Re^{it}$, $0 \leq t \leq \frac{\pi}{2}$, it turns out that

$$P(Re^{it}) = R^3 e^{3it}\left(1 - \frac{2}{Re^{it}} + \frac{4}{R^3 e^{it}}\right) = R^3 e^{3it}\left(1 + O\left(\frac{1}{R}\right)\right),$$

which reads that the term $z^3$ is the dominating one, if $R \to \infty$. If $R$ is big, $P(Re^{it})$ has then an argument which varies approximately between 0 and $\frac{3\pi}{2}$ until getting $P(iR) = -iR^3 + 2R^2 + 4$. Writing $P(iR) = R^3\left(-i + \frac{2}{R} + \frac{4}{R^3}\right) = R^3\left(-i + O\left(\frac{1}{R}\right)\right)$ we see that $P(iR)$ has argument $\frac{3\pi}{2} + O\left(\frac{1}{R}\right)$. Finally, when $z$ goes from $iR$ to 0 following the imaginary axis, $P(iy) = -iy^3 + 2y^2 + 4$ remains always in the fourth quadrant and ends up at $P(0)$. Conclusion: $\Gamma$ rotates once around zero and $P$ has a zero in the first quadrant (Figure 5.2). $\qquad\square$

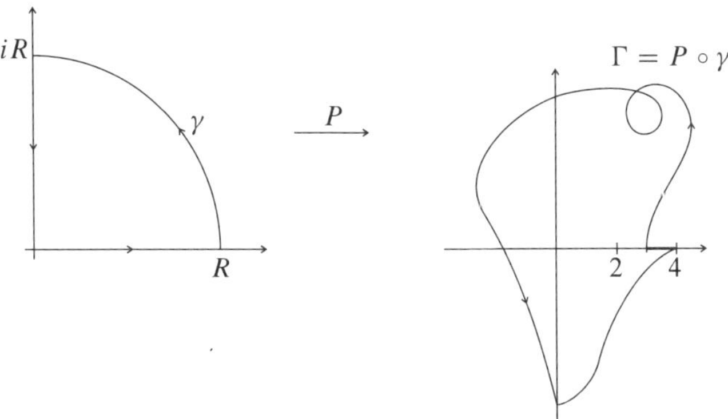


Figure 5.2

## 5.7 Dependence of the set of solutions of an equation with respect to parameters

Let $U$ be a bounded domain of the plane with positively oriented piecewise regular boundary. Suppose that the function $f$, holomorphic on a neighborhood $V$ of $\bar{U}$, is not a constant. If $w \in \mathbb{C}$, the set $f^{-1}\{w\}$ is formed by isolated points of $V$, and so the set

$$F(w) = f^{-1}\{w\} \cap \bar{U} \tag{5.10}$$

is finite.

If $w \notin \Gamma = f(\partial U)$, the set $F(w)$ is contained in $U$ and, by the argument principle, has a number of elements equal to $N(w) = \mathrm{Ind}(\Gamma, w)$. The function $N(w)$ is constant on each component of $\mathbb{C} \setminus \Gamma$.

Observe that Theorem 4.32 implies that if $f \in H(V)$ and $w_0 \in \mathbb{C}$ has a finite number $m$ of preimages by $f$ in $V$, then there is $\varepsilon > 0$ such that every point $w \in D(w_0, \varepsilon)$ has *at least* $m$ preimages (always counted according to multiplicities). With the present notation, this means that $N(w) \geq N(w_0)$ if $w$ is close to $w_0$. As said, the argument principle shows that $N(w)$ is locally constant.

The aim now is to prove that $F(w)$ depends continuously on $w$. In order to express that two sets $E$, $F$ are close it is convenient to use the *Hausdorff distance*, $d_H$, between $E$ and $F$ defined by

$$d_H(E, F) = \max_{z \in E} d(z, F) + \max_{z \in F} d(z, E).$$

Hence, $E = F$ implies $d_H(E, F) = 0$, and the converse holds if $E$, $F$ are closed.

**Theorem 5.33.** *Let $U$ be a bounded domain of the plane with piecewise regular positively oriented boundary. If $V$ is an open neighborhood of $\bar{U}$ and $f \in H(V)$ is not constant, then the function $w \mapsto F(w)$ defined by (5.10) is continuous, that is, if $w_0 \notin \Gamma = f(\partial U)$ and $\varepsilon > 0$, there is $\delta > 0$ such that $|w - w_0| < \delta$, $w \notin \Gamma$, implies $d_H(F(w), F(w_0)) < \varepsilon$.*

*Proof.* Take $N(w_0) = N$ and let $a_1, \ldots, a_m$ (with multiplicities $k_1, \ldots, k_m$) be the roots of $f(z) = w_0$ in $U$, $\sum_{j=1}^{m} k_j = N$. By Theorem 4.32, $f$ is $k_j$ to 1 in a neighborhood of $a_j$, that is, there is $\delta_j > 0$ such that $f(z) = w$ has $k_j$ different roots in a neighborhood of $a_j$, provided that $|w - w_0| < \delta_j$. Taking $\delta = \min \delta_j$, we get that every point $w$ with $|w - w_0| < \delta$ has at least $N$ preimages ($k_1$ on a neighborhood of $a_1$, $k_2$ on a neighborhood of $a_2, \ldots, k_m$ on a neighborhood of $a_m$). But, making $\delta$ smaller if needed, to get $D(w_0, \delta) \subset \mathbb{C} \setminus f(\partial U)$, it is known that each point $w \in D(w_0, \delta)$ has $N$ preimages, that is, $F(w)$ consists of exactly the roots that appear around each $a_j$. Since these roots collapse at the point $a_j$ when $w \to w_0$ (also by Theorem 4.32), one deduces that $d_H(F(w), F(w_0)) \to 0$ when $w \to w_0$. $\qquad\square$

With a similar process it will be shown now that the roots of a polynomial $P(z) = c_0 + c_1 z + c_2 z^2 + \cdots + c_n z^n$ depend continuously on the coefficients $c_0, c_1, \ldots, c_n \in \mathbb{C}$. One must be very careful with this kind of statements, because one deals with the *set* of zeros and not with a particular root determined by some additional criterion. For polynomials of degree 1 and 2 one has the explicit formulae

$$c_0 + c_1 z = 0, \quad c_1 \neq 0 \quad \Longrightarrow \quad z = -\frac{c_0}{c_1},$$

$$c_0 + c_1 z + c_2 z^2 = 0, \quad c_2 \neq 0 \quad \Longrightarrow \quad z = \frac{-c_1 + (c_1^2 - 4 c_0 c_2)^{\frac{1}{2}}}{2 c_2}$$

which prove continuous dependence, with respect to the coefficients, of the set of roots. Now, Galois theory shows that it is not possible, in general, to write a formula for the roots of a polynomial in terms of the coefficients. Nevertheless, it will be seen that the set

$$\mathbb{Z}(P) = \mathbb{Z}(c_0, c_1, \ldots, c_n) = \{z : P(z) = 0\}$$

depend continuously on $\{c_0, c_1, \ldots, c_n\}$ in the sense of the Hausdorff distance.

**Theorem 5.34.** *The function* $\{c_0, c_1, \ldots, c_n\} \mapsto \mathbb{Z}(P)$ *is continuous. That is, if* $P(z) = c_0 + c_1 z + \cdots + c_n z^n$, *given* $\varepsilon > 0$ *there is a* $\delta = \delta(\varepsilon, P) > 0$ *such that for* $Q(z) = d_0 + d_1 z + \cdots + d_n z^n$ *with* $|d_i - c_i| < \delta$, $i = 0, \ldots, n$, *one has* $d_H(\mathbb{Z}(Q), \mathbb{Z}(P)) < \varepsilon$.

*Proof.* As we know, $\mathbb{Z}(P)$ and $\mathbb{Z}(Q)$ each consist of $n$ points (counted according to multiplicities); let $a_1, \ldots, a_m$ be the zeros of $P$, and $k_1, \ldots, k_m$ their multiplicities, $k_1 + \cdots + k_m = n$. Let $\varepsilon > 0$ be such that the discs $\bar{D}(a_i, \varepsilon)$ are disjoint and $m = \min_{i=1,\ldots,m} \min\{|P(z)| : |z - a_i| = \varepsilon\}$; let $\delta > 0$ such that

$$\delta(1 + |z| + \cdots + |z|^n) < m$$

if $|z - a_i| = \varepsilon$ for some $i$ (for example, if $\alpha = \max |a_i|$, it is enough to impose $\delta(1 + (\alpha + \varepsilon) + \cdots + (\alpha + \varepsilon)^n) < m$). Then, if $|d_i - c_i| < \delta, i = 0, \ldots, n$, Rouché's theorem gives that $Q$ has $k_i$ zeros in $D(a_i, \varepsilon)$; therefore, $\mathbb{Z}(Q) \subset \bigcup_{i=1}^m D(a_i, \varepsilon)$ and $d_H(\mathbb{Z}(Q), \mathbb{Z}(P)) < \varepsilon$. $\qquad\square$

In particular, every well-defined and continuous function of the set of zeros is a continuous function of the coefficients, for example, the functions $\min\{|z| : z \in \mathbb{Z}(P)\}$ or $\max\{|z| : z \in \mathbb{Z}(P)\}$.

If $g$ is an entire function, it is known by Theorem 5.25 that

$$\sum_{\alpha \in \mathbb{Z}(P)} g(\alpha) = \sum_{i=1}^m k_i g(a_i) = \frac{1}{2\pi i} \int_{C(0,R)} g(z) \frac{P'(z)}{P(z)} \, dz,$$

for $R$ big enough, supposing that $a_1, \ldots, a_m$ are the zeros of $P$, with multiplicities $k_1, \ldots, k_m$. Hence $\sum_{\alpha \in Z(P)} g(\alpha)$ is a continuous function of the zeros, and therefore of $P$. The continuity of $\sum_{z \in Z(P)} g(\alpha)$ with respect to $P$ in this case is also a consequence of the previous formula. Observe that writing

$$P(z) = c_0 + c_1 z + \cdots + c_n z^n = c_n \prod_{j=1}^{n} (z - \alpha_j),$$

where now $\alpha_1, \alpha_2, \ldots, \alpha_n$ are the zeros of $P$ counted according to multiplicities, and equating coefficients one obtains

$$c_0 = c_n(-1)^n \prod_{j=1}^{n} \alpha_j = c_n(-1)^n \prod_{i=1}^{m} a_i^{k_i} = (-1)^n c_n \prod_{\alpha \in Z(P)} \alpha,$$

$$c_1 = c_n \sum_{j=1}^{n} (-1)^{n-1} \prod_{l \neq j} \alpha_l = (-1)^{n-1} c_n \sum_{i=1}^{m} k_i a_i^{k_i - 1} \prod_{l \neq i} a_l^{k_l},$$

$$\vdots$$

$$c_{n-1} = -c_n \sum_{j=1}^{n} \alpha_j = -c_n \sum_{i=1}^{m} k_i a_i.$$

These formulae show that the so-called *symmetric functions* of the zeros of $P$ depend continuously on the coefficients of $P$.


## 5.8  Calculus of real integrals

One of the typical and more important applications of the residue theorem is the calculus of some integrals of functions defined on the real line. There is no unique method to deal with these integrals, and the best way to check the possibilities of the calculus of residues is considering several kinds of examples.


**A)**   First consider an integral of the form

$$I = \int_0^{2\pi} R(\cos \theta, \sin \theta) \, d\theta,$$

where $R$ is a rational function of two real variables $x$, $y$, that is, $R(x, y) = \frac{P(x,y)}{Q(x,y)}$ with $P$ and $Q$ polynomials. Assume $R$ is continuous on the unit circle $\mathbb{T}$. The value of $I$ is $2\pi$ times the mean value of $R$ on $\mathbb{T}$. If $z = e^{i\theta}$, one has

$$\cos\theta = \frac{1}{2}(e^{i\theta} + e^{-i\theta}) = \frac{1}{2}\left(z + \frac{1}{z}\right),$$

$$\sin\theta = \frac{1}{2i}(e^{i\theta} - e^{-i\theta}) = \frac{1}{2i}\left(z - \frac{1}{z}\right)$$

and $dz = i e^{i\theta}\, d\theta$. Therefore,

$$I = \frac{1}{i}\int_{\mathbb{T}} R\left(\frac{1}{2}\left(z + \frac{1}{z}\right), \frac{1}{2i}\left(z - \frac{1}{z}\right)\right)\frac{dz}{z}.$$

Now the residue theorem yields

$$I = 2\pi \sum_{|\alpha|<1} \operatorname{Res}\left(\frac{1}{z}R\left(\frac{1}{2}\left(z + \frac{1}{z}\right), \frac{1}{2i}\left(z - \frac{1}{z}\right)\right), \alpha\right).$$

**Example 5.35.**

$$I = \int_0^{2\pi} \frac{d\theta}{a + \cos\theta}, \quad a > 1.$$

In this case,

$$I = \frac{1}{i}\int_{\mathbb{T}} \frac{1}{a + \frac{1}{2}\left(z + \frac{1}{z}\right)}\frac{dz}{z} = \frac{2}{i}\int_{\mathbb{T}} \frac{dz}{z^2 + 2az + 1}.$$

The zeros of $z^2 + 2az + 1$ are $-a \pm \sqrt{a^2 - 1}$, and only $-a + \sqrt{a^2 - 1}$ belongs to $\mathbb{D}$. One then obtains

$$I = 4\pi \operatorname{Res}\left(\frac{1}{z^2 + 2az + 1}, -a + \sqrt{a^2 - 1}\right) = \frac{2\pi}{\sqrt{a^2 - 1}}. \qquad \square$$

**B)**  Now consider integrals of the form

$$I = \int_{-\infty}^{+\infty} f(x)\, dx,$$

where $f$ is a meromorphic function on a neighborhood of the half plane $\{z : \operatorname{Im} z \geq 0\}$, with a finite number of poles in $\{z : \operatorname{Im} z \geq 0\}$, none of them being real. Assume that $|f(z)| = o\left(\frac{1}{|z|}\right)$ when $|z| \to \infty$, that is, $\lim_{|z|\to\infty} |z||f(z)| = 0$, ($\operatorname{Im} z > 0$).

Consider the path $\gamma_r$ formed by the segment going from $-r$ to $r$, $r > 0$, followed by the semicircle $C_r : re^{it}$, $0 \leq t \leq \pi$. Suppose that $r$ is big enough so that $\gamma_r$ contains in its interior all the poles of $f$ located on $\{z : \operatorname{Im} z > 0\}$ (Figure 5.3).

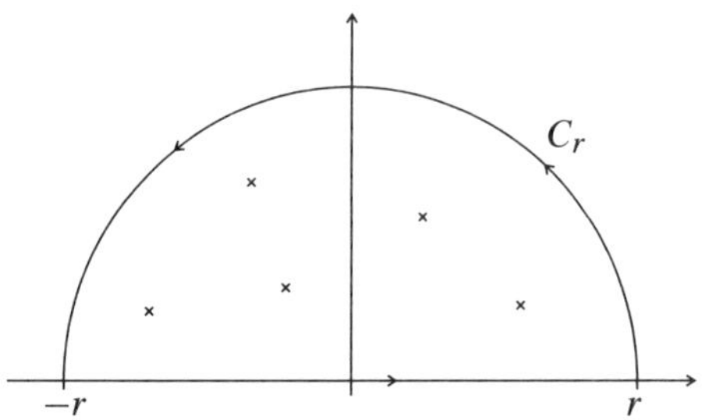

Figure 5.3

By the residue theorem,

$$\int_{-r}^{r} f(x)\, dx + \int_{C_r} f(z)\, dz = 2\pi i \sum_{\operatorname{Im}\alpha > 0} \operatorname{Res}(f(z), \alpha).$$

Now, the quantity $\left|\int_{C_r} f(z)\, dz\right| \le \int_{C_r} |f(z)||dz| \le \pi r \sup_{|z|=r} |f(z)|$ tends to zero when $r \to +\infty$, by the hypothesis on $f$. So, $\lim_{r\to\infty} \int_{-r}^{r} f(x)\, dx$ exists and its value is

$$\lim_{r\to\infty} \int_{-r}^{r} f(x)\, dx = 2\pi i \sum_{\operatorname{Im}\alpha > 0} \operatorname{Res}(f(z), \alpha).$$

Typically, $|f(z)|$ will have a bound, for $|z| \to \infty$, which will imply $|f(z)| = o\left(\frac{1}{|z|}\right)$ and the absolute convergence of the integral $I$. This is the case, for example, if $f$ is a rational function, $R = \frac{P}{Q}$ without real poles and $\deg Q \ge \deg P + 2$, because then $|R(z)| = O(|z|^{-2})$.

**Example 5.36.**

$$I = \int_{0}^{\infty} \frac{dx}{(1 + x^2)^2} = \frac{1}{2} \int_{-\infty}^{+\infty} \frac{dx}{(1 + x^2)^2}.$$

The rational function $R(z) = (1 + z^2)^{-2}$ has double poles at $z = \pm i$. It is enough to consider $z = i$ a point at which the residue is

$$\operatorname{Res}(R(z), i) = \operatorname{Res}\left(\frac{1}{(z - i)^2}\frac{1}{(z + i)^2}, i\right) = \frac{d}{dz}\frac{1}{(z + i)^2}\bigg|_{z=i} = \frac{1}{4i}.$$

Therefore, $I = \frac{1}{2} \cdot 2\pi i \cdot \frac{1}{4i} = \frac{\pi}{4}$. $\qquad\qquad\square$

**Example 5.37.**

$$I = \int_{-\infty}^{+\infty} \frac{e^{ix}}{(1+x^2)^2} \, dx = 2\pi i \operatorname{Res}\left(\frac{e^{iz}}{(1+z^2)^2}, i\right)$$

$$= 2\pi i \frac{d}{dz} \frac{e^{iz}}{(z+i)^2}\bigg|_{z=i} = 2\pi i \left(-\frac{i}{2e}\right) = \frac{\pi}{e}$$

(the function $f(z) = \frac{e^{iz}}{(1+z^2)^2}$ satisfies $|f(z)| = \frac{e^{-\operatorname{Im} z}}{|1+z^2|^2} \leq c\frac{1}{|z|^4}$, if $|z|$ is big enough).

Taking real parts, it turns out that

$$\operatorname{Re} I = \int_{-\infty}^{+\infty} \frac{\cos x}{(1+x^2)^2} \, dx = \frac{\pi}{e}. \qquad \square$$

Consider now the function $f(z) = \operatorname{Log}(z + ia)R(z)$, where $R$ is a rational function as before ($R = P/Q$ without real poles and $\deg P \geq \deg P + 2$), $a > 0$ and Log denotes the principal branch of the logarithm. Since $|\operatorname{Log}(z + ia)|^2 = (\operatorname{Log}|z + ia|)^2 + (\operatorname{Arg}(z + ia))^2$, it keeps being true that

$$\lim_{|z|\to\infty} |z||f(z)| = 0.$$

Therefore,

$$\int_{-\infty}^{\infty} f(x) \, dx = \int_{-\infty}^{+\infty} \operatorname{Log}(x + ia)R(x) \, dx = 2\pi i \sum_{\operatorname{Im}\alpha>0} \operatorname{Res}(f(z), \alpha).$$

**Example 5.38.**

$$I = \int_{-\infty}^{+\infty} \frac{\operatorname{Log}(x + ia)}{1 + x^2} \, dx = 2\pi i \operatorname{Res}\left(\frac{\operatorname{Log}(z + ia)}{1 + z^2}, i\right)$$

$$= 2\pi i \frac{\operatorname{Log}(z + ia)}{z + i}\bigg|_{z=i} = 2\pi i \left(\frac{\operatorname{Log}[(1 + a)i]}{2i}\right)$$

$$= \pi\left(\operatorname{Log}(1 + a) + i \operatorname{Arg}[(1 + a)i]\right) = \pi\left(\operatorname{Log}(1 + a) + i\frac{\pi}{2}\right).$$

Equating real and imaginary parts, it turns out that

$$\int_{-\infty}^{+\infty} \frac{\operatorname{Log}(x^2 + a^2)}{1 + x^2} \, dx = \pi \operatorname{Log}(1 + a), \qquad \int_{-\infty}^{+\infty} \frac{\operatorname{Arg}(x + ia)}{1 + x^2} \, dx = \frac{\pi^2}{2}.$$

Since $\operatorname{Arg}(x + ia) + \operatorname{Arg}(-x + ia) = \pi$, if $x > 0$, the last equality is equivalent to

$$\int_0^{\infty} \frac{dx}{1 + x^2} = \frac{\pi}{2}. \qquad \square$$

**C)**   In all the examples in item B), the hypothesis about the decrease of the function $f$, $\lim_{|z|\to\infty} |z|\,|f(z)| = 0$, implies that the integrals considered are absolutely convergent, that is, $\int_{-\infty}^{+\infty} |f(x)|\,dx < +\infty$. In turn it gives the existence of the following limit:

$$\lim_{\substack{a\to-\infty\\b\to+\infty}} \int_a^b f(x)\,dx = \int_{-\infty}^{+\infty} f(x).$$

But, there are integrals which are convergent but not absolutely convergent. For example, if $g(x)$ is real and decreasing for $x > 0$ and $\lim_{x\to+\infty} g(x) = 0$, the integral

$$I = \int_0^\infty g(x)e^{ix}\,dx$$

is convergent. Actually, integrating by parts (suppose $g$ differentiable), the partial integral

$$\int_a^b g(x)e^{ix}\,dx = -i\,ge^{ix}\Big|_a^b + \int_a^b i e^{ix} g'(x)\,dx$$

$$= i(g(a)e^{ia} - g(b)e^{ib}) + i\int_a^b g'(x)e^{ix}\,dx$$

has modulus smaller than or equal to $g(a)+g(b)+\int_a^b |g'(x)|\,dx = g(a)+g(b)-\int_a^b g'(x)\,dx = 2g(a)$. Now by Cauchy's criterion, $\int_0^\infty g(x)e^{ix}$ is convergent (this fact is analogous to the convergence of alternating series $\sum(-1)^n a_n$ with $a_n \searrow 0$). Hence, the integrals $\int_{-\infty}^\infty \frac{\sin x}{x}\,dx$, $\int_{-\infty}^\infty \frac{\cos x}{1+|x|}\,dx$ are convergent, but it is not difficult to see that they are not absolutely convergent.

   For integrals extended over all $\mathbb{R}$, there is still a third concept of convergence, more general, which is given by the existence of *Cauchy's principal value*, defined as

$$\lim_{r\to+\infty} \int_{-r}^r f(x)\,dx = \text{p.v.} \int_{-\infty}^\infty f(x)\,dx.$$

It is clear that for an integral, $\int_{-\infty}^\infty f(x)\,dx$, the following implications hold:

$$\text{absolutely convergent} \implies \text{convergent} \implies \text{existence of p.v.}$$

   In this subsection it will be seen that the residue theorem allows to prove the existence and to compute Cauchy's principal value of some integrals, not necessarily absolutely convergent.

**Proposition 5.39.** *Suppose that $f$ is meromorphic on a neighborhood of $\{z: \operatorname{Im} z \geq 0\}$, with a finite number of poles in $\{z: \operatorname{Im} \geq 0\}$, none of them being* *real, and that* $\lim_{|z|\to\infty} f(z) = 0$. *Then p.v.* $\int_{-\infty}^{\infty} f(x)e^{ix}\,dx$ *exists and*

$$p.v. \int_{-\infty}^{\infty} f(x)e^{ix}\,dx = 2\pi i \sum_{\alpha} \mathrm{Res}(f(z)e^{iz}, \alpha),$$

*where the sum is taken over the poles of* $f$ *located at* $\{z: \mathrm{Im}\, z > 0\}$.

*Proof.* Considering the same path $\gamma_r$ as in item B), it is enough to show that the contribution of the half circle $C_r$ tends to zero, when $r \to \infty$. Now,

$$\left| \int_{C_r} f(z)e^{iz}\,dz \right| = \left| \int_0^\pi f(re^{i\theta})e^{ire^{i\theta}} ire^{i\theta}\,d\theta \right|$$

$$\leq \int_0^\pi |f(re^{i\theta})|e^{-r\sin\theta}r\,d\theta$$

$$\leq \left( \max_{|z|=r} |f(z)| \right) \int_0^\pi e^{-r\sin\theta}r\,d\theta.$$

But the last integral is $2\int_0^{\frac{\pi}{2}} e^{-r\sin\theta}r\,d\theta$ and using inequalities $\frac{2}{\pi} \leq \frac{\sin\theta}{\theta} \leq 1$, if $0 < \theta < \pi/2$, it turns out that it is bounded by $2\int_0^{\frac{\pi}{2}} e^{-\frac{2}{\pi}r\theta}r\,d\theta \leq 2\int_0^\infty e^{-\frac{2}{\pi}r\theta}r\,d\theta = \pi$. $\qquad\square$

**Example 5.40.**

$$I = \int_0^\infty \frac{x\sin x}{1+x^2}\,dx.$$

In this case the principal value

$$\lim_{r\to\infty} \int_0^r \frac{x\sin x}{1+x^2}\,dx = \frac{1}{2}\lim_{r\to\infty} \int_{-r}^r \frac{x\sin x}{1+x^2}\,dx = \frac{1}{2}\,\mathrm{Im}\lim_{r\to\infty} \int_{-r}^r \frac{xe^{ix}}{1+x^2}\,dx.$$

exists. Actually, here $f(z) = \frac{z}{1+z^2}$ has a pole at the point $z = i$ with residue $\frac{z}{z+i}\big|_{z=i} = \frac{1}{2}$ and it turns out that $\lim_{r\to\infty} \int_0^r \frac{x\sin x}{1+x^2}\,dx = \frac{1}{2}\,\mathrm{Im}\,\pi i = \frac{\pi}{2}$. $\qquad\square$

**D)** The notion of principal value may be formulated also for the so-called improper integrals of second kind (the integrand $f$ tends to infinity at a finite point). Suppose, for example, that $f$ is a continuous function on the interval $(a - R, a + R)$, $R > 0$, except at the point $a$, and $f$ is not bounded on $(a - R, a + R)$. The *principal value* of $\int_{a-R}^{a+R} f(x)\,dx$ is, by definition,

$$\lim_{\varepsilon\to 0} \left\{ \int_{a-R}^{a-\varepsilon} + \int_{a+\varepsilon}^{a+R} \right\} f(x)\,dx.$$

The (improper) integral $\int_{a-R}^{a+R} f(x)\,dx$ is said to be *convergent* if the integrals $\int_{a-R}^{a}$ and $\int_{a}^{a+R}$ are convergent, that is, if there exist separately the two limits: $\lim_{\varepsilon\to 0}\int_{a-R}^{a-\varepsilon}$ and $\lim_{\varepsilon\to 0}\int_{a+\varepsilon}^{a+R}$. The integral is *absolutely convergent* if $\int_{a-R}^{a+R}|f(x)|\,dx < +\infty$. As before, absolutely convergent $\Rightarrow$ convergent $\Rightarrow$ existence of p.v.

Suppose now that the function $f$ of items B) or C), meromorphic on a neighborhood of $\{z\colon \operatorname{Im} z \geq 0\}$, has a finite number of real simple poles: $a_1,\ldots,a_m \in \mathbb{R}$. In this case, the integral $\int_{-\infty}^{\infty} f(x)\,dx$ cannot be absolutely convergent because close to each point $a_j$ one has $|f(x)| \sim c_j|x-a_j|^{-1}$ with $c_j$ constant, and so

$$\int_{a_j-\alpha}^{a_j+\alpha} |f(x)|\,dx = +\infty, \quad \text{if } \alpha > 0.$$

However, if the pole $a_j$ is of order 1, there exists p.v. $\int_{a_j-\alpha}^{a_j+\alpha} f(x)\,dx$ because $f(x) = \frac{c_j}{x-a_j} + O(1)$ and, trivially, $\left\{\int_{a_j-\alpha}^{a_j-\varepsilon} + \int_{a_j+\varepsilon}^{a_j+\alpha}\right\}\frac{dx}{x-a_j} = 0$, $(x-a_j)^{-1}$ being an odd function with respect to the point $a_j$.

In order to compute the p.v. $\int_{-\infty}^{\infty} f(x)\,dx$, modify the circuit $\gamma_r$, which has been used in B) and C), in such a way that it travels, in the negative sense, the semicircles centered at $a_j$ with radius $\varepsilon > 0$ which go from $a_j - \varepsilon$ to $a_j + \varepsilon$, for $j = 1,\ldots,m$ (Figure 5.4). If $\gamma_j(\varepsilon)$ are these semicircles, then one has

$$\int_{\gamma_j(\varepsilon)} f(z)\,dz = \int_{\gamma_j(\varepsilon)} \frac{c_j}{z-a_j}\,dz + \int_{\gamma_j(\varepsilon)} O(1)\,dz$$

and

$$\lim_{\varepsilon\to 0}\int_{\gamma_j(\varepsilon)} f(z)\,dz = \lim_{\varepsilon\to 0}\int_{\gamma_j(\varepsilon)} \frac{c_j}{z-a_j}\,dz$$

$$= -c_j \lim_{\varepsilon\to 0}\int_0^{\pi} \frac{\varepsilon i e^{i\theta}\,d\theta}{\varepsilon e^{i\theta}} = -i\pi c_j = -i\pi \operatorname{Res}(f,a_j).$$

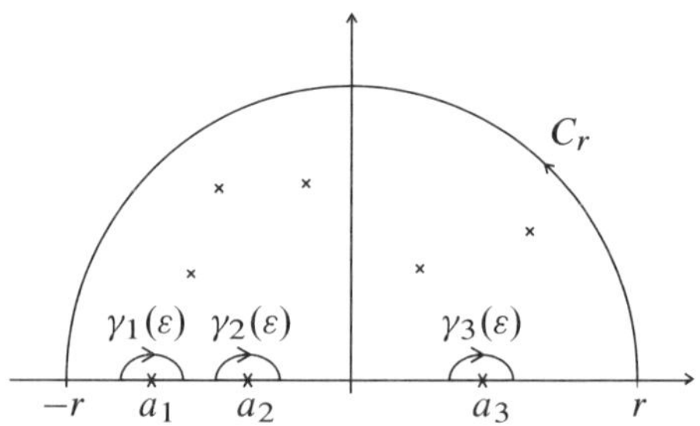


Figure 5.4

Supposing that $a_1 < \cdots < a_m$, that $r$ is big enough and $\varepsilon$ small enough and on account of $\int_{c_r} f(z)\,dz \to 0$, $r \to +\infty$ (because it is supposed either $|f(z)| = O\left(\frac{1}{|z|}\right)$, or $\lim_{|z|\to\infty} f(z) = 0$ and one considers the integrand $f(x)e^{ix}$) the residue theorem leads to

$$\text{p.v.} \int_{-\infty}^{\infty} f(x)\,dx$$

$$= \lim_{\substack{r\to+\infty \\ \varepsilon\to 0}} \left\{ \int_{-r}^{a_1-\varepsilon} + \int_{a_1+\varepsilon}^{a_2-\varepsilon} + \int_{a_2+\varepsilon}^{a_3-\varepsilon} + \cdots + \int_{a_{n-1}+\varepsilon}^{a_n-\varepsilon} \mid \int_{a_n+\varepsilon}^{r} \right\} f(x)\,dx$$

$$= 2\pi i \sum_{\operatorname{Im} z>0} \operatorname{Res}(f,\alpha) + \pi i \sum_{j=1}^{m} \operatorname{Res}(f,a_j).$$

**Example 5.41.**

$$\text{p.v.} \int_{-\infty}^{+\infty} \frac{e^{ix}}{x}\,dx = \pi i \operatorname{Res}\left(\frac{e^{iz}}{z}, 0\right) = \pi i.$$

Equating imaginary parts, one finds p.v. $\int_{-\infty}^{+\infty} \frac{\sin x}{x}\,dx = \pi$. Since the function $\frac{\sin x}{x}$ is continuous at the point 0, one may remove p.v. and the integral $\int_{-\infty}^{+\infty} \frac{\sin x}{x}\,dx$, which is convergent but not absolutely convergent, equals $\pi$. $\qquad\square$

The previous process lets one also compute integrals of the form

$$\int_{-\infty}^{+\infty} f(x) \begin{cases} \cos^n x \\ \sin^n x \end{cases} dx,$$

expressing $\sin^n x$, $\cos^n x$ as a linear combination of the functions $\cos nx$, $\sin nx$ and writing $\int_{-\infty}^{+\infty} f(x)e^{imx}\,dx$ as $\frac{1}{m} \int_{-\infty}^{+\infty} f(\frac{x}{m})e^{ix}\,dx$, after making a change of variables.

**Example 5.42.** The integral

$$I = \int_{-\infty}^{+\infty} \frac{\sin^3 x}{x}\,dx$$

is convergent (not absolutely) and one has

$$\sin^3 x = \left(\frac{1}{2i}(e^{ix} - e^{-ix})\right)^3 = \frac{1}{8i^3}(e^{3ix} - 3e^{ix} + 3e^{-ix} - e^{-3ix})$$

$$= -\frac{1}{8i}(2i \sin 3x - 6i \sin x)$$

$$= \frac{3}{4} \sin x - \frac{1}{4} \sin 3x.$$

The change $3x = t$ leads to $\int_{-\infty}^{+\infty} \frac{\sin 3x}{x}\,dx = \int_{-\infty}^{+\infty} \frac{\sin t}{t}\,dt = \pi$, and we obtain $I = \frac{\pi}{2}$. $\qquad\square$

**E)**  Consider now integrals of the form

$$I = \int_0^\infty \frac{R(x)}{x^\alpha}\,dx, \quad 0 < \alpha < 1,$$

where $R$ is a rational function, $R = \frac{P}{Q}$ with $\deg Q \geq \deg P + 1$, without poles on $[0, +\infty)$. This fact guarantees the absolute convergence of $I$.

Apply the residue theorem to a branch of $\frac{R(z)}{z^\alpha}$ on the domain $\mathbb{C} \setminus [0, \infty)$: the one given by $\arg z \in (0, 2\pi)$. For $r$ big and $\delta$ small, the path $\gamma_{r,\delta}$ of integration is the one in Figure 5.5.

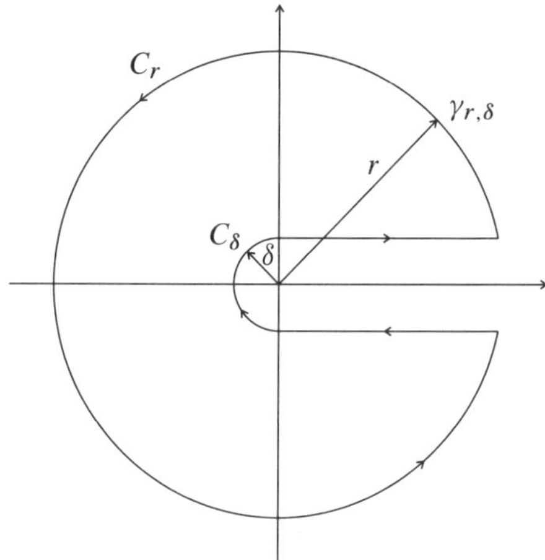


Figure 5.5

The two horizontal lines give

$$\int_0^r \frac{R(x + i\delta)}{(x + i\delta)^\alpha}\,dx - \int_0^r \frac{R(x - i\delta)}{(x - i\delta)^\alpha}\,dx$$

$$= \int_0^r \left( \frac{R(x) + O(\delta)}{|x + i\delta|^\alpha e^{i\alpha \arg(x+i\delta)}} - \frac{R(x) + O(\delta)}{|x - i\delta|^\alpha e^{i\alpha \arg(x-i\delta)}} \right) dx.$$

When $\delta \to 0$ the limit of the expression above is $\int_0^r \frac{R(x)}{x^\alpha}(1 - e^{-2\pi i\alpha})\,dx$. For the limit when $\delta \to 0$ of the integral over the semicircle of radius $\delta$ we get

$$\left| \int_{C_\delta} \frac{R(z)}{z^\alpha}\,dz \right| \leq k\delta \sup_{|z|=\delta} \frac{1}{|z^\alpha|} = k\delta^{1-\alpha} \xrightarrow[\delta \to 0]{} 0, \quad k \text{ constant.}$$

In the arc of the circle with radius $r$ one has

$$\left| \int_{C_r} \frac{R(z)}{z^\alpha}\,dz \right| \leq r^{1-\alpha} \sup_{|z|=r} |R(z)| = r^{1-\alpha} O\!\left(\frac{1}{r}\right) \longrightarrow 0 \quad \text{if } r \longrightarrow \infty.$$

All together, plus the residue theorem, give

$$(1 - e^{-2\pi i\alpha})I = 2\pi i \sum_{p \notin [0,+\infty)} \mathrm{Res}\left(\frac{R(z)}{z^\alpha}, p\right).$$

**Example 5.43.**

$$I = \int_0^\infty \frac{x^{\beta-1}}{1+x}\,dx, \quad 0 < \beta < 1.$$

Here one has $\alpha = 1 - \beta$, $R(z) = (1+z)^{-1}$. Since $\mathrm{Res}\left(\frac{z^{\beta-1}}{1+z}, -1\right) = (-1)^{\beta-1} = e^{i\pi(\beta-1)}$ it follows that

$$I = 2\pi i \frac{e^{i\pi(\beta-1)}}{1 - e^{-2\pi i(1-\beta)}} = \frac{2\pi i}{e^{i\pi(1-\beta)} - e^{-i\pi(1-\beta)}} = \frac{\pi}{\sin \pi(1-\beta)}. \qquad \square$$

This method would also work if one replaces $R$ by a function $f$, meromorphic on $\mathbb{C}$ with a finite number of poles, none of them in $[0, \infty)$ and satisfying $\lim_{|z|\to\infty} |z|^{1-\alpha}|f(z)| = 0$. Since this implies that $f$ has a removable singularity at the point $\infty$ (because $\lim_{|z|\to\infty} f(z) = 0$) $f$ is, indeed, a rational function, by Proposition 5.24.

**F)** Now integrals of the type $\int_0^\infty R(x)\,\mathrm{Log}\,x\,dx$ and $\int_0^\infty R(x)\,dx$ will be computed. Here the function $R = \frac{P}{Q}$ is rational with $\deg Q \geq \deg P + 2$, without poles in $[0, \infty)$; so the integrals are absolutely convergent. Suppose also that $R$ takes real values on $\mathbb{R}$. We use the same path as in item E) but now integrating the function $R(z)\log^2 z$, where the branch of $\log z$ is the one given on $\mathbb{C} \setminus [0, \infty)$ by $\arg z \in (0, 2\pi)$. With an argument similar to the one in E), one finds

$$\int_0^\infty R(x)\,\mathrm{Log}^2\,x\,dx - \int_0^\infty R(x)(\mathrm{Log}\,x + 2\pi i)^2\,dx = 2\pi i \sum_{\alpha \neq 0} \mathrm{Res}(R(z)\log^2 z, \alpha).$$

Equating real an imaginary parts, we get

$$\int_0^\infty R(x)\,\mathrm{Log}\,x\,dx = -\frac{1}{2}\,\mathrm{Re}\sum_{\alpha \neq 0} \mathrm{Res}(R(z)\log^2 z, \alpha),$$

$$\int_0^\infty R(x)\,dx = -\frac{1}{2\pi}\,\mathrm{Im}\sum_{\alpha \neq 0} \mathrm{Res}(R(z)\log^2 z, \alpha).$$

**Example 5.44.**

$$I = \int_0^\infty \frac{\mathrm{Log}\,x}{(x^2 + 1)^2}.$$

There is a double pole at $z = \pm i$, and the corresponding residues are

$$\operatorname{Res}\left(\frac{\log^2 z}{(z^2 + 1)^2}, i\right) = \left(\frac{\log^2 z}{(z + i)^2}\right)'(i) = \frac{2\log i}{i(2i)^2} - \frac{2(\log i)^2}{(2i)^3}$$

$$= \frac{2i\frac{\pi}{2}}{-4i} + \frac{2\left(i\frac{\pi}{2}\right)^2}{8i} = -\frac{\pi}{4} + i\frac{\pi^2}{16},$$

and similarly,

$$\operatorname{Res}\left(\frac{\log^2 z}{(z^2 + 1)^2}, -i\right) = \frac{3\pi}{4} - \frac{9}{16}i\pi^2.$$

Then it turns out that

$$\int_0^\infty \frac{\operatorname{Log} x}{(x^2 + 1)^2}\, dx = -\frac{\pi}{4}; \quad \int_0^\infty \frac{dx}{(x^2 + 1)^2} = \frac{\pi}{4}. \qquad \Box$$

Since the factor $\operatorname{Log} x$ appearing in the integrand of $\int_0^\infty R(x)\operatorname{Log} x\, dx$, vanishes at the point $x = 1$, one may allow $R$ to have a simple pole at the point 1 and the integral will be still absolutely convergent. In order to see the influence of this fact on the previous computation, replace the part of the integral given by

$$-\int_0^r R(x - i\delta)\log^2(x - i\delta)\, dx$$

by the integrals

$$-\int_0^{1-c(\delta)} R(x - i\delta)\log^2(x - i\delta)\, dx - \int_{\beta_\delta} R(z)\log^2 z\, dz$$

$$-\int_{1+c(\delta)}^r R(x - i\delta)\log^2(x - i\delta)\, dx.$$

The points $(1 - c(\delta), -\delta)$, $(1 + c(\delta), -\delta)$ are the intersection points of the line $y = -\delta$ with the circle centered at 1 and with radius $2\delta$, and $\beta_\delta$ is the arc of this circle determined by these two points (Figure 5.6).

Taking the limit when $\delta \to 0$, the imaginary parts of the first and third integrals converge, both together, to

$$-\operatorname{Im}\int_0^r R(x)(\operatorname{Log} x + 2\pi i)^2\, dx = -4\pi\int_0^r R(x)\operatorname{Log} x\, dx.$$

The integral in the middle term is $\int_{\beta_\delta} R(z)(2\pi i + O(\delta))^2\, dz$ which, for $\delta \to 0$, converges to $\lim_{\delta \to 0}\int_{\beta_\delta} R(z)(2\pi i)^2\, dz = -4\pi^2\pi i\operatorname{Res}(R, 1)$. Hence

$$\int_0^\infty R(x)\operatorname{Log} x\, dx = \pi^2\operatorname{Re}\left(\operatorname{Res}(R, 1)\right) - \frac{1}{2}\operatorname{Re}\sum_\alpha \operatorname{Res}(R(z)(\log z)^2, \alpha).$$

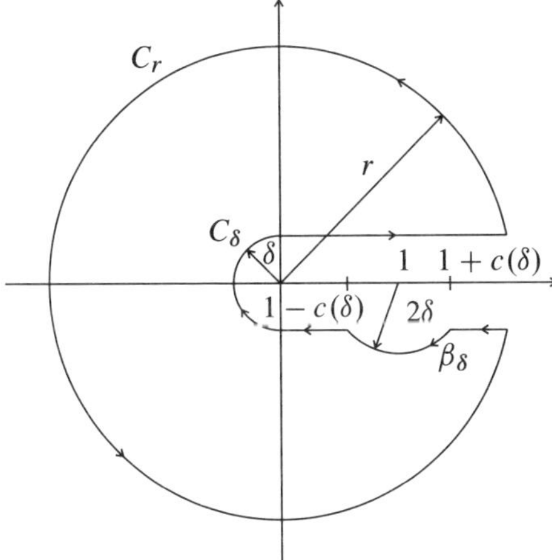

Figure 5.6

Using real parts one obtains now

$$\mathrm{p.v.} \int_0^\infty R(x)\,dx = -\frac{1}{2\pi}\,\mathrm{Im}\sum_\alpha \mathrm{Res}(R(z)(\log z)^2,\alpha) + \pi\,\mathrm{Im}\,\mathrm{Res}(R,1),$$

where $\mathrm{p.v.} \int_0^\infty R(x)\,dx = \lim_{\delta\to 0}\left(\int_0^{1-\delta} + \int_{1+\delta}^\infty\right)R(x)\,dx$ (in this case the integral $\int_0^\infty R(x)\,dx$ is not absolutely convergent). On the other hand, since $R$ is real on $\mathbb{R}$, the quantity $\mathrm{Res}(R,1) = \lim_{x\to 1} R(x)(x-1)$ is also real.

**Example 5.45.**

$$I = \int_0^{+\infty} \frac{\log x}{x^3 - 1}\,dx.$$

The residue of $R(z) = \frac{1}{z^3-1}$ at the point 1 is $\lim_{z\to 1} R(z)(z-1) = \frac{1}{3}$. At the other two poles $e^{i\frac{2\pi}{3}}, e^{i\frac{4\pi}{3}}$ the residues of $R(z)(\log z)^2$ are

$$\lim_{z\to e^{i\frac{2\pi}{3}}} \frac{z - e^{i\frac{2\pi}{3}}}{z^3 - 1}(\log z)^2 = \frac{\left(i\frac{2\pi}{3}\right)^2}{3\left(e^{i\frac{2\pi}{3}}\right)^2} = -\frac{4\pi^2}{27}e^{-i\frac{4\pi}{3}},$$

$$\lim_{z\to e^{i\frac{4\pi}{3}}} \frac{z - e^{i\frac{4\pi}{3}}}{z^3 - 1}(\log z)^2 = \frac{\left(i\frac{4\pi}{3}\right)^2}{3\left(e^{i\frac{4\pi}{3}}\right)^2} = -\frac{16\pi^2}{27}e^{-i\frac{2\pi}{3}}.$$

It turns out, therefore, that

$$I = \frac{\pi^2}{3} + \frac{1}{2}\frac{\pi^2}{27}\left(4\cos\frac{4\pi}{3} + 16\cos\frac{2\pi}{3}\right) = \frac{\pi^2}{3} - \frac{5\pi^2}{27} = \frac{4\pi^2}{27}. \qquad \square$$

**G)**   Let us compute, finally, some integrals which are often used in real or complex analysis

Start with the integral $I = \int_0^{2\pi} \mathrm{Log}\,|1 - re^{i\theta}|\,d\theta$, $0 < r < 1$.

Since $I = \mathrm{Re}\,\int_0^{2\pi} \mathrm{Log}(1 - re^{i\theta})\,d\theta$, one has

$$I = \mathrm{Re}\,\frac{1}{i}\int_{C(0,r)} \mathrm{Log}(1 - z)\frac{dz}{z} = 0$$

because the function $\frac{\mathrm{Log}(1-z)}{z}$ is holomorphic on the unit disc $\mathbb{D}$.

For $r = 1$ the integral $\int_0^{2\pi} \mathrm{Log}\,|1 - e^{i\theta}|\,d\theta$ is still absolutely convergent since

$$\begin{aligned}
\mathrm{Log}\,|1 - e^{i\theta}| &= \frac{1}{2}\,\mathrm{Log}\,|1 - e^{i\theta}|^2 \\
&= \frac{1}{2}\,\mathrm{Log}((1 - \cos\theta)^2 + \sin^2\theta) \\
&= \frac{1}{2}\,\mathrm{Log}(2(1 - \cos\theta))
\end{aligned}$$

has the same order as $|\mathrm{Log}\,\theta|$ for $\theta$ near 0 and the same one as $|\mathrm{Log}(2\pi - \theta)|$ for $\theta$ near $2\pi$. In order to compute the integral, take the path of Figure 5.7. It follows that

$$i\int_\delta^{2\pi-\delta} \mathrm{Log}(1 - e^{i\theta})\,d\theta - \int_{C_\delta} \frac{1}{z}\,\mathrm{Log}(1 - z)\,dz = 0$$

because $\frac{\mathrm{Log}(1-z)}{z}$ is holomorphic on $\mathbb{C} \setminus [1, \infty)$. Now,

$$\begin{aligned}
\left|\int_{C_\delta} \frac{\mathrm{Log}(1 - z)}{z}\,dz\right| &\le 2\pi\delta \cdot \sup_{\substack{|z-1|=\delta \\ |z|\le 1}} \frac{|\mathrm{Log}\,(1 - z)|}{|z|} \\
&\le \frac{2\pi\delta}{1 - \delta}\left(\mathrm{Log}^2\,|1 - z| + \mathrm{Arg}^2(1 - z)\right)^{\frac{1}{2}} \\
&\le \frac{2\pi\delta}{1 - \delta}((\mathrm{Log}\,\delta)^2 + \pi^2)^{\frac{1}{2}} \longrightarrow 0, \quad \delta \longrightarrow 0.
\end{aligned}$$

Therefore, $\int_0^{2\pi} \mathrm{Log}\,|1 - e^{i\theta}|\,d\theta = \mathrm{Re}\,\int_0^{2\pi} \mathrm{Log}(1 - e^{i\theta})\,d\theta = 0$.

When $r > 1$, the integral $I$ may be easily computed since

$$I = \int_0^{2\pi} \mathrm{Log}\,|1 - re^{i\theta}|\,d\theta = \int_0^{2\pi} \mathrm{Log}\,r\left|\frac{1}{r}e^{-i\theta} - 1\right|d\theta$$

$$= \int_0^{2\pi}\left(\mathrm{Log}\,r + \mathrm{Log}\left|\frac{1}{r}e^{i\theta} - 1\right|\right)d\theta = 2\pi\,\mathrm{Log}\,r.$$

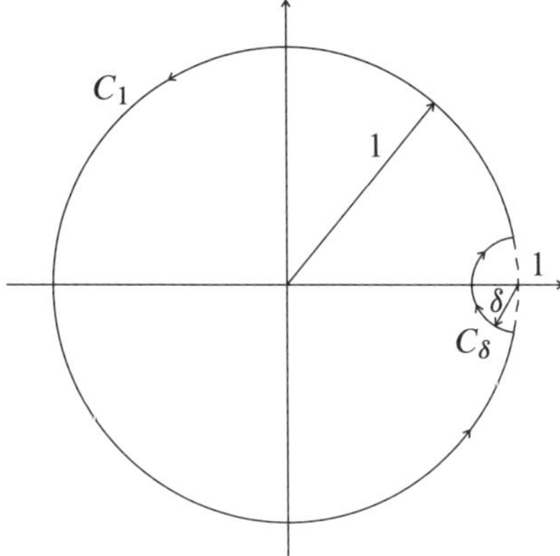

Figure 5.7

Summarizing, the following formula has been shown:

$$\int_0^{2\pi} \mathrm{Log}\,|1 - re^{i\theta}|\,d\theta = 2\pi\,\mathrm{Log}^+ r, \quad r > 0,$$

where $\mathrm{Log}^+ r$ is the positive part of $\mathrm{Log}\,r$.

Finally, combining the equality

$$\int_0^{\pi} \mathrm{Log}\,|1 - e^{2i\theta}|\,d\theta = \frac{1}{2}\int_0^{2\pi} \mathrm{Log}\,|1 - e^{it}|\,dt = 0$$

with the relation $|1 - e^{2i\theta}| = |e^{-i\theta} - e^{i\theta}| = 2|\sin\theta|$, one finds

$$\int_0^{\pi} \mathrm{Log}\,\sin\theta\,d\theta = -\pi\log 2.$$

Notice that throughout this section proper or improper integrals have been computed, without using antidifferentiation, which, thanks to the fundamental theorem of calculus, is the traditional process for computing definite integrals.

## 5.9 Exercises

**1.** Find the Laurent series of the function $f(z) = \frac{z^2}{z-1}\,e^{1/z}$ on each annulus: $C = \{z \in \mathbb{C}: 0 < |z| < 1\}$ and $C^* = \{z \in \mathbb{C}: |z| > 1\}$.

**2.** Let $f$ be a holomorphic function on the punctured disc $D'(a, r)$, $a \in \mathbb{C}$, $r > 0$, which takes all its values in the upper half plane $\Pi^+ = \{z: \mathrm{Im}\,z > 0\}$. Show that $f$ has a removable singularity at the point $a$.

3. Suppose that $f \in H(D'(a,r))$, $a \in \mathbb{C}$, $r > 0$, has a pole at the point $a$ and $g$ is an entire function such that the function $g(f(z))$, $z \in D'(a,r)$, has also a pole at the point $a$. Show that $g$ must be a polynomial.

4. Show that the Laurent series of the function $\frac{1}{e^z - 1}$ around the origin is of the form
$$\frac{1}{z} - \frac{1}{2} + \sum_{n=1}^{\infty} (-1)^{n-1} \frac{B_n}{(2n)!} z^{2n-1}.$$

The numbers $B_n$, $n = 1, 2, \ldots$ are *Bernoulli numbers*. Prove the recurrence formula
$$\sum_{k=1}^{n} \frac{(-1)^{k-1} B_k}{(2k)!(2n - 2k + 1)!} = \frac{1}{2(2n)!} - \frac{1}{(2n + 1)!}$$

and compute $B_n$ for $n = 1, 2, 3$.

5. Show that the Laurent series of the function $e^{\frac{t(z-1)}{2z}}$, for $t \in \mathbb{R}$, around the origin is of the form
$$\sum_{n=-\infty}^{\infty} J_n(t) z^n$$

with $J_n(t) = \frac{1}{2\pi} \int_0^{2\pi} \cos(n\theta - t \sin \theta) d\theta$, $n \in \mathbb{Z}$. The functions $J_n(t)$ are called *Bessel functions*.

6. Suppose that the series $\sum_{n=0}^{\infty} c_n z^n$ has radius of convergence 1 and let $f$ be the holomorphic function that it defines on $\mathbb{D}$. Show that $f$ has a meromorphic extension to a neighborhood of $\overline{\mathbb{D}}$ with a unique simple pole at a point $z_0 \in \mathbb{T}$ if and only if $\overline{\lim}_n |c_{n-1} - c_n z_0| < 1$. Prove that this last condition implies $\lim_n \frac{c_n}{c_{n+1}} = z_0$.

Show also that the condition $\lim_n c_n/c_{n+1} = z_0$, holds if $f$ has a meromorphic extension to a neighborhood of $\overline{\mathbb{D}}$ which has a pole at the point $z_0 \in \mathbb{T}$ of order greater than the orders of all the other poles located in $\mathbb{T}$.

7. Show that any function $f$ holomorphic on the half plane $\Pi^+ = \{z : \operatorname{Im} z > 0\}$ and $2\pi$-periodic (that is, $f(z + 2\pi) = f(z)$, $z \in \Pi^+$) has a complex Fourier series expansion of the form
$$f(z) = \sum_{n=-\infty}^{\infty} c_n e^{inz}, \quad z \in \Pi^+.$$

Find integral expressions for the coefficients $c_n$.

8. Using the argument principle give an alternative proof of Theorem 4.32 with a more geometrical character.

9. Let $f$ be a holomorphic one-to-one function on an open set $U$. For fixed $a \in U$ and $r > 0$ such that $\bar{D}(a,r) \subset U$, show the formula

$$f^{-1}(w) = \frac{1}{2\pi i} \int_{C(a,r)} \frac{zf'(z)}{f(z) - w}\, dz, \quad \text{if } w \in f(D(a,r)).$$

10. Prove that the zeros of a non-identically zero meromorphic function on a domain are isolated points.

11. Let $f$ be a meromorphic function on $\mathbb{C}$. Prove that $f$ is the logarithmic derivative of a meromorphic function on $\mathbb{C}$ if and only if all the poles of $f$ are simple poles with integer residue. Similarly $f$ is the logarithmic derivative of an entire function if and only if all the poles of $f$ are simple with natural residue.

12. Show that for $\alpha > 0$ and $w \in \mathbb{C}$ with $|w| < \alpha^p$, $p \in \mathbb{N}$, the equation $(z-\alpha)^p e^z = w$ has exactly $p$ different roots in the half plane $\{z : \operatorname{Re} z > 0\}$. Compute the smallest radius $r < \alpha$ so that all the roots of this equation are contained in the disc $D(\alpha, r)$.

13. Let $U$ be a bounded domain of the plane with regular positively oriented boundary. Let $f$ be a meromorphic function on a neighborhood of $\bar{U}$ without poles in $\partial U$, satisfying $|f(z)| \leq 1$ for $z \in \partial U$. Show that $f$ takes on $U$ every value $w \in \mathbb{C}$ with $|w| > 1$, as many times as the sum of the multiplicities of the poles of $f$ in $U$.

14. Let $f$ be an entire function satisfying $\operatorname{Im} f(z) = (\operatorname{Re} f(z))^3$ if and only if $\operatorname{Im} z = (\operatorname{Re} z)^2$. Prove that $f$ has at most a zero in $\mathbb{C}$.

15. Let $f$ be an entire one-to-one function. Show that $\lim_{|z| \to \infty} |f(z)| = \infty$ and deduce that $f$ must be a polynomial of degree 1. (See Exercise 20 of Section 4.7).

16. For $a \in \mathbb{C}$ fixed with $|a| \neq 1$, prove that $\left| \frac{z-a}{1-\bar{a}z} \right| = 1$ when $|z| = 1$. Use this fact to show that a rational function $R$, satisfying $|R(z)| = 1$ when $|z| = 1$, is of the form

$$R(z) = c\, z^m \prod_{k=1}^{n} \frac{z - a_k}{1 - \bar{a}_k z}$$

with $c, a_k \in \mathbb{C}$, $|a_k| \neq 1$ and $m \in \mathbb{Z}$.

17. Let $u$ be a real harmonic function on the annulus $C(a, R_2, R_1)$. Prove that there are constants $\alpha, \beta \in \mathbb{R}$ such that

$$\frac{1}{2\pi} \int_0^{2\pi} u(a + re^{i\theta})\, d\theta = \alpha \operatorname{Log} r + \beta, \quad R_2 < r < R_1.$$

If, in addition, $u$ is bounded then $\alpha = 0$ and $\beta = u(a)$.

**18.** Show the equality

$$\int_0^\infty \frac{dx}{1 + x^n} = \frac{\pi/n}{\sin(\pi/n)} \quad \text{for } n = 1, 2, \ldots .$$

**19.** Compute

$$\int_0^\infty \frac{x^2}{\operatorname{ch} x} \, dx \quad \text{and} \quad \int_0^\infty \frac{\operatorname{Log}^2 x}{1 + x^4} \, dx.$$

**20.** Compute

$$\int_0^\infty \frac{\sqrt{x} \operatorname{Log} x}{1 + x^2} \, dx \quad \text{and} \quad \int_0^\infty \frac{\operatorname{Log} x}{\sqrt{x}(1 + x^2)} \, dx.$$

# Chapter 6

# Homology and holomorphic functions

The aim of this chapter is to give the homological versions of Cauchy's theorem and Cauchy's integral formula. Cauchy's theorem asserts that if the function $f$ is holomorphic on a domain $U$, then the integral $\int_\gamma f(z)\,dz$ is zero for certain cycles $\gamma$ in $U$. The description of these cycles is a topological question that has to be properly stated. Here this formulation is done by means of the notion of homology, based on the index of a closed curve. It may be done also by means of the concept of homotopy of curves, and the relationship between both points of view is briefly commented.

On the other hand, homological versions of Green's formula and Cauchy–Green's formula are given. They may be considered as the multiplicity version of the classical Green's formulae and have some interesting applications, among these the Cauchy theorem for differential forms. The obtained results may be applied also to give several characterizations of simply connected domains in the plane.

## 6.1 Homology of chains and simply connected domains

In order to state properly the homological versions of both Cauchy's theorem and Cauchy's integral formula, it is suitable to extend the notion of line integral to a more general setting. Consider $\gamma_1, \gamma_2, \ldots, \gamma_k$ paths, that is, piecewise $C^1$ curves, inside a domain $U$ of the plane, and let $n_1, n_2, \ldots, n_k$ be integer numbers. The formal linear combination

$$\Gamma = \sum_{j=1}^{k} n_j \gamma_j$$

is, by definition, a *chain* in the domain $U$. Hence, the set of chains in $U$ is formally the free abelian group generated by the set of paths. In particular, if $\Gamma_1$, $\Gamma_2$ are chains in $U$, other chains $n\Gamma_1 + m\Gamma_2$ with $n, m \in \mathbb{Z}$ may be defined. The range of $\Gamma$ is $\Gamma^* = \gamma_1^* \cup \gamma_2^* \cup \cdots \cup \gamma_k^*$.

If $f \in C(\Gamma^*)$, we write, by definition,

$$\int_\Gamma f(z)\,dz = \sum_{j=1}^{k} n_j \int_{\gamma_j} f(z)\,dz.$$

Therefore, if $f \in C(U)$ and $\Gamma_1$, $\Gamma_2$ are chains in $U$, one has

$$\int_{n\Gamma_1+m\Gamma_2} f(z)\,dz = n \int_{\Gamma_1} f(z)\,dz + m \int_{\Gamma_2} f(z)\,dz, \quad n,m \in \mathbb{Z}.$$

One may have chains $\Gamma$, $\Gamma'$ with different representations,

$$\Gamma = \sum_{j=1}^{k} n_j \gamma_j, \quad \Gamma' = \sum_{j=1}^{k'} n'_j \gamma'_j,$$

but $\int_\Gamma f(z)\,dz = \int_{\Gamma'} f(z)\,dz$ holds for every function $f \in C(\Gamma^* \cup \Gamma'^*)$. This is the case, for example, if $\Gamma'$ is obtained from $\Gamma$ doing permutations of the paths $\gamma_j$, subdivisions, aggregations of paths, reparametrizations, etc. All throughout this chapter, given a chain $\Gamma$ in $U$ one will be primarily interested in the correspondence

$$f \longmapsto \int_\Gamma f(z)\,dz, \quad f \in C(U).$$

If $\int_\Gamma f(z)\,dz = \int_{\Gamma'} f(z)\,dz$, for every function $f \in C(U)$, it is said that $\Gamma$ and $\Gamma'$ *are equivalent chains* and $\Gamma$ will be identified with $\Gamma'$. With this convention, a statement such as the one in the following example may be made.

**Example 6.1.** Consider the unit square $Q = [0,1] \times [0,1]$ divided into four equal parts $Q_i, i = 1, 2, 3, 4$, and let $\Gamma_i = \partial Q_i$ be the boundary of $Q_i$ positively oriented. Then

$$\Gamma_1 + \Gamma_2 + \Gamma_3 + \Gamma_4 = \partial Q,$$

because there are cancellations of the contributions of the inner sides to the integral of any continuous function (Figure 6.1). $\qquad\qquad\square$

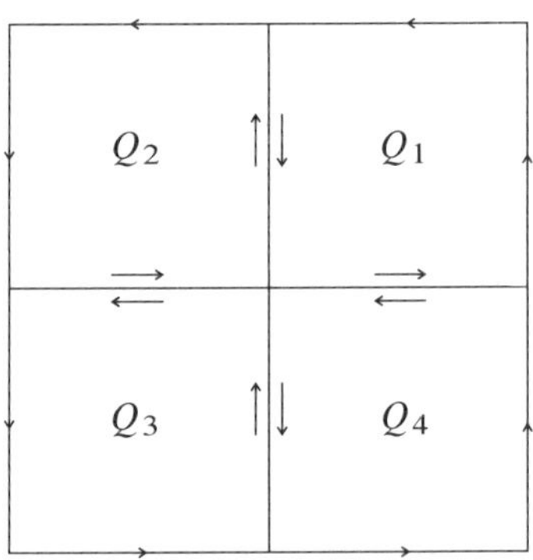


Figure 6.1

Observe the equality $\overline{\int_\Gamma f(z)\,dz} = \int_\Gamma \overline{f(z)}\,d\bar{z}$ and the fact that every differential 1-form $\omega = P\,dx + Q\,dy$ with real coefficients may be written as $\omega = \frac{1}{2}(f(z)\,dz + \overline{f(z)}\,d\bar{z})$ with $f = (P - iQ)$. It turns out, then, that two chains $\Gamma$ and $\Gamma'$ are equivalent in $U$ if and only if $\int_\Gamma \omega = \int_{\Gamma'} \omega$ for all differential 1-forms $\omega$ with continuous coefficients on $U$. One considers only forms of type $f(z)\,dz$ just for convenience.

A chain $\Gamma$ on $U$ is said to be a *cycle* provided it may be represented in the form $\Gamma = \sum_{j=1}^k n_j \gamma_j$, where each $\gamma_j$ is a closed path in $U$ and each $n_j$ an integer. Then the index of $\Gamma$ with respect to a point $z \notin \Gamma^*$ is defined as

$$\mathrm{Ind}\,(\Gamma, z) = \frac{1}{2\pi i} \int_\Gamma \frac{dw}{w - z} = \sum_{j=1}^k n_j \,\mathrm{Ind}(\gamma_j, z).$$

Obviously, $\mathrm{Ind}\,(\Gamma, z)$ is an integer number and the relation

$$\mathrm{Ind}(n_1\Gamma_1 + n_2\Gamma_2, z) = n_1\,\mathrm{Ind}(\Gamma_1, z) + n_2\,\mathrm{Ind}(\Gamma_2, z), \quad z \notin \Gamma_1^* \cup \Gamma_2^*, \; n_1, n_2 \in \mathbb{Z},$$

holds if $\Gamma_1$ and $\Gamma_2$ are cycles in $U$.

**Definition 6.2.** A cycle $\Gamma$, contained in a domain $U$, is said to be homologous to zero with respect to $U$ and it is written $\Gamma \approx 0\,(U)$, if $\mathrm{Ind}\,(\Gamma, z) = 0$ for every point $z \notin U$. Two chains $\Gamma_1$, $\Gamma_2$ contained in $U$ are homologous with respect to $U$, $\Gamma_1 \approx \Gamma_2\,(U)$, if $\Gamma_1 - \Gamma_2$ is a cycle homologous to zero with respect to $U$ (if $\Gamma_1$, $\Gamma_2$ are cycles, this is equivalent to the condition $\mathrm{Ind}\,(\Gamma_1, z) = \mathrm{Ind}\,(\Gamma_2, z)$, for every point $z \notin U$).

It is worthy noting that the definitions of index of a cycle and homologous cycles are done according to line integrals of type $\int_\Gamma (w - z)^{-1}\,dz$, and therefore they do not depend on the way that $\Gamma$ is represented as a combination of closed paths.

**Example 6.3.** In Section 1.6 it has been seen that if $U$ is a bounded domain with positively oriented piecewise regular boundary then $\partial U = \gamma_1^* \cup \gamma_2^* \cup \cdots \cup \gamma_N^*$ with $\gamma_i$ closed Jordan curves such that for $z \in U$,

$$\mathrm{Ind}\,(\gamma_1, z) = 1, \quad \mathrm{Ind}\,(\gamma_i, z) = 0, \quad i = 2, \ldots, N.$$

If $\tilde{U}$ is any domain with $\tilde{U} \supset \bar{U}$, then the cycle $\partial U = \gamma_1 + \gamma_2 + \cdots + \gamma_N$ is homologous to zero with respect to $\tilde{U}$.  $\qquad\square$

Actually, $\Gamma \approx 0\,(U)$ means that the components of $\mathbb{C} \setminus \Gamma^*$ with non-zero index lie in $U$. In Section 1.3 the concept of simply connected domain has been introduced as a domain $U$ of $\mathbb{C}$ such that $\mathbb{C}^* \setminus U$ is connected. Intuitively this condition has the same meaning as $\Gamma \approx 0\,(U)$ for any closed path $\Gamma$ in $U$, as the following result confirms.

**Proposition 6.4.** *A domain $U$ of the plane is simply connected if and only if every cycle contained in $U$ is homologous to zero with respect to $U$.*

*Proof.* If $U$ is simply connected and $\Gamma$ is a cycle with $\Gamma^* \subset U$, the connected set $\mathbb{C}^* \setminus U$ is inside the component of $\mathbb{C}^* \setminus \Gamma^*$ which contains the point at infinity and, therefore, $\mathrm{Ind}(\Gamma, z) = 0$, for $z \notin U$.

Conversely, suppose $U$ is not simply connected; then $\mathbb{C}^* \setminus U$ has at least two connected components $V_1, V_2$, which are disjoint closed sets of $\mathbb{C}^* \setminus U$. So they are disjoint closed sets of $\mathbb{C}^*$ and one of them, say $V_2$, contains the point at infinity. Then $V_1$ has $\infty$ as an exterior point and it must be compact. Construct a grid of the plane formed by squares with sides parallel to the axis, of size $\varepsilon/2$, with $\varepsilon = d(V_1, V_2)$, so that a point $P \in V_1$ will be within one of the squares (Figure 6.2). Consider now all the squares of the grid intersecting $V_1$; there are a finite quantity of them,

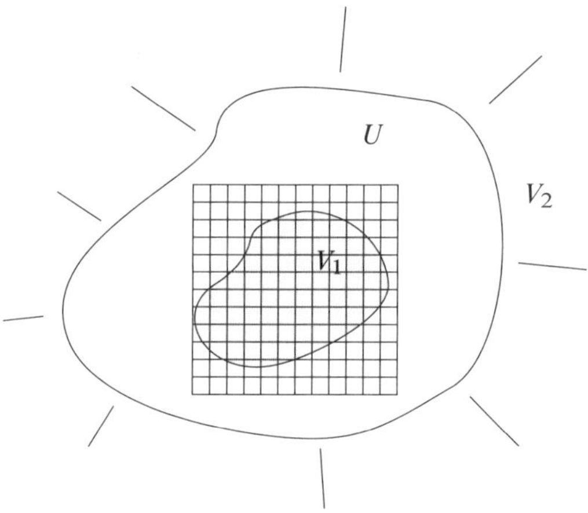


Figure 6.2

say $Q_i, i = 1, \ldots, N$, and none of them intersects $V_2$. Let $\Gamma$ be the cycle defined by $\Gamma = \sum_{i=1}^{N} \partial Q_i$, where $\partial Q_i$ is the boundary of $Q_i$, positively oriented. Since $V_1 \subset \bigcup_i Q_i$ it is clear that $\mathrm{Ind}(\Gamma, P) = \sum_{i=1}^{N} \mathrm{Ind}(\partial Q_i, P) = 1$. Once all possible cancellations are done one gets an equivalent expression of $\Gamma$ such that $\Gamma^*$ does not intersect either $V_2$ nor $V_1$. Indeed $\Gamma^*$ consists of some sides within $\bigcup_i \partial Q_i$. If $S$ is a side meeting $V_1$ then both adjacent squares having $S$ as a common side are among the $Q_i, i = 1, \ldots, N$, and hence $S$ is cancelled. $\qquad\square$

## 6.2 Homological versions of Green's formula and Cauchy's theorem

### 6.2.1 The homological version of Cauchy's theorem

A general version of Cauchy's theorem is the following.

**Theorem 6.5.** *If $U$ is a domain of the plane, $\Gamma$ a cycle homologous to zero with respect to $U$ and $f \in H(U)$, then*

$$\int_\Gamma f(z)\, dz = 0.$$

*As a consequence, if $\Gamma_1$ and $\Gamma_2$ are homologous chains with respect to $U$ and $f \in H(U)$, then*

$$\int_{\Gamma_1} f(z)\, dz = \int_{\Gamma_2} f(z)\, dz.$$

The shortest proof of this fundamental result is likely the one given below (due to A. F. Beardon), based on an integral representation formula by means of Cauchy's kernel.

**Lemma 6.6.** *Let $U$ be a domain of the plane, $K \subset U$ a compact set and $f \in H(U)$. Then there exist a finite number of segments $\gamma_1, \gamma_2, \ldots, \gamma_N$ contained in $U \setminus K$ such that if $\gamma = \gamma_1 + \gamma_2 + \cdots + \gamma_N$, one has*

$$f(z) = \frac{1}{2\pi i} \int_\gamma \frac{f(w)}{w - z}\, dw, \quad z \in K.$$

*Proof.* Let $\varepsilon$ be the distance from $K$ to the complement of $U$. Consider a grid of the plane composed by squares of diagonal smaller than $\varepsilon$ and consider the squares $Q_1, \ldots, Q_m$ of this grid intersecting $K$. If $\partial Q_i$ is the boundary of $Q_i$, positively oriented as always, consider $\gamma = \sum_i \partial Q_i$. Once all possible cancellations are done, it yields

$$\gamma = \sum_{j=1}^{N} \gamma_j$$

where each $\gamma_j$ is a side of a certain $Q_i$ and $\gamma_j \subset U \setminus K$.

Hence, for every continuous function $\varphi$ on $\bigcup_j \partial Q_j$, one has

$$\sum_{i=1}^{m} \int_{\partial Q_i} \varphi(w)\, dw = \sum_{j=1}^{N} \int_{\gamma_j} \varphi(w)\, dw = \int_\gamma \varphi(w)\, dw.$$

Suppose now that the point $z \in K$ is not in any $\partial Q_i$. Then $z$ belongs to a unique square, $Q_{i_0}$, of the set $Q_1, \ldots, Q_m$, and applying Corollary 4.3 to the function $f$

and Cauchy–Goursat's theorem (Remark 3.3 a)) to the function $\omega \to \frac{f(\omega)}{\omega - z}$, it turns out that

$$\frac{1}{2\pi i} \int_\gamma \frac{f(\omega)}{\omega - z} d\omega = \frac{1}{2\pi i} \sum_{j=1}^m \int_{\partial Q_j} \frac{f(\omega)}{\omega - z} d\omega$$

$$= \frac{1}{2\pi i} \int_{\partial Q_{i_0}} \frac{f(\omega)}{\omega - z} d\omega + \frac{1}{2\pi i} \sum_{j \neq i_0} \int_{\partial Q_j} \frac{f(\omega)}{\omega - z} d\omega = f(z).$$

Hence, the lemma is proved if $z$ is not in any $\partial Q_i$, and for the points of $\bigcup_i \partial Q_i$ it follows by continuity. $\qquad\square$

**Remark 6.1.** In fact, under the hypotheses of Lemma 6.6 it may be asserted that the segments $\gamma_1, \ldots, \gamma_N$ form a system of *closed Jordan polygonal curves* $\Gamma_1, \ldots, \Gamma_n$ contained in $U \setminus K$ such that

$$f(z) = \frac{1}{2\pi i} \sum_{j=1}^n \int_{\Gamma_j} \frac{f(w)}{w - z} dw, \quad z \in K, \text{ if } f \in H(U).$$

Actually, one only needs to remember that segments $\gamma_1, \ldots, \gamma_N$ are the topological boundary of the compact set $Q = \bigcup_{j=1}^m Q_j$. Then each connected component of $\partial Q$ is a compact and connected 1-manifold and by Theorem 1.31 it is a Jordan curve which in this case must be a polygonal curve. Moreover, Lemma 6.6 applied to $f = 1$ gives that every point of $K$ belongs to one and only one of the polygons $\text{Int}(\Gamma_i), i = 1, \ldots, n$.

This fact may be also proved, in a more direct way, using a combinatorial argument (see [3], p. 260, or [11], p. 155).

*Proof of Theorem* 6.5. Suppose that $\Gamma \approx 0\,(U)$ and let $f \in H(U)$. Let further $K = \Gamma^* \cup \{z \notin \Gamma^*\colon \text{Ind}\,(\Gamma, z) \neq 0\}$, which is the complement in $\mathbb{C}$ of the set $\{z \notin \Gamma^*\colon \text{Ind}\,(\Gamma, z) = 0\}$. This last set is the union of some connected components of $\mathbb{C} \setminus \Gamma^*$, among which there is the unbounded one that contains $\infty$; therefore, $K$ is a compact set and since $\Gamma \approx 0\,(U)$, one has $K \subset U$. Let now $\gamma$ be as in Lemma 6.6; one has

$$\int_\Gamma f(z)\,dz = \int_\Gamma \left\{ \frac{1}{2\pi i} \int_\gamma \frac{f(w)}{w - z}\,dw \right\} dz = \int_\gamma \frac{f(w)}{2\pi i} \left\{ \int_\Gamma \frac{dz}{w - z} \right\} dw.$$

But for $w \in \gamma^*$ one has $w \in U \setminus K$, so that

$$\int_\Gamma \frac{dz}{w - z} = -2\pi i \,\text{Ind}\,(\Gamma, w) = 0.$$

Notice that the application of Fubini's theorem is correct because the function $(z, w) \to \frac{f(w)}{w-z}$ is continuous on the compact set $\Gamma^* \times \gamma^*$. $\qquad\square$

## 6.2.2 Green's formula with multiplicities

We have just given a direct proof of the homological version of Cauchy's theorem. The classical version of Cauchy's theorem (Theorem 3.24) was obtained in Chapter 3 as a particular case of Green's formula (Theorem 3.22) when applied to the form $f(z)\,dz$, with $f$ holomorphic. So one may ask if there is also a homological version of Green's formula which gives rise to Theorem 6.5 when applied to $f(z)\,dz$.

The answer is yes; this more general version of Green's formula is based on the following integral representation given by Corollary 4.4, which is quoted here:

**Lemma 6.7.** *If $f$ is a differentiable function on $\mathbb{C}$ with compact support and $\bar{\partial} f$ is continuous on $\mathbb{C}$, then*

$$f(z) = -\frac{1}{\pi} \int_{\mathbb{C}} \frac{\partial f}{\partial \bar{w}} \frac{1}{w - z}\, dm(w), \quad z \in \mathbb{C}.$$

**Theorem 6.8** (Homological version of Green's formula). *Let $U$ be a domain of the complex plane, $f$ a differentiable function with $\bar{\partial} f$ continuous on $U$ and $\Gamma$ a cycle homologous to zero with respect to $U$. Then one has*

$$\int_{\Gamma} f(z)\,dz = 2i \iint_{\mathbb{C}} \frac{\partial f}{\partial \bar{z}} \operatorname{Ind}(\Gamma, z)\, dm(z). \tag{6.1}$$

*More generally, if $\omega = P\,dx + Q\,dy$ is a 1-form in $U$ with differentiable coefficients, the function $Q_x - P_y$ is continuous on $U$ and $\Gamma \approx 0$ with respect to $U$, then*

$$\int_{\Gamma} \omega = \int_{\Gamma} P\,dx + Q\,dy = \iint_{\mathbb{C}} \left( \frac{\partial Q}{\partial x} - \frac{\partial P}{\partial y} \right) \operatorname{Ind}(\Gamma, z)\, dm(z). \tag{6.2}$$

*In particular, if $U$ is simply connected, formulae (6.1) and (6.2) hold for every cycle $\Gamma$ contained in $U$.*

Note that, in fact, the integrals of the right-hand term of (6.1) and of (6.2) are taken over a compact set of $U$.

*Proof.* Take $K = \Gamma^* \cup \{z \notin \Gamma^*;\ \operatorname{Ind}(\Gamma, z) \neq 0\}$, as in the proof of Theorem 6.5, so that $K$ is a compact set, $K \subset U$. Let $\psi \in C^{\infty}(U)$ be a function with compact support contained in $U$ such that $\psi$ is 1 on a neighborhood of $K$ and apply Lemma 6.7 to the function $g = \psi f$. It yields

$$g(z) = -\frac{1}{\pi} \iint_{\mathbb{C}} \frac{\partial g}{\partial \bar{w}} \frac{1}{w - z}\, dm(w).$$

Since $g = f$ on $\Gamma^*$, it turns out that

$$\int_{\Gamma} f(z)\,dz = \int_{\Gamma} g(z)\,dz = -\frac{1}{\pi} \int_{\Gamma} \left( \iint_{\mathbb{C}} \frac{\partial g}{\partial \bar{w}} \frac{1}{w - z}\, dm(w) \right) dz.$$

In order to apply now Fubini's theorem it is enough to see that an iterated integral of the absolute value of the integrand is finite,

$$\int_{\Gamma} \iint_{F} \left| \frac{\partial g}{\partial \bar{w}} \right| \left| \frac{1}{w - z} \right| dm(w)|dz| < +\infty.$$

The inner integral is taken over the compact set $F = \mathrm{spt}(g)$ where $\left| \frac{\partial g}{\partial \bar{w}} \right|$ is bounded; writing $\left\| \frac{\partial g}{\partial \bar{w}} \right\|_{\infty} = \sup_{F} \left| \frac{\partial g}{\partial \bar{w}} \right|$ it turns out by Lemma 4.1 that the iterated integral is bounded above by

$$\int_{\Gamma} \left\| \frac{\partial g}{\partial \bar{w}} \right\|_{\infty} (m(F) + 2\pi)|dz| = L(\Gamma) \left\| \frac{\partial g}{\partial \bar{w}} \right\|_{\infty} (m(F) + 2\pi) < +\infty.$$

Now apply Fubini's theorem to obtain

$$\int_{\Gamma} f(z)\, dz = -\frac{1}{\pi} \iint_{\mathbb{C}} \frac{\partial g}{\partial \bar{w}}(w) \left( \int_{\Gamma} \frac{dz}{w - z} \right) dm(w)$$

$$= 2i \iint_{\mathbb{C}} \frac{\partial g}{\partial \bar{w}}(w) \, \mathrm{Ind}\,(\Gamma, w)\, dm(w).$$

But $\mathrm{Ind}\,(\Gamma, w) \neq 0$ gives $w \in K$ and $g = f$ on a neighborhood of $w$; therefore, $\frac{\partial g}{\partial \bar{w}} = \frac{\partial f}{\partial \bar{w}}$ at this point and this proves (6.1).

In order to show (6.2) one may suppose that $P$ and $Q$ are real. If the form $\omega = P\, dx + Q\, dy$ is $C^1$ applying (6.1) to the function $f = (P - iQ)$ and taking real parts, one has

$$\int_{\Gamma} \omega = \mathrm{Re} \int_{\Gamma} f(z)\, dz$$

$$= 2 \iint_{\mathbb{C}} \mathrm{Ind}\,(\Gamma, w) \, \mathrm{Re}\left( i \frac{\partial f}{\partial \bar{w}} \right) dm(w)$$

$$= \iint_{\mathbb{C}} \mathrm{Ind}\,(\Gamma, w) \left( \frac{\partial Q}{\partial x} - \frac{\partial P}{\partial y} \right) dm(z),$$

which is (6.2).

Assuming now only that $P$ and $Q$ are differentiable and $Q_x - P_y$ is continuous, and replacing $P$ and $Q$ by $P\psi$ and $Q\psi$ where $\psi$ is the function of the first part of the proof, we may assume that $P$ and $Q$ have compact support in $U$. Next apply a regularization process to $P$ and $Q$ using an approximation to the identity $\Phi_{\varepsilon}$, defined by $\Phi_{\varepsilon}(x, y) = \varepsilon^{-2}\Phi\left(\frac{x}{\varepsilon}, \frac{y}{\varepsilon}\right)$, $\varepsilon > 0$, with $\Phi \in C^{\infty}(\mathbb{C})$, $\mathrm{spt}\,(\Phi) \subset \bar{\mathbb{D}}$, $\iint_{\mathbb{C}} \Phi\, dm = 1$. That is, define

$$P^{\varepsilon}(x, y) = (P * \Phi_{\varepsilon})(x, y) = \iint_{\mathbb{C}} P(t, s)\Phi_{\varepsilon}(x - t, y - s)\, ds\, dt,$$

$$Q^{\varepsilon}(x, y) = (Q * \Phi_{\varepsilon})(x, y) = \iint_{\mathbb{C}} Q(t, s)\Phi_{\varepsilon}(x - t, y - s)\, ds\, dt$$

so that $P^\varepsilon$, $Q^\varepsilon$ are $C^\infty$ with compact support in $U_\varepsilon = \{z \in U : d(z, U^c) > \varepsilon\}$ and $P^\varepsilon \to P$, $Q^\varepsilon \to Q$ uniformly when $\varepsilon \to 0$.

Accordingly one has

$$\int_\Gamma P^\varepsilon dx + Q^\varepsilon dy = \iint_{\mathbb{C}} \left( \frac{\partial Q^\varepsilon}{\partial x} - \frac{\partial P^\varepsilon}{\partial y} \right) \operatorname{Ind}(\Gamma, z) \, dm(z)$$

and if we check the equality

$$\frac{\partial Q^\varepsilon}{\partial x} - \frac{\partial P^\varepsilon}{\partial y} = \left( \frac{\partial Q}{\partial x} - \frac{\partial P}{\partial y} \right) * \Phi_\varepsilon,$$

it will suffice to let $\varepsilon \to 0$ to finish the proof.

Hence, one must show that

$$\frac{\partial (Q * \Phi_\varepsilon)}{\partial x} - \frac{\partial (P * \Phi_\varepsilon)}{\partial y} = Q * \frac{\partial \Phi_\varepsilon}{\partial x} - P * \frac{\partial \Phi_\varepsilon}{\partial y} = \left( \frac{\partial Q}{\partial x} - \frac{\partial P}{\partial y} \right) * \Phi_\varepsilon.$$

To this end, take a bounded open set $\tilde{U} \subset U$ with piecewise regular boundary such that $\operatorname{spt}(P)$, $\operatorname{spt}(Q) \subset \tilde{U}$ (cf. Lemma 6.6 and Remark 6.1) and apply Green's formula (Theorem 3.22) to $\tilde{U}$ and to the form

$$P(x - t, y - s)\Phi_\varepsilon(t, s)dt + Q(x - t, y - s)\Phi_\varepsilon(t, s)ds,$$

with $\varepsilon > 0$ small enough. We get

$$0 = \int_{\partial \tilde{U}} P(x - t, y - s)\Phi_\varepsilon(t, s)dt + Q(x - t, y - s)\Phi_\varepsilon(t, s)ds$$

$$= -\iint_{\mathbb{C}} \left[ \frac{\partial Q}{\partial x}(x - t, y - s) - \frac{\partial P}{\partial y}(x - t, y - s) \right] \Phi_\varepsilon(t, s)$$

$$+ \iint_{\mathbb{C}} \left[ Q(x - t, y - s)\frac{\partial \Phi_\varepsilon}{\partial x} - P(x - t, y - s)\frac{\partial \Phi_\varepsilon}{\partial y} \right] dt\,ds,$$

which is the desired result. $\qquad\qquad\qquad\qquad\qquad\qquad\qquad\qquad\qquad\square$

**Example 6.9.** If $\Gamma$ is a cycle such that $\int_\Gamma f(z)dz = 0$ for each $f \in C(\Gamma^*)$, then $\operatorname{Ind}(\Gamma, z) = 0$, for every point $z \notin \Gamma^*$, because $w \to \frac{1}{w - z} \in C(\Gamma^*)$. Formula (6.1) leads to the converse: if $\operatorname{Ind}(\Gamma, z) = 0$, for every point $z \notin \Gamma^*$, then we can apply (6.1) to $U = \mathbb{C}$ and $f \in C^1(\mathbb{C})$ to obtain $\int_\Gamma f(z)dz = 0$. Since every continuous function on $\Gamma^*$ may be approximated by functions in $C^1(\mathbb{C})$, we get the expected conclusion. Hence, the closed curves that gives a line integral identically zero are exactly the curves with vanishing index. $\qquad\quad\square$

**Example 6.10.** By the previous example if $\Gamma_1$, $\Gamma_2$ are two cycles with $\Gamma_1^* = \Gamma_2^*$ and $\mathrm{Ind}(\Gamma_1, z) = \mathrm{Ind}(\Gamma_2, z)$ for every point $z \notin \Gamma_1^* = \Gamma_2^*$, then one has $\int_{\Gamma_1} f(z)dz = \int_{\Gamma_2} f(z)dz$ for each function $f$ continuous on the compact set $\Gamma_1^* = \Gamma_2^*$.

This is the case when $\Gamma_2$ is a closed curve obtained from $\Gamma_1$ by an orientation preserving change of parameter; that is, $\Gamma_1 \colon [a, b] \to \mathbb{C}$, $\Gamma_2 \colon [c, d] \to \mathbb{C}$, $\varphi \colon [c, d] \to [a, b]$ with $\varphi'(t) > 0$ and $\Gamma_2 = \Gamma_1 \circ \varphi$ (see Section 3.1). $\qquad\square$

**Example 6.11.** Formula (6.1) may also be applied to compute complex integrals. For example, if $f \in C^1(U)$, where $U$ is a domain that contains the segment $[-1, 1]$, $\Gamma$ is a cycle with $\Gamma^* \subset U \setminus [-1, 1]$, $\Gamma \approx 0(U)$ and $g(z)$ is the holomorphic branch of $\sqrt{z^2 - 1}$ on $\mathbb{C} \setminus [-1, 1]$ that is positive on $(1, \infty)$, by (6.1), it turns out that

$$\frac{1}{2i} \int_\Gamma f(z)g(z)dz - \iint_\mathbb{C} \bar{\partial} f(z)g(z)\,\mathrm{Ind}(\Gamma, z)dm(z)$$
$$= -\,\mathrm{Ind}(\Gamma, 0) \int_{-1}^{1} f(t)\sqrt{1 - t^2}dt.$$

Taking, in particular, $f(z) = z^2$, one has

$$4 \int_\Gamma z^2 \sqrt{z^2 - 1}dz = -\pi i\,\mathrm{Ind}(\Gamma, 0). \qquad\square$$

### 6.2.3 Cauchy–Green's formula with multiplicities and Cauchy's integral formula

In this subsection the versions of Cauchy–Green's formula and Cauchy's integral formula which correspond to the homological version of Green's formula (Theorem 6.8) are given.

The first one is a decomposition formula that gives any function $f$ as the sum of a holomorphic one and a term which depends only on $\bar{\partial} f$.

**Theorem 6.12** (Homological version of Cauchy–Green's formula). *Let $U$ be a domain of the plane, $\Gamma$ a cycle homologous to zero with respect to $U$ and $f$ a differentiable function with $\bar{\partial} f$ continuous on $U$. Then for $z \in U \setminus \Gamma^*$ one has*

$$\mathrm{Ind}\,(\Gamma, z)\,f(z) = \frac{1}{2\pi i} \int_\Gamma \frac{f(w)}{w - z}\,dw - \frac{1}{\pi} \iint_\mathbb{C} \frac{\partial f}{\partial \bar{w}} \frac{1}{w - z}\,\mathrm{Ind}\,(\Gamma, w)\,dm(w).$$

*Proof.* Fix a point $z \in U \setminus \Gamma^*$ and take $\varepsilon > 0$ small enough such that $D(z, \varepsilon) \cap \Gamma^* = \emptyset$ and $\bar{D}(z, \varepsilon) \subset U$. If $\gamma_\varepsilon(t) = z + \varepsilon e^{it}$, $0 \le t \le 2\pi$ (the circle centered at $z$ and radius $\varepsilon$ travelled in the positive sense), the cycle $\Gamma_\varepsilon = \Gamma - \mathrm{Ind}\,(\Gamma, z)\gamma_\varepsilon$ is homologous to zero with respect to the open set $U \setminus \{z\}$. Actually, if $w \notin U$, since $\mathrm{Ind}\,(\gamma_\varepsilon, w) = 0$ and $\mathrm{Ind}\,(\Gamma, w) = 0$ (as $\Gamma \approx 0\,(U)$) one also has $\mathrm{Ind}\,(\Gamma_\varepsilon, z) = 0$

and for $w = z$, $\text{Ind}\,(\Gamma_\varepsilon, z) = \text{Ind}\,(\Gamma, z) - \text{Ind}\,(\Gamma, z) = 0$. Applying equality (6.1) to the function $\frac{f(w)}{w-z}$ and to the cycle $\Gamma_\varepsilon$, it turns out that

$$\int_\Gamma \frac{f(w)}{w-z}\,dw - \text{Ind}\,(\Gamma, z)\int_{\gamma_\varepsilon} \frac{f(w)}{w-z}\,dw = 2i\iint_{\mathbb{C}} \frac{\partial f}{\partial \bar{w}}\,\frac{1}{w-z}\,\text{Ind}\,(\Gamma_\varepsilon, w)\,dm(w)$$

$$= 2i\iint_{\mathbb{C}} \frac{\partial f}{\partial \bar{w}}\,\frac{1}{w-z}\,\text{Ind}\,(\Gamma, w)\,dm(w) - \text{Ind}\,(\Gamma, z)\iint_{D(z,\varepsilon)} \frac{\partial f}{\partial \bar{w}}\,\frac{1}{w-z}\,dm(w).$$

Letting $\varepsilon \to 0$ and using the fact that $w \to \frac{\partial f}{\partial \bar{w}}\,\frac{1}{w-z}$ is integrable, one gets

$$\lim_{\varepsilon \to 0} \iint_{D(z,\varepsilon)} \frac{\partial f}{\partial \bar{w}}\,\frac{1}{w-z}\,dm(z) = 0,$$

and the result follows. $\qquad\square$

When $f \in H(U)$, one has as a particular case, a homological version of Cauchy's integral formula.

**Theorem 6.13** (Homological version of Cauchy's integral formula). *If $U$ is a domain of the plane, $\Gamma$ a cycle in $U$ homologous to zero with respect to $U$ and $f \in H(U)$, one has*

$$\text{Ind}\,(\Gamma, z)f(z) = \frac{1}{2\pi i}\int_\Gamma \frac{f(w)}{w-z}\,dw, \quad z \in U \setminus \Gamma^*, \tag{6.3}$$

*and also*

$$\text{Ind}\,(\Gamma, z)f^{(n)}(z) = \frac{n!}{2\pi i}\int_\Gamma \frac{f(w)}{(w-z)^{n+1}}\,dw, \quad z \in U \setminus \Gamma^*, \, n = 1, 2, \ldots.$$

The second equality is obtained from the first one by differentiating with respect to $z$ under the integral and bearing in mind that $\text{Ind}\,(\Gamma, \cdot)$ is constant on the neighborhood of a point $z \notin \Gamma^*$.

The formula (6.3) may be obtained directly from Cauchy's theorem (Theorem 6.5) applied to the holomorphic function

$$g(w) = \frac{f(w) - f(z)}{w - z} \quad \text{if } w \neq z \quad \text{and} \quad g(z) = f'(z).$$

Next some examples of applications of Cauchy's theorem and Cauchy's integral formula are given.

**Example 6.14.** If $U$ is simply connected, $f \in H(U)$, $a \in U$ and $\Gamma$ is a cycle of $U$ not containing $a$, the possible values of $\frac{1}{2\pi i}\int_\Gamma \frac{f(w)}{w-a}\,dw$ are $\text{Ind}\,(\Gamma, a)f(a)$, that is, an integer multiplied by $f(a)$. For example, $\frac{1}{2\pi i}\int_C \frac{e^w}{w-1}\,dw$, where $C$ is a circle not passing through 1, is 0 if 1 is in the exterior of $C$ and $e$ if 1 is in the interior of $C$. $\qquad\square$

**Example 6.15.** Suppose that $U$ is simply connected, $f \in H(U)$ (for example, $U = \mathbb{C}$ and $f$ an entire function), $a, b \in U$, $a \neq b$, and $C$ is a circle inside $U$ which does not pass through $a, b$ (always positively oriented). Compute all possible values of

$$I = \frac{1}{2\pi i} \int_C \frac{f(z)}{(z-a)(z-b)}\, dz.$$

If $a, b$ are in the exterior of $C$ one has $I = 0$, by Cauchy's theorem applied to the function $g(z) = \frac{f(z)}{(z-a)(z-b)}$, which is holomorphic on the interior of $C$. If $a$ is inside $C$ and $b$ is outside, then $I = \frac{f(a)}{a-b}$, by Cauchy's integral formula applied to $g(z) = \frac{f(z)}{z-b}$, which is holomorphic inside $C$. Similarly, if $b$ is inside $C$ and $a$ is outside, $I = \frac{f(b)}{b-a}$, by Cauchy's integral formula applied to $g(z) = \frac{f(z)}{z-a}$. Consider finally the case both $a$ and $b$ are in the interior of $C$. Let $C_1$, $C_2$ be two small circles centered respectively at $a,b$, inside $C$ and positively oriented. In the open set $U \setminus \{a, b\}$, where $\frac{f(z)}{(z-a)(z-b)}$ is holomorphic, the cycle $C$ is homologous to $C_1 + C_2$; thus,

$$I = \frac{1}{2\pi i} \int_{C_1} \frac{f(z)}{(z-a)(z-b)}\, dz + \frac{1}{2\pi i} \int_{C_2} \frac{f(z)}{(z-a)(z-b)}\, dz$$

$$= \frac{f(a)}{a-b} + \frac{f(b)}{b-a} = \frac{f(a) - f(b)}{a-b}. \qquad \square$$

**Example 6.16.** If $\gamma$ is the path of Figure 6.3, compute $I = \int_\gamma \frac{\sin z}{z^4}\, dz$. Using Cauchy's integral formula for derivatives, it yields

$$I = \frac{\mathrm{Ind}(\gamma, 0)(\sin)^{(3)}(0)}{3!} 2\pi i = \frac{-2(-\cos 0)}{6} 2\pi i = \frac{2\pi}{3} i. \qquad \square$$

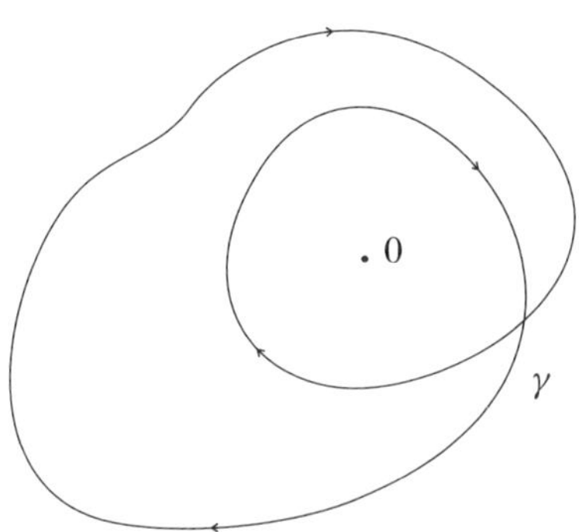


Figure 6.3

## 6.3 The residue theorem and the argument principle in a homological version

There is also a homological version of the residue theorem that is, indeed, a generalization of Cauchy's theorem (Theorem 6.5) when $f$ has singularities.

**Theorem 6.17** (Homological version of the residue theorem). *Let $U$ be a domain of the complex plane and let $A$ be a discrete and closed subset of $U$. Let $f \in H(U \setminus A)$ be such that each point $a \in A$ is an isolated singularity of $f$, and let $\Gamma$ be a cycle inside $U \setminus A$, homologous to zero with respect to $U$. Then one has*

$$\frac{1}{2\pi i} \int_{\Gamma} f(z)dz = \sum_{a \in A} \mathrm{Ind}\,(\Gamma, a)\,\mathrm{Res}\,(f, a).$$

**Remark 6.2.** Both terms of the previous equality are well defined: the one on the left-hand side, because $f \in C(\Gamma^*)$, and the one on the right-hand side is, actually, a finite sum because each bounded component of $\mathbb{C} \setminus \Gamma^*$ is inside a compact subset of $U$ that has at most a finite number of points of $A$.

*Proof.* Let $a_1, \ldots, a_N$ be the points of $A$ with $\mathrm{Ind}\,(\Gamma, a_j) \neq 0$. For each $j = 1, \ldots, N$ let $\gamma_j$ be a small circle centered at $a_j$, travelled in a direct sense, not containing in its interior any other point $a_i$, $i \neq j$, and not intersecting $\Gamma^*$ (Figure 6.4).

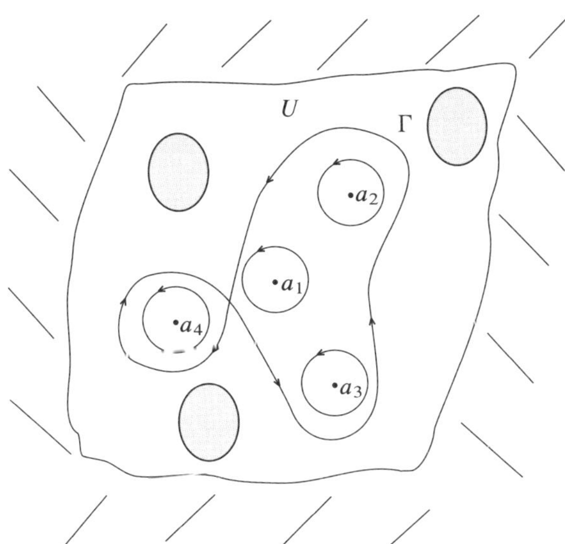


Figure 6.4

Then the cycle

$$\Gamma - \sum_{j=1}^{N} \operatorname{Ind}(\Gamma, a_j)\gamma_j$$

is homologous to zero with respect to the open set $U \setminus A$. By Cauchy's theorem it turns out that

$$\frac{1}{2\pi i} \int_{\Gamma} f(z)dz = \sum_{j=1}^{N} \operatorname{Ind}(\Gamma, a_j) \frac{1}{2\pi i} \int_{\gamma_j} f(z)dz.$$

Since each of the integrals of the right-hand side term is $2\pi i \operatorname{Res}(f, a_j)$, the theorem follows. $\qquad\square$

Cauchy's theorem is the particular case $A = \emptyset$ in the previous theorem. Cauchy's formula is also a particular case: if $f \in H(U)$, $\Gamma \sim 0$ with respect to $U$ and $z \in U \setminus \Gamma^*$, apply the residue theorem to $g(w) = \frac{f(w)}{w-z}$, with $A = \{z\}$ and $\operatorname{Res}(g, z) = f(z)$, to obtain

$$\frac{1}{2\pi i} \int_{\Gamma} \frac{f(w)}{w-z} dw = \frac{1}{2\pi i} \int_{\Gamma} g(w)dw = \operatorname{Ind}(\Gamma, z) f(z).$$

Finally, the corresponding version of the argument principle may be stated.

**Theorem 6.18** (Homological version of the argument principle). *Let $U$ be a domain of $\mathbb{C}$ and $f$ a meromorphic function on $U$. Let $\{a_j\}$ be the zeros of $f$ in $U$, with $n_j$ the multiplicity of $a_j$, and let $\{b_j\}$ the poles of $f$ in $U$, with $m_j$ the multiplicity of $b_j$. Let $\Gamma$ be a cycle inside $U$ not passing through any pole nor any zero of $f$ such that $\Gamma \approx 0$ with respect to $U$. Consider also a function $h$ holomorphic on $U$. Then one has*

$$\frac{1}{2\pi i} \int_{\Gamma} h(z) \frac{f'(z)}{f(z)} \, dz = \sum_{j} h(a_j)n_j \operatorname{Ind}(\gamma, a_j) - \sum_{j} h(b_j)m_j \operatorname{Ind}(\gamma, b_j). \tag{6.4}$$

*Proof.* Proceed as in the proof of Theorem 5.25, but using now the homological version of the residue theorem (Theorem 6.17). $\qquad\square$

Homological versions of Corollary 5.26 and Theorem 5.27, can be properly stated by the reader.

## 6.4 Cauchy's theorem for locally exact differential forms

Cauchy's theorem asserts that the integral of the form $\omega = f(z)\, dz$, with $f$ holomorphic on $U$, along a cycle homologous to zero with respect to $U$, is zero. It is natural to ask if this result also holds when $\omega$ is a general 1-form $\omega = P\,dx + Q\,dy$, locally exact. Note that $\omega = f(z)\,dz$ is locally exact if and only if $f$ is holomorphic. The answer is affirmative and it is given by the following result (see [4]).

**Theorem 6.19** (Cauchy's theorem for differential forms). *If $\omega = P\,dx + Q\,dy$ is a locally exact 1-form in a domain $U \subset \mathbb{C}$, with continuous coefficients $P, Q$ on $U$ and $\Gamma$ is a cycle, $\Gamma \approx 0$ with respect to $U$, then one has*

$$\int_\Gamma \omega = 0.$$

*Consequently, if $\Gamma_1, \Gamma_2$ are homologous chains with respect to $U$, one has $\int_{\Gamma_1} \omega = \int_{\Gamma_2} \omega$.*

This result may be found in [1], Theorem 16. In this book, the author gives a geometric proof and asks if it is possible to modify the proof of Cauchy's theorem so that it covers this more general case. Imposing the form $\omega$ has coefficients $P, Q \in C^1(U)$, then the answer to this question is Theorem 6.8. Actually, this theorem is a proof of Cauchy's theorem (formula (6.1) when $\frac{\partial f}{\partial \bar{z}} = 0$) and formula (6.2) gives Theorem 6.19 when $\omega$ is locally exact and $C^1$, because then $\omega$ is closed, that is, $P_y = Q_x$. The general case of Theorem 6.19, for $\omega$ only continuous, may be proved with a standard process of regularization of the form $\omega$, making convolutions with an approximation to the identity. This process, already used in the proof of Theorem 6.8, appears in the proof of the following statement, stressed here because it is a characterization, by duality, of locally exact forms.

**Proposition 6.20.** *In a domain $U$ of the plane, a continuous 1-form $\omega = P\,dx + Q\,dy$ is locally exact if and only if*

$$\iint_U \left( P\frac{\partial\varphi}{\partial y} - Q\frac{\partial\varphi}{\partial x} \right) dx\,dy = 0 \tag{6.5}$$

*for every $C^\infty$ function $\varphi$ with compact support contained in $U$.*

*Proof.* If $\omega$ is exact in a disc $\bar{D} \subset U$ then $\omega = dh$, for some function $h$, and if $\varphi$ has support in $D$ it follows by Green's formula that

$$\iint_U \left( \frac{\partial h}{\partial x}\frac{\partial\varphi}{\partial y} - \frac{\partial h}{\partial y}\frac{\partial\varphi}{\partial x} \right) dx\,dy = \int_{\partial D} h\frac{\partial\varphi}{\partial x}\,dx + h\frac{\partial\varphi}{\partial y}\,dy = 0.$$

Using partitions of unity we see that every $C^\infty$ function with compact support in $U$ is a finite sum of functions of the same kind which have support in discs and then we obtain (6.5) for $\varphi \in C^\infty$ with $\mathrm{spt}(\varphi) \subset U$.

Conversely, if condition (6.5) holds, let $\Phi_\varepsilon$ with $\varepsilon > 0$ be an approximation of the identity, that is, $\Phi_\varepsilon(x, y) = \varepsilon^{-2}\Phi\left(\frac{x}{\varepsilon}, \frac{y}{\varepsilon}\right)$ with $\Phi \in C^\infty(\mathbb{C})$, $\mathrm{spt}(\Phi) \subset \bar{\mathbb{D}}$ and

$\iint_{\mathbb{C}} \Phi dx dy = 1$. Consider now forms $\omega^{\varepsilon} = P^{\varepsilon} dx + Q^{\varepsilon} dy$ with

$$P^{\varepsilon}(x, y) = \iint_{\mathbb{C}} P(t, s) \Phi_{\varepsilon}(x - t, y - s) \, dt ds,$$

$$Q^{\varepsilon}(x, y) = \iint_{\mathbb{C}} Q(t, s) \Phi_{\varepsilon}(x - t, y - s) \, dt ds. \tag{6.6}$$

Then $\omega^{\varepsilon}$ is $C^{\infty}$ on $U_{\varepsilon} = \{z \in U : d(z, U^{C}) > \varepsilon\}$ and $\omega^{\varepsilon} \xrightarrow[\varepsilon \to 0]{} \omega$ (i.e. $P^{\varepsilon} \to P$, $Q^{\varepsilon} \to Q$ if $\varepsilon \to 0$), uniformly on compact sets of $U$.

Now, if $z = x + iy \in U_{\varepsilon}$, one has

$$\frac{\partial Q^{\varepsilon}}{\partial x}(x, y) - \frac{\partial P^{\varepsilon}}{\partial y}(x, y) = \iint_{\mathbb{C}} Q(t, s) \frac{\partial \Phi_{\varepsilon}}{\partial x}(x - t, y - s) \, dt ds$$

$$- \iint_{\mathbb{C}} P(t, s) \frac{\partial \Phi_{\varepsilon}}{\partial y}(x - t, y - s) \, dt ds$$

$$= -\left[ \iint_{\mathbb{C}} Q(t, s) \frac{\partial \Phi_{\varepsilon}}{\partial t}(x - t, y - s) \, dt ds \right.$$

$$\left. - \iint_{\mathbb{C}} P(t, s) \frac{\partial \Phi_{\varepsilon}}{\partial s}(x - t, y - s) \, dt ds \right] = 0,$$

by (6.5). Thus, $\omega^{\varepsilon}$ is closed and of class $C^{1}$ and, by (6.2), $\int_{\partial \Delta} \omega^{\varepsilon} = 0$ for each triangle $\Delta$ contained in $U$ and $\varepsilon > 0$ small enough. Since $\omega^{\varepsilon} \to \omega$ uniformly on $\Delta$, one has $\int_{\partial \Delta} \omega = 0$ and $\omega$ is locally exact (Corollary 3.12). $\qquad\square$

**Remark 6.3.** Proposition 6.20 may extend to continuous 1-forms $\omega = \sum_{i=1}^{n} P_i dx_i$ in open sets $U$ of $\mathbb{R}^n$. Then condition (6.5) must be replaced by $\int_{U} \omega \wedge d\eta = 0$ for each $(n - 2)$-form $\eta$ of class $C^{\infty}$ with compact support inside $U$. This condition, which is (6.5) when $n = 2$, replaces the condition $d\omega = 0$ that $\omega$ satisfies when it is $C^{1}$ and locally exact. When it is satisfied we say that the equality $d\omega = 0$ holds *in the weak sense* (see Subsection 7.7.2).

*Proof of Theorem* 6.19.  Take the compact set $K$ and the function $\psi$ as in the proof of Theorem 6.8 and change $P$ and $Q$ by $P\psi$ and $Q\psi$, so $\omega$ is now $\psi P dx + \psi Q dy$. Define, for $\varepsilon > 0$, the forms $\omega^{\varepsilon}$, as in the proof of Proposition 6.20; that is, $\omega^{\varepsilon} = P^{\varepsilon} dx + Q^{\varepsilon} dy$ with $P^{\varepsilon}$, $Q^{\varepsilon}$ given by (6.6) from $P\psi$ and $Q\psi$.

Then one has

$$\int_{\Gamma} \omega = \lim_{\varepsilon \to 0} \int_{\Gamma} \omega^{\varepsilon}.$$

Now the forms $\omega^{\varepsilon}$ are $C^{1}$ and, according to the proof of Proposition 6.20, satisfy $Q^{\varepsilon}_x = P^{\varepsilon}_y$ on a neighborhood of $K$ if $\omega$ is locally exact. Hence, we can apply (6.2) which yields $\int_{\Gamma} \omega^{\varepsilon} = 0$ and, consequently, $\int_{\Gamma} \omega = 0$. $\qquad\square$

## 6.5 Characterizations of simply connected domains

There are several ways to characterize, from a topological point of view, simply connected domains. The aim is to rigorously state the idea that the domain has no "hole". A result in this sense is Proposition 6.4, which uses the concept of a cycle homologous to zero. There is another characterization which only requires us to consider Jordan curves contained in the domain, and even another one taking account of the boundary. All of them are given in the following statement:

**Theorem 6.21.** *For a domain $U$ of the complex plane, the following are equivalent:*

a) *$U$ is simply connected, that is, $\mathbb{C}^* \setminus U$ is connected.*

b) *Every cycle contained in $U$ is homologous to zero with respect to $U$.*

c) *For every closed Jordan curve $\gamma$ with $\gamma^* \subset U$, one has $\mathrm{Int}(\gamma) \subset U$.*

d) *The boundary of $U$ in $\mathbb{C}^*$ is connected.*

*Proof.* The equivalence between a) and b) is Proposition 6.4. The equivalence between b) and c) is a consequence of Remark 6.1 and the proof of Proposition 6.4. In order to prove the equivalence with d) suppose, without loss of generality, that $U$ contains the point at infinity and so $K = \mathbb{C}^* \setminus U$ is compact. If $C$ is a connected component of $K$, one has $\partial C = C \cap \partial K$; thus, each non-trivial component of $K$ contains at least a non trivial component of $\partial K = \partial U$, a fact which proves that d) implies a). In order to show the converse, suppose $U$ is simply connected, that is $K$ is connected, and $\partial K$ is not connected. Then $\partial K$ is the union of two disjoint compact sets $E_1, E_2$. Let $\varepsilon > 0$ be the distance between $E_1, E_2$ and consider, as in Lemma 6.6, a grid of the plane formed by closed squares of diagonal strictly smaller than $\varepsilon$; let $Q$ be the union of the squares which intersect $E_1$ in such a way that $E_2$ is in the exterior of $Q$. Remark 6.1 shows that the boundary of $Q$ is composed by closed Jordan polygonal curves $\Gamma_1, \ldots, \Gamma_n$, so that every point of $E_1$ is inside one of the polygons $\mathrm{Int}(\Gamma_i)$ and, by construction, these polygonal curves do not meet $\partial K$. Consequently, one of them, say $\Gamma$, has a point of $E_1$ in its interior, $\mathrm{Int}(\Gamma)$. Since $E_1$ is a part of the boundary of $U$, $\mathrm{Int}(\Gamma)$ intersects $U$ and also $K$. The same may be said about the exterior of $\Gamma$, $\mathrm{Ext}(\Gamma)$, because it contains $E_2$. Hence, $U$ intersects both $\mathrm{Int}\,\Gamma$ and $\mathrm{Ext}(\Gamma)$ and, being connected, will meet the polygonal curve $\Gamma$; similarly, the connected set $K$ intersects $\mathrm{Int}(\Gamma)$ and $\mathrm{Ext}(\Gamma)$ and, therefore, $K \cap \Gamma \neq \emptyset$. Summarizing, the polygonal curve $\Gamma$ intersects $U$ and also $K$ and, therefore, $\partial K \cap \Gamma \neq \emptyset$ which is a contradiction. $\qquad\square$

Generically, one can say that local properties of holomorphic functions hold globally on simply connected domains. Moreover, the validity of these properties characterizes analytically these domains, as the following result specifies:

**Theorem 6.22.** *For a domain $U$ of the complex plane the following are equivalent:*

a) *$U$ is simply connected.*

b) *For every function $f \in H(U)$ and every cycle $\Gamma$ contained in $U$, one has $\int_\Gamma f(z)dz = 0$.*

c) *Every holomorphic function on $U$ has a holomorphic antiderivative on $U$.*

d) *Every real harmonic function on $U$ has a conjugated harmonic function on $U$ and it is, thus, the real part of a holomorphic function.*

e) *Every non-vanishing holomorphic function on $U$ admits a branch of the logarithm on $U$.*

f) *Every non-vanishing holomorphic function on $U$ admits a branch of the square root on $U$.*

g) *For every function $u$, real and harmonic on $U$, there is a function $f$, holomorphic and non-vanishing on $U$, such that $u = \mathrm{Log}\,|f|$ on $U$.*

h) *Every locally exact 1-form with continuous coefficients on $U$ is exact on $U$.*

*Proof.* If $U$ is simply connected, the conclusion of Cauchy's theorem, $\int_\Gamma f(z)dz = 0$, holds for all $f \in H(U)$ and for every cycle $\Gamma \subset U$ and, therefore, a) $\Rightarrow$ b). On the other hand, b) $\Rightarrow$ c) by Theorem 3.13.

The implication c) $\Rightarrow$ d) has already appeared in Subsection 3.7.3; if $u$ is real harmonic on $U$, consider the holomorphic function $f = u_x - iu_y$. If $F' = f$ and $v = \mathrm{Re}\,F$, then $v_x - iv_y = F' = f = u_x - iu_y$; hence, $\vec{\nabla}u = \vec{\nabla}v$ and, thus, $u - v$ is a constant $c$. Then $F + c$ has real part $u$, and its imaginary part is a conjugated function of $u$.

If d) holds and $f \in H(U)$ does not have zeros, consider $h = \mathrm{Log}\,|f|$. This function is harmonic, by Corollary 4.31. If $F$ is a holomorphic function with real part $h = \mathrm{Log}\,|f|$, then $e^F$ is holomorphic and has modulus $e^{\mathrm{Re}\,F} = |f|$; this means that $e^{-F}f$ has modulus 1, and so it is constant. If $e^{-F}f = e^{i\varphi}$, $F + i\varphi$ is a branch of $\log f$ and e) holds.

If e) holds and $f \in H(U)$, $f(z) \neq 0$ for $z \in U$, there exists $g \in H(U)$ with $f = e^g$. Then the function $h = e^{1/2g}$ satisfies $h^2 = f$ and f) follows.

The implication f) $\Rightarrow$ a) also holds. Actually if a) is false, there would exist, by Theorem 6.21, a closed Jordan curve $\gamma \subset U$ and a point $a \notin U$ with $\mathrm{Ind}(\gamma, a) = 1$. Now, the function $z \to z - a$ is holomorphic and non-vanishing on $U$ and it cannot have any branch of the square root, because if $f \in H(U)$ satisfies $f^2(z) = z - a$, one would have $\frac{1}{z-a} = \frac{2f'}{f}$, and so

$$1 = \mathrm{Ind}(\gamma, a) = \frac{1}{2\pi i}\int_\gamma \frac{dz}{z - a} = 2\frac{1}{2\pi i}\int_\gamma \frac{f'(z)}{f(z)}\,dz.$$

This yields $\frac{1}{2\pi i}\int_\gamma \frac{f'}{f}\,dz = \frac{1}{2}$, but taking $\Gamma = f \circ \gamma$, this last integral is $\mathrm{Ind}(\Gamma, 0)$, which is an integer. Hence, f) would not be true.

So far it has been proved that the first six assertions of the theorem are equivalent.

Assertion g) is a consequence of d). Actually, if $u$ is real harmonic, one will have by d) that $u = \mathrm{Re}\, g$ with $g \in H(U)$. The function $f = e^g$ is holomorphic and non-vanishing, and $|f| = e^u$, that is, $u = \mathrm{Log}\,|f|$.

Furthermore, if g) holds, then f) may be proved. Indeed, if $g \in H(U)$ has no zeros, the function $\frac{1}{2}\mathrm{Log}\,|g|$ is harmonic and there will be $f \in H(U)$, without zeros, with $\frac{1}{2}\mathrm{Log}\,|g| = \mathrm{Log}\,|f|$. This means $\frac{g}{f^2}$ is holomorphic on $U$ and has modulus 1. Therefore, it is a constant $e^{i\alpha}$ and the function $e^{i\alpha/2} f$ is a square root of $g$.

Finally, the equivalence a) $\Leftrightarrow$ h) is proved as follows. Let $\omega$ be continuous and locally exact. If $U$ is simply connected, every cycle $\Gamma$ is homologous to 0 with respect to $U$ and, by Theorem 6.19, one has $\int_\Gamma \omega = 0$, which means that $\omega$ is exact on $U$. Conversely for every point $a \notin U$, the form $\frac{1}{z-a}\,dz$ is closed, and if h) holds, it will be exact. Consequently $\int_\Gamma \frac{dz}{z-a} = 0$, that is, $\mathrm{Ind}(\Gamma, a) = 0$ whatever the cycle $\Gamma$ of $U$ be, which means that $U$ is simply connected.     $\square$

**Remark 6.4.** It is clear that statement e) is equivalent to saying that every holomorphic non-vanishing function on $U$ has a branch of the argument on $U$, an assertion that, by f), is equivalent to the existence of a branch of the square root.

Instead, in a non-simply connected domain there may be functions with a branch of its square root but without a branch of its argument (Example 1.21).

## 6.6 The first homology group of a domain and de Rham's theorem. Homotopy

Let $U$ be a domain of the complex plane. The relation $\Gamma_1 \approx \Gamma_2\,(U)$ ($\Gamma_1, \Gamma_2$ are cycles homologous with respect to $U$) states an equivalence relation in the set of all cycles in $U$. Denote by $\widetilde{\Gamma}$ the equivalence class of $\Gamma$ and by $H^1(U)$ the quotient set; $H^1(U)$ is called the *first homology group of $U$*.

Consider now the space of real 1 forms of class $C^1$ that are closed in $U$; this space will be represented as $T(U)$. By Green's theorem (6.2), if $\Gamma_1 \approx \Gamma_2\,(U)$ and $\omega \in T(U)$, one has $\int_{\Gamma_1} \omega = \int_{\Gamma_2} \omega$. This means that

$$\langle \widetilde{\Gamma}, \omega \rangle = \int_\Gamma \omega$$

is a good definition, that is, it does not depend on the chosen representative of the class $\widetilde{\Gamma}$.

Within $T(U)$ there are the exact forms, namely the ones of type $\omega = dh$ with $h \in C^2(U)$, which form a subspace of $T(U)$, represented by $E(U)$. The equivalence relation $\omega_1 \sim \omega_2$ when $\omega_1 - \omega_2$ is exact, has as quotient set the quotient space $T(U)/E(U)$, called *first differentiable cohomology group of $U$* and denoted by $E^1(U)$.

If $\omega_1 - \omega_2$ is exact, then $\int_\Gamma \omega_1 = \int_\Gamma \omega_2$, for any cycle $\Gamma$. Hence, $\langle \widetilde{\Gamma}, \widetilde{\omega} \rangle = \int_\Gamma \omega$ just depends on the equivalence class of $\omega$ in $T(U)/E(U) = E^1(U)$, denoted by $\widetilde{\omega}$. In short,

$$\langle \widetilde{\Gamma}, \widetilde{\omega} \rangle = \int_\Gamma \omega$$

is a well-defined duality between $H^1(U)$ and $E^1(U)$. Each $\widetilde{\omega} \in E^1(U)$ gives a mapping

$$H^1(U) \xrightarrow{I(\widetilde{\omega})} \mathbb{R},$$

$$\widetilde{\Gamma} \longmapsto \langle \widetilde{\Gamma}, \widetilde{\omega} \rangle = \int_\Gamma \omega,$$

which is a group homomorphism:

$$I(\widetilde{\omega})(n_1\Gamma_1 + n_2\Gamma_2) = n_1 I(\widetilde{\omega})(\Gamma_1) + n_2 I(\widetilde{\omega})(\Gamma_2), \quad n_1, n_2 \in \mathbb{Z}.$$

The space of all homomorphisms $H^1(U) \to \mathbb{R}$ is called the *first cohomology group of $U$* and is denoted by $H^1(U)^*$.

**Theorem 6.23** (de Rham's theorem). *The map $\widetilde{\omega} \mapsto I(\widetilde{\omega})$ is a bijection between $E^1(U)$ and $H^1(U)^*$.*

Injectivity here means that if $\omega_1$, $\omega_2$ are closed forms and $\int_\Gamma \omega_1 = \int_\Gamma \omega_2$, for every cycle $\Gamma$ inside $U$, then $\omega_1 - \omega_2$ is exact. This is already known.

Surjectivity means that if one associates a real number $a(\Gamma)$ to each cycle $\Gamma$ of $U$ with $a(\Gamma_1) = a(\Gamma_2)$ if $\Gamma_1 \approx \Gamma_2$ $(U)$ and $a(n_1\Gamma_1 + n_2\Gamma_2) = n_1 a(\Gamma_1) + n_2 a(\Gamma_2)$, then there is a closed form $\omega$ such that $\int_\Gamma \omega = a(\Gamma)$ for each $\Gamma$.

In the case $U$ is simply connected, one has $H^1(U)^* = \{0\}$ and de Rham's theorem implies $E^1(U) = \{0\}$, that is, every closed form is exact.

It may be proved that $H^1(U)$ is a free abelian group for every domain $U$. So if $\widetilde{\Gamma}_i$, $i \in I$, are the generators, the numbers $a(\widetilde{\Gamma}_i)$, $i \in I$, may be freely assigned. Thus, a closed form $\omega$ is exact if and only if $\int_{\Gamma_i} \omega = 0$ for every $i \in I$ and moreover for any family $(a_i)_{i \in I}$ of real numbers, there is a closed form $\omega$ in $U$ such that $\int_{\Gamma_i} \omega = a_i$, $i \in I$.

The proof of Theorem 6.23 may be found in [13], p. 154. Here de Rham's theorem will be justified only in the particular case that $U$ is an $n$-connected domain. Recall this means that $\mathbb{C}^* \setminus U$ has $n$ connected components: the component $C_\infty$, which contains $\infty$, and $n-1$ bounded components $C_1, \ldots, C_{n-1}$, which are compact connected sets of $\mathbb{C}$. In this case it may be proved that $H^1(U)$ is the free abelian group $\mathbb{Z}^{n-1}$, generated by some closed paths $\gamma_i$, $i = 1, 2, \ldots, n-1$ that wind once around each "hole", $C_1, \ldots, C_{n-1}$ of $U$ (Figure 6.5). It is enough to consider $\varepsilon < \min d(C_j, C_l)$, $j \neq l$, cover each component $C_j$ by a finite number of squares with diagonal smaller than $\varepsilon$ and take as $\gamma_j$ the boundary of the union of these squares.

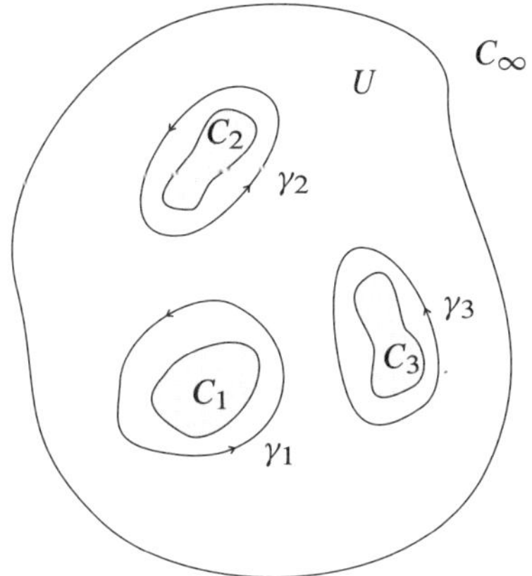


Figure 6.5

Since $\gamma_1, \ldots, \gamma_{n-1}$ are generators of $H^1(U)$, a closed form $\omega$ on $U$ is exact if and only if $\int_{\gamma_i} \omega = 0$, $j = 1, \ldots, n-1$. In other words, de Rham's mapping $I(\tilde{\omega})$ may be interpreted as

$$\gamma_j \mapsto \left( \int_{\gamma_j} \omega \right)_{j=1,\ldots,n-1}.$$

It is easy to show that this mapping is onto. Actually, choose points $\alpha_j \in C_j$, $j = 1, \ldots, n-1$; given $a_1, \ldots, a_{n-1} \in \mathbb{C}$, the form

$$\omega = \frac{1}{2\pi i} \sum_{j=1}^{n-1} a_j \frac{dz}{z - \alpha_j}$$

is closed in $U$ and satisfies

$$\int_{\gamma_k} \omega = \sum_{j=1}^{n-1} a_j \frac{1}{2\pi i} \int_{\gamma_k} \frac{dz}{z - \alpha_j} = \sum_{j=1}^{n-1} a_j \delta_{jk} = a_k.$$

When $\omega$ is of the form $f(z)dz$ with $f$ holomorphic on $U$, this also shows that $E^1(U)$ is isomorphic to $H(U)/H(U)'$, where $H(U)$ denotes, as always, the space of holomorphic functions on $U$ and $H(U)'$ is the subspace of functions with holomorphic antiderivative. This holds for any domain $U$, but it will not be proved here. Moreover, in the general case, in the definition of $T(U)$ and $E(U)$ the degree of differentiability of the forms may be chosen. So, it is possible to work with $C^\infty$ forms in $U$ or to consider that $T(U)$ is the space of locally exact continuous 1-forms in $U$ and $E(U)$ the subspace of exact ones (that is, which are the differential of a function in $C^1(U)$).

A concept related with the homology of paths is the one of *homotopy*. Two closed curves $\gamma_0, \gamma_1 : [0, 1] \to U$ *are homotopic in $U$* if there is a continuous map

$$H : [0, 1] \times [0, 1] \longrightarrow U$$

such that $H(t, 0) = \gamma_0(t)$, $H(t, 1) = \gamma_1(t)$, for $t \in [0, 1]$, and $H(0, s) = H(1, s)$, for $s \in [0, 1]$. The function $H$ is called a *homotopy* between $\gamma_0$ and $\gamma_1$. One may think of $H$ as a family of closed curves $\gamma_s(t) = H(t, s)$ that change continuously from $\gamma_0$ to $\gamma_1$. It is not difficult to show that two homotopic cycles in $U$, are homologous with respect to $U$. The converse is not true; for example, in the open set $U = \mathbb{C} \setminus \{-1, 1\}$, let $\alpha$, $\beta$ be the closed paths given by $\alpha(t) = -1 + e^{2\pi i t}$, $\beta(t) = 1 - e^{2\pi i t}$, $0 \le t \le 1$ and consider the cycles $\Gamma_1 = \alpha\beta\alpha^{-1}\beta^{-1}$, $\Gamma_2 = \beta\alpha\beta^{-1}\alpha^{-1}$ (here $\alpha^{-1}$ denotes the opposite path to $\alpha$ and $\alpha\beta$ the composition of paths, all of them parameterized on $[0, 1]$). Then $\Gamma_1$, $\Gamma_2$ are homologous with respect to $U$, but they are not homotopic in $U$.

**Example 6.24.** Let $U$ be an open set of the plane, starlike with respect to a point. Suppose, without loss of generality, that $0 \in U$ and $U$ is starlike with respect to the origin (this means that for each point $z \in U$, the segment $[0, z]$ is contained in $U$). Then, for each $0 < \lambda < 1$, the dilation $z \to \lambda z$ transforms $U$ into itself, and when $\lambda$ goes from 0 to 1, defines a homotopy between any closed curve $\gamma \subset U$ and the null curve reduced to the origin. Therefore, every closed curve of $U$ is homologous to 0 and $U$ is simply connected. A particular case is when $U$ is convex, because then $U$ is starlike with respect to all the points of $U$. $\qquad\square$

Fixing a point $a \in U$ and restricting all the considerations to closed paths starting and ending at $a$, the relation $\gamma_1 \sim \gamma_2$ if $\gamma_1, \gamma_2$ are homotopic in $U$ is an equivalence relation. The quotient set is called the *first homotopy group* or *fundamental group with base at the point $a$*, and it is denoted by $\Pi(U, a)$. For two points $a, b \in U$, $\Pi(U, a)$ and $\Pi(U, b)$ are isomorphic and one speaks, simply, about the *fundamental group* $\Pi(U)$. This group is not abelian in general. A basic theorem of algebraic topology states that $H^1(U)$ is the abelianization of $\Pi(U)$ (the quotient of $\Pi(U)$ by its commutator).

Of course, if $\Pi(U)$ is trivial, then so is $H^1(U)$. The example given above shows that two homologous cycles are not necessarily homotopic. But it is true that if $H^1(U)$ is trivial (all the cycles are homologous to zero), then $\Pi(U)$ is also trivial (all the cycles are homotopic to a point). In particular, $U$ is simply connected if and only if $\Pi(U)$ is trivial.

**Example 6.25.** If $U$ is simply connected and $\gamma \colon \mathbb{T} \to U$ is a continuous mapping, then $\gamma(\mathbb{T})$ is a closed curve in $U$ that, as said, must be homotopic to a point. The function $H(t, s)$ that defines the homotopy gives, for each $s \in [0, 1]$, a closed curve which may be parameterized by the circle with center $0$ and radius $s$. Hence, a continuous extension of $\gamma$ to the closed disc $\overline{\mathbb{D}}$ is obtained. That is, every continuous mapping from $\mathbb{T}$ to $U$, $U$ being simply connected, is the restriction of a continuous mapping from $\overline{\mathbb{D}}$ to $U$. $\qquad\square$

In summary, it has been shown that the equation $dh = \omega = P\,dx + Q\,dy$ or, equivalently, the system of equations

$$\frac{\partial h}{\partial x} = P, \qquad \frac{\partial h}{\partial y} = Q,$$

with $P, Q \in C^1(U)$ satisfying $\frac{\partial P}{\partial y} = \frac{\partial Q}{\partial x}$, has as many obstructions as $H^1(U)$ has generators. Simply connected open sets are the only domains where these equations may be solved without restrictions.

However, the same question for 2-forms, that is, solving $d\omega = \varphi\,dx \wedge dy$, where now $\omega = P\,dx + Q\,dy$ is the unknown (see Subsection 3.6.4 for the definition of $d\omega$), is equivalent to the equation

$$\frac{\partial Q}{\partial x} - \frac{\partial P}{\partial y} = \varphi$$

with $\varphi$ a given function on $U$, and here there is no obstruction for a domain $U$ of the plane. This is equivalent to the cancellation of a second homology group, $H^2(U)$.

As an ending remark, it is important to say that the definition of $H^1(U)$ given in this section is specific for plane domains since it is based on the winding number. In general, homology groups in any topological space are defined using other concepts.

## 6.7 Harmonic functions on $n$-connected domains

According to Theorem 6.22, a domain $U$ is simply connected if and only if every real harmonic function on $U$ is the real part of a holomorphic function on $U$, that is, it has a conjugated harmonic on $U$. If $U$ is not simply connected and $\alpha \in \mathbb{C}$ is a point of a bounded component of $\mathbb{C} \setminus U$, the function $\operatorname{Log}|z - \alpha|$ is harmonic on $U$ and does not have a conjugated harmonic on $U$.

Once again by Theorem 6.22, one knows that $U$ is simply connected when every function $u$, real and harmonic on $U$, may be written as, $u = \operatorname{Log}|f|$, with $f \in H(U)$ non-vanishing on $U$. Theorem 5.20 shows that the general expression of a real harmonic function on an annulus $U$ centered at the origin, the typical example of a 2-connected domain, is

$$u(z) = a \operatorname{Log}|z| + \operatorname{Re} f$$

with $f$ holomorphic on $U$ and $a \in \mathbb{R}$. Setting

$$u = a \operatorname{Log} \left| z \exp\left(\frac{f}{a}\right) \right|$$

it follows that every real harmonic function $u$ on $U$ is written as $a \operatorname{Log}|g|$ with $g$ holomorphic on $U$, non-vanishing, and $a$ a real constant. Next the ideas of Section 6.6 will be used to generalize this fact to all $n$-connected domains.

In the previous argumentation, if the constant $a$ is an integer, one may write $u = a \operatorname{Log}|g| = \operatorname{Log}|g^a|$ with $g^a$ holomorphic (see Theorem 6.27).

**Theorem 6.26.** *Let $U$ be an $n$-connected domain, $C_1, \ldots, C_{n-1}$ the bounded components of $\mathbb{C}^* \setminus U$, $\gamma_1, \gamma_2, \ldots, \gamma_{n-1}$ a system of regular closed curves which generate $H^1(U)$ and $\alpha_j \in C_j$, for $j = 1, \ldots, n-1$. Then the general expression of a real harmonic function on $U$ is*

$$u(z) = \frac{1}{2\pi} \sum_{j=1}^{n-1} a_j \operatorname{Log}|z - \alpha_j| + \operatorname{Re} f$$

*where $f$ is holomorphic on $U$ and $a_j = \int_{\gamma_j} \frac{\partial u}{\partial \vec{N}}\, ds$ ($\vec{N}$ is the exterior unit normal vector to $\gamma_j$). Thus, $u$ is the real part of a holomorphic function if and only if $a_j = 0$ for $j = 1, \ldots, n-1$.*

*Also*

$$u(z) = \sum_{j=1}^{n-1} b_j \operatorname{Log}|f_j|,$$

*with $f_j \in H(U)$ non-vanishing and $b_j \in \mathbb{R}$, is the general expression of real harmonic functions on $U$.*

*Proof.* Recall that the function $u$, harmonic on $U$, is the real part of a holomorphic function if and only if the holomorphic function $h = u_x - i u_y = 2\frac{\partial u}{\partial z}$ has a holomorphic antiderivative on $U$; that is, if for every closed path $\gamma$ contained in $U$, one has

$$\int_\gamma h(z)dz = i \int_\gamma d^*u = 0, \tag{6.7}$$

with $d^*u = -u_y dx + u_x dy$ and $\int_\gamma d^*u = \int_\gamma \frac{\partial u}{\partial \vec{N}}\, ds$, where $\vec{N}$ is the exterior unit normal vector to $\gamma$ (see Subsection 3.7.3).

As seen in Section 6.6, since $\gamma_1, \ldots, \gamma_{n-1}$ are closed paths generating $H^1(U)$, condition (6.7) is equivalent to

$$\int_{\gamma_j} h(z)dz = 0 \quad j = 1, \ldots, n-1.$$

Now, given $u$ a harmonic function on $U$, consider the associated holomorphic function $h = u_x - i u_y$, define the constants $a_j$, $j = 1, \ldots, n-1$, by

$$\int_{\gamma_j} h(z)\,dz = i \int_{\gamma_j} d^*u = i a_j$$

and write

$$v(z) = u(z) - \frac{1}{2\pi} \sum_{j=1}^{n-1} a_j \operatorname{Log} |z - \alpha_j|.$$

The harmonic function $u_j(z) = \operatorname{Log}|z - \alpha_j|$ has the associated holomorphic function $h_j = (u_j)_x - i(u_j)_y = \frac{1}{z-\alpha_j}$, so that $h_0 = v_x - i v_y = h - \frac{1}{2\pi} \sum_{j=1}^{n-1} a_j h_j$ is the associated one to the harmonic function $v$.

Since

$$\int_{\gamma_k} h_0(z)\,dz = i a_k - \frac{1}{2\pi} \sum_{j=1}^{n-1} a_j \int_{\gamma_k} \frac{dz}{z - \alpha_j}$$

$$= i a_k - \sum_{j=1}^{n-1} i a_j \delta_{jk} = 0, \quad k = 1, 2, \ldots, n-1,$$

it turns out that $h_0$ satisfies condition (6.7) and, hence, $v = \operatorname{Re} f$ with $f$ holomorphic on $U$. $\qquad\square$

Finally, one may ask which functions $u$, real and harmonic on a domain $U$, may be written as $u = \operatorname{Log}|g|$ with $g$ holomorphic and non-vanishing on $U$. If $u = \operatorname{Log}|g|$, one has

$$h = u_x - i u_y = \frac{g'}{g},$$

and, therefore,

$$\frac{1}{2\pi} \int_\gamma d^*u = \frac{1}{2\pi i} \int_\gamma h(z)\,dz = \frac{1}{2\pi i} \int_\gamma \frac{g'(z)}{g(z)} = \operatorname{Ind}(g \circ \gamma, 0),$$

which is an integer number, for every closed curve $\gamma$ contained in $U$. This necessary condition is also sufficient.

**Theorem 6.27.** *If $U$ is a domain of the plane, a real harmonic function $u$ on $U$ is of the form $u = \operatorname{Log}|g|$ with $g$ non-vanishing holomorphic on $U$ if and only if the integral*

$$\frac{1}{2\pi} \int_\gamma d^*u$$

*is integer for any closed path $\gamma$ contained in $U$.*

*Proof.* The proof of the sufficiency will be done only in the case of $n$-connected domains. Using the notations of the proof of Theorem 6.26, one has $a_j = 2\pi k_j$ with $k_j \in \mathbb{Z}$. Then, by the same theorem, it turns out that

$$u(z) = \sum_{j=1}^{n-1} k_j \operatorname{Log} |z - \alpha_j| + \operatorname{Re} f = \operatorname{Log} \left| \prod_{j=1}^{n-1} (z - \alpha_j)^{k_j} \exp(f) \right|$$

and $g = \prod_{j=1}^{n-1} (z - \alpha_j)^{k_j} \exp(f)$ is holomorphic and without zeros in $U$.     $\square$

## 6.8 Exercises

1. Let $U$ be a domain of the plane. Show that if the boundary of $U$ in $\mathbb{C}$ does not have any bounded component, then $U$ is simply connected.

2. Show that a domain $U$ is simply connected if and only if for every bounded open set $V$ with $\partial V \subset U$, one has $V \subset U$.

3. Let $U_1$, $U_2$ be two simply connected domains of $\mathbb{C}$ such that $U_1 \cap U_2$ is connected and non-empty. Prove that $U_1 \cup U_2$ is a simply connected domain.

4. Show that the union of an increasing sequence of simply connected domains is simply connected.

5. Consider the cycle $\Gamma = \gamma_1 - \gamma_2$ with $\gamma_j(t) = r_j e^{it}$ for $j = 1, 2, 0 \le t \le 2\pi$ and $0 \le r_2 < r_1$, and let $f \in C^1(U)$ where $U$ is an open set and $\Gamma^* \cup \{z : \operatorname{Ind}(\Gamma, z) \ne 0\} \subset U$. Show directly that Green's theorem holds for $\Gamma$ and $f$ (formula (6.1)). Use this result to provide a proof of Lemma 6.7.

6. Show that if $\gamma$ is a closed curve in $\mathbb{C}$, then the bounded components of $\mathbb{C} \setminus \gamma^*$ are simply connected, while the unbounded component is doubly connected.

7. Prove that if $K$ is a compact set of the plane and $f$ a holomorphic function on a neighborhood of $K$, then $f$ may be approximated, uniformly on $K$, by linear combinations of functions of the form $z \to \frac{1}{z-w}$ with $w \notin K$.

8. Prove that two closed curves which are homotopic in a domain $U$ are also homologous in $U$. Show that the converse is true for the domain $U = \mathbb{C} \setminus \{0\}$.

9. Let $f : \bar{\mathbb{D}} \to \mathbb{C}$ be continuous and put $\gamma(t) = f(e^{it}), 0 \le t \le 2\pi$. If $w \in \mathbb{C}$ and $\operatorname{Ind}(\gamma, w) \ne 0$, show that $f$ takes the value $w$ in the unit disc $\mathbb{D}$.

10. Let $f : \bar{\mathbb{D}} \to \bar{\mathbb{D}}$ be a continuous function. Show that $f$ has a fix point in $\bar{\mathbb{D}}$.

**11.** Prove the fundamental theorem of algebra, that is, every polynomial $P$ with degree greater than or equal to 1 has at least a complex root. Do this by computing Ind $(\gamma_r, 0)$, where $\gamma_r$ is the image of the circle $C(0, r)$ by the polynomial $P$.

**12.** Show that in every domain $U$ of the plane that has the points $1$ and $-1$ in the same connected component of $\mathbb{C} \setminus U$, a branch of the function $\sqrt{1 - z^2}$ may be defined. Determine all the possible values of the integral

$$\int_\gamma \frac{dz}{\sqrt{1 - z^2}}$$

when $\gamma$ is a piecewise regular closed curve of $U$.

**13.** State and prove a version of Rouché's theorem (Theorem 5.31) for two functions $f$, $g$ continuous on a compact set $K \subset \mathbb{C}$ and holomorphic on the interior of $K$.

**14.** Let $f(z) = \sum_{n=0}^\infty c_n z^n$ in the unit disc $\mathbb{D}$ and let $F \subset \mathbb{D}$ be a closed set which contains the origin. Write $m = \inf\{|f(z)|: z \in \partial F\}$ and let $N$ be the number of zeros of $f$ that belong to $F$. Prove that

$$m \leq |c_0| + |c_1| + \cdots + |c_N|.$$

Show also that this estimate improves when applied to the sequence of powers of $f$.

**15.** Let $U$ be an annulus centered at the origin. Prove the following statements:

a) If $\Gamma_1$ and $\Gamma_2$ are two cycles of $U$ and $f \in H(U)$, then

$$\text{Ind} \, (\Gamma_1, 0) \int_{\Gamma_2} f(z) \, dz = \text{Ind}(\Gamma_2, 0) \int_{\Gamma_1} f(z) \, dz.$$

b) If $f \in H(U)$ does not vanish on $U$, then $f$ has a branch of the logarithm in $U$ if and only if $\text{Ind}(f \circ \Gamma, 0) - 0$ for every cycle $\Gamma$ of $U$ such that $\text{Ind}(\Gamma, 0) \neq 0$. Show that in this case one has indeed, $\text{Ind}(f \circ \Gamma, 0) = 0$ for each cycle $\Gamma$ of $U$.

**16.** Let $\Gamma$ be a cycle homologous to zero with respect to the domain $U$ and $f \in H(U)$. Prove the following formulae:

$$\frac{1}{2\pi i} \int_\Gamma \overline{f(z)} f'(z) dz = \frac{1}{\pi} \iint_\mathbb{C} |f'(z)|^2 \, \text{Ind} \, (\Gamma, z) \, dm(z)$$

$$= \frac{1}{\pi} \iint_\mathbb{C} \sum_{z \in f^{-1}(w)} \text{Ind} \, (\Gamma, z) \, dm(w).$$

If $\mathrm{Ind}\,(\Gamma, z) \in \{0, 1\}$ for every point $z \notin \Gamma^*$, setting $G = \{z : \mathrm{Ind}\,(\Gamma, z) = 1\}$, the last integral is

$$\frac{1}{\pi} \iint_{\mathbb{C}} \#\{z \in G : f(z) = w\}dm(w)$$

and it represents the area of $f(G)$ counting multiplicities (the notation $\#A$ is used to denote the number of elements of a finite set $A$).

17. Let $\Gamma$ be a regular cycle homologous to zero with respect to the domain $U$ and $f, g \in C^2(U)$. Prove the following formulae:

i)

$$\int_{\Gamma} \left\langle f \frac{\partial g}{\partial \bar z}, \vec T \right\rangle |dz| + i \int_{\Gamma} \left\langle f \frac{\partial g}{\partial \bar z}, \vec N \right\rangle |dz|$$

$$= \frac{i}{2} \iint_{\mathbb{C}} \bar f(z) \Delta \bar g(z)\, \mathrm{Ind}\,(\Gamma, z)\, dm(z).$$

$$+ 2i \iint_{\mathbb{C}} \frac{\partial \bar f}{\partial \bar z}(z) \frac{\partial \bar g}{\partial z}(z)\, \mathrm{Ind}\,(\Gamma, z)\, dm(z).$$

ii)

$$\int_{\Gamma} f \cdot \frac{\partial g}{\partial \vec N} |dz| = \iint_{\mathbb{C}} (f \Delta g + \langle \vec\nabla f, \vec\nabla g\rangle)(z)\, \mathrm{Ind}(\Gamma, z)\, dm\,(z)$$

iii)

$$\int_{\Gamma} f \cdot \frac{\partial g}{\partial \vec T} |dz| = \iint_{\mathbb{C}} \left( \frac{\partial f}{\partial x} \cdot \frac{\partial g}{\partial y} - \frac{\partial f}{\partial y} \cdot \frac{\partial g}{\partial x} \right)(z)\, \mathrm{Ind}(\Gamma, z)\, dm\,(z).$$

As usual, $\vec T$, $\vec N$ denote the unit tangent vector and the exterior unit normal vector to $\Gamma$, respectively. In formula i) a complex number and the corresponding vector of $\mathbb{R}^2$ are identified.

18. Compute

$$\int_{C(0,3/2)} \frac{\mathrm{sh}\,5z}{(1 + z^2)z^2}\, dz.$$

19. Let $\gamma$ be the ellipse centered at the origin with semi-axis $a, b > 0$, travelled in the positive sense. Compute

$$\int_0^{2\pi} \frac{dt}{a^2 \cos^2 t + b^2 \sin^2 t}$$

using $\mathrm{Ind}\,(\gamma, 0)$.

**20.** Let $U$ be an $n$-connected domain of $\mathbb{C}$ and $C_1, C_2, \ldots, C_{n-1}, C_\infty$ the connected components of $\mathbb{C}^* \setminus U$ ($\infty \in C_\infty$). Prove that there exist cycles $\Gamma_1, \Gamma_2, \ldots, \Gamma_{n-1}$ in $U$ such that

$$\begin{aligned} \mathrm{Ind}(\Gamma_j, a) &= 1 \quad \text{if } a \in C_j, \\ \mathrm{Ind}(\Gamma_j, a) &= 0 \quad \text{if } a \in (\mathbb{C} \setminus U) \setminus U_j, \end{aligned} \qquad j = 1, \ldots, n-1.$$

Show now that for any cycle $\Gamma$ in $U$ there exists a unique linear combination $a_1 \Gamma_1 + \cdots + a_{n-1} \Gamma_{n-1}$ with $a_1, \ldots, a_{n-1} \in \mathbb{Z}$ such that

$$\Gamma \approx a_1 \Gamma_1 + \cdots + a_{n-1} \Gamma_{n-1} \quad \text{with respect to } U.$$

# Chapter 7

# Harmonic functions

With the results of Section 3.7 it has been established that, in the case of domains of the complex plane, there is an important link between complex analysis and the theory of harmonic functions. Just recall, for example, that every real harmonic function is locally the real part of a holomorphic function. More significantly, holomorphic functions correspond locally with vector fields that are both conservative and solenoidal, and these vector fields are exactly the gradients of the harmonic functions.

In this chapter, harmonic functions and the Laplace operator will be studied systematically in the context of real variables. The main problems to deal with are the Dirichlet problem, the Neumann problem and the solution of the Poisson equation. The Poisson equation, with boundary conditions, leads to non-homogeneous Dirichlet and Neumann problems.

Since much of the development is in terms of real variables, we have chosen to work on domains of $\mathbb{R}^n$, highlighting explicitly the specifities of the case $n = 2$ and their relation with the theory of holomorphic functions.

The Laplace operator appears in most of the equations of classical mathematical physics, a major reason why the relation between harmonic or holomorphic functions and problems of physics is of great interest. The chapter starts by describing in detail some examples of this relationship.

## 7.1 Problems of classical physics and harmonic functions

### 7.1.1 Distribution of heat in the stationary case

Suppose that a substance fills up a body $C$ in the space $\mathbb{R}^3$, on which a heat flow is distributed. Denote by $T(x, t)$ the temperature at the point $x \in C$ at the instant $t$. When the temperature does not depend on $t$, one says that the distribution is *stationary*. Consider also the heat flow vector field $\vec{H}(x, t)$ which, at any point and at any instant, indicates the direction and the quantity of the heat flow. This way, if $S$ is a closed surface contained in $C$ and $\vec{N}$ is the unit normal vector exterior to $S$, $\int_S \langle \vec{H}, \vec{N} \rangle \, dA$ represents the amount of heat flow that leaves $C$ through $S$. *Fourier's law* states that, at any instant, $\vec{H} = -k\vec{\nabla}T$ holds, where $k > 0$ is the *thermal conductivity*. This equation translates the intuitive fact that the heat flow goes from the warmer areas to the colder ones, with an intensity proportional to the difference of temperatures. Other variables that take part here are the *density* $\rho$ and the *heat capacity* $c$ of the substance that fills up the body $C$; one may think of

$c(x)$ as the amount of heat required to increase by one degree the temperature one gram of a substance located at the point $x$. Suppose that $B$ is a ball inside $C$ with boundary $S$; the heat flow that enters through $S$ between two instants $t_1 < t_2$ is

$$\int_{t_1}^{t_2} \int_S k \langle \vec{\nabla} T, \vec{N} \rangle \, dA \, dt = \int_{t_1}^{t_2} \int_S k \frac{\partial T}{\partial \vec{N}} \, dA \, dt.$$

If there are heat sources with known density $F(x,t)$, the quantity of heat that has entered inside $B$ during a time interval $(t_1, t_2)$ is

$$\int_{t_1}^{t_2} \int_B F \, dV \, dt + \int_{t_1}^{t_2} \int_S k \frac{\partial T}{\partial \vec{N}} \, dA \, dt.$$

This quantity of heat has been invested in passing from the distribution $T$ at the moment $t_1$ to the distribution $T$ at the moment $t_2$; therefore, it is equal to

$$\int_B c(x)\rho(x)(T(x,t_2) - T(x,t_1)) \, dV(x) = \int_B \int_{t_1}^{t_2} c(x)\rho(x) \frac{\partial T}{\partial t} \, dt \, dV(x).$$

Equating and applying the divergence theorem to the integral over $S$, it turns out that

$$\int_B \int_{t_1}^{t_2} \left( c(x)\rho(x) \frac{\partial T}{\partial t} - \mathrm{div}(k(x)\vec{\nabla} T) \right) dt \, dV(x) = \int_B \int_{t_1}^{t_2} F \, dt \, dV.$$

Since this holds for every ball and every time interval, this leads to the partial differential equation that controls the heat diffusion,

$$c(x)\rho(x) \frac{\partial T}{\partial t} - \mathrm{div}(k(x)\vec{\nabla} T) = F.$$

When $c$, $\rho$ and $k$ are constants one finds

$$a \frac{\partial T}{\partial t} - \Delta T = F,$$

with $a$ constant.

In general, in order to find the distribution of heat $T(\cdot, t)$ at every instant $t > 0$, one needs to know the initial distribution $T(\cdot, 0)$ and the boundary conditions on $C$. In the stationary case, one gets $\Delta T = -F$ and so the distribution of the temperature is a harmonic function on the domains where there is no heat source. The boundary conditions might be to know $T$ (room temperature) or $\frac{\partial T}{\partial \vec{N}}$ (insulation meaning that this derivative is zero) at the boundary. The problem

$$\Delta T = 0, \quad T(x) = \varphi(x), \quad x \in \partial C,$$

with a function $\varphi$ defined on $\partial C$, is a first example of *Dirichlet's problem*, and the problem

$$\Delta T = 0, \quad \frac{\partial T}{\partial \vec{N}}(x) = \varphi(x), \quad x \in \partial C,$$

with a function $\varphi$ given on $\partial C$, is an example of *Neumann's problem*.

The level surfaces of $T$, $T(x) = c$, are the *isothermal* surfaces. The problem of the distribution of heat may be also stated on a plane domain. In this case $T(x) = c$ defines the isothermal lines. If the harmonic function $T$ has a conjugated harmonic function $\widetilde{T}$, the level lines of $\widetilde{T}$, which are perpendicular to the isothermal lines (Section 3.7.3), are the ones followed by the flow.

## 7.1.2 Newtonian vector fields

A *newtonian vector field* on $\mathbb{R}^3$ is a vector field given by a mass distribution, according to Newton's law of universal gravitation. In the case of a point mass located at the origin, the vector field is, except for some constant,

$$\vec{X}(x) = -\frac{x}{|x|^3}, \quad x \in \mathbb{R}^3.$$

If the mass distribution is given by a density $\rho$ in a body $C$, one has to add the vector fields given by all infinitesimal masses $\rho(y)\, dV(y)$ and so the field is given by

$$\vec{X}(x) = -\int_C \frac{x-y}{|x-y|^3}\rho(y)\, dV(y).$$

A newtonian vector field is always conservative: just observe that, in the case of a point mass, one has

$$-\frac{x}{|x|^3} = \vec{\nabla}\frac{1}{|x|}.$$

In the case of a density $\rho$, summing up it turns out that

$$-\int_C \frac{x-y}{|x-y|^3}\rho(y)\, dV(y) = \vec{\nabla}\int_C \frac{1}{|x-y|}\rho(y)\, dV(y),$$

so that

$$\Phi(x) = \int_C \frac{1}{|x-y|}\rho(y)\, dV(y)$$

is the potential function of the vector field. This potential is called *newtonian potential*. The same considerations hold for electric fields, since Coulomb's law is formally equal to Newton's law.

An easy computation shows that for the unit point mass the equality

$$\operatorname{div}\frac{x}{|x|^3} = 0 \quad \text{if } x \neq 0,$$

holds, and hence also

$$\operatorname{div}\int_C \frac{x-y}{|x-y|^3}\rho(y)dV(y) = 0 \quad \text{if } x \notin C.$$

This means that the flow of the vector field $\vec{X}$ is zero on every surface that contains no mass in its interior. Now, newtonian vector fields are not solenoidal; for a point mass, if $S$ is the sphere of radius $r$ centered at the origin, the scalar product $\langle \vec{X}, \vec{N} \rangle$ equals $-\frac{1}{r^2}$. Therefore, the flow through $S$ is

$$\int_S \langle \vec{X}, \vec{N} \rangle dA = -4\pi.$$

This will be also the flow through any closed surface having the point mass inside, as it may be seen applying the divergence theorem to the difference between the surface and a sphere. If $S = \partial C$ is a surface that wraps a body $C$ and there is no mass on $S$, it yields *Gauss' law*, asserting that the flow of a vector field through $S$ is $-4\pi M$, where $M$ is the total mass contained in $C$. It may be formulated with the equality

$$\int_S \langle \vec{X}, \vec{N} \rangle \, dA = -4\pi \int_C \rho \, dV.$$

Over domains where the vector field $\vec{X}$ and the potential $\Phi$ are regular enough, the divergence theorem gives $\int_S \langle \vec{X}, \vec{N} \rangle \, dA = \int_C \operatorname{div} \vec{X} \, dV$ and one has

$$\operatorname{div} \vec{X} = -4\pi\rho, \quad \text{or} \quad \operatorname{div}(\operatorname{grad} \Phi) = \Delta\Phi = -4\pi\rho.$$

This last equation is an example of *Poisson's equation*. In particular, the potential $\Phi$ is a harmonic function on the regions without mass or charge. So far, the setting has been $\mathbb{R}^3$; now, if $\rho$ is independent from one of the three variables, so is $\Phi$, and hence it is also interesting to study Poisson's equation in dimension 2. In this chapter it will be seen in detail when the vector field or the potential are regular enough and how Poisson's equation must be interpreted in general.

### 7.1.3 Flow of an ideal fluid

**a) The continuity equation.**   Consider a region of the space where a fluid flows, a liquid for example. How may the motion of the fluid be described? At first one must consider a velocity vector field $\vec{X}(x,t)$ which indicates the velocity of the particle that at the instant $t$ is at $x$ and also the density of the fluid, $\rho(x,t)$, so that

$$\int_C \rho(x,t) \, dV(x)$$

is the total mass of fluid contained inside the domain $C$, at the instant $t$. Furthermore, we can consider the trajectories $\phi_t(x)$ that denote the position, at the instant $t$, of the particle that at the instant $t = 0$ is located at $x$; the trajectories are the solutions of the system of differential equations

$$\frac{dy}{dt} = \vec{X}(y, t), \quad y(0) = x.$$

Impose, now, the law of conservation of mass. Consider a region $C$ at the instant $t$ and compute in two different ways the rate of change of the mass of fluid that it contains: on one hand, it is

$$\frac{d}{dt} \int_C \rho(x, t)\, dV(x) = \int_C \frac{\partial \rho}{\partial t}(x, t)\, dV(x).$$

On the other hand, the mass that leaves $C$ through the boundary, does it with a rate equal to

$$\int_{\partial C} \rho(x, t)(\langle \vec{X}(x, t), \vec{N} \rangle)\, dA(x) = \int_C \mathrm{div}(\rho \vec{X})\, dV(x).$$

Changing the sign of the second rate and equating to the first one, it yields the *continuity equation*:

$$\frac{\partial \rho}{\partial t} + \mathrm{div}(\rho \vec{X}) = 0. \tag{7.1}$$

This equation may be reached also by imposing that the mass of fluid contained in $\phi_t(C)$ at the instant $t$ is constant over time, and it will be useful to reobtain (7.1) from this point of view: if $J_t(x)$ denotes the Jacobian of the transformation $\phi_t$, one has

$$\int_{\phi_t(C)} \rho(x, t)\, dV(x) = \int_C \rho(\phi_t(x), t) J_t(x)\, dV(x).$$

Imposing the conservation of mass means that the equation

$$0 = \frac{d}{dt}(\rho(\phi_t(x), t) J_t(x)) = J_t\left(\frac{\partial \rho}{\partial t} + \mathrm{grad}(\rho)\vec{X}\right) + \rho\frac{\partial}{\partial t}(J_t(x)).$$

must hold. A computation shows that $\frac{\partial}{\partial t}(J_t(x)) = J_t(x)\,\mathrm{div}(\vec{X})$ and dividing by $J_t$ the continuity equation is found again.

The same calculus is useful to differentiate integrals as

$$\frac{d}{dt} \int_{\phi_t(C)} \rho(x, t) F(x, t)\, dV(x), \tag{7.2}$$

where $F$ is a scalar or a vector function. Actually, doing the same change of variables as above, differentiating under the integral sign and using that $\rho J_t$ has null derivative with respect to $t$, one obtains for (7.2) the expression

$$\frac{d}{dt} \int_C \rho(\phi_t(x,t),t) J_t(x) F(\phi_t(x),t) \, dV(x)$$

$$= \int_C \rho(\phi_t(x),t) J_t(x) \left( \frac{\partial F}{\partial t}(\phi_t(x),t) + \langle \vec{X}, \vec{\nabla} F(\phi_t(x),t) \rangle \right) dV(x).$$

Changing the variables again, it leads to

$$\frac{d}{dt} \int_{\phi_t(C)} \rho(x,t)\, F(x,t) \, dV(x)$$

$$= \int_{\phi_t(C)} \rho(x,t) \left( \frac{\partial F}{\partial t} + \langle \vec{X}, \vec{\nabla} F(x,t) \rangle \right) dV(x).$$

Here $\vec{\nabla} F$ acts componentwise if $F$ is vectorial. It is common to use the notation $\langle \vec{X}, \vec{\nabla} \rangle$ for the operator $\sum X_i \frac{\partial}{\partial x_i}$, so that the last term may be also written as $\langle \vec{X}, \vec{\nabla} \rangle F$.

**b) Perfect fluids: Euler's equations.**   When $F = \vec{X}$, the expression (7.2) is the variation of momentum, which, according to Newton's second law, must be equal to the net force that acts at the instant $t$ over $\phi_t(C)$. This force has two components, an external one and an internal one. It is supposed that external forces are described by means of a density of external force by unit of mass $f(x,t)$, so that

$$\int_{\phi_t(C)} \rho(x,t) f(x,t) \, dV(x)$$

is the external net force acting on $\phi_t(C)$. *Cauchy's principle* establishes that the internal forces act on the surface that limits $\phi_t(C)$ and are described by an area density of force $\vec{\Sigma}(x, \vec{N}, t)$, so that

$$\int_{\partial \phi_t(C)} \vec{\Sigma}(x, \vec{N}, t) \, dA(x)$$

is the internal net force. As always, $\vec{N}$ denotes the normal vector to $\partial \phi_t(C)$. Moreover, $\vec{\Sigma}(x, \vec{N}, t)$ is linear in $\vec{N}$, that is, $\vec{\Sigma}(x, \vec{N}, t) = \langle \mathbf{S}(x,t), \vec{N} \rangle$, where $\mathbf{S}$ is a matrix, called the *stress tensor*. Using the divergence theorem one may write the internal net force in the form

$$\int_{\phi_t(C)} \mathrm{div}(\mathbf{S}(x,t)) \, dV(x),$$

where the divergence operator is interpreted as acting on each row of $\mathbf{S}$. Applying Newton's law, one obtains *Euler's equations*

$$\rho(x,t)\left(\frac{\partial \vec{X}}{\partial t} + \langle \vec{X}, \vec{\nabla}\rangle \vec{X}(x,t)\right) = \rho(x,t)f(x,t) + \operatorname{div}\mathbf{S}(x,t).$$

A fluid is said to be *perfect* if the internal forces acting on each region do it only in the normal direction to the boundary, at each point. That is, if $\mathbf{S}(x,t) = -p(x,t)\mathbf{I}$ holds for a certain function $p$, called the *internal pressure* of the fluid, $\mathbf{I}$ representing the identity matrix. In this case, Euler's equations may be written as

$$\rho(x,t)\left(\frac{\partial \vec{X}}{\partial t} + \langle \vec{X}, \vec{\nabla}\rangle \vec{X}(x,t)\right) = \rho(x,t)f(x,t) - \vec{\nabla}p(x,t).$$

Observe that so far there are four equations, one establishing continuity, the other three obtained by taking components in Euler's equations. But there are five unknowns: the three components of $\vec{X}$, $\rho$ and $p$. Therefore, the problem of describing the motion of the fluid is not to be determined.

**c) Incompressible fluids.** A fluid is said to be *incompressible* if the volume of $\phi_t(C)$ remains constant with respect to $t$, for any region $C$; this is equivalent to imposing $J_t(x) = 1$, that is, $\operatorname{div}(\vec{X}) = 0$; by the continuity equation, it is also equivalent to $\frac{d}{dt}\rho(\phi_t(x), x) = 0$, which means that the density is constant along the trajectories. For a perfect and incompressible fluid there is one more equation, altogether five, and these five equations should suffice to determine the motion, for fixed initial or boundary conditions. If the fluid is, moreover, homogeneous, in the sense that the density $\rho$ is constant with respect to $x$, then it is incompressible when $\rho$ is constant also with respect to $t$. Euler's equations which control the motion of the fluid are in this case, supposing $\rho = 1$,

$$\frac{\partial \vec{X}}{\partial t} + \langle \vec{X}, \vec{\nabla}\rangle \vec{X}(x,t) = f(x,t) - \vec{\nabla}p(x,t), \quad \operatorname{div}(\vec{X}) = 0. \qquad (7.3)$$

In order to determine the motion completely, one will need initial conditions on $p$, $\vec{X}$ and also boundary conditions on $\vec{X}$. For example, if the motion takes place inside a tube with an impermeable wall, a natural boundary condition is $\langle \vec{X}, \vec{N}\rangle = 0$, where $\vec{N}$ is the normal vector to the boundary.

An important concept is the *vorticity*, $\vec{\xi}$, of a fluid. This quantity measures the tendency to spin of the velocity vector field $\vec{X}$ and it is defined as the rotational of $\vec{X}$, that is, $\vec{\xi} = \operatorname{rot}(\vec{X}) = \vec{\nabla} \times \vec{X}$. The fluid is called *irrotational* if $\vec{\xi} = 0$. For perfect, incompressible and homogenous fluids (suppose $\rho = 1$) the equations of motion may be expressed in terms of the vorticity in the following way:

In order to simplify, suppose that there are no external forces acting, that is, $f(x,t) = 0$. The identity

$$\frac{1}{2}\vec{\nabla}|X|^2 = \vec{X} \times (\vec{\nabla} \times \vec{X}) + \langle \vec{X}, \vec{\nabla} \rangle \vec{X}$$

substituted in the first equation of (7.3) gives

$$\frac{\partial \vec{X}}{\partial t} + \frac{1}{2}\vec{\nabla}(|\vec{X}|^2) - \vec{X} \times (\vec{\nabla} \times \vec{X}) = -\vec{\nabla} p(x,t).$$

Taking rotationals the gradients disappear and it turns out that

$$\frac{\partial \vec{\xi}}{\partial t} - \vec{\nabla} \times (\vec{X} \times \vec{\xi}) = 0,$$

or

$$\frac{\partial \vec{\xi}}{\partial t} - \left\{ \langle \vec{\xi}, \vec{\nabla} \rangle \vec{X} - \vec{\xi} \langle \vec{\nabla}, \vec{X} \rangle - \langle \vec{X}, \vec{\nabla} \rangle \vec{\xi} + \vec{X} \langle \vec{\nabla}, \vec{\xi} \rangle \right\} = 0.$$

The last term is zero (divergence of a rotational) and using $\operatorname{div}(\vec{X}) = \langle \vec{\nabla}, \vec{X} \rangle = 0$, Euler's equations are written as

$$\frac{\partial \vec{\xi}}{\partial t} - \langle \vec{\xi}, \vec{\nabla} \rangle \vec{X} + \langle \vec{X}, \vec{\nabla} \rangle \vec{\xi} = 0, \quad \operatorname{div}(\vec{X}) = 0. \tag{7.4}$$

**d) Fluids on a plane.**   In the case of planc fluids, the only effective component of the vorticity $\vec{\xi}$ is the third one, which is denoted by $\xi$, and the second term of the first equation of (7.4) vanishes. Euler's equations are then

$$\frac{\partial \xi}{\partial t} + \langle \vec{X}, \vec{\nabla} \rangle \xi = 0, \quad \operatorname{div}(\vec{X}) = 0.$$

Suppose that the region $U$ of the plane where the motion takes place is simply connected; then the incompressibility condition on the vector field $\vec{X} = (u, v)$ is $u_x + v_y = 0$, which says that $-v\,dx + u\,dy$ is a closed form and, hence, exact. There exists, then, a function $\psi(x, y, t)$ such that $u = \psi_y$, $v = -\psi_x$ (it is common to write $\vec{X} = \vec{\nabla}^\perp \psi$). For $t$ fixed, the streamlines are the level curves of $\psi$. In terms of $\psi$, the vorticity is $\xi = -\Delta\psi$. If the condition $\vec{X} \cdot \vec{N} = 0$ holds on the boundary of $U$, then this boundary must be a level curve and, adding a constant, one may suppose $\psi = 0$ holds on the boundary. With all this, the motion of a perfect, incompressible and homogeneous plane fluid is described by the equations

$$\frac{\partial \xi}{\partial t} + \langle \vec{X}, \vec{\nabla} \rangle \xi = 0, \quad \Delta\psi = -\xi, \quad \psi(x) = 0, \quad x \in \partial U, \quad \vec{X} = \vec{\nabla}^\perp \psi.$$

Knowing $\xi$ at an instant $t$, then $\psi$ is also known and, therefore, also $\vec{X}$ at the same instant. Hence, $\xi$ determines $\frac{\partial \xi}{\partial t}$, so one just needs to know the initial vorticity.

If the motion of the fluid takes place in an open set $U$ of $R^3$, simply connected, there a is a potential function $\Phi$ such that $\vec{X} = \vec{\nabla}\Phi$ and one gets Laplace's equation

$$\Delta\Phi = \vec{\nabla}^2\Phi = 0$$

on $U$. In this case, the natural boundary conditions consist of prescribing the value of $\langle \vec{X}, \vec{N} \rangle = \frac{\partial \Phi}{\partial \vec{N}}$ on the boundary of $U$ and this poses a Neumann's problem.

## 7.2  Harmonic functions on domains of $\mathbb{R}^n$

### 7.2.1  The Laplacian

The presence of the Laplace operator in mathematical physics equations may be explained by means of laws of physics, as it has been done in the previous section. But it can be also justified from a mathematical perspective.

Laws of classical physics use measures of position and time referred to a coordinate system and an origin of time. Let the coordinates $(x_1, x_2, \ldots, x_n)$ denote the position of a point in $\mathbb{R}^n$ with respect to certain cartesian axes and an origin of coordinates. Considering other cartesian axes and another origin means passing from the coordinates $(x_1, \ldots, x_n)$ to coordinates $(y_1, \ldots, y_n)$; the relation between the column vectors $\vec{X} = \begin{pmatrix} x_1 \\ \vdots \\ x_n \end{pmatrix}$ and $\vec{Y} = \begin{pmatrix} y_1 \\ \vdots \\ y_n \end{pmatrix}$ is

$$\vec{Y} = \vec{A} + O\vec{X}$$

where $\vec{A}$ is a translation vector and $O$ is a unitary transformation, that is, it preserves the scalar product and so satisfies $O^t = O^{-1}$ ($O^t$ is the transpose of $O$). The transformation $\vec{X} \to \vec{Y}$ is called *isometry* of $\mathbb{R}^n$ and the set of all isometries is a group of transformations of $\mathbb{R}^n$.

The linear differential operators are written, in a system of coordinates $x = (x_1, x_2, \ldots, x_n)$, as

$$L = \sum_{|\alpha| \leq N} a_\alpha(x) D^\alpha.$$

Here, $\alpha = (\alpha_1, \ldots, \alpha_n)$ with $\alpha_i \in \mathbb{N}$ is a multi-index, $|\alpha| = \alpha_1 + \cdots + \alpha_n$ is its length,

$$D^\alpha = \frac{\partial}{\partial x_1^{\alpha_1}} \cdots \frac{\partial}{\partial x_n^{\alpha_n}}$$

and $a_\alpha(x)$ are functions. The natural number $N$ is the order of the operator. Differential operators act on functions $u$ according to $u \mapsto Lu = \sum_{|\alpha| \leq N} a_\alpha(x) D^\alpha u(x)$.

If $\varphi : \mathbb{R}^n \to \mathbb{R}^n$ is a diffeomorphism (a general change of coordinates), $\varphi^* L$ will be the operator defined by the equation

$$(\varphi^* L)(v) \circ \varphi = L(v \circ \varphi).$$

If $L = \sum_{|\alpha| \leq N} a_\alpha(x) D_x^\alpha$, then $\varphi^* L$ will have an expression $\sum b_\beta(y) D_y^\beta$, obtained by iterating the chain rule and identifying coefficients in the equality

$$\sum b_\beta(y) D_y^\beta v = \sum a_\alpha(x) D_x^\alpha (v \circ \varphi).$$

We say that $L$ is *invariant under* $\varphi$ if $\varphi^* L = L$, that is, if $(Lv) \circ \varphi = L(v \circ \varphi)$ for any function $v$.

Clearly, the laws of classical physics do not depend on an arbitrary selection of the axes and the origin of coordinates. In other words, if one describes phenomena invariant under isometries, the operators that one uses must also be invariant.

**Theorem 7.1.** *The differential operators $L$ which are invariant under the group of isometries are, exactly, the ones of the form $L = P(\Delta)$ where $P$ is a polynomial in one variable and $\Delta$ the Laplacian.*

*Proof.* First observe that the natural rules

$$(\varphi_1^* \circ \varphi_2^*)(L) = (\varphi_1 \circ \varphi_2)^*(L),$$
$$\varphi^*(L_1 \circ L_2) = \varphi^* L_1 \circ \varphi^* L_2$$

hold for $\varphi_1$, $\varphi_2$ diffeomorphisms of $\mathbb{R}^n$ and $L_1$, $L_2$ linear differential operators. Therefore, in order to see that every operator $P(\Delta)$ is invariant under the group of isometries, it is enough to show that $\Delta$ is invariant under translations and unitary transformations. The invariance under translations is obvious. If $\varphi$ is a unitary transformation with matrix $O = (u_{ij})$, then $y_i = \varphi_i(x) = \sum_{j=1}^n u_{ij} x_j$ and one has

$$\Delta(v \circ \varphi)(x) = \Delta[v(\varphi_1 x, \ldots, \varphi_n x)]$$

$$= \Delta\left[v\left(\sum_j u_{1j} x_j, \ldots, \sum_j u_{nj} x_j\right)\right]$$

$$= \sum_{k=1}^n \frac{\partial^2}{\partial x_k^2}\left[v\left(\sum_j u_{1j} x_j, \ldots, \sum_j u_{nj} x_j\right)\right]$$

$$= \sum_{k=1}^n \frac{\partial}{\partial x_k} \sum_{\rho=1}^n \frac{\partial v}{\partial y_\rho} u_{\rho k} = \sum_{k,\rho=1}^n \sum_{j=1}^n \frac{\partial^2 v}{\partial y_\rho \partial y_j} u_{\rho k} u_{jk}$$

$$= \sum_{\rho,j=1}^n \frac{\partial^2 v}{\partial y_\rho \partial y_j} \sum_{k=1}^n u_{\rho k} u_{jk} = \sum_{\rho,j=1}^n \frac{\partial^2 v}{\partial y_\rho \partial y_j} \delta_{\rho j} = \sum_{\rho=1}^n \frac{\partial^2 v}{\partial y_\rho^2}$$

$$= (\Delta v)(y).$$

Conversely, assume that $L = \sum a_\alpha(x)D^\alpha$ is invariant under isometries. The invariance under translations forces each $a_\alpha$ to be constant. Put the function $v_\xi(x) = e^{\langle \xi, x\rangle}$ in the equality $(Lv) \circ O = L(v \circ O)$, valid for every unitary transformation $O$. Since $v_\xi(Ox) = e^{\langle \xi, Ox\rangle} = e^{\langle O^t\xi, x\rangle}$, we get

$$L(v_\xi \circ O)(x) = \sum_\alpha a_\alpha D^\alpha \big[e^{\langle O^t\xi, x\rangle}\big] = \Big[\sum_\alpha a_\alpha (O^t\xi)_1^{\alpha_1} \ldots (O^t\xi)_n^{\alpha_n}\Big] e^{\langle O^t\xi, x\rangle},$$

which may be written as $\big[\sum_\alpha a_\alpha (O^t\xi)^\alpha\big] e^{\langle O^t\xi, x\rangle}$. On the other hand, $Lv_\xi(x) = \big(\sum_\alpha a_\alpha\xi^\alpha\big) e^{\langle \xi, x\rangle}$ and $Lv_\xi(Ox) = \big(\sum_\alpha a_\alpha\xi^\alpha\big) e^{\langle \xi, Ox\rangle} = \big(\sum_\alpha a_\alpha\xi^\alpha\big) e^{\langle O^t\xi, x\rangle}$. Consequently, one must have

$$\sum_\alpha a_\alpha\xi^\alpha = \sum_\alpha a_\alpha (O^t\xi)^\alpha$$

for every unitary matrix $O$. This is the same as saying that the polynomial $\sum_\alpha a_\alpha\xi^\alpha$ is radial, that is, it takes constant values on spheres, which forces it to be of the form $P(|\xi|^2)$. $\qquad\square$

In this chapter we will often use *Green's identities*, which are integral formulae involving the Laplacian that follow from the divergence theorem. Let $U$ be a bounded domain in $\mathbb{R}^n$ with oriented regular boundary, that is, $\partial U$ is a regular hypersurface oriented with the exterior unit normal $\vec{N}$. If $u$, $v$ are functions twice differentiable on a neighborhood of $\bar{U}$ with $\Delta u$, $\Delta v$ continuous on $\bar{U}$ and if we apply Theorem 3.32 to the vector field $\vec{X} = u\vec{\nabla}v$, we obtain the *first Green's identity*:

$$\int_{\partial U} u\frac{\partial v}{\partial \vec{N}}\, dA = \int_U \langle \vec{\nabla}u, \vec{\nabla}v\rangle\, dV + \int_U u\Delta v\, dV. \tag{7.5}$$

Permuting $u$ and $v$ in (7.5) and subtracting the two equalities we get the *second Green's identity*:

$$\int_{\partial U} \left(u\frac{\partial v}{\partial \vec{N}} - v\frac{\partial u}{\partial \vec{N}}\right) dA = \int_U (u\Delta v - v\Delta u)\, dV. \tag{7.6}$$

In particular, taking $u = v$ in (7.5) and $v = 1$ in (7.6), it turns out that

$$\int_{\partial U} u\frac{\partial u}{\partial \vec{N}}\, dA = \int_U |\vec{\nabla}u|^2\, dV + \int_U u\Delta u\, dV,$$

$$\int_{\partial U} \frac{\partial u}{\partial \vec{N}}\, dA = \int_U \Delta u\, dV. \tag{7.7}$$

These formulae hold when both members make sense and are finite. For example, when $u$ is harmonic on $U$ and continuous on $\bar{U}$ it yields

$$\int_{\partial U} u \frac{\partial u}{\partial \vec{N}} \, dA = \int_U |\vec{\nabla} u|^2 \, dV, \tag{7.8}$$

$$\int_{\partial U} \frac{\partial u}{\partial \vec{N}} \, dA = 0. \tag{7.9}$$

As a particular case, one finds again a result in Section 3.7: if $u$ is harmonic, the vector field $\vec{\nabla} u$ is conservative and solenoidal.

**Example 7.2.** Green's identities, in the case $n = 1$ and $U = (a, b)$, become

$$u(b)v'(b) - u(a)v'(a) = \int_{\partial [a,b]} u \frac{\partial v}{\partial n} = \int_a^b (u'v' + uv'') \, dx,$$

$$u(b)v'(b) - u(a)v'(a) - v(b)u'(b) + v(a)u'(a) = \int_a^b (uv'' - vu'') \, dx$$

which are consequences of the fundamental theorem of calculus.     $\square$

From now on, if $U$ is an open set in $\mathbb{R}^n$, the following notations will be used:

$$C^r(U) = \{f : U \to \mathbb{R} : f \text{ has continuous partial derivatives up to order } r\},$$
$$0 \le r \le +\infty, \text{ with } C^0(U) = C(U);$$
$$C_c^r(U) = \{f \in C^r(U) : \text{spt}(f) \text{ is compact}\}, \quad 0 \le r \le +\infty;$$
$$L_{\text{loc}}^1(U) = \{f : U \to \mathbb{R} : f \text{ is integrable on each compact set of } U\};$$
$$L_c^p(U) = \{f \in L^p(U) : \text{spt}(f) \text{ is compact}\}, \quad 1 \le p \le +\infty,$$

where $L^p(U)$ are the usual Lebesgue spaces.

## 7.2.2 The mean value property

Recall the definition of harmonic function given in Subsection 3.7.2.

**Definition 7.3.** A real-valued function $u$, twice differentiable on an open set $U$ of $\mathbb{R}^n$, is harmonic if the equation

$$\Delta u = \sum_{i=1}^n \frac{\partial^2 u}{\partial x_i^2} = 0 \quad \text{on } U \tag{7.10}$$

holds.

Throughout this chapter the case $n = 1$ is also considered, and, in fact, will help sometimes to better understand the situation when $n \geq 2$. What is a harmonic function on an interval $I$ of $\mathbb{R}$? In this case, equation (7.10) is $u'' = 0$, which is equivalent to $u' = a$, $a$ constant, hence to $u(x) = ax + b$, with $a$, $b$ constants. That is, harmonic functions in one variable are exactly linear functions.

**Example 7.4.** Let us analyze when a polynomial $P(x) = \sum_\alpha c_\alpha x^\alpha$ is harmonic. Here $\alpha = (\alpha_1, \ldots, \alpha_n)$ is a multi-index and $x^\alpha = x_1^{\alpha_1} x_2^{\alpha_2} \ldots x_n^{\alpha_n}$. One may write $P = \sum_k P_k$, where $P_k$ gathers all the terms with $|\alpha| = \alpha_1 + \alpha_2 + \cdots + \alpha_n = k$; $P_k$ is a *homogeneous* polynomial of degree $k$. Observe that $\Delta P_k$ is a homogeneous polynomial of degree $k - 2$, and consequently $P$ is harmonic if and only if each term $P_k$ is so. When $k = 1$ (linear terms), $P_1$ is always harmonic. A homogeneous polynomial of degree 2,

$$P_2(x) = \sum_{i=1}^n a_i x_i^2 + \sum_{\substack{i,j=1 \\ i \neq j}}^n a_{ij} x_i x_j,$$

is harmonic if and only if $a_1 + a_2 + \cdots + a_n = 0$. For $k \geq 3$ it is not easy to characterize when a homogeneous polynomial of degree $k$ is harmonic, in terms of its coefficients. The coefficients must satisfy a system of linear equations, which number depends on $k$ and $n$. For example, if $n = 2$, $k = 3$ the homogeneous polynomial $P(x, y) = ax^3 + by^3 + cx^2 y + dxy^2$ is harmonic if and only if $3a + d = 0$ and $3b + c = 0$, that is, if and only if it is a linear combination of the polynomials $x^3 - 3xy^2$, $y^3 - 3yx^2$. If $n = 3$, $k = 3$ the number of coefficients (the dimension of the space of all the homogeneous polynomials) is 10 and the number of equations is 6, so that the dimension of the space of harmonic homogeneous polynomials of degree 3 on $\mathbb{R}^3$ is 4.     $\square$

The term harmonic is rather associated to solutions of the equation

$$\Delta u = \lambda u, \quad \lambda \in \mathbb{R} \tag{7.11}$$

(eigenfunctions of the Laplacian). In one variable, the solutions of (7.11) are known: if $\lambda > 0$, the solution is

$$u(x) = A e^{\sqrt{\lambda} x} + B e^{-\sqrt{\lambda} x},$$

and if $\lambda < 0$,

$$u(x) = A \sin \sqrt{-\lambda} x + B \cos \sqrt{-\lambda} x,$$

where $A$, $B$ are constants. The unique solutions that remain bounded when $x \in \mathbb{R}$ are the functions $\sin \sqrt{-\lambda} x$, $\cos \sqrt{-\lambda} x$, which are the ones that appear in the expansion in harmonics of periodic functions. In dimension $n > 1$ it is more complicated to study equation (7.11), called *Helmholtz's equation*. Among its solutions there are *spherical harmonics*, functions that, to some extent, play the same role as the sine and the cosine in one variable.

Recall that the notation $|x|$ is used for the norm of $x$ in $\mathbb{R}^n$ and $B(a,r) = B_r(a) = \{x \in \mathbb{R}^n : |x - a| < r\}$ is the ball with center $a \in \mathbb{R}^n$ and radius $r > 0$. Also, $S(a,r) = S_r(a) = \partial B(a,r) = \{x \in \mathbb{R}^n : |x - a| = r\}$ is the corresponding sphere. We will write $\mathbb{B} = B(0,1)$ and $\mathbb{S} = S(0,1)$.

In order to understand what condition $\Delta u \equiv 0$ imposes on a twice differentiable function $u$ on the domain $U$, consider the Taylor expansion of $u$ around a point $a \in U$:

$$u(x) = u(a) + \langle \vec{\nabla} u(a), x - a \rangle + \frac{1}{2} Hu(a)(x - a, x - a) + o(|x - a|^2).$$

Here $\vec{\nabla} u(a)$ is, as always, the gradient of $u$ at the point $a$ and $Hu(a)$ is the Hessian, so that the homogeneous term of order 2 is

$$\frac{1}{2} Hu(a)(x - a, x - a) = \frac{1}{2} \sum_{i,j=1}^{n} \frac{\partial^2 u}{\partial x_i \partial x_j}(a)(x_i - a_i)(x_j - a_j).$$

With the aim of isolating in this expansion the terms of second order with $i = j$, take a ball centered at $a$, with radius $r > 0$ small enough to get $\overline{B}(a,r) \subset U$ and integrate the Taylor expansion of $u$ on the sphere $S(a,r)$. In other words, put $x = a + rw$ with $|w| = 1$ and integrate with respect to $dA(x) = r^{n-1} d\sigma(w)$, where $d\sigma$ *is the measure on the unit sphere*, $\mathbb{S} = S(0,1)$, of $\mathbb{R}^n$. One then has

$$\int_{S(a,r)} u(x)\, dA(x) = r^{n-1} \int_{\mathbb{S}} u(a + rw)\, d\sigma(w).$$

Denote by $c_n = d\sigma(\mathbb{S})$ ($2\pi$ if $n = 2$, $4\pi$ if $n = 3,...$) the area of $\mathbb{S}$. The mean value of $u$ on $S(a,r)$ is, then,

$$\frac{1}{c_n r^{n-1}} \int_{S(a,r)} u(x)\, dA(x) = \frac{1}{c_n} \int_{\mathbb{S}} u(a + rw)\, d\sigma(w)$$

$$= u(a) + \frac{1}{2c_n} \sum_{i=1}^{n} \frac{\partial^2 u}{\partial x_i^2}(a) r^2 \int_{\mathbb{S}} w_i^2\, d\sigma(w) + o(r^2).$$

Here we have used the fact that $\int_{\mathbb{S}} w_i\, d\sigma(w) = 0$ and $\int_{\mathbb{S}} w_i w_j\, d\sigma(w) = 0$ if $i \neq j$, because we are dealing with functions that have integrals of opposite sign on each hemisphere. Now, it is clear that $\int_{\mathbb{S}} w_i^2\, d\sigma(w)$ does not depend on $i$ and its value is, therefore, $\frac{c_n}{n}$. Hence, denoting by $M(u,a,r)$ the mean value of $u$ on $S(a,r)$, it turns out that

$$\frac{1}{2n} \Delta u(a) = \lim_{r \to 0} \frac{M(u,a,r) - u(a)}{r^2}.$$

This equality implies the following fact: if $M(u, a, r) = u(a)$ for $a \in U$ and $r$ arbitrarily small, then $\Delta u(a) = 0$.

**Definition 7.5.** If $u$ is a continuous function on the domain $U$ of $\mathbb{R}^n$, it is said that $u$ satisfies the mean value property on $U$ if

$$u(a) = \frac{1}{c_n r^{n-1}} \int_{S(a,r)} u(x) \, dA(x) = \frac{1}{c_n} \int_{S} u(a + rw) \, d\sigma(w),$$

whenever $S(a, r) \subset U$.

**Example 7.6.** In dimension $n = 1$ the mean value property for a continuous function $u$ on $\mathbb{R}$ is written

$$u\left(\frac{a + b}{2}\right) = \frac{1}{2}(u(a) + u(b)), \quad a, b \in \mathbb{R}.$$

Functions having this property are exactly the linear ones: $u(x) = m + nx$, with $m$ and $n$ constants. Indeed, first it is immediate to check that a linear function $u(x) = m + nx$ has the mean value property. Conversely, suppose that $u$ has the mean value property and let $v$ be the linear function which coincides with $u$ at $x = 0$ and $x = 1$, so that $h = u - v$ has the mean value property and vanishes at the points $0$ and $1$. Then $h(\frac{1}{2}) = 0$ and iterating, it turns out that $h(\frac{1}{4}) = h(\frac{3}{4}) = 0$. Applying successively this argument, it follows that $h$ vanishes at all the dyadic points $\frac{k}{2^n}, k = 0, 1, \ldots, 2^n$, and since these points are dense in $[0, 1]$ and $h$ is continuous, we get that $h$ is identically zero on $[0, 1]$. Taking $a = 0$ and $b \in [1, 2]$ the mean value property implies that $h$ vanishes on $[1, 2]$ and so on. This gives $h = 0$ and $u = v$ is linear. $\qquad\square$

It has just been seen that on the line the functions that fulfill the mean value property are the linear ones, that is, the harmonic functions. The aim now is to prove the same fact in any other dimension. Before Definition 7.5 it had been observed that if $u$ is twice differentiable and has the mean value property, then $u$ is harmonic. Now, it turns out that the mean value property for $u$ implies that $u$ is infinitely differentiable. Actually, let $\chi$ be a $C^\infty$ function with compact support contained in the unit ball $\mathbb{B}$, radial (that is, $\chi(x) = \psi(|x|)$ for some function $\psi$) and normalized by $\int \chi(x) \, dV(x) = 1$. Take $\chi_\varepsilon(x) = \varepsilon^{-n} \chi(x/\varepsilon)$. If $d(x, U^c) > \varepsilon$, the function $y \mapsto \chi_\varepsilon(x - y)$ has support contained in $B(x, \varepsilon) \subset U$ and one has

$$(u * \chi_\varepsilon)(x) = \int_{B(x,\varepsilon)} u(y)\chi_\varepsilon(x - y) \, dV(y) = \int_{B(0,\varepsilon)} u(x - y)\chi_\varepsilon(y) \, dV(y)$$

$$= \int_{B(0,\varepsilon)} u(x-y)\varepsilon^{-n}\chi(y/\varepsilon)\,dV = \int_{\mathbb{B}} u(x-\varepsilon y)\chi(y)\,dV(y)$$

$$= \int_0^1 \int_{\mathbb{S}} u(x-\varepsilon ry)\,\psi(r)r^{n-1}\,d\sigma(y)\,dr$$

$$= c_n u(x)\int_0^1 \psi(r)r^{n-1}\,dr = u(x)\int_{\mathbb{B}} \chi(x)\,dV(x) = u(x).$$

So $(u * \chi_\varepsilon)$ – the convolution of $u$ and $\chi_\varepsilon$ – coincides with $u$ on $\{x: d(x,U)^c > \varepsilon\}$ if $u$ has the mean value property. Since $\varepsilon$ is arbitrary and $u * \chi_\varepsilon$ is $C^\infty$ (see Proposition 7.36), it yields $u \in C^\infty(U)$.

Therefore, a part of the following theorem has been proved:

**Theorem 7.7.** *A continuous function on the domain $U$ fulfill the mean value property if and only if it is harmonic on $U$.*

*Proof.* It only remains to show that every harmonic function satisfies the mean value property. Looking at the derivative with respect to $r$ of the mean value $M(u,a,r)$ of the harmonic function $u$ on $S(a,r)$, one has

$$\frac{d}{dr} M(u,a,r) = \frac{d}{dr}\frac{1}{c_n}\int_{\mathbb{S}} u(a+rw)\,d\sigma(w)$$

$$= \frac{1}{c_n}\int_{\mathbb{S}}\sum_i D_i u(a+rw)w_i\,d\sigma(w)$$

$$= \frac{1}{c_n}\int_{\mathbb{S}}\frac{\partial u}{\partial \vec{N}}(a+rw)\,d\sigma(w)$$

$$= \frac{1}{c_n r^{n-1}}\int_{S(a,r)}\frac{\partial u}{\partial \vec{N}}(x)\,dA(x) = 0$$

according to (7.9) applied to the domain $B(a,r)$. Therefore, $M(u,a,r)$ is constant with respect to $r$, and since $\lim_{r\to 0} M(u,a,r) = u(a)$, it turns out that $M(u,a,r) = u(a)$. Hence, the theorem is proved. $\qquad\square$

Observe that it has been also proved that every harmonic function is of class $C^\infty$.

If a function $u$ has the mean value property on the domain $U$ in the sense of Definition 7.5, it also has the mean value property with respect to balls, in the following sense:

$$u(a) = \frac{n}{c_n r^n}\int_{B(a,r)} u(x)\,dV(x) \quad \text{if } \bar{B}(a,r) \subset U. \tag{7.12}$$

The coefficient before the integral is justified since the volume of the ball $B(a,r)$ is $\int_0^r t^{n-1}c_n\,dt = \frac{c_n r^n}{n}$. To prove (7.12) integrate in polar coordinates and use the mean value property with respect to spheres:

$$\int_{B(a,r)} u(x)\,dV(x) = \int_0^r \int_S u(a+sw)s^{n-1}\,d\sigma(w)\,ds$$

$$= \int_0^r s^{n-1}c_n u(a)\,ds = \frac{r^n}{n}c_n u(a).$$

A useful consequence of the mean value property is the following:

**Theorem 7.8** (Liouville's theorem for harmonic functions). *A bounded harmonic function on $\mathbb{R}^n$ is constant.*

*Proof.* If the function $u$ satisfies the hypothesis of the theorem, applying (7.12) for each $a \in \mathbb{R}^n$ and each $r > 0$ big enough, it turns out that

$$u(a) - u(0) = \frac{n}{c_n r^n}\left\{\int_{B(a,r)} u(x)\,dV(x) - \int_{B(0,r)} u(x)\,dV(x)\right\}.$$

If $\|u\|_\infty = \sup\{|u(x)| : x \in \mathbb{R}^n\}$ one has

$$|u(a) - u(0)| \le \frac{n}{c_n r^n}\|u\|_\infty\left\{\int_{B(a,r)\setminus B(0,r)} dV(x) + \int_{B(0,r)\setminus B(a,r)} dV(x)\right\}$$

$$\le \frac{n}{c_n r^n}\|u\|_\infty\, dV(\{x : r - |a| < |x| < r + |a|\})$$

$$= \frac{\|u\|_\infty}{r^n}\left((r+|a|)^n - (r-|a|)^n\right) = O(r^{-1}).$$

Letting $r \to +\infty$ we get $u(a) = u(0)$, for every point $a \in \mathbb{R}^n$.  $\square$

In the case $n = 2$, the relation between holomorphic and harmonic functions may be used to give a stronger version of the previous theorem.

**Theorem 7.9.** *If $u$ is a harmonic function on the plane and satisfies*

$$u(z) = c\,\mathrm{Log}\,|z| + O(1) \text{ when } |z| \to \infty, \text{ with } c \text{ constant,}$$

*then $u$ is constant. In the same conditions, if $u(z) = c\,\mathrm{Log}\,|z| + o(1)$, then $u$ is identically zero.*

*Proof.* If $f$ is an entire function with $\mathrm{Re}\,f = u$, then $F = e^f$ is also entire and

$$|F(z)| = e^{\mathrm{Re}\,f(z)} = e^{u(z)} = O(|z|^N)$$

for some natural number $N$. By Exercise 11 of Section 4.7, $F$ is a polynomial and, being of the form $F = e^f$, $F$ does not vanish; consequently $F$ is constant. This implies that $f$ and $u$ are also constant.  $\square$

## 7.3 Newtonian and logarithmic potentials. Riesz' decomposition formulae

Which examples of harmonic functions on a domain $U$ of $\mathbb{R}^n$ are known at present? When $n = 2$, there are a lot of them: just take $u = \operatorname{Re} f$ with $f$ holomorphic. But if $n > 2$, the only obvious harmonic functions are the linear ones, $u(x) = \sum_{i=1}^{n} a_i x_i + b$ with $a_i, b \in \mathbb{R}$.

Since the Laplacian commutes with rotations (the unitary transformations), it seems natural to look also at the radial functions. How must a function $\varphi \colon (0, \infty) \to \mathbb{R}$ be so that $u(x) = \varphi(|x|)$ is harmonic? Setting $r = |x|$ it turns out that, by a direct calculation,

$$\frac{\partial r}{\partial x_i} = \frac{x_i}{r}, \quad \frac{\partial u}{\partial x_i} = \varphi'(r)\frac{x_i}{r}, \quad \frac{\partial^2 u}{\partial x_i^2} = \varphi''(r)\frac{x_i^2}{r^2} + \varphi'(r)\frac{1}{r} - \varphi'(r)x_i\frac{x_i}{r^3},$$

$$\Delta u = \varphi''(r) + n\varphi'(r)\frac{1}{r} - \varphi'(r)\frac{1}{r} = \varphi''(r) + \frac{n-1}{r}\varphi'(r). \qquad (7.13)$$

It yields, then, the equation $\varphi''(r) + \frac{n-1}{r}\varphi'(r) = 0$, which gives $\varphi'(r) = Cr^{1-n}$, $C$ constant, and

$$\varphi(r) = k_1 r^{2-n} + k_2 \quad \text{if } n > 2,$$

$$\varphi(r) = k_1 \operatorname{Log} r + k_2 \quad \text{if } n = 2,$$

with $k_1, k_2$ constants. These functions have a singularity for $r = 0$, that is, they tend to $\infty$ when $r \to 0$. Therefore, there are no radial harmonic functions on a disc different from a constant. There is also a direct relation between the fact that a radial harmonic function on a ball is constant and the mean value property. On one hand, it is clear that if $u$ is radial on a ball centered at $a$ and has the mean value property (that is, it is harmonic), then $u$ must be constant, $u = u(a)$. On the other hand, if $u$ is harmonic, the invariance of the Laplacian with respect to the group $\Sigma$ of rotations implies that the function

$$v = \frac{1}{c_n} \int_{\Sigma} (u \circ g)\, dg \quad \left( (v(z) = \frac{1}{2\pi} \int_0^{2\pi} u(e^{i\theta}z)\, d\theta \ \text{if } n = 2 \right)$$

is harmonic too and, being radial, it must be constant.

The function

$$G(x) = \begin{cases} d_n |x|^{2-n} & \text{if } n > 2, \\ d_2 \operatorname{Log} |x| & \text{if } n = 2, \end{cases}$$

is called the *fundamental solution of the Laplacian with pole at the origin.* The constant $d_n$ is a constant of normalization which will be chosen later. It is known that $G(x)$ is harmonic on $\mathbb{R}^n \setminus \{0\}$. Moving to a point $a \in \mathbb{R}^n$, $G(x - a)$ is called the *fundamental solution of the Laplacian with pole a*.

**Example 7.10.** Clearly, if $m < n$ and $u$ is a harmonic function of $m$ variables taken among $x_1, x_2, \ldots, x_n$, the same function looked at on $\mathbb{R}^n$ as a function independent of the $n - m$ remaining variables is also harmonic wherever it is well defined. For example, if $i, j$ are different arbitrary indexes, the function $\mathrm{Log}(x_i^2 + x_j^2)$ is harmonic on $U = \{x \in \mathbb{R}^n : x_i^2 + x_j^2 \neq 0\}$. If $2 < k < n$ and one takes $k$ coordinates of $x$, for example the first $k$ ones, the function $(x_1^2 + x_2^2 + \cdots + x_k^2)^{1-\frac{k}{2}}$ is harmonic on $U = \{x \in \mathbb{R}^n : x_1^2 + x_2^2 + \cdots + x_k^2 \neq 0\}$.     $\square$

Since $G(x - a)$ is harmonic for $x \neq a$, it turns out that the function

$$u(x) = \sum_{i=1}^{N} c_i\, G(x - a_i)$$

is harmonic on $\mathbb{R}^n \setminus \{a_1, \ldots, a_N\}$ if $c_i$ are constants and $a_1, \ldots, a_N \in \mathbb{R}^n$.

One may also consider "infinite sums", that is integrals, instead of finite sums and put

$$u(x) = \int_K \phi(y) G(x - y)\, dV(y),$$

where $K$ is a compact set of $\mathbb{R}^n$ and $\phi$ is continuous on $K$.

To give sense to these integrals one needs to prove the local integrability of $G$.

**Lemma 7.11.** *For fixed $x \in \mathbb{R}^n$, the function $y \mapsto G(x - y)$ is integrable on each compact set of $\mathbb{R}^n$.*

*Proof.* It is enough to prove that $G(y)$ is integrable on a ball $B(0, R)$. Using polar coordinates, if $n > 2$, one has

$$\int_{B(0,R)} G(y)\, dV(y) = \int_{B(0,R)} d_n |y|^{2-n}\, dV(y)$$

$$= d_n \int_0^R r^{2-n} r^{n-1} c_n\, dr = c\, R^2 < +\infty, \quad c \text{ constant.}$$

If $n = 2$ the computation is a little bit different,

$$\int_{D(0,R)} G(y)\, dm(y) = \int_{D(0,R)} d_2 \,\mathrm{Log}\,|y|\, dm(y) = 2\pi d_2 \int_0^R r\,\mathrm{Log}\, r\, dr$$

$$= 2\pi d_2 \left[ \frac{r^2}{2}\,\mathrm{Log}\, r - \frac{r^2}{4} \right]_0^R = 2\pi d_2 \left[ \frac{R^2}{2}\,\mathrm{Log}\, R - \frac{R^2}{4} \right]$$

$$< +\infty. \qquad \square$$

If $\phi$ is a continuous function on a compact set $K$ of positive measure, $\phi$ is bounded and Lemma 7.11 implies that the function

$$u(x) = G(\phi)(x) = \int_K \phi(y) G(x-y)\, dV(y) \tag{7.14}$$

is well defined for all $x \in \mathbb{R}^n$. Outside $K$, $u$ is harmonic because

$$\Delta u(x) = \int_K \phi(y) \Delta_x G(x-y)\, dV(y) = 0.$$

Moreover, by the dominated convergence theorem, $G(\phi)$ vanishes at infinity if $n > 2$. The function $u$ defined by (7.14) is called the *Riesz potential* of $\phi$.

When $n = 3$, the function

$$u(x) = G(\phi)(x) = d_3 \int_K \frac{\phi(y)}{|x-y|}\, dV(y)$$

is the potential created by the distribution of charge (or mass) given by the density function $\phi$. Actually, an easy computation shows that $\vec{\nabla}_x \frac{1}{|x-y|} = \frac{\vec{r}(x,y)}{|x-y|^3}$, $\vec{r}(x,y)$ being the vector $y - x$. Therefore, on $\mathbb{R}^3 \setminus K$, one has

$$\vec{\nabla} u(x) = d_3 \int_K \phi(y) \frac{\vec{r}(x,y)}{|x-y|^3}\, dV(y),$$

which, by Newton's law, is the vector field of forces created by the density $\phi$. For this reason $u$ is called *newtonian potential*. When $n = 2$, the corresponding potentials

$$u(x) = G(\phi)(x) = d_2 \int_U \phi(y) \operatorname{Log} |x-y|\, dm(y),$$

are called *logarithmic potentials*.

Every potential in dimension $k$ may be interpreted as the restriction to $\mathbb{R}^k$ of a potential on $\mathbb{R}^n, n > k$. Write $\mathbb{R}^n$ as the product $\mathbb{R}^k \times \mathbb{R}^{n-k}$ and use the notation $x = (x', x'')$, with $x' = (x_1, \ldots, x_k)$, $x'' = (x_{k+1}, \ldots, x_n)$. Consider a potential corresponding to a function $\phi(y)$ independent from $y''$, that is,

$$u(x) = \int_{K' \times \mathbb{R}^{n-k}} \phi(y') |x-y|^{2-n}\, dV(y)$$

$$= \int_{K'} \phi(y') \left( \int_{\mathbb{R}^{n-k}} |x-y|^{2-n}\, dV(y'') \right) dV(y'),$$

where $K'$ is a compact set of $\mathbb{R}^k$. Suppose $k > 2$; evaluating at a point $x \in \mathbb{R}^k$ ($x'' = 0$) the previous integral, it yields a constant multiple of $|x' - y'|^{2-k}$ and one has that $u(x', 0)$ is, except for constants, the $k$-dimensional potential of $\phi$. In order to obtain logarithmic potentials ($k = 2$) consider the $n$-dimensional potential of $\phi(y')\mathbb{1}_N(y'')$, where $\mathbb{1}_N$ stands for the characteristic function of $\{y'' \in \mathbb{R}^{n-k} : |y''| \leq N\}$. Then the previous integral becomes

$$\int_{B(0,N)} |x - y|^{2-n} \, dV(y'').$$

Now with the change $y'' = tr'\omega$, $|\omega| = 1$, where $r' = |x' - y'|$, it is recognized as a multiple of

$$\int_0^{\frac{N}{r'}} \frac{t^{n-3}dt}{(1 + t^2)^{\frac{n}{2}-1}}.$$

With $N$ fixed, this last integral is of the order of $\mathrm{Log}\, r'$ for $r'$ small and one finds the logarithmic potential of $\phi(y')$. This calculus explains the interest in logarithmic potentials, although the emphasis is on newtonian potentials.

**Example 7.12.** Riesz potentials may just be computed explicitly in situations where there is a lot of symmetry. Suppose, for example, that $\phi$ is a radial function of the form $\phi(x) = h(|x|)$, where $h$ is continuous and supported in the interval $[1, 2]$. Then the potential $G(\phi)$ is also radial on the unit ball $\mathbb{B}$; indeed, if $T$ is a linear transformation given by a unitary matrix, one has

$$G(\phi)(Tx) = \int_{B_2(0)\setminus\bar{B}_1(0)} G(Tx - y)h(|y|) \, dV(y)$$

$$= \int_{B_2(0)\setminus\bar{B}_1(0)} G(x - T^{-1}y)h(|y|) \, dV(y).$$

We now change variables in this integral, introducing $z = T^{-1}(y)$ as a new variable, to get $G(\phi)(x)$. Since $G(\phi)$ is harmonic and radial on the unit ball $\mathbb{B}$, it must be constant:

$$\int_{B_2(0)\setminus\bar{B}_1(0)} G(x - y)h(|y|)dV(y) = C, \quad |x| < 1.$$

Outside $B(0, 2)$, for the same reason, $G(\phi)$ must satisfy

$$G(\phi)(x) = k_1 G(x) + k_2, \quad k_1, k_2 \text{ constants}.$$

If $n > 2$, letting $|x| \to \infty$ we get $k_2 = 0$, and multiplying by $|x|^{n-2}$ and letting again $|x| \to \infty$ one finds $k_1 = \int \phi(y) \, dV(y)$. The same conclusion is reached in the case $n = 2$ looking at the behavior of $G(\phi)(x)$ for $|x|$ big enough. $\qquad\square$

In Section 7.7 Riesz potentials will be studied in detail and it will be proved that they satisfy the equation $\Delta G(\phi) = \phi$, in a certain sense, on the compact set $K$ supporting the function $\phi$. For the moment we will see that these potentials appear in a natural way in the Riesz decomposition formula, which is the analogue of Cauchy–Green's formula for the Laplacian. This is the following theorem, which includes the choice of the constant $d_n$ in the definition of $G(x)$. The notation $\dfrac{\partial}{\partial \vec{N}_y}$ indicates that the derivative is taken in the normal direction to the boundary of the domain, with respect to the variable $y$.

**Theorem 7.13** (Riesz decomposition formula). *Let $U \subset \mathbb{R}^n$ be a bounded domain with regular boundary oriented by the exterior normal $\vec{N}$. If $u$ is a function, twice differentiable on a neighborhood of $\bar{U}$ with $\Delta u$ continuous on $\bar{U}$, then, for $x \in U$, one has*

$$u(x) = \int_{\partial U} \left\{ u(y) \frac{\partial}{\partial \vec{N}_y} G(x-y) - G(x-y) \frac{\partial u}{\partial \vec{N}}(y) \right\} dA(y)$$
$$+ \int_U G(x-y) \Delta u(y) \, dV(y).$$
$$(7.15)$$

*Proof.* Proceed as in the proof of Cauchy–Green's formula. Fix $x \in U$, consider $\bar{B}(x, \varepsilon) \subset U$ and apply the second Green's identity to the domain $U \setminus \bar{B}(x, \varepsilon)$ and to the functions $u$, $v(y) = G(x-y)$. We obtain

$$\left( \int_{S(x,\varepsilon)} + \int_{\partial U} \right) \left\{ u(y) \frac{\partial}{\partial \vec{N}_y} G(x-y) - G(x-y) \frac{\partial u}{\partial \vec{N}}(y) \right\} dA(y)$$

$$= \int_{U \setminus \bar{B}(x,\varepsilon)} \{ u(y) \Delta_y G(x-y) - G(x-y) \Delta u(y) \} \, dV(y)$$

$$= - \int_{U \setminus \bar{B}(x,\varepsilon)} G(x-y) \Delta u(y) \, dV(y).$$

For $y \in S(x, \varepsilon)$, the unit exterior normal derivative is the opposite of the radial derivative, $\frac{\partial}{\partial r}$, where $r$ is the distance to $x$. Since $G(x-y) = d_n r^{2-n}$ if $n > 2$ and $G(x-y) = d_2 \operatorname{Log} r$ if $n = 2$, one has $\frac{\partial}{\partial r} G(x-y) = d_n(2-n) r^{1-n}$ if $n > 2$ and $\frac{\partial}{\partial r} G(x-y) = d_2 r^{-1}$ if $n = 2$. Therefore,

$$\int_{S(x,\varepsilon)} \left\{ u(y) \frac{\partial}{\partial \vec{N}_y} G(x-y) - G(x-y) \frac{\partial u}{\partial \vec{N}}(y) \right\} dA(y)$$

$$= - \left\{ \begin{array}{l} d_n(2-n)\varepsilon^{1-n} \\ d_2 \varepsilon^{-1} \end{array} \right\} \!\!\int_{S(x,\varepsilon)} u(y) \, dA(y) + \left\{ \begin{array}{l} d_n \varepsilon^{2-n} \\ d_2 \operatorname{Log} \varepsilon \end{array} \right\} \!\!\int_{S(x,\varepsilon)} \frac{\partial u}{\partial r}(y) \, dA(y).$$

The second term is $O(\varepsilon^{2-n}) O(\varepsilon^{n-1}) = O(\varepsilon)$ if $n > 2$, and $O(\operatorname{Log} \varepsilon) O(\varepsilon) = O(\varepsilon \operatorname{Log} \varepsilon)$ if $n = 2$, and in both cases it tends to 0 when $\varepsilon \to 0$. The first term, choosing $d_n(2-n) = \frac{1}{c_n}$ if $n > 2$ and $d_2 = \frac{1}{2\pi}$ if $n = 2$, is the mean of $-u$ on $S(x, \varepsilon)$ and, hence, it tends to $-u(x)$ when $\varepsilon \to 0$.

On the other hand, since $\Delta u(y)G(x-y)$ is integrable with respect to $y$ on $U$ by Lemma 7.11, the limit of its integral on $U \setminus \bar{B}(x, \varepsilon)$ is the integral on the whole domain $U$. $\qquad\square$

As has been said in the proof of the previous theorem, $d_n = 1/2\pi$ if $n = 2$ and $d_n = 1/(2-n)c_n$ if $n > 2$ have been fixed. Recall $c_n$ is the $(n-1)$-dimensional measure of the unit sphere.

An integral of the kind

$$\int_{\partial U} \phi(y)G(x-y)\,dA(y)$$

is called a *simple layer potential* and it is, clearly, a harmonic function on $\mathbb{R}^n \setminus \partial U$.

**Example 7.14.** Once again, simple layer potentials may only be explicitly computed when there are symmetries. In a similar way to Example 7.12, it is easy to see that if $\phi$ is constant and equal to $C$ on the sphere $S(0, R)$, the simple layer potential $C \int_{S(0,R)} G(x-y)\,dA(y)$ is constant on the ball $B(0, R)$ and has value $k_1 G(x)$, with $k_1$ constant, on the exterior of this ball. The value of the constant on $B(0, R)$ is the value at $x = 0$, that is,

$$C \int_{S(0,R)} G(y)\,dA(y) = C \frac{R}{2-n}$$

if $n > 2$ and $CR \operatorname{Log} R$ if $n = 2$. The value of $k_1$ is

$$\lim_{|x|\to\infty} \frac{1}{G(x)} C \int_{S(0,R)} G(x-y)\,dA(y) = Cc_n R^{n-1},$$

so that the potential for $|x| > R$ is $C \frac{R^{n-1}}{(2-n)|x|^{n-2}}$ if $n > 2$ and $CR \operatorname{Log}|x|$ if $n = 2$. This means, in particular, that an electrical field created by a homogenously charged sphere is zero inside the sphere. Observe also that this potential is a continuous function on $\mathbb{R}^n$. $\qquad\square$

**Example 7.15.** Think once again of $\mathbb{R}^n$ as the product $\mathbb{R}^k \times \mathbb{R}^{n-k}$ and use the notation $x = (x', x'')$, with $x' = (x_1, \ldots, x_k)$, $x'' = (x_{k+1}, \ldots, x_n)$. Consider the potential

$$u(x) = \int_{\mathbb{R}^k} \phi(y')G(x-(y',0))\,dV(y'),$$

defined for $x \notin \mathbb{R}^k$, that is $x'' \neq 0$, provided that

$$\int_{\mathbb{R}^k} |\phi(y')||y'|^{2-n}\,dV(y') < \infty.$$

When $k = n-1$ it is formally a simple layer potential on the boundary of a half-space of $\mathbb{R}^n$. The function $u$ is harmonic in $x$ and, by definition, depends on $x'$ and $|x''|$. $\qquad\square$

An integral of the kind

$$\int_{\partial U} \phi(y) \frac{\partial}{\partial \vec{N}_y} G(x - y) \, dA(y)$$

is called a *double layer potential*. This is because when setting

$$\frac{\partial}{\partial \vec{N}_y} G(x - y) = \lim_{\varepsilon \to 0} \frac{G(x - y - \varepsilon \vec{N}(y)) - G(x - y)}{\varepsilon},$$

then, formally, the previous integral may be written as

$$\lim_{\varepsilon \to 0} \frac{1}{\varepsilon} \left\{ \int_{\partial U} \phi(y) G(x - y - \varepsilon \vec{N}(y)) \, dA(y) - \int_{\partial U} \phi(y) G(x - y) \, dA(y) \right\}$$

$$= \lim_{\varepsilon \to 0} \frac{1}{\varepsilon} \left\{ \int_{(\partial U)_\varepsilon} \phi(y) G(x - z) \, dA(z) - \int_{\partial U} \phi(y) G(x - y) \, dA(y) \right\},$$

where $(\partial U)_\varepsilon$ consists of points $y + \varepsilon \vec{N}(y)$, $y \in \partial U$. Double layer potentials are also harmonic functions on $\mathbb{R}^n \setminus \partial U$; to see this it is sufficient to set that the function $\frac{\partial}{\partial \vec{N}_y} G(x - y)$ is harmonic in $x$, for $y \in \partial U$. Actually, if $\vec{N}_y$ has components $(A_1(y), \ldots, A_n(y))$, then

$$\frac{\partial}{\partial \vec{N}_y} G(x - y) = \sum_{i=1}^{n} A_i(y) \frac{\partial}{\partial y_i} G(x - y)$$

and each term $\frac{\partial}{\partial y_i} G(x - y)$ is a harmonic function of $x$, because $\Delta_x \frac{\partial}{\partial y_i} G(x - y) = \frac{\partial}{\partial y_i} \Delta_x G(x - y) = 0$.

Hence, Riesz' decomposition formula expresses $u$ as a sum of a harmonic function on $U$, which at the same time is the sum of a simple layer potential and a double layer potential, plus another term that is a Riesz potential and only depends on $\Delta u$. Recall that, analogously, Cauchy–Green's formula decomposes a function $u$ into the sum of a holomorphic function plus another term depending on $\bar{\partial} u$ and Cauchy's kernel.

**Corollary 7.16.** *If $U$ is a bounded domain with oriented regular boundary and $u$ is harmonic on a neighborhood of $\bar{U}$, then*

$$u(x) = \int_{\partial U} \left\{ u(y) \frac{\partial}{\partial \vec{N}_y} G(x - y) - G(x - y) \frac{\partial u}{\partial \vec{N}}(y) \right\} dA(y), \quad \text{if } x \in U.$$

We can see now the form this formula takes in dimension $n = 2$. Write $w = u + iv$ instead of $y$, $z = x + iy$ instead of $x$, $G(w) = \frac{1}{2\pi} \text{Log} |w|$ and

$$\vec{N}_w = A_1(w) + i A_2(w), \qquad \frac{\partial}{\partial \vec{N}_w} = A_1(w) \frac{\partial}{\partial u} + A_2(w) \frac{\partial}{\partial v},$$

where $A_1(w)$, $A_2(w)$ are two real functions. If follows that

$$\frac{\partial}{\partial \vec{N}_w} \operatorname{Log} |z - w| = A_1(w)\frac{u - x}{|z - w|^2} + A_2(w)\frac{v - y}{|z - w|^2}$$

$$= \operatorname{Re} \frac{A_1(w) + i A_2(w)}{w - z} = \operatorname{Re} \frac{\vec{N}_w}{w - z}.$$

If $U$ is a bounded domain of $\mathbb{C}$ with positively oriented regular boundary and $w = \gamma(t)$ a local parametrization of $\partial U$, then $\vec{N}_w = -i\gamma'(t)/|\gamma'(t)|$, $ds = |\gamma'(t)|dt$ and $dw = \gamma'(t)dt$. It turns out that

$$\frac{1}{2\pi}\int_{\partial U} u(w)\frac{\partial}{\partial \vec{N}_w} \operatorname{Log} |z - w|\, ds(w) = \frac{1}{2\pi}\int_{\partial U} u(w) \operatorname{Re} \frac{-i\gamma'(t)}{w - z}\, dt$$

$$= \operatorname{Re} \frac{1}{2\pi i}\int_{\partial U} \frac{u(w)}{w - z}\, dw.$$

This proves:

**Corollary 7.17.** *If $U$ is a bounded domain of $\mathbb{C}$ with positively oriented regular boundary and $u$ is harmonic on a neighborhood of $\overline{U}$, then $u$ is the sum $u = u_1 + u_2$, where $u_1 = \operatorname{Re} f$ with $f$ holomorphic on $U$ and $u_2$ is a harmonic logarithmic potential on $U$.*

Suppose now that, in Theorem 7.13, $U$ is the ball with center $x$ and radius $R$. The same computation done in the proof of this theorem but changing the sense of the exterior normal derivative, gives

$$\frac{\partial}{\partial \vec{N}_y}G(x - y) = \frac{1}{c_n R^{n-1}},$$

where

$$G(x - y) = \begin{cases} d_n R^{2-n}, & d_n < 0 \text{ if } n > 2, \\ d_2 \operatorname{Log} R & \text{if } n = 2, \end{cases} \qquad \text{for } y \in S(x, R). \qquad (7.16)$$

If $u$ is twice differentiable on $\overline{B}(x, R)$, formula (7.7) yields

$$\int_{\partial U} G(x - y)\frac{\partial u}{\partial \vec{N}}(y)\, dA(y) = \left\{ \begin{array}{c} d_n R^{2-n} \\ d_2 \operatorname{Log} R \end{array} \right\} \int_{\partial U} \frac{\partial u}{\partial \vec{N}}(y)\, dA(y)$$

$$= \left\{ \begin{array}{c} d_n R^{2-n} \\ d_2 \operatorname{Log} R \end{array} \right\} \int_{U} \Delta u\, dV(y).$$

With all this, Theorem 7.13 gives the following result:

**Corollary 7.18.** *If $u$ is twice differentiable on a neighborhood of the closed ball $\bar{B}(x, R)$ with $\Delta u$ continuous on this ball, one has*

$$u(x) = \frac{1}{c_n R^{n-1}} \int_{S(x,R)} u(y)\, dA(y) + \int_{B(x,R)} G_{B(x,R)}(x, y)\Delta u(y)\, dV(y),$$

*with $G_{B(x,R)}(x, y) = d_n(|x - y|^{2-n} - R^{2-n})$ if $n > 2$ and $G_{B(x,R)}(x, y) = d_2 \log \frac{|x-y|}{R}$ if $n = 2$.*

Observe that the previous corollary proves again that a function $u$, with the noted regularity conditions, has the mean value property if and only if it is harmonic.

The function $G_{B(x,R)}(x, y)$ of Corollary 7.18 is always less than or equal to zero when $y \in B(x, R)$. As a consequence if $\Delta u \geq 0$ on an open set $U$ and $\bar{B}(x, R) \subset U$, then

$$u(x) \leq \frac{1}{c_n R^{n-1}} \int_{S(x,R)} u(y)\, dA(y),$$

that is, the function $u$ is always below its mean over spheres. When a continuous function on $U$ has this property, it is said to be *subharmonic on $U$*. The twice differentiable functions $u$ that are subharmonic are exactly the ones that satisfy $\Delta u \geq 0$. In dimension $n = 1$, subharmonic functions are those satisfying the condition

$$u(x) \leq \frac{1}{2}(u(x + h) + u(x - h)) \quad \text{for } x, h \in \mathbb{R},$$

that is, the convex functions.

Later on, in Section 9.4 subharmonic functions will be dealt with again.

**Remark 7.1.** It is appropriate to explain the meaning of previous considerations in the case $n = 1$. First, the fundamental solution is $G(x) = c|x|$, $c$ constant. The analog of (7.15) is, for $u \in C^2([a, b])$,

$$\frac{1}{c} u(x) = u(b) + u(a) - (|x - b|u'(b) - |x - a|u'(a)) + \int_a^b |x - y|u''(y)\, dy.$$

This formula may be obtained by integrating by parts. The last integral is

$$\int_a^x u''(y)(x - y)\, dy + \int_x^b u''(y)(y - x)\, dy$$

$$= [u'(y)(x - y)]_a^x + \int_a^x u'(y)\, dy + [u'(y)(y - x)]_x^b - \int_x^b u'(y)\, dy$$

$$= u'(b)(b - x) - u(b) + u(x) - u'(a)(x - a) + u(x) - u(a).$$

Therefore, $c = \frac{1}{2}$ and $G(x) = \frac{1}{2}|x|$ is the fundamental solution when $n = 1$.

## 7.4 Maximum principle. Dirichlet and Neumann homogeneous problems

Recall that when talking about harmonic functions, one refers to real-valued functions. If the function $u$ is harmonic on a neighborhood of the closure of a bounded domain $U$ with regular boundary oriented by the exterior normal $\vec{N}$, one has, by (7.8),

$$\int_{\partial U} u \frac{\partial u}{\partial \vec{N}} \, dA = \int_U |\vec{\nabla} u|^2 \, dV.$$

This implies that if $u$ is zero on $\partial U$, then $u$ vanishes on $U$, because $\int_U |\vec{\nabla} u|^2 \, dV = 0$ means $\vec{\nabla} u = 0$, that is, $u$ is constant and being zero on $\partial U$, this constant is zero. Hence, every harmonic function on $\bar{U}$ is determined by its values on $\partial U$. Equivalently, every twice differentiable function $u$ with $\Delta u$ continuous is determined on $U$ by its values on $\partial U$ and by $\Delta u$: if $u_1 = u_2$ on $\partial U$ and $\Delta u_1 = \Delta u_2$ on $U$, then $u = u_1 - u_2$ is 0 on $\partial U$ and it is harmonic on $U$, so then $u_1 = u_2$.

Similar considerations apply when replacing the value of $u$ on $\partial U$ by its derivative with respect to the normal, $\frac{\partial u}{\partial \vec{N}}$, also on $\partial U$. Now, it turns out that if $u$ is harmonic on $\bar{U}$ and $\frac{\partial u}{\partial \vec{N}} = 0$ on $\partial U$, then $u$ is constant. Every harmonic function on $\bar{U}$ is determined, except for constants, by $\frac{\partial u}{\partial \vec{N}}$ on $\partial U$. Equivalently, every twice differentiable function $u$ with $\Delta u$ continuous is determined on $U$, except for constants, by the values of $\frac{\partial u}{\partial \vec{N}}$ on $\partial U$ and $\Delta u$ on $U$.

What has been just said shows that there is necessarily some redundance in the statement of Corollary 7.16. This result expresses the value of the function $u$ at the points of the domain $U$ in terms of $u$ and $\frac{\partial u}{\partial \vec{N}}$ on $\partial U$. However it should be possible to express the values of $u$ on $U$ only in terms of $u$ on $\partial U$ or only in terms of $\frac{\partial u}{\partial \vec{N}}$ on $\partial U$, except for constants.

Since to consider the values of $u$ on $\partial U$ no condition on the regularity of $\partial U$ is needed, *Dirichlet's problem* (homogeneous) is stated for an arbitrary bounded domain $U$: if $U$ is such a domain of $\mathbb{R}^n$ and $\varphi$ is a function continuous on the boundary of $U$, Dirichlet's problem consists in finding a function $u$, harmonic on $U$, such that for each point $y \in \partial U$ the equality

$$\lim_{\substack{x \to y \\ x \in U}} u(x) = \varphi(y)$$

holds. Another way of expressing the properties of the function $u$ is to say that $u \in C(\bar{U})$, $u$ is harmonic on $U$ and $u\big|_{\partial U} = \varphi$. In particular, $u$ must be bounded.

There is a more general version of the problem without assuming $U$ is bounded; in this case, one must impose $\varphi \in C(\partial U)$ is bounded and look for a solution $u$, bounded on $U$ too.

Dirichlet's problem models, for example, the stationary distribution of temperatures (independent of time) on $U$ when $\partial U$ has a distribution, also stationary, given by $\varphi$ (for example, the outside room temperature). Another situation modelled by Dirichlet's problem is given in electrostatics. Suppose that $U \subset \mathbb{R}^3$ represents a region filled with a charged conductor material and one puts an electrical charge $q$ at a point $P \in U$. Then there is a redistribution of the charges on $\partial U$ to achieve a new equilibrium. Let $u$ be the electrostatic potential caused by the charge distribution on $\partial U$; as explained in Subsection 7.1.2, $u$ is a harmonic function on $U$. The total potential on $U$ is $u + \frac{q}{R}$, where $R$ is the distance function to $P$ ($\frac{q}{R}$ is the potential caused by the charge at the point $P$). The equilibrium of charge over $\partial U$ means that the potential $u + \frac{q}{R}$ is zero on $\partial U$. Therefore, $u$ is the solution of the Dirichlet problem on $U$ with data $-\frac{q}{R}$ on $\partial U$.

When $\partial U$ is regular, formula (7.8) gives the uniqueness of solution of the Dirichlet problem. In order to prove it in general one needs the maximum principle.

**Theorem 7.19** (Maximum principle). *Let $U$ be a domain of $\mathbb{R}^n$ and $u$ a continuous function on $U$ satisfying the mean value property, that is, $u$ is harmonic on $U$. Suppose that there is a constant $M \geq 0$ such that $u(x) \leq M$ for all $x \in U$ and that there is a point $x_0 \in U$ such that $u(x_0) = M$. Then one has $u(x) = M$ for all $x \in U$.*

*Proof.* Consider $A = \{x \in U : u(x) = M\}$ which is a closed subset of $U$; it is non-empty because $x_0 \in A$. It is enough to prove that $A$ is open: if $a \in A$ and $\overline{B}(a, r) \subset U$, by the mean value property on balls, one has

$$M = u(a) = \frac{n}{c_n r^n} \int_{B(a,r)} u(x)\, dV(x),$$

that is,

$$\frac{1}{V(B(a,r))} \int_{B(a,r)} (u(x) - u(a))\, dV(x) = 0.$$

Since $u(x) - u(a) = u(x) - M$ is a continuous function, less than or equal to zero and has integral zero on $B(a, r)$, it must be identically zero; therefore, $u(x) = M$ on $B(a, r)$, that is, $B(a, r) \subset A$ and $A$ is open. $\qquad\square$

Clearly, there is also a corresponding minimum principle replacing $u$ by $-u$.

**Corollary 7.20.** *If $U$ is a bounded domain of $\mathbb{R}^n$, $u \in C(\overline{U})$ and $u$ is harmonic on $U$, then the maximum and minimum values of $u$ in $\overline{U}$ are achieved in $\partial U$. Consequently, if $u = 0$ on $\partial U$, then $u = 0$ on $U$. If $u_1, u_2 \in C(\overline{U})$ are harmonic on $U$ and $u_1 = u_2$ on $\partial U$, then $u_1 = u_2$ on $U$.*

*Proof.* Since $\overline{U}$ is compact and $u \in C(\overline{U})$, $u$ has a global maximum: $u(x) \leq u(x_0) = M$, for all $x \in U$, with $x_0 \in \overline{U}$. If $x_0 \in \partial U$ there is nothing to prove.

If $x_0 \in U$, then, by the maximum principle, $u = M$ on $U$, and the conclusion is obvious. Arguing with $-u$, a similar result for the minimum is found. $\qquad\square$

**Corollary 7.21.** *The solution of the Dirichlet problem, if it exists, is unique.*

*Proof.* Let $\varphi \in C(U)$ and suppose there are $u_1, u_2 \in C(\bar{U})$ with $\Delta u_1 = \Delta u_2 = 0$ on $U$ and $u_1 = u_2 = \varphi$ on $\partial U$. Then $u = u_1 - u_2$ is $0$ on $\partial U$ and $\Delta u = 0$ on $U$. By Corollary 7.20, $u$ must be identically zero in $U$. $\qquad\square$

**Example 7.22.** Consider in $\mathbb{R}^3$ the harmonic polynomial $P(x, y, z) = x^3 - 3xy^2 + x^2 + y^2 - 2z^2$. The global maximum and minimum of $P$ in the closed unit ball are reached in the sphere and may be calculated by the Lagrange multipliers method. The solutions are $(\pm 1, 0, 0)$, $(\pm\frac{1}{2}, \pm\frac{\sqrt{3}}{2}, 0)$, $(0, 0, \pm 1)$, the global maximum is $2$ and the global minimum is $-2$. $\qquad\square$

*Neumann's problem* (homogeneous) is stated in a bounded domain $U$ with $\partial U$ regular and oriented. If $\vec{N}$ denotes the unit exterior normal vector field to $\partial U$, and $\varphi \in C(\partial U)$ is a given function, Neumann's problem consists in finding a function $u \in C^1(\bar{U})$, harmonic on $U$, such that $\frac{\partial u}{\partial \vec{N}} = \varphi$ in $\partial U$. By (7.9), the data $\varphi \in C(\partial U)$, must satisfy the necessary condition

$$\int_{\partial U} \varphi \, dA = 0.$$

This problem models, for example, a stationary distribution of temperatures on a domain $U$ when there is a controlled heat flow, $\frac{\partial u}{\partial \vec{N}} = \varphi$, on $\partial U$. The solution of Neumann's problem, if it exists, is unique except for constants, because if $u_1$, $u_2$ are solutions, $u = u_1 - u_2$ is harmonic on $U$ and $\frac{\partial u}{\partial \vec{N}} = 0$ on $\partial U$. Therefore $\vec{\nabla} u \equiv 0$ by (7.8) and $u$ is constant.

Evidently, Dirichlet's problem is trivial if $n = 1$, but however it should be mentioned. Recall that, in this case, harmonic functions are the linear functions. In a connected open set $U = (a, b)$, giving $\varphi \in C(\partial U)$ is just giving two values at $a$, $b$, say $A$, $B$. Then the solution of Dirichlet's problem is the line that joins $(a, A)$ with $(b, B)$, that is,

$$y - A = \frac{B - A}{b - a}(x - a).$$

## 7.5  Green's function. The Poisson kernel

Let $U$ be a bounded domain of $\mathbb{R}^n$ with regular boundary oriented by the unit exterior normal vector field $\vec{N}$. As we have seen in the previous section, every $C^2$ function $u$ on a neighborhood of $\bar{U}$ is determined by the values of $\Delta u$ in $U$ and by the values of $u$ in $\partial U$. In particular, if $u$ is harmonic, it is determined by $u_{|\partial U}$.

As a strategy to study the existence of solutions of Dirichlet's problem one should find, first, in which way $u$ is determined by $\Delta u$ and $u_{|\partial U}$. In other words, suppose there is a solution of Dirichlet's problem and try to express it in terms of the data at the boundary. To do this it is necessary to remove the term $\frac{\partial u}{\partial \vec{N}}$ in the Riesz decomposition formula (Theorem 7.13).

Let $u \in C^2(\bar{U})$ be given and let $v$ be any function, also of class $C^2$ on a neighborhood of $\bar{U}$, harmonic on $U$. By the second Green's identity, one has

$$0 - \int_U u \Delta v \, dV(x) = \int_{\partial U} \left( u(y) \frac{\partial v}{\partial \vec{N}}(y) - v(y) \frac{\partial u}{\partial \vec{N}}(y) \right) dA(y)$$

$$+ \int_U v(y) \Delta u(y) \, dV(y).$$

Subtracting this equality from (7.15), it yields

$$u(x) = \int_{\partial U} \left( u(y) \frac{\partial}{\partial \vec{N}_y}(G_x - v) - (G_x - v) \frac{\partial u}{\partial \vec{N}} \right) dA(y)$$

$$+ \int_U (G_x - v)(y) \Delta u(y) \, dV(y),$$

with $G_x(y) = G(x - y)$. The idea is now, for $x \in U$ fixed, choosing $v_x$ such that $v_x(y) = G_x(y)$ if $y \in \partial U$. In this case, the function $G_U(x, y) = G(x-y) - v_x(y)$ would satisfy

$$u(x) = \int_{\partial U} u(y) \frac{\partial}{\partial \vec{N}_y} G_U(x, y) \, dA(y) + \int_U G_U(x, y) \Delta u(y) \, dV(y). \quad (7.17)$$

Hence one would find $v_x$, harmonic, such that $v_x(y) = G(x - y)$ if $y \in \partial U$. This is a *particular* Dirichlet's problem, one for each point $x \in U$. If this particular problem is solved the function $G_U$ would be defined and one might expect that the solution of Dirichlet's problem $\Delta u = 0$ on $U$, $u_{|\partial U} = \varphi$, with $\varphi \in C(\partial U)$ is given by

$$u(x) = \int_{\partial U} \varphi(y) \frac{\partial}{\partial \vec{N}_y} G_U(x, y) \, dA(y).$$

The function $G_U(x, y)$ is called the *first Green's function of $U$ with pole $x$* or, simply, *Green's function of $U$ with pole $x$*. If $G_U(x, y)$ exists and everything makes sense, the function $P_U(x, y) = \frac{\partial}{\partial \vec{N}_y} G_U(x, y)$, $x \in U$, $y \in \partial U$ is called the *Poisson kernel* of $U$. The solution of Dirichlet's problem $\Delta u = 0$ on $U$, $u = \varphi$ on $\partial U$ would be then

$$u(x) = \int_{\partial U} P_U(x, y) \varphi(y) \, dA(y).$$

The existence of Green's function may be considered for any bounded domain $U$, not necessarily with regular boundary, because it is about solving a Dirichlet's problem with data $G(x - y)$ at the boundary; if it exists, it is the only function $G_U(x, y)$ harmonic on $U \setminus \{x\}$, as a function of $y$, that satisfies $G_U(x, y) = 0$ if $y \in \partial U$ and $G_U(x, y) = G(x - y) - v_x(y)$ with $v_x$ harmonic.

**Proposition 7.23.** *Let $U$ be a bounded domain of $\mathbb{R}^n$; if Green's function $G_U(x, y)$ exists, then it satisfies $G_U(x, y) = G_U(y, x)$ for $x, y \in U$, $x \neq y$.*

*Proof.* The proof of this result for any bounded domain is subtle and goes beyond the level of this book. Here we will only give an idea of the proof. In the special case that the domain has a regular boundary, a rigorous proof is provided.

In general, if $v_x$ is the solution of Dirichlet's problem with data $\varphi(y) = G(x-y)$, one needs to check that $v_x(y) = v_y(x)$. At first, for $x_1, x_2 \in U$ the maximum principle gives

$$\left| v_{x_1}(y) - v_{x_2}(y) \right| \leq \max_{z \in \partial U} \left| G(x_1 - z) - G(x_2 - z) \right|, \quad y \in U,$$

an inequality implying that $v_x(y)$ is continuous with respect to $x$ on $U$. One proves now that this function is harmonic in $x$, checking the validity of the mean value property. If $B(x, r)$ is a ball with center $x$ contained in $U$, the mean

$$f(y) = \frac{1}{c_n r^{n-1}} \int_{S(x,r)} v_z(y) dA(z)$$

is a harmonic function in $y$ and is continuous on $\bar{U}$. Their boundary values are

$$\frac{1}{c_n r^{n-1}} \int_{S(x,r)} G(z - y) dA(z), \quad y \in \partial U.$$

Since $G(z - y)$ is harmonic in $z$, these boundary values are $G(x - y)$. Therefore, $f$ is harmonic with boundary values $G(x - y)$, that is, $f = v_x$. This proves that $v_x(y)$ is harmonic in $x$. In order to show that $v_x(y) = v_y(x)$, thanks to the uniqueness of solution of Dirichlet's problem, it is enough to show that $v_x(y)$ has a continuous extension to $x \in \partial U$ and that it is $G(x - y)$. When $y \in \partial U$, this is true by construction, and when $y \in U$, one needs a more elaborated version of the maximum principle than the one in Theorem 7.19.

In order to have a rigorous proof supposing now that $U$ has a regular boundary, we just need to apply the second Green's identity. Taking the domain $U \setminus (\bar{B}_\varepsilon(x) \cup \bar{B}_\varepsilon(y))$ for $x, y \in U$, $x \neq y$ and $\varepsilon > 0$ small enough and the functions $u(z) = G_U(x, z)$, $v(z) = G_U(y, z)$, we get

$$\int_{S(x,\varepsilon)} \left( u \frac{\partial v}{\partial \vec{N}} - v \frac{\partial u}{\partial \vec{N}} \right) dA(z) = - \int_{S(y,\varepsilon)} \left( u \frac{\partial v}{\partial \vec{N}} - v \frac{\partial u}{\partial \vec{N}} \right) dA(z).$$

The limit, when $\varepsilon \to 0$, of the left-hand side term is $-v(x) = -G_U(y, x)$. The limit of the right-hand side term is $-u(y) = -G_U(x, y)$, as the proof of Theorem 7.13 shows, and this gives the stated result. $\qquad\square$

**Proposition 7.24.** *If Green's function $G_U(x, y)$ exists, then it satisfies $G_U(x, y) \le 0$, for $x, y \in U$, $x \ne y$.*

*Proof.* As said, $G_U(x, y)$ is harmonic with respect to $y$ on $U \setminus \{x\}$. Consider $U \setminus \bar{B}_\varepsilon(x)$, $\varepsilon > 0$. When $y \in \partial U$, $G_U(x, y)$ is zero, and if $y \in S_\varepsilon(x)$, $G_U(x, y) = G(x, y) - v_x(y)$ tends to $-\infty$ when $\varepsilon \to 0$. The maximum principle gives $G(x, y) \le 0$. $\qquad\square$

In order to solve Neumann's problem, one could try, similarly, to find a harmonic function $v_x$ such that $\frac{\partial}{\partial \vec{N}_y} v_x = \frac{\partial}{\partial \vec{N}_y} G(x - y)$ if $y \in \partial U$, for $x \in U$ fixed. In general, this is inconsistent, because it is known that $\int_{\partial U} \frac{\partial}{\partial \vec{N}} v_x \, dA = 0$ since $v_x$ is harmonic. Put, then,

$$m = \frac{1}{A(\partial U)} \int_{\partial U} \frac{\partial}{\partial \vec{N}_y} G(x - y) \, dA(y),$$

with $A(\partial U) = \int_{\partial U} dA$.

Since $G(x - y)$ is harmonic on $U \setminus \{x\}$, one has, for $\varepsilon > 0$,

$$\int_{\partial U} \frac{\partial}{\partial \vec{N}_y} G(x - y) \, dA(y) = -\int_{S(x,\varepsilon)} \frac{\partial}{\partial \vec{N}_y} G(x - y) \, dA(y) = 1,$$

the last equality following from (7.16). Therefore, $m = 1/A(\partial U)$ is independent of $x$. Now state the particular Neumann's problem: $\Delta v_x = 0$ on $U$, $\frac{\partial}{\partial \vec{N}_y} v_x(y) = \frac{\partial}{\partial \vec{N}_y} G(x - y) - m$ on $\partial U$, so that now the data at the boundary has integral zero. If one could find the solution $v_x$ of this problem, the function $H_U(x, y) = v_x(y) - G(x - y)$ would verify

$$u(x) = \frac{1}{A(\partial U)} \int_{\partial U} u(y) \, dA(y) + \int_{\partial U} H_U(x, y) \frac{\partial u}{\partial \vec{N}}(y) \, dA(y)$$
$$- \int_U H_U(x, y) \Delta u \, dV(y),$$

and one might expect that the solution of Neumann's problem $\Delta u = 0$ on $U$, $\frac{\partial u}{\partial \vec{N}} = \varphi$ on $\partial U$ (with $\varphi \in C(\partial U)$ and $\int_{\partial U} \varphi \, dA = 0$) is given by

$$u(x) = \int_{\partial U} H_U(x, y) \varphi(y) \, dA(y) + c, \quad c \text{ constant.} \qquad (7.18)$$

The function $H_U(x, y)$ is called the *second Green's function of $U$ with pole $x$.*

**Example 7.25.** Once again it is worth considering the case $n = 1$. Calculate first Green's function of $U = (a, b)$ with pole $x \in (a, b)$. Here $G(x - y) = \frac{1}{2}|x - y|$. Therefore, $v_x$ is the linear function on $(a, b)$ taking the values $\frac{1}{2}|a - x|$ at $a$ and $\frac{1}{2}|b - x|$ at $b$; that is,

$$
\begin{aligned}
v_x(y) &= \frac{1}{2}|a - x| + \frac{1}{2}\frac{|b - x| - |a - x|}{b - a}(y - a) \\
&= \frac{1}{2}(x - a) + \frac{1}{2}\frac{b - x - (x - a)}{b - a}(y - a) \\
&= \frac{1}{2}(x - a) + \frac{1}{2}\frac{b + a - 2x}{b - a}(y - a)
\end{aligned}
$$

and it turns out that

$$
G_U(x, y) = \frac{1}{2}|x - y| - v_x(y) =
\begin{cases}
\dfrac{(a - y)(b - x)}{b - a} & \text{if } y < x, \\[2mm]
\dfrac{(x - a)(y - b)}{b - a} & \text{if } x < y,
\end{cases}
$$

making clear that $G_U(x, y) \le 0$. $\qquad\square$

## 7.6 Plane domains: specific methods of complex variables. Dirichlet and Neumann problems in the unit disc

The relation between harmonic and holomorphic functions makes it possible, in dimension 2, to apply specific methods of complex variables to solve Dirichlet and Neumann problems. First of all, it will be seen how to find Green's function of the unit disc in a direct way and, consequently, to solve Dirichlet's problem in the disc. Later it will be shown that holomorphic transformations leave harmonic functions invariant, as well as Dirichlet and Neumann problems. This fact will give the solution of these problems in other plane domains by means of conformal mappings.

### 7.6.1 First Green's function and Poisson kernel in the unit disc

According to the previous section, to obtain the first Green's function of the unit disc $\mathbb{D}$ one needs to find, for $z \in \mathbb{D}$ fixed, a harmonic function $v$ on $\overline{\mathbb{D}}$ such that $v(w) = G(z - w) = \frac{1}{2\pi} \mathrm{Log}\,|z - w|$ if $w \in \partial\mathbb{D} = \mathbb{T}$, that is, if $|w| = 1$. For $z = 0$, evidently $v \equiv 0$ works. Now, if $|w| = 1$, one has $|z - w| = |1 - w\bar{z}|$, and the function

$$
v(w) = \frac{1}{2\pi} \mathrm{Log}\,|1 - w\bar{z}|,
$$

actually, satisfies $v(w) = G(z - w)$ if $|w| = 1$. Since $1 - w\bar{z}$ is holomorphic with respect to $w$ with a unique zero at the point $\frac{1}{\bar{z}}$, which is outside the disc, it follows that $\mathrm{Log}\,|1 - w\bar{z}|$ is harmonic on $\mathbb{D}$. Hence, one finds that the first Green's function of the disc is

$$G_{\mathbb{D}}(z, w) = \frac{1}{2\pi}\,\mathrm{Log}\left|\frac{z - w}{1 - w\bar{z}}\right|.$$

Compute now the Poisson kernel, $P_{\mathbb{D}}(z, w) = \frac{\partial}{\partial \vec{N}_w} G_{\mathbb{D}}(z, w)$. Write $w = re^{i\theta}$ and observe that $r\frac{\partial}{\partial r} = w\frac{\partial}{\partial w} + \bar{w}\frac{\partial}{\partial \bar{w}}$. If $r = |w| = 1$, one has

$$P_{\mathbb{D}}(z, w) = \frac{\partial}{\partial r} G_{\mathbb{D}}(z, w) = \left(w\frac{\partial}{\partial w} + \bar{w}\frac{\partial}{\partial \bar{w}}\right) G_{\mathbb{D}}(z, w)$$

$$= \left(w\frac{\partial}{\partial w} + \bar{w}\frac{\partial}{\partial \bar{w}}\right)\frac{1}{4\pi}\,\mathrm{Log}\left|\frac{z - w}{1 - w\bar{z}}\right|^2$$

$$= \frac{1}{4\pi}\left(w\frac{\partial}{\partial w} + \bar{w}\frac{\partial}{\partial \bar{w}}\right)\left(\log\frac{z - w}{1 - w\bar{z}} + \log\frac{\bar{z} - \bar{w}}{1 - \bar{w}z}\right)$$

$$= \frac{1}{2\pi}\frac{1 - |z|^2}{|1 - w\bar{z}|^2} = \frac{1}{2\pi}\frac{1 - |z|^2}{|w - z|^2}.$$

So formula (7.17) may be written, in the case of the unit disc, as

$$u(z) = \frac{1}{2\pi}\int_{\mathbb{T}}\frac{1 - |z|^2}{|z - w|^2}u(w)\,d\sigma(w) + \frac{1}{2\pi}\int_{\mathbb{D}}\mathrm{Log}\left|\frac{z - w}{1 - w\bar{z}}\right|\Delta u(w)\,dm(w).$$

Recall that $d\sigma$ is the length element on the unit circle, that is, $d\sigma(w) = d\theta$ if $w = e^{i\theta}$.

The discussion in Section 7.5 shows that a harmonic function on $\mathbb{D}$, $u$, continuous on $\bar{\mathbb{D}}$, is determined by its values on $\mathbb{T}$ according to the formula

$$u(z) = \frac{1}{2\pi}\int_{\mathbb{T}}\frac{1 - |z|^2}{|z - w|^2}u(w)\,d\sigma(w). \tag{7.19}$$

As a consequence, in order to solve Dirichlet's problem in $\mathbb{D}$ with data $\varphi \in C(\mathbb{T})$, one must consider the function

$$u(z) = \frac{1}{2\pi}\int_{\mathbb{T}}\frac{1 - |z|^2}{|z - w|^2}\varphi(w)\,d\sigma(w), \quad z \in \mathbb{D}. \tag{7.20}$$

To prove that, in fact, the function defined by (7.20) is the solution of Dirichlet's problem, some properties of the kernel $P_{\mathbb{D}}(z, w)$ are needed:

1) For every point $w \in \mathbb{T}$, the kernel $P_{\mathbb{D}}(z, w)$ is a harmonic function in $z$ for $z \in \mathbb{D}$ and satisfies $P_{\mathbb{D}}(z, w) \geq 0$.

2) $\displaystyle\int_{\mathbb{T}} P_{\mathbb{D}}(z,w)\,d\sigma(w) = 1$, for $z \in \mathbb{D}$.

3)  For fixed $w_0 \in \mathbb{T}$ and $\delta > 0$, one has

$$\lim_{z \to w_0} \int_{\mathbb{T} \setminus D_\delta(w_0)} P_{\mathbb{D}}(z,w)\,d\sigma(w) = 0.$$

Property 1) may be directly checked, by calculating $\Delta_z P_{\mathbb{D}}(z,w)$, but it will be more useful to show that the Poisson kernel is the real part of a holomorphic function in $z$. Write $G_{\mathbb{D}}(z,w) = \frac{1}{2\pi}\,\mathrm{Log}\left|\frac{z-w}{1-\bar{w}z}\right|$ and observe that $G_{\mathbb{D}}$ is the real part of $\frac{1}{2\pi}\log\frac{z-w}{1-\bar{w}z}$. This last function, for $w$ fixed with $|w| \leq 1$, is holomorphic on $D(0,|w|)$. Therefore, $P_{\mathbb{D}}(z,w)$ is the real part of the holomorphic function in $z$,

$$\left(w\frac{\partial}{\partial w} + \bar{w}\frac{\partial}{\partial\bar{w}}\right)\Big|_{\mathbb{T}}\left(\frac{1}{2\pi}\log\frac{z-w}{1-\bar{w}z}\right) = \frac{1}{2\pi}\left(\frac{w}{w-z} + \frac{\bar{w}z}{1-z\bar{w}}\right)$$

$$= \frac{1}{2\pi}\frac{w-\bar{w}z^2}{(1-z\bar{w})(w-z)} = \frac{1}{2\pi}\frac{w^2-z^2}{(w-z)^2} = \frac{1}{2\pi}\frac{w+z}{w-z} \qquad (7.21)$$

$$= \frac{1}{2\pi}\frac{1+z\bar{w}}{1-z\bar{w}} = \frac{1}{2\pi}\frac{2}{1-z\bar{w}} - 1.$$

Property 2) is obtained from (7.19) taking $u \equiv 1$. Property 3) is immediate: if $|w-w_0| \geq \delta$ and $|z-w_0| \leq \frac{\delta}{2}$, then $|w-z| \geq \frac{\delta}{2}$ and $P_{\mathbb{D}}(z,w) \leq \frac{1}{2\pi}\frac{4}{\delta^2}(1-|z|^2)$; therefore

$$\int_{\mathbb{T}\setminus D_\delta(w_0)} P_{\mathbb{D}}(z,w)\,d\sigma(w) \leq \frac{4}{\delta^2}(1-|z|^2) \xrightarrow[z\to w_0]{} 0.$$

**Theorem 7.26** (Solution of Dirichlet's problem in the unit disc). *If $\varphi \in C(\mathbb{T})$, the function $u$ defined by equality (7.20) is harmonic on $\mathbb{D}$ and satisfies*

$$\lim_{z\to w} u(z) = \varphi(w), \quad \text{for each } w \in \mathbb{T}.$$

*Proof.* First,

$$\Delta u(z) = \Delta\int_{\mathbb{T}} P_{\mathbb{D}}(z,w)\varphi(w)\,d\sigma(w) = \int_{\mathbb{T}} \Delta_z P_{\mathbb{D}}(z,w)\varphi(w)\,d\sigma(w) = 0,$$

by property 1) and so $u$ is harmonic on $\mathbb{D}$.

Now one must show that $\lim_{z\to w_0} u(z) = \varphi(w_0)$, if $|w_0| = 1$. Using property 2) and $P_{\mathbb{D}}(z,w) \geq 0$ one has, for $\delta > 0$,

$$u(z) - \varphi(w_0) = \int_{\mathbb{T}} \{\varphi(w) - \varphi(w_0)\} P_{\mathbb{D}}(z,w)\,d\sigma(w),$$

$$|u(z) - \varphi(w_0)| \leq \int_{\mathbb{T}} |\varphi(w) - \varphi(w_0)| P_{\mathbb{D}}(z, w) \, d\sigma(w)$$

$$\leq \int_{\mathbb{T} \setminus D_\delta(w_0)} + \int_{\mathbb{T} \cap D_\delta(w_0)} = \mathrm{I} + \mathrm{II}.$$

Part II can be estimated as follows:

$$\mathrm{II} \leq \sup_{|w - w_0| \leq \delta} |\varphi(w) - \varphi(w_0)| \int_{\mathbb{T} \cap D_\delta(w_0)} P_{\mathbb{D}}(z, w) \, d\sigma(w)$$

$$\leq \sup_{|w - w_0| \leq \delta} |\varphi(w) - \varphi(w_0)|,$$

so that given $\varepsilon > 0$ we get $\mathrm{II} \leq \varepsilon/2$ for every $z$, if $\delta > 0$ is small enough, thanks to uniform continuity of $\varphi$. Now, with this $\delta$ fixed, property 3) yields

$$\mathrm{I} \leq 2\|\varphi\|_\infty \int_{\mathbb{T} \setminus D_\delta(w_0)} P_{\mathbb{D}}(z, w) \, d\sigma(w) \leq \varepsilon/2$$

if $z$ is close enough to $w_0$ and this ends the proof. $\qquad\square$

The solution $u$ – called the *Poisson transform* of $\varphi$ and represented by $P[\varphi]$ – is expressed sometimes in a slightly different way. Write $w = e^{it}, 0 \leq t \leq 2\pi, z = re^{i\theta}$; then $|z - w|^2 = |re^{i\theta} - w|^2 = 1 + r^2 - 2\operatorname{Re} r\bar{w}e^{i\theta} = 1 + r^2 - 2r\cos(\theta - t)$. Hence, defining functions $u_r$ by $u_r(e^{i\theta}) = u(re^{i\theta}), 0 \leq r < 1$, it turns out that

$$u_r(e^{i\theta}) = \frac{1}{2\pi} \int_0^{2\pi} \frac{1 - r^2}{1 + r^2 - 2r\cos(\theta - t)} \varphi(e^{it}) \, dt.$$

That is, $u_r$ is the *convolution* of the data $\varphi$ with the function

$$P_r(t) = P(r, e^{it}) = \frac{1}{2\pi} \frac{1 - r^2}{1 + r^2 - 2r\cos t}$$

as functions defined on the multiplicative group $\mathbb{T}$. In summary, one has

$$u_r(e^{i\theta}) = u(re^{i\theta}) = (\varphi * P_r)(e^{i\theta}).$$

The fact that the solution is expressed in this way comes also from the *rotation invariance* of Dirichlet's problem: if there are two data $\varphi_1, \varphi_2$ and one of them is obtained from the other by means of a rotation, $\varphi_2(w) = \varphi_1(\lambda w)$ with $|\lambda| = 1$, then the respective solutions $u_1, u_2$ satisfy the same relation, $u_2(w) = u_1(\lambda w)$. The relevant fact is that every operator $\varphi \mapsto u$ with this property can be written as a convolution of $\varphi$ by a certain kernel.

Now one may calculate the expansion of the harmonic function $u$ in power series of $z$ and $\bar{z}$ (Theorem 4.19) and see how coefficients of this expansion depend on Fourier coefficients of $\varphi$. First observe that the equality

$$\frac{2}{1 - z\bar{w}} = 2\sum_{0}^{\infty}(z\bar{w})^n$$

holds uniformly for $w \in \mathbb{T}$, if $|z| < 1$. Therefore, by (7.21), one has

$$2\pi P_{\mathbb{D}}(z, w) = \operatorname{Re}\left(\frac{2}{1 - z\bar{w}} - 1\right) = 2\operatorname{Re}\sum_{0}^{\infty}(z\bar{w})^n - 1$$

$$= \sum_{0}^{\infty}(z\bar{w})^n + \sum_{0}^{\infty}(\bar{z}w)^n - 1 = \sum_{0}^{\infty}(z\bar{w})^n + \sum_{1}^{\infty}(\bar{z}w)^n,$$

also uniformly on $\mathbb{T}$. Now, recalling that the $n$-th Fourier coefficient of $\varphi$ is $\hat{\varphi}(n) = \frac{1}{2\pi}\int_0^{2\pi}\varphi(e^{it})e^{-int}\,dt = \frac{1}{2\pi}\int_0^{2\pi}\varphi(e^{it})\bar{w}^n\,d\sigma(w)$, one finds

$$u(z) = \sum_{0}^{\infty}z^n\hat{\varphi}(n) + \sum_{1}^{\infty}\bar{z}^n\hat{\varphi}(-n).$$

In other words, $P_r(t) = P(r, e^{it}) = \sum_{-\infty}^{+\infty}r^{|n|}e^{int}$ and assuming that $\varphi(t) \sim \sum\hat{\varphi}(n)e^{int}$, then

$$u(z) = (\varphi * P_r)(e^{it}) = \sum_{-\infty}^{+\infty}r^{|n|}\hat{\varphi}(n)e^{int}, \quad z = re^{it}. \tag{7.22}$$

The solution corresponding to $\varphi(t) = e^{int}$ is $z^n = r^n e^{int}$ if $n \geq 0$ and $\bar{z}^n = r^{|n|}e^{int}$ if $n < 0$.

**Example 7.27.** Let us compute the solution of Dirichlet's problem in the unit disc when the data is $\varphi(e^{it}) = \cos^2 t$. We find first the Fourier expansion of $\varphi$, expressing $\cos t$ in terms of complex exponentials:

$$\varphi(e^{it}) = \left(\tfrac{1}{2}(e^{it} + e^{-it})\right)^2 = \tfrac{1}{4}(e^{2it} + 2 + e^{-2it}).$$

The solution is, therefore, $u(z) = \tfrac{1}{4}(z^2 + 2 + \bar{z}^2) = \tfrac{1}{2}(1 + x^2 - y^2)$.

More generally if $\varphi$ is a polynomial in $\cos t$ and $\sin t$, then the solution $u$ is the sum of a polynomial in $z$ and a polynomial in $\bar{z}$. $\qquad\square$

Tacitly, in the formulation and the solution of Dirichlet's problem in the unit disc it has been supposed that one deals with real functions. Considering separately real and imaginary parts, a complex Dirichlet's problem can be solved: given any continuous function $\varphi\colon \mathbb{T} \to \mathbb{C}$, there is a unique harmonic function $u\colon \mathbb{D} \to \mathbb{C}$, $u \in C(\bar{\mathbb{D}})$, with $u_{|\mathbb{T}} = \varphi$, also given by (7.20) and (7.22). Hence, representing by $h(\mathbb{D})$ the space of harmonic functions on $\mathbb{D}$, one has a one-to-one correspondence

$$C(\mathbb{T}) \longrightarrow C(\bar{\mathbb{D}}) \cap h(\mathbb{D}),$$
$$\varphi \longmapsto u = P[\varphi],$$

that associates to $\varphi$ the function $u = P[\varphi]$, the only one continuous on $\bar{\mathbb{D}}$ and harmonic on $\mathbb{D}$ such that $u_{|\mathbb{T}} = \varphi$.

A subspace of $h(\mathbb{D})$ is the space $H(\mathbb{D})$ of holomorphic functions on $\mathbb{D}$. The class $C(\bar{\mathbb{D}}) \cap H(\mathbb{D})$ which, therefore, consists of holomorphic functions on $\mathbb{D}$ that have a continuous extension to $\bar{\mathbb{D}}$ is called the *disc algebra* and is denoted by $A(\mathbb{D})$. In the correspondence $\varphi \to u = P[\varphi]$, the subspace $A(\mathbb{D}) \subset C(\bar{\mathbb{D}}) \cap h(\mathbb{D})$ corresponds to the subspace $A(\mathbb{T}) \subset C(\mathbb{T})$ formed by functions $\varphi \in C(\mathbb{T})$ such $P[\varphi] \in H(\mathbb{D})$; in view of (7.22), one has

$$A(\mathbb{T}) = \{\varphi \in C(\mathbb{T})\colon \hat{\varphi}(n) = 0,\ n \in \mathbb{Z},\ n < 0\}.$$

One should be aware that, by Weierstrass' theorem, every function of $C(\mathbb{T})$ is the uniform limit on $\mathbb{T}$ of a sequence of polynomials in $x, y$, and so, of a sequence of polynomials in $z, \bar{z}$. Observe, however, that if

$$\sideset{}{'}\sum a_{nm} z^n \bar{z}^m$$

is a polynomial in $z, \bar{z}$, using the fact that $z\bar{z} = 1$ if $|z| = 1$, it turns out that the restriction to $\mathbb{T}$ of previous polynomial is the sum of a polynomial in $z$ and a polynomial in $\bar{z}$. Hence, when solving the Dirichlet problem $\Delta u = 0$ on $\mathbb{D}$, $u = \varphi$ on $\mathbb{T}$ for a data $\varphi \in C(\mathbb{T})$ by means of the function $u = P[\varphi]$, polynomials in $z, \bar{z}$ which approximate $\varphi$ uniformly on $\mathbb{T}$ are explicitly provided. If one looks for a polynomial $P(z, \bar{z})$ such that $|\varphi(z) - P(z, \bar{z})| < \varepsilon$ if $|z| = 1$, write

$$\varphi(e^{it}) = \lim_{r \to 1} u(re^{it}) = \lim_{r \to 1} \sum_{-\infty}^{\infty} \hat{\varphi}(n) r^{|n|} e^{int}$$

with uniform convergence on $\mathbb{T}$. Now choose $r$ such that $|\varphi(e^{it}) - u(re^{it})| < \varepsilon/2$ and after that, a partial sum $\sum_{|n| \le N} \hat{\varphi}(n) r^{|n|} e^{int}$ with $N$ big enough such that $|u(re^{it}) - \sum_{|n| \le N} \hat{\varphi}(n) r^{|n|} e^{int}| < \varepsilon/2$, using uniform convergence on the circle $|z| = r$. Hence, it is not true that

$$\varphi(e^{it}) = \lim_{N \to \infty} \sum_{|n| \le N} \hat{\varphi}(n) e^{int} \quad \text{uniformly in } t,$$

but, instead, it is true that

$$\varphi(e^{it}) = \lim_{r \to 1} \lim_{N \to \infty} \sum_{|n| \leq N} \widehat{\varphi}(n) r^{|n|} e^{int} \quad \text{uniformly in } t.$$

If $\varphi \in A(\mathbb{T})$, the polynomials that approximate $\varphi$ – the previous partial sums – are polynomials only in $z$. It is worth mentioning that, as said before, *it is not true* in general that trigonometric polynomials

$$S_N(\varphi) = \sum_{|n| \leq N} \widehat{\varphi}(n) e^{int}$$

converge to $\varphi$ uniformly on $\mathbb{T}$ if $\varphi \in C(\mathbb{T})$ (but it is true if in addition $\varphi$ has bounded variation). Another classical way of constructing polynomials that approximate uniformly a function $\varphi \in C(\mathbb{T})$ is using Fejér sums $\sigma_N(\varphi)$, arithmetic means of the polynomials $S_N(\varphi)$, defined by

$$\sigma_N(\varphi) = \frac{1}{N+1}(S_0(\varphi) + \cdots + S_N(\varphi)).$$

In the following statement, which describes the space $A(\mathbb{T})$, the item a) says that $A(\mathbb{T})$ is the "half" of $C(\mathbb{T})$, the part generated by polynomials in $z$.

**Proposition 7.28.** a) *$A(\mathbb{T})$ is the subspace of $C(\mathbb{T})$ formed by functions that can be approximated uniformly on $\mathbb{T}$ by polynomials in $z$.*

b) *A function $\varphi \in C(\mathbb{T})$ belongs to $A(\mathbb{T})$ if and only if $\int_{\mathbb{T}} \varphi(z) z^n dz = 0$, for $n \geq 0$.*

c) *A function $\varphi \in C(\mathbb{T})$ belongs to $A(\mathbb{T})$ if and only if*

$$\int_{\mathbb{T}} \frac{\varphi(z)}{z - w} \, dz = 0, \quad \text{for } |w| > 1.$$

d) *The functions $\varphi \in C(\mathbb{T})$ that can be approximated uniformly on $\mathbb{T}$ by polynomials in $\bar{z}$ are the ones of $\overline{A(\mathbb{T})}$ (conjugates of functions of $A(\mathbb{T})$) and are characterized by the condition*

$$\int_{\mathbb{T}} \varphi(z) \bar{z}^n d\bar{z} = 0, \quad \text{for } n \geq 0,$$

*or by*

$$\frac{1}{2\pi i} \int_{\mathbb{T}} \frac{\varphi(z)}{z - w} \, dz = \frac{1}{2\pi} \int_0^{2\pi} \varphi(e^{it}) \, dt, \quad |w| < 1.$$

e) *Let $A_0(\mathbb{T})$ denote the subspace of functions $\varphi \in A(\mathbb{T})$ with $\widehat{\varphi}(0) = 0$ (a condition which means that the function $u \in A(\mathbb{D})$ with boundary values $\varphi$*

*satisfies $u(0) = 0$). Then a function $\varphi \in C(\mathbb{T})$ belongs to the space $\overline{A_0(\mathbb{T})}$ if and only if $\varphi$ is perpendicular to $A(\mathbb{T})$ by the scalar product of $L^2(\mathbb{T})$ given by*

$$\langle \varphi, \psi \rangle = \frac{1}{2\pi} \int_0^{2\pi} \varphi(e^{it})\overline{\psi(e^{it})}\, dt.$$

*Proof.* It has been already seen that $\varphi \in A(\mathbb{T})$ if and only if $\hat{\varphi}(m) = 0$ for $m < 0$; if $m = -n, n > 0$, then

$$\hat{\varphi}(m) - \frac{1}{2\pi} \int_0^{2\pi} \psi(e^{it})e^{\,imt}\, dt = \frac{1}{2\pi} \int_T \varphi(z)z^{n-1}\, dz$$

and b) is proved. In the considerations made before the statement of this proposition, it was observed that every function $\varphi \in A(\mathbb{T})$ can be approximated uniformly by polynomials in $z$ on $\mathbb{T}$. The converse is immediate, since every polynomial in $z = e^{it}$ gives effective Fourier coefficients only for $n \geq 0$. The integral in c) may be written, expanding in geometric series the integrand, as

$$\int_{\mathbb{T}} \frac{\varphi(z)}{z - w}\, dz = -\int_{\mathbb{T}} \varphi(z)\frac{1}{w\left(1 - \frac{z}{w}\right)}\, dz$$

$$= -\int_{\mathbb{T}} \varphi(z) \sum_{n=0}^{\infty} \frac{z^n}{w^{n+1}}\, dz = -\sum_{n=0}^{\infty} w^{-n-1} \int_{\mathbb{T}} \varphi(z)z^n\, dz,$$

and then c) follows from b). Conjugating in b) one has $\varphi \in \overline{A(\mathbb{T})}$ if and only if $\int \varphi(z)\bar{z}^n\, d\bar{z} = 0, n \geq 0$ or, equivalently, $\int \varphi(z)z^m\, dz = 0, m \leq -2$. The second integral in d) is then

$$\frac{1}{2\pi i} \int_{\mathbb{T}} \varphi(z) \sum_{n=0}^{\infty} \frac{w^n}{z^{n+1}}\, dz = \frac{1}{2\pi i} \int_{\mathbb{T}} \varphi(z) \frac{dz}{z} = \frac{1}{2\pi} \int_0^{2\pi} \varphi(e^{it})\, dt.$$

Item e) is obvious, because $\overline{A_0(\mathbb{T})}$ is generated by the functions $\bar{z}^n$ with $n \geq 1$ and $A(\mathbb{T})$ by the functions $z^m$ with $m \geq 0$ and

$$\langle z^m, \bar{z}^n \rangle - \frac{1}{2\pi} \int_0^{2\pi} e^{i(m+n)t}\, dt = 0. \qquad \square$$

The calculation done to prove item d) of the proposition above shows that the Cauchy's integral of a function $\varphi \in C(\mathbb{T})$ is

$$\frac{1}{2\pi i} \int_{\mathbb{T}} \frac{\varphi(z)}{z - w}\, dz = \sum_{n=0}^{\infty} \hat{\varphi}(n)w^n.$$

Now, this function is not necessarily continuous on $\overline{\mathbb{D}}$.

In Chapter 10 a more general version of Proposition 7.28 will be given.

### 7.6.2 Relationship between the Cauchy kernel and the Poisson kernel

If $f$ is a holomorphic function on a neighborhood of the closed unit disc $\overline{\mathbb{D}}$, then the Cauchy integral formula may be written as

$$
f(z) = \frac{1}{2\pi i} \int_{\mathbb{T}} \frac{f(w)}{w - z}\, dw
$$
$$
= \frac{1}{2\pi} \int_{\mathbb{T}} \frac{f(w)}{1 - \bar{w}z}\, d\sigma(w) = \frac{1}{2\pi} \int_{-\pi}^{+\pi} \frac{f(e^{i\theta})}{1 - e^{-i\theta}z}\, d\theta, \quad z \in \mathbb{D},
\tag{7.23}
$$

with $w = e^{i\theta}$ if $|w| = 1$ and $d\sigma(w) = d\theta$.

A holomorphic function is harmonic, and, therefore, one also has

$$
f(z) = \frac{1}{2\pi} \int_{\mathbb{T}} f(w)\frac{1 - |z|^2}{|1 - \bar{w}z|^2}\, d\sigma(w), \quad z \in \mathbb{D}.
\tag{7.24}
$$

Hence, both Cauchy kernel and Poisson kernel are reproducing kernels for *holomorphic functions* from their values on $\mathbb{T} = \partial\mathbb{D}$. The difference between both kernels

$$
\frac{1}{2\pi}\left\{ \frac{1 - |z|^2}{|1 - \bar{w}z|^2} - \frac{1}{1 - \bar{w}z} \right\} = \frac{1}{2\pi}\frac{1}{1 - \bar{w}z}\left\{ \frac{1 - |z|^2}{1 - w\bar{z}} - 1 \right\}
$$
$$
= \frac{1}{2\pi}\frac{w\bar{z} - |z|^2}{|1 - \bar{w}z|^2} = \frac{\bar{z}(w - z)}{2\pi(1 - \bar{w}z)(1 - w\bar{z})} = \frac{\bar{z}}{2\pi\bar{w}(1 - w\bar{z})} = \frac{1}{2\pi}\frac{\bar{z}w}{1 - \bar{z}w}
$$

is, therefore, a kernel that cancels holomorphic functions in the sense that

$$
\frac{1}{2\pi} \int_{\mathbb{T}} f(w)\frac{\bar{z}w}{1 - \bar{z}w}\, d\sigma(w) = \frac{1}{2\pi i} \int_{\mathbb{T}} f(w)\frac{\bar{z}\, dw}{1 - \bar{z}w} = 0.
$$

This equality is also a consequence of Cauchy's theorem because the function $f(w)\frac{\bar{z}}{1-\bar{z}w}$ is holomorphic on a neighborhood of $\overline{\mathbb{D}}$.

Another way to obtain Poisson's formula for the holomorphic function $f$ is applying Cauchy's formula, to the function $h(w) = f(w)/(1 - \bar{z}w)$, for fixed $z$. Then it yields

$$
\frac{f(z)}{1 - |z|^2} = h(z) = \frac{1}{2\pi} \int_{\mathbb{T}} \frac{h(w)}{1 - \bar{w}z}\, d\sigma(w) = \frac{1}{2\pi} \int_{\mathbb{T}} \frac{f(w)}{|1 - \bar{w}z|^2}\, d\sigma(w).
$$

It is worth observing that the Cauchy kernel is a *universal* kernel, that is, it is the same for all domains, while the Poisson kernel depends on the domain (likewise, in the first case the integral is with respect to $dw$ – also universal – and in the second case, with respect to $d\sigma(w)$, which depends on the domain). The Poisson kernel is only explicit in domains with a lot of symmetries, as it is the case of the disc.

Writing the Cauchy kernel, $C(z, w) = \frac{1}{2\pi i}\frac{dw}{w-z}$, as before in the way $C(z, w) = \frac{1}{2\pi}\frac{d\sigma(w)}{1-\bar{w}z}$, and using (7.21), the following relation between the Poisson kernel and the Cauchy kernel is found:

$$P_{\mathbb{D}}(z, w) = \frac{1}{2\pi}\,\mathrm{Re}\left\{\frac{2}{1-z\bar{w}} - 1\right\}$$

$$= \frac{1}{2\pi}\,\mathrm{Re}\,\frac{1+z\bar{w}}{1-z\bar{w}}, \quad z \in \mathbb{D}, \; |w| = 1.$$

This formula has an interesting consequence related to the following observation. It is known that a function $f$ holomorphic on $\bar{\mathbb{D}}$ is determined by its real part except for an imaginary constant. This real part is a harmonic function determined by its values on $\mathbb{T}$. In conclusion, $f$ is determined, except for an imaginary constant, by its real part on $\mathbb{T}$. The kernel $\frac{1+z\bar{w}}{1-z\bar{w}}$ makes the reconstruction of $f$ in $\mathbb{D}$ explicit from the values of $\mathrm{Re}(f)$ on $\mathbb{T}$.

**Theorem 7.29.** *If the function $f$ is holomorphic on a neighborhood of $\bar{\mathbb{D}}$ and $u = \mathrm{Re}\, f$, then one has*

$$f(z) = \frac{1}{2\pi}\int_{\mathbb{T}}\frac{1+z\bar{w}}{1-z\bar{w}}u(w)\,d\sigma(w) + i\,\mathrm{Im}\, f(0), \quad z \in \mathbb{D}.$$

*Proof.* The integral defines a holomorphic function on $\mathbb{D}$ that, as it has just been seen, has real part $u$. Its difference with $f$ has real part zero and, therefore, is an imaginary constant. $\qquad\square$

Consequently, the conjugated harmonic function of $u$, say $v = \mathrm{Im}\, f$, satisfying $v(0) = 0$ is given by

$$v(z) = \frac{1}{2\pi}\int_{\mathbb{T}}\mathrm{Im}\,\frac{1+z\bar{w}}{1-z\bar{w}}u(w)\,d\sigma(w) = \frac{1}{\pi}\int_{\mathbb{T}}\frac{\mathrm{Im}\,z\bar{w}}{|1-z\bar{w}|^2}u(w)\,d\sigma(w).$$

The kernel

$$\frac{1+z\bar{w}}{1-z\bar{w}}$$

appearing in Theorem 7.29 is called the *Herglotz kernel*, and the formula representing $f$ in this theorem is also written

$$f(z) = \frac{1}{2\pi}\int_0^{2\pi}\frac{e^{it}+z}{e^{it}-z}u(e^{it})\,dt + i\,\mathrm{Im}\, f(0), \quad z \in \mathbb{D}.$$

**Corollary 7.30.** *Every function $f$, holomorphic on a neighborhood of $\overline{\mathbb{D}}$ and non-vanishing on $\overline{\mathbb{D}}$, may be expressed in terms of the values of $|f|$ on $\mathbb{T}$ by means of the formula*

$$f(z) = \lambda \exp \frac{1}{2\pi} \int_{\mathbb{T}} \frac{1 + z\overline{w}}{1 - z\overline{w}} \operatorname{Log}|f(w)|\, d\sigma(w), \quad z \in \mathbb{D}$$

*with $\lambda \in \mathbb{C}$, $|\lambda| = 1$.*

*Proof.* Just apply Theorem 7.29 to a branch of the function $\log f$. $\qquad\square$

**Example 7.31.** Apply the formula of Theorem 7.29 to the principal branch of the logarithm of $f(z) = z - a$, with $a$ real and $a < -1$:

$$\operatorname{Log}(z - a) = \frac{1}{2\pi} \int_0^{2\pi} \frac{e^{it} + z}{e^{it} - z} \operatorname{Log}|e^{it} - a|\, dt.$$

Taking $z = 0$ one finds

$$\operatorname{Log}|a| = \frac{1}{4\pi} \int_0^{2\pi} \operatorname{Log}(1 + a^2 - 2a \cos t)\, dt.$$

Since $\operatorname{Log}|e^{it} + 1|$ is still integrable, one may $a$ tend to $-1$ to obtain

$$\int_0^{2\pi} \operatorname{Log}(1 + \cos t)\, dt = -2\pi \operatorname{Log} 2. \qquad\square$$

### 7.6.3 The solution of Neumann's problem in the unit disc

In order to compute the second Green's function of the unit disc $\mathbb{D}$ one has to find, for fixed $z \in \mathbb{D}$, a harmonic function, $v = v_z$, on $\overline{\mathbb{D}}$ such that $\frac{\partial v}{\partial \vec{N}} = \frac{\partial G}{\partial \vec{N}} - \frac{1}{2\pi}$ if $|w| = 1$. At first, calculating, one has

$$\frac{\partial G}{\partial \vec{N}}(w) = w\frac{\partial}{\partial w} + \overline{w}\frac{\partial}{\partial \overline{w}}\Big|_{\mathbb{T}} \left( \frac{1}{2\pi} \operatorname{Log}|z - w| \right)$$

$$= \frac{1}{2\pi} \operatorname{Re} \frac{w}{w - z} = \frac{1}{2\pi} \operatorname{Re} \frac{1}{1 - \overline{w}z},$$

$$\frac{\partial G}{\partial \vec{N}}(w) - \frac{1}{2\pi} = \frac{1}{2\pi}\left\{ \operatorname{Re} \frac{1}{1 - \overline{w}z} - 1 \right\} = \frac{1}{2\pi} \operatorname{Re} \frac{\overline{w}z}{1 - \overline{w}z}.$$

Now it is clear that the function $v(w) = -\frac{1}{2\pi} \operatorname{Log}|1 - \overline{w}z|$ works, because

$$\frac{\partial v}{\partial \vec{N}} = w\frac{\partial}{\partial w} + \overline{w}\frac{\partial}{\partial \overline{w}}\Big|_{\mathbb{T}} \left( -\frac{1}{2\pi} \operatorname{Log}|1 - \overline{w}z| \right) = \frac{1}{2\pi} \operatorname{Re} \frac{\overline{w}z}{1 - \overline{w}z}.$$

Hence, $H_{\mathbb{D}}(z, w) = v(w) - G(z - w) = -\frac{1}{2\pi} \operatorname{Log}(|z - w|\,|1 - \overline{w}z|)$ is the second Green's function of $\mathbb{D}$.

**Theorem 7.32.** *Suppose that $\varphi \in C(\mathbb{T})$ satisfies the compatibility condition $\int_{\mathbb{T}} \varphi \, d\sigma = 0$. Then the unique solution of the problem $\Delta u = 0$ on $\mathbb{D}$, $\frac{\partial u}{\partial \vec{N}} = \varphi$ on $\mathbb{T}$, that vanishes at the origin, is the function*

$$u(z) = \int_{\mathbb{T}} H_{\mathbb{D}}(z, w)\varphi(w) \, d\sigma(w), \quad z \in \mathbb{D}.$$

*Proof.* According to (7.18) this is the expected solution. Let us check that it is so. When $|w| = 1$, one has $H_{\mathbb{D}}(z, w) = \frac{1}{\pi} \mathrm{Log}\, |z - w|$, which is harmonic in $z$, and therefore $u$ is harmonic. By the same computation done above, it turns out that

$$\frac{\partial}{\partial \vec{N}} H_{\mathbb{D}}(z, w) = \frac{1}{\pi} \, \mathrm{Re}\, \frac{z}{w - z} = \frac{1}{\pi} \, \mathrm{Re}\, \frac{\bar{w} z - |z|^2}{|w - z|^2} = \frac{1}{2\pi}\left(\frac{1 - |z|^2}{|w - z|^2} - 1\right).$$

This means

$$\frac{\partial u}{\partial \vec{N}}(z) = \frac{1}{2\pi} \int_{\mathbb{T}} \frac{1 - |z|^2}{|z - w|^2} \varphi(w) \, d\sigma(w) - \frac{1}{2\pi} \int_{\mathbb{T}} \varphi(w) \, d\sigma(w) = P[\varphi](z).$$

That is, $\frac{\partial u}{\partial \vec{N}}$ is the Poisson transform of $\varphi$ and, then, has value $\varphi$ on $\mathbb{T}$. $\qquad\square$

**Remark 7.2.** Speaking properly, the function $u$ is not of class $C^1$ on $\bar{\mathbb{D}}$. It has just been proved that $\frac{\partial u}{\partial \vec{N}}$ is continuous on $\bar{\mathbb{D}}$, and in order to set the continuity of the whole gradient, $\vec{\nabla} u$, $\varphi$ must fulfill some additional properties.

## 7.7 The Poisson equation in $\mathbb{R}^n$

### 7.7.1 The Riesz potential of a measure

The Poisson equation is

$$\Delta u = \phi,$$

where, to start with, $\phi$ is supposed to be a continuous function with compact support in $\mathbb{R}^n$. If this equation has a twice differentiable solution $u$, then it has infinitely many, because adding to $u$ any harmonic function, another solution is obtained. In order to distinguish a particular solution one must impose some other condition. The maximum principle suggests one such condition, since every function $u$ is determined on a domain by $\Delta u$ and the values of $u$ on the boundary.

**Proposition 7.33.** a) *If $u$ is a twice differentiable function on $\mathbb{R}^n$, vanishing at infinity (that is: $\lim_{|x| \to +\infty} u(x) = 0$) and such that $\Delta u$ is continuous with compact support (in particular if $u \in C_c^2(\mathbb{R}^n)$), then one has*

$$u(x) = \int_{\mathbb{R}^n} G(x - y)\Delta u(y) dV(y), \quad x \in \mathbb{R}^n.$$

b) *Given a function $\phi \in C_c(\mathbb{R}^n)$, if there is a twice differentiable solution $u$ of the equation $\Delta u = \phi$, vanishing at infinity, this solution is unique and it is the Riesz potential*

$$u = G(\phi)(x) = \int_{\mathbb{R}^n} G(x - y)\phi(y)dV(y).$$

*When $n = 2$, a necessary condition for the existence of the solution $u$ in item b) is $\int \phi\, dV = 0$.*

*Proof.* Suppose, for the moment, $n > 2$ and apply Corollary 7.18. Denoting by $K$ the support of $\Delta u$ and using that $u(x) \to 0$ when $|x| = R \to +\infty$, it turns out that

$$u(x) = \lim_{R \to +\infty} \int_{B(x,R)} G_{B(x,R)}(x, y)\Delta u(y)dV(y)$$

$$= \lim_{R \to +\infty} \int_K G_{B(x,R)}(x, y)\Delta u(y)dV(y)$$

$$= \int_K G(x - y)\Delta u(y)dV(y) - \lim_{R \to +\infty} d_n R^{2-n} \int_K \Delta u(y)dV(y)$$

$$= \int_{\mathbb{R}^n} G(x - y)\Delta u(y)dV(x).$$

If $n = 2$, $R^{2-n}$ is replaced by $\mathrm{Log}\, R$. In this case, the existence of the limit at infinity implies $\int_K \Delta u(y)dV(y) = 0$ and a) is proved. Obviously b) is a consequence of a) (observe that the uniqueness is also a consequence of the maximum modulus principle). The necessary condition when $n = 2$ follows from the fact that $\int_K \Delta u\, dV = 0$, as said. $\qquad\square$

The particularity of the case $n = 2$ in item b) is explained by the following remark: if $n > 2$, every Riesz potential $u = G(\phi)$ of a continuous function $\phi$ with compact support $K$ vanishes at infinity because $\sup_{y \in K} |x - y|^{2-n} \to 0$ if $|x| \to \infty$. If $n = 2$, this is not true, but assuming $\int_K \phi(y)dV(y) = 0$, then

$$\int_K \mathrm{Log}\,|x - y|\phi(y)dV(y) = \int_K (\mathrm{Log}\,|x - y| - \mathrm{Log}\,|x|)\,\phi(y)dV(y),$$

and now $\mathrm{Log}\,\dfrac{|x-y|}{|x|} \xrightarrow[|x| \to \infty]{} 0$ uniformly for $y \in K$.

Item b) of Proposition 7.33 does not show that given a function $\phi \in C_c(\mathbb{R}^n)$ (with $\int \phi\, dV = 0$ if $n = 2$) the Riesz potential $u = G(\phi)$ is a solution of $\Delta u = \phi$, vanishing at infinity; it only says that if there is a solution, it is this one. In order to study the existence of solutions of the Poisson equation, one needs to analyze the properties of Riesz potentials. To this end, and also for the applications, it is advisable to extend the definition of Riesz potential given in (7.14) changing $\phi\, dV$ by a real measure $\mu$ defined on $\mathbb{R}^n$ with compact support $K$ and finite total variation, $|\mu|(K) < +\infty$. Recall that the *support* of a measure $\mu$, $\mathrm{spt}(\mu)$, is the complement of the biggest open set on which $\mu$ vanishes. One may think about $\mu$ as a charge or mass distribution, which may go from being absolutely continuous, as in the case $\mu = \phi\, dV$, to be point charges $\mu = \sum c_i \delta_{a_i}$, with $\sum |c_i| < +\infty$ and $\delta_{a_i}$ denoting the mass 1 at the point $a_i$ (Dirac delta at $a_i$). In an intermediate situation, $\mu$ may be the measure of integration on a curve or a hyper-surface or, for example, a charge distribution on a sphere.

Given, then, a real measure $\mu$ on $\mathbb{R}^n$ with compact support $K$ and $|\mu|(K) < +\infty$, consider the *Riesz potential* of $\mu$:

$$u(x) = \int_K G(x - y)\, d\mu(y) \overset{\text{def}}{=} G(\mu)(x). \tag{7.25}$$

Hence the simple layer potentials

$$\int_{\partial U} \phi(y) G(x - y)\, dA(y),$$

that appear in the Riesz decomposition formula, are a particular case of Riesz potentials. If $\mu = \sum_i c_i \delta_{a_i}$ is a discrete measure, then $u(x) = \sum_i c_i G(x - a_i)$.

One may start specifying the definition of the function $G(\mu)$.

**Proposition 7.34.** *The potential $G(\mu)$ is defined at every point not belonging to $K = \mathrm{spt}(\mu)$ and is a harmonic function on $\mathbb{R}^n \setminus K$.*

*Proof.* The proof is done in the case $n > 2$. If $x \notin K$, one has $|x-y| \geq d(x, K) > 0$ for $y \in K$; then $G(x - y)$ is bounded on $K$ and $G(\mu)(x)$ is well defined. Moreover $G(x-y)$ is of class $C^1$ on a neighborhood of $x$ and continuous for $y \in K$; therefore,

$$|G(x - y) - G(z - y)| \leq C(x)|x - z|, \quad y \in K$$

for $|x - z| < \frac{1}{2} d(x, K)$ and some function $C(x)$. Then $|G(\mu)(x) - G(\mu)(z)| \leq |\mu|(K)C(x)|z - x|$; this proves that $G(\mu)$ is continuous on $\mathbb{R}^n \setminus K$. Now one can check that $G(\mu)$ has the mean value property. If $B$ is a ball with center $x_0$ not intersecting $K$ one has, by Fubini's theorem,

$$\frac{1}{A(\partial B)} \int_{\partial B} G(\mu)(x)\, dA(x) = \int_K \left\{ \frac{1}{A(\partial B)} \int_{\partial B} G(x - y)\, dA(x) \right\} d\mu(y).$$

Now, $G(x - y)$ is harmonic in $x$, for $y \in K$; hence, the inner integral is $G(x_0 - y)$ and one gets $G(\mu)(x_0)$. $\qquad\square$

Does the function $u(x) = G(\mu)(x)$ make sense at the points $x$ of $K$? In general, the answer is no. The clearest case is when one takes as $\mu$ a point mass at $y$, $\mu = \delta_y$; then $u(x) = G(x - y)$ tends to $\infty$ when $x \to y$.

In some special cases the potential is defined at the points of the support of the measure; for example, this happens for simple layer potentials. If $\mu$ is the measure of integration on a manifold $M$ of dimension $k$, in the space $\mathbb{R}^n$, $n > 2$, the function $|x - y|^{2-n}$ is locally integrable on $M$ if and only if $2 - n + k > 0$, that is, if $k > n - 2$. If $k \le n - 2$, (for example $n = 3$ and $M$ is a line) $u(x)$ becomes infinite on $M$. If $n = 2$, $\mathrm{Log}\,|x - y|$ is integrable on lines and curves. When $G(x - y)$ is integrable with respect to $\mu$, in general $u$ will be continuous; otherwise it will have discontinuities. Hence, in general, simple layer potentials are continuous functions.

**Example 7.35.** When $\mu$ is the measure of integration on $S(0, R)$, $R > 0$, it has been seen in Example 7.14 that the corresponding (simple layer) potential has value $\frac{1}{2-n} R$ ($R \,\mathrm{Log}\, R$ if $n = 2$) for $|x| < R$ and $\frac{R^{n-1}}{(2-n)|x|^{n-2}}$ ($R \,\mathrm{Log}\, |x|$ if $n = 2$) for $|x| > R$. In this case the potential is a continuous function at the points $x$ with $|x| = R$, but the derivatives have discontinuities. $\qquad\square$

Next, potentials of type

$$G(\phi)(x) = \int_{\mathbb{R}^n} G(x - y)\phi(y)dV(y),$$

where $\phi$ is a bounded function with compact support, will be analyzed in detail. They are the *convolution* of the functions $G$ and $\phi$, and that is why it is convenient to know some general properties of convolutions. Recall that $L^1_{\mathrm{loc}}(\mathbb{R}^n)$ is the space of measurable functions on $\mathbb{R}^n$ which are integrable on each compact set of $\mathbb{R}^n$ and $L^1_c(\mathbb{R}^n)$ is the space of integrable functions on $\mathbb{R}^n$ with compact support. Analogously, $L^\infty_{\mathrm{loc}}(\mathbb{R}^n)$ denotes the space of measurable functions that are essentially bounded on compact sets and $L^\infty_c(\mathbb{R}^n)$ the space of essentially bounded measurable functions with compact support.

**Proposition 7.36.** *Consider a convolution*

$$(k * \phi)(x) = \int_{\mathbb{R}^n} k(x - y)\phi(y)dV(y)$$

*with $k \in L^1_{\mathrm{loc}}(\mathbb{R}^n)$ and $\phi \in L^\infty_c(\mathbb{R}^n)$. Then one has:*

a) *The function $k * \phi$ is defined and continuous on $\mathbb{R}^n$.*

b) *If $\phi \in C^1_c(\mathbb{R}^n)$, then $k * \phi \in C^1(\mathbb{R}^n)$ and*

$$\frac{\partial(k * \phi)}{\partial x_i}(x) = \int_{\mathbb{R}^n} \frac{\partial\phi}{\partial x_i}(x - y)k(y)dV(y),$$

*that is, $D_i(k * \phi) = k * D_i\phi$, $i = 1, 2, \ldots, n$.*

c) *If the partial derivatives $\frac{\partial k}{\partial x_i}$ exist at every point of $\mathbb{R}^n$ and $\frac{\partial k}{\partial x_i} \in L^1_{\mathrm{loc}}(\mathbb{R}^n)$, for $i = 1, \ldots, n$, then $k * \phi \in C^1(\mathbb{R}^n)$ and one has*

$$\frac{\partial(k * \phi)}{\partial x_i}(x) = \int_{\mathbb{R}^n} \phi(y) \frac{\partial k}{\partial x_i}(x - y) dV(y),$$

*that is, $D_i(k * \phi) = (D_i k) * \phi$, $i = 1, 2, \ldots, n$.*

The hypotheses $k \in L^1_{\mathrm{loc}}(\mathbb{R}^n)$, $\phi \in L^\infty_c(\mathbb{R}^n)$ in item a) may be replaced by $k \in L^1(\mathbb{R}^n)$, $\phi \subset L^\infty(\mathbb{R}^n)$ or by $k \in L^1_c(\mathbb{R}^n)$, $\phi \in L^\infty_{\mathrm{loc}}(\mathbb{R}^n)$. Likewise, in b) one may suppose $k \in L^1(\mathbb{R}^n)$, $\phi \in C^1(\mathbb{R}^n)$, $\vec{\nabla}\phi \in L^\infty(\mathbb{R}^n)$ and in c): $\frac{\partial k}{\partial x_i} \in L^1(\mathbb{R}^n)$, $\phi \in L^\infty(\mathbb{R}^n)$.

The meaning of items b), c) is that in equalities

$$(k * \phi)(x) = \int \phi(y) k(x - y) dV(y) = \int k(y) \phi(x - y) dV(y) = (\phi * k)(x),$$

one may differentiate under the integral sign in any of both integrals provided that resulting integral is absolutely convergent.

Before proving Proposition 7.36, recall the general rule of differentiation of integrals depending on a parameter. Denote by $(x, y)$ a point of $\mathbb{R}^n \times \mathbb{R}^n$ and let $f(x, y)$ be a function defined on $\mathbb{R}^n \times \mathbb{R}^n$ that is integrable in $y$ for each fixed point $x \in \mathbb{R}^n$. Take $F(x) = \int_{\mathbb{R}^n} f(x, y) dV(y)$ and suppose that there exist partial derivatives $\frac{\partial f}{\partial x_i}(x, y)$, $i = 1, \ldots, n$, for almost every point $(x, y)$ and, furthermore, for each point $x_0 \in \mathbb{R}^n$ there is a neighborhood $U(x_0)$ in $\mathbb{R}^n$ and an integrable function $H$ such that

$$\left| \frac{\partial f}{\partial x_i}(x, y) \right| \leq H(y), \quad y \in \mathbb{R}^n, \ x \in U(x_0), \ i = 1, 2, \ldots, n.$$

Then the partial derivatives $\frac{\partial F}{\partial x_i}(x)$ exist and $\frac{\partial F}{\partial x_i}(x) = \int_{\mathbb{R}^n} \frac{\partial f}{\partial x_i}(x, y) dV(y)$, $i = 1, \ldots, n$, holds. This is a consequence of the dominated convergence theorem (see [12], p. 721).

*Proof of Proposition 7.36.* All conclusions are of local character and, for $x_0 \in \mathbb{R}^n$ fixed, we just need to analyze $(k * \phi)(x)$ in the ball $B(x_0, 1)$. If $k \in L^1_{\mathrm{loc}}(\mathbb{R}^n)$ and $\phi \in L^\infty_c(\mathbb{R}^n)$ has support in $B(0, r)$, then in $B(x_0, 1)$ we may change $k$ by $k\chi$, where $\chi$ is a function in $C^\infty_c(\mathbb{R}^n)$ that has value 1 on an appropriate neighborhood of $x_0$, without changing $k * \phi$ in $B(x_0, 1)$. Suppose, then, that $k \in L^1(\mathbb{R}^n)$. The convolution is continuous because, translations being continuous in the space $L^1(\mathbb{R}^n)$, one has

$$|(k * \phi)(x + h) - (k * \phi)(x)| \leq \int_{\mathbb{R}^n} |k(x + h - y) - k(x - y)||\phi(y)| \, dV(y)$$

$$\leq \|\phi\|_\infty \int_{\mathbb{R}^n} |k(z + h) - k(z)| dV(z) \xrightarrow[h \to 0]{} 0.$$

Under the hypothesis of b) it is

$$\left| \frac{\partial}{\partial x_i} \phi(x-y)k(y) \right| = \left| \frac{\partial \phi}{\partial x_i}(x-y) \right| |k(y)| \le C|k(y)|, \quad C \text{ constant,}$$

and the equality in b) is justified by the rule of differentiation of integrals depending on a parameter. Now by a) it turns out that $\frac{\partial}{\partial x_i}(k * \phi)$ is a continuous function for $i = 1 \ldots n$, that is $k * \phi \in C^1(\mathbb{R}^n)$.

The proof of c) is similar. One needs to see

$$\int_{\mathbb{R}^n} \phi(y) \left\{ \frac{k(x_0 + he_i - y) - k(x_0 - y)}{h} - \frac{\partial k}{\partial x_i}(x_0 - y) \right\} dV(y) \xrightarrow[h \to 0]{} 0$$

and it is enough to check

$$\int_{\mathbb{R}^n} \left| \frac{k(x_0 + he_i - y) - k(x_0 - y)}{h} - \frac{\partial k}{\partial x_i}(x_0 - y) \right| dV(y) \xrightarrow[h \to 0]{} 0.$$

Writing $k(x_0 + he_i - y) - k(x_0 - y) = h \int_0^1 \frac{\partial k}{\partial x_i}(x_0 + the_i - y)dt$, for almost every point $y \in \mathbb{R}^n$, it yields

$$\int_{\mathbb{R}^n} \left| \int_0^1 \left( \frac{\partial k}{\partial x_i}(x_0 + the_i - y) - \frac{\partial k}{\partial x_i}(x_0 - y) \right) dt \right| dV(y)$$

$$\le \int_{\mathbb{R}^n} \int_0^1 \left| \frac{\partial k}{\partial x_i}(x_0 + the_i - y) - \frac{\partial k}{\partial x_i}(x_0 - y) \right| dt\, dV(y)$$

$$= \int_0^1 \int_{\mathbb{R}^n} \left| \frac{\partial k}{\partial x_i}(x_0 + the_i - y) - \frac{\partial k}{\partial x_i}(x_0 - y) \right| dV(y)\, dt.$$

Finally, the continuity of the translation in $L^1(\mathbb{R}^n)$ applied to the integrable function $\frac{\partial k}{\partial x_i}$ allows one to ensure that this last expression is arbitrarily small, if $h \to 0$.  $\square$

**Proposition 7.37.** *If $\phi \in C_c^2(\mathbb{R}^n)$, then the potential $G(\phi)$ is a function of class $C^2$ on $\mathbb{R}^n$ and satisfies $\Delta G(\phi) = \phi$. If $\phi \in C_c^k(\mathbb{R}^n)$, then $G(\phi)$ is also of class $C^k$ on $\mathbb{R}^n$.*

*Proof.* The function $G$ is locally integrable, by Lemma 7.11. Applying twice item b) of Proposition 7.36 and summing up with respect to $i = 1, \ldots, n$, it turns out that

$$\Delta(G * \phi)(x) = G * \Delta\phi(x) = \int_{\mathbb{R}^n} G(y)\Delta\phi(x - y)dV(y).$$

By item a) of Proposition 7.33, this integral equals $\phi(x)$.  $\square$

This result may be considerably improved using the fact that partial derivatives of $G$,

$$\frac{\partial G}{\partial x_i}(x) = \frac{1}{c_n} \frac{x_i}{|x|^n}, \qquad c_n = A(\partial B(0,1)), \; i = 1, 2, \ldots, n$$

are still locally integrable, because

$$\int_{B(0,R)} \frac{|x_i|}{|x|^n} dV(x) \le \int_{B(0,R)} |x|^{1-n} dV(x) = c \cdot \int_0^R dr = c \cdot R,$$

with $c$ constant. Hence, applying items b) and c) of Proposition 7.36 one gets that $G(\phi)$ is of class $C^2(\mathbb{R}^n)$ if $\phi \in C_c^1(\mathbb{R}^n)$. This is so because one can differentiate once each factor in the integral defining $G(\phi)$:

$$\frac{\partial^2 G(\phi)}{\partial x_i \partial x_j}(x) = \int_{\mathbb{R}^n} \frac{\partial G}{\partial x_i}(y) \frac{\partial \phi}{\partial x_j}(x - y) dV(y), \quad i, j = 1, 2, \ldots, n.$$

In particular, one has

$$\Delta G(\phi)(x) = \int_{\mathbb{R}^n} \sum_{i=1}^n \frac{\partial G}{\partial x_i}(x - y) \frac{\partial \phi}{\partial x_i}(y) dV(y).$$

One may check that this expression coincides with $\phi(x)$ using Green's identities; later it will be seen that if $G(\phi)$ is of class $C^2$, then $\Delta G(\phi) = \phi$ holds automatically.

If $\phi \in C_c(\mathbb{R}^n)$, the potential $G(\phi)$ is not necessarily of class $C^2$, but it is really close to it:

**Proposition 7.38.** *If $\phi \in L_c^\infty(\mathbb{R}^n)$, then the potential $G(\phi)$ is of class $C^1$ on $\mathbb{R}^n$ and all its first-order derivatives satisfy a Lipschitz condition of type*

$$|\vec{\nabla} G(\phi)(x) - \vec{\nabla} G(\phi)(z)| \le C_L |x - z| \operatorname{Log} \frac{1}{|x - z|}, \tag{7.26}$$

*on every compact set $L \subset \mathbb{R}^n$, with a constant $C_L$ depending on $L$.*

*Proof.* It has been observed that $\vec{\nabla} G \in L^1_{\mathrm{loc}}(\mathbb{R}^n)$. Let $K$ be the support of $\phi$; then, by Proposition 7.36 c), one has

$$\frac{\partial G(\phi)}{\partial x_i}(x) = \frac{1}{c_n} \int_K \phi(y) \frac{x_i - y_i}{|x - y|^n} dV(y), \quad i = 1, 2, \ldots, n.$$

Now, if $L$ is a compact set and $x, z \in L$, it yields

$$D_i G(\phi)(x) - D_i G(\phi)(z) = \frac{1}{c_n} \int_K \phi(y) \left\{ \frac{x_i - y_i}{|x - y|^n} - \frac{z_i - y_i}{|z - y|^n} \right\} dV(y)$$

and it is enough to prove the inequality

$$\int_K \left| \frac{x_i - y_i}{|x-y|^n} - \frac{z_i - y_i}{|z-y|^n} \right| dV(y) \le C(L,K)\,|x-z|\,\mathrm{Log}\,\frac{1}{|x-z|}, \quad x,z \in L,$$

where $C(K,L)$ is a constant depending on $L$ and $K$.

Let $\delta = |x-z|$ and break the integral over $K$ into the integrals corresponding to the four regions (Figure 7.1):

$$I = \left\{ y : |y-x| < \frac{\delta}{2} \right\},$$

$$II = \left\{ y : |y-x| < |y-z|, |y-x| > \frac{\delta}{2} \right\},$$

$$III = \left\{ y : |y-z| < \frac{\delta}{2} \right\},$$

$$IV = \left\{ y : |y-x| > |y-z|, |y-x| > \frac{\delta}{2} \right\}.$$

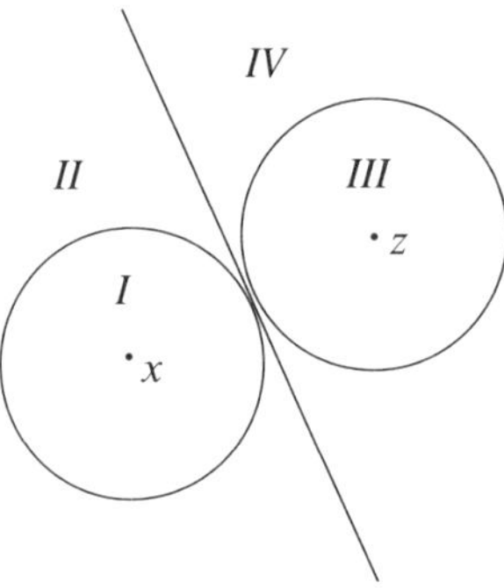


Figure 7.1

On region $I$ the integrand is bounded by the sum of moduli:

$$\int_{B_{\delta/2}(x)} \left( |x-y|^{1-n} + |z-y|^{1-n} \right) dV(y).$$

The integral of the first summand is $0(\delta)$, and the integral of the second one is bounded by

$$\left( \frac{\delta}{2} \right)^{1-n} V\left( B\left( x, \frac{\delta}{2} \right) \right) = 0(\delta).$$

Similarly, the contribution of region $III$ is of the order $0(\delta)$. On region $II$ the estimate

$$\left| \frac{x_i - y_i}{|x-y|^n} - \frac{z_i - y_i}{|z-y|^n} \right| = \frac{|(x_i - y_i)\,|z-y|^n - (z_i - y_i)\,|x-y|^n|}{|x-y|^n\,|z-y|^n}.$$

holds. The numerator of this expression is $|(x_i - z_i)|z - y|^n + (z_i - y_i)(|z - y|^n - |x - y|^n)|$. Using the inequality

$$|A^n - B^n| = |(A - B)(A^{n-1} + A^{n-2}B + \cdots + AB^{n-2} + B^{n-1})|$$
$$\leq n\,|A - B|\,[\max(A, B)]^{n-1},$$

for $A, B > 0$, we get

$$||z - y|^n - |x - y|^n| \leq n|z - x| \max\left(|z - y|, |x - y|\right)^{n-1} = n|z - x|\,|y - z|^{n-1}.$$

So the numerator of the integrand is dominated by $|x - z|\,|y - z|^n$ on region $II$, and the contribution of the corresponding integral is bounded by

$$|x - z| \int_{K \setminus B_{\delta/2}(x)} |x - y|^{-n} dV(y) = 0\left(\delta \log \frac{1}{\delta}\right).$$

A similar estimate holds for the integral on region $IV$.    $\square$

**Example 7.39.** Let the function $\phi$ be given by $\phi(x) = h(|x|)$, where $h$ is continuous on $(0, +\infty)$ with $\int_0^{+\infty} t|h(t)|dt < +\infty$ if $n > 2$ and $\int_0^{+\infty} t|\operatorname{Log} t|\,|h(t)|dt < +\infty$ if $n = 2$. According to Example 7.12, the potential $G(\phi)$ is a radial function, $G(\phi)(x) = H(|x|)$. The calculus of the function $H$ in the case $n > 2$ is the following: if $|x| = s$, one has

$$H(s) = \int_{\mathbb{R}^n} h(|y|)G(x-y)\,dV(y) = \int_0^{+\infty} h(r)\left(\int_{S(0,r)} G(x - y)\,dA(y)\right) dr.$$

The integral on the sphere $S(0, r)$ has been computed in Example 7.14; its value is $\frac{r}{2-n}$ if $s < r$ and $\frac{r^{n-1}}{(2-n)s^{n-2}}$ if $s > r$. This yields

$$H(s) = \frac{1}{2 - n}\left(s^{2-n}\int_0^s r^{n-1}h(r)\,dr + \int_s^{+\infty} rh(r)\,dr\right).$$

The hypothesis on $h$ guarantees that $H$ and, hence, $G(\phi)$ is continuous. One may also differentiate $H$, twice,

$$H'(s) = s^{1-n}\int_0^s r^{n-1}h(r)\,dr,$$

$$H''(s) = (1 - n)s^{-n}\int_0^s r^{n-1}h(r)\,dr + h(s).$$

The Laplacian of a radial function appears in equation (7.13), and it turns out that

$$\Delta G(\phi)(x) = H''(s) + \frac{n - 1}{s}H'(s).$$

Changing $H'$ and $H''$ by the expressions found above equality $\Delta G(\phi)(x) = h(s) = \phi(x)$ is proved.    $\square$

## 7.7.2 Weak solutions

Proposition 7.34 may be improved because, indeed, the Riesz potential of a measure is defined at almost every point, with respect to the Lebesgue measure of $\mathbb{R}^n$:

**Proposition 7.40.** *The integral giving the potential $u(x) = G(\mu)(x)$ of a measure $\mu$ with compact support in (7.25) is absolutely convergent at almost every point $x \in \mathbb{R}^n$ and defines a locally integrable function on $\mathbb{R}^n$.*

*Proof.* If $L$ is a compact set of $\mathbb{R}^n$ and $K = \mathrm{spt}(\mu)$, by Fubini's theorem one has

$$\int_L |u(x)| \, dV(x) \le \int_L \int_K |G(x-y)| \, d|\mu|(y) \, dV(x)$$

$$= \int_K \left\{ \int_L |G(x-y)| \, dV(x) \right\} d|\mu|(y)$$

$$\le \int_K \left\{ \int_{L-K} |G(x)| \, dV(x) \right\} d|\mu|(y).$$

Here $L - K = \{x - y \colon x \in L, \ y \in K\}$. Since $L - K$ is compact, it is contained in a ball $B(0, R)$ and Lemma 7.11 implies that the integral on $L - K$ is dominated by some constant $C(R)$. Finally, it holds that

$$\int_K |u(x)| \, dV(x) \le C(R)|\mu|(K) < +\infty$$

and, in particular, $|u(x)| < +\infty$ for almost every $x$. $\qquad\square$

Since in general the potential $G(\mu)$ of a measure or even the potential $G(\phi)$ of a function is not defined at every point, and in general not twice differentiable if defined everywhere, it makes no sense to apply the operator $\Delta$ to it. To know if $G(\phi) \in C^2(\mathbb{R}^n)$ and the equality $\Delta G(\phi) = \phi$ holds (besides the case $\phi \in C_c^2(\mathbb{R}^n)$ considered at Proposition 7.37), it is very convenient to generalize the concept of solution of the equation $\Delta u = \phi$, or $\Delta u = \mu$, so that it does not require the differentiability of $u$ beforehand.

**Definition 7.41.** Given a measure $\mu$ of locally finite mass in a domain $U$ of $\mathbb{R}^n$, a function $u \in L_{\mathrm{loc}}^1(U)$ is said to be a solution of the equation $\Delta u = \mu$ in $U$ in the weak sense, if for every function $\varphi \in C_c^2(U)$ one has

$$\int_U u(x) \Delta\varphi(x) \, dV(x) = \int_U \varphi(x) \, d\mu(x). \tag{7.27}$$

**Remark 7.3.** The hypothesis on the measure $\mu$ means that $|\mu|(K) < +\infty$ for every compact set $K \subset U$. Observe that both members of (7.27) make sense.

A measure $\mu$ on $U$ is determined by the integrals $\int_U \varphi(x)\,d\mu(x)$ with $\varphi \in C_c^2(U)$; hence, if $u \in L_{\mathrm{loc}}^1(U)$ and $\Delta u = \mu$, in the weak sense, the measure $\mu$ is unique and is well determined by $u$. An important remark – which justifies the definition – is the following: if $u \in C^2(U)$ and $v = \Delta u \in C(U)$ is the classical Laplacian, then one also has $\Delta u = v\,dV$ in the weak sense, that is,

$$\int_U u(x)\Delta\varphi(x)\,dV(x) = \int_U \varphi(x)v(x)\,dV(x), \quad \varphi \in C_c^2(U).$$

This follows from the second Green's identity (7.6) applied to a domain $\widetilde{U}$ with oriented regular boundary such that $\mathrm{spt}(\varphi) \subset \widetilde{U} \subset U$. As a consequence, if $u \in C^2(U)$ and $\Delta u = \mu$ in the weak sense then $\mu = v\,dV$, because $v$ as well as $\mu$ are weak Laplacians of $u$. In other words, the weak Laplacian is a concept that, extending the concept of the classical Laplacian for $u \in C^2(U)$, allows to consider $\Delta u$ for non-differentiable functions $u \in L_{\mathrm{loc}}^1(U)$.

**Theorem 7.42.** *Let $u$ be the Riesz potential of a measure $\mu$ with compact support on $\mathbb{R}^n$ and finite total mass. Then $\Delta u = \mu$ on $\mathbb{R}^n$ in the weak sense. Consequently, if $\phi \in C_c(\mathbb{R}^n)$ and $G(\phi)$ is of class $C^2$, one has $\Delta G(\phi) = \phi$ in the classical sense.*

*Proof.* By Fubini's theorem, if $\varphi \in C_c^2(\mathbb{R}^n)$, then one has

$$\int_{\mathbb{R}^n} u(x)\Delta\varphi(x)\,dV(x) = \int_{\mathbb{R}^n} \left\{ \int_{\mathbb{R}^n} G(x-y)\,d\mu(y) \right\} \Delta\varphi(x)\,dV(x)$$

$$= \int_{\mathbb{R}^n} \left\{ \int_{\mathbb{R}^n} G(x-y)\Delta\varphi(x)\,dV(x) \right\} d\mu(y).$$

Now just note that the inner integral equals $\varphi(y)$ by item a) of Proposition 7.33.
$\square$

**Example 7.43.** As a particular case of Theorem 7.42 taking $\mu = \delta_0$, the Dirac mass at the origin, it turns out that the function $G(x)$ satisfies $\Delta G(x) = \delta_0$ in the weak sense. This is why $G(x)$ is called the fundamental solution of the Laplacian, as it has been said in Section 7.3. Similarly, Green's function $G_U(x, y)$ of a domain $U$ with pole $x \in U$ satisfies $\Delta_y G_U(x, y) = \delta_x$ in the weak sense, and vanishes on the boundary of $U$.
$\square$

**Example 7.44.** In general the weak Laplacian of a function in $L_{\mathrm{loc}}^1(U)$ is not a measure. Consider, for example, $u(x) = |x|^\alpha$ in $\mathbb{R}^n$ with $\alpha > -n$ so that $u \in L_{\mathrm{loc}}^1(\mathbb{R}^n)$. Since $u$ is $C^\infty$ on $\mathbb{R}^n \setminus \{0\}$, if there was a measure $\mu$ such that $\Delta u = \mu$ in $\mathbb{R}^n$, in the weak sense, then $\mu$ should necessarily be the classical Laplacian outside zero; taking $\varphi(r) = r^\alpha$ in (7.13) yields that this Laplacian is

$\Delta u = \alpha(\alpha - 1)|x|^{\alpha - 2} + \frac{n-1}{|x|}\alpha|x|^{\alpha - 1} = |x|^{\alpha - 2}[\alpha(\alpha - 1) + n - 1]$. In particular, it should be

$$\int_{\mathbb{R}^n} |x|^{\alpha - 2}[\alpha(\alpha - 1) + n - 1]\varphi(x)\, dV(x) = \int_{\mathbb{R}^n} \varphi(x)\, d\mu(x)$$

whenever $\varphi$ has compact support in $\mathbb{R}^n \setminus \{0\}$. This implies that $|x|^{\alpha - 2}[\alpha(\alpha-1)+n-1]$ must be also locally integrable, an impossible fact if $-n < \alpha < -n + 3$. For example, if $n > 2$ and $\alpha = 1 - n$, the function $u = |x|^{1-n}$ is in $L^1_{\mathrm{loc}}(\mathbb{R}^n)$, but there cannot be any measure $\mu$ with compact support such that $\Delta u = \mu$, in the weak sense. $\qquad\square$

Observe that it does not follow from Definition 7.41 that the equation $\Delta u = 0$ in the weak sense, implies $u$ is harmonic. This fact is the content of the following statement.

**Theorem 7.45** (Weyl's lemma). *If $u \in L^1_{\mathrm{loc}}(U)$ and $\Delta u = 0$ in the weak sense, then $u$ is harmonic, that is, $\Delta u = 0$ in the classical sense in $U$.*

*Proof.* It will be proved that $u$ is continuous and satisfies the mean value property in $U$. Then Theorem 7.7 gives the harmonicity of $u$.

Take a function $\phi \in C_c^\infty(B(0, 1))$ with $\int \phi\, dV = 1$ and put $\phi_\varepsilon(x) = \varepsilon^{-1}\phi(x/\varepsilon)$ and $U_\varepsilon = \{x \colon d(x, U^c) \ge \varepsilon\}$. If $\psi \in C_c^\infty(U_\varepsilon)$, one has $\psi * \phi_\varepsilon \in C_c^\infty(U)$ and $u * \phi_\varepsilon \in C^\infty(U_\varepsilon)$. Therefore, by the second Green's identity (7.6) applied to an open set with regular boundary contained between $U_\varepsilon$ and spt $(\psi)$, one has, considering all integrals extended to $\mathbb{R}^n$,

$$\int \Delta(u * \phi_\varepsilon)\psi\, dV = \int (u * \phi_\varepsilon)\Delta\psi\, dV$$

$$= \iint u(x - y)\phi_\varepsilon(y)\, dV(y)\Delta\psi(x)\, dV(x)$$

$$= \iint u(y)\phi_\varepsilon(x - y)\, dV(y)\Delta\psi(x)\, dV(x)$$

$$= \int u(y) \int \phi_\varepsilon(x - y)\Delta\psi(x)\, dV(x)\, dV(y)$$

$$= \int u(y) \int \phi_\varepsilon(z)\Delta\psi(y + z)\, dV(z)\, dV(y)$$

$$= \int u(y)\Delta\left(\int \phi_\varepsilon(z)\psi(y + z)\, dV(z)\right) dV(y).$$

The hypothesis is that $\int u(y)\Delta\chi(y)\, dV(y) = 0$ if $\chi \in C_c^\infty(U)$. This is the case when $\chi(y) = \int \phi_\varepsilon(z)\psi(y + z)\, dV(z)$. Therefore, $\Delta(u * \phi_\varepsilon) = 0$, that is, $u * \phi_\varepsilon$ is harmonic on $U_\varepsilon$. Writing the mean value property on balls for the function $u * \phi_\varepsilon$ yields

$$(u * \phi_\varepsilon)(x) = \frac{n}{r^n c_n} \int_{B(0,r)} (u * \phi_\varepsilon)(x + y)\, dV(y), \tag{7.28}$$

for $x \in U$, $r > 0$ small enough. Due to the fact that $u \in L^1_{\mathrm{loc}}(U)$, it follows that $u * \phi_\varepsilon \to u$ in $L^1_{\mathrm{loc}}(U)$ when $\varepsilon \to 0$ and, in particular, in $L^1(B)$ for every ball $B \subset U$. Formula (7.28) applied to $u * \phi_\varepsilon - u * \phi_\delta$ shows that $u * \phi_\varepsilon$ is uniformly convergent on compact sets. Hence, $u * \phi_\varepsilon$ tends to $u$ uniformly on compact sets and, therefore, $u$ is continuous. Now each $u * \phi_\varepsilon$ being harmonic has the mean value property on spheres in $U_\varepsilon$. Letting $\varepsilon \to 0$ if follows that $u$ has the mean value property on spheres and so it is harmonic. $\qquad\square$

Weyl's lemma implies, for instance, that the two properties $\Delta_y G_U(x, y) = \delta_x$ and $G_U(x, y) = 0$ if $y \in \partial U$ characterize Green's function of a domain $U$, if it exists.

**Remark 7.4.** Throughout this section it has been tacitly supposed that $n \geq 2$, but it is interesting to consider also the case $n = 1$. Recall that in this case $G(x) = \frac{1}{2}|x|$ ($d_1 = \frac{1}{2}$); the analog of Proposition 7.33 a) is

$$\varphi(x) = \frac{1}{2} \int_{-\infty}^{+\infty} \varphi''(y)|x - y|\, dy \quad \text{if } \varphi \in C_c^2(\mathbb{R}^n). \tag{7.29}$$

Theorem 7.42 now asserts that the function

$$u(x) = \frac{1}{2} \int_{-\infty}^{+\infty} |x - y|\, d\mu(y)$$

has weak Laplacian $\Delta u = \mu$, if $\mu$ is a measure with compact support. An elementary classical version of this fact is obtained taking $\mu = f\, dx$ with $f$ continuous and with compact support. Then

$$u(x) = \frac{1}{2} \int_{-\infty}^{+\infty} |x - y| f(y)\, dy$$

satisfies $u'' = f$. Actually, one has

$$2u(x) = \int_{-\infty}^{x} + \int_{x}^{+\infty} = \int_{-\infty}^{x} (x - y) f(y)\, dy + \int_{x}^{+\infty} (y - x) f(y)\, dy$$

and differentiating twice,

$$2u'(x) = \int_{-\infty}^{x} f(y)\, dy - \int_{x}^{+\infty} f(y)\, dy, \quad 2u''(x) = f(x) - (-f(x)) = 2f(x).$$

Now go back to the case $n > 1$.

**Theorem 7.46.** *Let $\phi \in C_c(\mathbb{R}^n)$ be locally Lipschitz, that is, satisfying a local Lipschitz condition with positive exponent:*

$$|\phi(y) - \phi(x)| \le c(x)|y - x|^\alpha, \ \alpha > 0, \ c(x) \ge 0, \ \text{whenever } |y - x| \le \delta(x).$$

*Then the potential $u = G(\phi)$ is of class $C^2$ on $\mathbb{R}^n$ and satisfies $\Delta u = \phi$.*

*Proof.* In the proof of Proposition 7.38 it has been seen that $u \in C^1(\mathbb{R}^n)$ and that the equalities

$$\frac{\partial u}{\partial x_i}(x) = \frac{1}{c_n} \int_{\mathbb{R}^n} \phi(y) \frac{x_i - y_i}{|x - y|^n} \, dV(y), \quad i = 1, 2, \ldots, n$$

hold. Now one cannot differentiate again under the integral sign as before, because

$$\frac{\partial}{\partial x_j} \frac{x_i - y_i}{|x - y|^n} = \frac{\delta_{ij}}{|x - y|^n} - n \frac{(x_i - y_i)(x_j - y_j)}{|x - y|^{n+2}}$$

behaves like $|x - y|^{-n}$ and so is not locally integrable. Instead we will use that the integral over any sphere centered at $x$ is zero,

$$\int_{S(x,\varepsilon)} \frac{\partial}{\partial x_j} \frac{x_i - y_i}{|x - y|^n} \, dA(y) = 0 \tag{7.30}$$

(if $i \ne j$ the integrand is odd with respect to the $i$-th axis, and if $i = j$ both resulting terms from differentiation have the same integral). In particular, one has

$$\int_{\bar{B}_R(x) \setminus B_\varepsilon(x)} \frac{\partial}{\partial x_j} \frac{x_i - y_i}{|x - y|^n} \, dV(y) = 0 \quad \text{if } 0 < \varepsilon < R.$$

Fix $x$ and let $R$ be such that the support of $\phi$ is inside the ball $B(x, R)$. Take a function $\chi \in C^\infty(\mathbb{R})$ such that $\chi(t) = 0$ in $(-1, 1)$ and $\chi(t) = 1$ if $|t| \ge 2$ and consider

$$v_\varepsilon^i(x) = \frac{1}{c_n} \int_{\mathbb{R}^n} \phi(y) \frac{x_i - y_i}{|x - y|^n} \chi\left(\frac{|x - y|}{\varepsilon}\right) dV(y),$$

so that $v_\varepsilon^i(x) \to \frac{\partial u}{\partial x_i}(x)$ pointwise, if $\varepsilon \to 0$ (uniformly on compact sets indeed). We can apply Proposition 7.36 to the function $v_\varepsilon^i(x)$ to obtain

$$\frac{\partial v_\varepsilon^i}{\partial x_j}(x) = \frac{1}{c_n} \int_{B(x,R)} \phi(y) \frac{\partial}{\partial x_j} \left(\frac{x_i - y_i}{|x - y|^n}\right) \chi\left(\frac{|x - y|}{\varepsilon}\right) dV(y)$$

$$+ \frac{1}{c_n} \int_{B(x,R)} \phi(y) \frac{x_i - y_i}{|x - y|^n} \frac{1}{\varepsilon} \frac{x_j - y_j}{|x - y|} \chi'\left(\frac{|x - y|}{\varepsilon}\right) dV(y).$$

Using (7.30) and the fact that $\chi$ and $\chi'$ are radial, it turns out that

$$\frac{\partial v_\varepsilon^i}{\partial x_j}(x) = \frac{1}{c_n} \int_{B(x,R)} (\phi(y) - \phi(x)) \frac{\partial}{\partial x_j}\left(\frac{x_i - y_i}{|x-y|^n}\right) \chi\left(\frac{|x-y|}{\varepsilon}\right) dV(y)$$

$$+ \frac{1}{c_n} \int_{B(x,R)} \phi(y) \frac{1}{\varepsilon} \frac{(x_i - y_i)(x_j - y_j)}{|x-y|^{n+1}} \chi'\left(\frac{|x-y|}{\varepsilon}\right) dV(y)$$

$$= A_\varepsilon(x) + C_\varepsilon(x).$$

In polar coordinates, $y = x + rw$, one has

$$C_\varepsilon(x) = \frac{1}{c_n} \int_{\mathbb{S}} \int_\varepsilon^{2\varepsilon} \phi(x + rw) \frac{1}{\varepsilon} \frac{w_i w_j}{r^{n-1}} r^{n-1} \chi'\left(\frac{r}{\varepsilon}\right) dr \, d\sigma(w)$$

$$= \frac{1}{c_n} \int_{\mathbb{S}} \int_1^2 \phi(x + t\varepsilon w) w_i w_j \chi'(t) dt \, d\sigma(w).$$

If $i \neq j$, then $\int_{\mathbb{T}} w_i w_j d\sigma(w) = 0$ and

$$C_\varepsilon(x) = \frac{1}{c_n} \int_{\mathbb{S}} \int_1^2 (\phi(x + t\varepsilon w) - \phi(x)) w_i w_j) \chi'(t) dt \, d\sigma(w)$$

is $0(\varepsilon^\alpha)$, uniformly for $x$ on a compact set. If $i = j$, then $\int_{\mathbb{T}} w_i^2 d\sigma(w) = c_n/n$, which yields

$$C_\varepsilon(x) = \frac{1}{c_n} \int_{\mathbb{S}} \int_1^2 (\phi(x + t\varepsilon w) - \phi(x)) w_i^2 \chi'(t) dt \, d\sigma(w)$$

$$+ \frac{1}{c_n} \phi(x) \left(\int_1^2 \chi'(t) dt\right) \left(\int_{\mathbb{T}} w_i^2 d\sigma(w)\right) = 0(\varepsilon^\alpha) + \frac{1}{n}\phi(x).$$

Hence, $C_\varepsilon(x) \to 0$ if $i \neq j$ and $C_\varepsilon(x) \to \frac{1}{n}\phi(x)$ if $i = j$, uniformly on compact sets when $\varepsilon \to 0$. On the other hand, from $|\phi(y) - \phi(x)| = 0(|y-x|^\alpha)$ and

$$\frac{\partial}{\partial x_j}\left(\frac{x_i - y_i}{|x-y|^n}\right) = 0(|x-y|^{-n}),$$

it follows that the integral

$$A(x) = \int_{B(x,\varepsilon)} (\phi(y) - \phi(x)) \left(\frac{\partial}{\partial x_j}\left(\frac{x_i - y_i}{|x-y|^n}\right)\right) dV(y)$$

is absolutely convergent because the integrand is $0\,(|x-y|^{\alpha-n})$. In addition,

$$|A_\varepsilon(x) - A(x)| \leq c \int_{B(x,2\varepsilon)} |x-y|^{\alpha-n} dV(y) = 0(\varepsilon^\alpha), \quad c \text{ constant.}$$

In summary, it has been shown that

$$\frac{\partial v^i_\varepsilon}{\partial x_j}(x) \longrightarrow \int_{B(x,R)} (\phi(y) - \phi(x)) \left( \frac{\partial}{\partial x_j} \left( \frac{x_i - y_i}{|x - y|^n} \right) \right) dV(y) + C(x)$$

for $\varepsilon \to 0$, where $C(x) = 0$ if $i \neq j$ and $C(x) = \frac{1}{n}\phi(x)$ if $i = j$ with uniform convergence on compact sets. This implies $u$ is of class $C^2$ on $\mathbb{R}^n$. Equality $\Delta u = \phi$ comes then from Theorem 7.42; however, it follows also from previous computation, since

$$\Delta u(x) = \lim_{\varepsilon \to 0} \sum_{i=1}^{n} \frac{\partial v^i_\varepsilon}{\partial x_i}(x)$$

$$= \int_{B(x,R)} (\phi(y) - \phi(x)) \left( \sum_{i=1}^{n} \frac{\partial}{\partial x_i} \left( \frac{x_i - y_i}{|x - y|^n} \right) \right) dV(y)$$

$$+ \sum_{i=1}^{n} \frac{1}{n}\phi(x) = \phi(x),$$

because $\sum_{i=1}^{n} \frac{\partial}{\partial x_i} \left( \frac{x_i - y_i}{|x-y|^n} \right) = 0$ if $y \neq x$ (it is the Laplacian of $G(x - y)$). $\quad\square$

Summarizing the results in this section, one may state:

**Theorem 7.47.** *If $\mu$ is a measure with compact support and finite total mass, the potential $u = G(\mu)$ is a function in $L^1_{\mathrm{loc}}(\mathbb{R}^n)$ such that $\Delta u = \mu$ in $\mathbb{R}^n$ in the weak sense. If $n > 2$, it is the unique solution of $\Delta u = \mu$ in the weak sense which vanishes at infinity; if $n = 2$, it is the unique solution of $\Delta u = \mu$ in the weak sense which satisfies $u(x) = c \operatorname{Log}|x| + o(1)$, $|x| \to +\infty$, and, in this case, one has $c = \frac{1}{2\pi}\mu(K)$ if $\mu$ is supported on $K$. If $\mu = \phi dV$ and $\phi$ is bounded, then $u$ is of class $C^1$ on $\mathbb{R}^n$ and $\vec{\nabla}u$ locally satisfies a condition of type*

$$|\vec{\nabla}u(x) - \vec{\nabla}u(z)| = 0\left( |z - x| \operatorname{Log} \frac{1}{|z - x|} \right).$$

*If $\phi \in C_c(\mathbb{R}^n)$ is locally Lipschitz, then $u \in C^2(\mathbb{R}^n)$ and $\Delta u = \phi$ in the classical sense.*

## 7.8  The Poisson equation and the non-homogeneous Dirichlet and Neumann problems in a domain of $\mathbb{R}^n$

The Poisson equation $\Delta u = \phi$ may also be stated in a bounded domain $U \subset \mathbb{R}^n$. Given, for example, a function $\phi \in C(U)$, one wants to solve the equation $\Delta u = \phi$

with a function $u \in C^2(U)$. One possible solution is simply to consider the measure $\mu = \phi \, dV$ and the corresponding potential

$$G(\phi)(x) = \int_U G(x - y)\phi(y)dV(y)$$

only at the points $x \in U$; for this purpose it is better to assume $\phi \in L^1(U)$. According to the results of Subsection 7.7.2, one will have $G(\phi) \in L^1(U)$ and $\Delta G(\phi) - \phi$ in the weak sense in $U$. Observe that in this situation one cannot write $G(\phi)(x) = \int_U G(y)\phi(x - y)dV(y)$.

In order to apply the results obtained in Section 7.7 it will be supposed that $\phi$ is a bounded function on $U$. In this case, Proposition 7.38 and Theorem 7.46 give directly:

**Theorem 7.48.** *Let $\phi$ be a bounded function on a bounded domain $U$ of $\mathbb{R}^n$. Then the potential $G(\phi)$ is a $C^1$ function on $U$ and $\vec{\nabla}G(\phi)$ satisfies a Lipschitz condition of type (7.26). If $\phi$ is locally Lipschitz, then $G(\phi) \in C^2(U)$ and $\Delta G(\phi) = \phi$ in $U$ in the classical sense.*

On the other hand, one can prove that for $\phi \in C(U)$ (possibly unbounded), there always exists a function $u \in C^1(U)$ that satisfies $\Delta u = \phi$ in the weak sense and in addition $u \in C^2(U)$, provided $\phi$ locally satisfies a Lipschitz condition. For the case $n = 2$, this will be proved in Chapter 10.

Unlike the case $U = \mathbb{R}^n$, when $U$ is an arbitrary domain, the Riesz potential $G(\phi)$ is not a distinguished solution of the equation $\Delta u = \phi$ in $U$. The general solution is obtained by adding to $G(\phi)$ a harmonic function on $U$, and to determine a particular solution one needs to impose some *boundary condition*. For example, since in $\mathbb{R}^n$, $G(\phi)$ is the only solution vanishing at infinity, it would be natural to consider among all the solutions of $\Delta u = \phi$ on $U$ that one which vanishes on the boundary, if it exists. More generally, in a *non-homogeneous Dirichlet's problem* or general Dirichlet's problem, one starts from two data in the bounded domain $U \subset \mathbb{R}^n$: a function $\varphi \subset C(\partial U)$ and a function $\phi \in L^\infty(U)$. The problem consists in finding a function $u \in C(\overline{U})$ that satisfies

$$\begin{cases} \Delta u = \phi & \text{in } U, \\ u_{|\partial U} = \varphi & \text{in } \partial U, \end{cases} \tag{D}$$

where the equality $\Delta u = \phi$ is taken in the weak sense.

The solution to this problem, if one exists, is unique. Indeed if $u_1$, $u_2$ are solutions, then $u = u_1 - u_2$ is harmonic, by Weyl's lemma, and has value 0 in $\partial U$. Therefore $u = 0$ in $U$ and $u_1 = u_2$.

Problem (D) splits into two problems corresponding to $\phi = 0$ and to $\varphi = 0$, respectively:

$$(D_1) \quad \begin{cases} \Delta u = 0 & \text{in } U, \\ u = \varphi & \text{in } \partial U, \end{cases} \qquad\qquad (D_2) \quad \begin{cases} \Delta u = \phi & \text{in } U, \\ u = 0 & \text{in } \partial U, \end{cases}$$

where equations for the Laplacian are in the weak sense.

The first problem or homogeneous Dirichlet's problem has been considered in Section 7.4. Problem $(D_2)$ reduces to problem $(D_1)$ considering the potential

$$v(x) = G(\phi)(x) = \int_U G(x-y)\phi(y)\,dV(y).$$

By Theorem 7.47 one has $v \in C^1(\mathbb{R}^n)$ and $\Delta u = \phi$ in the weak sense in $U$. Let $\varphi \in C(\partial U)$ be the restriction of $v$ to $\partial U$ and let $u_0$ be the solution of $\Delta u_0 = 0$ in $U$, $u_0 = \varphi$ in $\partial U$. Then $u = v - u_0$ satisfies $\Delta u = \Delta v - \Delta u_0 = \phi$ in $U$ and $u_{|\partial U} = 0$ in $\partial U$. With this the following statement is proved:

**Proposition 7.49.** *If the homogeneous Dirichlet's problem $(D_1)$ has a solution for every function $\varphi \in C(\partial U)$, then the general Dirichlet's problem $(D)$ has a solution for every pair of functions $\varphi \in C(\partial U)$, $\phi \in L^\infty(U)$.*

If the domain $U$ admits a first Green's function, then the solution of $(D_2)$ exists and is explicit:

**Theorem 7.50.** *If the bounded domain $U$ admits a Green's function, $G_U(x, y)$, then the solution of the problem $(D_2)$ for $\phi \in L^\infty(U)$ is*

$$u(x) = \int_U G_U(x, y)\phi(y)dV(y), \quad x \in U.$$

*If $\phi$ is locally Lipschitz, then $u \in C^2(U)$ and $\Delta u = \phi$ in the classical sense.*

*Proof.* Recall that $G_U(x, y) = G(x-y) - v_x(y)$ where $v_x \in C(\bar{U})$ satisfies $v_x(y) = G(x-y)$, $y \in \partial U$ and $v_x$ is harmonic on $U$. With this, $u$ has the structure described above,

$$u(x) = \int_U G(x-y)\phi(y)\,dV(y) - \int_U v_x(y)\phi(y)\,dV(y) = G(\phi)(x) - u_0(x).$$

Now one only has to check that $u_0$ is the solution of problem $(D_1)$ with data $\varphi = G(\phi)$ in $\partial U$. Let us first prove that $u_0 \in C(\bar{U})$. By Proposition 7.24,

$$G(x-y) - v_x(y) = G_U(x-y) \le 0;$$

therefore, one has $|v_x(y)| \le |G(x-y)|$ if $n > 2$ and $|v_x(y)| \le |G(x, y)| + K(U)$ if $n = 2$, where $K(U)$ is a constant which just depends on the domain $U$. Moreover, by Proposition 7.23 one has $v_x(y) = v_y(x)$, which is continuous on $\bar{U}$ in one of the variables when the other is fixed. This allows us to prove that $u_0 \in C(\bar{U})$. Actually, one has

$$u_0(x) - u_0(z) = \int_U (v_x(y) - v_z(y))\phi(y) dV(y).$$

Choose $\delta > 0$, $x, z \in \bar{U}$ with $|z - x| < \frac{\delta}{2}$ and break the integral into two parts I, II corresponding, respectively, to the regions of integration $U \cap B(x, \delta)$ and $U \setminus B(x, \delta)$. When $z \to x$, part II converges to zero by the dominated convergence theorem; using the estimate of $v_x$ just seen before, part I is bounded above by

$$\int_{U \cap B_\delta(x)} (|G(y - x)| + |G(y - z)|) \, dV(y),$$

which also converges to zero when $\delta \to 0$. Since $v_x(y) = v_y(x)$ is harmonic in $x$, the mean value property and Fubini's theorem imply that $u_0$ is harmonic on $U$. Finally, when $x \in \partial U$, $v_x(y) = G(x - y)$ and, therefore, $u_0 = G(\phi)$ in $\partial U$.

If $\phi$ is locally Lipschitz on $U$, $\phi$ is automatically continuous and Theorem 7.47 gives that $G(\phi)$ is of class $C^2$ on $U$. Then $u$ also is $C^2$ ($u_0$ is harmonic and, hence, $u_0 \in C^\infty(U)$) and $\Delta u = \Delta G(\phi) = \phi$ in the classical sense.   $\square$

The function $u$ of Theorem 7.50 is called *Green's potential of the function $\phi$ in $U$.*

Assume now $U$ has regular boundary (oriented by the vector field $\vec{N}$) and Poisson kernel associated to $G_U$, $P_U(x, y) = \frac{\partial}{\partial \vec{N}_y} G_U(x, y)$, solves problem (D$_1$) in $U$. Then as a consequence of Theorem 7.50, the solution of the general Dirichlet's problem (D) is given by

$$u(x) = \int_{\partial U} \varphi(y) P_U(x, y) \, dA(y) + \int_U G_U(x, y)\phi(y) \, dV(y).$$

**Example 7.51.** In the particular case that $U$ is the unit disc, $\mathbb{D}$, Theorem 7.26 and Theorem 7.50 give the function

$$u(z) = \frac{1}{2\pi} \int_{\mathbb{T}} \varphi(w) \frac{1 - |z|^2}{|z - w|^2} \, d\sigma(w) + \frac{1}{2\pi} \int_{\mathbb{D}} \mathrm{Log} \left| \frac{z - w}{1 - w\bar{z}} \right| \phi(w) \, dm(w)$$

$$\tag{7.31}$$

as a solution of problem (D).

Once it is known how to solve Dirichlet's problem in the disc $\mathbb{D}$, it is also possible to solve it in any other disc $D(a, R)$. One simply goes from one disc to another by means of transformations $z \mapsto \zeta = a + Rz$, $\zeta \mapsto z = \frac{\zeta - a}{R}$. If $u$ is of class $C^2$ on $\bar{D}(a, R)$, then the function $v(z) = u(a + Rz)$ is of class $C^2$ on $\bar{\mathbb{D}}$ and one has the formula

$$v(z) = \frac{1}{2\pi} \int_{\mathbb{T}} v(w) \frac{1 - |z|^2}{|z - w|^2} \, d\sigma(w)$$

$$+ \frac{1}{2\pi} \int_{\mathbb{D}} \mathrm{Log} \left| \frac{z - w}{1 - \bar{w}z} \right| \Delta v(w) \, dm(w)$$

$$= \frac{1}{2\pi} \int_{\mathbb{T}} u(a + Rw) \frac{1 - |z|^2}{|z - w|^2} \, d\sigma(w)$$

$$+ \frac{1}{2\pi} \int_{\mathbb{D}} \mathrm{Log} \left| \frac{z - w}{1 - \bar{w}z} \right| R^2 \Delta u(a + Rw) \, dm(w).$$

When $|w| = 1$, $\xi = a + Rw$ describes $\{\xi : |\xi - a| = R\}$ and $ds(\xi) = R \, d\sigma(w)$; when $|w| < 1$, $\xi = a + Rw$ describes $\{\xi : |\xi - a| \leq R\}$ and $dm(\xi) = R^2 \, dm(w)$. The change of variables $\xi = a + Rw$ in the previous integrals yields

$$u(a + Rz) = \frac{1}{2\pi R} \int_{C(a,R)} u(\xi) \frac{1 - |z|^2}{\left| z - \frac{\xi - a}{R} \right|^2} \, ds(\xi)$$

$$+ \frac{1}{2\pi} \int_{D(a,R)} \mathrm{Log} \left| \frac{z - \frac{\xi - a}{R}}{1 - \frac{\bar{\xi} - \bar{a}}{R} z} \right| \Delta u(\xi) \, dm(\xi).$$

In terms of $\zeta = a + Rz$, it is written as

$$u(\zeta) = \frac{1}{2\pi R} \int_{C(a,R)} u(\xi) \frac{R^2 - |\zeta - a|^2}{|\xi - \zeta|^2} \, ds(\xi)$$

$$+ \frac{1}{2\pi} \int_{D(a,R)} \mathrm{Log} \left| \frac{(\xi - \zeta) R}{R^2 - (\bar{\xi} - \bar{a})(\zeta - a)} \right| \Delta u(\xi) \, dm(\zeta). \tag{7.32}$$

Hence, the Green's function of $D(a, R)$ is

$$G_{D(a,R)}(\xi, \zeta) = \frac{1}{2\pi} \mathrm{Log} \left| \frac{R(\xi - \zeta)}{R^2 - (\bar{\xi} - \bar{a})(\zeta - a)} \right|.$$

Observe that this is no more than composition of the Green's function of $\mathbb{D}$ with the transformation $\zeta \to \frac{\zeta - a}{R}$. In particular, one has $G_{D(a,R)} \leq 0$. Making $\zeta = a$, one finds

$$u(a) = \frac{1}{2\pi R} \int_{C(a,R)} u(\xi) \, ds(\xi) + \frac{1}{2\pi} \int_{D(a,R)} \mathrm{Log} \left| \frac{\xi - a}{R} \right| \Delta u(\xi) \, dm(\xi).$$

As well the Poisson kernel of the disc $D = D(a, R)$ is

$$P_D(\xi, \zeta) = \frac{1}{2\pi R} \frac{R^2 - |\zeta - a|^2}{|\xi - \zeta|^2}, \quad |\xi - a| = R, \ |\zeta - a| < R. \qquad \square$$

The potential $u$ of Theorem 7.50 makes sense for functions that are not necessarily bounded on $U$; one may even change $\phi\, dV$ by a measure $\mu$ of locally finite mass, as done with Riesz potentials, and define

$$u(x) = \int_U G_U(x, y)d\mu(y)$$

whenever the integral is convergent. For example, in the case of the unit disc $\mathbb{D}$, where the function $G_{\mathbb{D}}$ is explicit, this analysis may be done in detail.

**Proposition 7.52.** *Let $\mu$ be a measure of locally finite mass in $\mathbb{D}$ such that*

$$\int_{\mathbb{D}} (1 - |w|^2)d|\mu|(w) < +\infty.$$

*Then the potential*

$$u(z) = \frac{1}{2\pi} \int_{\mathbb{D}} \mathrm{Log}\left|\frac{z - w}{1 - w\bar{z}}\right| d\mu(w)$$

*defines a function $u \in L^1(\mathbb{D})$ that satisfies $\Delta u = \mu$ on $\mathbb{D}$ in the weak sense.*

*Proof.* One has

$$\int_{\mathbb{D}} |u(z)|dm(z) \le \frac{1}{2\pi} \int_{\mathbb{D}} d|\mu|(w) \int_{\mathbb{D}} \left|\mathrm{Log}\left|\frac{z - w}{1 - w\bar{z}}\right|\right| dm(z).$$

Applying formula (7.31) to the function $u(z) = |z|^2$ for which $\phi = \Delta u = 4$, one finds

$$\frac{1}{2\pi} \int_{\mathbb{D}} \mathrm{Log}\left|\frac{z - w}{1 - w\bar{z}}\right| dm(z) = \frac{1}{4}(|w|^2 - 1),$$

so that

$$\int_{\mathbb{D}} |u(z)|\, dm(z) \le \frac{1}{4} \int (1 - |w|^2)\, d|\mu|(w),$$

a finite quantity by hypothesis. To check $\Delta u = \mu$ in the weak sense, repeat the proof of Theorem 7.42. For $\varphi \subset C_c^2(\mathbb{D})$ one has

$$\int_{\mathbb{D}} u(z)\Delta\varphi(z)dm(z) = \frac{1}{2\pi} \int_{\mathbb{D}} d\mu(w) \int_{\mathbb{D}} \mathrm{Log}\left|\frac{z - w}{1 - w\bar{z}}\right| \Delta\varphi(z)\, dm(z)$$

$$= \int_{\mathbb{D}} \varphi(w)\, d\mu(w),$$

because formula (7.31), bearing in mind that $\varphi(z) = 0$ if $|z| = 1$, gives

$$\varphi(w) = \frac{1}{2\pi} \int_{\mathbb{D}} \mathrm{Log}\left|\frac{z - w}{1 - w\bar{z}}\right| \Delta\varphi(z)dw(z). \qquad \square$$

**Example 7.53.** Suppose that $\phi$ is radial on the disc $\mathbb{D}$ with $\phi(w) = h(|w|)$, where $h$ satisfies

$$\int_0^1 (1 - r)|h(r)|\, dr < +\infty.$$

Then Green's potential of the function $\phi$ in $\mathbb{D}$ is:

$$u(z) = \int_0^1 h(r)r \left( \frac{1}{2\pi} \int_0^{2\pi} \mathrm{Log} \left| \frac{z - re^{it}}{1 - re^{it}\bar{z}} \right| dt \right) dr.$$

The inner integral only depends on $|z| = s$; since the function $\mathrm{Log}\,|1 - rs\eta|$ is harmonic with respect to $\eta$ for $|\eta| \leq 1$ and has value $0$ at the origin, thanks to the mean value property, the integral of $\mathrm{Log}\,|1 - rse^{it}|$ is zero. Hence, the inner integral is

$$\frac{1}{2\pi} \int_0^{2\pi} \mathrm{Log}\,|s - re^{it}|\, dt.$$

If $s > r$, by the mean value property the integral equals the value of $\mathrm{Log}\,|s - z|$ at the origin, which is $\mathrm{Log}\,s$. Since $|s - re^{it}| = |r - se^{it}|$, the integral equals $\mathrm{Log}\,r$ for $s < r$, and so

$$u(z) = \mathrm{Log}\,s \int_0^s rh(r)\, dr + \int_s^1 rh(r) \log r\, dr.$$

In this case, one may directly check that for continuous $h$, the function $u$ is of class $C^2$ and has Laplacian equal to $\phi$.     $\square$

Similar considerations may arise for a general Neumann's problem or *non-homogenous Neumann's problem*. In a bounded domain $U$ with regular boundary oriented by the exterior normal $\vec{N}$, the data are now $\varphi \in C(\partial U)$ and $\phi \in L^\infty(U)$. The problem consists in finding a function $u \in C^1(U)$ such that

$$\Delta u = \phi \text{ in } U \text{ in the weak sense} \quad \text{and} \quad \frac{\partial u}{\partial \vec{N}} = \varphi \text{ in } \partial U. \tag{N}$$

The data $\varphi, \phi$ must fulfill the compatibility condition

$$\int_{\partial U} \varphi\, dA = \int_{\partial U} \frac{\partial u}{\partial \vec{N}}\, dA = \int_U \Delta u\, dV = \int_U \phi\, dV.$$

As in the homogeneous case, this problem, if it has a solution, has only one except for constants. Indeed if $u_1$, $u_2$ are solutions, $u = u_1 - u_2$ is harmonic on $U$ and $\frac{\partial u}{\partial \vec{N}} = 0$ on $\partial U$. Therefore (7.8) implies

$$\int_U |\vec{\nabla} u|^2\, dV = 0,$$

and $u$ is constant on $U$.

Neumann's problem splits into two corresponding problems

$$\Delta u = 0 \quad \text{in } U, \qquad \frac{\partial u}{\partial \vec{N}} = \varphi \quad \text{in } \partial U, \tag{$N_1$}$$

$$\Delta u = \phi \quad \text{in } U, \qquad \frac{\partial u}{\partial \vec{N}} = 0 \quad \text{in } \partial U. \tag{$N_2$}$$

The way the equation $\Delta u = \phi$ is dealt with is similar to the case of Dirichlet's problem. First, one looks for solutions in the weak sense, and afterwards one analyzes how properties of $\phi$ influence regularity of $u$. Analogously to Proposition 7.49, the following result holds:

**Proposition 7.54.** *If problem* $(N_1)$ *has a solution for every function* $\varphi \in C(\partial U)$ *satisfying* $\int_{\partial U} \varphi \, dA = 0$, *then problem* $(N_2)$ *has a solution for every function* $\phi \in L^\infty(U)$ *such that* $\int_U \phi \, dV = 0$. *If* $\phi$ *is locally Lipschitz, then there is a solution of problem* $(N_2)$ *in the classical sense.*

*Proof.* Consider the Riesz potential

$$G(\phi)(x) = \int_U G(x - y)\phi(y) \, dV(y).$$

By Theorem 7.47, $G(\phi)$ is of class $C^1$ on $\mathbb{R}^n$, it is harmonic outside $\bar{U}$ and satisfies $\Delta G(\phi) = \phi$ in $U$, in the weak sense. Since $\int_U \phi \, dV = 0$, one has

$$G(\phi)(x) = \int_U (G(x - y) - G(x))\phi(y) \, dV(y),$$

$$\vec{\nabla} G(\phi)(x) = \int_U (\vec{\nabla}_x G(x - y) - \vec{\nabla} G(x))\phi(y) \, dV(y).$$

Now, a computation shows that $|\vec{\nabla}_x G(x - y) - \vec{\nabla} G(x)| \le c\,|x - y|^{-n}$, $c$ constant, uniformly in $y \in U$ for $|x|$ big and, therefore, $|\vec{\nabla} G(\phi)(x)| = O(|x|^{-n})$ for $|x| \to +\infty$. Apply now (7.7) to the function $G(\phi)$ and the domain $B(0, R) \setminus U$ for $R$ big. It turns out that

$$\int_{S(0,R)} \frac{\partial G(\phi)}{\partial \vec{N}} \, dA - \int_{\partial U} \frac{\partial G(\phi)}{\partial \vec{N}} \, dA = \int_{B(0,R)\setminus \bar{U}} \Delta G(\phi) \, dV = 0.$$

Since $\left| \int_{S(0,R)} \frac{\partial G(\phi)}{\partial \vec{N}} \, dA \right| \le c \frac{R^{n-1}}{R^n} \xrightarrow[R\to\infty]{} 0$, $c$ constant, we get that $\varphi = \frac{\partial G(\phi)}{\partial \vec{N}}$ satisfies $\int_{\partial U} \varphi \, dA = 0$. If now $u_1$ is a solution of $(N_1)$ with data $\varphi$, it follows that $u = G(\phi) - u_1$ is a solution of $(N_2)$ with data $\phi$.

If $\phi$ is locally Lipschitz, then $G(\phi)$ is of class $C^2$ and $\Delta G(\phi) = \phi$ in the classical sense. So, $\Delta u = \phi$ also holds in the classical sense, since $u_1$ is harmonic. $\qquad \square$

Assume $U$ admits a second Green's function, $H_U$, so that the function

$$u(x) = \int_{\partial U} H_U(x, y)\varphi(y)\, dA(y) + c, \quad c \text{ constant}$$

solves problem $(N_1)$, according to (7.18). Then the function

$$u(x) = -\int_U H_U(x, y)\phi(y)\, dV(y)$$

will be the solution of $\Delta u = \phi$, $\frac{\partial u}{\partial \vec{N}} = 0$ in $\partial U$ if $\int \phi\, dV = 0$.

Consequently, the solution of the general Neumann's problem (N) will be given by the function

$$u(x) = \int_{\partial U} H_U(x, y)\varphi(y)\, dA(y) - \int_U H_U(x, y)\phi(y)\, dV(y) + c, \quad c \text{ constant.}$$

In the case of the unit disc, the previous formula with

$$H_{\mathbb{D}}(z, w) = -\frac{1}{2\pi} \text{Log}(|z - w||1 - \bar{w}z|)$$

(Subsection 7.6.3) gives, in fact, the solution of the general Neumann's problem.

## 7.9 The solution of the Dirichlet and Neumann problems in the ball

In the case $n > 2$, it is also possible to use the symmetry of the unit ball $\mathbb{B} = B(0, 1)$ of $\mathbb{R}^n$ to calculate Green's function and the Poisson kernel. In this case, for $x \in \mathbb{B}$ one needs to find a harmonic function $v_x$ such that

$$v_x(y) = G(x - y) = d_n|x - y|^{2-n} \quad \text{for } |y| = 1.$$

If $x = 0$, obviously one has $v_x = d_n$; for $x \neq 0$, let $x'$ be the point that corresponds to $x$ in the inversion with respect to $\mathbb{S} = \partial\mathbb{B}$, that is, $x' = \frac{x}{|x|^2}$. Then, if $|y| = 1$, one has

$$|x' - y|^2 = \left|\frac{x}{|x|^2} - y\right|^2 = \frac{|x|^2}{|x|^4} + |y|^2 - \frac{2\langle x, y\rangle}{|x|^2}$$

$$= \frac{1}{|x|^2} + 1 - \frac{2\langle x, y\rangle}{|x|^2} = \frac{1 + |x|^2 - 2\langle x, y\rangle}{|x|^2} = \frac{|x - y|^2}{|x|^2}.$$

It follows that $|x - y| = |x||x' - y|$ if $|y| = 1$ and one may take $v_x(y) = d_n|x|^{2-n}|x' - y|^{2-n}$ which is harmonic on $\mathbb{R}^n \setminus \{x'\}$. Hence, Green's function of the ball $\mathbb{B}$ is

$$G_{\mathbb{B}}(x, y) = d_n(|x - y|^{2-n} - |x|^{2-n}|x' - y|^{2-n}).$$

Finally, one may compute the Poisson kernel of $\mathbb{B}$. Observe that $\frac{\partial}{\partial \vec{N}_y} = \frac{\partial}{\partial r} = \sum y_i \frac{\partial}{\partial y_i}$, for $r = |y|$. It yields, then,

$$\frac{\partial}{\partial r}|x - y|^{2-n} = \frac{\partial}{\partial r}\left(|x|^2 + r^2 - \sum_i 2x_i y_i\right)^{1-\frac{n}{2}}$$

$$= \left(1 - \frac{n}{2}\right)(2r - 2x \cdot y)|x - y|^{-n}$$

$$= (2 - n)(r - x \cdot y)|x - y|^{-n}$$

$$= (2 - n)(1 - x \cdot y)|x - y|^{-n}.$$

In a similar way

$$\frac{\partial}{\partial r}|x' - y|^{2-n} = (2 - n)(1 - x' \cdot y)|x' - y|^{-n},$$

and one arrives at

$$P_\mathbb{B}(x, y) = d_n(2 - n)|x - y|^{-n}\{1 - x \cdot y - (1 - x' \cdot y)|x|^2\} = \frac{1}{c_n}\frac{1 - |x|^2}{|x - y|^n},$$

recalling $c_n$ is the area of the unit sphere of $\mathbb{R}^n$. Observe that this kernel has the same structure as for the unit disc. Analogously to Theorem 7.26, the following result is proved.

**Theorem 7.55.** *If $\varphi \in C(\mathbb{S})$, the solution of Dirichlet's problem $u = \varphi$ in $\mathbb{S}$ and $\Delta u = 0$ in $\mathbb{B}$ is the function*

$$u(x) = \frac{1}{c_n}\int_\mathbb{S} \frac{1 - |x|^2}{|x - y|^n}\varphi(y)\, dA(y).$$

**Example 7.56.** Calculate the solution $u_i$ corresponding to the restriction of the function $\varphi_i(x) = x_i^2$ to $\mathbb{S}$, $i = 1, 2, \ldots, n$. Since $x_i^2 - x_j^2$ is harmonic, one has $u_i - u_j = x_i^2 - x_j^2$, that is, $u_i - x_i^2$ is independent of $i$. The sum of the functions $u_i$ is the solution of Dirichlet's problem corresponding to $\sum_i x_i^2$, which has value 1 in $\mathbb{S}$; therefore, $\sum_i u_i(x) = 1$ for $x \in \mathbb{B}$, and, consequently, $\sum_i(u_i - x_i^2) = 1 - |x|^2$. Hence, it turns out that $u_i(x) = x_i^2 + \frac{1}{n}(1 - |x|^2)$. $\square$

As an application of the solution of Dirichlet's problem on balls one may prove the symmetry principle. A domain $U$ of $\mathbb{R}^n$ is *symmetric with respect to* $\mathbb{R}^{n-1}$ if $(x_1, x_2, \ldots, x_n) \in U$ implies $(x_1, x_2, \ldots, x_{n-1}, -x_n) \in U$, and one writes $U^+ = \{(x_1, x_2, \ldots, x_n) \in U : x_n > 0\}$ to denote the upper part of $U$.

**Proposition 7.57** (Symmetry principle for harmonic functions). *Let $U$ be a domain of $\mathbb{R}^n$ symmetric with respect to $\mathbb{R}^{n-1}$ and let $u$ be a harmonic function on $U^+$, continuous on $\overline{U}^+$ such that $u = 0$ in $U \cap \mathbb{R}^{n-1}$. Then the odd extension $v$ of $u$ to $U$ defined by*

$$v(x_1, x_2, \ldots, x_n) = \begin{cases} u(x_1, x_2, \ldots, x_n) & \text{if } x_n \geq 0, \\ -u(x_1, x_2, \ldots, -x_n) & \text{if } x_n < 0, \end{cases}$$

*is harmonic on $U$.*

*Proof.* Clearly the function $v$ is continuous on $U$ and harmonic on $U \setminus \mathbb{R}^{n-1}$. Fix a point $a \in U \cap \mathbb{R}^{n-1}$ and let $B(a, r) \subset U$ be a ball with center $a$ contained in $U$. Consider the harmonic function $h$ on $B(a, r)$ with $v$ as boundary value (on $\partial B(a, r)$). Since the data $v$ satisfies $v(x_1, x_2, \ldots, x_n) = -v(x_1, x_2, \ldots, -x_n)$ on $\partial B(a, r)$, the function $h$ satisfies the same property on $B(a, r)$ (this is a consequence of the Poisson formula or the uniqueness of solution of Dirichlet's problem in the ball). The anti-symmetry of $h$ implies $h = 0$ in $B(a, r) \cap \mathbb{R}^{n-1}$. Then $h$ and $v$ are harmonic functions on $B^+(a, r)$ and $B^-(a, r)$ (the upper and the lower parts of $B(a, r)$), continuous on $\overline{B}^+(a, r)$ and $\overline{B}^-(a, r)$ and coincide on the boundaries of these half-balls. The uniqueness of solution of Dirichlet's problem yields $h = v$ on $B(a, r)$, and $v$ is harmonic on this ball. The point $a \in U \cap \mathbb{R}^{n-1}$ is any one and $v$ is harmonic on $U$. $\qquad\square$

Regarding the non-homogeneous Dirichlet problem $\Delta u = \phi$ in $\mathbb{B}$, $u_{|\mathbb{S}} = \varphi$ with $\phi \in L^\infty(\mathbb{B})$ and $\varphi \in C(\mathbb{S})$, the solution will be, according to (7.17),

$$u(x) = \frac{1}{c_n} \int_{\mathbb{S}} \frac{1 - |x|^2}{|x - y|^n} \varphi(y) \, dA(y)$$

$$+ \frac{1}{c_n(n-2)} \int_{\mathbb{B}} (|x - y|^{2-n} - |x|^{2-n}|x' - y|^{2-n}) \phi(y) \, dV(y).$$

In the case of the ball of radius $R$, rescaling, we obtain as the solution of the homogeneous Dirichlet problem the function

$$u(x) = \frac{1}{c_n R} \int_{S(0,R)} \frac{R^2 - |x|^2}{|x - y|^n} \varphi(y) \, dA(y). \qquad (7.33)$$

The following result is the version of Cauchy's inequalities for harmonic functions:

**Proposition 7.58.** *Let $u$ be harmonic on a ball $B(0, R)$ satisfying $|u(x)| \leq M$ for $|x| \leq R$. Then one has, for every multi-index $\alpha$,*

$$\left| \frac{\partial^\alpha}{\partial x^\alpha} u(0) \right| \leq C_\alpha \frac{M}{R^{|\alpha|}}$$

*with a constant $C_\alpha$ which only depends on the multi-index $\alpha$.*

*Proof.* One may suppose that $u$ is harmonic on $\bar{B}(0, R)$ so that (7.33) holds with $\varphi(y) = u(y)$. Differentiating under the integral sign and evaluating at $x = 0$, the result follows.      $\square$

As an application one may deal with isolated singularities of harmonic functions.

**Proposition 7.59.** *Suppose that $u$ is harmonic on $B(0, R) \setminus \{0\}$ and*

$$|u(y)| = o(G(y)), \quad y \to 0,$$

*$G$ being the fundamental solution of the Laplacian. Then $u$ is harmonic on $B(0, R)$, that is, the origin is a removable singularity of the function $u$.*

*Proof.* The proof is done first for the case $n > 2$. Since $|u(y)| = o(|y|^{2-n})$, applying Proposition 7.58 to the ball $B(y, \frac{|y|}{2})$, we obtain

$$|\vec{\nabla} u(y)| = o(|y|^{1-n}).$$

For fixed $x \neq 0$ with $|x| < R$, apply Corollary 7.16 on $U = B(0, R) \setminus \bar{B}(0, \varepsilon)$. This yields

$$u(x) = \int_{S(0,R)} \left( u(y) \frac{\partial}{\partial \vec{N}_y} G(x - y) - G(x - y) \frac{\partial u}{\partial \vec{N}} (y) \right) dA(y)$$

$$- \int_{S(0,\varepsilon)} \left( u(y) \frac{\partial}{\partial \vec{N}_y} G(x - y) - G(x - y) \frac{\partial u}{\partial \vec{N}} (y) \, dA(y) \right)$$

$$= v_1(x) - v_2(x).$$

Now, for some constant $C(x)$ one has

$$|v_2(x)| \leq C(x)\varepsilon^{n-1}\left( \sup_{|y|=\varepsilon} |u(y)| + \sup_{|y|=\varepsilon} |\vec{\nabla} u(y)| \right) \to 0, \quad \text{if } \varepsilon \to 0.$$

Therefore, $v_2(x) \equiv 0$ and $u(x) = v_1(x)$ is harmonic on $B(0, R)$.

In the case $n = 2$, one has $|u(y)| = o(|\log|y||)$ and $|\vec{\nabla} u(y)| = o\left(\frac{|\text{Log}\,|y||}{|y|}\right)$. With this, the term $\int_{C(0,\varepsilon)} u(y) \frac{\partial}{\partial \vec{N}_y} G(x - y) \, dA(y)$ tends to zero. For the other term $\int_{C(0,\varepsilon)} G(x - y) \frac{\partial u}{\partial \vec{N}} (y) \, dA(y)$, one would only obtain the estimate $o(\text{Log}\,\varepsilon)$ and one must work a little more. Write

$$\int_{C(0,\varepsilon)} G(x - y) \frac{\partial u}{\partial \vec{N}} (y) \, dA(y) = G(x) \int_{C(0,\varepsilon)} \frac{\partial u}{\partial \vec{N}} (y) \, dA(y)$$

$$+ \int_{C(0,\varepsilon)} [G(x - y) - G(x)] \frac{\partial u}{\partial \vec{N}} (y) \, dA(y).$$

Since $|G(x - y) - G(x)| = o(\varepsilon)$, the last integral tends to zero with $\varepsilon$. Now we will show that $\int_{C(0,\varepsilon)} \frac{\partial u}{\partial \vec{N}}(y)\, dA(y) = 0$ (as it corresponds *a posteriori* if $u$ must be harmonic on $B(0, R)$), that is, we will prove that $\int_{C(0,\varepsilon)} u(y)\, dA(y)$ is constant with respect to $\varepsilon$. Actually, consider the function

$$v(x) = \frac{1}{2\pi} \int_0^{2\pi} u(xe^{i\theta})\, d\theta,$$

which is also harmonic on $B(0, R) \setminus \{0\}$, radial and satisfies $v(x) = o(\mathrm{Log}\,|x|)$. Since every radial harmonic function is of the form $a\,\mathrm{Log}\,|x| + b$ and $v(x) = o(\mathrm{Log}\,|x|)$, constant $a$ must be 0, $v$ is constant and $\int_{C(0,\varepsilon)} u(y)\, dA(y)$ is also.   $\square$

The computation of the second Green's function $H_{\mathbb{B}}(x, y)$ of the unit ball of $\mathbb{R}^n$ is more complicated. A not completely explicit expression will be given that is sufficient to set the existence and properties of this Green's function. Instead of seeking, for each $x \in \mathbb{B}$, the harmonic function $v = v_x$ such that $\frac{\partial v}{\partial \vec{N}_y} = \frac{\partial}{\partial \vec{N}_y} G(x - y)$ for $|y| = 1$, one will look for $H_{\mathbb{B}}(x, y) = v_x(y) - G(x - y)$ directly. Recall that $H_{\mathbb{B}}(x, y)$ must satisfy the following equality whenever $u$ is harmonic on $\bar{\mathbb{B}}$:

$$u(x) = \frac{1}{c_n} \int_{\mathbb{S}} u(y)\, dA(y) + \int_{\mathbb{S}} H_{\mathbb{B}}(x, y)\frac{\partial u}{\partial \vec{N}}(y)\, dA(y).$$

We will use a fact that is specific to the ball and the derivative with respect to the normal to its boundary: if $u$ is harmonic, the function $\sum_{i=1}^{n} x_i \frac{\partial u}{\partial x_i}$ is also, because

$$\Delta\left(\sum_{i=1}^{n} x_i \frac{\partial u}{\partial x_i}\right) = \sum_{j=1}^{n} \frac{\partial}{\partial x_j}\left(\sum_{i=1}^{n} \delta_{ij} \frac{\partial u}{\partial x_i} + x_i \frac{\partial^2 u}{\partial x_i \partial x_j}\right)$$

$$= 2 \sum_{i,j=1}^{n} \delta_{ij} \frac{\partial^2 u}{\partial x_i \partial x_j} + x_i \frac{\partial^3 u}{\partial x_i \partial x_j^2} = 2\Delta u + \sum_{i=1}^{n} x_i \frac{\partial \Delta u}{\partial x_i}.$$

In spherical coordinates, $x = rw$, $w \in \mathbb{S}$, one has $\sum_{i=1}^{n} x_i \frac{\partial u}{\partial x_i} = r \frac{\partial u}{\partial r}(rw)$, and, applying the Poisson representation formula to this harmonic function, it turns out that

$$r \frac{\partial u}{\partial r}(rw) = \int_{\mathbb{S}} P_{\mathbb{B}}(rw, y)\frac{\partial u}{\partial \vec{N}}(y)\, dA(y).$$

Now one would like to divide by $r$ and integrate radially to obtain $u(rw)$, but $\frac{1}{r}$ is not integrable around the origin; to avoid this difficulty one uses the fact $\int_{\mathbb{S}} \frac{\partial u}{\partial \vec{N}}\, dA(y) = 0$. Hence, one has

$$r \frac{\partial u}{\partial r}(rw) = \int_{\mathbb{S}} (P_{\mathbb{B}}(rw, y) - 1)\frac{\partial u}{\partial \vec{N}}(y)\, dA(y)$$

and

$$u(rw) = \int_{\mathbb{S}} \left\{ \int_0^r \frac{1}{s} \left( P_{\mathbb{B}}(sw, y) - 1 \right) ds \right\} \frac{\partial u}{\partial \vec{N}}(y) \, dA(y).$$

For $x \in \mathbb{B}$, $y \in \mathbb{S}$, write $H_{\mathbb{B}}(x, y)$ for the kernel:

$$H_{\mathbb{B}}(x, y) = \int_0^r \frac{1}{s} \{ P_{\mathbb{B}}(sw, y) - 1 \} \, ds = \int_0^1 \{ P_{\mathbb{B}}(tx, y) - 1 \} \frac{dt}{t}.$$

The Poisson kernel $P_{\mathbb{B}}(x, y)$ is harmonic in $x$; therefore, $P_{\mathbb{B}}(tx, y)$ is also harmonic in $x$ for all $t$, and so is $H_{\mathbb{B}}(x, y)$. By construction,

$$\frac{\partial H_{\mathbb{B}}}{\partial r}(rw, y) = \frac{1}{r} (P_{\mathbb{B}}(x, y) - 1)$$

holds. With this one may prove that Neumann's problem in the ball,

$$\Delta u = 0 \text{ in } \mathbb{B}, \quad \frac{\partial u}{\partial r} = \varphi \text{ in } \mathbb{S}, \quad \text{with } \int_{\mathbb{S}} \varphi \, dA = 0$$

has a unique solution (unique except for constants). Simply define

$$u(x) = \int_{\mathbb{S}} H_{\mathbb{B}}(x, y)\varphi(y) \, dA(y).$$

Since $H_{\mathbb{B}}(x, y)$ is harmonic in $x$, the function $u$ is harmonic. One also has

$$r\frac{\partial u}{\partial r}(x) = \int_{\mathbb{S}} (P_{\mathbb{B}}(x, y) - 1)\varphi(y) \, dA(y) = \int_{\mathbb{S}} P_{\mathbb{B}}(x, y)\varphi(y) \, dA(y).$$

The right-hand side term, the solution of Dirichlet's problem with data $\varphi$, has boundary value $\varphi$, and as a consequence one gets $\frac{\partial u}{\partial r} = \varphi$ in $\mathbb{S}$.

Finally, according to Proposition 7.54 the problem

$$\Delta u = \phi \text{ in } \mathbb{B}, \quad \frac{\partial u}{\partial \vec{N}} = 0 \text{ in } \mathbb{S}$$

also has a solution in the classical sense provided $\phi$ is locally Lipschitz.

## 7.10 Decomposition of vector fields

Recall that in an arbitrary domain $U$ of $\mathbb{R}^3$ a continuous vector field is conservative if and only if it is a gradient, and it is solenoidal if and only if it is a rotational. With differentiability assumptions, locally conservative (respectively, locally solenoidal) vector fields $\vec{X}$ are characterized by the condition rot $\vec{X} = 0$ (respectively, div $\vec{X} = 0$). Using the language of solutions in the weak sense, we will now make considerations for the equations $\operatorname{div} \vec{X} = h$ and $\operatorname{rot} \vec{X} = \vec{Y}$ with given $h$ and $\vec{Y}$, similar to those for the Poisson equation $\Delta u = \phi$.

A continuous vector field $\vec{X}$ on a domain $U$ of $\mathbb{R}^3$ is determined by the scalar products,

$$\langle \vec{X}, \vec{Y} \rangle_U \underset{\text{def}}{=} \int_U \langle \vec{X}(x), \vec{Y}(x) \rangle \, dV(x) = \int_U \left\{ \sum_{i=1}^{3} X_i(x) Y_i(x) \right\} dV(x),$$

involving $\vec{X}$ and all the vector fields $\vec{Y}$ of class $C^\infty$ with compact support in $U$. Assume $\vec{X}$ is of class $C^1$ on $U$ and $\vec{Y} = \vec{\nabla}\phi$ with $\phi \in C_c^\infty(U)$. Then

$$\langle \vec{X}, \vec{\nabla}\phi \rangle_U = \int_U \left\{ \sum_{i=1}^{3} \vec{X}_i(x) \frac{\partial \phi}{\partial x_i}(x) \right\} dV(x)$$

$$= - \int_U \left( \sum_{i=1}^{3} \frac{\partial \vec{X}_i}{\partial x_i}(x) \right) \phi(x) \, dV(x) = -\langle \operatorname{div} \vec{X}, \phi \rangle_U.$$

This is the reason to give the following definition:

**Definition 7.60.** If $\vec{X}$ is a continuous vector field on $U$ and $h \in L^1_{\text{loc}}(U)$, it is said that equality $\operatorname{div} \vec{X} = h$ holds in the weak sense in $U$ if

$$\langle \vec{X}, \vec{\nabla}\phi \rangle_U = \int_U \left\{ \sum_{i=1}^{3} \vec{X}_i(x) \frac{\partial \phi}{\partial x_i}(x) \right\} dV(x) = - \int_U h(x)\phi(x) \, dV(x),$$

for every $\phi \in C_c^\infty(U)$.

Analogously, if $\vec{X}$ is of class $C^1$ on $U$ and $\vec{X} = \operatorname{rot} \vec{Z}$, $\vec{Z} = (Z_1, Z_2, Z_3)$ with $Z_i \in C_c^\infty(U)$, then one has

$$\langle \vec{X}, \operatorname{rot} \vec{Z} \rangle_U = \int_U \left[ \vec{X}_1 \left( \frac{\partial Z_3}{\partial y} - \frac{\partial Z_2}{\partial z} \right) + \vec{X}_2 \left( \frac{\partial Z_1}{\partial z} - \frac{\partial Z_3}{\partial x} \right) \right.$$

$$\left. + \vec{X}_3 \left( \frac{\partial Z_2}{\partial x} - \frac{\partial Z_1}{\partial y} \right) \right] dV(x)$$

$$= - \int_U \left[ \left( Z_3 \frac{\partial \vec{X}_1}{\partial y} - Z_2 \frac{\partial \vec{X}_1}{\partial z} \right) + \left( Z_1 \frac{\partial \vec{X}_2}{\partial z} - Z_3 \frac{\partial \vec{X}_2}{\partial x} \right) \right.$$

$$\left. + \left( Z_2 \frac{\partial \vec{X}_3}{\partial x} - Z_1 \frac{\partial \vec{X}_3}{\partial y} \right) \right] dV(x)$$

$$= -\int \left[ Z_1 \left( \frac{\partial \vec{X}_2}{\partial z} - \frac{\partial \vec{X}_3}{\partial y} \right) + Z_2 \left( \frac{\partial \vec{X}_3}{\partial x} - \frac{\partial \vec{X}_1}{\partial z} \right) \right.$$

$$\left. + Z_3 \left( \frac{\partial \vec{X}_1}{\partial y} - \frac{\partial \vec{X}_2}{\partial x} \right) \right] dV(x) = -\langle \mathrm{rot}\, \vec{X}, \vec{Z} \rangle_U .$$

**Definition 7.61.** If $\vec{X}$ is a continuous vector field on $U$ and $\vec{Y}$ is a locally integrable vector field on $U$, it is said that equality rot $\vec{X} = \vec{Y}$ holds in the weak sense in $U$ if

$$\langle \vec{X}, \mathrm{rot}\, \vec{Z} \rangle_U = -\langle \vec{Y}, \vec{Z} \rangle_U ,$$

for every vector field $\vec{Z}$ with $Z_i \in C_c^\infty(U)$, $i = 1, 2, 3$.

**Proposition 7.62.** *In an arbitrary domain $U$ of $\mathbb{R}^3$, a continuous vector field $\vec{X}$ is locally conservative if and only if* rot $\vec{X} = 0$ *in the weak sense, that is,*

$$\langle \vec{X}, \mathrm{rot}\, \vec{Z} \rangle_U = 0, \quad \text{for } \vec{Z} \text{ with } Z_i \in C_c^\infty(U), \ i = 1, 2, 3.$$

*A continuous vector field $\vec{X}$ is locally solenoidal on $U$ if and only if* div $\vec{X} = 0$ *in the weak sense, that is,*

$$\langle \vec{X}, \vec{\nabla}\phi \rangle_U = 0 \quad \text{for } \phi \in C_c^\infty(U).$$

*If $U$ is simply connected (respectively, without holes), locally conservative may be changed to conservative (respectively, locally solenoidal to solenoidal).*

For the notions of simply connected domain and domain without holes see Section 3.7.1.

*Proof.* It is quite similar to Proposition 6.20 (Remark 6.3). If $\vec{X}$ is of class $C^1$, the result is a consequence of previous considerations. In general, we approach $\vec{X}$ by $C^\infty$ vector fields, uniformly on compact sets, which still verify the same hypotheses. $\square$

Hence, considering, in order to make it simpler, that $U$ is a simply connected domain of $\mathbb{R}^3$ without holes and assuming, in addition, that $\partial U$ is regular, one has these two classes of continuous vector fields on $U$:

A) $\vec{X}$ conservative $\iff$ $\vec{X} = \vec{\nabla}\psi$ $\iff$ rot $\vec{X} = 0$, in the weak sense.

B) $\vec{X}$ solenoidal $\iff$ $\vec{X} = \mathrm{rot}\, \vec{Y}$ $\iff$ div $\vec{X} = 0$, in the weak sense.

**Theorem 7.63.** *If a continuous vector field $\vec{X}$ is simultaneously conservative and solenoidal on a domain $U$, then there is a harmonic function $\psi$ on $U$ such that $\vec{X} = \vec{\nabla}\psi$. In particular, $\vec{X}$ is of class $C^\infty$. If $U$ has a regular boundary oriented by the exterior normal vector field $\vec{N}$ and $\langle \vec{X}, \vec{N} \rangle = 0$ holds on $\partial U$, then one has $\vec{X} \equiv 0$ on $U$.*

*Proof.* Since $\vec{X}$ is conservative, one has $\vec{X} = \vec{\nabla}\psi$, with $\psi \in C^1(U)$. Moreover, $\mathrm{div}(\vec{\nabla}\psi) = 0$ in the weak sense, that is,

$$\int_U \left( \sum_{i=1}^{3} \frac{\partial \psi}{\partial x_i}(x) \frac{\partial \phi}{\partial x_i} \right) dV(x) = 0$$

for $\phi \in C_c^\infty(U)$. This is equivalent to

$$\int_U \psi(x) \Delta \phi(x)\, dV(x) = 0 \quad \text{for } \phi \in C_c^\infty(U),$$

which means $\Delta\psi = 0$ in the weak sense. By Weyl's lemma (Theorem 7.45) $\psi$ is harmonic. Since $\langle \vec{X}, \vec{N} \rangle = \langle \vec{\nabla}\psi, \vec{N} \rangle = \frac{\partial \psi}{\partial \vec{N}}$, if the normal component of $\vec{X}$ is zero, then $\psi$ is constant and $\vec{X} = \vec{\nabla}\psi = 0$. $\qquad\square$

Suppose one wants to split a vector field $\vec{X}$, regular enough on $\bar{U}$, as a sum of a vector field of type A) plus a vector field of type B):

$$\vec{X} = \vec{X}_1 + \vec{X}_2.$$

Then, applying the divergence operator, one would have

$$\mathrm{div}\, \vec{X}_1 = \mathrm{div}\, \vec{X} \overset{\mathrm{def}}{=} h$$

in the weak sense. Writing $\vec{X}_1 = \vec{\nabla}\psi$, it follows that

$$\Delta\psi = h$$

in the weak sense. In order to determine $\psi$ it is enough to give a condition of Neumann's type on $\partial U$. One may impose, for example,

$$\frac{\partial \psi}{\partial \vec{N}} = \langle \vec{X}, \vec{N} \rangle$$

so that $\langle \vec{X}_1, \vec{N} \rangle = \langle \vec{\nabla}\psi, \vec{N} \rangle = \frac{\partial \psi}{\partial \vec{N}} = \langle \vec{X}, \vec{N} \rangle$. Hence, in any possible decomposition $\vec{X} = \vec{X}_1 + \vec{X}_2$ with $\vec{X}_1$ conservative and $\vec{X}_2$ solenoidal, $\vec{X}_1$ is completely determined if $\langle \vec{X}_1, \vec{N} \rangle = \langle \vec{X}, \vec{N} \rangle$. Then the vector field $\vec{X}_2 = \vec{X} - \vec{X}_1$ satisfies $\mathrm{div}\, \vec{X}_2 = 0$, so it is solenoidal, and in addition $\langle \vec{X}_2, \vec{N} \rangle = 0$. On the other hand, this decomposition turns out to be orthogonal because

$$\langle \vec{X}_1, \vec{X}_2 \rangle_U = \int_U \langle \vec{X}_1(x), \vec{X}_2(x) \rangle\, dV(x) = \int_U \langle \vec{\nabla}\psi(x), \vec{X}_2 \rangle\, dV(x).$$

But $\langle \vec{\nabla}\psi, \vec{X}_2 \rangle = \mathrm{div}(\psi\vec{X}_2) - \psi(\mathrm{div}\,\vec{X}_2) = \mathrm{div}(\psi\vec{X}_2)$ and

$$\langle \vec{X}_1, \vec{X}_2 \rangle_U = \int_U \mathrm{div}(\psi\vec{X}_2)\,dV = \int_{\partial U} \psi(\langle \vec{X}_2, \vec{N}\rangle\,dA = 0.$$

By construction, this decomposition is unique. Indeed if $\vec{X}_1 + \vec{X}_2 = \vec{X}'_1 + \vec{X}'_2$ were two decompositions of this kind, then $\vec{X}_1 - \vec{X}'_1 = \vec{X}'_2 - \vec{X}_2$ would be a vector field simultaneously conservative and solenoidal with normal component zero. Therefore, it is identically zero by Theorem 7.63. The conditions on $\vec{X}$ so that $\vec{X}_1$, $\vec{X}_2$ be $C^1(\bar{U})$ vector fields are delicate, but the condition $\vec{X} \in C^{1+\alpha}(\bar{U})$, $\alpha > 0$, suffices; then the function $h = \mathrm{div}\,\vec{X}$ is locally Lipschitz on $U$ and according to Proposition 7.54 the problem $\Delta\psi = h$ in $U$, $\frac{\partial\psi}{\partial\vec{N}} = \langle \vec{X}, \vec{N}\rangle$ in $\partial U$, has a solution $\psi \in C^2(\bar{U})$ and it turns out that $\vec{X}_1 = \vec{\nabla}\psi \in C^1(\bar{U})$. With all this, the following theorem has been shown.

**Theorem 7.64** (Helmholtz). *Assume $U$ is a bounded domain with regular boundary oriented by the exterior normal vector field $\vec{N}$, in which Neumann's problem can be solved, and $U$ is simply connected without holes. Then every vector field $\vec{X}$ regular enough on $\bar{U}$ (for example $\vec{X} \in C^{1+\alpha}(\bar{U})$, $\alpha > 0$) has a unique decomposition*

$$\vec{X} = \vec{X}_1 + \vec{X}_2,$$

*with $\vec{X}_1$, $\vec{X}_2$ of class $C^1$ on $\bar{U}$, where $\vec{X}_1$ is conservative and satisfying $\langle \vec{X}_1, \vec{N}\rangle = \langle \vec{X}, \vec{N}\rangle$ and $\vec{X}_2$ is solenoidal and satisfying $\langle \vec{X}_2, \vec{N}\rangle = 0$. Furthermore, this decomposition is orthogonal with respect to the scalar product of vector fields in $U$.*

It is possible as well to get other decompositions: choosing $\psi$ with $\Delta\psi = h = \mathrm{div}\,\vec{X}$ and $\psi = 0$ on $\partial U$, then the decomposition is also orthogonal and $\vec{X}_1$ has only a component normal to $\partial U$.

Let us now consider the previous decomposition on the whole space $\mathbb{R}^3$, this time solving the equation $\Delta\psi = h = \mathrm{div}\,\vec{X}$ by means of the newtonian potential

$$\psi(x) = \int_{\mathbb{R}^3} h(y)G(x - y)\,dV(y).$$

**Theorem 7.65.** *Suppose that $\vec{X}$ is a vector field of class $C^1$ on $\mathbb{R}^3$ with $|\mathrm{div}\,\vec{X}(x)| = O(|x|^{-1-\varepsilon})$ when $|x| \to +\infty$ for some $\varepsilon > 0$. Then $\vec{X}$ has a unique decomposition $\vec{X} = \vec{X}_1 + \vec{X}_2$, with $\vec{X}_1$ conservative, $\vec{X}_2$ solenoidal and both vector fields continuous and vanishing at infinity.*

*Proof.* If $\vec{X}_1 + \vec{X}_2 = \vec{X}'_1 + \vec{X}'_2$, then the vector field $\vec{X}_1 - \vec{X}'_1 = \vec{X}'_2 - \vec{X}_2$ is both conservative and solenoidal and, therefore, is of the form $\vec{\nabla}u$ with $u$ harmonic and

$\vec{\nabla}u(x)$ vanishing at infinity. Since the function $\frac{\partial u}{\partial x_j}$ is harmonic and vanishing at infinity, by Liouville's theorem (Theorem 7.8) it must be identically zero. Hence, the decomposition is unique. In order to prove the existence, define $\vec{X}_1 = \vec{\nabla}\psi$, where $\psi$ is the solution of $\Delta\psi = \operatorname{div}\vec{X}$ given by the newtonian potential, that is,

$$\vec{X}_1 = \vec{\nabla}\int_{\mathbb{R}^3} \operatorname{div}\vec{X}(y)\, G(x-y)\, dV(y) = \frac{1}{c_3}\int_{\mathbb{R}^3} \operatorname{div}\vec{X}(y)\frac{x-y}{|x-y|^3}\, dV(y).$$

Observe that the hypothesis $|\operatorname{div}\vec{X}(y)| = O(|y|^{-1-\varepsilon})$ guarantees the convergence of the integrals. We just need to show that the vector field $\vec{X}_1$ vanishes at infinity (it is clear that $\vec{X}_2 = \vec{X} - \vec{X}_1$ is solenoidal, because $\operatorname{div}\vec{X}_2 = 0$ and vanishes at infinity as well).

It suffices to check that

$$\int_{\mathbb{R}^3} |y|^{-1-\varepsilon}|x-y|^{-2}\, dV(y) = O(|x|^{-\varepsilon}).$$

By symmetry, it is enough to consider the contribution of the half-space $\{y : |y| \le |y-x|\}$. The contribution of $\{y : |y| \le \frac{|x|}{2}\}$ (where $|x-y|$ is about $|x|$) is estimated by

$$|x|^{-2}\int_{|y|\le\frac{|x|}{2}} |y|^{-1-\varepsilon}\, dV(y) = |x|^{-2}O(|x|^{2-\varepsilon}) = O(|x|^{-\varepsilon})$$

and the contribution of $\{y : \frac{|x|}{2} \le |y| \le |y-x|\}$ (where $|y-x|$ is about $|y|$) is bounded above by

$$\int_{|y|\ge\frac{|x|}{2}} |y|^{-3-\varepsilon}\, dV(y) = O(|x|^{-\varepsilon}). \qquad\qquad \square$$

One can go further and make the vector field $\vec{X}_2$ explicit. Since $\vec{X}_2$ is solenoidal, it has a potential vector, that is, $\vec{X}_2 = \operatorname{rot}\vec{Y}$. The vector field $\vec{Y}$ is not uniquely determined, because $\operatorname{rot}(\vec{X} + \vec{\nabla}f) = \operatorname{rot}\vec{Y}$ for any conservative vector field $\vec{\nabla}f$, with $f$ a function. A way of normalizing $\vec{Y}$ is to impose that $\vec{Y}$ be solenoidal and vanishing at infinity: actually, if $\operatorname{rot}\vec{Y}_1 = \operatorname{rot}\vec{Y}_2$ and $\vec{Y}_1$, $\vec{Y}_2$ are both solenoidal and vanishing at infinity, then $\vec{Y}_1 - \vec{Y}_2 = \vec{\nabla}u$ with $u$ harmonic and $\vec{\nabla}u$ vanishing at infinity, which gives $\vec{Y}_1 = \vec{Y}_2$.

Hence we look for a solenoidal vector field $\vec{Y}$ such that $\vec{X}_2 = \operatorname{rot}\vec{Y}$. Then

$$\operatorname{rot}\vec{X} = \operatorname{rot}\vec{X}_1 + \operatorname{rot}\vec{X}_2 = \operatorname{rot}\vec{X}_2 = \operatorname{rot}(\operatorname{rot}\vec{Y}).$$

Now,

$$\operatorname{rot}(\operatorname{rot}\vec{Y}) = \vec{\nabla}(\operatorname{div}\vec{Y}) - \Delta\vec{Y},$$

where $\Delta \vec{Y}$ is interpreted vectorially. If $\vec{Y}$ is solenoidal, then $\operatorname{div} \vec{Y} = 0$ and we conclude that

$$\Delta \vec{Y} = -\operatorname{rot} \vec{X}.$$

If $|\operatorname{rot} \vec{X}| = O(|y|^{-2-\varepsilon})$, the only solution vanishing at infinity is again

$$\vec{Y}(x) = -\int_{\mathbb{R}^3} \operatorname{rot} \vec{X}(y) G(x - y) \, dV(y),$$

by a computation similar to the one in the proof of Theorem 7.65. Moreover, if one also has $|\operatorname{div} \vec{X}(y)| = O(|y|^{-2-\varepsilon})$, the potential function $\psi$ defined before by

$$\psi(x) = \int_{\mathbb{R}^3} \operatorname{div} \vec{X}(y) \, G(x - y) \, dV(y)$$

will be vanishing at infinity as well. All this proves the following statement:

**Theorem 7.66.** *Let $\vec{X}$ be a vector field of class $C^1$ on $\mathbb{R}^3$ with $|\operatorname{div} \vec{X}(x)| = O(|x|^{-2-\varepsilon})$ and $|\operatorname{rot} \vec{X}(x)| = O(|x|^{-2-\varepsilon})$ when $|x| \to \infty$, for some $\varepsilon > 0$. Then the decomposition*

$$\vec{X} = \vec{\nabla}\psi + \operatorname{rot} \vec{Y}$$

*with*

$$\psi(x) = \int_{\mathbb{R}^3} \operatorname{div} \vec{X}(y) G(x - y) \, dV(y),$$

$$\vec{Y}(x) = -\int_{\mathbb{R}^3} \operatorname{rot} \vec{X}(y) G(x - y) \, dV(y)$$

*is the unique decomposition of $\vec{X}$ as a sum of a gradient, $\vec{\nabla}\psi$, and a rotational, $\operatorname{rot} \vec{Y}$, with $\psi, \vec{Y}$ vanishing at infinity.*

Finally we find an expression for the vector field $\vec{X}_1 = \vec{\nabla}\psi$ in terms of $\vec{X}$, which is useful in fluid mechanics. Applying the divergence theorem to the field $\frac{x_i - y_i}{|x-y|^3} \vec{X}$ and to the domain $B(x, R) \setminus \bar{B}(x, \varepsilon)$, it turns out that

$$\frac{1}{c_3} \int_{S(x,R)} \frac{x_i - y_i}{|x - y|^3} \langle \vec{X}, \vec{N} \rangle \, dA(y) - \frac{1}{c_3} \int_{S(x,\varepsilon)} \frac{x_i - y_i}{|x - y|^3} \langle \vec{X}, \vec{N} \rangle \, dA(y)$$

$$= \frac{1}{c_3} \int_{B_R(x) \setminus \bar{B}_\varepsilon(x)} \operatorname{div} \left( \frac{x_i - y_i}{|x - y|^3} \vec{X} \right) dV(y)$$

$$= \frac{1}{c_3} \int_{B_R(x) \setminus \bar{B}_\varepsilon(x)} (\operatorname{div} \vec{X}) \frac{x_i - y_i}{|x - y|^3} \, dV(y)$$

$$+ \frac{1}{c_3} \int_{B_R(x) \setminus \bar{B}_\varepsilon(x)} \left\langle \vec{\nabla}_y \left( \frac{x_i - y_i}{|x - y|^3} \right), \vec{X} \right\rangle dV(y).$$

If $\vec{X}$ vanishes at infinity, one has

$$\lim_{R\to\infty} \frac{1}{c_3} \int_{S(x,R)} \frac{x_i - y_i}{|x-y|^3} \langle \vec{X}, \vec{N} \rangle \, dA(y) = 0.$$

Moreover,

$$\frac{1}{c_3} \int_{S(x,\varepsilon)} \frac{x_i - y_i}{|x-y|^3} \langle \vec{X}, \vec{N} \rangle \, dA(y)$$
$$= \frac{1}{c_3} \int_{S(x,\varepsilon)} \frac{x_i - y_i}{|x-y|^3} \sum_j X_j(y) \frac{y_j - x_j}{|y-x|} \, dA(y).$$

The limit when $\varepsilon \to 0$ of this last integral does not change when replacing $X_j(y)$ with $X_j(x)$; then the terms with $j \neq i$ are zero and the term with $j = i$ has limit $-\frac{1}{n} X_i(x)$, when $\varepsilon \to 0$. Hence one has

$$\vec{X}_1(x) = \frac{1}{n} \vec{X}(x) - \lim_{\varepsilon\to 0} \frac{1}{c_3} \int_{\varepsilon < |x-y|} \langle \vec{\nabla}_y \left( \frac{x-y}{|x-y|^3} \right), \vec{X} \rangle \, dV(y),$$

where $\vec{\nabla}_y \left( \frac{x-y}{|x-y|^3} \right)$ must be interpreted as the matrix

$$\left( \frac{\partial}{\partial y_j} \left( \frac{x_i - y_i}{|x-y|^3} \right) \right)_{i,j=1,2,3}.$$

## 7.11 Dirichlet's problem and conformal transformations

In order to solve Dirichlet's problem or Neumann's problem in more general domains than the disc or the ball, it is natural to analyze how these problems behave when a domain is transformed into another. First we look for transformations $\Psi$ that preserve harmonic functions, that is, those giving a harmonic function $u = v \circ \Psi$ whenever $v$ is harmonic. This analysis is done here in arbitrary dimension $n$.

It is worth remarking that one does not ask that equality $\Delta(v \circ \Psi) = (\Delta v) \circ \Psi$ holds. In Subsection 7.2.1 it has been seen that $L = P(\Delta)$, where $P$ is a polynomial, is the general expression of operators satisfying $L(v \circ \Psi) = Lv \circ \Psi$ for every isometry $\Psi$. In particular, one has $\Delta(v \circ \Psi) = \Delta v \circ \Psi$ for every isometry $\Psi$. However $\Psi$ may preserve harmonic functions under more general conditions. For example, there could exist a factor $E$ such that

$$\Delta(v \circ \Psi) = [\Delta v \circ \Psi] \cdot E.$$

Let $\Psi = (\Psi_1, \Psi_2, \dots, \Psi_n)$ be a diffeomorphism of class $C^2$ between two open sets $U$, $V$ of $\mathbb{R}^n$, $v \in C^2(V)$ and compute $\Delta(v \circ \Psi)$:

$$\frac{\partial}{\partial x_i}(v \circ \Psi) = \sum_{k=1}^{n} \frac{\partial v}{\partial y_k}(\Psi(x)) \frac{\partial \Psi_k}{\partial x_i}(x),$$

$$\frac{\partial^2}{\partial x_i^2}(v \circ \Psi) = \sum_{k,\ell=1}^{n} \frac{\partial^2 v}{\partial y_\ell \partial y_k}(\Psi(x)) \frac{\partial \Psi_\ell}{\partial x_i}(x) \frac{\partial \Psi_k}{\partial x_i}(x) + \sum_{k=1}^{n} \frac{\partial v}{\partial y_k}(\Psi(x)) \frac{\partial^2 \Psi_k}{\partial x_i^2}(x),$$

$$\Delta(v \circ \Psi)(x) = \sum_{i,k,\ell=1}^{n} \frac{\partial^2 v}{\partial y_\ell \partial y_k}(\Psi(x)) \frac{\partial \Psi_\ell}{\partial x_i}(x) \frac{\partial \Psi_k}{\partial x_i} + \sum_{k=1}^{n} \frac{\partial v}{\partial y_k}(\Psi(x)) \Delta \Psi_k(x).$$

$$(7.34)$$

Suppose now this expression is zero whenever $\Delta v = 0$. Taking $v \equiv y_k$, it follows that $\Delta \Psi_k = 0$, that is, each component $\Psi_k$ of $\Psi$ must be harmonic. Taking $v = y_\ell^2 - y_k^2$ one gets

$$\sum_{i=1}^{n} \left( \frac{\partial \Psi_\ell}{\partial x_i} \right)^2 = \sum_{i=1}^{n} \left( \frac{\partial \Psi_k}{\partial x_i} \right)^2,$$

and taking $v = y_k y_\ell$ with $k \neq \ell$,

$$\sum_{i=1}^{n} \frac{\partial \Psi_\ell}{\partial x_i} \frac{\partial \Psi_k}{\partial x_i} = 0, \quad k \neq \ell.$$

This means $J_\Psi(x) = \Psi'(x) \in \mathbb{R}O(n)$. So $J_\Psi(x)$ is a scalar multiple of an orthogonal matrix, that is, the linear mapping $\Psi'(x)$ conserves the size of angles and is therefore a linear similarity for each $x \in U$. In other words, $\Psi$ is conformal or anticonformal at each point of $U$.

**Theorem 7.67.** *A diffeomorphism $\Psi$ of class $C^2$ between two open sets of $\mathbb{R}^n$ preserves harmonic functions if and only if it is conformal or anticonformal at each point and the components of $\Psi$ are harmonic functions.*

*Proof.* We have already seen that $\Psi$ is necessarily of this kind. For the converse, notice that formula (7.34) reads

$$\Delta(v \circ \Psi) = \mathrm{trace}[\Psi'(x)^t H \Psi'(x)],$$

where $H$ is the Hessian matrix of $v$ at the point $\Psi(x)$. Since $\Psi'(x)$ is conformal, $\Psi'(x)^t$ is a multiple of $\Psi'(x)^{-1}$ and then it follows that $\Delta(v \circ \Psi)$ is a multiple of $\mathrm{trace}\,[\Psi'(x)^{-1} H \Psi'(x)] = \mathrm{trace}\,[H] = 0$. $\qquad\square$

In dimension $n \geq 3$, Liouville's theorem (Theorem 2.29) asserts that if $\Psi$ is conformal or anticonformal at each point of $U$, then $\Psi$ is the restriction to $U$ of a similarity in $\mathbb{R}^n$ (composition of isometries, dilations and inversions). However, in dimension $n = 2$, Theorem 7.67 asserts that all transformations $w = \Psi(z)$ that are holomorphic or antiholomorphic preserve the class of harmonic functions.

In the case $n = 2$, when $w = \Psi(z)$ is holomorphic, repeating the calculus of $\Delta(v \circ \Psi)$ in terms of operators $\partial$, $\bar{\partial}$ yields

$$\partial(v \circ \Psi) = \frac{\partial v}{\partial w} \Psi'(z) + \frac{\partial v}{\partial \bar{w}} \frac{\partial \bar{\Psi}}{\partial z} = \frac{\partial v}{\partial w}(\Psi(z))\Psi'(z),$$

$$\bar{\partial}\partial(v \circ \Psi)(z) = \frac{\partial^2 v}{\partial w \partial \bar{w}}(\Psi(z))|\Psi'(z)|^2,$$

$$\Delta(v \circ \Psi)(z) = \Delta v(\Psi(z))|\Psi'(z)|^2.$$

All that has been said can be expressed, more precisely, in the following result.

**Theorem 7.68** (Invariance of the Dirichlet problem for holomorphic transformations). *Let $U$ and $V$ be two bounded domains of $\mathbb{C}$ such that there exists a holomorphic one-to-one function from $U$ onto $V$ which extends to a homeomorphism from $\bar{U}$ onto $\bar{V}$. Suppose one knows how to solve the non-homogeneous Dirichlet problem in $U$. Then, if $\varphi \in C(\partial V)$ and $\phi \in L^\infty(V)$, the Dirichlet problem $\Delta v = \phi$ in $V$, $v = \varphi$ in $\partial V$ has a unique solution (in the weak sense). If $\phi$ is locally Lipschitz, then the solution $v$ is of class $C^2$ on $V$ and satisfies $\Delta v = 0$ in the classical sense.*

*Proof.* Let $\Psi: U \to V$ be bijective and holomorphic extending to a homeomorphism from $\bar{U}$ onto $\bar{V}$. Recall that $\Psi'(z) \neq 0$, for all $z \in U$ and $\Psi^{-1}: V \to U$ is holomorphic with $(\Psi^{-1})'(\Psi(z)) = \Psi'(z)^{-1}$. Consider the Dirichlet problem in $U$ with data $\varphi_0(z) = \varphi(\Psi(z))$, $\phi_0(z) = \phi(\Psi(z))|\Psi'(z)|^2$. One has

$$\int_U |\Psi'(z)|^2 \, dm(z) = \int_{\Psi(U)} dm(w) = m(V) < +\infty$$

and, therefore, $\phi_0$ is integrable and locally bounded on $U$. Let $u$ be the solution of the problem $\Delta u_0 = \phi_0$ in $U$, $u_0 = \varphi_0$ in $\partial(U)$. Then $v = u \circ \Psi^{-1}$ is continuous on $\bar{V}$ and satisfies $v_{|\partial V} = \varphi$ and $\Delta v = \phi$ in $V$. $\qquad\square$

Furthermore, one has a formula for the solution in $V$ if one has it in $U$. In the particular case $U = \mathbb{D}$ and $\Psi: \mathbb{D} \to V$ one-to-one and holomorphic, one can write, by (7.31), using the previous notation:

$$u(z) = \frac{1}{2\pi} \int_{\mathbb{T}} \frac{1 - |z|^2}{|z - \zeta|^2} \varphi_0(\zeta) \, d\sigma(\zeta)$$

$$+ \frac{1}{2\pi} \int_{\mathbb{D}} \mathrm{Log} \left| \frac{z - \zeta}{1 - \bar{z}\zeta} \right| \phi_0(\zeta) \, dm(\zeta) = u^{\mathrm{I}}(z) + u^{\mathrm{II}}(z).$$

The second term $u^{\mathrm{II}}(z)$ is the solution of $\Delta u = \phi_0$ in $\mathbb{D}$, $u = 0$ in $\mathbb{T}$. Making the change of variable $\Psi(\zeta) = \eta$, the measure $\phi_0(\zeta)\,dm(\zeta) = \phi(\Psi(\zeta))|\Psi'(\zeta)|^2\,dm(\zeta)$ becomes $\phi(\eta)\,dm(\eta)$ and the function

$$v^{\mathrm{II}}(w) = u^{\mathrm{II}}(\Psi(w)) = \frac{1}{2\pi}\int_V \mathrm{Log}\,\frac{|\Psi(w) - \Psi(\eta)|}{|1 - \overline{\Psi(w)}\Psi(\zeta)|}\phi(\eta)\,dm(\eta)$$

is the solution of the problem $\Delta v = \phi$ in $V$, $v = 0$ in $\partial V$. The function

$$G_V(w,\eta) = \mathrm{Log}\,\frac{|\Psi(w) - \Psi(\eta)|}{|1 - \overline{\Psi(w)}\Psi(\zeta)|}$$

is thus the Green's function of $V$. When $\partial V$ is regular and $\Psi$ is differentiable along $\mathbb{T}$ with $\Psi'(z) \neq 0$ one may also transform the integral that defines $u^{\mathrm{I}}(z)$ with the change $\Psi(\zeta) = \eta$. Then the measure $\varphi_0(\zeta)\,d\sigma(\zeta)$ becomes $|(\Psi)'(\eta)|\varphi(\eta)\,ds(\eta)$ and the function

$$v^{\mathrm{I}}(w) = u^{\mathrm{I}}(\Psi(w)) = \frac{1}{2\pi}\int_{\partial V}\frac{1 - |\Psi(w)|^2}{|\Psi(w) - \Psi(\eta)|^2}|(\Psi)'(\eta)|\varphi(\eta)ds(\eta)$$

is the solution of the problem $\Delta v = 0$ in $V$, $v = \varphi$ in $\partial V$. The function

$$P(w,\eta) = \frac{1}{2\pi}\frac{1 - |\Psi(w)|^2}{|\Psi(w) - \Psi(\eta)|^2}|(\Psi)'(\eta)|$$

is the Poisson kernel of $V$.

The same idea may be used for the Neumann problem. Suppose that $U$ and $V$ are two bounded domains of $\mathbb{C}$ with positively oriented regular boundary and $\Psi$ is a one-to-one and holomorphic transformation from a neighborhood of $\overline{U}$ onto a neighborhood of $\overline{V}$ so that it is a homeomorphism from $\overline{U}$ onto $\overline{V}$. In all the examples one may check that this hypothesis holds.

First observe that $\Psi$ satisfies $J_\Psi(z) = |\Psi'(z)|^2 > 0$ and so preserves orientation. Hence, $\Psi$ transforms $\partial U$ into $\partial V$ and when $z$ describes $\partial U$ in the positive sense, the point $\Psi(z) = w$ describes $\partial V$ in the positive sense. Let $\vec{N}_1(z)$ denote the unit exterior normal vector to $\partial U$ at the point $z \in \partial U$. Since $\Psi'$ preserves angles at the points $z \in \partial U$ and $\Psi$ transforms $\partial U$ into $\partial V$, this yields that the vector $(d\Psi)(z)(\vec{N}_1(z)) = \Psi'(z)\vec{N}_1(z)$ is an exterior normal vector to $\partial V$ at the point $w = \Psi(z)$. So, denoting by $\vec{N}_2(w)$ the unit exterior normal vector at $w \in \partial V$, one has

$$\vec{N}_2(w) = \frac{\Psi'(z)}{|\Psi'(z)|}\vec{N}_1(z).$$

Assume now that $v$ is a solution of the problem $\Delta v = \phi$ in $V$, $\frac{\partial v}{\partial \vec{N}_2} = \varphi$ in $\partial V$. Then the function $u = v \circ \Psi$ satisfies $\Delta u(z) = \phi(\Psi(z))|\Psi'(z)|^2$ in $U$ and

$$\frac{\partial u}{\partial \vec{N}_1}(z) = \frac{\partial v}{\partial \vec{N}_2}(w)|\Psi'(z)| = \varphi(w)|\Psi'(z)|.$$

So, if one wants to solve the problem $\Delta v = \phi$ in $V$, $\frac{\partial u}{\partial \vec{N}_2} = \varphi$ in $\partial V$ and knows how to find a solution $u$ of

$$\Delta u(z) = \phi(\Psi(z))|\Psi'(z)|^2 \text{ in } U, \qquad \frac{\partial u}{\partial \vec{N}_1}(z) = \varphi(\Psi(z))|\Psi'(z)| \text{ in } \partial U,$$

then $v(w) = u(\Psi(w))$ is the desired solution.

## 7.12  Dirichlet's principle

Dirichlet originally solved the problem that bears his name, namely $\Delta u = 0$ in a domain $U \subset \mathbb{R}^n$, $u = \varphi$ in $\partial U$ with $\varphi \in C(\partial U)$, as a problem of *calculus of variations*. He did so on the basis of the observation that Laplace's equation is Euler's equation associated to the functional $I$ acting on the function $u$, according to

$$I(u) = \int_U |\vec{\nabla} u|^2 \, dV. \tag{7.35}$$

This functional is called the *Dirichlet integral*.

Given $\varphi \in C(\partial U)$, consider the class of functions

$$\Sigma = \{u \in C^2(\bar{U}) : u = \varphi \text{ in } \partial U\}$$

and the functional $I : \Sigma \to \mathbb{R}^+$ defined by (7.35). Suppose that $u_0 \in \Sigma$ is a global minimum of $I$ on $\Sigma$. Then, fixing $\psi \in C_c^2(U)$, the function $u_0 + \varepsilon \psi$ is in $\Sigma$ for all $\varepsilon > 0$ and $I(u_0 + \varepsilon \psi) \geq I(u_0)$ holds. The function $\varepsilon \mapsto I(u_0 + \varepsilon \psi)$ has a minimum for $\varepsilon = 0$, and so has vanishing derivative at $\varepsilon = 0$, whenever this derivative exists. One may calculate it:

$$\frac{d}{d\varepsilon} I(u_0 + \varepsilon \psi) = \frac{d}{d\varepsilon} \int_U |\vec{\nabla} u_0 + \varepsilon \vec{\nabla} \psi|^2 \, dV(x)$$

$$= \frac{d}{d\varepsilon} \int_U \sum_{i=1}^{n} \left( \frac{\partial u_0}{\partial x_i} + \varepsilon \frac{\partial \psi}{\partial x_i} \right)^2 dV(x)$$

$$= 2 \int_U \sum_{i=1}^{n} \left( \frac{\partial u_0}{\partial x_i} + \varepsilon \frac{\partial \psi}{\partial x_i} \right) \frac{\partial \psi}{\partial x_i} \, dV(x).$$

Letting $\varepsilon \to 0$, it yields

$$\int_U \langle \vec{\nabla} u_0, \vec{\nabla} \psi \rangle \, dV(x) = 0.$$

By the first Green's identity, and the fact that $\psi$ has compact support in $U$, we get

$$\int_U \Delta u_0(x) \cdot \psi(x) \, dV(x) = 0.$$

Since this equality holds for any $\psi \in C_c^2(U)$, one has $\Delta u_0 \equiv 0$. Therefore, if $u_0$ is a global minimum of $I$ in $\Sigma$, then $u_0$ is *the* solution of Dirichlet's problem. This is the statement of *Dirichlet's principle*. Since $I$ takes positive values, it could seem clear that this global minimum always exists. This (mistakenly) led Dirichlet to regard the previous argument as a proof of the existence of a solution for the problem. One may try, however, to keep the argument up and then some questions on the completeness of spaces of functions appear. These questions were not solved in Dirichlet's time and motivate further study of the nowadays well-known theory of Hilbert spaces.

First, change the definition of $\Sigma$ and take now

$$\Sigma = \{u \in C^1(\bar{U}) : u = \varphi \text{ a } \partial U\}.$$

If $u_0 \in \Sigma$ minimizes $I$ on $\Sigma$ one will have, as before,

$$\int_U \left( \sum_{i=1}^n \frac{\partial u_0}{\partial x_i} \frac{\partial \psi}{\partial x_i} \right) dV = \int_U \langle \vec{\nabla} u_0, \vec{\nabla} \psi \rangle \, dV(x) = 0.$$

Integrating by parts yields

$$\int_U u_0 \Delta \psi \, dV(x) = 0 \quad \text{for } \psi \in C_c^2(U).$$

So, $\Delta u_0 = 0$ in the weak sense and by Weyl's lemma, $u_0$ is a harmonic function. Consider

$$\alpha = \inf\{I(u), u \in \Sigma\} \geq 0$$

and a sequence of functions $u_n \in \Sigma$ such that $(I(u_n)) \to \alpha$, when $n \to \infty$. Since $\frac{1}{2}u_n + \frac{1}{2}u_m \in \Sigma$, one has

$$0 \leq \frac{1}{4} \int_U |\vec{\nabla} u_n - \vec{\nabla} u_m|^2 \, dV + \alpha$$

$$< \frac{1}{4} \int_U |\vec{\nabla} u_n - \vec{\nabla} u_m|^2 \, dV + \frac{1}{4} \int |\vec{\nabla} u_n + \vec{\nabla} u_m|^2 \, dV$$

$$= \frac{1}{2} \int_U |\vec{\nabla} u_n|^2 \, dV + \frac{1}{2} \int_U |\vec{\nabla} u_m|^2 \, dV \longrightarrow \alpha \quad \text{if } n, m \to \infty.$$

Therefore, $\int_U |\vec{\nabla} u_n - \vec{\nabla} u_m|^2 \to 0$, for $n, m \to \infty$. Now notice that defining $\rho(u, v) = I(u - v)$ for $u, v \in \Sigma$, $\rho$ is a distance. Indeed it satisfies the triangle inequality and $\rho(u, v) = 0$, implies $\vec{\nabla} u = \vec{\nabla} v$ and since $u = v = \varphi$ in $\partial U$, one has $u = v$. If the space $\Sigma$ was complete with this distance, the conclusion just obtained, saying that $(u_n)$ is a Cauchy sequence in $(\Sigma, \rho)$, would imply that $(u_n) \to u_0$ with $u_0 \in \Sigma$ and $I(u_0) = \alpha$ and would prove the existence of solutions of Dirichlet's problem. But $\Sigma$ is not complete with the distance $\rho$. However, one could redevelop what has been said, on the basis of similar arguments and the projection theorem for Hilbert spaces and prove that a variant of Dirichlet's problem (in the sense of Sobolev spaces) has a solution.

The most general way of solving Dirichlet's problem in open sets of $\mathbb{R}^n$ is *Perron's method*. It is based on subharmonic functions and the concept of "subsolution". In Section 9.4 this method will be explained in the case of domains of the complex plane.

## 7.13 Exercises

1. Show that a continuous function $u$ on a domain $U \subset \mathbb{R}^n$ is harmonic if and only if it has the mean value property with respect to balls, that is,

$$u(a) = \frac{n}{c_n r^n} \int_{B(a,r)} u(x)\, dV(x),$$

whenever $B(a,r) \subset U$, where $c_n = d\sigma(\mathbb{S})$.

2. Find the dimension of the vector space formed by homogeneous polynomials on $\mathbb{R}^n$ of degree $k$, $k \in \mathbb{N}$, that are harmonic.

3. Recall (Section 7.3) that a continuous function $v$ on a domain $U \subset \mathbb{R}^n$ is said to be subharmonic if it has the sub-mean property with respect to spheres, that is,

$$v(a) \leq \frac{1}{c_n r^{n-1}} \int_{S(a,r)} v(x)\, dA(x),$$

whenever $\bar{B}(a,r) \subset U$. Show that in the case $n = 1$ subharmonic functions are the convex functions.

a) Prove that a function $v \in C^2(U)$ is subharmonic on $U$ if and only if $\Delta v \geq 0$ on $U$ (see Exercise 13 of Section 3.8).

b) Let $\psi \geq 0$, $\psi \in C_c^\infty(\mathbb{R}^n)$, with support in the unit ball $\mathbb{B}$ and

$$\int_{\mathbb{R}^n} \psi\, dV = 1.$$

Put $\psi_\varepsilon(x) = \varepsilon^{-n} \psi(x/\varepsilon)$ and consider the convolution $v_\varepsilon = v * \psi_\varepsilon$ in the open set $U_\varepsilon = \{x : d(x, U^c) > \varepsilon\}$. Prove that $v$ is subharmonic on $U$ if and only if $v_\varepsilon$ is subharmonic on $U_\varepsilon$, for all $\varepsilon > 0$.

c) Show that $v$ is subharmonic on $U$ if and only if

$$\int_U v(x)\Delta\phi(x)\,dV(x) \geq 0,$$

for every function $\phi \in C_c^\infty(U)$, $\phi \geq 0$.

d) Show that $v$ is subharmonic on $U$ if and only if has the sub-mean value property with respect to balls, that is,

$$v(a) \leq \frac{n}{c_n r^n}\int_{B(a,r)} v(x)\,dV(x),$$

whenever $B(a,r) \subset U$.

4. Show that the maximum modulus principle holds for subharmonic functions: if $U$ is a bounded domain of $\mathbb{R}^n$ and $v \in C(\bar{U})$ is subharmonic on $U$ satisfying $v(x) \leq M$ for $x \in \partial U$, then one also has $v(x) \leq M$ for $x \in U$. Show that if $u \in C(\bar{U})$ is harmonic on $U$ and $v \in C(\bar{U})$ is subharmonic on $U$ and satisfies $v(x) \leq u(x)$ for $x \in \partial U$, then one also has $v(x) \leq u(x)$ for $x \in U$.

5. Starting from the solution of Dirichlet's problem in a ball given by the Poisson kernel, prove that every harmonic function on an open set of $\mathbb{R}^n$ is an analytic function of its real variables.

6. Suppose that the function $f$ is harmonic and complex-valued on a domain $U$ of $\mathbb{R}^n$, say $f = u + iv$. Prove that the following are equivalent:

   a) $f^2$ is harmonic on $U$.

   b) $\vec{\nabla}u, \vec{\nabla}v$ are perpendicular and have the same norm in $U$.

   If $n = 2$, prove that a) or b) are equivalent to saying that $f$ is holomorphic or antiholomorphic on $U$. In the general case $n \geq 2$, show that if $f$, harmonic, satisfies a) or b) and takes real values, then $f$ is constant.

7. Prove Harnack's inequality: if $u$ is harmonic and positive on a domain $U$ of $\mathbb{R}^n$ and $B(a,r) \subset U$, then one has

$$r^{n-2}\frac{r - |x - a|}{(r + |x - a|)^{n-1}}u(a) \leq u(x) \leq r^{n-2}\frac{r + |x - a|}{(r - |x - a|)^{n-1}}u(a)$$

for $x \in B(a,r)$. Use this inequality to prove that every harmonic function on $\mathbb{R}^n$ bounded above is constant.

8. a) Use Harnack's inequality (Exercise 7 of this section) to prove the following: let $U$ be a domain of $\mathbb{R}^n$, $K$ a compact set contained in $U$ and

$a \in U$; then there are constants $m, M$ depending on $U, K$ and $a$ such that

$$mu(a) \leq u(x) \leq Mu(a),$$

for every function $u$ harmonic and positive on $U$ and every point $x \in K$.

b) Let $(u_n)_{n \in \mathbb{N}}$ be a sequence of harmonic functions on a domain $U$ that converge uniformly on each compact subset of $U$ to the function $u$. Show that $u$ is harmonic on $U$ and the derivatives $(D^\alpha u_n)$ tend to $D^\alpha u$, uniformly on each compact subset of $U$, for each multi-index $\alpha$.

**9.** Let $(u_n)_{n \in \mathbb{N}}$ be a monotone sequence of harmonic functions on a domain $U \subset \mathbb{R}^n$. Assume that $(u_n(a))$ converges at a point $a \in U$. Show then that $(u_n)$ converges uniformly on each compact subset of $U$ to a harmonic function on $U$.

**10.**  a) Let $\Pi^+ = \{(x_1, x_2, \ldots, x_n) \in \mathbb{R}^n : x_n > 0\}$ be the upper half space of $\mathbb{R}^n$ and let $u$ be a bounded harmonic function on $\Pi^+$, continuous on $\overline{\Pi}^+$. Assume that $u(x) = 0$ if $x \in \mathbb{R}^{n-1} \simeq \{x = (x_1, x_2, \ldots, x_n) \in \mathbb{R}^n : x_n = 0\}$. Show that $u \equiv 0$ on $\Pi^+$.

b) Prove the same result changing the boundedness of $u$ on $\Pi^+$ by the condition: $|u(x)| = O(G(x))$, $|x| \to \infty$, where $G$ is the fundamental solution of the Laplacian.

*Hint:* For part b) use Proposition 7.40 and the fact that the composition of a harmonic function with the inversion with respect to the unit ball is harmonic.

**11.** Find the harmonic function on the unit disc $\mathbb{D}$ that has a continuous extension to $\overline{\mathbb{D}}$ with value 1 at the points $(x, y) \in \partial\mathbb{D}$ with $y > 0$ and with value 0 at the other points of $\partial\mathbb{D}$. Find also the harmonic function that has a continuous extension with value $|y|$ at every point $(x, y) \in \partial\mathbb{D}$.

**12.** Find the harmonic functions on the unit disc $\mathbb{D}$ extending continuously to each point $e^{i\theta} \in \partial\mathbb{D}$ with value a) $\sin\theta$, b) $\sin^2\theta$, c) $\cos^3\theta$. Find also the harmonic function on $\mathbb{D}$ extending continuously to the function $x^2 + y^3$ at the points $(x, y) \in \partial\mathbb{D}$.

**13.** Let $u$ be a harmonic function on a ball $B(a, r)$ of $\mathbb{R}^n$ such that

$$\int_{B(a,r)} |u(x)|^p \, dV(x) = M < +\infty, \quad p > 0.$$

Prove the inequality $|u(a)| \leq \left(\frac{n}{c_n r^n} M\right)^{\frac{1}{p}}$, where $c_n = d\sigma(\mathbb{S})$. As a consequence, show that if $u$ is harmonic on $\mathbb{R}^n$ and $|u|^p$ is integrable on $\mathbb{R}^n$, then $u$ vanishes on $\mathbb{R}^n$.

**14.** Consider in the unit disc $\mathbb{D}$ of the complex plane the mixed problem: look for $u \in C^1(\overline{\mathbb{D}})$ such that $\Delta u = 0$ in $\mathbb{D}$, $\lambda \frac{\partial u}{\partial \theta}(e^{i\theta}) + \lambda \frac{\partial u}{\partial N}(e^{i\theta}) = \varphi(e^{i\theta})$, if $e^{i\theta} \in \partial\mathbb{D}$, with $\varphi \in C(\partial\mathbb{D})$ given and $\lambda$ constant. Show that this problem has a solution if and only if the data $\varphi$ has zero integral on $\partial\mathbb{D}$ and find the solution of the form

$$u(z) = \int_0^{2\pi} \varphi(e^{i\theta}) Q(z, e^{i\theta}) \, d\theta,$$

with an explicit kernel $Q(z, e^{i\theta})$.

**15.** The aim of this exercise is to solve Dirichlet's problem in an annulus $C = \{z \in \mathbb{C} : R_2 < |z| < R_1\}$ of the complex plane. That is, given two continuous functions $\varphi_1, \varphi_2 \in C(\partial C)$, one wants to find $u$ harmonic on $C$, continuous on $\overline{C}$ such that $u(R_1 e^{i\theta}) = \varphi_1(R_1 e^{i\theta}), u(R_2 e^{i\theta}) = \varphi_2(R_2 e^{i\theta})$, for $e^{i\theta} \in \mathbb{T}$.

a) First show that the general expression of a harmonic function $u$ on $C$ is

$$u(re^{i\theta}) = \sum_{n\in\mathbb{Z}} (A_n r^{|n|} + B_n r^{-|n|}) e^{in\theta}, \quad A_n, B_n \in \mathbb{C}.$$

b) Calculate coefficients $A_n, B_n$ in terms of Fourier coefficients of $\varphi_1$ and $\varphi_2$.

c) Solve Dirichlet's problem in $C$.

**16.** Let $U$ be a non-bounded domain of $\mathbb{R}^n$ and $u$ a continuous function on $\overline{U}$, harmonic and bounded on $U$, vanishing on the boundary of $U$. Show that $u$ is identically zero on $U$.

**17.** Let $\Pi^+$ be the upper half plane $\{z = x + yi : y > 0\}$ and $z \in \Pi^+$ fixed. Show that the function

$$G(w, z) = \frac{1}{2\pi} \operatorname{Log} \frac{w - z}{w - \bar{z}}$$

is the Green's function of $\Pi^+$ with pole $z$. Use the function $G$ and Exercise 16 of this section to prove the following: for $\varphi$ bounded and continuous on $\mathbb{R}$, the function of $z = x + iy$,

$$u(z) = \frac{1}{\pi} \int_{-\infty}^{+\infty} \frac{\varphi(t)}{(t - x)^2 + y^2} \, dt,$$

is the only bounded harmonic function on $\Pi^+$ with value $\varphi$ at the boundary of $\Pi^+$.

**18.** Show that the Riesz potential $G(\phi)$ is a continuous function on $\mathbb{R}^n$ provided that $\phi \in L^\infty_{loc}(\mathbb{R}^n)$ satisfies the condition

$$\int_{\mathbb{R}^n} |G(x)||\phi(x)|dV(x) < \infty.$$

**19.** The aim of this exercise is to solve Dirichlet's problem in the square $Q = [0, 1] \times [0, 1]$. The data is a continuous function $\varphi$ on $\partial Q$.

    a) First show that one may assume that $\varphi$ vanishes at the vertices of $Q$. For this consider the harmonic polynomial $a + bx + cy + dxy$, with $a$, $b$, $c$, $d$ conveniently chosen.

    b) If $\varphi$ vanishes at the vertices, then it may be expressed as a sum of four functions, each of which is zero in three of the edges of $Q$.

    c) It is enough, now, to solve the problem $\Delta u = 0$ in $\overset{\circ}{Q}$ and

$$u(x, 0) = u(0, y) = u(1, y) = 0 \quad \text{for } 0 \le x, y \le 1,$$
$$u(x, 1) = \varphi(x) \quad \text{for } 0 \le x \le 1,$$

with $\varphi(0) = \varphi(1) = 0$. For this problem, use separation of variables to prove that the solution is

$$u(x, y) = \sum_{n=1}^{\infty} \frac{2}{\operatorname{sh} n\pi} \hat{\varphi}(n) \sin n\pi x \cdot \operatorname{sh} n\pi y,$$

where $\hat{\varphi}(n)$ denotes the $n$-th coefficient of $\varphi$ in the orthonormal basis $\{\sqrt{2} \sin n\pi t : n \in \mathbb{N}\}$ of $L^2([0, 1])$.

**20.** Do something similar to the previous exercise to solve Neumann's problem in the square $Q$.

**21.** This exercise provides an interpretation of the Poisson kernel of the unit disc $\mathbb{D}$. For $z \in \mathbb{D}$ fixed, associate to each point $w \in \partial\mathbb{D}$ the point $\xi = \xi(z, w) \in \partial\mathbb{D}$ such that $w, z$ and $\xi$ are aligned. Show that the solution of Dirichlet's problem in $\mathbb{D}$ with data $\varphi \in C(\partial\mathbb{D})$ is

$$u(z) = \frac{1}{2\pi} \int_{\partial\mathbb{D}} \varphi(\xi(z, w))d\sigma(w).$$

**22.** Let $U$ be a bounded domain of $\mathbb{R}^n$ with regular boundary oriented by the exterior normal vector field $\vec{N}$ and $\partial U = S_1 \cup S_2$ a partition of the boundary.

Let $\varphi_1$, $\varphi_2$ be continuous functions on $S_1$, $S_2$, respectively. Show that the problem

$$\Delta u = 0 \text{ in } U, \quad u = \varphi_1 \text{ in } S_1, \quad \frac{\partial u}{\partial \vec{N}} = \varphi_2 \text{ in } S_2$$

has at most one solution. Consider the case $U$ is the upper half disc of the plane, $U = \{(x, y) : x^2 + y^2 < 1, \ y > 0\}$, $S_1 = [-1, 1]$ and $S_2 = \partial U \setminus S_1$. Solve the problem

$$\Delta u = 0 \text{ in } U, \quad u = \varphi \text{ in } S_2, \quad \frac{\partial u}{\partial \vec{N}} = 0 \text{ in } S_1,$$

with $\varphi$ continuous on $S_2$, in two different ways: a) using the symmetry principle, b) by separation of variables.

# Chapter 8
# Conformal mapping

Holomorphic functions of one complex variable are conformal mappings in the sense that they preserve angles. This property makes them a very useful tool to study different problems in Physics, such as the motion of fluids, or in the solution of problems with boundary conditions.

In this chapter conformal mappings are studied, stressing geometrical aspects, examples and applications. The proof of Riemann's theorem or fundamental theorem of conformal mapping will be given in Chapter 9.

Besides linear transformations and conformal mappings provided by elementary functions, one deals with conformal mappings of polygons and with the boundary behavior of conformal mappings of domains limited by analytic arcs, both important questions in view of the applications. The chapter ends with some examples of the use of conformal mapping, one in cartography and the others in hydrodynamics.

## 8.1 Conformal transformations

We start with an overview of the relation between holomorphy and conformality, a question already considered in Subsection 2.4.2.

Suppose that $\gamma_1(t)$, $\gamma_2(t)$ are two regular curves passing through the point $z_0$, let $\gamma_1(0) = \gamma_2(0) = z_0$, and having, at this point, a non-zero tangent vector, $\gamma_1'(0) \neq 0$, $\gamma_2'(0) \neq 0$. The angle between $\gamma_1$ and $\gamma_2$ at the point $z_0$ is the angle between the tangent vectors:

$$\operatorname{Arg}\left[\gamma_1'(0) \cdot \overline{\gamma_2'(0)}\right].$$

Let now $f$ be a differentiable mapping on a neighborhood of the point $z_0$ with $df(z_0)$ invertible. The linear mapping $df(z_0)$ is called the *tangent linear mapping* to $f$ at the point $z_0$ for the following reason: considering the curve $\widetilde{\gamma}_1(t) = f(\gamma_1(t))$, the image of $\gamma_1$ by $f$, then the tangent vector to $\widetilde{\gamma}_1$ at the point $f(z_0)$ is the image by $df(z_0)$ of the tangent vector to $\gamma_1$ at the point $z_0$. That is,

$$\widetilde{\gamma}_1'(0) = (f \circ \gamma_1)'(0) = df(z_0)(\gamma_1'(0)).$$

Of course, the same holds for $\gamma_2$ and for any curve passing through the point $z_0$.

Hence, the curves $\widetilde{\gamma}_1 = f \circ \gamma_1$ and $\widetilde{\gamma}_2 = f \circ \gamma_2$ meet at the point $f(z_0)$ with an angle equal to

$$\operatorname{Arg}\left[df(z_0)(\gamma_1'(0)) \cdot \overline{df(z_0)(\gamma_2'(0))}\right].$$

Observe that the condition $df(z_0)$ to be invertible is necessary to consider the angle between $\tilde{\gamma}_1$ and $\tilde{\gamma}_2$. Otherwise it could happen that one of these curves has null tangent vector at the point $f(z_0)$.

Suppose now that $f$ is holomorphic around the point $z_0$ and $f'(z_0) \neq 0$. Then it is known (Theorem 2.25) that the tangent linear mapping $df(z_0)$ is the $\mathbb{C}$-linear one given by complex multiplication by $f'(z_0)$. Therefore,

$$\tilde{\gamma}_1'(0) = f'(z_0) \cdot \gamma_1'(0) \quad \text{and} \quad \tilde{\gamma}_2'(0) = f'(z_0) \cdot \gamma_2'(0).$$

This means that the action of the function $f$ makes its tangent vector to rotate by an angle $\operatorname{Arg} f'(z_0)$.

The angle between $\tilde{\gamma}_1$ and $\tilde{\gamma}_2$ is now

$$\operatorname{Arg}\left[|f'(z_0)|^2 \gamma_1'(0) \cdot \overline{\gamma_2'(0)}\right] = \operatorname{Arg}\left[\gamma_1'(0) \cdot \overline{\gamma_2'(0)}\right].$$

The function $f$ preserves, then, the angles between curves at the point $z_0$. We say that $f$ is *conformal* at this point. Observe that $f$ preserves the values of the angles and also their sense (which is the sign of $\operatorname{Arg}[\gamma_1'(0) \cdot \overline{\gamma_2'(0)}]$).

Now, instead of comparing the angle between tangent vectors to two curves with the angle between tangent vectors to their images by $f$, let us compare the length of these tangent vectors. That is, let us compare $|\gamma'(0)|$ with $|df(z_0)(\gamma'(0))|$.

In general, the relation between these lengths depends on the direction of the vector $\gamma'(0)$, but if $f$ is holomorphic, then

$$|df(z_0)(\gamma'(0))| = |f'(z_0)| \cdot |\gamma'(0)|$$

and it yields that the distortion of the length induced by $f$ is the same in all directions and equal to $|f'(z_0)|$.

Consider now the converse of the results that have just been stated. In Subsection 2.4.2 it has been seen that if $f$ is differentiable around the point $z_0$ (with $df(z_0)$ invertible) and $f$ is conformal, that is, preserves the angles between curves and their sense, then $f$ is holomorphic at the point $z_0$. Let us review the argument: with matrix notation one has $df(z_0) = \begin{pmatrix} a & h \\ c & d \end{pmatrix}$ where $a = \frac{\partial u}{\partial x}, b = \frac{\partial u}{\partial y}, c = \frac{\partial v}{\partial x}, d = \frac{\partial v}{\partial y}$, if $f = u + iv$ and $df(z_0)(z) = \alpha z + \beta \bar{z}$ with $\alpha + \beta = a + ic$ and $\alpha - \beta = d - ib$; the fact that $df(z_0)$ preserves angles means that $\frac{df(z_0)(z)}{z} = \alpha + \beta \frac{\bar{z}}{z}$ has constant argument, and this only can happen if $\beta = 0$, because $\alpha + \beta \frac{\bar{z}}{z}$ describes a circle with center $\alpha$ and radius $|\beta|$; finally, $\beta = 0$ means $a = d, b = -c$, which are Cauchy–Riemann equations.

Let us assume now that the differentiable mapping $f$, with $df(z_0)$ invertible, dilates lengths of vectors with the same rate in all directions. This means that

$$\frac{df(z_0)(z)}{z} = \alpha + \beta \frac{\bar{z}}{z}$$

has constant modulus, independently of $z$, which is only possible if $\alpha + \beta \frac{\bar{z}}{z}$ describes either a circle of radius zero or a circle with center at the origin. That is, either $\beta = 0$, or $\alpha = 0$. In the first case, it is known that $f$ is $\mathbb{C}$-derivable at the point $z_0$. In the second one, $\alpha = 0$ and one gets $a = -d$, $b = c$. Now $df(z_0)(z) = \beta \bar{z}$ is a linear mapping which preserves angles, but changes orientation. The equations corresponding to $a = -d$, $b = c$ tell now that $\bar{f}$ is a holomorphic function. We say that $f$ is *antiholomorphic* or *inversely conformal*.

Altogether the following statement holds:

**Proposition 8.1.** *If $f$ is a differentiable function around the point $z_0$, $f$ is conformal at $z_0$ if and only if $f$ has complex derivative at $z_0$ and $f'(z_0) \neq 0$. In this case, the tangent linear mapping $df(z_0)$ is a rotation of angle $\mathrm{Arg}\, f'(z_0)$ followed by a dilation of factor $|f'(z_0)|$.*

## 8.2 Conformal mappings

It has just been shown that if $f$ is a holomorphic function on an open set $U$ of the complex plane and $f'(z) \neq 0$, for each $z \in U$, then $f$ is conformal at each point of $U$. Now, $f$ does not need to be an injective mapping on $U$ (think, for example, about $U = \mathbb{C}$ and $f(z) = e^z$).

However, if $f$ is a holomorphic function on $U$ that, in addition, is injective on $U$, then necessarily $f'(z) \neq 0$, for $z \in U$, according to Theorem 4.32. So in this case, $f$ is a one-to-one mapping from $U$ onto the open set $f(U)$ that is conformal at each point of $U$ ($f(U)$ is open by Theorem 4.33). It is also clear that the inverse function $f^{-1}$ is conformal at each point of $f(U)$, because, $f'(z) \neq 0$ yields that $f^{-1}$ is holomorphic with non-vanishing derivative as well.

**Definition 8.2.** If $U$ is an open set of $\mathbb{C}$, a holomorphic and one-to-one function $f$ on $U$ is said to be a *conformal mapping* from $U$ onto the open set $f(U)$.

Two domains $U$, $U'$ of the complex plane are said to be *conformally equivalent* if there exists a conformal mapping from $U$ onto $U'$ (or from $U'$ onto $U$).

So conformal mappings are the mappings that are conformal at each point of its domain and, furthermore, are globally injective. For example, the function $e^z$ is a conformal mapping of the strip $U = \{z : 0 < \mathrm{Im}\, z < 2\pi\}$ onto $\mathbb{C} \setminus \{z : \mathrm{Re}\, z > 0, \mathrm{Im}\, z = 0\}$.

If $U$, $U'$ are conformally equivalent, they must also be topologically equivalent, because a conformal mapping from $U$ onto $U'$ is, in particular, a homeomorphism between $U$ and $U'$.

Hence, for example, the unit disc of the plane $\mathbb{D} = \{z : |z| < 1\}$ and an annulus $C(0, 1, 2) = \{z : 1 < |z| < 2\}$ cannot be conformally equivalent. Another negative example is the following.

**Proposition 8.3.** *The unit disc $\mathbb{D}$ and the complex plane $\mathbb{C}$ are not conformally equivalent even though they are topologically equivalent.*

*Proof.* If $f : \mathbb{C} \to \mathbb{D}$ is holomorphic, being also bounded, Liouville's theorem (Theorem 4.41) yields that $f$ would be constant and, therefore, not injective.

On the other hand, $f(z) = \frac{z}{1+|z|}$ is a homeomorphism from $\mathbb{C}$ onto $\mathbb{D}$.     $\square$

It is a natural question, then, to ask which domains can be conformally mapped onto the unit disc $\mathbb{D}$. The answer is the following:

**Theorem 8.4.** *A domain of the complex plane can be conformally mapped onto the unit disc if and only if it is simply connected and different from $\mathbb{C}$.*

One may easily see that the conditions on $U$ are necessary. First the condition $U \neq \mathbb{C}$ is necessary by Proposition 8.3. Secondly if $f : \mathbb{D} \to U$ is a conformal mapping, then for each piecewise regular closed curve $\gamma$ with $\gamma^* \subset U$ and each point $a \notin U$, one has

$$\int_{\gamma} \frac{dw}{w - a} = \int_{f^{-1}\circ\gamma} \frac{f'(z)}{f(z) - a}\, dz = 0,$$

by Cauchy's theorem since $f(z) \neq a$ if $z \in U$. So $\mathrm{Ind}(\gamma, a) = 0$ and $U$ is simply connected.

The fact that simply connected domains different from $\mathbb{C}$ can be conformally mapped on $\mathbb{D}$ is due to Riemann and it is known as the *fundamental theorem of conformal mapping*. More precisely, Riemann's theorem is the following.

**Theorem 8.5** (Fundamental theorem of conformal mapping). *Given a simply connected domain $U$, $U \neq \mathbb{C}$, and a point $z_0 \in U$, there exists a unique conformal mapping $f$ from $U$ onto the unit disc $\mathbb{D}$, normalized with the conditions $f(z_0) = 0$ and $f'(z_0) > 0$.*

**Corollary 8.6.** *Two domains of the plane, simply connected and different from the whole plane, are conformally equivalent to each other.*

Two proofs of Riemann's theorem will be given in Chapter 9. This theorem says that there are two models of simply connected plane domains: $\mathbb{C}$ and the unit disc $\mathbb{D}$. Any other simply connected domain is equivalent to one of these.

If two domains $U$, $U'$ are conformally equivalent, then it makes no difference if one studies holomorphic functions on $U$ or on $U'$, because if $f : U \to U'$ is conformal, the transformation $g \to g \circ f^{-1}$ gives an isomorphism between the rings of functions $H(U)$ and $H(U')$. That is, if we want to know the analytic functions on a domain without holes, it is enough to study entire functions and holomorphic functions on the unit disc.

In particular, conformal mappings from a domain $U$ onto itself may be considered. It is clear that these transformations form a group, called the *automorphism group* of $U$ and denoted by $\mathrm{Aut}(U)$. If $U$ and $U'$ are conformally equivalent by $f: U \to U'$, then $\mathrm{Aut}(U)$ and $\mathrm{Aut}(U')$ are isomorphic groups, through the correspondence $g \to g \circ f^{-1}$ from $\mathrm{Aut}(U)$ onto $\mathrm{Aut}(U')$.

If $f$ is a conformal mapping from $U$ onto $U'$, the effect of $f$ on harmonic functions has been analyzed in Section 7.11. It is known that the transformation $u \to u \circ f$ gives a bijection between harmonic functions on $U'$ and harmonic functions on $U$. Indeed, the converse also holds in the sense that $u \to u \circ f$ is a bijection between the spaces of harmonic functions only when $f$ is either holomorphic or antiholomorphic (Theorem 7.67). This fact suggests that conformal mappings might be useful in the solution of Dirichlet's problem, since it may be convenient to change the domain by means of a conformal mapping, before seeking for the appropriate harmonic function.

Recall that if $U$, $U'$ are two domains and $f: U \to U'$ is a conformal mapping, the relationship between Dirichlet's problem in $U$ and the corresponding one in $U'$ is the following (Section 7.11):

Suppose that one knows how to solve Dirichlet's problem in the domain $U'$ and wants to solve it in the domain $U$. If $\varphi \in C(\partial U)$ is the data at the boundary of $U$, define $\psi = \varphi \circ f^{-1}$. Now take $v$ to be the solution of Dirichlet's problem in $U'$ with data $\psi$, that is, $\Delta v = 0$ on $U'$ and $v\big|_{\partial U'} = \psi$. Then $u = v \circ f$ is the solution of the Dirichlet problem in $U$ (Figure 8.1).

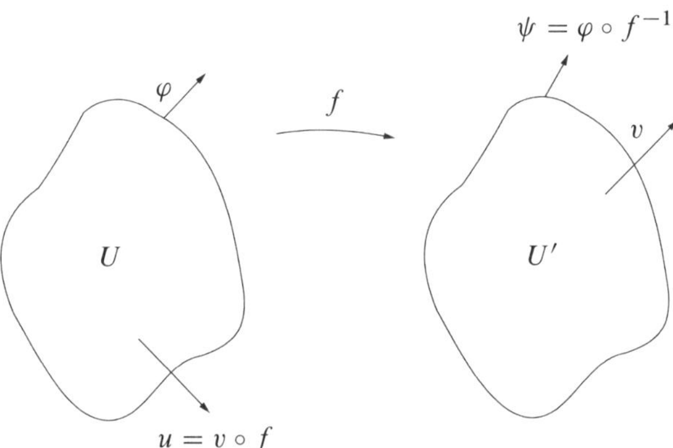


Figure 8.1

Remark that $f$ must be a bijective and bicontinuous transformation (homeomorphism) between boundaries of $U$ and $U'$. Therefore, it is very important to know what is the behavior at the boundary of a conformal mapping. In the case of a simply connected domain $U \neq \mathbb{C}$ it is known, by Riemann's theorem, that there are conformal mappings $f : U \to \mathbb{D}$, where $\mathbb{D}$ is the unit disc. The question is: when does $f$ extend to a homeomorphism $\tilde{f} : \bar{U} \to \bar{\mathbb{D}}$?

Of course, if this is possible, then $\tilde{f}$ will be also a homeomorphism from $\partial U$ onto $\partial \mathbb{D}$ and, therefore, $\partial U$ must be a closed Jordan curve that, by definition, is a space homeomorphic to $\partial \mathbb{D}$. What matters here is the fact that this condition on $\partial U$ is enough to guarantee existence of the extension $\tilde{f}$. Indeed, the following statement holds.

**Theorem 8.7** (Carathéodory). *If $U$ is a simply connected domain, $U \neq \mathbb{C}$, such that $\partial U$ is a closed Jordan curve, then any conformal mapping $f$ from $U$ onto the unit disc $\mathbb{D}$ extends to a homeomorphism from $\bar{U}$ onto $\bar{\mathbb{D}}$.*

**Corollary 8.8.** *A conformal mapping between two simply connected domains bounded by Jordan curves extends to a homeomorphism between the closures of the domains.*

The solution of Dirichlet's problem in the unit disc given in the previous chapter (Subsection 7.6.1) allows us, then, to solve Dirichlet's problem in a simply connected domain bounded by a Jordan curve whenever one is able to map explicitly this domain onto the unit disc. Later on some example will be given.

The proof of Theorem 8.7 when $\partial U$ is an arbitrary Jordan curve falls beyond the scope of this text and will not be given. It may be found in [9], p. 324. However, the proof will be done in the particular case that $\partial U$ consists of a finite number of analytic arcs. This is, essentially, the situation of Proposition 1.35, but with the function $\varphi$ analytic, that is, holomorphic. More precisely, an arc $\gamma(t), a \leq t \leq b$, is an *analytic arc* in the boundary of the domain $U$ if there exists a simply connected domain $\Delta$, symmetric with respect to the interval $(a, b)$, and a function $\varphi$ that maps conformally $\Delta$ onto an open set $\tilde{\Delta} = \varphi(\Delta)$ such that

$$\varphi(\Delta \cap \{z : \operatorname{Im} z > 0\}) = \tilde{\Delta} \cap U,$$
$$\varphi(t) = \gamma(t), \quad t \in (a, b), \tag{8.1}$$
$$\varphi(\Delta \cap \{z : \operatorname{Im} < 0\}) = \tilde{\Delta} \cap U^c.$$

First let us note a property of conformal mappings that is purely topological and so holds for homeomorphisms as well. Let $f : U \to U'$ be a continuous mapping from a domain $U$ on a domain $U'$. One says that $f(z)$ *tends to the boundary of $U'$ when $z$ tends to the boundary of $U$* if for any compact set $K' \subset U'$ there exists a compact set $K \subset U$ such that if $z \notin K$, then $f(z) \notin K'$.

**Proposition 8.9.** *If $f : U \to U'$ is a homeomorphism, then $f(z)$ tends to the boundary of $U'$ when $z$ tends to the boundary of $U$.*

*Proof.* Since $f^{-1}$ is continuous, given a compact set $K' \subset U'$ just take $K = f^{-1}(K')$ to have a compact set of $U$ satisfying the required condition. $\qquad \square$

This means that for a sequence $(z_n)$ of points in $U$ converging to a point of $\partial U$, the sequence $(f(z_n))$ gets close to $\partial U'$, but one cannot say that it tends to a point of $\partial U'$. The same happens with an arc of curve $\gamma(t) \in U$ which tends to a point of $\partial U$.

We will also need a *symmetry principle for holomorphic functions*, which is worth being stressed.

**Proposition 8.10.** *Let $U$ be a domain of the plane, symmetric with respect to the real axis, $f$ a holomorphic function on $U^+ = U \cap \{z : \mathrm{Im} > 0\}$ and suppose that the function $u = \mathrm{Im}\, f$ is continuous on $\overline{U^+}$ and $u = 0$ in $U \cap \mathbb{R}$. Then $f$ has a holomorphic extension to $U$ that satisfies $f(\bar{z}) = \overline{f(z)}$.*

*Proof.* Consider a disc $\Delta$ centered at a point of $U \cap \mathbb{R}$. It is known by the symmetry principle for harmonic functions (Proposition 7.57) that $u$ has a harmonic extension to $\Delta$ satisfying $u(\bar{z}) = -u(z)$. The function $u$ has in $\Delta$ a conjugated harmonic function which will be called $-u_0$ and one may suppose that $u_0 = \mathrm{Re}\, f$ in $\Delta^+$, that is, $f(z) = u_0(z) + iu(z)$, $z \in \Delta^+$. Now the function $v(z) = u_0(z) - u_0(\bar{z})$ satisfies $\frac{\partial v}{\partial x} = 0$ and $\frac{\partial v}{\partial y} = 2\frac{\partial u_0}{\partial y} = -2\frac{\partial u}{\partial x} = 0$ on the real axis. Therefore, the holomorphic function $\frac{\partial v}{\partial x} - i\frac{\partial v}{\partial y}$ vanishes on an interval and, consequently, is identically zero.

Hence, $v$ is constant and it is clear that $v = 0$, that is, $u_0(z) = u_0(\bar{z})$, which gives $f(\bar{z}) = \overline{f(z)}$ defining $f(z) = u_0(z) + iu(z)$ in $\Delta$. Now it is sufficient to repeat the construction on arbitrary discs and observe that $u_0$ must be the same in two intersecting discs. $\qquad\square$

**Remark 8.1.** The statement of Proposition 8.10 is very similar to the symmetry principle given after Theorem 4.27 (Remark 4.2). The difference is that there it was assumed that the given function, holomorphic on $U^+$, had a limit at points on the real axis, while now this hypothesis relates only to its imaginary part.

**Theorem 8.7′.** *Let $U$ be a simply connected domain whose boundary is a piecewise analytic closed Jordan curve, that is, formed by a finite number of analytic arcs, and let $f$ be a conformal mapping from $U$ onto the unit disc $\mathbb{D}$. Then $f$ can be extended to a homeomorphism from $\overline{U}$ onto $\overline{\mathbb{D}}$.*

*Proof.* Suppose that the boundary of $U$ is formed by the analytic arcs $\gamma_1, \gamma_2, \ldots, \gamma_n$ and let $z_1, z_2, \ldots, z_n$ be their vertices; that is, $z_k$ is the final point of $\gamma_{k-1}$ and the starting point of $\gamma_k$, $k = 2, \ldots, n$ and $z_1$ joins $\gamma_n$ with $\gamma_1$. Suppose also that $a_0 \in U$ is such that $f(a_0) = 0$ (Figure 8.2).

The proof is done in several steps.

A) The function $f$ can be extended continuously to the interior of each arc $\gamma_k$, that is, to $\gamma_k$ without its end points.

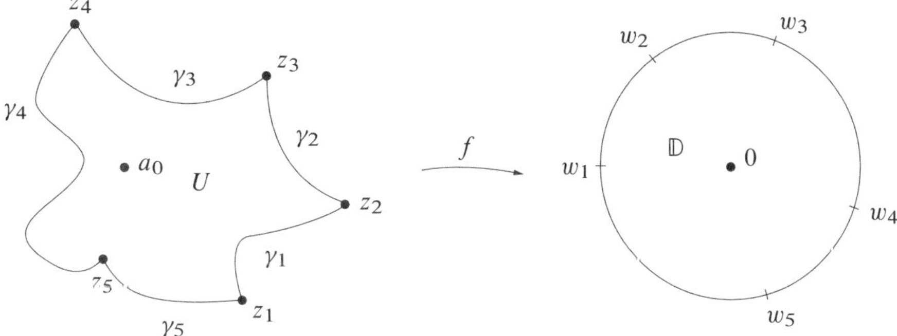

Figure 8.2

In order to prove this assertion take $\Delta$ and $\varphi$ like in (8.1), corresponding to the arc $\gamma_k$. If $\Delta$ is small enough such that $a_0 \notin \varphi(\Delta)$, the function $\log f(\varphi(z))$ has a holomorphic branch on $\Delta \cap \{z : \operatorname{Im} z > 0\}$. Its real part, $\operatorname{Log} |f(\varphi(z))|$ tends to 0 when $\operatorname{Im} z \to 0$, by Proposition 8.9. Therefore, by Proposition 8.10, $\log f(\varphi(z))$ has a holomorphic extension to $\Delta$, as well as the function $f(\varphi(z))$. For each $t_0 \in (a, b)$ one has $\varphi'(t_0) \neq 0$, so that $\varphi$ has a holomorphic inverse on a neighborhood of $\varphi(t_0)$. This means that $f$ also extends analytically around $\varphi(t_0)$.

B) The extension of $f$ is one-to-one on the union of interiors of the arcs $\gamma_1, \ldots, \gamma_n$.

Actually, let $z_0, z_0'$ be two different interior points of some arc of the boundary of $U$ and assume that $f(z_0) = f(z_0')$. Let $\Delta$ be a disc of center $z_0$ not containing $z_0'$ where the extension of $f$ is defined, and write $w_0 = f(z_0) = f(z_0')$. The image $f(\Delta)$ is an open set that contains $w_0$. Moreover if $z \in U$, $z \notin \Delta$, is a point close enough to $z_0'$, one will have $f(z) \in f(\Delta) \cap \mathbb{D}$, because $f$ is continuous at the point $z_0'$. But then there would be a point $z_1 \in \Delta \cap U$ with $f(z_1) = f(z)$, which is not possible, $f$ being one-to-one on $U$.

C) The function $f$ extends continuously to the vertices $z_1, \ldots, z_n$ and, writing $w_k = f(z_k)$, the images of the arcs $\gamma_{k-1}$ and $\gamma_k$ are two arcs of the unit circle which have $w_k$ as the only common point (Figure 8.3).

The image of the interior of each arc $\gamma_k$, given by the extension of $f$, is an open arc of $\partial \mathbb{D}$. Let $I_k = (a_k, b_k)$, $k = 1, 2, \ldots, n$, be the image of the interior of $\gamma_k$. Let us prove that if $\gamma_{k-1}$ and $\gamma_k$ meet at the vertex $z_k$, then $I_{k-1}$ and $I_k$ have a common end. Actually, otherwise there would be an arc between $I_{k-1}$ and $I_k$ which one may assume to be $(b_{k-1}, a_k)$. For each $\varepsilon > 0$, consider the part of the circle of center $z_k$ and radius $\varepsilon$ that is inside $U$ and let $C_\varepsilon$ be its image by $f$ (Figure 8.4). It is a Jordan arc that goes from a point of $I_{k-1}$ to a point of $I_k$.

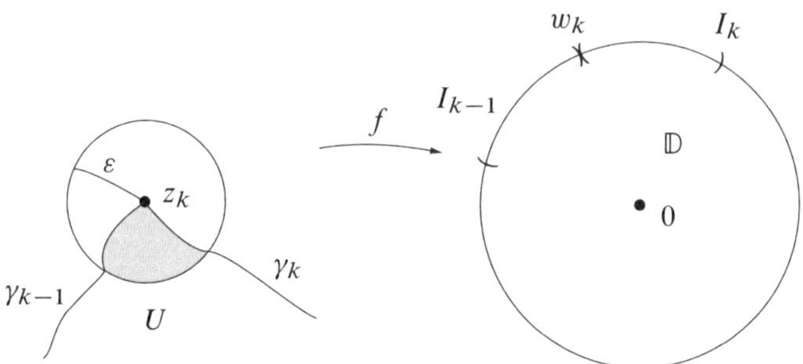



Figure 8.3

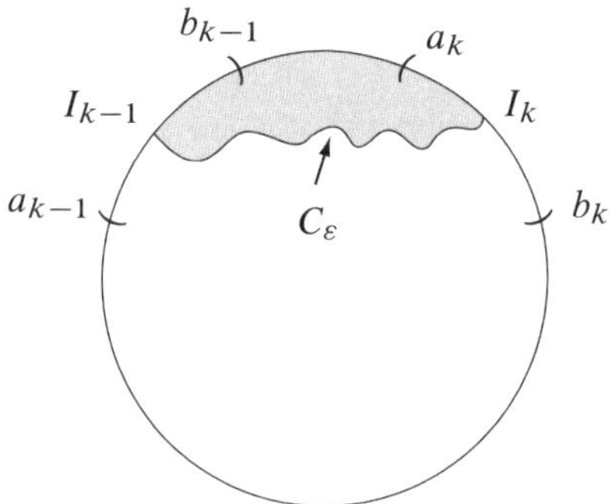



Figure 8.4

If $\varepsilon \to 0$, $C_\varepsilon$ tends to $\partial \mathbb{D}$ by Proposition 8.9, and the end points of $C_\varepsilon$ converge to $b_{k-1}$ and $a_k$, respectively. Consider now the inverse function of $f$, $f^{-1} \colon \mathbb{D} \to U$ and remark that, being injective, if two points of $\mathbb{D}$ are not separated by $C_\varepsilon$, their images on $U$ are not separated as well by the circle of center $z_k$ and radius $\varepsilon$. We conclude now that for a point $w \in \mathbb{D}$ tending to the arc $(b_{k-1}, a_k)$, $f^{-1}(w)$ converges to $z_k$ and, therefore, the function $f^{-1}(w) - z_k$ tends to $0$ when $w$ approaches $(b_{k-1}, a_k)$. Applying again Proposition 8.10, this yields that $f^{-1}(w) - z_k$ extends analytically with the value $0$ through $(b_{k-1}, a_k)$, and the principle of analytic continuation tells us that this function vanishes on an open set of $\mathbb{D}$ and, consequently, in the whole of $\mathbb{D}$. This is a contradiction, because $f^{-1}$ is not constant.

Hence, we must have $b_{k-1} = a_k$ and the argument shows that $f$ extends continuously to $z_k$ with $f(z_k) = w_k = b_{k-1} = a_k$.

D) Finally, what has been proved also shows that the images by $f$ of the arcs $\gamma_1, \ldots, \gamma_n$, now including the vertices, must cover the whole unit circle, and this finishes the proof of the theorem. $\qquad \square$

It is worth noting that the proof just given shows that the mapping $f$ satisfies a stronger condition than the one in Theorem 8.7′. Indeed, it proves that $f$ extends *analytically* through each open analytic arc which is a part of the boundary of $U$. This implies that $f$ is conformal at the points of $\partial U$, in the sense that it preserves the angle between two curves that meet at a point of $\partial U$ and their images. In particular, $f$ preserves orthogonality, that is, a curve passing through $z_0$ orthogonal to $\partial U$ becomes a curve that intersects the unit circle orthogonally (Figure 8.5).

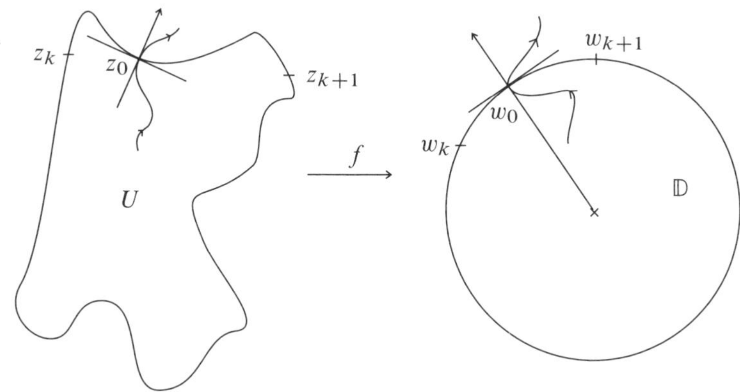

Figure 8.5

In general, if $U$ is bounded by an arbitrary Jordan curve, there is no preservation of angles at the boundary of $U$ (indeed, it makes no sense to talk about orthogonality with respect to the boundary); but if $\partial U$ is regular enough, there is conformality at the boundary. A sufficient condition is $\partial U$ being a Jordan curve of class $C^2$ (or even of class $C^{1+\varepsilon}$ with $\varepsilon > 0$). With these conditions the derivative of the conformal mapping extends continuously to the boundary with non-vanishing values, and then there is conformality at the points of $\partial U$. In particular, a curve normal to $\partial \mathbb{D}$ is mapped into a curve normal to $\partial U$. This property will be used later to solve Neumann's problem for a particular domain.

## 8.3 Homographic transformations

### 8.3.1 The linear group

Among all analytic functions, rational functions of first degree, also called *homographic transformations* or *linear fractional transformations* or even *linear transformations*, are of special interest. Indeed they are conformal mappings of the whole Riemann sphere onto itself and, in addition, they have very simple geometrical properties. It is worth studying them in detail.

A homographic transformation is a rational function of the form

$$Tz = T(z) = \frac{az+b}{cz+d} \quad \text{with } ad - bc \neq 0, \; a,b,c,d \in \mathbb{C}. \tag{8.2}$$

The condition $ad - bc \neq 0$ avoids $T$ being constant and $T$ is a holomorphic function on the whole plane, except at the point $z = -d/c$, if $c \neq 0$.

A homographic transformation $T$ is said to be *real* when it transforms the real axis on itself. It is clear that this happens if and only if the coefficients $a$, $b$, $c$, $d$ are all four real.

It is convenient to look at $T$ as a transformation from the Riemann sphere $S^2$ into itself, defining $T(-d/c) = \infty$ and $T(\infty) = a/c$. A homographic transformation is a bijective mapping from $S^2$ onto $S^2$, since if $T$ is defined by (8.2), then it has an inverse which is also homographic:

$$T^{-1}(w) = \frac{dw - b}{-cw + a}.$$

It is easy to check that the composition of two homographic transformations is another one. Indeed, associating to $T$ the matrix $\left(\begin{smallmatrix} a & b \\ c & d \end{smallmatrix}\right)$ of its coefficients and to another homographic transformation $S$ the corresponding matrix $\left(\begin{smallmatrix} a' & b' \\ c' & d' \end{smallmatrix}\right)$, then it yields that $S \circ T$ has associated the product matrix

$$\begin{pmatrix} a' & b' \\ c' & d' \end{pmatrix} \begin{pmatrix} a & b \\ c & d \end{pmatrix}.$$

Hence, homographic transformations form a group of transformations of $S^2$, called the *linear group*.

For example, the homographic transformations with matrices

$$\begin{pmatrix} 1 & b \\ 0 & 1 \end{pmatrix}, \quad \begin{pmatrix} a & 0 \\ 0 & 1 \end{pmatrix} \quad \text{and} \quad \begin{pmatrix} 0 & 1 \\ 1 & 0 \end{pmatrix}$$

are, respectively, the *translation* $z \to z + b$, the *dilation* $z \to az$ and the *inversion* $z \to 1/z$.

It is an interesting fact that every homographic transformation is the composition of these three kinds of transformations. Actually, if $c \neq 0$, we can write

$$Tz = \frac{az+b}{cz+d} = \frac{bc - ad}{c^2(z + d/c)} + \frac{a}{c}.$$

When $c = 0$, then $Tz = \frac{a}{d}z + \frac{b}{d}$, which is a dilation followed by a translation.

Observe that the homographic transformation $T$ is a conformal mapping of the open set $\mathbb{C} \setminus \{-d/c\}$ onto the open set $\mathbb{C} \setminus \{a/c\}$ if $c \neq 0$. If $c = 0$, it is a conformal mapping from $\mathbb{C}$ onto $\mathbb{C}$. Therefore, $T$ preserves the angles of any pair of curves which meet inside $\mathbb{C} \setminus \{-d/c\}$.

Consider now the set of straight lines and circles of the complex plane. If we look at them in the Riemann sphere, we can consider all of them as circles, the lines being circles passing through $\infty$. Recall from Subsection 1.2.1 that a straight line with direction vector $\alpha \in \mathbb{C}$ (that is, it is perpendicular to $\alpha$) satisfies the equation $\bar{\alpha} z + \alpha \bar{z} + m = 0$, with $m \in \mathbb{R}$. For a circle of center $\alpha \in \mathbb{C}$, its equation is

$$|z|^2 + \bar{\alpha} z + \alpha \bar{z} + m = 0 \tag{8.3}$$

with $m \in \mathbb{R}$ and $|\alpha|^2 - m > 0$.

**Proposition 8.11.** *Every homographic transformation sends the set of straight lines and circles into itself.*

*Proof.* It is obvious, geometrically or analytically, that a translation or a dilation transforms a line into a line, and a circle into a circle. An inversion, $Tz = 1/z$, transforms the line $\bar{\alpha} z + \alpha \bar{z} + m = 0$, $m \neq 0$ into the circle through the origin with equation $|z|^2 + \frac{\alpha}{m} z + \frac{\bar{\alpha}}{m} \bar{z} = 0$. If $m = 0$, the line passes through the origin and is transformed into the line $\alpha z + \bar{\alpha}\, \bar{z} = 0$. The circle (8.3) with $m \neq 0$ is mapped by $T$ onto the circle with equation $|z|^2 + \frac{\alpha}{m} z + \frac{\bar{\alpha}}{m} \bar{z} + \frac{1}{m} = 0$. If $m = 0$, the circle (8.3) contains the origin and its image by $T$ is the line $\alpha z + \bar{\alpha}\, \bar{z} + 1 = 0$. $\square$

What has been proved justifies not making differences between straight lines and circles. From now on, the term *circle* will be used to denote either a straight line or a circle. Hence, it is said that homographic transformations send circles into circles.

The remaining question is if, given two arbitrary circles, there is always a homographic transformation that takes one to the other. The affirmative answer is a consequence of the following result.

**Proposition 8.12.** *Given three different points $z_1$, $z_2$, $z_3$ of $S^2$ and three more points, also in $S^2$ and pairwise different, $w_1$, $w_2$, $w_3$, there exists a unique homographic transformation $T$ such that $T(z_i) = w_i$, $i = 1, 2, 3$.*

*Proof.* Being aware of the structure of the linear group, it suffices to see there exists a homographic transformation $T$ which sends $z_1, z_2, z_3$ to $0, 1, \infty$, respectively, and that a homographic transformation leaving the points $0, 1, \infty$ fixed is the identity.

In order to check the first assertion take

$$Tz = \frac{z - z_1}{z - z_3} \cdot \frac{z_2 - z_3}{z_2 - z_1}$$

if $z_1, z_2, z_3 \neq \infty$. If, say, $z_1 = \infty$, then take

$$Tz = \frac{z_2 - z_3}{z - z_3}.$$

Finally, if $Tz = \frac{az+b}{cz+d}$ fixes $0, 1, \infty$, one has necessarily $Tz = \frac{a}{d}z + \frac{c}{d}$, because $T(\infty) = \infty$. Now $T(0) = 0$ gives $Tz = \frac{a}{d}z$ and $Tz = z$, since $T(1) = 1$.    $\square$

Since every circle is determined by three different points, one obtains the following corollary.

**Corollary 8.13.** *Given two circles there always exists a homographic transformation passing from one to the other.*

One must observe that this homographic transformation is not unique; indeed, one can obtain infinitely many of them by taking groups of three different points on each circle.

**Example 8.14.** Given two circles $C_1$, $C_2$ and two points $z_1 \notin C_1$ and $z_2 \notin C_2$, there is a homographic transformation that sends $C_1$ into $C_2$ and $z_1$ into $z_2$. Indeed, one may pass, by two homographic transformations, $C_1$ and $C_2$ to the real axis and then it is enough to find a homographic transformation that fixes the real axis and sending $z_1'$ into $z_2'$, where $z_1'$, $z_2'$ are two points not on the real axis.

If $z_1'$, $z_2'$ are on a line parallel to the real axis, the translation $w = z + (z_2' - z_1')$ works. Otherwise the line joining $z_1'$ and $z_2'$ intersects the real axis at a point $z_0$ and one takes the transformation $w = z_0 + \frac{z_2' - z_0}{z_1' - z_0}(z - z_0)$, which is either a dilation with center $z_0$, or a dilation with center $z_0$ followed by a rotation of angle $\pi$, according to the position of $z_1'$ and $z_2'$ with respect to the real axis.    $\square$

**Remark 8.2.** In the proof of Proposition 8.12 it has been seen that a homographic transformation $T$ which satisfies $T(0) = 0$, $T(1) = 1$, $T(\infty) = \infty$ must be the identity. More generally one may assert that a homographic transformation with three fixed points is the identity. In other words, if $Tz = \frac{az+b}{cz+d}$ is not the identity, then $T$ has at most two fixed points. To see this, note that the equality $Tz = z$ is equivalent to

$$cz^2 + (d - a)z - b = 0,$$

a second-degree equation that has at most two different solutions.

**Example 8.15.** We look first for the homographic transformation $Tz = \frac{az+b}{cz+d}$ that sends $0, i, \infty$ into $-1, 0, 1$, respectively. Taking $a = 1$ and considering that $T(i) = 0$, we get $Tz = \frac{z-i}{cz+d}$. Now $T(\infty) = 1$ gives $c = 1$ and $T(0) = -1$ gives $d = i$. It turns out, then, that

$$Tz = \frac{z - i}{z + i}.$$

Look now for the image of both axes by $T$. The imaginary axis contains $0, i, \infty$, and, therefore, its image is a circle through $-1, 0, 1$. That is, it is the real axis. The real axis contains $0, \infty$. Its image will be a circle through $-1$ and $1$ perpendicular to the image of imaginary axis, that is, to the real axis. Therefore, it is the unit circle.

Consider now the upper half plane $\Pi^+ = \{z: \operatorname{Im} z > 0\}$; its image cannot meet the unit circle and so it is either the unit disc or its exterior. Since $T(i) = 0$, it must be the unit disc.     $\square$

**Example 8.16.** Consider the inverse mapping of the homographic transformation $T$ in Example 8.15. If

$$w = Tz = \frac{z - i}{z + i},$$

it turns out that $Sw = T^{-1}w = i\frac{1+w}{1-w}$ applies the unit disc conformally into the upper half plane $\Pi^+ : \{z: \operatorname{Im} z > 0\}$. Now we can use the transformation $S$ to solve Dirichlet's problem in $\Pi^+$, with boundary values given by $\varphi(x) = \frac{1}{1+x^2}$, $x \in \mathbb{R} = \partial\Pi^+$ (Figure 8.6).

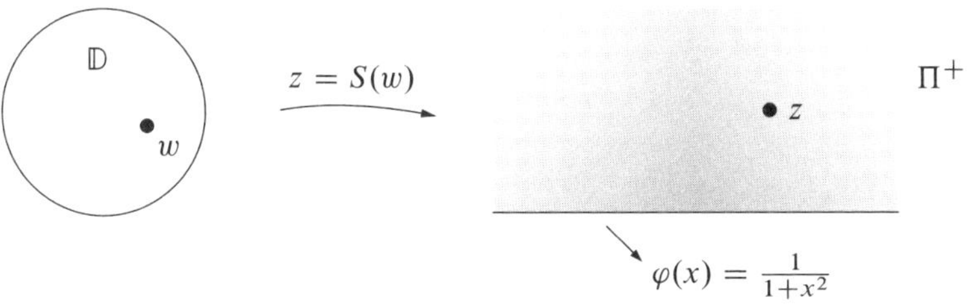


Figure 8.6

Start by transporting the function $\varphi$ to the boundary of the disc. If $w = e^{i\theta} \in \partial\mathbb{D}$, one has

$$\psi(e^{i\theta}) = \varphi(S(e^{i\theta})) = \frac{1}{1 - \left(\frac{1+e^{i\theta}}{1-e^{i\theta}}\right)^2}$$

$$= \frac{2 - e^{i\theta} - e^{-i\theta}}{4} = \frac{1}{2} - \frac{1}{2}\cos\theta.$$

Now we have to solve Dirichlet's problem in the disc with boundary values $\psi(e^{i\theta})$. Extend the function $\psi$ to the whole disc $\mathbb{D}$ defining

$$u(w) = u(re^{i\theta}) = \frac{1}{2} - \frac{1}{2}r\cos\theta = \frac{1}{2} - \frac{1}{2}\sigma$$

if $w = re^{i\theta} = \sigma + i\tau, 0 < r \leq 1, 0 \leq \theta \leq 2\pi$. This function clearly satisfies the equation $\Delta u = 0$ and, moreover, $u(e^{i\theta}) = \psi(e^{i\theta})$.

Now it is enough to transport the function $u$ to the upper half plane $\Pi^+$, writing

$$v(z) = u(T(z)) = \frac{1}{2} - \frac{1}{2} \operatorname{Re}(T(z)) = \frac{1}{2} - \frac{1}{2} \operatorname{Re} \frac{z-i}{z+i}$$

$$= \frac{y+1}{x^2 + y^2 + 2y + 1} \qquad \text{if } z = x + iy,$$

so that $v$ is harmonic on $\Pi^+$ and $v(x, 0) = \frac{1}{1+x^2} = \varphi(x)$. $\qquad\square$

### 8.3.2 The cross ratio

Consider four points of the Riemann sphere $z_1, z_2, z_3, z_4$ with the condition that $z_2$, $z_3$, $z_4$ are different. It is known that there is a unique homographic transformation $T$ such that $T(z_2) = 1$, $T(z_3) = 0$, $T(z_4) = \infty$. The *cross ratio* of the given four points, $(z_1, z_2, z_3, z_4)$, is defined as

$$(z_1, z_2, z_3, z_4) = T z_1.$$

Hence, if all the points are finite, we have

$$(z_1, z_2, z_3, z_4) = \frac{z_1 - z_3}{z_1 - z_4} \cdot \frac{z_2 - z_4}{z_2 - z_3}.$$

If, for example, $z_3 = \infty$, then one has

$$(z_1, z_2, \infty, z_4) = \frac{z_2 - z_4}{z_1 - z_4}.$$

The cross ratio of four points is invariant under homographic transformations.

**Proposition 8.17.** *If $z_1$, $z_2$, $z_3$, $z_4$ are different points of $S^2$ and $T$ is a homographic transformation, then one has*

$$(z_1, z_2, z_3, z_4) = (T z_1, T z_2, T z_3, T z_4).$$

*Proof.* Writing $Sz = (z, z_2, z_3, z_4)$, $S$ is the homographic transformation satisfying $Sz_2 = 1$, $Sz_3 = 0$, $Sz_4 = \infty$ and $Sz_1 = (z_1, z_2, z_3, z_4)$. Therefore, $ST^{-1}$ transforms $T z_2, T z_3, T z_4$ into $1, 0, \infty$, and this means that

$$(T z_1, T z_2, T z_3, T z_4) = ST^{-1}(T z_1) = (z_1, z_2, z_3, z_4). \qquad\square$$

As an application, if $z_1$, $z_2$, $z_3$ are three different points, the homographic transformation $T$ sending them into $w_1$, $w_2$, $w_3$, also different, is given by

$$(z, z_1, z_2, z_3) = (Tz, w_1, w_2, w_3).$$

It is obvious that if $z_1$, $z_2$, $z_3$, $z_4$ are on the real axis, then $(z_1, z_2, z_3, z_4)$ is a real number, and that if three of the previous points are real and $(z_1, z_2, z_3, z_4) \in \mathbb{R}$, then the forth point must be real as well.

Combining this observation with Proposition 8.11, the following result is obtained.

**Proposition 8.18.** *The cross ratio of four points $(z_1, z_2, z_3, z_4)$ is real if and only if they are located on a circle.*

### 8.3.3 Symmetry

The concept of symmetric points with respect to an axis is quite clear: $z$, $z^*$ are symmetric with respect to the line $L$ if $L$ is perpendicular to the segment $[z, z^*]$ at its middle point.

Now we consider the symmetry of a pair of points with respect to an arbitrary circle.

**Definition 8.19.** Two points $z$, $z^*$ are said to be symmetric with respect to a circle $C$ if and only if there exists a homographic transformation $T$ which sends $C$ to the real axis such that $Tz^* = \overline{Tz}$.

This definition does not depend on $T$, because if $S$ satisfies the same requirements than $T$, then $ST^{-1}$ is a real homographic transformation and $Sz^* = ST^{-1}(Tz^*)$ will be also the conjugate of $Sz = ST^{-1}(Tz)$, because it is obvious that real homographic transformations preserve conjugation.

The previous definition and Example 8.14 yields:

**Proposition 8.20.** *The points $z$, $z^*$ are symmetric with respect to the circle passing through the points $z_1$, $z_2$, $z_3$ if and only if*

$$(z^*, z_1, z_2, z_3) = \overline{(z, z_1, z_2, z_3)}. \tag{8.4}$$

So, given a circle $C$, for each point $z$ there is only one point $z^*$, symmetric to $z$ with respect to $C$. The mapping $z \rightarrow z^*$ is a bijection called *symmetry with respect to $C$*.

The homographic transformations preserve symmetry, as stated next.

**Proposition 8.21** (Symmetry principle). *If a homographic transformation $T$ sends a circle $C$ onto a circle $C'$, then it transforms each pair of symmetric points with respect to $C$ into a pair of symmetric points with respect to $C'$.*

*Proof.* If $C$ or $C'$ is the real axis, the symmetry principle is the definition of symmetric points. Otherwise take a homographic transformation $S$ sending $C$ into the real axis and then apply the previous case to $S$ and to $TS^{-1}$.    $\square$

Next, geometrical properties of symmetry with respect to $C$, $z \to z^*$, will be studied.

If $C$ is a straight line, one may take $z_3 = \infty$ and (8.4) becomes

$$\frac{z^* - z_2}{z_1 - z_2} = \frac{\bar{z} - \bar{z}_2}{\bar{z}_1 - \bar{z}_2},$$

which gives $|z^* - z_2| = |z - z_2|$. This means that $z$ and $z^*$ are equidistant from any point $z_2 \in C$, a condition that holds only if $C$ is the perpendicular bisector of the segment $[z, z^*]$.

If $C$ is a circle of the plane with center $a$ and radius $r$, then it is easy to see that

$$z^* = a + \frac{r^2}{\bar{z} - \bar{a}}. \tag{8.5}$$

Actually, if $z_1, z_2, z_3 \in C$, then $(z_j - a)(\bar{z}_j - \bar{a}) = r^2$, $j = 1, 2, 3$, holds and applying repeatedly Proposition 8.17, it follows that

$$\overline{(z, z_1, z_2, z_3)} = \overline{(z - a, z_1 - a, z_2 - a, z_3 - a)}$$

$$= \left( \overline{z - a}, \frac{r^2}{z_1 - a}, \frac{r^2}{z_2 - a}, \frac{r^2}{z_3 - a} \right)$$

$$= \left( \frac{r^2}{\bar{z} - \bar{a}}, z_1 - a, z_2 - a, z_3 - a \right) = \left( \frac{r^2}{\bar{z} - \bar{a}} + a, z_1, z_2, z_3 \right),$$

which, according to (8.4), gives (8.5).

From equality (8.5) written in the form

$$(z^* - a)(\bar{z} - \bar{a}) = r^2$$

one gets, $|z^* - a||z - a| = r^2$, and also that the quotient $\frac{z^* - a}{z - a}$ is a positive number.

Interpreting geometrically these results, it turns out that the points $a$, $z$ and $z^*$ are aligned and that the product of distances from $z$ to $a$ and from $z^*$ to $a$ is $r^2$. This means that symmetry with respect to $C$ is the *inversion* of center $a$ and power $r^2$.

The notion of symmetry with respect to a circle allows us to extend the reflection principle with respect to the real axis to any circle. If $C$ is a circle, a domain $U$ is said to be *symmetric with respect to* $C$ if $z \in U$ implies $z^* \in U$, where $z^*$ is the symmetric point of $z$ with respect to $C$. Using homographic transformations to carry a circle into the real axis and being aware of Proposition 8.21 and Remark 4.2, one obtains the following result.

**Proposition 8.22** (Reflection principle with respect to circles). *Let $U$ be a domain symmetric with respect to a circle $C$ and let $U^+$ be one of the regions in which $C$ divides $U$. Let $f$ be a holomorphic function on $U^+$ and let there exist a circle $C'$ such that $f(z)$ tends to a point of $C'$ when $z$ tends to a point of $C$. Then $f$ has a holomorphic extension to $U$ that carries symmetric points with respect to $C$ into symmetric points with respect to $C'$.*

## 8.4 Automorphisms of simply connected domains

According to Riemann's theorem, there are essentially two kinds of simply connected domains: the whole plane $\mathbb{C}$ and the unit disc $\mathbb{D}$. It is interesting to know the automorphism group of these two domains; then the automorphism group of any simply connected domain $U$ will be obtained, whenever a conformal mapping from $U$ onto $\mathbb{D}$ is known.

Let us start with automorphisms of $\mathbb{C}$. If one considers a homographic transformation $Tz = \frac{az+b}{cz+d}$ and wants it to be holomorphic on $\mathbb{C}$, it must necessarily be $c = 0$, that is, $T$ is linear in $z$ and not constant. These transformations form an automorphism group of $\mathbb{C}$. It turns out, however, that they are *all* the automorphisms of $\mathbb{C}$.

**Theorem 8.23.** *The automorphism group of $\mathbb{C}$ consists of linear transformations $Tz = az + b$ with $a, b \in \mathbb{C}$, $a \neq 0$.*

*Proof.* We need to show that $f : \mathbb{C} \to \mathbb{C}$ bijective and entire, implies $f$ linear in $z$. There are only two possibilities (see Example 5.22): either $f$ is a polynomial, or the point $\infty$ is an essential singularity for $f$.

The latter case is not possible. If it were so, the image by $f$ of the punctured neighborhood of $\infty$, $\{z \in \mathbb{C} : |z| > 1\}$ should be dense in $\mathbb{C}$, according to Casorati–Weierstrass's theorem (Theorem 5.6). But this is impossible because $f$ being injective, $f\{z : |z| > 1\}$ does not intersect the open set $f\{z : |z| < 1\}$ and so it cannot be a dense set.

If $f$ is a polynomial, it must be of first degree, because otherwise $f$ would not be bijective and, hence, $f(z) = az + b$ with $a, b \in \mathbb{C}$.                           $\square$

One may also find the automorphisms of the extended plane. Observe, first, that a homographic transformation $Tz = \frac{az+b}{cz+d}$, $ad - bc \neq 0$, is a bijective mapping $T : S^2 \to S^2$, holomorphic at every point of $S^2$. Actually, $T$ is holomorphic at every finite point $z \neq -d/c$, and since

$$T(1/z) = \frac{a + bz}{c + dz},$$

it follows that $T$ is also holomorphic at infinity, whenever $c \neq 0$.

In case $c = 0$, or in order to see that $T$ is holomorphic at the point $-d/c$ with value $\infty$, just remark that a function $f$ with $f(z_0) = \infty$ is holomorphic at $z_0$ when $g(z) = 1/f(z)$, with $g(z_0) = 0$, is holomorphic at $z_0$.

The interesting fact is that homographic transformations are *all* the automorphisms of $S^2$. In order to show this it is convenient to make some general considerations on transformation groups of a domain.

If $U$ is an open set of $S^2$ and $G \subset \mathrm{Aut}(U)$ is an automorphism group of $U$, it will be said that $G$ is *transitive on $U$* if for all pair of points $z, z' \in U$ there exists a transformation $\varphi \in G$ such that $\varphi(z) = z'$.

If $z \in U$, the *isotropy group* of $z$, $A_z$, is the set of automorphisms of $U$ that leave $z$ fixed, that is, $A_z = \{\varphi \in \mathrm{Aut}(U): \varphi(z) = z\}$. The following result will be useful.

**Lemma 8.24.** *Let $G \subset \mathrm{Aut}(U)$ be an automorphism group of the open set $U$ of $S^2$. Suppose that the following conditions hold:*

a) *$G$ is transitive on $U$.*

b) *There exists at least one point $z_0 \in U$ such that $A_{z_0} \subset G$.*

*Then $G = \mathrm{Aut}(U)$.*

*Proof.* Let $\varphi \in \mathrm{Aut}(U)$. Since $G$ is transitive, there is a transformation $\psi \in G$ so that $\psi(z_0) = \varphi(z_0)$. Since $(\psi^{-1} \circ \varphi)(z_0) = z_0$, it turns out that $\psi^{-1} \circ \varphi \in A_{z_0} \subset G$, and this implies that $\varphi \in G$ as well. $\qquad\square$

**Theorem 8.25.** *The automorphism group of $S^2$ coincides with the homographic transformations group.*

*Proof.* Apply the previous lemma, taking as $G$ the group of homographic transformations. It is obvious that $G$ is transitive on $S^2$. Moreover, if $z_0 = \infty$, the automorphisms of $S^2$ fixing $z_0$ are the automorphisms of the plane, which, according to Theorem 8.23, are homographic transformations. Thus, $A_\infty \subset G$. $\qquad\square$

Next, automorphisms of $\mathbb{D}$, the fundamental domain of the conformal mapping, are studied. It is easy to find homographic transformations mapping the disc $\mathbb{D}$ into itself. If $a \in \mathbb{D}$, define

$$\tau_a(z) = \frac{z - a}{1 - \bar{a}z},$$

which is a homographic transformation with determinant $1 - |a|^2 \neq 0$. If $z = e^{i\theta}$, one has

$$\tau_a(e^{i\theta}) = \frac{e^{i\theta} - a}{1 - \bar{a}e^{i\theta}} = e^{-i\theta} \cdot \frac{e^{i\theta} - a}{e^{-i\theta} - \bar{a}},$$

so that $|\tau_a(e^{i\theta})| = 1$. Thus, $\tau_a$ transforms the unit circle into itself. Since $\tau_a(a) = 0$, it turns out that $\tau_a$ transforms $\mathbb{D}$ into $\mathbb{D}$ and the domain $\{z: |z| > 1\}$ into itself.

Combining $\tau_a$ with a rotation, $z \to e^{i\alpha}z$, one finds the homographic transformation

$$\varphi_{a,\alpha}(z) = e^{i\alpha}\frac{z - a}{1 - \bar{a}z},$$

which is an automorphism of $\mathbb{D}$ for each $a \in \mathbb{D}$, $\alpha \in \mathbb{R}$. Let us check now that mappings of the type $\varphi_{a,\alpha}$ form a group. One may manage without the rotation and multiply $\tau_a$ by $\tau_b$. It follows that

$$(\tau_a \circ \tau_b)(z) = \frac{(1 + a\bar{b})z - (a + b)}{-(\bar{a} + \bar{b})z + (1 + \bar{a}b)} = \frac{Az + B}{Cz + D} \qquad (8.6)$$

with $A = 1 + a\bar{b}$, $B = -(a + b)$, $C = -(\bar{a} + \bar{b})$, $D = 1 + \bar{a}b$, so that $\bar{A} = D$ and $\bar{B} = C$. Hence,

$$(\tau_a \circ \tau_b)(z) = \frac{Az + B}{\bar{B}z + \bar{A}} = \frac{A(z + B/A)}{\bar{A}(1 + (\bar{B}/\bar{A})z)},$$

that is, $\tau_a \circ \tau_b = \varphi_{c,\alpha}$ with $c = -B/A$, $e^{i\alpha} = A/\bar{A}$, $|c| < 1$. Inequality $|c| < 1$ is a consequence of

$$|c| = \left| \frac{B}{A} \right| = \left| \frac{a + b}{1 + \bar{a}b} \right| = |\tau_{-b}(a)|, \quad |a|, |b| < 1,$$

and it has already been noted that $\tau_{-b}$ takes $\mathbb{D}$ into $\mathbb{D}$.

Taking $b = -a$ in (8.6) one finds $\tau_{-a} = \tau_a^{-1}$.

In turns out that the group $G = \{\varphi_{a,\alpha} : a \in \mathbb{D}, \alpha \in \mathbb{R}\}$ contains all the automorphisms of $\mathbb{D}$. In order to prove this the following useful statement, known as Schwarz's lemma, is needed.

**Lemma 8.26** (Schwarz's lemma). *If the function $f$ is analytic on the unit disc $\mathbb{D}$ and satisfies the conditions $|f(z)| \leq 1$, for all $z \in \mathbb{D}$, and $f(0) = 0$, then one has $|f(z)| \leq |z|$, for all $z \in \mathbb{D}$, and $|f'(0)| \leq 1$. Equality $|f(z)| = |z|$ at some point $z \in \mathbb{D}$ or $|f'(0)| = 1$ holds only when $f(z) = \alpha z$ with $|\alpha| = 1$.*

*Proof.* By hypothesis, $g(z) = f(z)/z$ is a holomorphic function on $\mathbb{D}$. On the circle $\{z : |z| = r\}$ one has $|g(z)| = \frac{|f(z)|}{|z|} \leq 1/r$, and the maximum principle gives $|g(z)| \leq 1/r$ for $|z| \leq r$. Letting $r \to 1$ we obtain $|g(z)| \leq 1$, that is, $|f(z)| \leq |z|$. Moreover, $|f'(0)| = |g(0)| \leq 1$.

If equality holds at some point, the function $|g(z)|$ reaches its maximum and, therefore, it must be constant. Thus, $g(z) = \alpha$ with $|\alpha| = 1$ and $f(z) = \alpha z$.    $\square$

**Theorem 8.27.** *The automorphism group of the unit disc is the set of homographic transformations of the form*

$$\varphi_{a,\alpha}(z) = e^{i\alpha} \frac{z - a}{1 - \bar{a}z}, \quad a \in \mathbb{D}, \alpha \in \mathbb{R}.$$

*Proof.* Apply Lemma 8.24 to the group $G$ of all transformations $\varphi_{a,\alpha}$. It is a transitive group because for $a, b \in \mathbb{D}$, one has $(\tau_{-b} \circ \tau_a)(a) = b$. Let us check that the isotropy group of the origin is contained in $G$.

If $\varphi : \mathbb{D} \to \mathbb{D}$ is bijective and holomorphic with $\varphi(0) = 0$, Schwarz's lemma gives $|\varphi(z)| \leq |z|, z \in \mathbb{D}$. But applying the same lemma to $\varphi^{-1}$ we get $|z| \leq |\varphi(z)|$, for $z \in \mathbb{D}$. Hence $|\varphi(z)| = |z|$, and once again Schwarz's lemma implies that $\varphi(z) = e^{i\alpha}z$ for some $\alpha \in \mathbb{R}$. That is, $\varphi = \varphi_{0,\alpha} \in G$.    $\square$

**Example 8.28.** Let $f : \mathbb{D} \to \mathbb{D}$ be any holomorphic function and fix a point $a \in \mathbb{D}$. Put $w = \tau_a(z) = \frac{z-a}{1-\bar{a}z}$ and $z = \tau_a^{-1}(w)$. Then the function in $w$, $g(w) = \frac{f(\tau_a^{-1}(w))-f(a)}{1-\overline{f(a)}f(\tau_a^{-1}(w))}$, is holomorphic on $\mathbb{D}$ and satisfies $|g(w)| \le 1$ and $g(0) = 0$, by Theorem 8.27. So Schwarz's lemma gives, $|g(w)| \le |w|$, that is,

$$\left| \frac{f(z) - f(a)}{1 - \overline{f(a)}f(z)} \right| \le \left| \frac{z - a}{1 - \bar{a}z} \right|, \quad z \in \mathbb{D}.$$

Letting $z \to a$ in the previous inequality, it yields

$$\frac{|f'(a)|}{1 - |f(a)|^2} \le \frac{1}{1 - |a|^2}, \quad a \in \mathbb{D}.$$

Now equality holds at some point if and only if $f$ is an automorphism of the unit disc. $\qquad\square$

**Example 8.29.** If $a_1, a_2, \ldots, a_N$ are points of $\mathbb{D}$ and $\alpha \in \mathbb{R}$, the function $B(z) = e^{i\alpha} \prod_{j=1}^{N} \frac{z-a_j}{1-\bar{a}_j z}$, obtained as a product of transformations $\tau_{a_j}(z)$ and $e^{i\alpha}$, is called a *Blaschke product* of degree $N$. The function $B$ has the following properties:

i) $B$ is holomorphic on $\mathbb{D}$ and continuous on $\overline{\mathbb{D}}$ (indeed, it is holomorphic on the disc of center 0 and radius $\min\{\frac{1}{|a_j|} : j = 1, \ldots, N\}$).

ii) $|B(z)| = 1$ if $|z| = 1$.

iii) $B$ vanishes exactly at the points $\{a_1, a_2, \ldots, a_N\}$.

If a function $f$ satisfies i), ii) and iii), then $f$ must be equal to $B$, except for the factor $e^{i\alpha}$. Actually, observe that $f/B$ as well as $B/f$ are holomorphic functions on $\mathbb{D}$, continuous on $\overline{\mathbb{D}}$; by ii) and the maximum principle they satisfy $|f(z)/B(z)| \le 1$ and $|B(z)/f(z)| \le 1$ for $z \in \mathbb{D}$. So, $f/B$ must be a constant of modulus 1. $\qquad\square$

In Example 8.15 it has been shown that the homographic transformation $Tz = \frac{z-i}{z+i}$ maps the half plane $\Pi^+ = \{z : \operatorname{Im} z > 0\}$ onto the unit disc $\mathbb{D}$. Let us now exhibit, explicitly, the automorphisms of $\Pi^+$.

Consider, first of all, a homographic transformation $Sz = \frac{az+b}{cz+d}, ad - bc \ne 0$, and impose to it the condition that takes the real axis into itself. This happens if $a, b, c, d$ are real numbers, that is, if the homographic transformation is real. But this condition is also necessary because coefficients $a, b, c, d$ will be determined by a system of linear equations with real coefficients if $Sx \in \mathbb{R}$, for all $x \in \mathbb{R}$. Without loss of generality one may suppose that $ad - bc = \pm 1$, and one finds that $S$ transforms the half plane $\Pi^+$ into itself when $ad - bc = 1$, a condition which implies $\operatorname{Im} \frac{ai+b}{ci+d} > 0$. It is then clear that the set of real homographic transformations $Sz = \frac{az+b}{cz+d}$ with $a, b, c, d \in \mathbb{R}, ad - bc = 1$ is an automorphism group $G$ of $\Pi^+$.

**Theorem 8.30.** *The automorphism group of the upper half space* $\Pi^+ = \{z : \operatorname{Im} z > 0\}$ *is the group $G$ of real homographic transformations with determinant 1.*

*Proof.* Apply once again Lemma 8.24: $G$ is transitive because for every $w = a + ib \in \Pi^+$ the homographic transformation of $G$, $Tz = a + bz$ sends $i$ into $w$.

Let us check now that the isotropy group of the point $i$ is contained in $G$. We need to show that each element of this isotropy group is a homographic transformation. Now, the mapping

$$S \longrightarrow T^{-1} \circ S \circ T \quad \text{with } Tz = \frac{z - i}{z + i}.$$

is an isomorphism between the isotropy group of the origin in $\operatorname{Aut}(\mathbb{D})$ and the isotropy group of the point $i$ in $\operatorname{Aut}(\Pi^+)$. The first group, as it has been seen in the proof of Theorem 8.27, consists of rotations, and if $S$ is a rotation, then $T^{-1} \circ S \circ T$ is a homographic transformation. $\square$

## 8.5 Dirichlet's problem and Neumann's problem in the half plane

In this section homographic transformations will be used to transfer Dirichlet's problem and Neumann's problem from the unit disc to the half space, using the invariance of both problems under conformal transformations (Section 7.11).

First of all, to illustrate this method one may find the solution of Dirichlet's problem in the disc using only the mean value property. Let $\varphi \in C(\mathbb{T})$ and look for a function $u \in C(\overline{\mathbb{D}})$ with $\Delta u = 0$ and $U|_{\mathbb{T}} = \varphi$. Fix a point $a \in \mathbb{D}$ and consider the homographic transformation

$$\tau(z) = \tau_{-a}(z) = \frac{z + a}{1 + \bar{a}z},$$

which is known (Section 8.4) to transform $\mathbb{D}$ into $\mathbb{D}$ and $\mathbb{T}$ into $\mathbb{T}$. If $u$ is the desired solution, the function $u \circ \tau$ must be harmonic on $\mathbb{D}$ and in $\mathbb{T}$ will take the values given by $\varphi(\tau(t))$. The mean value property yields

$$(u \circ \tau)(0) = u(a) = \frac{1}{2\pi} \int_{\mathbb{T}} \varphi(\tau(z)) \, |dz|.$$

Making the change of variable $w = \tau(z)$, $z = \tau^{-1}(w)$, it turns out that $|dz| = \frac{1 - |a|^2}{|1 - \bar{a}w|^2} |dw|$ and

$$u(a) = \frac{1}{2\pi} \int_{\mathbb{T}} \frac{1 - |a|^2}{|1 - \bar{a}w|^2} \varphi(w) \, |dw|$$

and one gets back the fact that $u$ must be the Poisson's integral of $\varphi$.

Let us consider now the upper half plane $\Pi^+ = \{z = x + iy\colon y > 0\}$. Start noting that $\Pi^+$ being unbounded, the maximum principle cannot be applied to show the uniqueness of a solution of Dirichlet's problem. For example, if $\varphi(x) = 0$, one has the two solutions $u_1 \equiv 0$ and $u_2(x, y) = y$.

However, there is uniqueness if one just looks for bounded solutions. That is, the problem $\Delta u = 0$ on $\Pi^+$, $u|_{\mathbb{R}} = \varphi$ has at most one *bounded* solution in $\Pi^+$, continuous on $\overline{\Pi^+}$, because if $u_1, u_2$ are two solutions, then $u_1 - u_2$ is harmonic and bounded on $\Pi^+$ and has value $0$ on $\mathbb{R}$; from this it follows that $u_1 - u_2 = 0$ on $\Pi^+$. This is a consequence of the symmetry principle (see Exercise 10 in Section 7.13) and Liouville's theorem for harmonic functions.

Let us see now how the solution of Dirichlet's problem in $\Pi^+$ is obtained.

In Subsection 7.6.2 it was shown that the Poisson kernel of the disc $P_{\mathbb{D}}(z, w)$ is the real part of $\frac{1}{2\pi} \frac{1+z\bar{w}}{1-z\bar{w}}$:

$$P_{\mathbb{D}}(z, w) = \frac{1}{2\pi} \operatorname{Re} \frac{1 + z\bar{w}}{1 - z\bar{w}} = \frac{1}{2\pi} \frac{1 - |z|^2}{|1 - z\bar{w}|^2}, \qquad |z| < 1, \ |w| = 1.$$

Take $w = 1$ and consider the function $\phi(z) = \frac{1+z}{1-z}$ that has positive real part for $|z| < 1$. If $\operatorname{Re} w \geq 0$, the point $z$ given by $\frac{1+z}{1-z} = w$, that is, $z = \frac{w-1}{w+1}$ satisfies $|z|^2 = \frac{1+|w|^2-2\operatorname{Re} w}{1+|w|^2+2\operatorname{Re} w}$, $1-|z|^2 = \frac{4\operatorname{Re} w}{|1+w|^2}$ and, therefore, $|z| < 1$. The transformation $\phi$ is, then, conformal from $\mathbb{D}$ into the half plane $\{w\colon \operatorname{Re} w > 0\}$. Considering $\phi$ defined on $\mathbb{C}^*$, it is a homeomorphism from $\overline{\mathbb{D}}$ onto $\{w\colon \operatorname{Re} w \geq 0\} \cup \infty$, the imaginary axis $w = it$ being the image of $\mathbb{T} \setminus \{1\}$.

The transformation $S(z) = i\frac{1+z}{1-z}$ maps, therefore, $\mathbb{D}$ into $\Pi^+$ and $\mathbb{T} \setminus \{1\}$ into $\mathbb{R}$. Its inverse transformation is $z = \frac{w-i}{w+i}$.

If $z = \frac{w-i}{w+i}$ and $e^{it} = \frac{x-i}{x+i}$, one has

$$1 - |z|^2 = \frac{4 \operatorname{Im} w}{|w + i|^2}, \qquad |e^{it} - z|^2 = \frac{4|w - x|^2}{|1 + iw|^2(1 + x^2)} \quad \text{and} \quad dt = \frac{2\,dx}{1 + x^2}.$$

This means that the Poisson formula in the disc becomes for the half plane

$$u(w) = \frac{1}{\pi} \int_{-\infty}^{+\infty} \frac{\operatorname{Im} w}{|w - x|^2} u(x)\,dx$$

and suggests that, given a function $\varphi$ continuous and bounded on $\mathbb{R}$, the solution of the problem $\Delta u = 0$, $u|_{\mathbb{R}} = \varphi$ should be

$$u(w) = \frac{1}{\pi} \int_{-\infty}^{+\infty} \frac{\operatorname{Im} w}{|w - x|^2} \varphi(x)\,dx. \tag{8.7}$$

Hence, the Poisson kernel of the half plane is $P_{\Pi^+}(w, x) = \frac{1}{\pi} \frac{\operatorname{Im} w}{|w-x|^2}$ and one can check that the function $u$, the Poisson transform of $\varphi$, given by (8.7) is harmonic on $\Pi^+$ and satisfies

$$\lim_{\substack{w \to x \\ w \in \Pi^+}} u(w) = \varphi(x), \quad x \in \mathbb{R}$$

(see Exercise 15 of Section 8.11).

One may use the same calculations to find the solution of Neumann's problem in the upper half plane; that is, to solve

$$\Delta u = 0, \quad -\frac{\partial u}{\partial y}(x,0) = \varphi(x), \quad x \in \mathbb{R}, \text{ with } \varphi \in C(\mathbb{R}).$$

Here $-\frac{\partial}{\partial y}$ is the derivative with respect to the unit normal vector to $\mathbb{R} = \partial \Pi^+$.

Define $v, \psi$ on $\mathbb{D}$ and $\mathbb{T}$ by means of the transformation $S$, that is,

$$v(z) = u(S(z)) = u(w), \quad \psi(e^{it}) = \varphi(S(e^{it})) = \varphi(x).$$

The inverse transformation $w \xrightarrow{S^{-1}} \frac{w-i}{w+i}$ has derivative $\frac{2i}{(w+i)^2}$ with modulus $\frac{2}{|w+i|^2} = \frac{2}{1+x^2}$, when $w = x$; moreover it preserves angles, so that

$$-\frac{\partial u}{\partial y}(x,0) = \frac{2}{1+x^2}\frac{\partial v}{\partial r}(e^{it}).$$

Therefore, $\frac{\partial v}{\partial r}(e^{it}) = \frac{1+x^2}{2}\psi(e^{it})$, and so by Theorem 7.32, the solution of Neumann's problem in the disc is

$$v(z) = -\frac{1}{\pi}\int_0^{2\pi} \text{Log}\,|z - e^{it}|\,\frac{\partial v}{\partial r}(e^{it})\,dt.$$

Changing variables one gets

$$u(w) = -\frac{1}{2\pi}\int_{-\infty}^{\infty} \text{Log}\left[\frac{4|w-x|^2}{|1+iw|^2(1+x^2)}\right]\varphi(x)\,dx$$

and, up to a constant, it follows that

$$u(w) = -\frac{1}{\pi}\int_{-\infty}^{\infty}\left(\text{Log}\,\frac{|w-x|}{|1+iw|}\right)\varphi(x)\,dx. \tag{8.8}$$

Now one may also check directly that function $u$, given by (8.8) is harmonic on $\Pi^+$ and satisfies $-\frac{\partial u}{\partial y}(x,0) = \varphi(x)$, $x \in \mathbb{R}$ (see Exercise 15 of Section 8.11).

## 8.6 Level curves

If $f$ is an analytic function and $z_0$ a point at which $f'(z_0) \neq 0$, then $w = f(z)$ is a conformal mapping from a neighborhood of $z_0$ into a neighborhood of $w_0 = f(z_0)$.

A way to understand the properties of $f$ is to display the behavior of the conformal mapping given by $f$, in a similar way as the graph of a function of a real variable gives information about the behavior of the function. This visualization can be obtained by considering the image curves by $f$ of lines parallel to the coordinate axes, and also the level curves of the functions Re $f(z)$ and Im $f(z)$ (Figure 8.7).

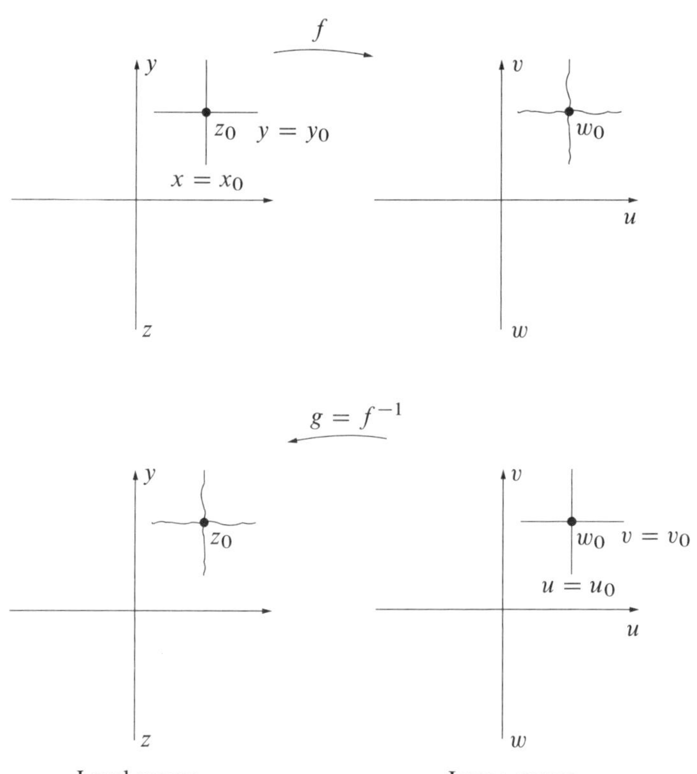

Level curves          Image curves

Figure 8.7

Suppose, then, that $w = f(z)$ is a conformal mapping from the domain $U$ of the plane $z = x + iy$ onto the domain $U'$ of the plane $w = u + iv$ and write $f(z) = u(z) + iv(z)$, $z_0 = x_0 + iy_0 \in U$ and $w_0 = f(z_0) = u_0 + iv_0$. Let $g = f^{-1}$ be the conformal mapping from $U'$ onto $U$, inverse of $f$.

The straight lines $x = x_0$, $y = y_0$ give an orthogonal coordinate system around the point $z_0$. Their images are the curves $y \to (u(x_0, y), v(x_0, y))$ and $x \to (u(x, y_0), v(x, y_0))$, respectively, which intersect each other orthogonally at the point $w_0$. These two curves, linked to the conformal mapping $f$, may be also considered an orthogonal coordinate system around $w_0$.

Starting now from the lines $u = u_0$, $v = v_0$ in the $w$-plane, their images by $g$ are the curves in the $z$-plane,

$$v \longrightarrow g(u_0, v) = \{z : u(z) = u_0\},$$
$$u \longrightarrow g(u, v_0) = \{z : v(z) = v_0\};$$

that is, they are the level curves of functions $u$ and $v$, which intersect each other orthogonally at the point $z_0$ as well, since $g = f^{-1}$ is conformal.

Some examples of determination of level curves are the following ones:

A) The function $w = \operatorname{Log} z$ is a conformal mapping from the $z$-plane up to the ray $(-\infty, 0]$ onto the strip $-\pi < v < \pi$ of the $w$-plane.

The level curves $u = \operatorname{Log}|z| = u_0$ are circles with center at the origin, and the curves $v = \operatorname{Arg} z = v_0$ are rays starting from the origin. These curves are the images of straight lines $u = u_0$ and $v = v_0$ by the function $z = e^w$ (Figure 8.8).

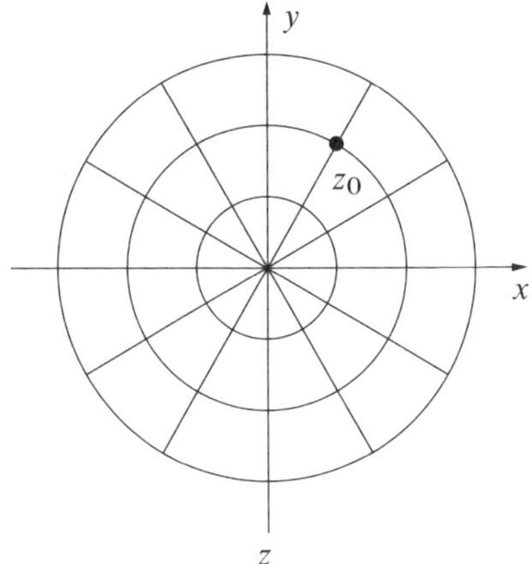


Figure 8.8

B) The level curves of the function $w = z^2$ are the curves $u = x^2 - y^2 = u_0$ and $v = 2xy = v_0$. Equations $x^2 - y^2 = u_0$ represent hyperbolas having as asymptotes the bisectors of quadrants, and equations $2xy = v_0$ correspond to hyperbolas with coordinate axes as asymptotes (Figure 8.9).

Regarding the level curves of $z = \sqrt{w}$, since they are the image curves by $w = z^2$ of $x = x_0$ and $y = y_0$, one obtains

$$u = x_0^2 - y^2, \quad v = 2x_0 y,$$
$$u = x^2 - y_0^2, \quad v = 2x y_0.$$

Equivalently

$$v^2 = 4x_0^2(x_0^2 - u) \quad \text{and} \quad v^2 = 4y_0^2(y_0^2 + u),$$

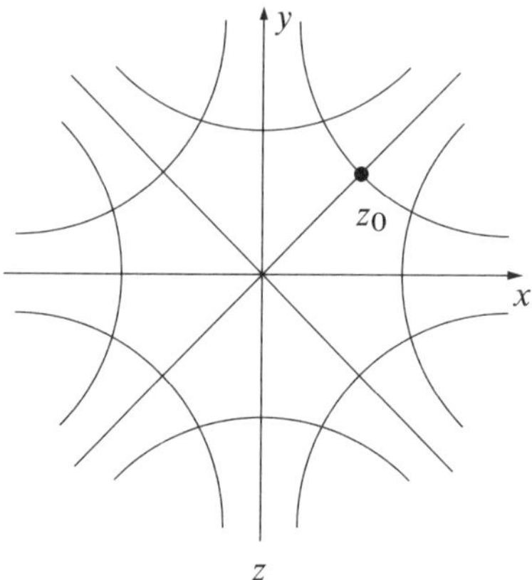



Figure 8.9

that describe parabolas of the $w$-plane, intersecting the $u$-axis at points $x_0^2$ and $-y_0^2$, respectively, and having their focus at the origin (Figure 8.10).

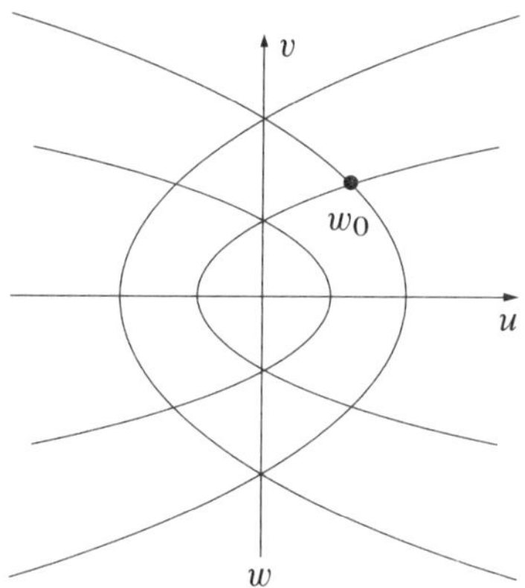



Figure 8.10

C) Let us now study the level curves of the transformation $w = \mathrm{Log}\,\frac{z-1}{z+1}$. They are, for each point $z_0$, the curves

$$\mathrm{Log}\left|\frac{z-1}{z+1}\right| = u_0, \quad \mathrm{Arg}\,\frac{z-1}{z+1} = v_0, \quad \text{with } u_0 + iv_0 = w_0 = \mathrm{Log}\,\frac{z_0-1}{z_0+1}.$$

Setting $\tau = \tau(z) = \frac{z-1}{z+1}$, the level curves are the preimages by $\tau$ of level curves of $\mathrm{Log}\,\tau$, that is, of the family of circles centered at the origin and the family of rays starting from the origin in the $\tau$-plane. Since $\tau(1) = 0$, $\tau(-1) = \infty$, the rays starting from the origin in the $\tau$-plane correspond to the family of circles of the $z$-plane centered at the origin and passing through points $-1$ and $1$.

The concentric circles with center at the origin of the $\tau$-plane correspond to the circles with equation

$$\left|\frac{z-1}{z+1}\right| = \rho, \quad \rho > 0, \tag{8.9}$$

which are orthogonal to the ones passing through $-1$ and $1$. They are called *Apollonian circles* with limit points $-1$ and $1$; according to equation (8.9) they are the locus of points whose ratio of distances to the points $-1$ and $1$ is constant (Figure 8.11).

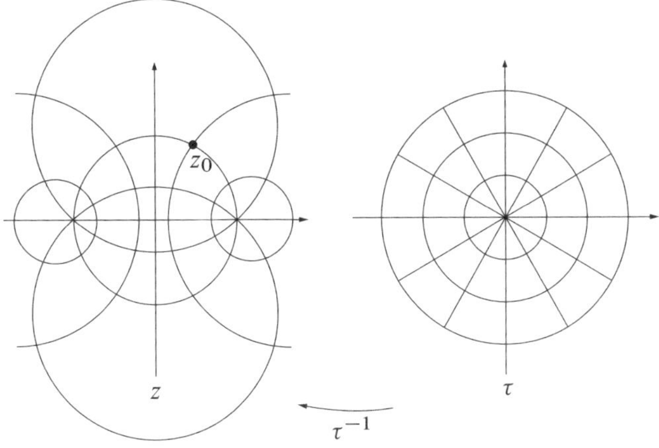

Figure 8.11

This double set of circles that, as said, are the level curves of $\operatorname{Log}\frac{z-1}{z+1}$, are called *Steiner's circles* with limit points $-1$ and $1$.

## 8.7 Elementary conformal transformations

According to Riemann's theorem, given two domains of the plane $U$ and $U'$, simply connected and different from $\mathbb{C}$, there is always a conformal mapping from $U$ onto $U'$. In order to find effectively a conformal mapping between these domains, one may proceed in the following way: look first for a conformal mapping from $U$ onto the unit disc $\mathbb{D}$ and then map conformally $\mathbb{D}$ onto $U'$. Hence, the problem is reduced to finding conformal mappings of simply connected domains onto $\mathbb{D}$. In order to achieve this we can use, at least, the homographic transformations that have been studied in detail, and the elementary functions $e^z$, $\log z$, $z^\alpha$ ($\alpha \in \mathbb{R}$ fixed), $\sin z$, etc. The aim of this section is to show, through several examples, how some elementary domains can be mapped onto $\mathbb{D}$ with the help of these functions.

First of all, it is convenient to recall that the disc $\mathbb{D}$ and the half plane $\Pi^+ = \{z \in \mathbb{C}: \operatorname{Im} z > 0\}$ are conformally equivalent by means of the homographic transformation $Tz = i\frac{1+z}{1-z}$. Actually, $T$ transforms $-1$, $i$, $1$ into the points $0$, $-1$, $\infty$, respectively. That is, it transforms the unit circle into the real axis. Since $T(0) = i$, $T$ maps $\mathbb{D}$ into $\Pi^+$. So, it makes no difference to transform a domain into the disc $\mathbb{D}$ or into the half plane $\Pi^+$.

Some examples of elementary mappings are the following ones:

A) The transformation $z \to z^p$ with $0 < p < 2$ maps the upper half plane $\Pi^+$ into the region $U = \{z = re^{i\theta}: 0 < r < \infty, 0 < \theta < p\pi\}$. In particular, one may go from any sector

$$S_\alpha = \{z = re^{i\theta}: 0 < r < \infty, 0 < \theta < \alpha\}$$

to the half plane $\Pi^+$ by means of the mapping $z \to z^{\pi/\alpha}$ (Figure 8.12).

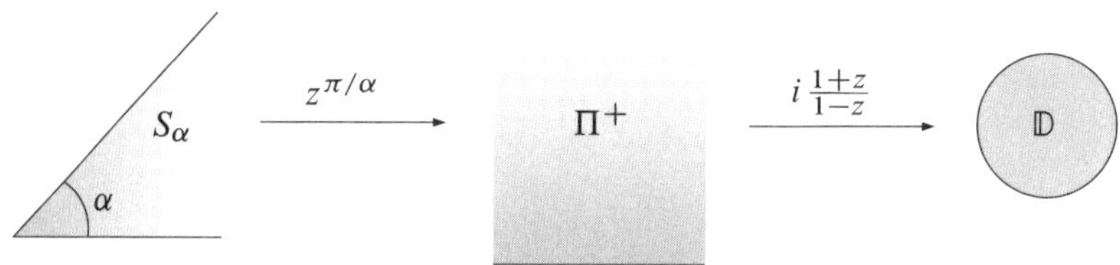


Figure 8.12

B) Another kind of elementary domains are the strips. Using rotations, dilations and translations one may assume that we are dealing with the strip

$$B = \{z = x + iy: -\pi < y < \pi\}.$$

The exponential function $z \to e^z$ maps $B$ onto the domain $\mathbb{C} \setminus \{(-\infty, 0]\}$, which in turn can be mapped onto the half plane $\{z: \operatorname{Re} z > 0\}$ with the transformation $z \to \sqrt{z}$. From this half plane one may rotate to get $\Pi^+$ and then go into the disc (Figure 8.13).

C) The fact that homographic transformations preserve the family of circles makes them helpful when transforming into the unit disc a domain bounded by arcs of a circle. For example, consider a half disc

$$U = \{z: |z| < 1, \operatorname{Im} z > 0\}$$

and let us see how it may be conformally mapped into a disc or a half plane. Observe that the mapping $z \to z^2$ does not work, since it only opens $U$ to the region $\mathbb{D} \setminus \{[0, \infty)\}$. Instead, the homographic transformation $Tz = \frac{z-1}{z+1}$, satisfying $T(-1) = \infty$, $T(1) = 0$, $T(i) = i$, $T(0) = -1$, maps $U$ into the second quadrant, and then one goes into $\Pi^- = \{z: \operatorname{Im} z < 0\}$ applying $z \to z^2$ (Figure 8.14).

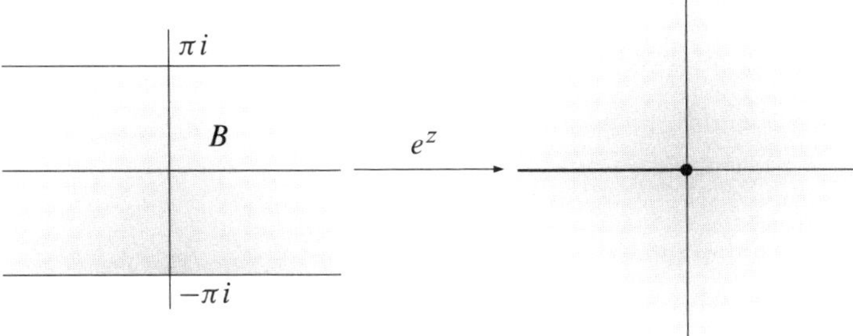

Figure 8.13

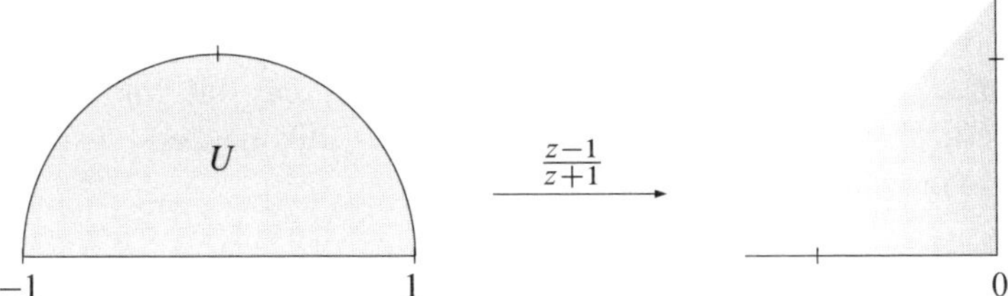

Figure 8.14

D) One can show now the usefulness of trigonometric functions in the conformal mapping. Starting with a half strip $S = \{z = x + iy : x > 0, \, -\pi < y < \pi\}$, not a full strip as in B), one cannot use the function $e^z$ because one would only obtain the part of $\mathbb{C} \setminus \{(-\infty, 0]\}$ that is outside $\mathbb{D}$. If, using a similarity, one starts with the half strip $S_0$,

$$S_0 = \{z = x + iy : 0 < x < \pi/2, \, y > 0\},$$

one may go to a quadrant thanks to the function $\sin z$ (Figure 8.15):

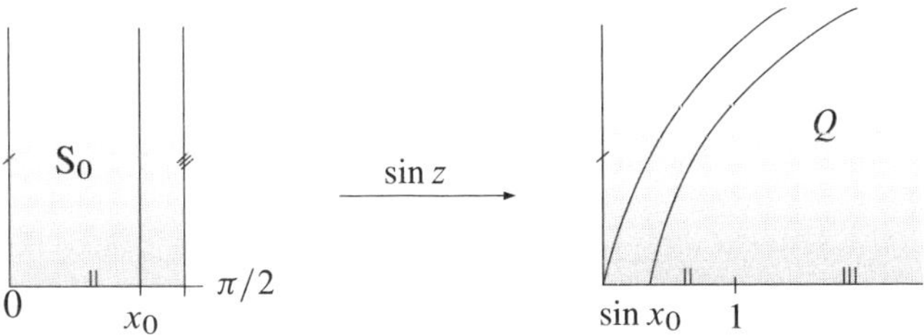

Figure 8.15

The function $\sin z$ maps conformally the half strip $S_0$ onto the first quadrant, $Q = \{z = x + iy, \ x > 0, \ y > 0\}$.

To prove this assertion, recall the formulae in Subsection 2.2.5:

$$\mathrm{Re}(\sin z) = \sin x \, \mathrm{ch}\, y, \quad \mathrm{Im}(\sin z) = \cos x \, \mathrm{sh}\, y$$

if $z = x + iy$, which tell us that $\sin z \in Q$ when $z \in S_0$. Analyze first the action of $\sin z$ on the boundary of $S_0$. When $x = 0$, $y > 0$, one has $\sin z = i\,\mathrm{sh}\,y$, and since $\mathrm{sh}\,y$ increases from 0 to $+\infty$ when $0 < y < +\infty$, $\sin z$ describes the positive imaginary axis on this piece of the boundary of $S_0$. If $y = 0$, $0 < x < \pi/2$, it is clear that $\sin z$ increases from 0 to 1. Finally, since $\sin(\pi/2 + iy) = \mathrm{ch}\,y$, the part of the boundary of $S_0$ given by $x = \pi/2$, $y > 0$ is mapped onto $(1, +\infty)$. So, the boundary of $S_0$ is transformed, by $\sin z$, into the boundary of $Q$.

We can see now that each point of $Q$ comes from only one point of $S_0$.

The vertical ray $x = x_0$, $0 < x_0 < \pi/2$, $y > 0$ is mapped by $\sin z = w = \sigma + i\tau$ into the piece of the hyperbola

$$\frac{\sigma^2}{\sin^2 x_0} - \frac{\tau^2}{\cos^2 x_0} = 1$$

that falls within $Q$, because $\sigma = \sin x_0 \cdot \mathrm{ch}\,y$ and $\tau = \cos x_0 \cdot \mathrm{sh}\,y$. Now, given a point $a + ib$, $a > 0$, $b > 0$ of $Q$, the equation $\sin z = a + ib$ must be solved. To this end, take $x$, $0 < x < \pi/2$ such that $a + ib$ is in the hyperbola of equation

$$\frac{\sigma^2}{\sin^2 x} - \frac{\tau^2}{\cos^2 x} = 1.$$

Now take as value of $\sin x$ the intersection point of the interval $(0, 1)$ and the hyperbola passing through $a + ib$; afterwards choose $y > 0$ such that $a = \sin x \cdot \mathrm{ch}\,y$, $b = \cos x \cdot \mathrm{sh}\,y$.

Therefore $z \to \sin z$ is bijective (and holomorphic) from $S_0$ onto $Q$. Since $\sin(-\bar{z}) = -\overline{\sin z}$, we obtain (see Figure 8.16) that $\sin z$ maps the half strip $S_1 = \{z = x + iy\colon \ -\pi/2 < x < \pi/2, \ y > 0\}$ onto the half plane $\Pi^+$. Observe that the function $\sin z$, unlike the ones considered so far, does not map a straight line into a line or a circle, but vertical lines go to hyperbolas.

E) Sometimes it may not be clear if a function $f$ provides a conformal mapping, but some changes in the expression of $f$ may reduce it to a combination of well-known transformations. An interesting example of this is the following one:

Take $f(z) = \sum_{n=1}^{\infty} nz^n = z + 2z^2 + 3z^3 + \cdots + nz^n + \cdots$ and consider it on the unit disc $\mathbb{D}$, the disc of convergence of this power series. Observe that $f(0) = 0$ and $f'(0) = 1$. The question is to know if $f$ is one-to-one in $\mathbb{D}$. One has

$$f(z) = z \sum_{1}^{\infty} nz^{n-1} = \frac{z}{(1-z)^2},$$

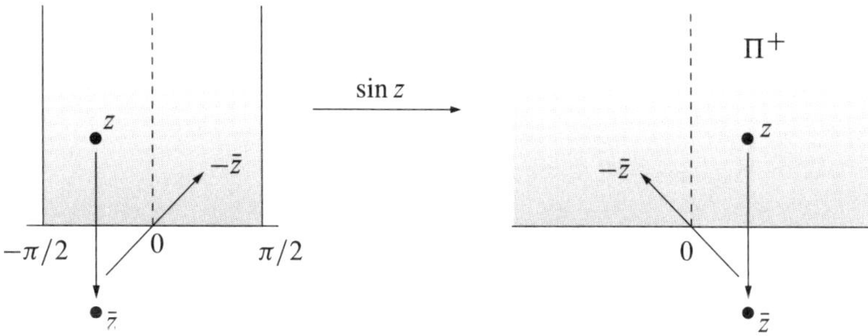


Figure 8.16

because $\sum_1^\infty n z^{n-1}$ is the derivative series of $\sum_0^\infty z^n = \frac{1}{1-z}$, if $|z| < 1$. So,

$$f(z) = \frac{z}{(1-z)^2} = \frac{1}{4}\left(\frac{1+z}{1-z}\right)^2 - \frac{1}{4}$$

and one obtains an expression for $f$ as a composition of well-known elementary transformations.

The transformation $z \to \frac{1+z}{1-z}$ maps $\mathbb{D}$ into the half plane $\{z: \operatorname{Re} z > 0\}$, and $z \to z^2/4$ opens this half plane to the whole plane except for the negative real axis; so it turns out that $f$ maps conformally the disc $\mathbb{D}$ onto the domain $\mathbb{C} \setminus (-\infty, -1/4]$ (Figure 8.17).

The function $f$ is called *Koebe's function*.

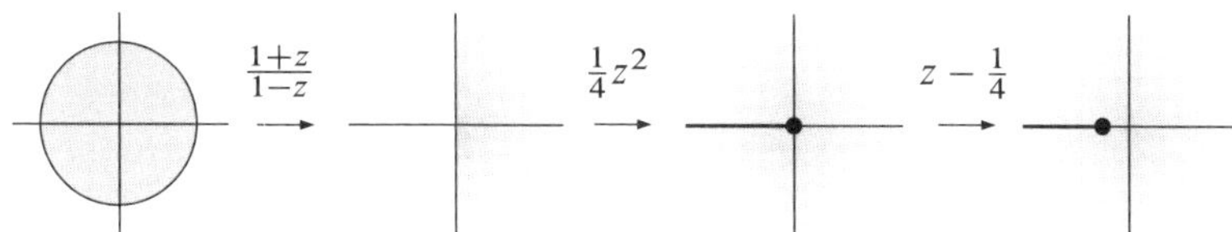


Figure 8.17

F) In the last example we will consider a region which is simply connected in the Riemann sphere $S^2$, that is, its complement in $S^2$ is connected.

Take as domain $U$ the complement of a segment in the Riemann sphere, for example $U = S^2 \setminus [-1, 1]$. The homographic transformation $z \to \frac{z+1}{z-1}$ maps $U$ into $\mathbb{C} \setminus (-\infty, 0]$; next, using $z \to \sqrt{z}$ take it to the right half plane $\{z: \operatorname{Re} z > 0\}$ and, finally, with the homographic transformation $z \to \frac{z-1}{z+1}$, to the unit disc $\mathbb{D}$ (Figure 8.18). Hence, $U$ is transformed into $\mathbb{D}$ by the function

$$\varphi(z) = w = \frac{\sqrt{\frac{z+1}{z-1}} - 1}{\sqrt{\frac{z+1}{z-1}} + 1}, \qquad \varphi(\infty) = 0,$$

which may be written also in the form

$$w = z - \sqrt{z^2 - 1}, \quad z \in U. \tag{8.10}$$

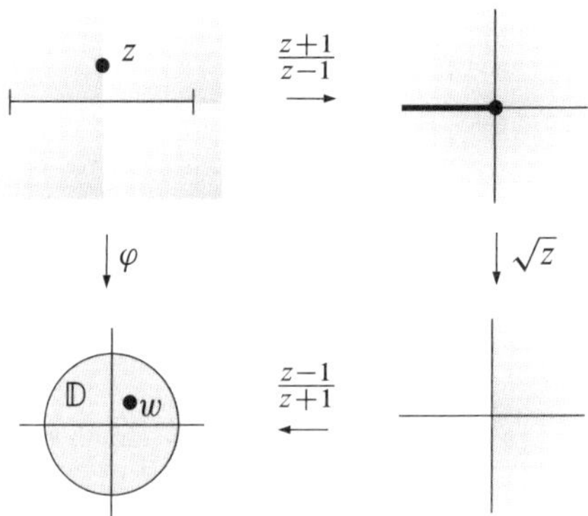


Figure 8.18

The inverse of this mapping, from $\mathbb{D}$ to $U$, is

$$z = \frac{1}{2}\left(w + \frac{1}{w}\right). \tag{8.11}$$

Setting $\mathbb{D}^* = \{z \in S^2 : |z| > 1\}$, one has $w \in \mathbb{D}$ if and only if $1/w \in \mathbb{D}^*$ and, therefore, the function in (8.11) also transforms $\mathbb{D}^*$ into $U$ ($w \in \mathbb{D}$ and $1/w \in \mathbb{D}^*$ both go to the same point of $U$ by (8.11)). Starting from a point $z \in U$, the corresponding point $w \in \mathbb{D}$ is obtained from (8.10) choosing the positive sign in $\sqrt{z^2 - 1}$. Changing the sign, we get the point $z + \sqrt{z^2 - 1}$, which is in $\mathbb{D}^*$, because $(z - \sqrt{z^2 - 1})(z + \sqrt{z^2 - 1}) = 1$.

Now look for the images by the transformation (8.11) of the family of circles $\{w : |w| = r < 1\}$, and also for those of radius of the unit circle, $\{w : w = re^{i\theta}, 0 \le r < 1\}$ (Figure 8.19).

Note that they are as well the images by (8.11) of the family of circles $\{w : |w| = r, r > 1\}$ and of rays $\{w : w = re^{i\theta}, r > 1\}$, $0 \le \theta \le 2\pi$.

We use polar coordinates, $w = re^{i\theta}$. The point $z = x + iy$ given by (8.11) is now written as

$$x = \frac{1}{2}\left(r + \frac{1}{r}\right)\cos\theta; \quad y = \frac{1}{2}\left(r - \frac{1}{r}\right)\sin\theta.$$

The cancellation of $\theta$ in these two equations yields

$$\frac{4x^2}{[(r + 1/r)]^2} + \frac{4y^2}{[(r - 1/r)]^2} = 1,$$

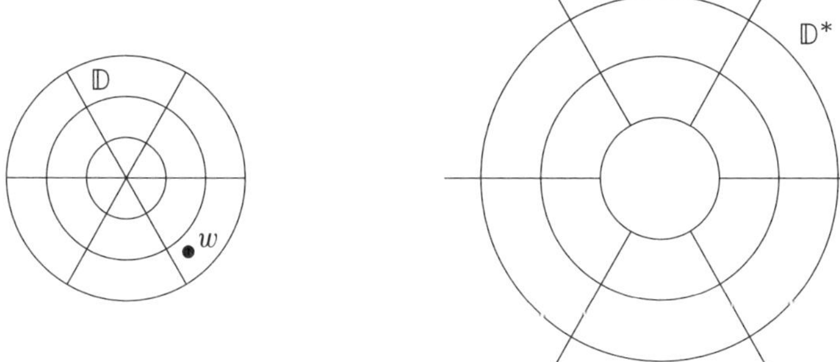

Figure 8.19

which is the equation of an ellipse in the plane $(x, y)$ with axes $r + 1/r$ and $r - 1/r$. Starting now from a radius, $\arg w = \theta$, and cancelling $r$, it turns out that

$$\frac{x^2}{\cos^2 \theta} - \frac{y^2}{\sin^2 \theta} = 1,$$

which is the equation of a hyperbola. Observe that the image of a radius $w = re^{i\theta}$, $0 \leq r < 1$ is the part of a branch of this hyperbola inside the circle $x^2 + y^2 = 1$; analogously the image of the ray $w = re^{i\theta}$, $r > 1$ is the part of the branch that falls outside (Figure 8.20).

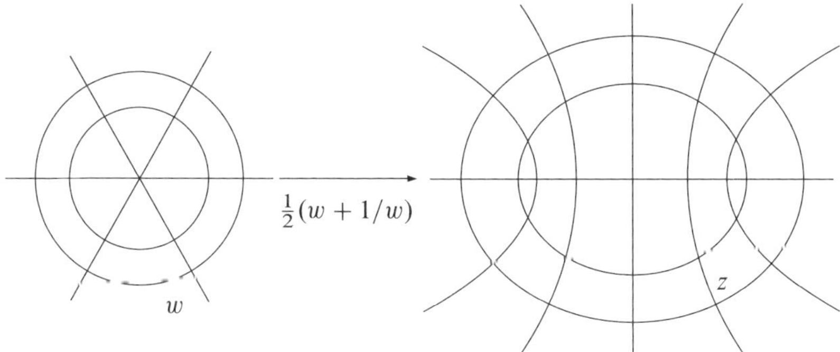

Figure 8.20

Transformation (8.11), called *Jowkowsky's mapping*, is very important in the applications of conformal mapping, especially in fluid mechanics.

## 8.8  Conformal mappings of polygons

Consider a simply connected domain $U$ bounded by a closed polygonal Jordan curve (Figure 8.21). One says that $U$ is a *polygon*. Assume it has $n$ sides and the vertices are the points $z_1, z_2, \ldots, z_n$. At each vertex $z_k$ suppose the interior angle of the polygon measures $\alpha_k \pi$ with $0 < \alpha_k < 2$. Then the corresponding exterior angle is $\beta_k \pi$ with $\beta_k = 1 - \alpha_k$ and $-1 < \beta_k < 1$. One may easily prove that $\beta_1 + \cdots + \beta_n = 2$.

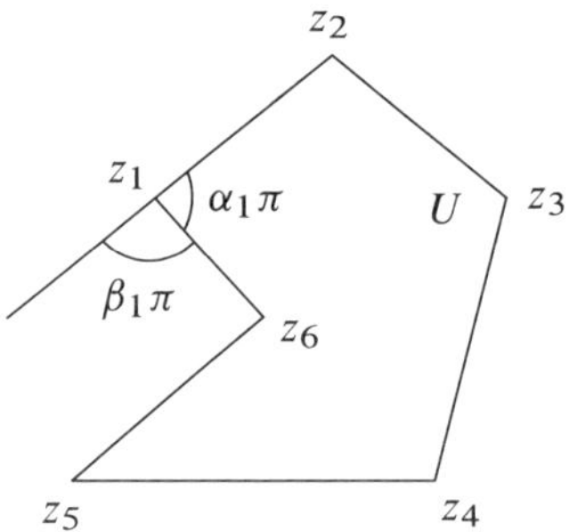

Figure 8.21

Consider now a conformal mapping $f$ of the polygon $U$ onto the unit disc $\mathbb{D}$. By Theorem 8.7′ it is known that $f$ extends continuously to $\bar{U}$ and transforms injectively each side of the polygon into an arc of $\partial\mathbb{D}$ (Figure 8.22).

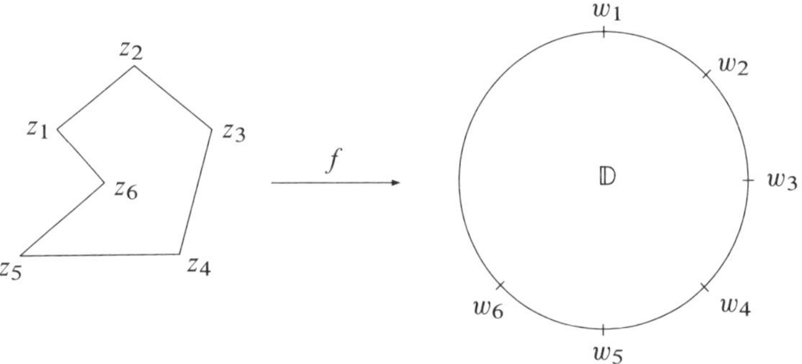

Figure 8.22

Now we will study in detail the mapping $f$ and find an explicit formula for its inverse. It will be easier to work in the half plane; so one goes from the disc $\mathbb{D}$ to the upper half plane $\Pi^+$ by means of a homographic transformation and denotes by $a_1, \ldots, a_n$ the points on the real line which correspond to points $w_k = f(z_k)$,

$k = 1, \ldots, n$. Remark that one of these could be the point at infinity. Let $g$ be the mapping from the half plane $\Pi^+$ onto $U$ (Figure 8.23).

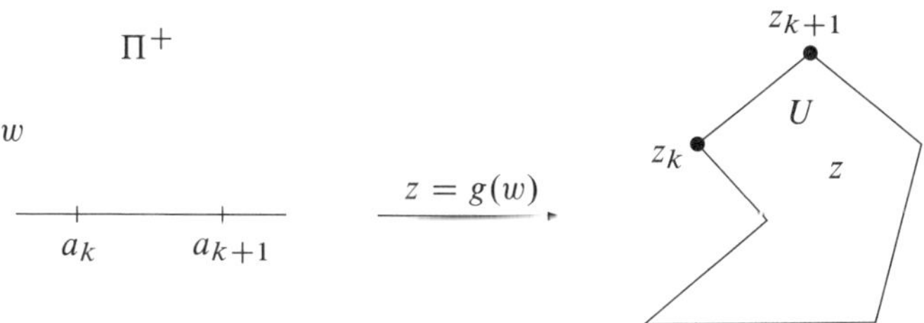


Figure 8.23

The explicit form of the function $g$ is given by the following result.

**Theorem 8.31** (Schwarz–Christoffel formula). *Let $g$ be a conformal mapping from the upper half plane $\Pi^+$ onto a polygon of $n$ sides which has interior angles with measures $\alpha_1 \pi, \ldots, \alpha_n \pi$ and let $a_1, a_2, \ldots, a_n$ be the points on the real axis mapped into the vertices of the polygon. Then, fixing a point $w_0 \in \overline{\Pi^+}$, there exist complex constants $A$, $B$ such that*

$$g(w) = A \int_{w_0}^{w} (w - a_1)^{\alpha_1 - 1} \cdots (w - a_n)^{\alpha_n - 1} \, dw + B, \quad w \in \Pi^+. \qquad (8.12)$$

**Remark 8.3.** The integral in (8.12) may be taken along the line segment going from $w_0$ to $w$, but, indeed, the integral is the same for any path from $w_0$ to $w$, not leaving $\Pi^+$, as a consequence of Cauchy's theorem.

Before starting the proof of this formula, observe that (8.12) is equivalent to

$$g'(w) = A(w - a_1)^{\alpha_1 - 1} \cdots (w - a_n)^{\alpha_n - 1}, \quad w \in \Pi^+ \qquad (8.13)$$

as well as to

$$\frac{g''(w)}{g'(w)} = \frac{\alpha_1 - 1}{w - a_1} + \cdots + \frac{\alpha_n - 1}{w - a_n}, \quad w \in \Pi^+. \qquad (8.14)$$

The equivalence of (8.12), (8.13) and (8.14) follows by differentiation and integration. Equality (8.14) tells us that the function $g''/g'$ has a simple pole with residue $\alpha_k - 1$ around the point $a_k$. This fact suggests how to proceed to prove the theorem.

*Proof.* As just said, we will study the behavior of the function $g''/g'$, which is holomorphic on $\Pi^+$, around a point $a_k$ with $g(a_k) = z_k$.

Consider the function $f_k(w) = (g(w) - z_k)^{1/\alpha_k}$, holomorphic on $\Pi^+$ (Figure 8.24). When $w$ tends to the real axis near the point $a_k$, $\operatorname{Im} f_k(w)$ tends to 0 and, by the symmetry principle (Proposition 8.10), $f_k(w)$ is holomorphic on a neighborhood of $a_k$ with $f_k(a_k) = 0$ and $f'_k(a_k) \neq 0$. Indeed if $f'_k(a_k) = 0$, then there would be two points $w_1, w_2 \in \Pi^+$ such that $f_k(w_1) = f_k(\overline{w}_2) = \overline{f_k(w_2)}$, whence $g(w_2)$ would not be inside the polygon.

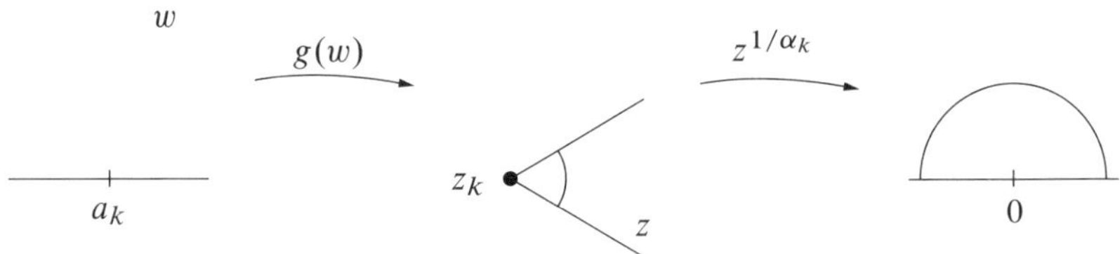


Figure 8.24

Therefore, $f_k(w) = (g(w) - z_k)^{1/\alpha_k} = (w - a_k)\varphi_k(w)$ with $\varphi_k(w)$ holomorphic on a neighborhood of $a_k$ and $\varphi_k(a_k) \neq 0$. Taking a branch of $\varphi_k(w)^{\alpha_k}$, denoted by $h_k(w)$, one has

$$g(w) - z_k = (w - a_k)^{\alpha_k} h_k(w), \quad w \in \Pi^+,$$

where $h_k(w)$ is holomorphic on a neighborhood of $a_k$ and $h_k(a_k) \neq 0$.

Now, computing, it turns out that $\frac{g''(w)}{g'(w)}$ equals

$$\frac{\alpha_k(\alpha_k - 1)(w - a_k)^{\alpha_k - 2} h_k(w) + 2\alpha_k(w - a_k)^{\alpha_k - 1} h'_k(w) + (w - a_k)^{\alpha_k} h''_k(w)}{\alpha_k(w - a_k)^{\alpha_k - 1} h_k(w) + (w - a_k)^{\alpha_k} h'_k(w)}$$

and

$$\frac{g''(w)}{g'(w)} - \frac{\alpha_k - 1}{w - a_k} = \frac{2\alpha_k h'_k(w) + (w - a_k)h''_k(w) - (\alpha_k - 1)h'_k(w)}{\alpha_k h_k(w) + (w - a_k)h'_k(w)}. \tag{8.15}$$

The right-hand side of (8.15) is a holomorphic function on the neighborhood of $a_k$ because $h_k(a_k) \neq 0$. Hence, $g''/g'$ has an isolated singularity at the point $a_k$ which is a simple pole with residue $\alpha_k - 1$.

Finally, let us study the function $g''/g'$ around the point at infinity. One may assume that infinity corresponds to a point $z_0$ of the boundary of the polygon which is not a vertex, which is the same as considering it to be a vertex with angle $\pi$, that is, with $\alpha_0 = 1$. Then, arguing as before, one finds $g$ will be of the form

$$g(w) = z_0 + \frac{1}{w} + h(w)$$

with $h(w)$ analytic on a neighborhood of infinity. Calculating, we get

$$\frac{g''(w)}{g'(w)} = -\frac{2}{w} + \cdots \qquad \text{if } w \in \Pi^+ \text{ and } |w| \text{ is big enough.}$$

Therefore, $g''/g'$ is analytic and vanishing at infinity.

Consider now the function $\varphi(z)$ defined on the lower half plane by the equality

$$\varphi(z) = \overline{g(\bar{z})},$$

so that $\varphi$ is holomorphic on $\Pi^-$. Fix an interval $(a_k, a_{k+1})$ and apply to $\varphi$ the rotation that transforms the side $[z_k, z_{k+1}]$ of the polygon into a segment of the real axis. Then this rotated function of $\varphi$ will be the reflection of $g$ through $(a_k, a_{k+1})$ in the sense that it will join $g$ continuously on this interval. If this is done for each $k = 1, \ldots, n$, one finds several "reflections" of $g$ to the lower half plane that are different and, therefore, several extensions to $\Pi^-$ of $g'$ and $g''$. However, the value of $g''/g'$ for all these extensions is the same, because applying a rotation is just to multiply by a constant with modulus 1. Hence, it turns out that $g''/g'$ is a meromorphic function on the Riemann sphere that has a simple pole at each $a_k$ with residue $\alpha_k - 1$ and, moreover, it is analytic and vanishing at infinity. It must be, therefore, of the form (8.14). $\qquad\square$

It is worth making some comments on the Schwarz–Christoffel formula.

a) Formula (8.12) holds also when $g$ is the conformal mapping of $\mathbb{D}$ onto $U$ replacing points $a_1, \ldots, a_n$ by points $w_1, \ldots, w_n$ in $\mathbb{T}$ such that $g(w_k) = z_k$. Just use a homographic transformation between the disc and the half plane to conclude that $g''/g'$ has a simple pole at each $w_k$ with residue $\alpha_k - 1$, and vanishes at infinity. After that one may argue by reflection on the arcs that $w_1, \ldots, w_n$ determine on the unit circle. Hence, we get the formula corresponding to (8.14) and then it is enough to integrate twice.

b) For better understanding formula (8.12), note that as $w = x$ describes the real axis, $g(x)$ must travel along the boundary of $U$, which is a polygonal line. Actually, formula (8.13) leads to the equality (Figure 8.25)

$$\arg g'(x) = \arg A + (\alpha_1 - 1)\arg(x - a_1) + \cdots + (\alpha_n - 1)\arg(x - a_n).$$

Since $\arg g'(x)$ is the value of the rotation needed to pass from the real axis to the boundary of the polygon, it follows that on each interval $(a_k, a_{k+1})$, $\arg g'(x)$ remains constant. Therefore, $g(x)$ describes a line segment, but when passing through $a_k$, $\arg g'(x)$ has an increase of $-(\alpha_k - 1)\pi = \beta_k \pi$, that is, $g(x)$ jumps to the next side of the polygon.

c) With the previous considerations it is easy to give now the proof of Riemann's theorem in the particular case that the simply connected domain $U$ is the interior of a triangle.

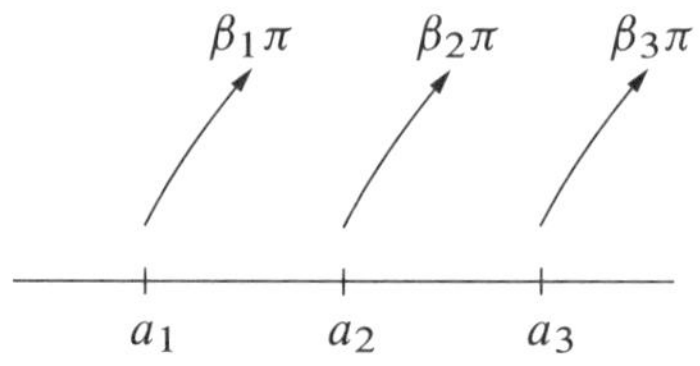

Figure 8.25

Suppose that $U$ is the interior of a triangle with vertices $z_1, z_2, z_3$ and interior angles $\alpha_k \pi$, $k = 1, 2, 3$ with $0 < \alpha_k < 2$ and $\sum_1^3 \alpha_k = 1$. Consider the function

$$g(w) = \int_{w_0}^{w} (w-a_1)^{\alpha_1 - 1}(w-a_2)^{\alpha_2 - 1}(w-a_3)^{\alpha_3 - 1}\, dw, \quad w_0, w \in \Pi^+, \quad (8.16)$$

where $a_1 < a_2 < a_3$ are three arbitrary points of $\mathbb{R}$. It is clear that $g$ is holomorphic on $\Pi^+$ and moreover it extends continuously to $\mathbb{R}$ because $1 - \alpha_k < 1, k = 1, 2, 3$, so the integral defining $g$ is convergent around $a_k$. But, in addition, this integral is convergent at the point at infinity, because changing $w$ to $1/w$, (8.16) becomes

$$-\int_{1/w_0}^{1/w} (1 - a_1 w)^{\alpha_1 - 1}(1 - a_2 w)^{\alpha_2 - 1}(1 - a_3 w)^{\alpha_3 - 1}\, dw,$$

which is continuous at the origin. In b) it has been shown that, when $w = x$ describes the real axis (including the point at infinity), $g(x)$ travels along the boundary of a triangle $T$, with vertices $g(a_1)$, $g(a_2)$ and $g(a_3)$ and interior angles $\alpha_1, \alpha_2, \alpha_3$. Moreover, the function $g$ must be one-to-one on $\Pi^+$ and it must send $\Pi^+$ to the interior of the triangle $T$.

In order to prove this last assertion, replace the half plane $\Pi^+$ by the unit disc $\mathbb{D}$ and apply the argument principle (Theorem 5.27). Even though $g$ is not holomorphic on $\overline{\mathbb{D}}$, but it is only holomorphic on $\mathbb{D}$ and continuous on $\overline{\mathbb{D}}$, Theorem 5.27 remains true (consider $g_r(w) = g(rw)$ and let $r \to 1$). Since $\mathrm{Ind}(\partial T, z) = 1$ for each $z \in T$, it turns out that $g$ takes the value $z$ exactly once in $\mathbb{D}$ (Figure 8.26).

Back to the half plane, it has been seen that the function $g$ maps conformally $\Pi^+$ into the interior of $T$. Now, the triangle $T$ and the starting triangle with vertices $z_1, z_2, z_3$ have the same angles and, therefore, are similar. Using, then, a similarity after $g$, one will go conformally from $\Pi^+$ to the given triangle, $U$.

**Remark 8.4.** Starting from four or more points $a_1 < a_2 < \cdots < a_n$ on the real axis, and angles $\alpha_1, \ldots, \alpha_n$ with $\sum \alpha_k = n - 2$, one can not guarantee that the function $g$ defined by (8.12) gives a conformal mapping of $\Pi^+$ onto the interior of a polygon of $n$ sides. Indeed it could happen that as $x$ describes the real axis, $g(x)$ travels along the boundary of a polygon which is not a Jordan curve, but instead

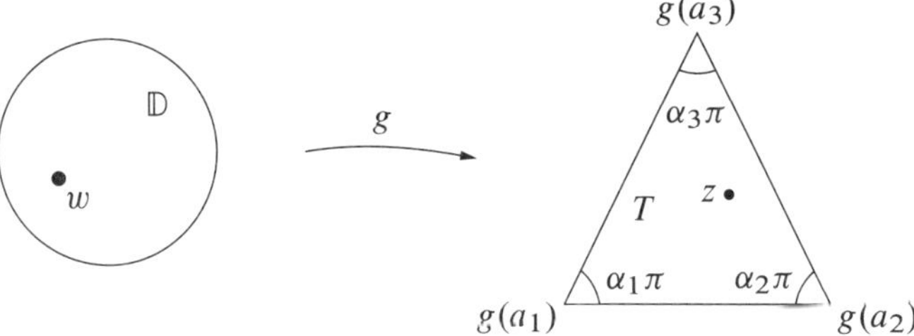

Figure 8.26

has auto-intersections (Figure 8.27). There is no easy way to know, on the basis of angles $\alpha_k$ and points $a_k$, whether the polygonal curve given by the function $g(x)$ of (8.12) for $x \in \mathbb{R}$ is simple or not.

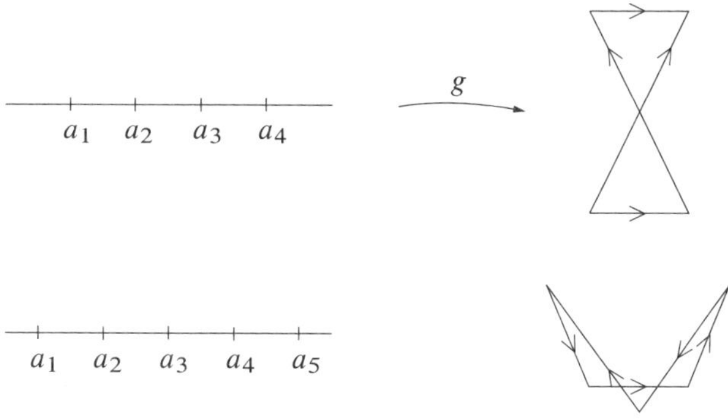

Figure 8.27

d) The Schwarz–Christoffel formula may be extended to an unbounded domain $U$ limited by a polygonal curve which, now, will not be closed. In applications one often deals with the case $z_0 = \infty$ is a vertex of the polygonal curve and the mapping $g \colon \Pi^+ \to U$ satisfies $g(\infty) = \infty$. Then formulae (8.12), (8.13) and (8.14) hold without any term corresponding to the point at infinity. This can be shown by letting one of the parameters $a_k$ tend to $\infty$.

**Example 8.32.** Suppose $\alpha_1, \alpha_2, \alpha_3$ are given with $0 < \alpha_k < 2$ and $\sum_1^3 \alpha_k = 1$. Fix the size and the position of a triangle with angles $\alpha_k \pi$, $k = 1, 2, 3$, choosing the segment $[0, 1]$ as one side and let $z_3$ be the third vertex. The conformal mapping $g$ of $\Pi^+$ onto the triangle satisfying $g(-1) = 1$, $g(0) = 0$, $g(1) = z_3$ will be of the form

$$g(w) = A \int_0^w (w + 1)^{\alpha_1 - 1} \cdot w^{\alpha_2 - 1} \cdot (w - 1)^{\alpha_3 - 1} \, dw + B.$$

Remark that, in this case, one may freely choose the points, $a_1, a_2, a_3$ of the real axis associated by $g$ to the vertices of the triangle, because there is a homographic transformation from $\Pi^+$ into $\Pi^+$ mapping a set of three real points into another one (Figure 8.28).

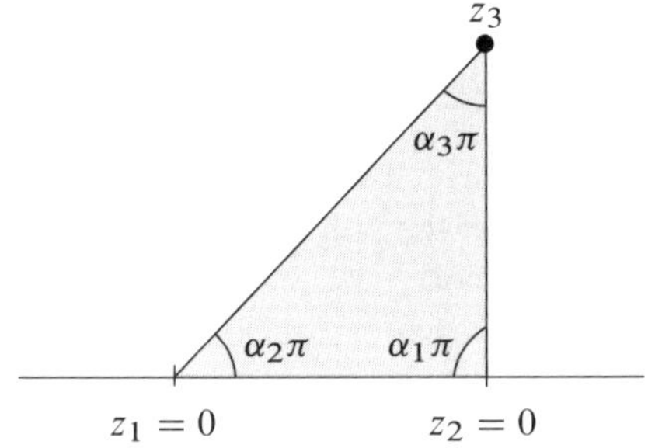


Figure 8.28

Since $g(0) = 0$, it must be $B = 0$ and $A$ is uniquely determined because $g(1) = z_3$; however the previous integral cannot be calculated explicitly.    $\square$

**Example 8.33.** Consider now the conformal mapping $g$ of $\Pi^+$ onto the domain $U$, the exterior region of a horizontal strip $B = \{z = x + iy : x > 0, -a < y < a\}$ with $a > 0$. The polygonal curve bounding $U$ has at the vertices $\pm ia$ an interior angle with measure $3\pi/2$ and at infinity, a vertex with angle 0. Take $g$ with $g^{-1}(-ia) = -1$, $g^{-1}(ia) = 1$, $g^{-1}(0) = 0$ (Figure 8.29).

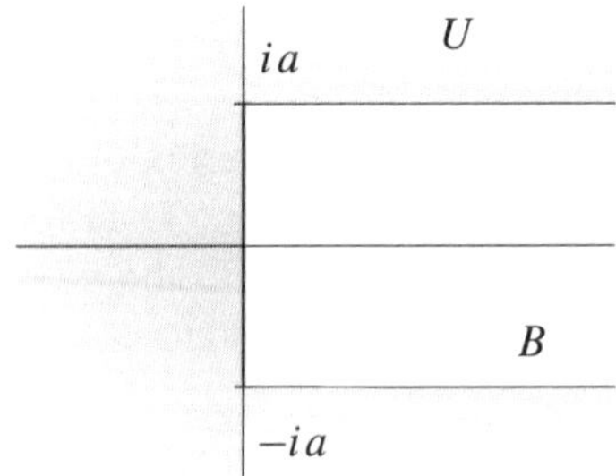


Figure 8.29

This yields

$$g(w) = A \int_0^w (w^2 - 1)^{1/2} \, dw.$$

This integral may be computed (exercise) and gives

$$g(w) = A(w\sqrt{1 - w^2} + \arcsin w).$$

Since $g(1) = A\pi/2 = ia$, we get

$$g(w) = i\frac{2a}{\pi}(w\sqrt{1 - w^2} + \arcsin w). \qquad \square$$

**Example 8.34.** Let us exhibit a conformal mapping of the half plane $\Pi^+$ into itself with the vertical segment $[0, ia]$, $a > 0$, removed.

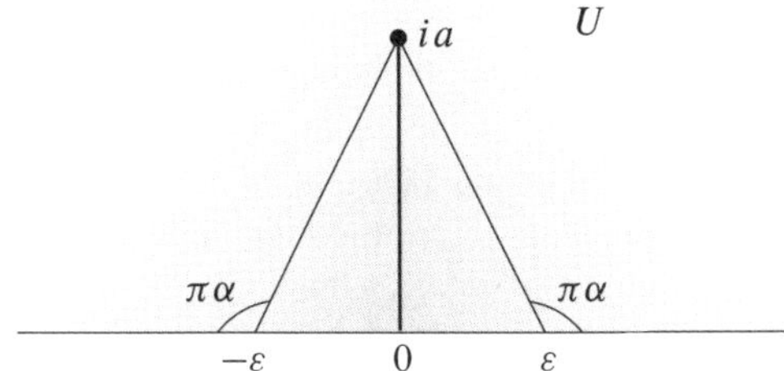


Figure 8.30

In this case, one looks at the domain $U$ as the limit of domains $U_\varepsilon$ obtained by removing from $\Pi^+$ a triangle with a small basis of length $2\varepsilon$, and height $a$, when $\varepsilon \to 0$ (Figure 8.30). For $U_\varepsilon$, the conformal mapping $g_\varepsilon$ such that $g_\varepsilon(-1) = -\varepsilon$, $g_\varepsilon(0) = ia$, $g_\varepsilon(1) = \varepsilon$ is given by

$$g_\varepsilon(w) = A \int_0^w (w + 1)^{\alpha - 1} \cdot w^{2 - 2\alpha} \cdot (w - 1)^{\alpha - 1} \, dw + B$$

(see the figure to understand the meaning of $\alpha$)   Now, when $\varepsilon \to 0$, one has $\alpha \to 1/2$ and it follows that

$$g(w) = A \int_0^w \frac{w}{\sqrt{w^2 - 1}} \, dw + B,$$

which yields

$$g(w) = A\sqrt{w^2 - 1} + B$$

with $g(-1) = B = 0$, $g(0) = iA = ia$. Therefore,

$$g(w) = a\sqrt{w^2 - 1}. \qquad \square$$

## 8.9 Conformal mapping of doubly connected domains

Riemann's theorem classifies simply connected domains, up to conformal mapping: the plane $\mathbb{C}$ and the unit disc $\mathbb{D}$.

What happens if one considers $n$-connected domains with $n \geq 2$ (Definition 1.14)? That is, what are the classes of conformally equivalent $n$-connected domains? This problem is much more complicated and will not be dealt with here. We will refer, basically, to the case $n = 2$ without going into detail.

Consider, then, doubly connected domains, that is, simply connected domains with a hole. The easiest examples are the annuli. If $0 < r < R$, let $C(r, R) = C(0, r, R) = \{z \in \mathbb{C} : r < |z| < R\}$ be an annulus centered at the origin. One may prove that every doubly connected domain is conformally equivalent to an annulus or to a punctured disc or to the punctures plane (which may be taken as annuli as well). Considering this result, which will not be proved here, one gets that in order to classify doubly connected domains it is enough, basically, to classify annuli centered at the origin. The fact is that, in general, two different annuli are not equivalent.

Dilating the annulus $C(r, R)$ by a factor $\alpha > 0$, that is, applying the transformation $z \to \alpha z$, one obtains the annulus $C(r', R')$ with $r' = \alpha r$, $R' = \alpha R$ (Figure 8.31).

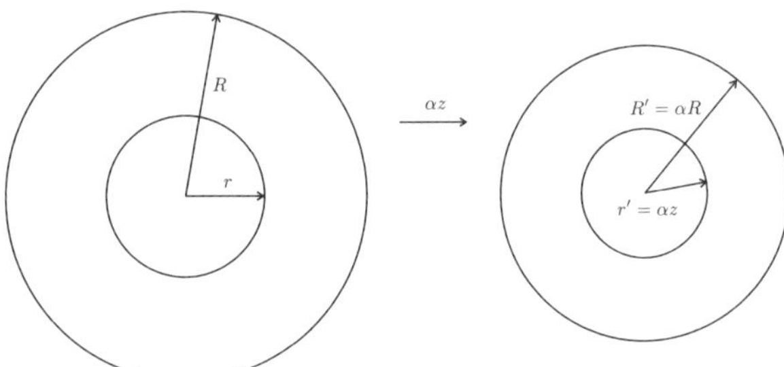


Figure 8.31

These two annuli are conformally equivalent and their radii satisfy

$$\frac{R}{r} = \frac{R'}{r'}. \tag{8.17}$$

It is clear that condition (8.17) is sufficient for $C(r, R)$ and $C(r', R')$ to be equivalent, since one needs only to take $z \to \alpha z$ with $\alpha = r'/r = R'/R$. The crucial fact is that it is also necessary, as the following statement shows.

**Theorem 8.35.** *Two annuli $C(r, R)$ and $C(r', R')$ are conformally equivalent if and only if the condition*

$$\frac{R}{r} = \frac{R'}{r'}$$

*holds.*

*Proof.* Applying dilations one may assume $r = r' = 1$. Let $w = f(z)$ be a conformal mapping from $C(1, R)$ onto $C(1, R')$ and let us show that $R = R'$.

By Proposition 8.9, when $|z| \to 1$ in $C(1, R)$, we must have either $|f(z)| \to 1$, or $|f(z)| \to R'$. Changing $f$ by $R'/f$ if necessary, one may get $|f(z)| \to 1$ when $|z| \to 1$ and, consequently, $|f(z)| \to R'$ when $|z| \to R$ (Figure 8.32).

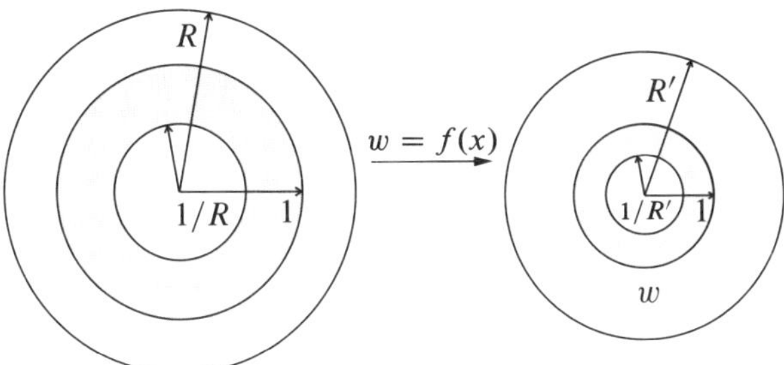


Figure 8.32

Now, applying the reflection principle (Proposition 8.22) with respect to the unit circle one may extend $f$ to a conformal mapping from the annulus $\{z : 1/R < |z| < 1\}$ onto the annulus $\{w : 1/R' < |w| < 1\}$. Applying once more this principle, now with respect to the circle with radius $1/R$, one will extend $f$ to a mapping of $\{z : \frac{1}{R^2} < |z| < \frac{1}{R}\}$ onto $\{w : \frac{1}{(R')^2} < |w| < \frac{1}{R'}\}$. Iterating this process one will obtain a conformal mapping from the punctured disc $D'(0, R)$ onto $D'(0, R')$, which has a removable singularity at the origin. So, one arrives at a conformal mapping from $D(0, R)$ onto $D(0, R')$, still denoted by $f$, satisfying $f(0) = 0$ and sending the circle $\{z : |z| = 1\}$ onto the circle $\{w : |w| = 1\}$.

Now, the function $z \to \frac{f(Rz)}{R'}$ is an automorphism of the unit disc $\mathbb{D}$ that sends the point 0 to 0. Therefore, by Theorem 8.27, it is a rotation,

$$\frac{f(Rz)}{R'} = e^{i\alpha} z, \quad \alpha \in \mathbb{R}, \ |z| < 1,$$

or

$$f(z) = R' e^{i\alpha} \frac{z}{R}, \quad |z| < R.$$

Taking $|z| = 1$, which implies $|f(z)| = 1$, it yields that $R = R'$.  $\square$

The theorem just proved states that every annulus of the plane is conformally equivalent to some annulus of the family $C(1, R)$ with $R > 0$, and two different annuli of this family are not equivalent to each other.

Other simple examples of doubly connected domains are the punctured disc, $\mathbb{D} \setminus \{0\}$, and the punctured plane, $\mathbb{C} \setminus \{0\}$. These two domains are not equivalent to each other and cannot be equivalent to any annulus, because the corresponding conformal mapping would extend analytically to the point 0, according to part a) of Theorem 5.5. As it has been said, one may prove that every doubly connected domain of the plane is conformally equivalent to the punctured plane, or to the punctured disc or to an annulus $C(1, R)$, $R > 1$. This result describes, up to conformal mapping, all domains with connection degree equal to 2.

**Example 8.36.** Consider the ellipse centered at the origin with semi-axes $5/2$ and $3/2$ and let $U$ be the interior of this ellipse with the segment $[-1, 1]$ removed. This way $U$ is a doubly connected domain, which must be, as just said, conformally equivalent to an annulus. Example F) of Section 8.7 shows that this annulus will be $C(1/2, 1)$ and that the mapping of $U$ onto the annulus is given by the inverse of Jowkowsky's mapping (Figure 8.33). $\qquad\square$

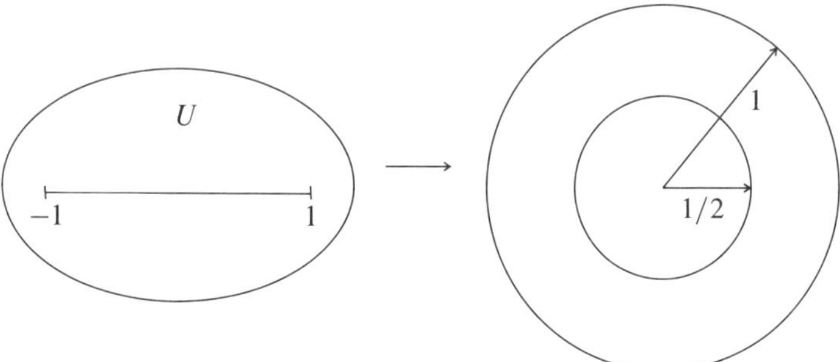


Figure 8.33

As it was said at the beginning of this section, conformal mappings between domains with connection degree greater than 2 are more complicated. One result that extends in some way what has been proved about doubly connected domains is the following one (see [1], p. 255):

*If $U$ is an $n$-connected domain, then $U$ is conformally equivalent to an annulus with $n - 2$ arcs located in circles centered at the origin removed (Figure 8.34).*

In this case there are two boundary contours of the components of $\mathbb{C} \setminus U$ that are mapped to the boundary of the annulus, while the other ones are mapped to the arcs.

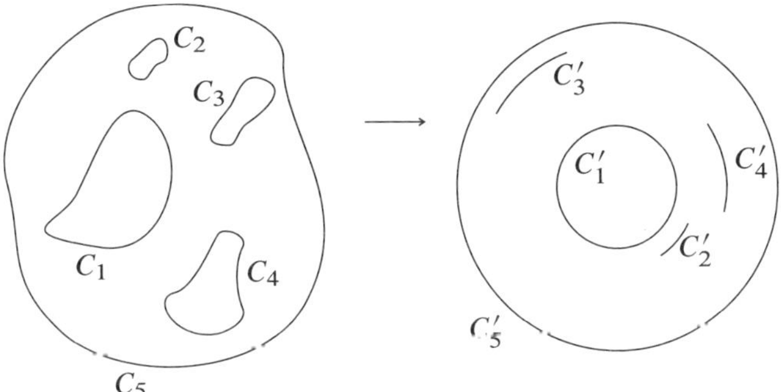

Figure 8.34

## 8.10  Applications of conformal mapping

In this section two applications of conformal mapping to quite different problems
are given.

### 8.10.1  Transverse Mercator projection

The problem considered here is the basic problem of cartography, that is, the mapping of the surface of the Earth, or a piece of it, on a plane. Ideally this representation
should preserve distances, so that, except for a scaling factor, the distance between
two points on the Earth coincides with the distance between the corresponding
points on the plane. But this is impossible, because no part of the sphere can be
isometrically parameterized by a plane. This fact is a consequence of a very important theorem in differential geometry, Gauss's Theorem Egregium, but, in the
case of the sphere, it may be also proved with elementary considerations.

What kind of mappings are used in cartography, then, if they do not preserve
distances? Remark that a mapping of the Earth on a plane preserving distances
would preserve angles too; otherwise, representations used in cartography preserve
angles, although they cannot preserve distances. The idea of a conformal projection
(that is, preserving angles) of the Earth on a plane was conceived by Mercator, and
the projection used nowadays is a kind of Mercator projection, called *transverse
Mercator projection*. In this projection the conformal mapping plays an important
role, as explained below.

The Mercator projection is a representation of the Earth on a plane, namely of
coordinates $(x, y)$, that satisfies the following conditions:

a) It is conformal.

b) It maps the equator onto the $y$-axis, preserving distances (on the equator).

c) It maps the points of the Earth of positive longitude onto the right half plane
   ($x > 0$) and the points of positive latitude onto the upper half plane ($y > 0$).

Mercator projection works because it is conformal and moreover parallels are transformed into horizontal straight lines, and meridians, into vertical ones. But it has the disadvantage that, as one moves far away from the equator, distances are much distorted. This is why it has been replaced by the so-called transverse Mercator projection. The idea behind this projection is to divide the region of the Earth between the two polar circles into spindles of $6°$ of width and apply to each of these spindles a different Mercator projection where the central meridian of the spindle plays the role of the equator.

Hence, consider the region $R$ of the Earth between the two polar circles and between two meridians giving a $6°$ spindle, and let $m_0$ be the central meridian of this region. We want to find a projection of $R$ on the plane $(x, y)$ such that the following conditions hold:

a$'$) It is conformal.

b$'$) It maps the meridian $m_0$ onto the $y$-axis.

c$'$) Distances are preserved on the meridian $m_0$.

Of course some scaling factors must be taken into account, which we do not consider here.

In order to construct this projection, consider the Earth as an ellipsoid with semi-axes $a, b$, so that $R$ may be parameterized in the form

$$x = a \cos u \cos \lambda,$$

$$y = a \cos u \sin \lambda, \quad (\lambda, u) \in V, \tag{8.18}$$

$$z = b \sin u,$$

where $\lambda$ is the longitude of the point $P(x, y, z)$ counted from $m_0$, $u$ is the angle represented in Figure 8.35 and $V$ is an open set of $\mathbb{R}^2$. Observe that $u$ is not the latitude, $\varphi$, of $P$, but it is really close. The latitude is the angle with the plane $x, y$ of the perpendicular to the ellipsoid through $P$, and one may easily set the relation between $u$ and $\varphi$. Remark also that in the case of a sphere, $a = b$, $u$ is the latitude, and the parametrization (8.18) corresponds to spherical coordinates.

The function $(x, y, z) = f(\lambda, u)$ given by (8.18), for $(\lambda, u) \in V$, exhibits $R$ as a surface parameterized by $f$. The parametrization $f$ is said to be conformal if the tangent linear mapping $f'(\lambda, u)$ of $\mathbb{R}^2$ into the tangent plane to $R$ at the point $f(\lambda, u)$ preserves angles, when the usual scalar product is taken in $\mathbb{R}^2$ and the scalar product induced by the one of $\mathbb{R}^3$ is taken in the tangent space. It is known that $f$ is conformal when the matrix of vectors $\frac{\partial f}{\partial \lambda}$ and $\frac{\partial f}{\partial u}$ is a multiple of an orthogonal matrix, that is, when the matrix of their scalar products is

$$h(\lambda, u) \begin{pmatrix} 1 & 0 \\ 0 & 1 \end{pmatrix}, \tag{8.19}$$

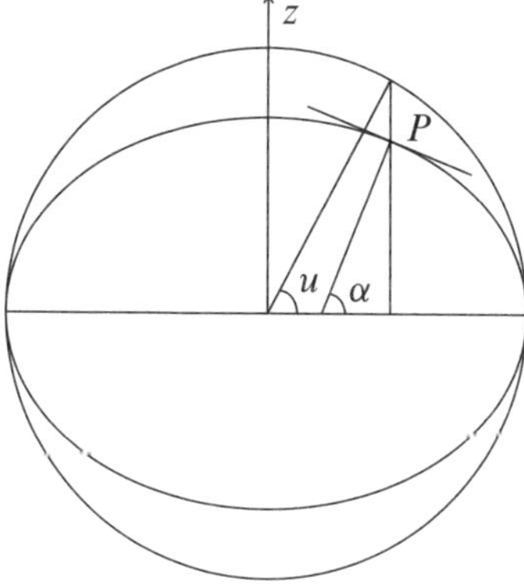

Figure 8.35

where $h(\lambda, u)$ is a differentiable function non-vanishing at any point.

Thus, one has

$$\frac{\partial f}{\partial \lambda} = (-a \cos u \sin \lambda, a \cos u \cos \lambda, 0),$$

$$\frac{\partial f}{\partial u} = (-a \sin u \cos \lambda, -a \sin u \sin \lambda, b \cos u),$$

and the matrix of scalar products is

$$\begin{pmatrix} a^2 \cos^2 u & 0 \\ 0 & a^2 \sin^2 u + b^2 \cos^2 u \end{pmatrix} = a^2 \cos^2 u \begin{pmatrix} 1 & 0 \\ 0 & \tan^2 u + \left(\frac{b}{a}\right)^2 \end{pmatrix},$$

which is not of the form in (8.19). Now, if one introduces a new variable $l$, defined by

$$l = l(u) = \int \sqrt{\tan^2 u + \left(\frac{b}{a}\right)^2}\, du,$$

and changes $u$ by $l$, it turns out that in coordinates $(\lambda, l)$ the new matrix of scalar products is

$$a^2 \cos^2 u(l) \begin{pmatrix} 1 & 0 \\ 0 & 1 \end{pmatrix},$$

which actually has the form of (8.19). This means that the new parametrization of $R$ given by

$$\begin{array}{ccc} V & \xrightarrow{f} & R, \\ (\lambda, l) & \longmapsto & f(\lambda, l), \end{array}$$

is conformal. Its inverse $f^{-1}\colon R \to V$ is also conformal and maps the meridian $m_0$, which corresponds to $\lambda = 0$, onto the $l$-axis. Hence, in order to make $f^{-1}$ be the transverse Mercator projection, condition b′) about preservation of distances between two points of $m_0$ and their corresponding points in the plane must hold.

To this end, introduce complex notation writing $w = \lambda + il$ for the point $(\lambda, l)$ and look for a function $F\colon V \to \mathbb{C}$ such that the plane representation of $R$ given by the function $F \circ f^{-1}\colon R \to \mathbb{C}$ satisfies a′), b′) and c′). This will be achieved by choosing $F$ holomorphic, hence conformal, and then imposing it to transform the imaginary axis into itself: $F(il) = iy$, if $F = x + iy$. Finally, the preservation of distances on $m_0$ may be obtained this way:

For each point $P$ of $m_0$ one has $f^{-1}(P) = il$, where $l$ is the parameter corresponding to $P$. Let $\tau$ be the distance from $P$ to the equator following $m_0$, so as $P$ changes, $\tau$ is a function of $l$, $\tau = \tau(l)$. This function is not explicitly known, but it is easy to argue that it is a real analytic function (Figure 8.36).

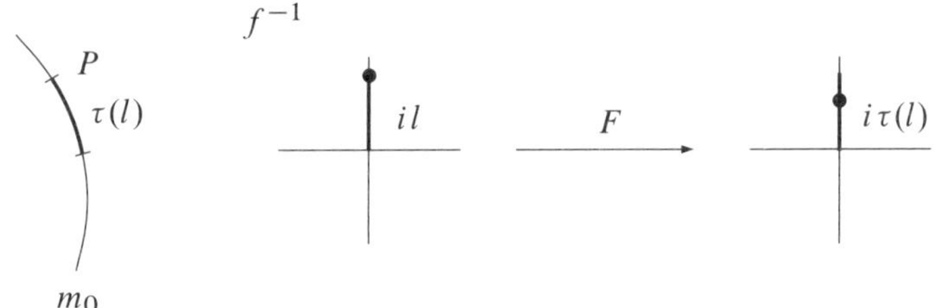


Figure 8.36

If $F$ can be chosen such that $F(il) = i\tau(l)$, that is, the ordinate of $F(il)$ coincides with the distance to the equator of the point of parameters $(0, l)$, property b′) will hold. The problem, from the point of view of complex variables, is then to find a holomorphic function $F$ on $V$ known on the imaginary axis through $F(il) = i\tau(l)$, where $\tau(l)$ is analytic. The solution of this problem, locally at least, is easy to find because $\tau(l)$ will be given by a power series, say $\tau(l) = \sum_0^\infty c_n l^n$. Now take as $F$ the analytic extension of $\tau$ obtained by substituting $l$ by the complex variable $w$, indeed $w/i$, and define

$$F(w) = i \sum_0^\infty c_n \left(\frac{w}{i}\right)^n,$$

so that $F$ is holomorphic and $F(il) = i\sum_0^\infty c_n l^n = i\tau(l)$.

More details on the previous construction may be found in Joan Girbau's paper: "Si no es pot representar cap terreny a escala, que fan els cartògrafs?" (*Les bases matemàtiques de la civilització tecnològica*, Fundació Caixa de Sabadell, 1999, pages 73–81), on which this exposition is based.

## 8.10.2 Conformal mapping and hydrodynamics

Suppose that one wants to study the motion of a perfect, incompressible, homogeneous and irrotational fluid that flows through a simply connected plane region $U$ (see Subsection 7.1.3 for general notions on fluids). Using complex notation, one will represent the velocity vector field of the fluid by a function $f(z)$, defined for $z \in U$, and will suppose that the velocity does not change with time.

The incompressibility means that $\mathrm{div}(f) = 0$, that is, writing $f = u + iv$, equation $u_x + v_y = 0$ holds. If $\gamma$ is a closed path inside $U$ with normal vector $\vec{N}$, Green's formula gives

$$\int_\gamma \langle f, \vec{N} \rangle \, ds = 0,$$

which means that the flow of the fluid through $\gamma$ is zero. One also says that *there are no sources in $U$*, that is, intakes or outtakes of fluid at any point of $U$. It could happen, however, that there are sources at points of the boundary of $U$.

Moreover the fluid being irrotational, that is, $\mathrm{rot}(f) = 0$, equation $v_x - u_y = 0$ holds and so the function $\overline{f(z)} = u - iv$, the conjugate of the velocity vector field, is holomorphic on $U$. Since $U$ is simply connected, there is a holomorphic antiderivative $F$ of $\bar{f}$ on $U$. Writing $F = \varphi + i\psi$, the relation $F' = \bar{f}$ becomes

$$\vec{\nabla}\varphi = \vec{\nabla}^{\perp}\psi = f.$$

The function $\varphi$ is, therefore, a *potential function* of the vector field $f$ and the holomorphic function $F$ is called a *complex potential* of this vector field. The function $\psi$ that appears in item d) of Subsection 7.1.3, but not depending on $t$ now, is harmonic because $\psi = \mathrm{Im}\, F$. This corresponds to Euler's equation $\Delta\psi = -\mathrm{rot}(f) = 0$. The function $\psi$ is called a *stream function* of the fluid. The reason for this name is the following: the level lines of $\varphi$ are perpendicular to the vector $\vec{\nabla}\varphi = f$. As remarked in Subsection 7.1.1, the level lines of a pair of conjugated harmonic functions are perpendicular to each other. Therefore, level lines of $\psi$ are tangent to $f$ and, consequently, they are the trajectories of the vector field $f$ which are the trajectories of the fluid or *stream lines*.

Now it is convenient to consider the behavior of the fluid at the boundary of the domain $U$. It has been supposed there were no sources in $U$, but there may be at some points of $\partial U$. Observe that the word *source* has a relative meaning because it may be an intake point of fluid as well as an outtake one: it depends on the sign of the flow of the vector field $f$ around the point.

Suppose that in case there is no source in a whole arc of $\partial U$, the velocity vector field $f$ is tangent to $\partial U$ along this arc, that is, $\langle f, \vec{N} \rangle = 0$ if $\vec{N}$ is the normal vector to $\partial U$. Then this arc of $\partial U$ may be considered as a stream line, and, therefore, the stream function $\psi = \mathrm{Im}\, F$ must be constant on this part of $\partial U$. An example is a fluid flowing through the upper half plane $\Pi^+ = \{w : \mathrm{Im}\, w > 0\}$ with an intake source at the origin. The stream function $\psi$ is a harmonic function on $\Pi^+$, constant on each ray $(-\infty, 0)$ and $(0, +\infty)$. This function is $c \, \mathrm{Arg}\, w$, with $c$ constant. The complex potential of the fluid is $F(w) = c \, \mathrm{Log}\, w$, $w \in \Pi^+$, and the velocity vector field

$$f(w) = \overline{F'(w)} = c\,\frac{1}{\overline{w}},$$

given in Example 3.6 (see Figure 8.37). This vector field has a source at the origin because if $\gamma$ is the circle with center 0 and radius 1, one has $\int_\gamma \langle f, \vec{N} \rangle \, ds = 2\pi c$. The flow of the vector field $f$ entering into $\Pi^+$ is given by the previous integral extended only to $\gamma \cap \Pi^+$ and equals $c\pi$. If $c > 0$, there is a fluid intake; if $c < 0$, a fluid outtake.

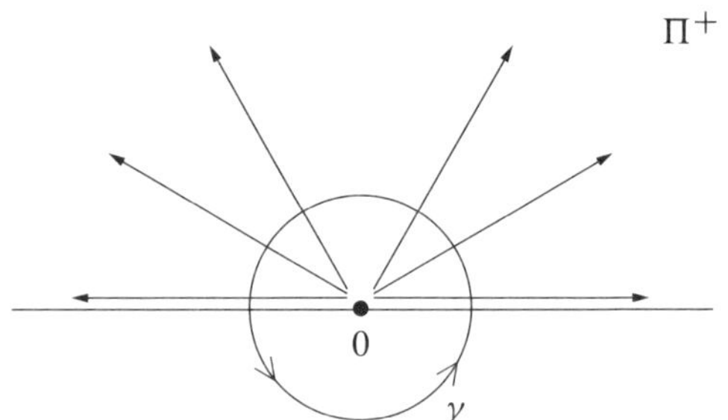


Figure 8.37

Even in the case when there is no source in the boundary of the domain, the fact that $\psi$ is harmonic on $U$ and constant on $\partial U$ does not determine the stream function. For example in the upper half plane $\Pi^+$ of the plane $w = u + iv$, the function $\psi_1(u, v) = Av$ with $A$ a real constant is harmonic and vanishing on $\partial \Pi^+$. It corresponds to the complex potential $F_1(w) = Aw$. But also the function $\psi_2(u, v) = Ae^u \sin v$ that corresponds to the complex potential $F_2(w) = Ae^w$ is harmonic and vanishing on the boundary of $\Pi^+$. In the first case the stream line $\psi_1 = 0$ consists of $\partial \Pi^+$ while in the second one the line $\psi_2 = 0$ consists of the real axis plus the straight lines $v = n\pi, n \in \mathbb{N}$, which are inside the domain, and $Ae^w$ is the complex potential for the flow in a strip of width $\pi$. In many cases, however, the physical conditions of the problem make clear that the stream function is uniquely determined, being harmonic on the domain and constant on its boundary. When the domain is unbounded and there are no sources at the boundary, an analytic condition that determines the stream function is a correct behavior at the point of infinity, given by a condition of the kind $\lim_{|w| \to \infty} (F(w) - Aw) = 0$, for some constant $A$ where $F$ is the complex potential.

Assume that in a simply connected domain $U'$ one has a fluid flowing with a velocity vector field $f$ which has a complex potential $F = \varphi + i\psi$. Let $g$ be a conformal mapping of another domain $U$ onto $U'$ sending the boundary of $U$ into the boundary of $U'$. Then harmonic functions $\varphi \circ g$ and $\psi \circ g$ may be interpreted as the potential of the velocity and as the stream function of a flow in the region $U$, having $F \circ g$ as complex potential. For example consider $U' = \Pi^+$ and the uniform flow given by the complex potential $F(w) = Aw$, $A$ a real constant, as done before and let $U$ be the first quadrant of the plane $z = x + iy$, $U = \{z : x, y > 0\}$; then the transformation $w = z^2$ maps conformally $U$ onto $\Pi^+$ and preserves the boundaries. The function $Az^2 = A(x^2 - y^2 + 2ixy)$ is the complex potential of a flow in $U$ that has as stream function $2Axy$. The stream lines are the hyperbolas $xy = c$, $c$ constant (Figure 8.38) and the velocity vector field is $2A\bar{z} = 2A(x - iy)$ of modulus $2A\sqrt{x^2 + y^2}$. Consider now in a domain $U$ a uniform flow, that is,

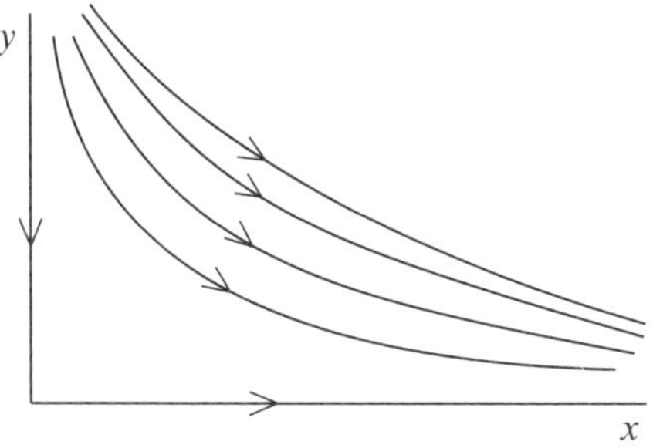

Figure 8.38

corresponding to the uniform flow in the upper half plane of complex potential $Aw$ by means of a conformal mapping $g : U \to \Pi^+$ which transforms $\partial U$ into $\partial \Pi'$. Then, the stream lines in $U$ may be described in terms of the inverse mapping $h = g^{-1} : \Pi^+ \to U$ by means of curves $\gamma(t) = h(t + ic)$, $t \in \mathbb{R}$, $c$ constant. This is what is done in the two following examples.

**Example 8.37.** Consider a fluid moving uniformly in the upper half plane with an obstacle given by a vertical stick $[0, ia]$, $a > 0$, so that there is no source on the real axis (Figure 8.39).

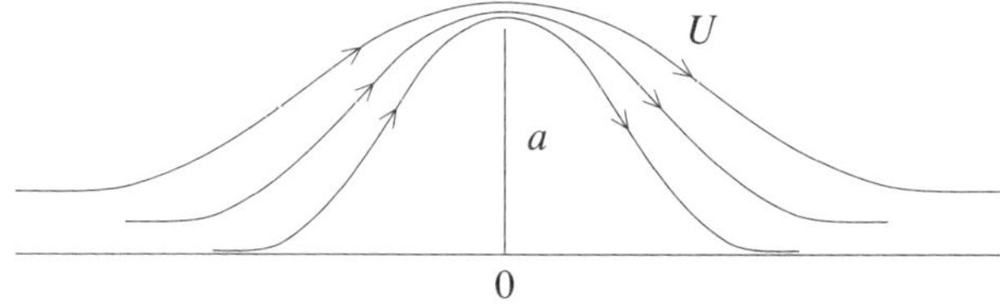


Figure 8.39

In Example 8.34 it was seen that the conformal mapping from $\Pi^+$ onto $U = \Pi^+ \setminus [0, ia]$ is given by the function

$$h(w) = a\sqrt{w^2 - 1},$$

sending $\infty$ to $\infty$. Therefore, the stream lines are the trajectories

$$\gamma(t) = h(t + ic) = a\sqrt{(t + ic)^2 - 1}, \quad t \in \mathbb{R},\ c > 0. \qquad \square$$

**Example 8.38.** Let us study now the uniform motion of a fluid that flows horizontally and jumps a step of height 1 (Figure 8.40). The corresponding domain is the upper half plane with a step:

$$U = \{z : \operatorname{Re} z > 0,\ \operatorname{Im} z > 0\} \cup \{z : \operatorname{Re} z < 0,\ \operatorname{Im} z > 1\}.$$

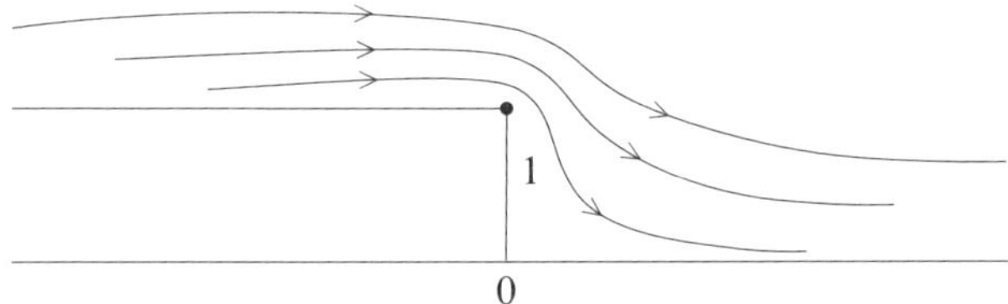


Figure 8.40

We look for the conformal mapping $h$ from $\Pi^+$ onto $U$ with $h(\infty) = \infty$, $h(-1) = i$, $h(1) = 0$. Applying the Schwarz–Christoffel formula with angles $\alpha_1 = 3\pi/2$, $\alpha_2 = \pi/2$ yields that

$$h(w) = A \int_0^w (w + 1)^{\frac{1}{2}} (w - 1)^{-\frac{1}{2}}\, dw + B, \quad A \text{ and } B \text{ constants.}$$

This integral may be computed, turning it into a rational one using the change of variable $\frac{w+1}{w-1} = u$, and this yields

$$h(w) = A(\sqrt{w^2 - 1} + \operatorname{Log}(w + \sqrt{w^2 - 1})) + B.$$

Using $h(-1) = i$, $h(1) = 0$ one obtains

$$Ai\pi + B = i, \quad B = 0,$$

that is, $A = 1/\pi$, $B = 0$,

Hence, the stream lines are the trajectories

$$\gamma(t) = h(t + ic)$$

$$= \frac{1}{\pi}(\sqrt{(t+ic)^2 - 1} + \operatorname{Log}(t + ic + \sqrt{(t+ic)^2 - 1})), \quad t \in \mathbb{R},\ c > 0. \qquad \square$$

**Example 8.39.** Consider, finally, the uniform flow of a liquid that flows horizontally and finds a semicylindrical obstacle. The model in this case is given by the domain $U = \{z \in \Pi^+ : |z| > 1\}$ (Figure 8.41). This domain is conformally mapped into $\Pi^+$ by the function

$$w = g(z) = \frac{1}{2}\left(z + \frac{1}{z}\right),$$

that transforms the boundary of $U$ into the real axis (Example F) of Section 8.7).

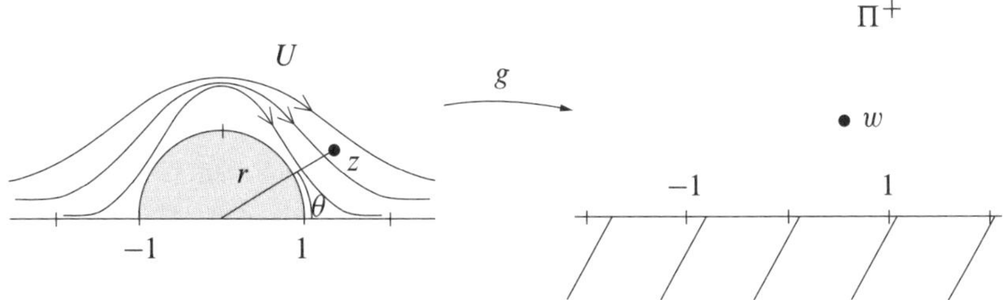

Figure 8.41

Taking polar coordinates in the plane $z$, $z = re^{i\theta}$, it turns out that

$$\operatorname{Re} g(r, \theta) = \frac{1}{2}\left(r + \frac{1}{r}\right)\cos\theta,$$

$$\operatorname{Im} g(r, e^{i\theta}) = \frac{1}{2}\left(r - \frac{1}{r}\right)\sin\theta,$$

so that the stream lines are the trajectories

$$\operatorname{Im} g(re^{i\theta}) = \frac{1}{2}\left(r - \frac{1}{r}\right)\sin\theta = c, \quad c > 0. \qquad \square$$

## 8.11 Exercises

1. Let $D_1$, $D_2$ be two discs of the plane and $z_0 \in D_1$, $w_0 \in D_2$. Find a conformal mapping $f : D_1 \to D_2$ such that $f(z_0) = w_0$.

2. Let $f : U \to U$ be holomorphic where $U = \{z = x + iy : |y| < \pi/2\}$; use Schwarz's lemma to prove that $|f'(x)| \leq 1$ for $x \in \mathbb{R}$.

3. Let $f \in H(\mathbb{D})$ with $f(0) = 1$ and $\operatorname{Re} f(z) \geq 0$, $z \in \mathbb{D}$. Show that

$$\frac{1 - |z|}{1 + |z|} \leq |f(z)| \leq \frac{1 + |z|}{1 - |z|}, \quad z \in \mathbb{D}.$$

What may be said when some one of the two previous inequalities is an equality at some point $z \in \mathbb{D}$?

4. Let $f$ be a continuous function on $\overline{\mathbb{D}}$ and meromorphic on $\mathbb{D}$ satisfying $|f(z)| = 1$ if $|z| = 1$. Let $a_1, a_2, \ldots, a_n$ and $b_1, b_2, \ldots, b_m$ be the zeros and the poles of $f$ in $\mathbb{D}$, respectively (justify that there are only a finite number of them). Writing $\tau_a(z) = \frac{z-a}{1-\bar{a}z}$ when $|a| < 1$, show that the function

$$f(z) \cdot \tau_{b_1}(z) \ldots \tau_{b_m}(z) \frac{1}{\tau_{a_1}(z)} \cdots \frac{1}{\tau_{a_n}(z)} \quad \text{for } |z| < 1,$$

is a constant of modulus 1.

5. Let $f : \mathbb{D} \to U$ be a conformal transformation such that $U = f(\mathbb{D})$ is a convex domain of the plane. Show that for each $0 < r < 1$ the domain $f(D(0, r))$ is also convex.

   *Hint:* For $z_1, z_2 \in D(0, r)$ with $|z_1| \le |z_2|$ and $0 < t < 1$ consider the function $\varphi(z) = tf(zz_1/z_2) + (1 - t)f(z)$.

6. Prove that a homographic transformation maps a pair of concentric circles into another pair of concentric circles and the ratio of the radii remains constant.

7. Let $f : \mathbb{D} \to \mathbb{D}$ be holomorphic with Taylor series $f(z) = \sum_{n=0}^{\infty} c_n z^n$ around the origin. Show, by induction, that for each $n \in \mathbb{N}$ there is a Blaschke product, $B_n$, of degree $n$ (Example 8.29), having as Taylor series at the origin

$$B_n(z) = c_0 + c_1 z + \cdots + c_{n-1} z^{n-1} + b_n z^n + \cdots, \quad b_n \in \mathbb{C}.$$

   That is, $f$ and $B_n$ have the same $n$ first Taylor coefficients.

8. Find a conformal mapping of the upper half plane onto the domain $U = \mathbb{D} \setminus \{x \in \mathbb{R} : 0 \le x \le 1\}$.

9. Let $U$ be a Jordan domain and $f \in C(\overline{U}) \cap H(U)$ such that $f$ is injective on $\partial U$. Show that $f$ is a conformal mapping of $U$ onto $f(U)$ and a homeomorphism of $\overline{U}$ onto $\overline{f(U)}$.

   *Hint:* Use the argument principle.

10. Find a conformal mapping of the upper half plane onto the exterior domain of an ellipse with focuses at the points 1 and $-1$.

11. Let $\Pi^+$ be the upper half plane and $U = \Pi^+ \setminus \{z = iy : 1 \le y < +\infty\}$. Prove that on $U$, continuous branches of the functions $\sqrt{1 + z^2}$ and $\sqrt{1 - z^2}$ can be defined and use them to show that the transformation $f(z) = \frac{z}{\sqrt{1-z^2}}$ maps conformally $\Pi^+$ onto $U$.

**12.** Let $Q$ be a square centered at the origin and $f : \mathbb{D} \to Q$ a conformal mapping of $\mathbb{D}$ onto $Q$. Put $f(z) = \sum_{n=0}^{\infty} c_n z^n$, if $|z| < 1$. Show that $c_n = 0$ if $n$ is not of the form $n = 4k + 1$, $k \in \mathbb{N}$.

*Hint:* The equality $f(iz) = if(z)$ must hold for $z \in \mathbb{D}$.

**13.** Find a conformal mapping from the unit disc onto the exterior region of a parabola with focus at the origin and vertex at the point 1.

**14.** Find a harmonic function $u$ on the unit disc $\mathbb{D}$ such that $\lim_{z \to e^{i\theta}} u(z) = 0$ if $-\pi/2 < \theta < \pi/2$ and $\lim_{z \to e^{i\theta}} u(z) = 1$ if $\pi/2 < \theta < 3\pi/2$. Show that this problem has more than one solution.

**15.**   a) Give precise hypotheses in order that the function $u$ defined by equality (8.7) is the solution of Dirichlet's problem in the half plane $\Pi^+$ with boundary values equal to $\varphi$ and prove it.

   b) Proceed analogously for Neumann's problem in $\Pi^+$ with the function defined by (8.8).

**16.** For $c > 0$, find the integral expression of a conformal mapping of the upper half plane onto the domain $U$ given by

$$\{z \in \mathbb{C} : 0 < \operatorname{Re} z < 1, \ \operatorname{Im} z > 0\} \cup \{z \in \mathbb{C} : \operatorname{Re} z > 0, \ 0 < \operatorname{Im} z < c\}.$$

In the particular case $c = 1$, find explicitly the conformal mapping $f : \Pi^+ \to U$ such that $f(0) = 0$, $f(1) = \infty$ and $f(\infty) = 1 + i$.

**17.** Show that the function

$$f(z) = \int_0^z \frac{d\xi}{\sqrt{(1 - \xi^2)(1 - k^2 \xi^2)}}, \qquad \text{where } 0 < k < 1,$$

gives a conformal mapping of the upper half plane onto a rectangle. Find the vertices of this rectangle by means of the parameters

$$K = \int_0^1 \frac{dt}{\sqrt{(1 - t^2)(1 - k^2 t^2)}}, \qquad K' = \int_0^{1/k} \frac{dt}{\sqrt{(t^2 - 1)(1 - k^2 t^2)}},$$

$t \in \mathbb{R}$. Prove that the transformation $f^{-1}$ extends to a meromorphic function on the whole complex plane and find its zeros and poles.

**18.** Show that functions

$$f_I(z) = \int_0^z (1 - \xi^n)^{-2/n} \, d\xi \quad \text{and} \quad f_E(z) = \int_0^z (1 - \xi^n)^{2/n} \frac{d\xi}{\xi}, \qquad n \in \mathbb{N},$$

apply conformally the unit disc onto the interior and exterior domains, respectively, of a regular polygon of $n$ sides.

**19.** Find a conformal mapping from the domain $U = \{z \in \mathbb{C} : |z| > 1, \operatorname{Im} z > 0\}$ onto the lower half plane which, when extended to $\bar{U}$, fixes the points $-1$ and $1$ and sends $i$ to the origin.

**20.** Find a conformal mapping from the oval $U = \{z \in \mathbb{C} : |z^2 - x_0^2| < r^2\}$ with $x_0 \in \mathbb{R}$ and $0 < x_0 < r$ onto the unit disc.

*Hint:* Use the reflection principle.

**21.** Find a conformal mapping from the upper half plane $\Pi^+$ onto the domain $U = \Pi^+ \setminus \{z = x + i : x \leq 0\}$.

**22.** Consider the domains

$$U_1 = \{z = x + iy : |y| < \pi\} \setminus \{z = x + iy : x = 0, \ |y| \geq \pi/2\},$$
$$U_2 = \{z = x + iy : |y| < \pi\} \setminus \{z = x + iy : |x| \geq 1, \ y = 0\}.$$

Find conformal mappings from both $U_1$, $U_2$ onto the unit disc.

# Chapter 9
# The Riemann mapping theorem and Dirichlet's problem

The aim of this chapter is to prove Riemann's theorem on conformal mapping. The usual proof of this statement (due to Koebe) is based on the properties of uniformly bounded sequences of holomorphic functions (normal families). Riemann's theorem may be also obtained from the existence of a solution of Dirichlet's problem in a simply connected domain. This way is longer, but it highlights the relationship between conformal mapping and Dirichlet's problem by means of Green's function, and corresponds to Riemann's point of view. Both proofs will be given, each requiring an appropriate preparation. Namely, the study of normal families and the construction of the Green's function of a simply connected domain, respectively. The properties of sequences of holomorphic or harmonic functions, which are dealt with in detail, have an interest on their own, beyond the proof of Riemann's theorem.

## 9.1 Sequences of holomorphic or harmonic functions

### 9.1.1 Local uniform convergence

In real variable function theory, functions are often defined as limits or sums of series of known functions. So, given the function sequences $(f_n)$, $(g_n)$, $n = 1, 2 \ldots$, defined on a common domain $U$, we can consider the functions

$$f(x) = \lim_{n \to \infty} f_n(x),$$
$$g(x) = \sum_{n=1}^{\infty} g_n(x), \qquad x \in U.$$

It is clear that this is possible only when the sequence $(f_n)$ or the series $\sum g_n$ converge in some sense. At least, it is necessary to have pointwise convergence, that is, for each $x \in U$ the sequence $(f_n(x))$ or the series $\sum_{n=1}^{\infty} g_n(x)$ should be convergent.

From now on, only function sequences will be considered.

The pointwise convergence of a sequence $(f_n)$ to $f$ does not guarantee that properties of the functions $f_n$ remain valid for $f$. For instance each $f_n$ could be continuous without $f$ being continuous. In order to ensure continuity, or other properties of the functions $f_n$, to be preserved, stronger ways of convergence must be imposed, the most important of which is the *uniform convergence*. Thus, if the functions $f_n$ are continuous and converge uniformly to $f$, then $f$ is also continuous.

Now we will consider functions defined on an open set $U$ of the complex plane. Suppose that functions $f_n$, $n = 1, 2, \ldots$ are continuous on $U$ and $f(z) = \lim_{n \to \infty} f_n(z)$ is its pointwise limit. Bearing in mind the local character of continuity, that is, a function is continuous on $U$ if and only if it is continuous on a neighborhood of each point of $U$, we see that in order to ensure that $f$ is continuous on $U$ it is enough that the following condition holds:

For each point $a \in U$ there exists a disc $D(a, r) \subset U$ such that $f_n(z) \to f(z)$ uniformly for $z \in D(a, r)$, when $n \to \infty$.

If the previous condition is satisfied, the sequence $(f_n)$ is said to converge to $f$ *locally uniformly on $U$* (from now on the reference *when $n \to \infty$* will be omitted). It is clear, as has been said, that if $f_n \to f$ locally uniformly on $U$ and the functions $f_n$ are continuous on $U$, then also $f$ is so. One has the following statement:

**Proposition 9.1.** *The functions $(f_n)$ converge locally uniformly to $f$ on $U$ if and only if for every compact set $K \subset U$, the functions $(f_n)$ converge uniformly to $f$ on $K$.*

*Proof.* In one direction this is clear just taking $K = \bar{D}(a, r) \subset U$. In the other direction, given $K \subset U$, cover it by a finite quantity of discs, $K \subset \bigcup_{i=1}^{N} D(a_i, r_i)$, $\bar{D}(a_i, r_i) \subset U$ and remark that if there is uniform convergence on each disc $D(a_i, r_i)$, then there is also on $K$. $\qquad\square$

Due to the previous proposition, when $f_n \to f$ locally uniformly on $U$, it is also said that $(f_n)$ converges to $f$ *uniformly on compact sets of $U$*.

**Example 9.2.** Let $f(z) = \sum_0^{\infty} c_n z^n$ be the sum of a power series with radius of convergence $R > 0$. It is known that the series $\sum c_n z^n$ converges uniformly to $f(z)$ on each compact disc $\bar{D}(0, r)$ with $r < R$. This means that $\sum c_n z^n$ converges to $f(z)$ uniformly on compact sets of the open disc $D(0, R)$ because each compact set of this disc is contained in a disc $\bar{D}(0, r)$ for some $r < R$. However, the series $\sum c_n z^n$ may not converge uniformly on the whole disc $D(0, R)$. As an example consider the geometric series $\sum_0^{\infty} z^n$ with radius of convergence $R = 1$. The convergence to the function $\frac{1}{1-z}$ is not uniform on $\mathbb{D}$, because if it was so, the general term $z^n$ would tend uniformly to $0$ on the disc $\mathbb{D}$ and this is false. $\qquad\square$

Here the interest is to consider sequences of functions $(f_n)$ that are holomorphic or harmonic on an open set $U$ such that $f(z) = \lim_n f_n(z)$ for $z \in U$, and to decide if $f$ is also holomorphic or harmonic on $U$. How must be the convergence of functions $(f_n)$ to $f$ in order to preserve holomorphicity or harmonicity?

Observe that being holomorphic or harmonic on $U$ is also a local property, like continuity. As a matter of fact, uniform convergence on compact sets is indeed appropriate to ensure the preservation of holomorphicity and of harmonicity.

**Theorem 9.3** (Weierstrass). *Let $f_n$, $n = 1, 2, \ldots$ be holomorphic functions on the open set $U$ of the complex plane and suppose that the sequence $(f_n)$ converges to a function $f$ uniformly on compact sets of $U$. Then $f$ is holomorphic on $U$ and, furthermore, the sequence of derivatives $(f_n')$ tends to $f'$ also uniformly on compact sets of $U$.*

*Proof.* We start by observing the following fact: if $\gamma(t), t \in [a, b]$ is a path contained in $U$, one has, under the hypotheses of the theorem,

$$\int_\gamma f(z)\, dz = \lim_n \int_\gamma f_n(z)\, dz. \tag{9.1}$$

Actually, since $f_n \to f$ uniformly on the compact set $\gamma([a, b])$, one has $f_n(\gamma(t)) \cdot \gamma'(t) \to f(\gamma(t)) \cdot \gamma'(t)$ uniformly on $[a, b]$, and integrating it yields (9.1).

Now, to prove that $f$ is holomorphic on $U$ one has two possibilities. The first one is using Morera's theorem (Theorem 4.11). We already know that $f$ is continuous on $U$; if $\Delta$ is any triangle contained in $U$, one has $\int_{\partial\Delta} f_n(z)\, dz = 0$ by Cauchy's theorem. Using (9.1) it turns out that

$$\int_{\partial\Delta} f(z)\, dz = \lim_n \int_{\partial\Delta} f_n(z)\, dz = 0.$$

So, according to Morera's theorem, $f$ is holomorphic on $U$.

The second possibility is using Cauchy's integral formula. Let $\bar{D}(a, r) \subset U$ be a disc and $\gamma$ its boundary travelled in the positive sense. One has

$$f_n(z) = \frac{1}{2\pi i} \int_\gamma \frac{f_n(w)}{w - z}\, dw, \quad z \in D(a, r). \tag{9.2}$$

Since $\dfrac{f_n(w)}{w - z} \to \dfrac{f(w)}{w - z}$ uniformly for $w$ in $\gamma$ and $z$ fixed, it turns out that

$$f(z) = \lim_n f_n(z) = \lim_n \frac{1}{2\pi i} \int_\gamma \frac{f_n(w)}{w - z}\, dw = \frac{1}{2\pi i} \int_\gamma \frac{f(w)}{w - z}, \quad z \in D(a, r)$$

and this shows that $f(z)$ is holomorphic on the disc $D(a, r)$.

To prove that $f_n'(z) \to f'(z)$ locally uniformly fix, as before, a disc $\bar{D}(a, r) \subset U$. Then (9.2) yields

$$f_n'(z) = \frac{1}{2\pi i} \int_\gamma \frac{f_n(w)}{(w - z)^2}\, dw, \quad z \in D(a, r),$$

which implies

$$\lim_n f_n'(z) = \lim_n \frac{1}{2\pi i} \int_\gamma \frac{f_n(w)}{(w - z)^2}\, dw = \frac{1}{2\pi i} \int_\gamma \frac{f(w)}{(w - z)^2}\, dw = f'(z)$$

if $z \in D(a,r)$.  Finally, the convergence of $(f_n')$ to $f'$ is uniform on the disc $\bar{D}(a,r')$ with $r' < r$ because, given $\varepsilon > 0$, one has

$$|f'(z) - f_n'(z)| \le \frac{1}{2\pi} \int_\gamma \frac{|f(w) - f_n(w)|}{|w - z|^2} |dw| \le \frac{\varepsilon}{2\pi} \int_\gamma \frac{|dw|}{(r' - r)^2} = \frac{\varepsilon r}{(r' - r)^2}$$

if $n \ge n_0$, $n_0$ big enough.     $\square$

**Remark 9.1.** a) The previous result may be iterated to get that $f_n \to f$ uniformly on compact sets of $U$ implies that $f_n^{(m)} \to f^{(m)}$ also uniformly on compact sets of $U$ for each $m = 1, 2, \ldots$.

b) Given a series of holomorphic functions uniformly convergent on compact sets

$$f(z) = \sum_{n=1}^{\infty} f_n(z),$$

Weierstrass' theorem states that one may differentiate term by term to obtain the series of derivatives

$$f'(z) = \sum_{n=1}^{\infty} f_n'(z),$$

also converging uniformly on compact sets.  In particular, starting from a power series

$$f(z) = \sum_{0}^{\infty} c_n z^n$$

with positive radius of convergence, we again get Theorem 2.31, stating that

$$f'(z) = \sum_{0}^{\infty} n c_n z^{n-1},$$

in the interior of the disc of convergence.

c) Dealing with convergence of sequences of holomorphic functions, it is convenient to bear the maximum principle in mind.  Indeed, if $f_n \in C(\bar{U}) \cap H(U)$, where $U$ is a bounded domain, the maximum principle ensures that if the sequence $(f_n)$ converges uniformly on the boundary of $U$, then it converges uniformly on the whole $U$; actually, just use the Cauchy criterion and observe that

$$|f_n(z) - f_m(z)| \le \varepsilon \text{ for } z \in \partial U \text{ implies } |f_n(z) - f_m(z)| \le \varepsilon \text{ for each } z \in U.$$

In Weierstrass' theorem, two proofs of the fact that the uniform limit on compact sets of a sequence of holomorphic functions is holomorphic were given.  The one based on Cauchy integral representation suggests an analogous result for harmonic functions for which one has Poisson integral representation.

**Theorem 9.4.** *Let $(u_n)$ be a sequence of harmonic functions on the open set $U$ and suppose that $u(z) = \lim_{n\to\infty} u_n(z)$ uniformly on compact sets of $U$. Then $u$ is harmonic on $U$.*

*Proof.* Take a disc $\bar{D}(a,r) \subset U$ and apply to each $u_n$ Poisson's formula (7.32) to get

$$u_n(z) = \frac{1}{2\pi r} \int_{C(a,r)} \frac{r^2 - |z-a|^2}{|w-z|^2} u_n(w)\, ds(w).$$

By the hypothesis on the uniform convergence of $(u_n)$, it yields

$$u(z) = \lim_n u_n(z) = \frac{1}{2\pi r} \int_{C(a,r)} \frac{r^2 - |z-a|^2}{|w-z|^2} u(w)\, ds(w),$$

proving that $u$ is harmonic, since so is Poisson's kernel,

$$z \longrightarrow \frac{r^2 - |w-a|}{|w-z|^2}.$$  $\square$

Coming back to the case of a sequence of holomorphic functions $(f_n)$ converging uniformly on compact sets to a function $f$, it is easy to prove that the univalence of the functions $f_n$ is a property preserved whenever $f$ is not constant.

**Theorem 9.5** (Hurwitz). *Let $U$ be a domain of the complex plane and $(f_n)$ a sequence of holomorphic functions on $U$ with $f_n(z) \neq 0$, $z \in U$, $n = 1, 2, \ldots$ converging uniformly on compact sets of $U$ to a function $f$. Then either $f$ is identically zero on $U$, or $f(z) \neq 0$ for every point $z \in U$.*

*Proof.* Suppose $f$ is not identically zero and fix a point $z_0 \in U$. Since the zeros of $f$ (if there are any) must be isolated, there is $r > 0$ such that $f(z) \neq 0$ if $0 < |z - z_0| \leq r$. In particular, $|f(z)| \geq \delta > 0$ if $z \in C(z_0, r)$, and, by the uniform convergence, one will also have $|f_n(z)| \geq \delta/2$ if $z \in C(z_0, r)$ and $n \geq n_0$. Hence, it turns out that $\frac{1}{f_n(z)} \to \frac{1}{f(z)}$ uniformly on $C(z_0, r)$ thanks to the above lower bounds. It is also known that $f_n'(z) \to f'(z)$ uniformly on $C(z_0, r)$, by Weierstrass's theorem. So we may conclude that

$$\lim_n \frac{1}{2\pi i} \int_{C(z_0,r)} \frac{f_n'}{f_n(z)}\, dz = \frac{1}{2\pi i} \int_{C(z_0,r)} \frac{f'(z)}{f(z)}\, dz.$$

Now, by the argument principle (Theorem 5.27) the integrals on the left-hand side are zero because no function $f_n$ has zeros. The fact that the right-hand side integral is zero means, by the same principle, that $f(z_0) \neq 0$.  $\square$

**Corollary 9.6.** *Let $U$ be a domain of the plane, $f_n \in H(U)$, $n = 1, 2, \ldots$ and $f(z) = \lim_n f_n(z)$, uniformly on compact sets of $U$. If each function $f_n$ is one-to-one on $U$, then the function $f$ is either one-to-one, or constant on $U$.*

*Proof.* Suppose that $f$ is not constant. If it was not one-to-one on $U$, one could find two points $z_1 \neq z_2$ of $U$ with $f(z_1) = f(z_2) = w$. Take disjoint discs $D(z_1, r)$, $D(z_2, r) \subset U$. On each one, $f_n(z) - w \to f(z) - w$ uniformly on compact sets. Since the function $f(z) - w$ is not identically zero and has a zero in $D(z_1, r)$ as well as in $D(z_2, r)$, according to Hurwitz's theorem the functions $f_n(z) - w$ must have a zero in $D(z_1, r)$ and in $D(z_2, r)$ for $n \geq n_0$. But this means that $f_n$, for $n$ big enough, will take the value $w$ in $D(z_1, r)$ and in $D(z_2, r)$, a contradiction. $\qquad\square$

## 9.1.2 Normal families

To obtain new functions as limits of sequences of known functions, some criterion stating when a given sequence has a limit is needed. Generally it is enough to guarantee that, given a sequence of functions $(f_n)$ of a certain class, there is a subsequence $(f_{k_n})$ having a limit, $f(z) = \lim_n f_{k_n}(z)$. Indeed, then $f$ is a limit of a sequence of functions of the class and will get the properties of the functions $f_n$ if the convergence is good enough.

To deal with the problem of the existence of convergent partial sequences, recall first what happens in the case of numerical sequences. If $(z_n)$ is a sequence of complex numbers which converges, then the sequence $(z_n)$ is necessarily bounded, that is, $\sup_n |z_n| < +\infty$. The converse is not true; but the Bolzano–Weierstrass theorem asserts that a bounded $(z_n)$ has at least a convergent subsequence. Hence, in this case, boundedness is the criterion for a sequence to have a convergent subsequence. The Bolzano–Weierstrass theorem is a basic principle of analysis which, indeed, is equivalent to the completeness of the real field. It is also equivalent to the statement that closed and bounded sets in the Euclidian space are exactly the compact ones.

What happens if one replaces the sequence of points $(z_n)$ by a sequence of functions $(f_n)$? Suppose that functions $f_n$ are defined on a compact set $K$ and are continuous on $K$. If $f(z) = \lim_{n \to \infty} f_n(z)$ uniformly on $K$, it follows easily that the sequence $(f_n)$ must be *uniformly bounded on $K$*, that is, there is a constant $M > 0$ such that $|f_n(z)| \leq M$, $n = 1, 2, \ldots$ and $z \in K$. Now, assuming that $(f_n)$ is uniformly bounded on $K$, is it true, as in the case of numerical sequences, that there is a subsequence $(f_{k_n})$ uniformly convergent on $K$? The answer is no; indeed it can happen that $(f_n)$ is uniformly bounded on $K$, but it has no subsequence pointwise convergent on $K$, as the following example shows.

**Example 9.7.** Let $K = [0, 2\pi]$, $f_n(x) = \sin nx$, $n = 1, 2, \ldots$. Then $|f_n(x)| \leq 1$, for $x \in K$, $n = 1, 2, \ldots$ but, there cannot exist integer numbers $(k_n)$ such that $\lim_n \sin k_n x$ exists for each $x \in K$. Indeed, this would imply

$$\lim_n (\sin k_n x - \sin k_{n+1} x)^2 = 0, \quad x \in [0, 2\pi],$$

and, by the dominated convergence theorem,

$$\lim_{n} \int_0^{2\pi} (\sin k_n x - \sin k_{n+1} x)^2 \, dx = 0.$$

But an easy computation shows that

$$\int_0^{2n} (\sin k_n x - \sin k_{n+1} x)^2 \, dx = 2\pi \quad \text{if } k_n \text{ is an integer,}$$

a contradiction. $\qquad\square$

What else is necessary, then, in order to ensure that a uniformly bounded sequence of continuous functions on $K$ has a uniformly convergent subsequence on $K$? The key notion is *equicontinuity*.

In a more general setting, consider a family $\mathcal{F}$ of continuous functions on a compact $K$. Recall that the family $\mathcal{F}$ is said to be *equicontinuous on $K$* if, given $\varepsilon > 0$, there exists $\delta > 0$ such that the condition

$$z, w \in K, \; |z - w| < \delta \implies |f(z) - f(w)| < \varepsilon, \quad \text{for all } f \in \mathcal{F}$$

holds. The point is that a uniformly convergent sequence of continuous functions on $K$ is equicontinuous on $K$. In the opposite direction, if a sequence of functions is pointwise bounded and equicontinuous on $K$, then it has a uniformly convergent subsequence on $K$. This is the content of the following classical result.

**Theorem 9.8** (Arzelà–Ascoli). *Let $\mathcal{F}$ be a family of continuous functions on a compact set $K$. The following conditions are equivalent:*

a) *Every sequence of functions of the family $\mathcal{F}$ has a subsequence converging uniformly on $K$.*

b) *The family $\mathcal{F}$ is pointwise bounded and equicontinuous on $K$.*

Pointwise bounded means that for each $z \in K$ one has $\sup_{f \in \mathcal{F}} |f(z)| < \infty$. The proof of Theorem 9.8 may be found in [10], p. 156–158. Moreover, $(f_n)$ being pointwise bounded and equicontinuous implies that $(f_n)$ is uniformly bounded.

It is convenient to remark that the definition given above is the one of *uniform equicontinuity* on $K$. There is also a notion of equicontinuity at a point $z_0 \in K$. The family $\mathcal{F}$ is said to be *equicontinuous at the point $z_0$* if for each $\varepsilon > 0$ there is $\delta > 0$ such that

$$z \in K, \; |z - z_0| < \delta \implies |f(z) - f(z_0)| < \varepsilon, \quad \text{for each } f \in \mathcal{F}.$$

Using a covering argument, one gets the equivalence of both concepts.

**Proposition 9.9.** *A family $\mathcal{F}$ of functions is equicontinuous (uniformly) on the compact set $K$ if and only if it is equicontinuous at each point of $K$.*

Since the interest here is in studying sequences of holomorphic and harmonic functions uniformly convergent on compact sets, a version of Arzelà–Ascoli's theorem in this context will be given.

**Theorem 9.10.** *Let $\mathcal{F}$ be a family of continuous functions on an open set $U$ of the plane. The following conditions are equivalent:*

a) *Every sequence of functions of the family $\mathcal{F}$ has a subsequence converging uniformly on compact sets of $U$.*

b) *The family $\mathcal{F}$ is pointwise bounded on $U$ and equicontinuous at each point of $U$.*

*Proof.* The proof of a) $\Rightarrow$ b) is the same as the corresponding one in Theorem 9.8.

To show that b) $\Rightarrow$ a) take an exhaustive sequence of compact sets of $U$, that is, a sequence of compact sets $K_n \subset U$ such that

1) $K_n \subset \overset{\circ}{K}_{n+1}, n = 1, 2, \ldots.$

2) If $K \subset U$ is compact, there exists a number $n \in \mathbb{N}$ with $K \subseteq K_n$,

according to Lemma 1.15.

Let now $(f_n)$ be a sequence of functions of the family $\mathcal{F}$. By b), Proposition 9.9 and Theorem 9.8 applied to the compact set $K_1$, we can find a subsequence of $(f_n)$, called $(f_n^1)$, uniformly convergent on $K_1$. By the same reason we can find a subsequence of $(f_n^1)$, say $(f_n^2)$, uniformly convergent on $K_2$, which will also converge on $K_1$. Continuing this process we find for each $m$ a sequence $(f_n^m)_{n=1,2,\ldots}$, uniformly convergent on $K_1, K_2, \ldots, K_m$. The diagonal sequence $(f_n^n)_{n=1,2,\ldots}$ is then uniformly convergent on each compact set $K_n$ and, by property 2), also on every compact set of $U$. $\qquad\square$

According to Theorem 9.10, to show that a family $\mathcal{F}$ of continuous functions on $U$ has the property that every sequence of $\mathcal{F}$ has a subsequence uniformly convergent on compact sets of $U$, it is enough to check that $\mathcal{F}$ is equicontinuous on $U$. In the case that the functions of $\mathcal{F}$ are differentiable, a criterion for equicontinuity is the uniform boundedness of the derivatives on compact sets of $U$ or, equivalently, their local uniform boundedness.

For example, suppose that we are dealing with a family of functions, $\mathcal{F}$, differentiable on the real line such that for each fixed $x_0 \in \mathbb{R}$, and each $r > 0$ there exists $M_r > 0$ with

$$|f'(x)| \leq M_r \quad \text{if} \ |x - x_0| < r, \ f \in \mathcal{F}.$$

The mean value theorem gives

$$|f(x) - f(x_0)| \le M_r |x - x_0| \quad \text{if } |x - x_0| < r, \ f \in \mathcal{F},$$

which proves that $\mathcal{F}$ is equicontinuous at the point $x_0$. As well, for functions of two real variables, the boundedness conditions

$$\left| \frac{\partial f}{\partial x}(x, y) \right| \le M_r, \quad \left| \frac{\partial f}{\partial y}(x, y) \right| \le M_r$$

if $d((x, y), (x_0, y_0)) < r$, for every $f \in \mathcal{F}$, imply $\mathcal{F}$ is equicontinuous at the point $(x_0, y_0)$.

It is an essential fact that a uniformly bounded family of holomorphic or harmonic functions is automatically equicontinuous.

**Proposition 9.11.** *Let $\mathcal{F}$ be a family of holomorphic functions on an open set $U$ of the plane that is uniformly bounded on each compact set of $U$. Then the family $\mathcal{F}$ is equicontinuous at each point of $U$.*

*The same holds if $\mathcal{F}$ is a family of harmonic functions on $U$.*

*Proof.* Assume that functions of $\mathcal{F}$ are holomorphic; take $z_0 \in U$ and $\bar{D}(z_0, r) \subset U$. By hypothesis there exists $M_r > 0$ such that

$$|f(z)| \le M_r \quad \text{if } |z - z_0| \le r, \ f \in \mathcal{F}.$$

The first Cauchy inequality (Theorem 4.38 for $n = 1$) gives

$$|f'(z)| \le 2\frac{M_r}{r} \quad \text{if } |z - z_0| \le \frac{r}{2}, \ f \in \mathcal{F},$$

because $\bar{D}(z, r/2) \subset \bar{D}(z_0, r)$ when $|z - z_0| \le r/2$.

Now, if $z \in D(z_0, r/2)$, integrating along the line segment joining $z_0$ with $z$, one has

$$|f(z) - f(z_0)| = \left| \int_{z_0}^{z} f'(w) \, dw \right| \le \int_{z_0}^{z} |f'(w)||dw| \le 2\frac{M_r}{r}|z - z_0|$$

whenever $|z - z_0| \le r/2$ and $f \in \mathcal{F}$. This proves that the family $\mathcal{F}$ is equicontinuous at $z_0$.

In the case that functions of $\mathcal{F}$ are harmonic, writing $z = x + iy$ they will be considered as functions of the real variables $x$, $y$. As said above, it is enough to show, for $z_0 \in U$ fixed, the following estimates, for some $r$:

$$\left| \frac{\partial u}{\partial x}(z) \right| \le M_r, \quad \left| \frac{\partial u}{\partial y}(z) \right| \le M_r, \quad \text{if } |z - z_0| \le r, \ u \in \mathcal{F}.$$

They are equivalent to

$$\left|\frac{\partial u}{\partial z}(z)\right| \le M_r, \quad \left|\frac{\partial u}{\partial \bar z}(z)\right| \le M_r, \quad |z - z_0| \le r, \ u \in \mathcal{F}.$$

Assume, in order to simplify the notation, $z_0 = 0$ and $\bar{\mathbb{D}} \subset U$. Represent $u$ by Poisson's formula (7.19):

$$u(z) = \int_0^{2\pi} P_{\mathbb{D}}(z, e^{i\theta}) u(e^{i\theta})\, d\theta = \frac{1}{2\pi} \int_0^{2\pi} \frac{1 - |z|^2}{|1 - e^{-i\theta}z|^2} u(e^{i\theta})\, d\theta, \quad |z| < 1.$$

According to the hypotheses one has $|u(e^{i\theta})| \le M$ for $u \in \mathcal{F}$. Differentiating the expression of $u(z)$ and estimating the result, one gets

$$\left|\frac{\partial u}{\partial z}(z)\right| \le M_{1/2} \frac{1}{2\pi} \int_0^{2\pi} \left|\frac{\partial}{\partial z} P_{\mathbb{D}}(z, e^{i\theta})\right| d\theta,$$

$$\left|\frac{\partial u}{\partial \bar z}(z)\right| \le M_{1/2} \frac{1}{2\pi} \int_0^{2\pi} \left|\frac{\partial}{\partial \bar z} P_{\mathbb{D}}(z, e^{i\theta})\right| d\theta$$

whenever $|z| < 1$, $u \in \mathcal{F}$.

Now we will show that both $\frac{\partial}{\partial z} P_{\mathbb{D}}(z, e^{i\theta})$ and $\frac{\partial}{\partial \bar z} P_{\mathbb{D}}(z, e^{i\theta})$ are uniformly bounded, with respect to $\theta$, when $|z| \le 1/2$.

To this end, just write Poisson's kernel in the form

$$2\pi\, P_{\mathbb{D}}(z, e^{i\theta}) = \frac{1 - |z|^2}{|1 - e^{-i\theta}z|^2} = 1 + \frac{ze^{-i\theta}}{1 - ze^{-i\theta}} + \frac{\bar z e^{i\theta}}{1 - \bar z e^{i\theta}},$$

from which it follows that

$$\frac{\partial P_{\mathbb{D}}(z, e^{i\theta})}{\partial z} = \frac{1}{2\pi} \frac{e^{-i\theta}}{(1 - ze^{-i\theta})^2}, \quad \frac{\partial P_{\mathbb{D}}(z, e^{i\theta})}{\partial \bar z} = \frac{1}{2\pi} \frac{e^{i\theta}}{(1 - \bar z e^{i\theta})^2},$$

and so

$$\left|\frac{\partial P_{\mathbb{D}}(z, e^{i\theta})}{\partial z}\right|, \left|\frac{\partial P_{\mathbb{D}}(z, e^{i\theta})}{\partial \bar z}\right| \le \frac{1}{2\pi} \frac{1}{|e^{i\theta} - z|^2} \le \frac{2}{\pi} \quad \text{if } |z| \le \frac{1}{2}. \qquad \square$$

Let us state now the basic result about holomorphic or harmonic function sequences.

**Theorem 9.12** (Montel). *Let $\mathcal{F}$ be a family of holomorphic functions on an open set $U$ of the plane. Then the following conditions are equivalent:*

  a) *Every sequence of functions of the family $\mathcal{F}$ has a subsequence that is uniformly convergent on compact sets of $U$.*

b) *The family $\mathcal{F}$ is uniformly bounded on compact sets of $U$.*

*The same statement holds if $\mathcal{F}$ is a family of harmonic functions on $U$.*

*Proof.* The implication a) $\Rightarrow$ b) holds for every family of continuous functions. Indeed condition a) implies that the family $\mathcal{F}$ is equicontinuous and pointwise bounded (Theorem 9.10), and then if $z_0 \in U$, there exists $\delta > 0$ with $|f(z) - f(z_0)| < 1$ if $|z - z_0| < \delta$, for $f \in \mathcal{F}$. Hence, $|f(z)| \leq |f(z_0)| + 1$, $|z - z_0| < \delta$, $f \in \mathcal{F}$, and the family $\mathcal{F}$ is locally uniformly bounded.

For b) $\Rightarrow$ a) just combine Theorem 9.10 with Proposition 9.11. $\quad\square$

A family of holomorphic functions uniformly bounded on compact sets of $U$ is also called a *normal family*. We point out, however, that in Montel's classical terminology the term normal family is applied to a class of holomorphic functions $\mathcal{F}$ such that every sequence of $\mathcal{F}$ has either a uniformly convergent subsequence, or a subsequence converging uniformly to $\infty$, on each compact set of $U$.

In the case of harmonic functions, there is another criterion of normality for a given family that guarantees that every sequence has a uniformly convergent subsequence on compact sets, either to a harmonic function, or to $\infty$. Namely, it is enough that the family consists of positive harmonic functions. This criterion, known as Harnack's principle, is based on an inequality already proposed in Exercise 7 in Section 7.13.

**Proposition 9.13** (Harnack's inequality). *Let $u$ be a harmonic and positive function on the disc $D(z_0, R)$. Then*

$$\frac{R - r}{R + r} u(z_0) \leq u(z) \leq \frac{R + r}{R - r} u(z_0) \quad if \ |z - z_0| = r < R. \tag{9.3}$$

*Proof.* It is enough to consider the case $z_0 = 0$, $R = 1$. We must show, then, that

$$\frac{1 - r}{1 + r} u(0) \leq u(re^{i\theta}) \leq \frac{1 + r}{1 - r} u(0) \quad if \ 0 \leq r < 1, \ 0 \leq \theta \leq 2\pi. \tag{9.4}$$

If $P_r(\tau - \theta) = \frac{1}{2\pi} \frac{1 - r^2}{1 - 2r\cos(\tau - \theta) + r^2}$, one has the inequalities

$$\frac{1 - r}{1 + r} = \frac{1 - r^2}{1 + 2r + r^2} \leq 2\pi P_r(\tau - \theta) \leq \frac{1 - r^2}{1 - 2r + r^2} = \frac{1 + r}{1 - r}, \tag{9.5}$$

$$0 \leq r < 1, \ 0 \leq \theta \leq 2\pi.$$

Since $\lim_{r \to 1} u(rz) = u(z)$, we may suppose that $u$ is continuous on $\overline{\mathbb{D}}$. Then Poisson's formula gives

$$u(re^{i\theta}) = \int_0^{2\pi} P_r(\tau - \theta) u(e^{i\tau}) \, d\tau.$$

Since $u(z) \geq 0$, multiplying (9.5) by $u(e^{i\tau})$, integrating and using (by the mean value property)

$$\frac{1}{2\pi} \int_0^{2\pi} u(e^{i\tau}) d\tau = u(0)$$

yields (9.4). $\qquad\square$

Harnack's principle can be given now.

**Theorem 9.14.** *Let $\mathcal{F}$ be a family of positive harmonic functions on a domain $U$ of the complex plane. Then every sequence of functions of $\mathcal{F}$ has a subsequence that converges, uniformly on compact sets of $U$, either to a harmonic function on $U$ or to infinity.*

*Proof.* Let $(u_n)$ be a sequence of functions of $\mathcal{F}$. If $(u_n(z))$ is bounded for every $z \in U$, then Harnack's inequality implies that $(u_n)$ is, in fact, locally uniformly bounded, whence uniformly bounded on compact sets and so, by Montel's theorem, it has a subsequence uniformly convergent on compact sets.

Otherwise there is a point $z_0 \in U$ and a subsequence $(u_{k_n})$ with $\lim_n u_{k_n}(z_0) = +\infty$. In this case we claim that $(u_{k_n})$ converges to $+\infty$ locally uniformly. Indeed consider

$$A = \{z \in U : \lim_n u_{k_n}(z) = +\infty\},$$

so that $A \neq \emptyset$; by Harnack's inequality $A$ is open and $u_{k_n}(z) \to +\infty$ locally uniformly on $A$. But $U \setminus A$ is also open by Harnack's inequalities again and so $A = U$. $\qquad\square$

**Corollary 9.15.** *Let $(u_n)$ be an increasing sequence of harmonic functions on a domain $U$. Then $(u_n)$ is uniformly convergent on compact sets of $U$, either to a harmonic function on $U$ or to infinity.*

*Proof.* The sequence $(v_n)$ with $v_n(z) = u_n(z) - u_1(z)$ consists of positive harmonic functions and is increasing. Theorem 9.14 gives the conclusion for the sequence $(v_n)$ and, therefore, also for the sequence $(u_n)$. $\qquad\square$

### 9.1.3 Topology of the space of continuous functions on an open set

In the previous subsections it has been highlighted how uniform convergence on compact subsets of an open set is suitable for sequences of continuous functions as well as for sequences of holomorphic or harmonic functions. We will show now that this convergence is associated to a complete metric.

Fix an open set $U$ of the complex plane and consider $C(U)$, the space of continuous functions on $U$. Let $(K_n)_{n=1,2,\ldots}$ be a sequence of compact sets of $U$ as in

Lemma 1.15 (in fact it is enough that the sets $K_n$ satisfy a) and b) of this lemma) and for each function $f \in C(U)$ write

$$M_n(f) = \sup\{|f(z)| : z \in K_n\},$$

$$p_n(f) = \min\{M_n(f), 1\},$$

$$p(f) = \sum_{n=1}^{\infty} \frac{p_n(f)}{2^n}.$$

Observe that $0 \le p(f) \le 1$, for every continuous function $f$. Now define a distance in the space $C(U)$ by

$$d(f, g) = p(f - g), \quad \text{if } f, g \in C(U).$$

Since $d(f, g) = 0$ if and only if $f = g$, to prove that $d$ is a distance in $C(U)$ it is enough to check that

$$p(f + g) \le p(f) + p(g),$$

an inequality which follows from the definition of the functional $p$ and the fact that $M_n(f + g) \le M_n(f) + M_n(g)$ for every $n \in \mathbb{N}$ and $f, g \in C(U)$.

Remark that the distance $d$ just defined is invariant under translations, that is,

$$d(f, g) = d(f + h, \ g + h), \quad \text{if } f, g, h \in C(U).$$

This means that it suffices to know the topology that $d$ defines around the zero of $C(U)$.

**Proposition 9.16.** *A sequence of functions* $(f_n)$, $f_n \in C(U)$, *has a limit* $f \in C(U)$ *in the metric* $d$, *that is,* $\lim_n d(f_n, f) = 0$ *if and only if* $(f_n)$ *converges to* $f$ *uniformly on compact sets of* $U$.

*Proof.* As said above, it is enough to consider the case $f = 0$.

Suppose $d(f_n, 0) = p(f_n) \to 0$ when $n \to \infty$. We must show that, given $K \subset U$ compact and $0 < \varepsilon < 1$, there exists $n_0 \in \mathbb{N}$ such that $\sup\{|f_n(x)| : x \in K\} < \varepsilon$ if $n \ge n_0$. Take $K_m$ with $K \subset K_m$ and $n_0$ such that $p(f_n) < \frac{\varepsilon}{2^m}$ if $n \ge n_0$. Then one has $p_m(f_n) \le 2^m p(f_n) < \varepsilon$ if $n \ge n_0$, giving $M_m(f_n) < \varepsilon$ and, *a fortiori*, $|f_n(x)| < \varepsilon$, for $x \in K$, $n \ge n_0$.

Conversely, suppose that $f_n \to 0$ uniformly on compact sets of $U$. Given $\varepsilon > 0$, choose $m$ with $\frac{1}{2^m} < \varepsilon/2$. Then there exists $n_0$ such that $M_m(f_n) < \varepsilon/2$ if $n \ge n_0$. Now, if $f \in C(U)$, one has $p_n(f) \le p_m(f)$ if $n \le m$ and, therefore,

$$p(f) \le \sum_{n=1}^{m} \frac{p_m(f)}{2^n} + \sum_{n>m} \frac{1}{2^m} \le p_m(f) + \frac{1}{2^m}.$$

Applying this inequality to $f_n$ for $n \ge n_0$ yields

$$d(f_n, 0) = p(f_n) \le M_m(f_n) + \frac{1}{2^m} < \varepsilon. \qquad \square$$

**Corollary 9.17.** *The space $C(U)$ with the distance $d$ is a complete metric space.*

*Proof.* If $(f_n)$ is a Cauchy sequence of $C(U)$ with respect to the metric $d$, then $(f_n)$ is a uniform Cauchy sequence on compact sets of $U$, by Proposition 9.16. So $(f_n)$ must have a uniform limit $f \in C(U)$ on compacts sets of $U$, and again by Proposition 9.16, $f$ is the limit of $(f_n)$ with respect to the metric $d$.    $\square$

Some results of Subsections 9.1.1 and 9.1.2 have more synthetic statements using the topology given by the metric $d$ of the space $C(U)$. For example, Weierstrass' theorem (Theorem 9.3) reads:

> The space $H(U)$ of holomorphic functions on an open set $U$ is a closed subspace of $C(U)$. The mapping $f \rightarrow f'$ from $H(U)$ into $H(U)$ is continuous.

Montel's theorem (Theorem 9.12) has also an interesting reformulation in terms of the topology of $H(U)$, induced by the one of $C(U)$. First, an appropriate definition of a bounded set of $C(U)$ must be given. Observe that defining a set $A \subset C(U)$ to be bounded when its diameter is finite does not work because the distance $d$ satisfies $d(f, g) \leq 1$ for every pair $f, g \in C(U)$, so every set would be bounded. The suitable definition is the following:

A set of continuous functions $A \subset C(U)$ is said to be *bounded* if for each $\varepsilon > 0$ there exists a constant $0 < \alpha_\varepsilon < \infty$ such that $d(\alpha_\varepsilon f, 0) = p(\alpha_\varepsilon f) \leq \varepsilon$, for every function $f \in A$.

Arguing as in Proposition 9.16, one can see that $A \subset C(U)$ is a bounded set in the above sense if and only if $A$ is uniformly bounded on compact sets of $U$. Now Theorem 9.12 reads as follows:

> A set of holomorphic functions $\mathscr{F} \subset H(U)$ is compact if and only if $\mathscr{F}$ is closed and bounded.

This result shows that compact sets in the topological space $H(U)$ have the same characterization as compact sets in $\mathbb{R}^n$. This assertion is not true for compact sets of $C(U)$ as shown in Subsection 9.1.2 and Arzelà–Ascoli's theorem is needed to characterize them.

Of course, equivalent statements for the space $h(U)$ of harmonic functions on $U$ hold: $h(U)$ is a closed subspace of $C(U)$ and the compact subsets of $h(U)$ are the close and bounded ones.

## 9.2 Riemann's theorem

The aim of this section is to prove Riemann's theorem on conformal mapping (Theorem 8.5), which is next recalled.

**Theorem 9.18** (Fundamental theorem of conformal mapping). *Given a simply connected domain of the plane $U$, $U \neq \mathbb{C}$, and a point $z_0 \in U$, there exists a unique conformal mapping $f$ from $U$ onto the unit disc $\mathbb{D}$, normalized with the conditions $f(z_0) = 0$, $f'(z_0) > 0$.*

The proof given below is due, essentially, to Koebe and is based on the theory of normal families. Later another proof will be given, closer to the original Riemann's spirit, based on the solution of Dirichlet's problem.

First, let us prove the uniqueness of $f$. If $f_1, f_2 \colon U \to \mathbb{D}$ are conformal, surjective and normalized, the function

$$T(w) = f_2(f_1^{-1}(w))$$

is an automorphism of the unit disc satisfying $T(0) = 0$, $T'(0) > 0$. According to Theorem 8.27, one has $T(w) = w$, for $w \in \mathbb{D}$, that is, $f_1 = f_2$.

To show the existence of a normalized conformal mapping from $U$ onto $\mathbb{D}$, consider the family $\mathcal{F}$ of holomorphic functions on $U$ defined as follows:

$$g \in \mathcal{F} \iff \begin{cases} \text{a) } g \text{ is holomorphic and one-to-one in } U, \\ \text{b) } |g(z)| < 1, \text{ for } z \in U, \\ \text{c) } g(z_0) = 0, \ g'(z_0) > 0, \end{cases}$$

where $z_0$ is the fixed point in $U$. The proof of Riemann's theorem is done in the three following steps:

A) $\mathcal{F} \neq \emptyset$.

B) There exists a function $f \in \mathcal{F}$ such that $g'(z_0) \leq f'(z_0)$, for every $g \in \mathcal{F}$.

C) If $f \in \mathcal{F}$ satisfies B), then $f$ is the Riemann mapping, that is, $f(U) = \mathbb{D}$, $f(z_0) = 0$ and $f'(z_0) > 0$.

*Proof of* A) Since $U \neq \mathbb{C}$, one may take a point $a \notin U$. Let $g(z) = \sqrt{z - a}$ be a branch of the square root of $z - a \neq 0$, $z \in U$ which exists by Theorem 6.22 f). Hence, $g \in H(U)$ and it is clear that $g$ is one-to-one in $U$. Moreover, if $g$ takes a value $w$ in $U$, it cannot take the value $-w$. It is known that $g(U)$ is open and, therefore, there exists $r > 0$ with $D(g(z_0), r) \subset g(U)$. So, one has $D(-g(z_0), r) \cap g(U) = \emptyset$, that is $|g(z) + g(z_0)| \geq r$, for $z \in U$. The function $g_1(z) = c/(g(z) + g(z_0))$ with $c$ constant is holomorphic and one-to-one in $U$ and satisfies $|g_1(z)| < 1$, $z \in U$, provided $|c|$ is small enough. Finally, composing $g_1$ with an automorphism $\tau$ of the unit disc, the function $g_0 = \tau \circ g_1$ satisfies, in addition, $g_0(z_0) = 0$, $g_0'(0) > 0$, that is, $g_0 \in \mathcal{F}$.

*Proof of* B). Remark that by condition b) above and Montel's theorem the family $\mathcal{F}$ is normal.

Let $M = \sup\{g'(z_0) : g \in \mathcal{F}\}$, $M \leq +\infty$. Consider a sequence of functions $g_n \in \mathcal{F}$ with $\lim_n g'_n(z_0) = M$. Since $\mathcal{F}$ is normal, by Theorem 9.12, there exists a subsequence $(g_{k_n})$ uniformly convergent on compact sets of $U$ to a function $f$. By Theorem 9.3, $f$ is holomorphic on $U$ and, in addition, $g'_{k_n}(z) \to f'(z)$, also uniformly on compact sets. In particular, $M < +\infty$ and $f'(z_0) = M$. Now, by Corollary 9.6, $f$ must be either one-to-one or constant, because each $g_n$ is one-to-one. But $f'(z_0) = M > 0$ since $\mathcal{F} \neq \emptyset$. So, $f$ is not constant and it is one-to-one. Since $f(z_0) = \lim_n g_{k_n}(z_0) = 0$ and $f'(z_0) > 0$, one has $f \in \mathcal{F}$ and, finally, $g'(z_0) \leq f'(z_0)$ if $g \in \mathcal{F}$, by the definition of the number $M$.

*Proof of* C). We show that if $f$ is the function in B), then $f(U) = \mathbb{D}$. Suppose there is a point $w_0 \in \mathbb{D}$, $w_0 \notin f(U)$. A function $g_1 \in \mathcal{F}$ such that $g'_1(z_0) > f'(z_0)$ will be constructed, in contradiction to B). The function $(f(z)-w_0)/(1-\bar{w}_0 f(z))$ is holomorphic and one-to-one from $U$ into $\mathbb{D}$ and, moreover, does not vanish on $U$. Therefore, there exists a branch of its square root

$$g(z) = \sqrt{\frac{f(z) - w_0}{1 - \bar{w}_0 f(z)}}$$

which is one-to-one and holomorphic on $U$ and satisfies $|g(z)| < 1$ as well. Now the function

$$g_1(z) = \frac{|g'(z_0)|}{g'(z_0)} \frac{g(z) - g(z_0)}{1 - \overline{g(z_0)}g(z)}$$

has the same properties as $g$ and, in addition, is normalized, that is, $g_1 \in \mathcal{F}$. Indeed, specifying the value of $g'_1(z_0)$ yields

$$g'_1(z_0) = \frac{|g'(z_0)|}{1 - |g(z_0)|^2}$$

and, by the relationship between $g(z)$ and $f(z)$, one has

$$g'_1(z_0) = \frac{1 + |w_0|}{2\sqrt{|w_0|}} \cdot f'(z_0) > f'(z_0),$$

because $\frac{1+|w_0|}{2\sqrt{|w_0|}} > 1$ is equivalent to $(1 - |w_0|)^2 > 0$. $\qquad\square$

## 9.3  Green's function and conformal mapping

Let $U$ be a bounded domain of the complex plane and let us recall what Green's function of $U$ is. As said in Section 7.5, to get *Green's function of $U$ with pole at* *the point* $z_0 \in U$ one must find a function $h$, harmonic on $U$ and continuous on $\bar{U}$, such that $h(z) = \frac{1}{2\pi} \mathrm{Log}\,|z - z_0|$ if $z \in \partial U$. Then the function

$$G_U(z_0, z) = \frac{1}{2\pi} \mathrm{Log}\,|z - z_0| - h(z)$$

is the sought Green's function.

The function $G_U(z_0, z)$, in the variable $z$, has the following properties:

a) $G_U(z_0, z)$ is continuous on $\bar{U} \setminus \{z_0\}$ and harmonic on $U \setminus \{z_0\}$.

b) $G_U(z_0, z) - \frac{1}{2\pi} \mathrm{Log}\,|z - z_0|$ is harmonic on $U$.

c) $G_U(z_0, z) = 0$ when $z \in \partial U$.

It is easy to show that these properties determine the function $G_U(z_0, z)$. Actually, if $\tilde{G}_U(z_0, z)$ also satisfies a), b) and c), then the function $G_U(z_0, z) - \tilde{G}_U(z_0, z)$ would be harmonic on $U$, continuous on $\bar{U}$ and vanishing on the boundary of $U$; hence by the maximum principle (Corollary 7.20) it should be identically zero in $U$.

Clearly the function $h$ used to construct Green's function is the solution of Dirichlet's problem in $U$ with boundary values $\varphi(z) = \frac{1}{2\pi} \mathrm{Log}\,|z - z_0|$. Therefore, knowing how to solve Dirichlet's problem in $U$ implies knowing how to construct Green's function of $U$ with pole at any point of $U$.

Green's function has several important properties. For instance, the symmetry. If $z$, $w$ are two points of $U$, $z \neq w$, the equality $G_U(z, w) = G_U(w, z)$ holds, according to Proposition 7.23. Another property is the *conformal invariance* of Green's function. This property is given in the following proposition. It states that a conformal mapping between two domains transforms Green's function of one of them into Green's function of the other.

**Proposition 9.19.** *Let $U$, $U'$ be two bounded domains of the plane, $z_0 \in U$, $w_0 \in U'$ and $f : U' \to U$ a conformal mapping from $U'$ onto $U$ such that $f(w_0) = z_0$. Suppose that the domain $U$ has Green's function with pole at $z_0$, $G_U(z_0, z)$. Then the function $G_U(z_0, f(w))$ is the Green's function of $U'$ with pole at $w_0$, that is,*

$$G_{U'}(w_0, w) = G_U(z_0, f(w)), \quad w \in \bar{U}' \setminus \{w_0\}.$$

*Proof.* It is enough to show that the function of $w$, $G_U(z_0, f(w))$ satisfies conditions a), b) and c) in the domain $U'$, for the point $w_0 \in U'$.

First, by Proposition 8.9, if $w$ approaches a point of $\partial U'$, then $f(w)$ approaches the boundary of $U$. Since $G_U(z_0, z)$ vanishes on $\partial U$, this proves that $G_U(z_0, f(w))$ is continuous up to $\partial U'$ and vanishes there. It is clear that this function is harmonic on $U' \setminus \{w_0\}$, because it is the composition of a holomorphic function with a harmonic one.

Finally, there is a function $h$ harmonic on $U$ with $G_U(z_0, z) - \frac{1}{2\pi} \operatorname{Log} |z - z_0| = -h(z)$, $z \in U$. Hence, $G_U(z_0, f(w)) - \frac{1}{2\pi} \operatorname{Log} |f(w) - z_0| = -h(f(w))$ is harmonic on $U'$. So,

$$G_U(z_0, f(w)) - \frac{1}{2\pi} \operatorname{Log} |w - w_0|$$
$$= -h(f(w)) - \frac{1}{2\pi} \operatorname{Log} |w - w_0| + \frac{1}{2\pi} \operatorname{Log} |f(w) - z_0|,$$

a harmonic function on $U'$, since

$$\operatorname{Log} |f(w) - z_0| - \operatorname{Log} |w - w_0| = \operatorname{Log} \left| \frac{f(w) - f(w_0)}{w - w_0} \right|$$

and $\frac{f(w) - f(w_0)}{w - w_0}$ is holomorphic and non-vanishing around $w_0$. $\qquad\square$

**Example 9.20.** Take as domain the unit disc $\mathbb{D}$. The Green's function of $\mathbb{D}$ with pole at the origin is

$$G_{\mathbb{D}}(0, z) = \frac{1}{2\pi} \operatorname{Log} |z|,$$

since it satisfies conditions a), b) and c). To find Green's function with a pole at another point $z_0 \in \mathbb{D}$, use the conformal mapping $\tau : \mathbb{D} \to \mathbb{D}$, which satisfies $\tau(z_0) = 0$. It is the function

$$\tau(z) = \frac{z - z_0}{1 - \bar{z}_0 z},$$

and, therefore, Proposition 9.19 gives

$$G_{\mathbb{D}}(z_0, z) = \frac{1}{2\pi} \operatorname{Log} \left| \frac{z - z_0}{1 - \bar{z}_0 z} \right|. \qquad\square$$

**Example 9.21.** Let now $U$ be a bounded simply connected domain of the plane, $z_0 \in U$ and look for Green's function of $U$ with pole at $z_0$. By now it is unknown if Dirichlet's problem can be solved in $U$ (later on it will be seen that it can). Anyway, by Riemann's theorem, there exists a conformal mapping $f$ from $U$ onto the unit disc $\mathbb{D}$ such that $f(z_0) = 0$. Proposition 9.19 yields

$$G_U(z_0, z) = \frac{1}{2\pi} \operatorname{Log} |f(z)|, \quad z \in U \setminus \{z_0\}. \qquad\square$$

So, according to the previous example, the knowledge of the conformal mapping of a simply connected domain onto the disc allows constructing Green's function of the domain. Now it will be shown that, conversely, the knowledge of Green's function allows constructing the conformal mapping.

**Proposition 9.22.** *Let $U$ be a bounded simply connected domain of the plane and $z_0 \in U$. If there exists the Green's function of $U$ with pole at $z_0$, then there is a conformal mapping $f$ from $U$ onto the unit disc $\mathbb{D}$ such that $f(z_0) = 0$ and $f'(z_0) > 0$.*

*Proof.* If $G_U(z_0, z)$ is the Green's function with pole at $z_0$, there is a function $h$, continuous on $\bar{U}$ and harmonic on $U$, such that

$$G_U(z_0, z) - \frac{1}{2\pi} \operatorname{Log}|z - z_0| = -h(z).$$

Since $U$ is simply connected, the function $h$ has a conjugated harmonic function on $U$, say $\tilde{h}$. Consider the function $f(z_0, z)$ defined by

$$f(z_0, z) = (z - z_0)e^{-2\pi(h(z) + i\tilde{h}(z))}.$$

Then $f(z_0, z)$ is a holomorphic function of $z$ on $U$ that has a simple zero at the point $z_0$ and $|f(z_0, z)| = |z - z_0|e^{-2\pi h(z)}$ has value 1 when $z \in \partial U$, because $h(z) = \frac{1}{2\pi} \operatorname{Log}|z - z_0|$, if $z \in \partial U$. Since $|f(z_0, z)|$ is continuous on $\bar{U}$, it turns out, by the maximum principle, that $|f(z_0, z)| \leq 1$, $z \in U$; that is, $f(z_0, z)$ maps $U$ into the unit disc and vanishes at the point $z_0$. We will now show that $f(z_0, z)$ is one-to-one in $U$ and takes every value $w \in \mathbb{D}$. Then the proposition will be proved taking $f(z) = f(z_0, z)$, since the condition $f'(z_0) > 0$ is obtained by changing $f(z)$ by $e^{i\theta} f(z)$, with a suitable $\theta \in \mathbb{R}$.

a) $f(z_0, z)$ is one-to-one in $U$.

If $z_1 \in U$ define the function of $z$,

$$\varphi(z_0, z_1, z) = \frac{f(z_0, z) - f(z_0, z_1)}{1 - \overline{f(z_0, z_1)} f(z_0, z)}$$

so that $\varphi$ is holomorphic on $U$, $|\varphi(z_0, z_1, z)| \leq 1$ and $\varphi(z_0, z_1, z_1) = 0$. Since the function $f(z_1, z)$, constructed similarly to $f(z_0, z)$ replacing $z_0$ by $z_1$, has a simple zero at the point $z_1$, it turns out that

$$h(z) = \frac{\varphi(z_0, z_1, z)}{f(z_1, z)}$$

is holomorphic on $U$ and $|h(z)| \to 1$ if $z$ tends to $\partial U$. By the maximum principle, one has $|h(z)| \leq 1$, $z \in U$. Hence,

$$|h(z_0)| = \left| \frac{f(z_0, z_1)}{f(z_1, z_0)} \right| \leq 1,$$

and by symmetry, $|f(z_0, z_1)| = |f(z_1, z_0)|$, which leads to $|h(z_0)| = 1$. Hence, the function $|h|$ reaches its maximum at a point of $U$ and so $|h(z)| = 1$, for $z \in U$.

The definition of $h$ gives now

$$|\varphi(z_0, z_1, z)| = |f(z_1, z)| \neq 0 \quad \text{if } z \neq z_1,$$

and, therefore, $f(z_0, z) \neq f(z_0, z_1)$ if $z \neq z_1$. This proves that $f(z_0, z)$ is one-to-one.

b) $f(z_0, z)$ is surjective.

Let $w \in \mathbb{D}$ and suppose that $f(z_0, z)$ does not take the value $w$ in $U$. Then the function

$$g(z) = \frac{f(z_0, z) - w}{1 - \overline{w} f(z_0, z)}$$

is holomorphic from $U$ into $\mathbb{D}$ with $g(z) \neq 0$, $z \in U$. However, $|g(z)| \to 1$ if $z$ tends to $\partial U$, and so the function $\frac{1}{g(z)}$ would be also holomorphic, with $\left|\frac{1}{g(z)}\right| \to 1$ if $z$ tends to $\partial U$. The maximum principle implies that $|g(z)| = 1$ for $z \in U$, and this is impossible because, taking $z = z_0$ yields $|g(z_0)| = |w| < 1$. This contradiction proves assertion b). $\qquad\qquad\square$

Proposition 9.22 allows us to prove Riemann's theorem avoiding the use of normal families on which the proof given in Section 9.2 is based. First, by what has been done in part A) of the proof of Riemann's theorem we may assume that the domain is bounded. Next, we need to show that every bounded simply connected domain $U$ has a Green's function. As known, this means we must solve Dirichlet's problem in $U$, with a particular boundary behavior.

This new strategy to show the fundamental theorem of the conformal mapping was, precisely, the one followed by Riemann, but his proof was incomplete due to the lack of rigor in the existence of a solution for the Dirichlet problem.

The existence of a solution of Dirichlet's problem was announced by Dirichlet himself in 1857, but neither Riemann when he established his theorem in the year 1851, nor Dirichlet provided a correct proof. This existence was, nevertheless, admitted to be true based on the so-called Dirichlet principle. This principle has been dealt with in Section 7.12, and now it is recalled here.

If $U$ is a bounded domain and a $\varphi$ continuous function on $\partial U$, one looks for a function $u$ such that $\Delta u = 0$ on $U$ and $u|_{\partial U} = \varphi$. Then the so-called Dirichlet integral is considered

$$I(u) = \int_U |\vec{\nabla} u|^2 \, dx \, dy.$$

Now, one minimizes the functional $I$ among all functions $u \in C^1(\overline{U})$ that satisfy $u|_{\partial U} = \varphi$. It turns out that if $u_0$ is a global minimum of this problem, $u_0$ is a harmonic function on $U$ and, therefore, $u_0$ is the solution of Dirichlet's problem in $U$ with boundary value $\varphi$. The mistake of Dirichlet was not proving the existence of a function minimizing the Dirichlet integral.

Therefore, to give Riemann's proof one has to show in a rigorous way that Dirichlet's problem has a solution in a bounded simply connected domain. As said at the end of Section 7.12, this can be done using Perron's method, explained in the next section.

## 9.4 Solution of Dirichlet's problem in an arbitrary domain

Perron's method for the solution of Dirichlet's problem is completely general and not limited, therefore, to the case of simply connected domains. It is based on subharmonic functions which are presented below.

### 9.4.1 Subharmonic functions

In one variable the Laplace equation is written $\frac{d^2 u}{dx^2} = 0$, and its solutions, that is, the harmonic functions, are of the form $u = ax + b$, with $a$ and $b$ constants. Imposing to a function $v$ that, for each interval of the line, it is below the linear function that equals $v$ at end-points of the interval, means that $v$ is convex. The same approach in two variables (and also in more variables) leads to the notion of subharmonic function. That is, a subharmonic function $v$ is a function that on any region of the plane is below the harmonic function that equals $v$ at its boundary. Since this formulation involves solving Dirichlet's problem, an alternative definition will be given.

**Definition 9.23.** A real function $v$, continuous on an open set $U$ of the complex plane, is said to be subharmonic on $U$ if it satisfies

$$v(z_0) \le \frac{1}{2\pi} \int_0^{2\pi} v(z_0 + re^{i\theta}) \, d\theta, \tag{9.6}$$

for each $z_0 \in U$ and $r > 0$ such that $\overline{D}(z_0, r) \subset U$.

Inequality (9.6) is called *mean value inequality*. Clearly every real harmonic function on $U$ is subharmonic on $U$ because, in this case, equality in (9.6) holds, Obviously if $u$ is harmonic on $U$, then $|u|$ is subharmonic on $U$. Hence, if $f$ is holomorphic, $|f|$ is subharmonic.

The following properties are direct consequences of the definition.

**Proposition 9.24.** *If $v_1$ and $v_2$ are subharmonic functions on $U$, then the following functions are also subharmonic:*

a) $v_1 + v_2$,

b) $\alpha v_1$ *if* $\alpha > 0$,

c) $v = \max\{v_1, v_2\}$.

*Proof.* Assertions a) and b) are evident from (9.6). For c) suppose that $\bar{D}(z_0, r) \subset U$. Then one has

$$v_j(z_0) \leq \frac{1}{2\pi} \int_0^{2\pi} v_j(z_0 + re^{i\theta}) \, d\theta \leq \frac{1}{2\pi} \int_0^{2\pi} v(z_0 + re^{i\theta}) \, d\theta, \quad j = 1, 2,$$

and, therefore, $v(z_0) \leq \frac{1}{2\pi} \int_0^{2\pi} v(z_0 + re^{i\theta}) \, d\theta$. $\qquad\square$

A remarkable property of subharmonic functions is that, like harmonic functions, they fulfill the maximum principle.

**Proposition 9.25** (Maximum principle for subharmonic functions). *Let $v$ be a subharmonic function on a domain $U$ such that $v(z) \leq M$, $z \in U$, where $M$ is a constant. If there exists a point $z_0 \in U$ such that $v(z_0) = M$, then $v(z) = M$, for all $z \in U$.*

*Proof.* We argue as in the proof of Theorem 7.19. If $\bar{D}(z_0, r) \subset U$, from (9.6) it follows that

$$v(z_0) \leq \frac{1}{\pi r^2} \int_{D(z_0, r)} v(z) \, dm(z).$$

Therefore, one has now

$$\frac{1}{\pi r^2} \int_{D(z_0, r)} [v(z) - v(z_0)] \, dm(z) \geq 0,$$

but the integrand satisfies $v(z) - v(z_0) \leq 0$ if $v(z_0) = M$ and so it must be identically zero on the disc $D(z_0, r)$. $\qquad\square$

**Corollary 9.26.** *Let $u$ be a harmonic function on the domain $U$, continuous on $\bar{U}$, and let $v$ be a subharmonic function on $U$ such that $\overline{\lim}_{z \to \xi} v(z) \leq u(\xi)$ for each $\xi \in \partial U$. Then one has $v(z) \leq u(z)$ for all $z \in U$.*

*Proof.* Since $v - u$ is subharmonic, one may assume $\overline{\lim}_{z \to \xi} v(z) \leq 0$, $\xi \in \partial U$ and to prove that $v(z) \leq 0$, $z \in U$. Now, the function $w(z) = \max(v(z), 0)$ is subharmonic and satisfies $\lim_{z \to \xi} w(\xi) = 0$, $\xi \in \partial U$. Therefore, defining $w(\xi) = 0$ when $\xi \in \partial U$, $w$ is continuous on $\bar{U}$, subharmonic on $U$ and zero on $\partial U$. Hence, the function $w$ has its maximum at some point $z_0 \in \bar{U}$. If $w(z_0) > 0$, then $z_0 \in U$, and, by Proposition 9.25, $w$ would be constant, a fact that contradicts $w(\xi) = 0$, if $\xi \in \partial U$. Therefore, $w(z_0) = 0$, which gives $w(z) = 0$, $z \in U$ and so $v(z) \leq 0$, $z \in U$. $\qquad\square$

This corollary shows that a subharmonic function is below a harmonic function that equals it at the boundary of a domain. As said, this is analogous to the behavior of convex functions of one real variable. In this case, it is also known that the function $v$ is convex if and only if $v'' \geq 0$, at least in the case that $v$ is twice differentiable. Analogously it may be proved that if $v$ has continuous partial derivatives of second order, then $v$ is subharmonic on $U$ if and only if $\Delta v(z) \geq 0$, $z \in U$ (see Exercise 13 in Section 3.8 and Exercise 3 in Section 7.13).

Recall that Dirichlet's problem in a disc $D$ has been already solved in Theorem 7.26 (see also Example 7.51). It is known that if $\varphi$ is continuous on $\partial D$, then the solution of Dirichlet's problem in $D$ with boundary values $\varphi$ is given by the Poisson integral of $\varphi$, denoted by $P[\varphi]$. So, $P[\varphi]$ is a harmonic function on $D$, satisfying $\lim_{z \to z_0} P[\varphi](z) = \varphi(z_0)$ for each $z_0 \in \partial D$.

Let now $v$ be a subharmonic function on the open set $U$ and $D$ a disc such that $\bar{D} \subset U$. Define the function $\tilde{v}$ on $U$ by

$$
\begin{cases}
\tilde{v}(z) = P[v](z) & \text{if } z \in D, \\
\tilde{v}(z) = v(z) & \text{if } z \in U \setminus D.
\end{cases}
\tag{9.7}
$$

That is, $\tilde{v}$ is the function $v$ outside the disc $D$ and inside $D$ it is the harmonic function which has value $v$ on $\partial D$.

The maximum principle has as a consequence the following proposition.

**Proposition 9.27.** *Let $v$ be a subharmonic function on $U$, $D$ a disc such that $\bar{D} \subset U$ and $\tilde{v}$ the function defined on $U$ by (9.7). Then $\tilde{v}$ is subharmonic on $U$ and $v(z) \leq \tilde{v}(z)$, $z \in U$.*

*Proof.* By the maximum principle (Corollary 9.26) one has $v(z) \leq \tilde{v}(z)$, if $z \in D$. Therefore, $v(z) \leq \tilde{v}(z)$, for $z \in U$. It is also clear that $\tilde{v}(z)$ is continuous on $U$.

Now, if $z_0 \in D$, one has $\tilde{v}(z_0) = \frac{1}{2\pi} \int_0^{2\pi} \tilde{v}(z_0 + re^{i\theta})\, d\theta$, if $r$ is small enough, because $\tilde{v}$ is harmonic on $D$. If $z_0 \notin D$ and $\bar{D}(z_0, r) \subset U$, then

$$
\tilde{v}(z_0) = v(z_0) \leq \frac{1}{2\pi} \int_0^{2\pi} v(z_0 + re^{i\theta})\, d\theta \leq \frac{1}{2\pi} \int_0^{2\pi} \tilde{v}(z_0 + re^{i\theta})\, d\theta
$$

and, so, $\tilde{v}$ is subharmonic on $U$. $\qquad\square$

## 9.4.2 Perron's method

Let $U$ be a bounded domain of the plane and let $\varphi$ be a real and continuous function on the boundary of $U$. Perron's method associates to the pair $U$, $\varphi$ a harmonic function $u$ on $U$ that is below $\varphi$ at the boundary of $U$. Under appropriate regularity conditions on $\partial U$, $u$ is the solution of Dirichlet's problem in $U$ with boundary values $\varphi$.

In order to construct the function $u$, define the class $\mathcal{P}(\varphi)$ of functions $v$ that satisfy the following conditions:

a) $v$ is subharmonic on $U$.

b) $\overline{\lim}_{z \to \xi} v(z) \leq \varphi(\xi)$, for each $\xi \in \partial U$.

Observe that the class $\mathcal{P}(\varphi)$ is not empty. Actually, since $\varphi$ is continuous on the compact set $\partial U$, there is a constant $M > 0$ such that $-M \leq \varphi(z) \leq M, z \in \partial U$ and the constant function equal to $-M$ belongs to $\mathcal{P}(\varphi)$. Applying Corollary 9.26 one gets that each function $v \in \mathcal{P}(\varphi)$ satisfies $v(z) \leq M$, for $z \in U$.

The harmonic function we are looking for is defined as follows:

$$u(z) = \sup\{v(z) : v \in \mathcal{P}(\varphi)\}, \quad z \in U. \tag{9.8}$$

As said above, one has $u(z) \leq M$ for $z \in U$, if $|\varphi(z)| \leq M, z \in \partial U$.

**Proposition 9.28.** *The function $u$ given by (9.8) is harmonic on $U$.*

*Proof.* It is enough to choose a disc $D$ with $\bar{D} \subset U$ and prove that $u$ is harmonic on $D$.

Fix a point $z_0 \in D$. By the definition of $u$ there exist functions $v_n \in \mathcal{P}(\varphi)$, $n = 1, 2, \ldots$, such that $\lim_n v_n(z_0) = u(z_0)$. Put $V_n = \max\{v_1, \ldots, v_n\}$ and for each $n$ let $\widetilde{V}_n$ be the function that equals $V_n$ on $U \setminus D$ and that is the harmonic function on $D$ with value $V_n$ at $\partial D$ (defined according to (9.7)). It is clear that the sequence $(V_n)$ is increasing and so is the sequence $(\widetilde{V}_n)$, by the maximum principle. Due to Propositions 9.25 and 9.27, $V_n$ as well as $\widetilde{V}_n$ belong to $\mathcal{P}(\varphi)$.

The inequalities

$$v_n(z_0) \leq V_n(z_0) \leq \widetilde{V}_n(z_0) \leq u(z_0)$$

prove that $\lim_n \widetilde{V}_n(z_0) = u(z_0) \leq M$. Now, by Corollary 9.15, the sequence $\widetilde{V}_n$ converges on $D$ to a harmonic function $\tilde{u}$. This function satisfies $\tilde{u}(z) \leq u(z)$, for $z \in D$ and $\tilde{u}(z_0) = u(z_0)$. If one shows that $u(z_1) = \tilde{u}(z_1)$ for any other point $z_1 \in D$, the proposition will be proved, because then $u \equiv \tilde{u}$ on $D$ and $u$ will be harmonic.

Fix now $z_1 \in D$ and repeat the previous argument. Take functions $w_n \in \mathcal{P}(\varphi)$, $n = 1, 2, \ldots$ with $\lim_n w_n(z_1) = u(z_1)$, replace $w_n$ by $w_n' = \max\{v_n, w_n\}$ and next write $W_n = \max\{w_1', \ldots, w_n'\}$. So $\widetilde{W}_n$ will be the function equal to $W_n$ on $U \setminus D$ and harmonic on $D$ with value $W_n$ on $\partial D$. Since $(\widetilde{W}_n)$ is an increasing sequence of functions in $\mathcal{P}(\varphi)$ which are harmonic on $D$, it will converge to a limit $\tilde{u}_1(z)$, a harmonic function on $D$ satisfying

$$\tilde{u}(z) \leq \tilde{u}_1(z) \leq u(z), \quad z \in D \text{ and } \tilde{u}_1(z_1) = u(z_1).$$

Therefore, the function $\tilde{u} - \tilde{u}_1$ takes its maximum value equal to zero at the point $z_0$ and, by the maximum principle, one has $\tilde{u}(z) = \tilde{u}_1(z), z \in D$. So, $\tilde{u}(z_1) = \tilde{u}_1(z_1) = u(z_1)$ as claimed. $\qquad\square$

In order to solve Dirichlet's problem in $U$ with boundary values $\varphi$, it remains to check that $\lim_{z \to z_0} u(z) = \varphi(z_0)$ for each point $z_0 \in \partial U$. We introduce the following definition.

**Definition 9.29.** A point $\xi_0$ of the boundary of a bounded domain $U$ is called a regular point of $\partial U$ if the condition

$$\lim_{\substack{z \to \xi_0 \\ z \in U}} u(z) = \varphi(\xi_0)$$

holds for every function $\varphi$ defined and continuous on the boundary of $U$, where $u$ is the harmonic function on $U$ defined by (9.8).

It is worth noting at this point that if Dirichlet's problem in $U$ with boundary values $\varphi$ has a solution, then this solution necessarily is Perron's, given by (9.8). To see this, just observe that if $h$ is a solution of Dirichlet's problem, then $h \in \mathcal{P}(\varphi)$, and hence, $h(z) \le u(z)$, $z \in U$. But also each function $v \in \mathcal{P}(\varphi)$ satisfies $\overline{\lim}_{z \to \xi} v(z) \le \varphi(\xi) = h(\xi)$ if $\xi \in \partial U$ and, by the maximum principle, it follows that $v(z) \le h(z)$, $z \in U$ and, consequently, $u(z) \le h(z)$, $z \in U$.

**Example 9.30.** To give an example of a non-regular point, consider the punctured disc $U = \mathbb{D} \setminus \{0\}$ and the point $0 \in \partial U$. Let $\varphi$ be the function with value $0$ on the unit circle and $\varphi(0) = 1$; then the corresponding Dirichlet's problem has no solution and, therefore, the origin is not a regular point of $U$. Actually, if $u$ was continuous on $\overline{U}$ with value $\varphi$ on $\partial U$ and harmonic on $U$, the origin would be a removable singularity of $u$ (Corollary 5.21), that is, $u$ would be harmonic on the whole disc $\mathbb{D}$ and zero on $\partial \mathbb{D}$. By the maximum principle $u$ should be identically zero on $\mathbb{D}$, but $u(0) = \varphi(0) = 1$. $\square$

In order to find conditions for a point $\xi_0 \in \partial U$ to be regular, Dirichlet's problem will be solved for some special boundary values. Fix $\xi_0 \in \partial U$ and assume that $\xi_0$ is a regular point. Take $\varphi(\xi) = |\xi - \xi_0|$, $\xi \in \partial U$, and let $u$ be the corresponding Perron's solution. Since the function $|z - \xi_0|$ is subharmonic, it belongs to $\mathcal{P}(\varphi)$ and, therefore, $u(z) \ge |z - \xi_0|$, $z \in U$. Writing now $h(z) = -u(z)$, $h$ is a harmonic function on $U$ satisfying

$$\lim_{z \to \xi_0} h(z) = -\varphi(\xi_0) = 0, \qquad \overline{\lim_{\substack{z \to \xi \\ z \in U}}} h(z) \le -|\xi - \xi_0| < 0, \qquad \text{for } \xi \in \partial U, \ \xi \ne \xi_0.$$

That is, the function $h$ is harmonic, vanishing at $\xi_0$ and has a strictly negative "value" at the points of $\partial U$ different from $\xi_0$.

The argument above justifies the following definition.

**Definition 9.31.** If $\xi_0$ is a point of the boundary of the bounded domain $U$, a barrier at the point $\xi_0$ is a harmonic function $h$ on $U$ such that

$$\lim_{z \to \xi_0} h(z) = 0 \quad \text{and} \quad \overline{\lim_{z \to \xi}} \, h(z) < 0, \quad \text{for } \xi \in \partial U, \ \xi \neq \xi_0.$$

**Example 9.32.** a) If $U = \mathbb{D}$ is the unit disc and $\xi_0 \in \partial \mathbb{D}$, a barrier at $\xi_0$ is the function $h(z) = \mathrm{Re}(z\bar{\xi_0}) - 1$. Just remark that $\mathrm{Re}(z\bar{\xi_0}) = 1$ is the equation of the tangent line to $\partial \mathbb{D}$ at the point $\xi_0$.

b) Assume now that $\xi_0 \in \partial U$ and there is a line segment $L$ having $\xi_0$ as an endpoint, with $L \setminus \{\xi_0\}$ contained in the exterior of $U$. Then one can construct a barrier at the point $\xi_0$ in the following way: if $\xi_1$ is the other endpoint of $L$, the function $(z - \xi_0)/(z - \xi_1)$ does not vanish on the simply connected domain $\mathbb{C}^* \setminus L$ and, therefore, there exists a holomorphic branch of $\sqrt{(z - \xi_0)/(z - \xi_1)}$ in this domain (we can assume the domain is in $\mathbb{C}$ applying the mapping $z^* = \frac{1}{z}$). This branch maps $\mathbb{C}^* \setminus L$ into a half-plane bounded by a straight line through the origin, which may in turn be mapped into the imaginary axis by a rotation of angle $\alpha$, for some $\alpha \in \mathbb{R}$. This means that the harmonic function

$$h(z) = \mathrm{Re}\left( e^{i\alpha} \sqrt{\frac{z - \xi_0}{z - \xi_1}} \right)$$

is negative in $\partial U \setminus \{\xi_0\}$ and satisfies $h(\xi_0) = 0$.

c) A particular case of b) is when the boundary of $U$ consists, around $\xi_0$, of a $C^1$ curve with non-null derivative. Then $\mathbb{C} \setminus U$ contains in its interior a line segment in the normal direction to $\partial U$. $\qquad\square$

The following result shows that the existence of barriers is not only necessary, but also sufficient for the regularity at a point of the boundary.

**Proposition 9.33.** *Let $\xi_0$ be a point of the boundary of the bounded domain $U$ and suppose there exists a barrier at the point $\xi_0$. Then $\xi_0$ is a regular point of $\partial U$.*

Before starting the proof of this result, let us make a useful remark on Perron's method. It associates to each continuous function $\varphi$ on $\partial U$ a harmonic function $u$ on $U$ according to (9.8) and Proposition 9.28. This correspondence is *superlinear* in the following sense:

Let $\varphi_1, \varphi_2 \in C(\partial U)$ and let $u_1, u_2$ be the harmonic functions associated to $\varphi_1$, $\varphi_2$ by (9.8). Then the harmonic function associated to $\varphi_1 + \varphi_2$ is greater than or equal to $u_1 + u_2$. Equivalently,

$$\sup\{v(z)\colon v \in \mathcal{P}(\varphi_1 + \varphi_2)\} \geq \sup\{v(z)\colon v \in \mathcal{P}(\varphi_1)\} + \sup\{v(z)\colon v \in \mathcal{P}(\varphi_2)\}.$$

This inequality follows from the definition of the class $\mathcal{P}(\varphi)$.

*Proof of Proposition* 9.33. Let $\varphi$ be continuous on $\partial U$ and assume $|\varphi(\xi)| \leq M$, $M > 0$, for $\xi \in \partial U$. Fix $\xi_0 \in \partial U$ and let $h$ be a barrier at the point $\xi_0$. If $u$ is the harmonic function associated to $\varphi$ by (9.8), one has to show that

$$\lim_{z \to \xi_0} u(z) = \varphi(\xi_0).$$

To this end it is enough to prove, for all $\varepsilon > 0$, the two inequalities

$$\overline{\lim_{z \to \xi_0}} \, u(z) \leq \varphi(\xi_0) + \varepsilon, \quad \underline{\lim_{z \to \xi_0}} \, u(z) \geq \varphi(\xi_0) - \varepsilon.$$

Given $\varepsilon > 0$, choose a $\delta > 0$ such that

$$\varphi(\xi_0) - \varepsilon \leq \varphi(\xi) \leq \varphi(\xi_0) + \varepsilon \quad \text{if } \xi \in \partial U \cap D_\delta(\xi_0).$$

Let $m = \sup\{h(z) \colon z \in U \setminus D_\delta(\xi_0)\}$. By the maximum principle and since $\overline{\lim}_{z \to \xi} h(z) < 0$ if $\xi \neq \xi_0$, one has $m < 0$. Write now

$$v(z) = \varphi(\xi_0) - \varepsilon - \frac{h(z)}{m}(M + \varphi(\xi_0))$$

so that $v$ is harmonic on $U$ and satisfies

$$\overline{\lim_{z \to \xi}} \, v(z) \leq \varphi(\xi_0) - \varepsilon \leq \varphi(\xi) \quad \text{if } \xi \in \partial U \cap D_\delta(\xi_0)$$

and

$$\overline{\lim_{z \to \xi}} \, v(z) \leq -M - \varepsilon \leq \varphi(\xi) \quad \text{if } \xi \in \partial U \setminus D_\delta(\xi_0).$$

Therefore, $v \in \mathcal{P}(\varphi)$, $v(z) \leq u(z)$, $z \in U$, and consequently

$$\underline{\lim_{z \to \xi_0}} \, u(z) \geq \lim_{z \to \xi_0} v(z) = \varphi(\xi_0) - \varepsilon.$$

Repeating now the previous argument replacing $\varphi$ by $-\varphi$ and $u$ by Perron's function associated to $-\varphi$, say $u_1$, we get

$$\underline{\lim_{z \to \xi_0}} \, u_1(z) \geq -\varphi(\xi_0) - \varepsilon.$$

Now, by the remark after Proposition 9.33 we know that $u(z) + u_1(z) \leq 0$. Hence,

$$u(z) \leq -u_1(z) \text{ and } \overline{\lim_{z \to \xi_0}} \, u(z) \leq \overline{\lim_{z \to \xi_0}} \, (-u_1(z)) = - \underline{\lim_{z \to \xi_0}} \, u_1(z) \leq \varphi(\xi_0) + \varepsilon,$$

which is the other needed inequality. $\qquad\square$

**Corollary 9.34.** *Dirichlet's problem can be solved in any bounded domain $U$ such that each point of $\partial U$ is the endpoint of a segment having all the other points in the exterior to $U$. This happens, in particular, if $\partial U$ consists of a finite number of regular curves.*

*Proof.* According to parts b) and c) of Example 9.32, there is a barrier at each point of $\partial U$ and now Proposition 9.33 applies. $\qquad\square$

### 9.4.3 Construction of barriers for simply connected domains and Riemann's theorem

As said in Section 9.3, to prove the fundamental theorem of conformal mapping following Riemann, one needs to show that Dirichlet's problem has a solution in every bounded simply connected domain. For this it suffices, according to Proposition 9.33, to prove the following result.

**Proposition 9.35.** *If $U$ is a bounded simply connected domain of $\mathbb{C}$, there is a barrier at every point of the boundary of $U$.*

*Proof.* Let us fix $\xi_0 \in \partial U$ and construct a barrier at $\xi_0$. If $\xi_1 \notin U$ is far enough from $U$, the set $A = \{z : |z - \xi_0| < \frac{1}{e}|z - \xi_1|\}$ is an open set containing $U$. Since $U$ is simply connected, the function $\operatorname{Log} \frac{z-\xi_0}{z-\xi_1}$ has a holomorphic branch on $U$, call it $f$. Now, if $z \in U \subset A$, one has $\operatorname{Re} f(z) = \operatorname{Log}\left|\frac{z-\xi_0}{z-\xi_1}\right| < -1$, that is, $f$ takes values in the half-plane $\{w : \operatorname{Re} w < -1\}$ (Figure 9.1).

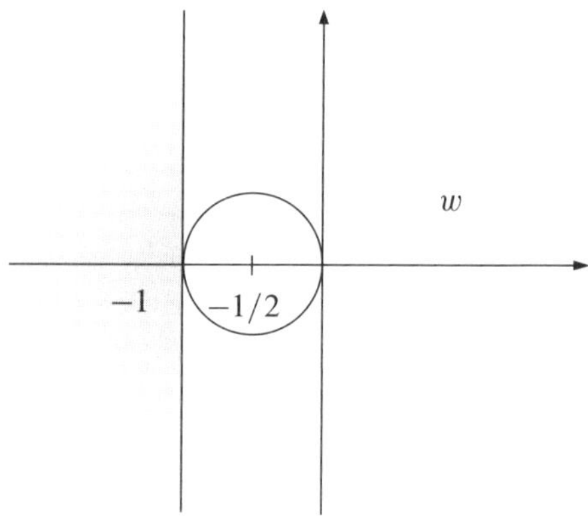


Figure 9.1

The mapping $w \to \frac{1}{w}$ transforms this half-plane into the disc $D(-1/2, 1/2)$ and so, the function $\frac{1}{f(z)}$, holomorphic on $U$, takes its values in the disc $D(-1/2, 1/2)$. The function $h(z) = \operatorname{Re} \frac{1}{f(z)}$ is harmonic on $U$ and satisfies $\overline{\lim}_{z \to \xi} h(z) \leq 0$, if $\xi \in \partial U$. But $\overline{\lim}_{z \to \xi} h(z) = 0$ only if $\lim_{z \to \xi} \operatorname{Re} f(z) = -\infty$, and this is only possible if $\xi = \xi_0$. Therefore, $h$ is a barrier at the point $\xi_0$. $\qquad\square$

**Corollary 9.36.** *Dirichlet's problem has a solution in every bounded simply connected domain of the plane.*

**Remark 9.2.** Proposition 9.35 does not follow from Corollary 9.34 because if $U$ is simply connected and $\xi_0 \in \partial U$, it may happen there is no segment with an endpoint at $\xi_0$ and contained in $\mathbb{C} \setminus U$. For example, consider a strip with decreasing width, rolled like a spiral, $U$, winding infinitely many times around the origin. Then $U$ is simply connected, $0 \in \partial U$ but, travelling from the origin in any direction, one finds points of $U$ (Figure 9.2).

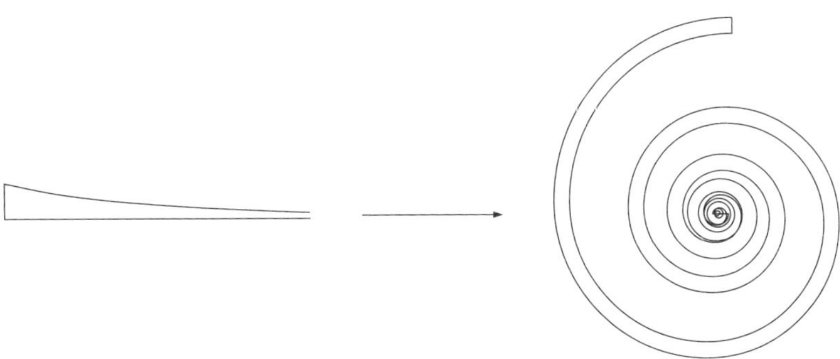

Figure 9.2

Summarizing, let us show how all these facts combine to give the second proof of Riemann's theorem (Theorem 9.18).

If $U \neq \mathbb{C}$ is a simply connected domain, it is easy to see that $U$ may be conformally mapped on a bounded simply connected domain. It is enough, following the proof of A) in Section 9.2, to take $g(z)$ as a branch of $\sqrt{z - a}$, $a \notin U$, which is holomorphic and one-to-one on $U$ and satisfies $|g(z) + g(z_0)| \geq r$ for some point $z_0 \in U$ and some $r > 0$. Then the function $g_1(z) = \frac{1}{g(z)+g(z_0)}$ sends $U$ conformally into a bounded domain.

Supposing, then, that $U$ is bounded and simply connected, one can solve Dirichlet's problem in $U$ with boundary values $z \to \frac{1}{2\pi} \mathrm{Log}\,|z - z_0|$, according to Corollary 9.36. Now, if the harmonic function $h$ is the solution of this Dirichlet problem, the function $G_U(z_0, z) = \frac{1}{2\pi} \mathrm{Log}\,|z - z_0| - h(z)$, $z \in U \setminus \{z_0\}$, is the Green's function of $U$ with pole at $z_0$ (Section 9.3), and one just has to apply Proposition 9.22 to finish the proof of Riemann's theorem.

## 9.5 Exercises

1. Show the space $H(\mathbb{D}) \cap L^1(\mathbb{D})$ is a closed subspace of $L^1(\mathbb{D})$.

2. Let $U$ be a domain of $\mathbb{C}$ and $\mathcal{F}$ a family of holomorphic functions on $U$. Assume for each point $z \in U$ there is a neighborhood $V(z) \subset U$ such that $\mathcal{F}$ is normal on $V(z)$. Prove that $\mathcal{F}$ is normal on $U$. (Normal family understood in Montel's sense.)

3. Let $U$ be a domain of $\mathbb{C}$ and $f_n \in H(U)$ with $\lim_n f_n(z) = f(z)$ uniformly on compact sets of $U$. Suppose $f \not\equiv 0$ and $f$ has $m$ different zeros in $U$. Show that, for $n$ big enough, the functions $f_n$ have at least $m$ different zeros on $U$.

4. Let $(c_n)_{n \in \mathbb{N}}$ be a bounded sequence of complex numbers. Prove that the function series $\sum_{n=1}^{\infty} \frac{c_n z^n}{1-z^n}$ converges uniformly on compact sets of $\mathbb{D}$. Show also that if the series $\sum_{n=1}^{\infty} c_n$ is convergent, then the same function series converges uniformly on compact sets of $\mathbb{C} \setminus \mathbb{T}$ to a holomorphic function $g$.

Prove in this last case that defining $f(z) = \sum_{n=1}^{\infty} c_n z^n$ on $\mathbb{D}$, the equalities

$$g(z) = \sum_{n=1}^{\infty} f(z^n) \text{ if } |z| < 1 \quad \text{and} \quad g(z) = -\sum_{n=0}^{\infty} f(z^{-n}) \text{ if } |z| > 1,$$

hold.

5. Let $U$ be a simply connected domain of the plane, $U \neq \mathbb{C}$ and $f : U \to U$ a holomorphic function. Assume there is a point $z_0 \in U$ with $f(z_0) = z_0$ and $|f'(z_0)| < 1$. Show that the sequence of iterates of $f$, $(f_n)$, where $f_n(z) = \underbrace{(f \circ \cdots \circ f)}_{\leftarrow \ n \ \text{times} \ \rightarrow}(z)$ for $n = 1, 2, \ldots$, is uniformly convergent on compact sets of $U$ and find the function $\lim_n (f_n)$.

Show also that this assertion does not hold if $U$ is any domain of $\mathbb{C}$.

6. Prove that a family of holomorphic functions on a domain $U$ taking no value in a fixed disc of positive radius is a normal family in $U$ (in Montel's sense).

7. Let $U$ be a domain of $\mathbb{C}$ and let $f_n \in H(U)$, $n = 1, 2, \ldots$, be such that $\sup_n |f_n(z)| < +\infty$, for each $z \in U$. Show there is a subset $\tilde{U} \subset U$, open and dense in $U$, and a function $f \in H(\tilde{U})$ with $f(z) = \lim_n f_n(z)$, uniformly on compact sets of $\tilde{U}$.

*Hint:* Use Baire's theorem.

8. Let $U$ be a domain and $f_n \in H(U)$, $n = 1, 2, \ldots$, $z_0 \in U$ with $f_n \to f$ uniformly on compact sets of $U$ and $f'(z_0) \neq 0$. Show that there exist a neighborhood $V$ of $z_0$, a neighborhood $W$ of $f(z_0)$ and $n_0 \in \mathbb{N}$ such that for $n \geq n_0$, $f_n$ and $f$ are one-to-one on $V$, $W \subset f_n(V)$ and $f_n^{-1} \to f^{-1}$ uniformly on $W$. Find out what happens in the case one assumes $f'(z_0) = f''(z_0) = f^{(m-1)}(z_0) = 0$, $f^{(m)}(z_0) \neq 0$, for some $m \in \mathbb{N}$, $m \geq 1$.

9. Let $(U_n)$ be a sequence of simply connected domains of $\mathbb{C}$ with $U_n \subset U_{n+1}$, $n = 1, 2, \ldots$ and $U = \bigcup_{n=1}^{\infty} U_n \neq \mathbb{C}$. Let $f_n : U_n \to \mathbb{D}$ be the conformal

mapping given by Riemann's theorem ($f_n(z_0) = 0$, $f_n'(z_0) > 0$, for a fixed point $z_0 \in U_1$). Show that $U$ is a simply connected domain and that functions $f_n$ converge uniformly on compact sets of $U$ to the conformal mapping $f : U \to \mathbb{D}$ with $f(z_0) = 0$, $f'(z_0) > 0$.

10. Let $f$ be a continuous function on a domain $U \subset \mathbb{C}$. For each $r > 0$ write $U_r = \{z \in U : d(z, U^c) > r\}$ and define on $U_r$ the function

$$f_r(z) = \frac{1}{\pi r^2} \int_0^{2\pi} \int_0^r f(z + \rho e^{i\theta})\rho\, d\rho\, d\theta.$$

Prove that $f$ is holomorphic on $U$ if and only if there is a sequence of positive numbers $(r_n) \to 0$ such that $f_{r_n}$ is holomorphic on $U_{r_n}$ for each $n = 1, 2, \ldots$.

11. Prove that a continuous function $f$ on a domain $U$ is holomorphic on $U$ if and only if it satisfies $\int_{\partial D} f(z)dz = 0$, for every disc $D$ such that $\bar{D} \subset U$.

   *Hint:* Use the previous exercise and Exercise 2 in Section 9.5.

12. Let $U$ be a domain of the plane, and $K$ a compact set, $K \subset U$. Show there exist two constants $m$, $M > 0$ with $m \leq u(z)/u(w) \leq M$ for every positive harmonic function $u$ on $U$ and every pair of points $z, w \in K$.

13. Let $U$ be a bounded domain. Define $d(z, w)$, if $z, w \in U$, as the infimum of the set of numbers $\mathrm{Log}\, C$, where $C$ is any constant $C > 1$ satisfying $1/C \leq u(z)/u(w) \leq C$ for every positive harmonic function $u$ on $U$. Show

   i) $d(z, w) \leq d(z, \tau) + d(\tau, w)$ if $z, w, \tau \in U$.

   ii) $\lim_n d(z_n, z_0) = 0$ if and only if $|z_n - z_0| \to 0$, $z_n, z_0 \in U$.

   iii) Compute $d(z, w)$ in the case $U$ is the unit disc.

14. Let $U$ be a simply connected domain such that $0 \in U \subset \mathbb{D}$ and let $\mathcal{F}$ be the family of functions $f : U \to \mathbb{D}$, holomorphic and one-to-one on $U$ with $f(0) = 0$. If $z_0 \in U$, $z_0 \neq 0$, let $S = \sup\{|f(z_0)| : f \in \mathcal{F}\}$. Prove that there is a function $g \in \mathcal{F}$ such that $|g(z_0)| = S$ and this function $g$ maps $U$ conformally onto $\mathbb{D}$.

15. Let $v : \mathbb{D} \to \mathbb{R}$ be a subharmonic function. Show the following conditions are equivalent:

   i) $\sup_{0 < r < 1} \int_0^{2\pi} v(re^{(i\theta)})d\theta < \infty$.

   ii) There exists a function $u$ harmonic on $\mathbb{D}$ with $v(z) \leq u(z)$, $z \in \mathbb{D}$.

**16.** Let $U$ be a bounded domain and $\varphi$ a function with continuous second-order derivatives on $\bar{U}$. Show that $\varphi$ may be written in $U$ as the difference of two subharmonic functions.

*Hint:* One can assume that $\varphi$ has compact support contained in a neighborhood of $\bar{U}$. Write $\Delta\varphi$ as the difference of two positive functions and apply Exercise 13 in Section 3.8.

**17.** Prove that every simply connected domain of $\mathbb{C}^*$ (that is, its complement is connected in $\mathbb{C}^*$) is conformally equivalent to one (and only one) of the three following domains:

    a) The compactified plane $\mathbb{C}^*$.

    b) The finite plane $\mathbb{C}$.

    c) The unit disc $\mathbb{D}$.

**18.** Show that the Green's function $G_U(z_0, z)$ of a domain $U$ is jointly continuous with respect to the two variables for $z \neq z_0$.

**19.** Let $U$ be a bounded domain of the plane whose boundary is an analytic Jordan curve $\gamma$ (see Section 8.2). Prove the following properties of Green's function of $U$ with pole at $z_0 \in U$, $G(z_0, z)$:

    i) $G(z_0, z) > 0$, if $z \in U$, $z \neq z_0$.

    ii) If $\widetilde{G}(z_0, z)$ is the harmonic reflection of $G$ through $\gamma$, then $\widetilde{G}(z_0, z) < 0$ if $z \notin \bar{U}$.

    iii) $\frac{\partial G}{\partial \vec{N}}(z_0, z) < 0$, for $z \in \partial U$, where $\vec{N}$ is the exterior normal unit vector to $\gamma$.

**20.** Let $U$ be a bounded domain of $\mathbb{C}$ such that $\mathbb{C}^* \setminus U$ consists of a finite number of compact connected sets with more than one point. Show that Dirichlet's problem has a solution in $U$. Show that, on the other hand, this problem may have no solution in $U$ if some of the compact connected sets in $\mathbb{C}^* \setminus U$ is reduced to a point.

# Chapter 10

# Runge's theorem and the Cauchy–Riemann equations

The first part of this chapter deals with the problem of approximating holomorphic functions by simpler ones, concretely by rational or polynomial functions. The basic result is Runge's theorem, asserting that it is possible to approximate every holomorphic function on an open set by rational functions with the poles located at prescribed points outside the open set. This theorem is applied to two classical problems in complex analysis. The first one is about constructing a meromorphic function with predetermined poles, as well as its principal part at these poles. The exact formulation is given by Mittag-Leffler type theorems.

The second problem deals with the solution of the non-homogeneous Cauchy–Riemann equations, that is, one looks for functions $f$ such that $\bar{\partial} f = \phi$, where $\phi$ is a given function. The treatment of this problem is parallel to the one done for the Poisson equation, $\triangle u = \phi$, in Section 7.7. When the data $\phi$ has compact support in $\mathbb{C}$, the solution of $\bar{\partial} f = \phi$ is given by the Cauchy integral of $\phi$, which plays a role similar to the Riesz potential in the equation with the Laplacian. For the case of a general $\phi$ on an open set of the plane, the solution of the Cauchy–Riemann equations is based on Runge's theorem. Finally, we deal with Dirichlet's problem corresponding to the operator $\bar{\partial}$, that is, the solution of the equation $\bar{\partial} f = \phi$ in a domain with conditions on $f$ at its boundary.

## 10.1  Runge's approximation theorems

Typical examples of holomorphic functions are rational functions and polynomials, and, to some extent, they are the simplest ones. In this section it will be proved that, in general, polynomials and rational functions approximate all holomorphic functions.

We start specifying what it is understood by approximation of functions. Two kinds of results will be obtained: the first one deals with approximation of functions on compact sets, and the second one with approximation of functions on open sets of the plane.

When dealing with continuous functions on a compact set $K \subset \mathbb{C}$, the suitable notion of approximation is *uniform* approximation on $K$. The fact that every function of a class $\mathcal{A}$ of continuous functions on $K$ can be uniformly approximated by functions of another class $\mathcal{B}$ on $K$ means that for every $f \in \mathcal{A}$ and every $\varepsilon > 0$

there is a function $g \in \mathcal{B}$ such that

$$|f(z) - g(z)| < \varepsilon, \quad z \in K.$$

Giving to $\varepsilon$ the values $1/n, n = 1, 2, \ldots$, and choosing the corresponding functions $g_n \in \mathcal{B}$, it turns out that the uniform approximation of the functions of $\mathcal{A}$ by functions of $\mathcal{B}$ is equivalent to the fact that every function $f \in \mathcal{A}$ is the uniform limit on $K$ of a sequence $(g_n)$ of functions of $\mathcal{B}$. A more abstract point of view is considering the metric space $C(K)$ of continuous functions on $K$ with the distance

$$d(f_1, f_2) = \max_{z \in K} |f_1(z) - f_2(z)|.$$

Then the approximation just described is expressed by $\mathcal{A} \subset \bar{\mathcal{B}}$ in $C(K)$.

When dealing with the approximation by continuous functions on an open set $U$ then, since these functions are in general not bounded on $U$, the appropriate notion of convergence is not the uniform convergence on $U$, but the *uniform convergence on compact sets* of $U$ (see Subsection 9.1.1). A function $f$ of a class $\mathcal{A}$ of continuous functions on an open set $U$ can be uniformly approximated on compact sets of $U$ by functions of the class $\mathcal{B}$, if for every compact set $K \subset U$ and every $\varepsilon > 0$ there is a function $g \in \mathcal{B}$ such that $|f(z) - g(z)| < \varepsilon$ for $z \in K$. It is easy to check that every function of $\mathcal{A}$ can be uniformly approximated on compact sets of $U$ by functions of $\mathcal{B}$ if and only if for every function $f \in \mathcal{A}$ there exists a sequence $(g_n)$ of functions of $\mathcal{B}$ such that $\lim_{n \to \infty} g_n = f$, uniformly on each compact set of $U$. Actually, we need only use Lemma 1.15 and take, for each compact set $K_n$, $\varepsilon_n = 1/n$ and $g_n \in \mathcal{B}$ the function that approximates $f$ on $K_n$ with an error smaller than $\varepsilon_n$.

We will often use the following *transitivity* property: if a class $\mathcal{B}$ approximates all the functions in the class $\mathcal{A}$ and a class $\mathcal{C}$ approximates all the ones in $\mathcal{B}$, then the class $\mathcal{C}$ approximates the class $\mathcal{A}$. The proof is immediate.

When dealing with the approximation by rational functions, the Cauchy integral formula suggests how to proceed; indeed an integral is a limit of Riemann sums and in the case of the Cauchy integral, each of these sums is a rational function. This is the contents of the following proposition, a warm-up for Runge's approximation theorem.

**Proposition 10.1.** *Let $\gamma$ be a path in the complex plane and $F$ the holomorphic function on $\mathbb{C} \setminus \gamma^*$ defined by the Cauchy integral*

$$F(z) = \frac{1}{2\pi i} \int_\gamma \frac{f(w)}{w - z} dw, \quad z \notin \gamma^*,$$

*where $f$ is a continuous function on $\gamma^*$. Let $K$ be a compact set of $\mathbb{C}$ disjoint from $\gamma^*$ and $\varepsilon > 0$ arbitrary. Then there is a rational function $R$ with simple poles on $\gamma^*$ such that*

$$|F(z) - R(z)| < \varepsilon, \quad z \in K.$$

*Proof.* Let $\gamma$ be defined by $\gamma(t)$, $a \le t \le b$. By the uniform continuity of the function $f(\gamma(t))/(\gamma(t) - z)$ for $t \in [a, b]$ and $z \in K$, there is a partition $a = t_0 < t_1 < \cdots < t_n = b$ of $[a, b]$ such that

$$\left| \frac{f(\gamma(t))}{\gamma(t) - z} - \frac{f(\gamma(t_i))}{\gamma(t_i) - z} \right| < \frac{2\pi\varepsilon}{M(b-a)}, \quad t_i \le t \le t_{i+1}, \ z \in K,$$

$M$ being an upper bound of $|\gamma'(t)|$, $a \le t \le b$. Consider now the rational function

$$R(z) = \frac{1}{2\pi i} \sum_{j=0}^{n-1} \frac{f(\gamma(t_j))}{\gamma(t_j) - z} \left( \gamma(t_{j+1}) - \gamma(t_j) \right).$$

Then one has

$$\left| \frac{1}{2\pi i} \int_\gamma \frac{f(w)}{(w-z)} dw - R(z) \right| = \left| \frac{1}{2\pi i} \sum_{j=0}^{n-1} \int_{t_j}^{t_{j+1}} \left( \frac{f(\gamma(t))}{\gamma(t) - z} - \frac{f(\gamma(t_j))}{\gamma(t_j) - z} \right) \gamma'(t) dt \right|$$

$$\le \frac{1}{2\pi} M \frac{2\pi\varepsilon}{M(b-a)} (b-a) = \varepsilon. \qquad \square$$

The following proposition explains the so-called method of *translation of poles*. It allows changing the location of poles of rational functions approximating a given function.

**Proposition 10.2.** *Let $K$ be a compact set of $\mathbb{C}$. Then one has:*

a) *If $V$ is a connected component of $\mathbb{C} \setminus K$ and $a, b \in V$, the function $1/(z-a)$ can be uniformly approximated on $K$ by polynomials in $1/(z-b)$ (that is, rational functions holomorphic at the point $\infty$ with a unique pole at the point $b$).*

b) *If $V_\infty$ is the unbounded component of $\mathbb{C} \setminus K$ and $a \in V_\infty$, the function $1/(z-a)$ can be uniformly approximated on $K$ by polynomials.*

c) *Conversely, if $a \notin K$ and $1/(z-a)$ can be uniformly approximated by polynomials on $K$, then $a \in V_\infty$.*

*Proof.* To prove a) consider, for fixed $b \in V$, the set

$$A = \{a \in V : 1/(z-a) \text{ is a uniform limit on } K \text{ of polynomials in } 1/(z-b)\}.$$

The set $A$ is closed in $V$ because if points $a_n \in V$ approach $a \in V$, then $d(a_n, K) \ge m > 0$, for some $m$. Hence

$$\left| \frac{1}{z-a} - \frac{1}{(z-a_n)} \right| = \frac{|a - a_n|}{|z-a||z-a_n|} \le m^{-2}|a - a_n|, \quad \text{for all } z \in K,$$

and so $1/(z - a_n)$ converges uniformly to $1/(z - a)$ on $K$. Transitivity yields $a \in A$. The set $A$ is also open in $V$; indeed, if $a \in A$ and $\overline{D}(a, r) \subset V$, let us show that every point $w \in D(a, r)$ is in $A$. Consider the expansion

$$\frac{1}{z - w} = \frac{1}{z - a - (w - a)} = \frac{1}{(z - a)\left(1 - \frac{w-a}{z-a}\right)} = \sum_{n=0}^{\infty} \frac{(w - a)^n}{(z - a)^{n+1}}, \quad z \in K,$$

uniformly convergent on $K$, since $\left|\frac{w-a}{z-a}\right| < \frac{r}{|z-a|} < 1$, $\overline{D}(a, r) \subset V$ and $z \notin V$. This means that $1/(z - w)$ is uniformly approximated on $K$ by polynomials in $1/(z - a)$; now, $1/(z - a)$ is uniformly approximated on $K$ by polynomials in $1/(z - b)$ and, therefore, also its powers. By transitivity, one gets $w \in A$. Hence, $A$ is open and closed in $V$, trivially $b \in A$ and, therefore, $A = V$.

To prove b) let $M = \max\{|z| : z \in K\}$ and $b \in \mathbb{C}$ with $|b| > M + 1$; clearly $b \in V_\infty$, and the expansion

$$\frac{1}{z - b} = \frac{1}{b\left(\frac{z}{b} - 1\right)} = -\frac{1}{b} \sum_{n=0}^{\infty} \left(\frac{z}{b}\right)^n$$

is uniformly convergent on $K$ because $\left|\frac{z}{b}\right| \leq \frac{M}{M+1} < 1$ if $z \in K$. This means that $1/(z - b)$ and, therefore, polynomials in $1/(z - b)$ can be uniformly approximated by polynomials on $K$. If $a \in V_\infty$, by part a), $1/(z - a)$ can be approximated by polynomials in $1/(z - b)$ and, by transitivity, $1/(z - a)$ is approximated by polynomials on $K$.

In order to prove c) let us show that if $a \notin K$ and $a$ belongs to a bounded component $V$ of $\mathbb{C} \setminus K$, then $1/(z - a)$ cannot be uniformly approximated by polynomials on $K$. It is clear that $\partial V \subset K$; hence, if $P_n(z) \to 1/(z - a)$ uniformly on $K$ with $P_n$ polynomials, we would have, in particular, that the sequence $(P_n)$ is uniformly convergent on $\partial V$. Then for $\varepsilon > 0$ we would get $\left|P_n(z) - \frac{1}{z-a}\right| < \varepsilon$ if $n$ is big enough and $z \in \partial V$. From here it turns out that

$$|(z - a)P_n(z) - 1| < \varepsilon|z - a| < 1/2, \quad z \in \partial V,$$

if $\varepsilon < \frac{1}{2\,\mathrm{diam}(\overline{V})}$. Now, by the maximum principle, we obtain $|(z - a)P_n(z) - 1| \leq 1/2$ for $z \in \overline{V}$ and this inequality is impossible if $z = a \in V$.     $\square$

**Theorem 10.3** (Runge's theorem for compact sets). a) *If $K$ is a compact set of $\mathbb{C}$ and $\mathbb{C} \setminus K$ is connected, then every holomorphic function on a neighborhood of $K$ can be uniformly approximated on $K$ by polynomials. Conversely, if this approximation is possible for every function holomorphic on a neighborhood of $K$, then $\mathbb{C} \setminus K$ is connected.*

b) *If $\mathbb{C} \setminus K$ is not connected and $A \subset \mathbb{C}$ is a set intersecting each bounded component of $\mathbb{C} \setminus K$, then every holomorphic function on a neighborhood of $K$*

*can be uniformly approximated on $K$ by rational functions that have their poles at points of $A$.*

*Proof.* Let $f$ be a holomorphic function on a neighborhood $U$ of $K$ and $\varepsilon > 0$. By Lemma 6.6 there is a chain $\gamma$ in $U \setminus K$ such that

$$f(z) = \frac{1}{2\pi i} \int_\gamma \frac{f(w)}{w - z} dw, \quad z \in K.$$

Then by Proposition 10.1 there is a rational function $R$ with poles on $\gamma^* \subset \mathbb{C} \setminus K$ such that

$$|f(z) - R(z)| < \varepsilon/2, \quad z \in K.$$

The function $R(z)$ is a linear combination of functions $1/(z - a_j)$, with $a_j \notin K$. If $\mathbb{C} \setminus K$ is connected, by Proposition 10.2 b), each of these functions can be uniformly approximated on $K$ by polynomials; hence, there is a polynomial $P$ satisfying $|R(z) - P(z)| < \varepsilon/2$ if $z \in K$, and so $|f(z) - P(z)| < \varepsilon, z \in K$. In other words, the poles that are in the unbounded component of $K$ may be translated to $\infty$, and then one obtains polynomials.

Suppose now that $\mathbb{C} \setminus K$ is not connected and $A$ intersects all the bounded components of $\mathbb{C} \setminus K$; write $R = R_1 + R_2$, where $R_1$ has poles in the unbounded component of $\mathbb{C} \setminus K$ and $R_2$ has all its poles in the bounded components. As shown, there is a polynomial $P$ such that $|R_1(z) - P(z)| < \varepsilon/4, z \in K$. To end with, it is enough to see that all the poles in bounded components may be translated to $A$, to obtain $R_3$ with poles in $A$ and $|R_2(z) - R_3(z)| < \varepsilon/4, z \in K$. The function $R_2(z)$ is a linear combination of functions $1/(z - a_j)$ with $a_j \notin K, a_j \notin V_\infty$, and therefore it suffices to show that $1/(z - a), a \notin K, a \notin V_\infty$ can be uniformly approximated on $K$ by rational functions with poles inside $A$. If $V$ is the bounded component of $\mathbb{C} \setminus K$ containing $a$, by hypothesis there is a point $b \in V \cap A$, and Proposition 10.2 a) finishes the proof.

The converse of item a) is a consequence of Proposition 10.2 c): if $\mathbb{C} \setminus K$ is not connected, it has a bounded component $V$, and if $a \in V$, the function $1/(z - a)$, which is holomorphic on a neighborhood of $K$, cannot be approximated by polynomials. $\qquad\square$

Rational functions with poles in the set $A$ of item b) of Theorem 10.3, approximating a given holomorphic function, are of the form

$$R = P + R_1,$$

where $P$ is a polynomial and $R_1(z)$ is a linear combination of polynomials in $1/(z - a), a \in A$. For example, if $K$ is the closed annulus $K = \{z : 0 < R_2 \le |z| \le R_1 < +\infty\}$, considering $A = \{0\}$ we find that every holomorphic function on a neighborhood of $K$ is uniformly approximated on $K$ by functions of type

$$P_1(z) + P_2(1/z)$$

with $P_1$ and $P_2$ polynomials. This fact is already known, and furthermore, it is explicitly known which the polynomials $P_1$ and $P_2$ are, since according to Theorem 5.10 they correspond to partial sums of the Laurent expansion of $f$ in an annulus $C(0, R_2 - \varepsilon, R_1 + \varepsilon)$, $\varepsilon > 0$.

If $K$ has two holes, for example,

$$K = \{z : 1 \leq |z|, \ 1 \leq |z - 2|, \ |z| \leq 5\},$$

with $A = \{0, 2\}$ we get that every holomorphic function on a neighborhood of $K$ is the uniform limit of functions of type $P_1(z) + P_2(1/z) + P_3(1/(z - 2))$ with $P_1, P_2, P_3$ polynomials.

**Theorem 10.4** (Runge's theorem for open sets). a) *An open set $U \subset \mathbb{C}$ has the property that every holomorphic function on $U$ is the uniform limit on compact sets of $U$ of a sequence of polynomials, if and only if $\mathbb{C} \setminus U$ has no bounded connected component.*

b) *If $\mathbb{C} \setminus U$ is not connected and $A \subset \mathbb{C}$ is such that $\bar{A}$ intersects all the bounded connected components of $\mathbb{C} \setminus U$, then every holomorphic function on $U$ is the uniform limit on compact sets of $U$ of a sequence of rational functions with their poles at points of $A$.*

*Proof.* Let $f$ be a holomorphic function on $U$. To prove that approximations in a) and b) are possible, we must show that, given a compact set $K \subset U$ and a number $\varepsilon > 0$, there exists a function $g$ of the desired kind, that is, a polynomial in case a) and a rational function with poles in $A$ in case b) such that $|f(z) - g(z)| < \varepsilon$, $z \in K$. To this end we can assume that $K$ is one of the compact sets $K_n$, of the exhaustive sequence of compact sets of $U$, given by Lemma 1.15. If $\mathbb{C} \setminus U$ has no bounded connected components, then neither has $\mathbb{C} \setminus K_n$, that is, it is connected and the result follows from Theorem 10.3 a). If $\mathbb{C} \setminus K_n$ has some bounded component $V$, $V$ is a bounded open set of $\mathbb{C}$ containing a bounded component $W$ of $\mathbb{C} \setminus U$. By hypothesis, $\bar{A} \cap W \neq \emptyset$; hence, $\bar{A} \cap V \neq \emptyset$ and being that $V$ is open, $A \cap V \neq \emptyset$. Thus $A$ intersects all bounded components of $\mathbb{C} \setminus K_n$, and the statement follows from Theorem 10.3 b).

It remains only to prove the converse in a); that is, assuming $\mathbb{C} \setminus U$ has some bounded component $C$, one has to find $f \in H(U)$ not approachable on compact sets by polynomials. As in Proposition 10.2 c), consider, for $a \in C$, the function $f(z) = 1/(z - a)$. If there is a bounded open set $V$ of $\mathbb{C}$, with $a \in V$ and $\partial V \subset U$, then one can repeat the proof of Proposition 10.2 c) and arrive at a contradiction in case that $f$ could be approximated by polynomials on $\partial V$. The existence of $V$ is a purely topological fact, which is a consequence of the following general result, known as Šura-Bura's theorem (see [3], p. 32):

If $F$ is a closed set of $\mathbb{C}$ and $C$ a bounded connected component of $F$, then $C$ is the intersection of all compact and relatively open subsets of $F$ containing $C$.

In our case, taking $F = \mathbb{C} \setminus U$ it turns out that if $\mathbb{C} \setminus U$ has bounded components, then in $\mathbb{C} \setminus U$ there are compact sets $A \neq \emptyset$ open in $\mathbb{C} \setminus U$. Consider then $B = (\mathbb{C} \setminus U) \setminus A$, which is closed in $\mathbb{C} \setminus U$ and, so, closed in $\mathbb{C}$. The set $\mathbb{C} \setminus B$ is an open set of $\mathbb{C}$ containing the compact set $A$; hence, there is a bounded open set $V$ with $A \subset V \subset \bar{V} \subset \mathbb{C} \setminus B$. Since $\partial V$ intersects neither $A$ nor $B$, one has $\partial V \subset \mathbb{C} \setminus (A \cup B) = U$. Summarizing, it has been shown that if $U$ has holes, there are bounded open sets $V$ of $\mathbb{C}$ with $\partial V \subset U$ and $V \cap (\mathbb{C} \setminus U) \neq \emptyset$; taking $a \in V \cap (\mathbb{C} \setminus U)$ and $f(z) = 1/(z - a)$ one gets that $f \in H(U)$ and $f$ is not uniformly approachable by polynomials on compact sets of $U$. $\qquad\square$

In the particular case that $U$ is a disc, the approximation by polynomials of a function $f \in H(U)$ given in Theorem 10.4 a) is already provided by the partial sums of the Taylor series of $f$.

In general, the easiest set $A$, when $\mathbb{C} \setminus U$ is not connected, consists of a point in each bounded component of $\mathbb{C} \setminus U$. Then every function $f \in H(U)$ is uniformly approximated on compact sets of $U$ by functions of the type

$$P(z) + {\sum_{a \in A}}' Q_a\left(\frac{1}{z - a}\right)$$

with $P$, $Q_a$ polynomials and the sum finite. When $U$ is a punctured disc $U = D'(0, R)$ and $A = \{0\}$, these functions are $P(z) + Q(1/z)$ with $P$, $Q$ polynomials; in this case, the Laurent expansion of $f$ in $D'(0, R)$ gives explicitly the result. In general, it is not easy to find explicitly rational functions approaching a given function. For instance, if $U = D(0, 2) \setminus [-1, 1]$, one can take $A = \{0\}$ and every function $f \in H(U)$ is the uniform limit on compact sets of $U$ of a sequence of functions of the kind $P_n(z) + Q_n(1/z)$, but it is not easy to specify the polynomials $P_n$, $Q_n$.

Remark that neither in Theorem 10.3 nor in Theorem 10.4 is it assumed that $K$ or $U$ is connected. This is the case in the following examples.

The characteristic function of any set $A$ will be denoted by $\mathbb{1}_A$.

**Example 10.5.** Using Runge's theorems, the existence of a sequence of polynomials $P_n$ such that

$$\lim_n P_n(z) = 0 \text{ if } z \notin \mathbb{R}, \quad \lim_n P_n(z) = 1 \text{ if } z \in \mathbb{R}$$

will be proved. Consider the compact sets

$$K_n = \{z = z + iy : |z| \leq n, \; y = 0 \text{ or } \tfrac{1}{n} \leq |y| \leq n\}$$

and the open sets $V_n = \{z = x + iy : |x| < n + 1, \ |y| < 1/3n\}$. The function $\mathbb{1}_{V_n}$ is holomorphic on a neighborhood of $K_n$ and $\mathbb{C} \setminus K_n$ is connected. Therefore, there are polynomials $P_n$ such that $|\mathbb{1}_{V_n}(z) - P_n(z)| \leq 1/n$ if $z \in K_n$, that is, $|P_n(z)| < 1/n$ if $|x| \leq n$, $1/n \leq |y| \leq n$ and $|1 - P_n(z)| \leq 1/n$ if $z \in [-n, n]$. Then, clearly $P_n(z) \to 1$ if $z \in \mathbb{R}$ and $P_n(z) \to 0$ if $z \notin \mathbb{R}$. $\qquad\square$

**Example 10.6.** We construct now a sequence of polynomials $P_n$ such that $P_n(0) = 0$ and $P_n(z) \to 1$ if $z \neq 0$. Writing $P_n(z) = z\, Q_n(z)$, it is enough to find polynomials $Q_n$ such that $Q_n(z) \to 1/z$ if $z \neq 0$, that is, $Q_n$ tends to $1/z$ pointwise in $\mathbb{C} \setminus \{0\}$. Notice that according to the argument used in Proposition 10.2 c) there cannot exist any sequence of polynomials $Q_n$ with $Q_n(z) \to 1/z$ uniformly on compact sets of $\mathbb{C} \setminus \{0\}$.

In order to get $Q_n$, consider the compact sets (Figure 10.1)

$$K_n = \left[\frac{1}{n}, n\right] \cup \left\{z : |z| \leq n, \ d(z, \mathbb{R}^+) \geq \frac{1}{n}\right\}.$$

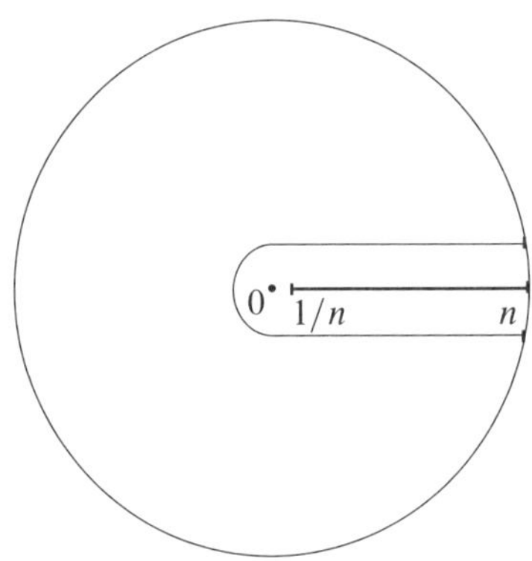


Figure 10.1

The function $1/z$ is holomorphic on a neighborhood of $K_n$ and $\mathbb{C} \setminus K_n$ is connected. Therefore, there is a polynomial $Q_n$ such that $|1/z - Q_n(z)| < 1/n$ if $z \in K_n$. Since $\bigcup_{n=1}^{\infty} K_n = \mathbb{C} \setminus \{0\}$ and $K_n \subset K_{n+1}$, it turns out that $Q_n(z) \to 1/z$ if $z \neq 0$. $\qquad\square$

A consequence of Runge's theorem is the following result:

**Theorem 10.7.** *Let $K \subset U$ with $K$ compact and $U$ open, then the following assertions are equivalent:*

a) *No connected component of $U \setminus K$ is relatively compact in $U$.*

b) *Every bounded component of $\mathbb{C} \setminus K$ intersects $\mathbb{C} \setminus U$.*

c) *Every holomorphic function on a neighborhood of $K$ can be approximated, uniformly on $K$, by functions holomorphic on $U$ (indeed, by rational functions with poles outside $U$).*

d) *For every point $z \in U \setminus K$ there is a holomorphic function $h \in H(U)$ such that $|h(z)| > \max_{w \in K} |h(w)|$.*

*Proof.* To see a) $\Rightarrow$ b), let $V$ be a bounded component of $\mathbb{C} \setminus K$. If $V \subset U$, $V$ would be a component of $U \setminus K$ with $\partial V \subset K \subset U$, and so relatively compact in $U$. Hence, $V$ intersects $\mathbb{C} \setminus U$. Theorem 10.3 gives b) $\Rightarrow$ c). Implication c) $\Rightarrow$ a) is like in Proposition 10.2 c). The fact that d) implies a) is based on the same idea: if $C$ is a component of $U \setminus K$ relatively compact in $U$, then $\partial C \subset K$, and, by the maximum modulus principle, one has

$$|h(z)| \le \max_{w \in K} |h(w)|, \quad z \in C,$$

contradicting a). Finally, suppose a) holds and let $z \in U \setminus K$. The compact set $K \cup \{z\}$ also satisfies a) and, therefore, c). Consider the function $f$ with value 0 on a neighborhood of $K$ and 1 on a neighborhood of $z$; by c), there is a function $h \in H(U)$ such that $|f(z) - h(z)| < 1/2$ in $K \cup \{z\}$. Then $|h(z)| > 1/2$ and $|h(w)| < 1/2$, $w \in K$. This proves d). $\qquad\square$

Notice that compact sets $K_n$ constructed in Lemma 1.15 satisfy the properties of Theorem 10.7 with respect to $U$.

## 10.2 Approximation of harmonic functions

In this section an approximation theorem for harmonic functions will be stated, analogous to the ones in the previous section. It is convenient to be aware of the results of Section 6.7. In particular, recall that if $U$ is an $n$-connected domain, every real harmonic function $u$ on $U$ can be written as

$$u(z) = \sum_{j=1}^{n-1} a_j \operatorname{Log} |z - \alpha_j| + \operatorname{Re} f$$

with $\alpha_j \notin U$, $f \in H(U)$ and $a_j \in \mathbb{R}$. For any domain $U$ it will be shown that every harmonic function can be approximated, uniformly on compact sets of $U$, by functions of this kind.

**Theorem 10.8.** a) *Every real-valued function $u$ harmonic on a neighborhood of a compact set $K$ can be written as $u(z) = \sum_{j=1}^{m} a_j \operatorname{Log} |z - \alpha_j| + \operatorname{Re} f$ for some $m \in \mathbb{N}$, with $a_j \in \mathbb{R}$, $\alpha_j \notin K$ and $f$ holomorphic on a neighborhood of $K$.*

b) *Let $U$ be an open neighborhood of the compact set $K$ such that every component of $\mathbb{C} \setminus K$ intersects $\mathbb{C} \setminus U$. Then every real-valued function $u$ harmonic on a neighborhood of $K$ can be approximated, uniformly on $K$, by harmonic functions on $U$ of the kind*

$$\sideset{}{'}\sum a_j \operatorname{Log} |z - \alpha_j| + \operatorname{Re} R$$

*with $\alpha_j \notin U$, $a_j \in \mathbb{R}$ and $R$ a rational function with simple poles outside $U$.*

c) *Every real-valued function $u$ harmonic on a domain $U$ is the uniform limit on compact sets of $U$ of a sequence of harmonic functions $(u_n)$, written as*

$$u_n(z) = \sideset{}{'}\sum a_j^{(n)} \operatorname{Log} |z - \alpha_j^{(n)}| + \operatorname{Re} R_n,$$

*with $\alpha_j^{(n)} \notin U$, $a_j^{(n)} \in \mathbb{R}$ and $R_n$ rational without poles in $U$.*

The notation $\sum'$ means that the sum only has a finite number of terms.

*Proof.* For part a) it is enough to note that if $U$ is an open set, $U \supset K$, by Remark 6.1 there is a polygonal domain $U_0$ with piecewise regular boundary, therefore, $n$-connected for some $n$, (see Section 6.6) such that $K \subset U_0 \subset \bar{U}_0 \subset U$. Afterwards, consider Theorem 6.26.

Let $K$ and $U$ satisfy the hypotheses of b) and let $u$ be harmonic on a neighborhood of $K$ written as in item a) with $f$ holomorphic on a neighborhood of $K$. Then by Runge's theorem, $f$ can be uniformly approximated on $K$ by rational functions $R_n$ with simple poles outside $U$ and, therefore, $\operatorname{Re} f$ is approximated by $\operatorname{Re} R_n$. Consequently, it is enough to show that $\operatorname{Log} |z - \beta|$, with $\beta \notin K$ fixed, satisfies the statement of b). Suppose first $\beta \in V$, being $V$ a bounded component of $\mathbb{C} \setminus K$, and let $\alpha \notin U$ be a point of this same component $V$. Define

$$W = \{\tau \in V : \operatorname{Log} |z - \tau| \text{ is uniformly approximated on } K \text{ by}$$
$$\text{functions of the kind } \operatorname{Log} |z - \alpha| + \operatorname{Re} f, \ f \text{ holomorphic}\},$$

so that $\alpha \in W$.

Arguing as in the proof of Proposition 10.2 we get that $W$ is closed in $V$. Now it will be shown that $W$ is open in $V$. If $\tau \in W$ and $D(\tau, r) \subset V$ with $r < \frac{d(\tau, K)}{2}$, let us prove that $D(\tau, r) \subset W$. If $w \in W$, any holomorphic branch of $\log \left( \frac{z - \tau}{z - w} \right)$ on a neighborhood of $K$ (which exists by Proposition 3.21) has real part $\operatorname{Log} |z - \tau| - \operatorname{Log} |z - w|$. If $|w - \tau| < r$, the expansion

$$\log \left( \frac{z - \tau}{z - w} \right) = \log \left( 1 + \frac{w - \tau}{z - w} \right) = \sum_{n=0}^{\infty} (-1)^n \frac{(w - \tau)^{n+1}}{n + 1} \frac{1}{(z - w)^{n+1}}$$

is uniformly convergent on $K$ because $|w - \tau| < r < |z - w|$ if $z \in K$. Taking real parts yields that $\operatorname{Log} |z - \tau| - \operatorname{Log} |z - w|$ can be uniformly approximated by $\operatorname{Re} g$ with $g$ holomorphic. In conclusion, $W = V$ and $\beta \in W$.

If $\beta$ belongs to $V_\infty$, the unbounded component of $\mathbb{C} \setminus K$, then $\mathrm{Log}\,|z - \beta|$ is uniformly approachable on $K$ by functions of the kind $\mathrm{Re}\,f$, with $f$ holomorphic. Indeed, repeating the previous argument one can replace $\beta$ by any point $\alpha \in V_\infty$ with $|\alpha|$ big enough. Now, in this case, any branch of $\log(\alpha - z)$ (which exists by Proposition 3.21) has real part $\mathrm{Log}\,|z - \alpha|$, and expanding in power series one has

$$\log(\alpha - z) = \log\alpha + \log\left(1 - \frac{z}{\alpha}\right) = \log\alpha - \sum_{n=0}^{\infty} \frac{z^{n+1}}{(n+1)\alpha^{n+1}},$$

with uniform convergence on $K$, if $|\alpha| > \max\{|z| : z \in K\}$.

Item c) follows from b) and Lemma 1.15. $\qquad\square$

## 10.3 Decomposition of meromorphic functions into simple elements

### 10.3.1 Mittag-Leffler's theorems

Let $f$ be a meromorphic function on $\mathbb{C}$ with a finite number of poles $z_1, \ldots, z_n$ and principal parts $P_1\left(\frac{1}{z-z_1}\right), \ldots, P_n\left(\frac{1}{z-z_n}\right)$, where $P_1, \ldots, P_n$ are polynomials without independent term. Then

$$g(z) = f(z) - \sum_{i=1}^{n} P_i\left(\frac{1}{z - z_i}\right)$$

is an entire function. If, further, $f$ has a pole at infinity, that is, $\lim_{|z|\to\infty} f(z) = \infty$, then $g$ is a polynomial, $P$, and $f$ is a rational function, being

$$f(z) = P(z) + \sum_{i=1}^{n} P_i\left(\frac{1}{z - z_i}\right)$$

the decomposition of $f$ into simple fractions.

This section is devoted to study the general case when the function $f$, meromorphic on $\mathbb{C}$, has infinitely many poles $(z_n)_{n\in\mathbb{N}}$. For convenience it will be assumed that $f$ has a pole at the origin $z_0 = 0$ and that the points $z_n$ are all different, with $\lim_n |z_n| = +\infty$. The principal parts are written as $P_0\left(\frac{1}{z}\right)$, $P_n\left(\frac{1}{z-z_n}\right)$, where $P_0$, $P_n$ are polynomials without a constant term.

First we will show that there is no restriction for the points $z_n$ nor for the principal parts of $f$ at these points.

**Theorem 10.9** (Mittag-Leffler). *Let $z_0 = 0$, $(z_n)_{n\in\mathbb{N}}$ be a sequence of points in the plane with $|z_n| \to +\infty$ when $n \to \infty$ and $(P_n)_{n=0}^{\infty}$ any sequence of polynomials without constant term. Then there is a meromorphic function on $\mathbb{C}$ having its poles exactly at the points $z_n$ with principal parts $P_n\left(\frac{1}{z-z_n}\right)$, for $n \in \mathbb{N} \cup \{0\}$.*

The most obvious choice would be, simply, to take the function

$$f(z) = \sum_n P_n\left(\frac{1}{z - z_n}\right)$$

whenever this series is convergent in the sense that will be specified now. For each compact set $K \subset \mathbb{C}$, only a finite number of points $z_n$ are in $K$, so that

$$\sum_{z_n \notin K} P_n\left(\frac{1}{z - z_n}\right)$$

is a series of holomorphic functions on a neighborhood of $K$. If this series converges uniformly on $K$, then it is said that the complete series, $\sum_n P_n\left(\frac{1}{z-z_n}\right)$, *converges uniformly on compact sets.* Of course, one may consider this kind of convergence for any series of meromorphic functions $\sum_n g_n(z)$, with each $g_n$ having a unique pole at the point $z_n$.

Under the above hypothesis, the function $f$ defined before has the desired properties in Theorem 10.9, but in general the series $\sum_n P_n\left(\frac{1}{z-z_n}\right)$ will not converge uniformly on compact sets and one needs to introduce some correction terms.

*Proof of Theorem* 10.9. The idea is to introduce polynomials $Q_n$ such that the series

$$\sum_n g_n(z) \overset{\text{def}}{=} P_0\left(\frac{1}{z}\right) + \sum_{n=1}^{\infty}\left(P_n\left(\frac{1}{z - z_n}\right) - Q_n(z)\right) \tag{10.1}$$

converges uniformly on compact sets. Since $Q_n$ are polynomials, the principal part of each term of the series is $P_n\left(\frac{1}{z-z_n}\right)$ and the function $f(z) = \sum_n g_n(z)$ will have the right behavior.

The choice of polynomials $Q_n$ is not unique and it is easy to do. The function $P_n\left(\frac{1}{z-z_n}\right)$ is holomorphic on the disc $D(0, |z_n|)$. Write

$$P_n\left(\frac{1}{z - z_n}\right) = \sum_{m=0}^{\infty} c_{n,m} z^m$$

for its Taylor expansion. On the compact set $\bar{D}(0, |z_n|/2)$ this expansion converges uniformly and, therefore, we can consider a partial sum $Q_n(z) = \sum_{m=0}^{N_n} c_{n,m} z^m$ with $N_n$ big enough such that

$$\left|P_n\left(\frac{1}{z - z_n}\right) - Q_n(z)\right| \le 2^{-n} \quad \text{if } |z| \le |z_n|/2.$$

These polynomials $Q_n$ will work because, fixing a disc $\bar{D}(0, R)$, one will have $R < |z_n|/2$ for $n$ big enough and (10.1) will converge uniformly on $\bar{D}(0, R)$.     $\square$

In Theorem 10.9 a meromorphic function with prescribed principal parts at given points has been *constructed*. The result may be reinterpreted as a theorem about the structure of all meromorphic functions.

**Corollary 10.10.** *The general expression of a function $F$ meromorphic on $\mathbb{C}$ with principal parts $P_n\left(\frac{1}{z-z_n}\right)$, $n = 0, 1, 2, \ldots$ at the points $z_0 = 0$, $(z_n)_{n\in\mathbb{N}}$ is*

$$F(z) = h(z) + P_0\left(\frac{1}{z}\right) + \sum_{n=1}^{\infty}\left(P_n\left(\frac{1}{z - z_n}\right) - Q_n(z)\right),$$

*where $h$ is an entire function and the functions $Q_n$ are polynomials such that the series converges uniformly on compact sets.*

*Proof.* Just take the meromorphic function $f$ constructed in the proof of Theorem 10.9 and observe that another meromorphic function $F$ has the same poles and the same principal parts as $f$ if and only if $F - f$ is entire. $\qquad\square$

In the case all the poles are simple one has, $P_n\left(\frac{1}{z-z_n}\right) = \frac{\alpha_n}{z-z_n}$, $\alpha_n \in \mathbb{C}$, and the expansion of this fraction in $D(0, |z_n|)$ is

$$\frac{\alpha_n}{z - z_n} = -\frac{\alpha_n}{z_n - z} = -\frac{\alpha_n}{z_n\left(1 - \frac{z}{z_n}\right)} = -\alpha_n \sum_{m=0}^{\infty} \frac{z^m}{z_n^{m+1}}.$$

If $Q_n$ is the $\lambda_n$-th partial sum, $Q_n(z) = -\alpha_n \sum_{m=0}^{\lambda_n} \frac{z^m}{z_n^{m+1}}$, then the difference $\frac{\alpha_n}{z-z_n} - Q_n(z)$ is

$$-\alpha_n \sum_{m=\lambda_n+1}^{\infty} \frac{z^m}{z_n^{m+1}} = -\frac{\alpha_n}{z_n} \frac{\left(\frac{z}{z_n}\right)^{\lambda_n+1}}{1 - \frac{z}{z_n}} = \frac{\alpha_n}{z - z_n}\left(\frac{z}{z_n}\right)^{\lambda_n+1}.$$

**Corollary 10.11.** *The general expression of a function $F$ meromorphic on $\mathbb{C}$ with simple poles at the points $z_0 = 0$, $(z_n)_{n\in\mathbb{N}}$ and residue $\alpha_n$ at the pole $z_n$ is*

$$F(z) = h(z) + \frac{\alpha_0}{z} + \sum_{n=1}^{\infty} \frac{\alpha_n}{z - z_n}\left(\frac{z}{z_n}\right)^{\lambda_n+1},$$

*where the numbers $\lambda_n$ are positive integers such that the series converges uniformly on compact sets and $h$ is an entire function.*

If the residues $\alpha_n$ are uniformly bounded, the conditions of the previous corollary hold if the numbers $\lambda_n$ make the series

$$\sum_n \frac{|z|^{\lambda_n+1}}{|z_n|^{\lambda_n+2}}$$

converge uniformly on compact sets. For instance, if $\sum_n \frac{1}{|z_n|^{\lambda+2}} < +\infty$ for some positive integer $\lambda$, then one can take $\lambda_n = \lambda$, for every $n \in \mathbb{N}$.

**Example 10.12.** Consider the meromorphic function $f(z) = \frac{\pi^2}{\sin^2 \pi z}$ that has a double pole at each integer point $z_n = n$, $n \in \mathbb{Z}$. To find the principal part around each pole consider first the expansion of $\frac{\pi}{\sin \pi z}$ around the point $n$,

$$\frac{\pi}{\sin \pi z} = \frac{(-1)^n}{z - n} + c_1(z - n) + \cdots .$$

Now, squaring yields

$$\frac{\pi^2}{(\sin \pi z)^2} = \frac{1}{(z - n)^2} + \cdots .$$

In this case the series of principal parts already converges and one has

$$\frac{\pi^2}{\sin^2 \pi z} = \sum_{n=-\infty}^{+\infty} \frac{1}{(z - n)^2} + h(z)$$

with $h$ an entire function. To determine the function $h$, consider the square $Q_N$ centered at the origin and with side $2N + 1$; one has $|\sin \pi z|^2 = \sin^2 \pi x + \text{sh}^2 \pi y$, a quantity bounded below by 1 for $|x| = N + 1/2$ as well as for $|y| = N + 1/2$, that is, bounded below if $z \in \partial Q_N$, independently of $N$. Also if $z \in \partial Q_N$, one has

$$\left| \sum_n \frac{1}{(z - n)^2} \right| \le \sum_n \frac{1}{|z - n|^2} \le \sum_n \frac{1}{(n - 1/2)^2} < +\infty.$$

So, by the maximum modulus principle, the function $h$ is bounded on each square $Q_N$ by a bound which does not depend on $N$ and, by Liouville's theorem, it must be equal to a constant $C$. Letting $z = iy$ and $y \to +\infty$, we get $C = 0$. Hence, it turns out that

$$\frac{\pi^2}{\sin^2 \pi z} = \sum_{n=-\infty}^{+\infty} \frac{1}{(z - n)^2} = \frac{1}{z^2} + \sum_{n \ne 0} \frac{1}{(z - n)^2} .$$

In particular,

$$2 \sum_{n=1}^{\infty} \frac{1}{n^2} = \lim_{z \to 0} \left( \frac{\pi^2}{\sin^2 \pi z} - \frac{1}{z^2} \right) = \frac{\pi^2}{3} .$$

Similarly, the decomposition

$$\pi^3 \frac{\cot \pi z}{\sin^2 \pi z} = \sum_{n=-\infty}^{+\infty} \frac{1}{(z - n)^3} .$$

can be obtained. $\qquad\qquad\qquad\qquad\qquad\qquad\qquad\qquad\qquad\qquad\qquad\qquad\quad \square$

**Example 10.13.** Consider now the function $f(z) = \pi \cot \pi z = \pi \frac{\cos \pi z}{\sin \pi z}$. It has simple poles at the integers, $z_n = n \in \mathbb{Z}$, with residue 1. Since

$$\sum_{z_n \neq 0} \frac{1}{|z_n|^2} = \sum_{n \neq 0} \frac{1}{n^2} < +\infty,$$

one can consider $\lambda_n = 0$ for every $n$, that is, $Q_n(z) = -\frac{1}{n}$ and it turns out that

$$\pi \frac{\cos \pi z}{\sin \pi z} = h(z) + \frac{1}{z} + \sum_{\substack{n \neq 0 \\ n \in \mathbb{Z}}} \left( \frac{1}{z - n} + \frac{1}{n} \right) = h(z) + \frac{1}{z} + \sum_{n=1}^{\infty} \frac{2z}{z^2 - n^2}.$$

Now the function $\pi \cot \pi z$ is also bounded on the boundary of squares $Q_N$ with center 0 and side $2N + 1$ considered in Example 10.12. Actually, one has

$$\pi^2 |\cot \pi z|^2 = \pi^2 \frac{\cos^2 \pi x + \operatorname{sh}^2 \pi y}{\sin^2 \pi x + \operatorname{sh}^2 \pi y}$$

so that for $|x| = N + 1/2$,

$$\pi^2 |\cot \pi z|^2 = \pi^2 \frac{\operatorname{sh}^2 \pi y}{1 + \operatorname{sh}^2 \pi y} \leq \pi^2$$

and for $|y| = N + 1/2$,

$$\pi^2 |\cot \pi z|^2 \leq \pi^2 \frac{1 + \operatorname{sh}^2 \pi (N + 1/2)}{\operatorname{sh}^2 \pi (N + 1/2)} \leq \pi^2 \left( 1 + \frac{1}{\operatorname{sh}^2 3\pi/2} \right), \quad \text{for every } N.$$

A similar argument to the one of Example 10.12 gives $h \equiv 0$ and, finally, we obtain

$$\pi \frac{\cos \pi z}{\sin \pi z} = \frac{1}{z} + \sum_{n-1}^{\infty} \frac{2z}{z^2 - n^2} = \frac{1}{z} + \sum_{n \neq 0} \left( \frac{1}{z - n} + \frac{1}{n} \right).$$

If instead the function $\frac{\pi}{\sin \pi z}$ is considered, the poles are the same points $z_n = n$ and the residues $(-1)^n$. One has then

$$\frac{\pi}{\sin \pi z} = \frac{1}{z} + 2z \sum_{n=1}^{\infty} \frac{(-1)^n}{z^2 - n^2}. \qquad \square$$

Theorem 10.9 holds on any domain $U$. The proof is similar to the case $U = \mathbb{C}$; just choose the correction terms $Q_n$ – in this case, rational functions with poles outside $U$ – using Runge's theorem.

**Theorem 10.14.** *Let $U$ be a domain of the plane and $A \subset U$ a discrete closed set in $U$. For each point $a \in A$, suppose a polynomial $P_a$ without constant term is given. Then there is a meromorphic function on $U$ having its poles at the points $a \in A$ with principal part $P_a\left(\frac{1}{z-a}\right)$.*

*Proof.* Let $(K_n)_{n=1}^{\infty}$ be a sequence of compact sets as in Lemma 1.15. Let

$$Q_n(z) = \sum_{a \in A \cap (K_n \setminus K_{n-1})} P_a\left(\frac{1}{z-a}\right),$$

where the sum is finite because $A$ has a finite number of points in every compact set. Each function $Q_n$ is holomorphic on a neighborhood of $K_{n-1}$; as in the proof of Theorem 10.4, there is a rational function $R_n$ with poles in $\mathbb{C} \setminus U$ such that $|Q_n(z) - R_n(z)| < 2^{-n}$, $z \in K_{n-1}$. Then the function

$$Q_1(z) + \sum_{n=2}^{\infty} (Q_n(z) - R_n(z))$$

works. $\qquad\square$

### 10.3.2  Cauchy's method

One of the difficulties that appear when applying Corollary 10.10 is how to find the entire function $h$. For this reason it is convenient to consider besides Mittag-Leffler's theorem another method, due to Cauchy, which allows us to develop a meromorphic function on $\mathbb{C}$ into simple elements. It is explained below (see [11]).

Start with a meromorphic function on $\mathbb{C}$, $f$, with poles at points $(z_n)_{n \in \mathbb{N}}$ and principal parts $P_n\left(\frac{1}{z-z_n}\right)$. One can assume $f$ is holomorphic around the origin because if $z_0 = 0$ was a pole with principal part $P_0\left(\frac{1}{z}\right)$, we would consider the function $f(z) - P_0(1/z)$.

Let $C = \partial D(0, R)$ be a circle centered at the origin not passing through any pole of $f$ ($C$ could also be the boundary of a square centered at the origin) and consider the integral

$$\frac{1}{2\pi i} \int_C \frac{f(w)}{w - z}\, dw \quad \text{with } |z| < R,\ z \neq z_n,\ \text{for all } n \in \mathbb{N}.$$

By the residue theorem this integral equals $f(z)$ plus the sum of the residues of the function $\frac{f(w)}{w-z}$ at each point $z_n$ with $|z_n| < R$. Let us show that the residue of this function at $z_n$ is $-P_n\left(\frac{1}{z-z_n}\right)$. Write $P_n\left(\frac{1}{w-z_n}\right) = \sum_{l=1}^{r} \frac{c_l}{(w-z_n)^l}$ and $f(w) =$

$P_n\left(\frac{1}{w-z_n}\right) + f_n(w - z_n)$ with $f_n$ holomorphic around $z_n$. Then one has

$$\frac{f(w)}{w - z} = -\frac{1}{z - z_n}\frac{f(w)}{1 - \frac{w-z_n}{z-z_n}} = -f(w)\sum_{k=1}^{\infty}\frac{(w - z_n)^{k-1}}{(z - z_n)^k}$$

$$= -\left[f_n(w - z_n) + \sum_{l=1}^{r}\frac{c_l}{(w - z_n)^l}\right]\sum_{k=1}^{\infty}\frac{(w - z_n)^{k-1}}{(z - z_n)^k},$$

and the coefficient of $\frac{1}{w-z_n}$ is $-\sum_{l=1}^{r}\frac{c_l}{(z-z_n)^l} = -P_n\left(\frac{1}{z-z_n}\right)$. Using now

$$\frac{1}{w - z} = \frac{1}{w(1 - z/w)} = \frac{1}{w} + \frac{z}{w^2} + \cdots + \frac{z^{k-1}}{w^k} + \frac{z^k}{w^k(w - z)},$$

which holds for any natural $k$, it follows that

$$\frac{1}{2\pi i}\int_C \frac{f(w)}{w - z}\,dw = \frac{1}{2\pi i}\int_C f(w)\left(\frac{1}{w} + \frac{z}{w^2} + \cdots + \frac{z^{k-1}}{w^k}\right)dw$$

$$+ \frac{1}{2\pi i}\int_C \frac{z^k f(w)}{w^k(w - z)}\,dw.$$

The value of the first integral on the right-hand side is the sum of the residues of its integrand at the points of the set $\{0\} \cup \{z_n : |z_n| < R\}$. These residues are

$$\frac{1}{2\pi i}\int_{C_n} f(w)\left(\frac{1}{w} + \frac{z}{w^2} + \cdots + \frac{z^{k-1}}{w^k}\right)dw$$

if $C_n$ is a small circle around $z_n$ if $n \geq 1$ and $C_0$ surrounds the origin (Figure 10.2). Denoting these residues by $Q_0(z)$ at the origin and $-Q_n(z)$ at the point $z_n$, $Q_0(z)$ is the sum of the $k$ first terms of the Taylor series of $f$ around the origin (recall that $f$ is holomorphic at the origin) and $Q_n(z)$ is a polynomial in $z$ of degree smaller than $k$.

Summarizing, we get

$$f(z) = \sum_{|z_n|<R} P_n\left(\frac{1}{z - z_n}\right) + \frac{1}{2\pi i}\int_C \frac{f(w)}{w - z}\,dw$$

$$= \sum_{l=0}^{k-1}\frac{f^{(l)}(0)}{l!}z^l + \sum_{|z_n|<R}\left[P_n\left(\frac{1}{z - z_n}\right) - Q_n(z)\right] \qquad (10.2)$$

$$+ \frac{1}{2\pi i}\int_C \frac{z^k f(w)}{w^k(w - z)}\,dw.$$

We now claim that $Q_n(z)$ is the sum of the first $k$ terms of the Taylor series of $P_n\left(\frac{1}{z-z_n}\right)$ around the origin. Actually, $-Q_n(z)$ is the residue of the function

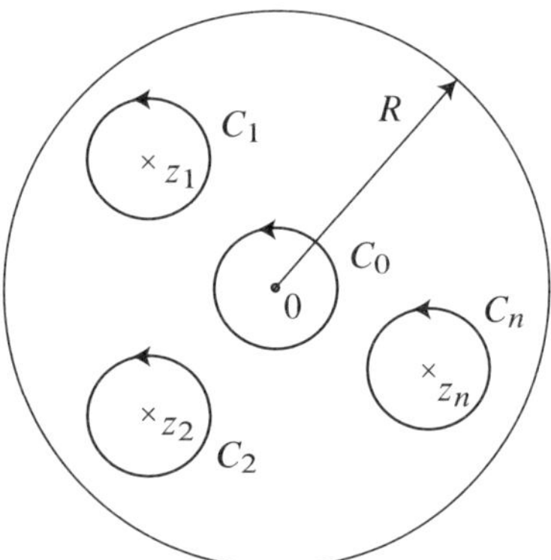


Figure 10.2

$P_n\!\left(\frac{1}{w-z_n}\right)\!\left(\frac{1}{w} + \frac{z}{w^2} + \cdots + \frac{z^{k-1}}{w^k}\right)$ at the point $z_n$. This function is rational and has one pole at the origin and one pole at the point $z_n$; moreover it is holomorphic at infinity. Then, by Proposition 5.24, the sum of its residues must be zero and the claim is proved because the residue at the origin is just the Taylor polynomial of degree $n - 1$ of $P_n\!\left(\frac{1}{z-z_n}\right)$.

Finally, assume $f$ satisfies the following growth condition:

> There exist circles $C(0, R_n)$, $R_n \to +\infty$, and numbers $\varepsilon_n > 0$ with $\varepsilon_n \to 0$ such that
>
> $$|f(z)| \le \varepsilon_n |z|^k, \quad \text{if } z \in C(0, R_n) \ (k \text{ a fixed natural number}).$$

In this case one can write formula (10.2) for each $C(0, R_n)$ and let $R_n \to +\infty$. The last integral in (10.2) tends to zero because it is bounded in absolute value by

$$\frac{|z|^k}{2\pi}\,\frac{\varepsilon_n R_n^k}{R_n^k(R_n - |z|)}\,2\pi R_n = |z|^k \varepsilon_n \frac{R_n}{R_n - |z|} \to 0, \quad \text{when } n \to \infty.$$

Hence we get the following decomposition of $f$:

$$f(z) = \sum_{l=0}^{k-1} \frac{f^{(l)}(0)}{l!} z^l + \sum_{n=1}^{\infty}\left[ P_n\!\left(\frac{1}{z - z_n}\right) - Q_n(z)\right],$$

where, recall, $Q_n(z)$ is the Taylor polynomial of degree $k - 1$ of $P_n\!\left(\frac{1}{z-z_n}\right)$ around the origin.

If all the poles $z_n$ are simple, then $P_n\!\left(\frac{1}{z-z_n}\right) = \frac{\alpha_n}{z-z_n}$ with $\alpha_n = \operatorname{Res}(f, z_n)$ and so $Q_n(z) = -\frac{\alpha_n}{z_n}\sum_{l=0}^{k-1}\left(\frac{z}{z_n}\right)^l$.

Assume, in addition, that $f$ is uniformly bounded on a sequence of circles $C(0, R_n)$, $R_n \to \infty$, say $|f(z)| \le M$, $z \in C(0, R_n)$, $n = 1, 2, \ldots$; then

$$\frac{f(z)}{|z|} \le \frac{M}{R_k} \to 0 \quad \text{if } z \in C(0, R_k)$$

and we may take $k = 1$, that is, $Q_n(z) = -\frac{\alpha_n}{z_n}$.

**Example 10.15.** The expansion of the function $\pi \cot \pi z$ of Example 10.13 is immediately obtained with Cauchy's method. Actually, since the function $f(z) = \pi \cot \pi z - \frac{1}{z}$ is bounded on the boundary of the squares $Q_N$ of Example 10.13, we can take $k = 1$. Then we compute

$$f(0) = 0, \quad P_n\left(\frac{1}{z - n}\right) = \frac{1}{z - n} \quad \text{if } n \in \mathbb{Z}, \ n \neq 0,$$

and since the poles are simple with residue 1, we get $Q_n(z) = -\frac{1}{n}$. So, $\pi \cot \pi z = \frac{1}{z} + \sum_{n \neq 0} \left(\frac{1}{z-n} + \frac{1}{n}\right)$. $\qquad\qquad\square$

**Example 10.16.** Let $w, w'$ be two non-zero complex numbers such that $w/w' \notin \mathbb{R}$ and consider the set of points of the plane

$$\Omega = \{mw + nw' : m, n \in \mathbb{Z}\}.$$

Considering at each point of $\Omega$ the straight lines directed by $w$ and $w'$, a division of the plane by a grid of parallelograms is obtained.

Let us now look for a meromorphic function on $\mathbb{C}$ having a simple pole with residue 1 at each point of $\Omega$. To this end enumerate the points of $\Omega$ by $z_0 = 0$, $z_1, z_2, \ldots, z_n, \ldots$. It is easy to check that $\sum_{n=1}^{\infty} \frac{1}{|z_n|^3} < +\infty$. Indeed, for $k \ge 1$ let $G_k$ be the set of points of $\Omega$ obtained with $m = \pm k$ and $-k \le n \le k$, or $n = \pm k$ and $-k \le m \le k$. There are $8k$ of these points. If $S_k = \sum_{z_n \in G_k} \frac{1}{|z_n|^3}$ and $\delta$ is the distance from the origin to the grid of parallelograms (see Figure 10.3), one has $|z_n| \ge k\delta$ if $z_n \in G_k$ and

$$S_n \le 8k \, \frac{1}{(\delta k)^3} = \frac{8}{\delta^3} \frac{1}{k^2}$$

so that $\sum_n \frac{1}{|z_n|^3} = \sum_k S_k < +\infty$.

In this case, $P_n\left(\frac{1}{z - z_n}\right) = \frac{1}{z - z_n}$ and we can take as $Q_n(z)$ the two first terms in the Taylor expansion of $\frac{1}{z - z_n}$ around the origin. That is,

$$Q_n(z) = -\frac{1}{z_n} - \frac{z}{z_n^2}.$$

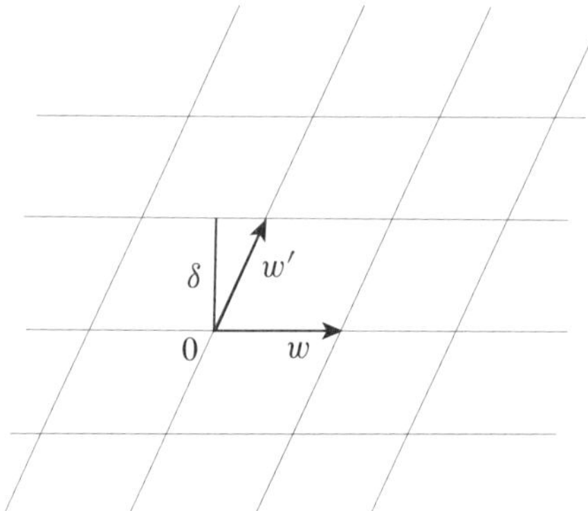


Figure 10.3

So, the function we are looking for is

$$\xi(z) = \frac{1}{z} + \sum_{n=1}^{\infty}\left(\frac{1}{z - z_n} + \frac{1}{z_n} + \frac{1}{z_n^2}\right),$$

where $z_n$, $n \geq 1$ are, as said above, the non-zero points of the set $\Omega$. The function $\xi(z)$ is called *Weierstrass' $\xi$-function*.    $\square$

### 10.3.3  The method of residues

Another way to expand a meromorphic function into simple elements and, also, of summing series is based on the residue theorem.

To illustrate this method fix a meromorphic function on the plane, $\varphi$, having only simple poles at a sequence of points of the kind $z_n = z_0 + n$, $z_0 \in \mathbb{C}$, $n \in \mathbb{Z}$, with residues $\alpha_n$, $n = 0, \pm 1, \pm 2, \dots$. Let $f$ be a meromorphic function on an open set $U$ and let $\gamma$ be a closed path of $U$ such that $\mathrm{Ind}(\gamma, z) = 0$ or 1, for every $z \in \mathbb{C}$, not passing through any singularity of $\varphi$ nor $f$. Then

$$\frac{1}{2\pi i}\int_{\gamma} f(z)\varphi(z)dz = {\sum_{n}}' \alpha_n f(z_0 + n) + {\sum_{w}}' \mathrm{Res}(f\varphi, w),$$

with the first sum taken at the poles of $\varphi$ at which $f$ is regular and the second one taken at the singular points of $f$, considering, for $\varphi$ as well as for $f$, just the singularities that are in the interior of $\gamma$. Suppose now $f$ is a rational function such that $|zf(z)| \to 0$ if $|z| \to \infty$ and take as $\gamma$ the boundary, $\gamma_N$, of a square of sides $x = \pm(a + N)$, $y = \pm(a + N)$ with $a > 0$ fixed and $N$ natural. If, in addition, the function $\varphi$ is uniformly bounded on $\gamma_N$, for all $N$, then $\lim_{N \to \infty} \frac{1}{2\pi i}\int_{\gamma_N} f(z)\varphi(z)dz = 0$. Actually, $|z| \geq N$ if $z \in \gamma_N$ and the length of $\gamma_N$ is $8N + 8a$ so that for $|\varphi(z)| \le M$ on $\gamma_N$, one gets

$$\left| \int_{\gamma_N} f(z)\varphi(z)dz \right| = \left| \int_{\gamma_N} zf(z)\varphi(z)\frac{dz}{z} \right| \le \frac{M(8N + 8a)}{N} \cdot \sup_{|z|=N} |zf(z)| \xrightarrow[N \to \infty]{} 0.$$

Summarizing, we obtain the formula

$$\sum_{n=-\infty}^{\infty} \alpha_n f(z_0 + n) = -\sum_{w} \mathrm{Res}(f\varphi, w), \tag{10.3}$$

with the sum on the left-hand side taken at the points $z_0 + n$ where $f$ is regular, and the sum on the right, taken at the poles of $f$.

**Example 10.17.** Considering $\varphi(z) = \pi \cot \pi z$, the poles are the points $z_n = n \in \mathbb{Z}$ and the residues are $\alpha_n = 1$; one can use $\gamma_N = \partial Q_N$, where $Q_N$ are the squares in Example 10.12. Then if $f$ is rational with $|zf(z)| \to 0$, $|z| \to \infty$, one has

$$\sum_{-\infty}^{\infty} f(n) = -\sum_{w} \mathrm{Res}(\pi f(z) \cot \pi z, w)$$

with the sum on the left-hand side extended to the integers $n$ that are not poles of $f$ and the sum on the right-hand side extended to the poles of $f$.

If $f(z) = \frac{1}{z^2}$, it turns out, in particular, that

$$2 \sum_{1}^{\infty} \frac{1}{n^2} = - \mathrm{Res}\left( \pi \frac{1}{z^2} \cot \pi z, 0 \right).$$

The origin is a pole of order 3 of the function $\pi z^{-2} \cot \pi z$ and we must compute the coefficient of $z$ in the expansion of $\cot \pi z$. Writing $\cot z = \frac{a_{-1}}{z} + a_0 + a_1 z + \cdots$ and identifying coefficients in the equality $\cos z = \sin z \cot z$, we obtain $a_1 = -\frac{1}{3}$. Therefore, $\mathrm{Res}(\pi z^{-2} \cot \pi z, 0) = -\pi^2/3$ and

$$\sum_{n=1}^{\infty} \frac{1}{n^2} = \frac{\pi^2}{6}. \qquad \square$$

Using the functions $\pi \operatorname{cosec} \pi z$, $\pi \tan \pi z$ and $\pi \sec \pi z$ one can sum up series of the form $\sum (-1)^n f(n)$, $\sum f(n + 1/2)$ and $\sum (-1)^n f(n + 1/2)$, where $f$ is of the kind considered above.

**Example 10.18.** Another remarkable application of formula (10.3) is the decomposition of the meromorphic function $\varphi$ into simple fractions. Suppose, for example, that the simple poles of $\varphi$ are the integer numbers with residue $\alpha_n$ and that $\varphi$ is uniformly bounded on $\partial Q_N$ (the squares of Example 10.12). Take $f(z) = \frac{1}{w^2 - z^2}$ with $w \in \mathbb{C}$, $w$ not integer; then (10.3) yields

$$\frac{\alpha_0}{w^2} + \sum_{n=1}^{\infty} \frac{\alpha_n + \alpha_{-n}}{w^2 - n^2} = -\sum \operatorname{Res}\left(\frac{1}{w^2 - z^2}\varphi(z), z = \pm w\right)$$

$$= \frac{1}{2w}\left(\varphi(w) - \varphi(-w)\right).$$

That is, changing now $w$ to $z$,

$$\frac{\varphi(z) - \varphi(-z)}{2} = \frac{\alpha_0}{z} + \sum_{n=1}^{\infty} \frac{(\alpha_n + \alpha_{-n})z}{z^2 - n^2}.$$

If $\varphi$ is an odd function, this expansion will be the one of $\varphi$ and if, in addition, the residues all have value 1, one gets $\varphi(z) = \pi \cot \pi z$. If the residues are $\alpha_n = (-1)^n$, then it turns out that $\varphi(z) = \frac{\pi}{\sin \pi z}$.     $\square$

## 10.4 The non-homogeneous Cauchy–Riemann equations in the plane. The Cauchy integral

Harmonic functions are the solutions of the homogeneous Laplace equation $\Delta u = 0$. In Chapter 7 it has been shown that it is also important to solve the non-homogeneous Laplace equation $\Delta u = \phi$, with $\phi$ a given function. In a similar way, holomorphic functions are solutions of the homogeneous Cauchy-Riemann equations $\bar{\partial} f = 0$. We consider now the non-homogeneous Cauchy–Riemann equations $\bar{\partial} f = \phi$, first in the whole plane and afterwards in any domain.

The equation $\bar{\partial} f = \phi$ is quite different from the equation $\Delta u = \phi$; first of all, $\bar{\partial}$ is a first-order operator and $\Delta$ a second-order one. Moreover the Laplacian is a real operator, and so the study of the equation $\Delta u = \phi$ may be done, without loss of generality, for real functions. Instead, the operator $\bar{\partial} = \frac{1}{2}\left(\frac{\partial}{\partial x} + i\frac{\partial}{\partial y}\right)$ transforms real functions into complex functions. Writing $f = u + iv$, $\phi = \alpha + i\beta$ and separating real and imaginary parts, the equation $\bar{\partial} f = \phi$ is equivalent to

$$\frac{1}{2}(u_x - v_y) = \alpha, \quad \frac{1}{2}(v_x + u_y) = \beta.$$

These are two connected real equations of first order, that is, they cannot be taken separately. If one is only interested in real solutions ($v = 0$), the above equations become

$$\frac{1}{2}u_x = \alpha, \quad \frac{1}{2}u_y = \beta$$

that is, $du = 2(\alpha dx + \beta dy)$. As it is known, this equation, saying that $\alpha dx + \beta dy$ is an exact form, does not always have a solution. It is necessary for $\alpha$ and $\beta$ to satisfy a compatibility condition; for a simply connected domain and $\alpha, \beta$ differentiable this condition is $\alpha_y = \beta_x$. Analogously, if the data $\phi$ is real ($\beta = 0$), then simply take $f = u$ real ($v = 0$) with $u$ depending only on $x$ and $u'(x) = 2\alpha$.

If the equation $\bar{\partial} f = \phi$, has a solution $f_0$, it has infinitely many because adding any holomorphic function to $f_0$ another solution is obtained. So to obtain uniqueness for the solution it will be necessary, as usual, to add some condition to the problem.

Next a parallel development to the one in Section 7.7 will be done, replacing the operator $\wedge$ by the operator $\bar{\partial}$.

**Proposition 10.19.** a) *If $f$ is a differentiable function on $\mathbb{C}$ such that $\bar{\partial} f$ is a continuous function with compact support and $\lim_{|z|\to\infty} f(z) = 0$, then one has*

$$f(z) = -\frac{1}{\pi} \int_{\mathbb{C}} \frac{\bar{\partial} f(w)}{w - z} \, dm(w), \quad z \in \mathbb{C}.$$

*In particular, this equality holds if $f \in C_c^1(\mathbb{C})$ (cf. Lemma 6.7).*

b) *Given a function $\phi \in C_c(\mathbb{C})$, if there is a solution $f$ differentiable on $\mathbb{C}$ of the equation $\bar{\partial} f = \phi$, with $\lim_{|z|\to\infty} f(z) = 0$, then this solution is unique and is given by*

$$f(z) = -\frac{1}{\pi} \int_{\mathbb{C}} \frac{\phi(w)}{w - z} \, dm(w), \quad z \in \mathbb{C}. \tag{10.4}$$

*Proof.* Applying Theorem 4.2 to a disc $D(0, R)$ big enough so that it contains the support of $\bar{\partial} f$, we get

$$f(z) = \frac{1}{2\pi i} \int_{C(0,R)} \frac{f(w)}{w - z} \, dw - \frac{1}{\pi} \int_{\mathbb{C}} \frac{\bar{\partial} f(w)}{w - z} \, dm(w), \quad |z| < R.$$

Now $f$ is holomorphic for $|w| > R$ and, since $\lim_{|z|\to\infty} f(z) = 0$, it has a removable singularity at infinity; therefore it can be expanded as $f(w) = \sum_{n=-\infty}^{-1} c_n w^n$ for $|w|$ big enough. With a fixed $z$, the function $\frac{f(w)}{w-z}$ is holomorphic at infinity as well, with an expansion of the kind $d_{-2} w^{-2} + O(w^{-3})$; that is, it has residue $0$ at the point $\infty$. Then (Subsection 5.5.2) one has

$$\int_{C(0,R)} \frac{f(w)}{w - z} \, dw = 0$$

for $R$ big enough and a) is proved. Obviously, b) is a reformulation of a). $\qquad\square$

Part b) in the proposition above does not prove that the function $f$ defined by (10.4) is a solution of $\bar{\partial} f = \phi$ for $\phi \in C_c(\mathbb{C})$. Observe that, since the integral is extended only to $K = \mathrm{spt}(\phi)$, $f$ is holomorphic outside $K$ and, therefore, the equality $\bar{\partial} f = \phi = 0$ holds outside $K$. As for the Laplacian, it is convenient to extend the integral of (10.4) to any measure.

**Definition 10.20.** Given a real or complex measure $\mu$ with compact. support $K \subset \mathbb{C}$ and $|\mu|(K) < +\infty$, the integral

$$C(\mu)(z) = \frac{1}{\pi} \int_{\mathbb{C}} \frac{d\mu(w)}{z - w}$$

is called the Cauchy integral of $\mu$. If $\mu = \phi\, dm$, we will write $C(\mu) = C(\phi)$

**Proposition 10.21.** *The Cauchy integral $C(\mu)(z)$ is defined at all points $z \notin K = \mathrm{spt}(\mu)$ and gives a holomorphic function on $\mathbb{C} \setminus K$.*

*Proof.* It is totally analogous to the proof of Proposition 7.34. $\qquad\square$

**Proposition 10.22.** *If $\phi \in C_c^1(\mathbb{C})$, the Cauchy integral $C(\phi)$ is of class $C^1$ on $\mathbb{C}$ and satisfies $\bar{\partial}C(\phi) = \phi$. If $\phi \in C_c^k(\mathbb{C})$, then $C(\phi) \in C^k(\mathbb{C})$.*

*Proof.* With a change of variable one has

$$C(\phi)(z) = \frac{1}{\pi} \int_{\mathbb{C}} \frac{\phi(z - w)}{w}\, dm(w).$$

Applying Proposition 7.36, on differentiation of convolutions, it turns out that

$$\bar{\partial}C(\phi)(z) = \frac{1}{\pi} \int_{\mathbb{C}} \frac{\bar{\partial}\phi(z - w)}{w}\, dm(w) = \frac{1}{\pi} \int_{\mathbb{C}} \frac{\bar{\partial}\phi(w)}{z - w}\, dm(w) = \phi(z),$$

where the last equality is a consequence of Proposition 10.19. $\qquad\square$

**Example 10.23.** Let us compute $C(\phi)$ for a radial function $\phi(z) = h(|z|)$ with $h$ locally integrable on $[0, +\infty)$. Integrating in polar coordinates, one has

$$C\phi(z) = \frac{1}{\pi} \int_0^\infty h(r)r \left\{ \int_0^{2\pi} \frac{d\theta}{z - re^{i\theta}} \right\} dr.$$

Using, for example, residues it can be easily checked that the integral with respect to $\theta$ has value 0 for $|z| < r$ and $2\pi/z$ for $|z| > r$, so that

$$C\phi(z) = \frac{2}{z} \int_0^{|z|} h(r)r\, dr.$$

If $g(r) = \int_0^r h(s)s\, ds$, one has therefore, $C\phi(z) = \frac{2}{z}g(|z|)$. Now, since $\bar{\partial}(|z|) = \frac{1}{2}\left(\frac{z}{\bar{z}}\right)^{1/2}$, it turns out that

$$\bar{\partial}C\phi(z) = \frac{2}{z}g'(|z|)\frac{1}{2}\left(\frac{z}{\bar{z}}\right)^{1/2} = \frac{1}{|z|}h(|z|)|z| = h(|z|) = \phi(z). \qquad\square$$

There is a close relation between Cauchy integral $C(\phi)$ and Riesz potential $G(\phi)$, which on $\mathbb{C}$ is the logarithmic potential,

$$G(\phi)(z) = \frac{1}{2\pi} \int \phi(w) \operatorname{Log} |z - w| dm(w).$$

Differentiating yields

$$\partial G(\phi)(z) = \frac{1}{2\pi} \int \phi(w) \partial_z \operatorname{Log} |z - w| dm(w).$$

Now,

$$\partial_z \operatorname{Log} |z - w| = \frac{1}{2} \partial_z \operatorname{Log} |z - w|^2 = \frac{1}{2} \partial_z \left( \operatorname{Log}(z - w) + \operatorname{Log}(\bar{z} - \bar{w}) \right)$$
$$= \frac{1}{2} \frac{1}{z - w}.$$

That is, if $\phi \in C_c(\mathbb{C})$, one has

$$C(\phi) = 4\partial G(\phi).$$

In general, $C(\mu)$ is not defined everywhere, but it exists at almost every point with respect to Lebesgue measure.

**Proposition 10.24.** *The Cauchy integral, $C(\mu)(z)$, of a measure $\mu$ with compact support $K$ and $|\mu|(K) < +\infty$ is defined for almost all $z \in \mathbb{C}$ and $C(\mu) \in L^1_{\mathrm{loc}}(\mathbb{C})$.*

*Proof.* Since $1/z$ is locally integrable (Lemma 4.1), one has, for $R > 0$,

$$\int_{D(0,R)} |C(\mu)|(z) dm(z) \leq \frac{1}{\pi} \int_{D(0,R)} \int_K \frac{d|\mu|(w)}{|z - w|} dm(z)$$
$$\leq \frac{1}{\pi} \int_K d|\mu|(w) \int_{D(0,R)} \frac{1}{|z - w|} dm(z)$$
$$\leq M |\mu|(K) < +\infty,$$

where $M$ is a constant depending on $R$. $\qquad\qquad\square$

**Definition 10.25.** Let $\mu$ be a measure of locally finite mass on a domain $U$ of $\mathbb{C}$. A function $u \in L^1_{\mathrm{loc}}(U)$ is said to be a solution of the equation $\bar{\partial} u = \mu$ in $U$ in the weak sense if for every function $\varphi \in C^1_c(U)$ one has

$$\int_U u(z) \bar{\partial} \varphi(z) dm(z) = - \int_U \varphi(z) d\mu(z).$$

The same comments made in Subsection 7.7.2 apply here. Namely, if $u \in C^1(U)$ and $\bar{\partial} u = \phi$ in the weak sense, then $\bar{\partial} u = \phi$ in the usual pointwise sense.

**Theorem 10.26.** *If $\mu$ is a measure with compact support and finite mass on $\mathbb{C}$, then $\bar{\partial}C(\mu) = \mu$ in the weak sense on $\mathbb{C}$.*

*Proof.* As in Theorem 7.42, one has for $\varphi \in C_c^1(\mathbb{C})$,

$$
\int_{\mathbb{C}} C(\mu)(z)\bar{\partial}\varphi(z)\,dm(z) = \int_{\mathbb{C}} \frac{1}{\pi} \int_{\mathbb{C}} \frac{d\mu(w)}{z-w}\bar{\partial}\varphi(z)\,dm(z)
$$

$$
= \int_{\mathbb{C}} \left\{ \frac{1}{\pi} \int_{\mathbb{C}} \frac{\bar{\partial}\varphi(z)}{z-w}\,dm(z) \right\} d\mu(w)
$$

$$
= - \int \varphi(w)\,d\mu(w),
$$

thanks to Proposition 10.19. $\qquad\square$

Analogously one may deal with weak solutions of the equation $\partial v = \mu$, in the domain $U$. They are functions $v \in L_{\text{loc}}^1(U)$ such that

$$
\int_U v(z)\partial\varphi(z)\,dm(z) = - \int_U \varphi(z)\,d\mu(z)
$$

for every $\varphi \in C_c^1(U)$.

The definition of weak solution for the operators $\partial$, $\bar{\partial}$, $\Delta$ does not depend on hypotheses of regularity and it can be applied to any locally integrable function. It is easy to check from the definitions that $u_1, u_2, u_3 \in L_{\text{loc}}^1(U)$ and

$$
\bar{\partial}u_1 = u_2 \quad \text{and} \quad \partial u_2 = u_3,
$$

in $U$ in the weak sense, imply $\Delta u_1 = 4\partial\bar{\partial}u_1 = 4\partial u_2 = 4u_3$ in the weak sense in $U$.

**Proposition 10.27.** *Let $U$ be a domain of the plane and let the function $u \in L_{\text{loc}}^1(U)$ satisfy in $U$ the equation $\bar{\partial}u = 0$ in the weak sense. Then $u$ is holomorphic on $U$. For every measure $\mu$ with compact support and finite mass on $\mathbb{C}$ one has $\partial G(\mu) = \frac{1}{4}C(\mu)$ in the weak sense.*

*Proof.* If $\bar{\partial}u = 0$ in the weak sense, then $\Delta u = 4\partial\bar{\partial}u = 0$ in the weak sense. By Weyl's lemma (Theorem 7.45) it turns out that $u \in C^\infty(U)$ and, therefore, $\bar{\partial}u = 0$ in the pointwise sense and $u$ is holomorphic on $U$. Also, $\bar{\partial}\partial G(\mu) = \frac{1}{4}\Delta G(\mu) = \frac{1}{4}\mu$ in the weak sense. Thus, $\bar{\partial}(4\partial G(\mu) - C(\mu)) = 0$ in the weak sense and $4\partial G(\mu) - C(\mu)$ is an entire function. Since this function vanishes at infinity, it must be identically zero and it turns out that $4\partial G(\mu) = C(\mu)$. $\qquad\square$

Combining the previous proposition with Proposition 7.38 and Theorem 7.46 one obtains the following result.

**Theorem 10.28.** a) *If $\phi \in L_c^\infty(\mathbb{C})$, then $C(\phi)$ satisfies a Lipschitz condition of type $|C(\phi)(z) - C(\phi)(w)| = O\big(|z - w| \operatorname{Log} \frac{1}{|z-w|}\big)$, on compact sets.*

b) *If $\phi \in C_c(\mathbb{C})$ is locally Lipschitz, that is, satisfies a local Lipschitz condition with positive exponent:*

$$|\phi(z) - \phi(w)| \leq c(z)|z - w|^\alpha, \quad \alpha > 0, \ c(z) \geq 0, \ if \ |z - w| \leq \delta(z),$$

*then $C(\phi)$ is of class $C^1$ on $\mathbb{C}$ and satisfies $\bar{\partial} C(\phi) = \phi$.*

**Example 10.29.** If $\phi$ is radial and bounded, $\phi(z) = h(|z|)$, the function $g(r) = \int_0^r h(s)s\,ds$ in Example 10.23 satisfies

$$|g(r_1) - g(r_2)| \leq \int_{r_1}^{r_2} |h(s)|s\,ds \leq \frac{1}{2}\|h\|_\infty(r_2^2 - r_1^2), \quad 0 < r_1 < r_2.$$

Since $C(\phi)(z) = \frac{2}{z}g(|z|)$, assuming $|z_1| \leq |z_2|$, we get

$$
\begin{aligned}
|C(\phi)(z_1) - C(\phi)(z_2)| &= 2\left| \frac{g(|z_1|)}{z_1} - \frac{g(|z_2|)}{z_2} \right| \\
&\leq \frac{2}{|z_2|}\left| g(|z_1|) - g(|z_2|) \right| + 2|g(|z_1|)|\left| \frac{1}{z_1} - \frac{1}{z_2} \right| \\
&\leq \|h\|_\infty\left( \frac{|z_2|^2 - |z_1|^2}{|z_2|} + \frac{|z_1|^2|z_1 - z_2|}{|z_1||z_2|} \right) \\
&= \|h\|_\infty|z_1 - z_2|\left( \frac{|z_2| + |z_1|}{|z_2|} + \frac{|z_1|}{|z_2|} \right) \\
&\leq 3\|h\|_\infty|z_1 - z_2|.
\end{aligned}
$$

Therefore when $\phi$ is bounded and radial, the Cauchy transform, $C(\phi)$, satisfies a stronger condition than one given in a) of Theorem 10.28.  $\square$

## 10.5 The non-homogeneous Cauchy–Riemann equations in an open set. Weighted kernels

So far the properties of the Cauchy integral $C(\mu)$, where $\mu$ is a measure on $\mathbb{C}$ with compact support $K$ and $|\mu|(K) < \infty$, have been considered, as well as the solutions in the weak sense of the equation $\bar{\partial} u = \mu$. To solve the equation $\bar{\partial} f = \phi$ with $\phi \in C(U)$ in a bounded domain $U \subset \mathbb{C}$ one may take, simply, the measure $\mu = \phi \, dm_{|U}$, whenever $\phi \in L^1(U)$.

**Proposition 10.30.** *Let $U$ be a bounded domain of $\mathbb{C}$ and $\phi \in L^1(U) \cap C(U)$. Then the Cauchy integral*

$$C(\phi)(z) = \frac{1}{\pi} \int_U \frac{\phi(w)}{z - w} dm(w), \quad z \in U$$

*defines a continuous function on $U$ satisfying $\bar{\partial}C(\phi) = \phi$ in the weak sense in $U$. If $\phi$ is locally Lipschitz on $U$, then $C(\phi) \in C^1(U)$ and $\bar{\partial}C(\phi) = \phi$ in the classical sense in $U$.*

*Proof.* It is known that $C(\phi) \in L^1_{\text{loc}}(\mathbb{C})$ and $\bar{\partial}C(\phi) = \phi\,dm_{|U}$ in the weak sense on $\mathbb{C}$ and, therefore, also on $U$. To see that $\phi$ is continuous on $U$, fix a disc $\bar{D}(z_0, r) \subset U$, consider a function $\chi \in C_c^\infty(U)$ such that $\chi = 1$ on $\bar{D}(z_0, r)$ and define $\phi_1 = \phi\chi$, $\phi_2 = \phi(1 - \chi)$. Then $C(\phi) = C(\phi_1) + C(\phi_2)$; $C(\phi_1)$ is continuous on $\mathbb{C}$ (because $\phi_1$ is bounded and Theorem 10.28 a) applies) and $C(\phi_2)$ is holomorphic on $D(z_0, r)$; thus $C(\phi)$ is continuous on $D(z_0, r)$. Since this disc is arbitrary, $C(\phi) \in C(U)$. If in addition $\phi$ satisfies a local Lipschitz condition, it is so for $\phi_1$ and, by Theorem 10.28 b), $C(\phi_1)$ is in $C^1(\mathbb{C})$, and $C(\phi) \in C^1(U)$. $\square$

If $\phi \notin L^1(U)$, one has, a priori, no way to solve the equation $\bar{\partial}f = \phi$ on $U$. Analogously, one has, at the moment, no way to solve the equation $\bar{\partial}f = \phi$ on $\mathbb{C}$ if $\phi$ has no compact support. Other methods for solving this equation, based on the following result, will be undertaken now.

**Theorem 10.31.** *Let $U$ be any domain of $\mathbb{C}$ and $H(z, w)$ a continuous function on $U \times U$ such that $H(z, z) = 1$ if $z \in U$ and $H(z, w)$ is holomorphic in $z$ for all $w \in U$. Suppose that $\phi$ is a continuous function on $U$ and that*

$$A(z) = \int_U |H(z, w)||\phi(w)|dm(w) \tag{10.5}$$

*is a locally bounded function on $U$. Then,*

$$C_H(\phi)(z) = \frac{1}{\pi} \int_U H(z, w)\frac{\phi(w)}{z - w}dm(w), \quad z \in U$$

*defines a continuous function on $U$ such that $\bar{\partial}C_H(\phi) = \phi$ in the weak sense. If $\phi$ is locally Lipschitz on $U$, then $C_H(\phi) \in C^1(U)$ and $\bar{\partial}C_H(\phi) = \phi$ holds in the classical sense.*

*Proof.* The hypotheses on $H$ imply this function may be written as

$$H(z, w) = 1 + (z - w)G(z, w)$$

with $G(z, w)$ holomorphic in $z$. Hence, formally,

$$C_H(\phi)(z) = C(\phi)(z) + \frac{1}{\pi} \int_U G(z, w)\phi(w)dm(w)$$

is the sum of $C(\phi)$ with a holomorphic function and, thus, $\bar{\partial}C_H(\phi) = \phi$. However, the hypotheses do not guarantee the convergence of the two terms on the right-hand side of the equality and another argumentation is needed. Fix a disc $\bar{D}(z_0, r) \subset U$; let $\chi$ be a function in $C_c^\infty(U)$ such that $\chi = 1$ on $\bar{D}(z_0, r)$ and take $\phi_1 = \chi\phi$, $\phi_2 = (1 - \chi)\phi$. Since $\phi_1 \in C_c(U)$, $C(\phi_1)$ is continuous on $\mathbb{C}$ by Theorem 10.28 a) and $G(z, w)$ is also jointly continuous in $z$, $w$ on $U \times U$. Therefore, if $K = \mathrm{spt}(\chi)$ one has

$$|G(z, w)| \le C, \quad \text{if } z \in \bar{D}(z_0, r) \text{ and } w \in K.$$

This proves the continuity of the function

$$z \to \frac{1}{\pi} \int_U G(z, w)\phi_1(w)dm(w),$$

which is holomorphic on $D(z_0, r)$ by Morera's theorem. On the other hand, we can prove now that the function

$$C_H(\phi_2)(z) = \frac{1}{\pi} \int_{|w - z_0| \ge r} H(z, w)\frac{\phi_2(w)}{z - w}dm(w)$$

is also holomorphic on $D(z_0, r)$. Once again, by Morera's theorem, it is enough to show that it is continuous and, according to the dominated convergence theorem, it suffices to prove the inequality

$$\int_U \sup_{|z - z_0| \le \frac{r}{2}} |H(z, w)||\phi(w)|dm(w) < +\infty. \tag{10.6}$$

By the Cauchy integral formula, one has

$$\sup_{|z - z_0| \le \frac{r}{2}} |H(z, w)| \le C \int_0^{2\pi} |H(z_0 + re^{i\theta}, w)|d\theta,$$

and (10.6) is dominated by

$$\int_U \int_0^{2\pi} |H(z_0 + re^{i\theta}, w)||\phi(w)|d\theta dm(w) = \int_0^{2\pi} A(z_0 + re^{i\theta})d\theta < +\infty,$$

where $A$ is the function defined in (10.5)

So, in the disc $D(z_0, r)$, $C_H(\phi)$ is the sum of $C(\phi_1)$ and a holomorphic function, a fact that implies the continuity of $C_H(\phi)$ and the equality $\bar{\partial}C_H(\phi) = \phi$. Furthermore, $C_H(\phi)$ will have the same local regularity as $C(\phi)$ and this proves the second part of the theorem. $\qquad\square$

**Example 10.32.** With suitable hypotheses on a measure $\mu$, Theorem 10.31 can be generalized so that writing

$$C_H(\mu)(z) = \frac{1}{\pi} \int_U H(z, w)\frac{d\mu(w)}{z - w}, \quad z \in U,$$

one may conclude that $C_H(\mu) \in L^1_{\mathrm{loc}}(U)$ and $\bar{\partial}C_H(\mu) = \mu$ in the weak sense. For example, when $\mu = \delta_a$ (Dirac's delta) with $a \in U$, we get

$$C_H(\delta_a)(z) = \frac{1}{\pi}\frac{H(z,a)}{z-a}.$$

The fact that $\bar{\partial}C_H(\delta_a) = \delta_a$ in the weak sense is shown, formally, in the following way: since $H$ is holomorphic in $z$, one has

$$\bar{\partial}C_H(\delta_a) = H(z,a)\bar{\partial}\left(\frac{1}{\pi}\frac{1}{z-a}\right).$$

Now $\bar{\partial}\left(\frac{1}{\pi}\frac{1}{z-a}\right) = \delta_a$, by Theorem 10.26, and

$$\bar{\partial}C_H(\delta_a) = H(z,a)\delta_a = H(a,a)\delta_a = \delta_a. \qquad \square$$

A more restrictive hypothesis, but quite general to be applied in many interesting cases, is that the function $H(z,w)$ of Theorem 10.31 satisfies the condition

$$|H(z,w)| \le \Psi_1(z)\Psi_2(w) \tag{10.7}$$

with $\Psi_1$, $\Psi_2$ continuous on $U$. Then the function $A(z)$ of (10.5) is locally bounded on $U$ whenever

$$\int_U \Psi_2(w)|\phi(w)|\,dm(w) < +\infty$$

holds. This condition allows us to solve the equation $\bar{\partial}f = \phi$ for functions $\phi \in C(U)$, not necessarily integrable on $U$, and the larger the class of admissible functions $\phi$ will be the faster the decrease to zero of $\Psi_2(w)$ when $w \to \partial U$.

For example, if $U$ is the unit disc, one may take

$$H(z,w) = \left(\frac{1-|w|^2}{1-z\bar{w}}\right)^k, \quad k \in \mathbb{N},$$

so that

$$|H(z,w)| \le \left(\frac{1-|w|^2}{1-|z|}\right)^k, \quad |z|,|w| < 1,$$

which is (10.7) with $\Psi_1(z) = (1-|z|)^{-k}$ and $\Psi_2(w) = (1-|w|^2)^k$. One has then the following result:

**Proposition 10.33.** *If $\phi \in C(\mathbb{D})$ and $\int_{\mathbb{D}}|\phi(w)|(1-|w|^2)^k\,dm(w) < +\infty$ for some $k \in \mathbb{N}$, then the function*

$$f_k(z) \stackrel{\mathrm{def}}{=} C_k(\phi)(z) = \frac{1}{\pi}\int_{\mathbb{D}}\left(\frac{1-|w|^2}{1-z\bar{w}}\right)^k\frac{\phi(w)}{z-w}\,dm(w)$$

*is continuous on $\mathbb{D}$ and satisfies the equation $\bar{\partial}f_k = \phi$ in the weak sense; if $\phi$ is locally Lipschitz, then $f_k \in C^1(\mathbb{D})$ and $\bar{\partial}f_k = \phi$ in the classical sense on $\mathbb{D}$.*

The solution $f_k$ of the proposition above is related to a decomposition formula similar to the Cauchy–Green formula.

**Proposition 10.34.** *If $f \in C^1(\mathbb{D})$ and, for $k \in \mathbb{N}$,*

$$\int_{\mathbb{D}} (1 - |w|^2)^{k-1} |f(w)| dm(w) < +\infty,$$

$$\int_{\mathbb{D}} (1 - |w|^2)^{k} |\bar{\partial} f(w)| dm(w) < +\infty,$$

*then one has, for $z \in \mathbb{D}$,*

$$f(z) = \frac{k}{\pi} \int_{\mathbb{D}} \frac{(1 - |w|^2)^{k-1}}{(1 - z\bar{w})^{k+1}} f(w) dm(w) + \frac{1}{\pi} \int_{\mathbb{D}} \frac{(1 - |w|^2)^{k}}{(1 - z\bar{w})^{k}} \frac{\bar{\partial} f(w)}{z - w} dm(w).$$

*Proof.* Consider the form $\eta = \frac{(r - |w|^2)^k}{(1 - z\bar{w})^k} \frac{f(w)}{w - z} dw$ in the domain $D(0, r) \setminus \bar{D}(z, \varepsilon)$ for $r < 1$, $|z| < r$ and $\varepsilon$ small enough. Computing yields

$$d\eta = \frac{(r - |w|^2)^{k}}{(1 - z\bar{w})^{k}} \frac{\bar{\partial} f(w)}{w - z} d\bar{w} \wedge dw + k \frac{f(w)}{w - z} \frac{(r - |w|^2)^{k-1}(rz - w)}{(1 - z\bar{w})^{k+1}} d\bar{w} \wedge dw$$

and, applying Stoke's theorem,

$$\int_{D(0,r)\setminus\bar{D}(z,\varepsilon)} d\eta = \left( \int_{C(0,r)} - \int_{C(0,r)} \right) \eta = - \int_{C(z,\varepsilon)} \frac{(r - |w|^2)^{k}}{(1 - \bar{w}z)^{k}} \frac{f(w)}{w - z} dw.$$

Letting $r \to 1$, $\varepsilon \to 0$ and using the dominated convergence theorem the proof is finished. $\square$

Using Proposition 10.34 one can also check that the function $f_k$ given by Proposition 10.33 satisfies $\bar{\partial} f_k = \phi$. Actually, if $f \in C^1(\mathbb{D})$ is any solution of $\bar{\partial} f = \phi$, then Proposition 10.34 and the definition of $f_k$ give

$$f(z) - f_k(z) = \frac{k}{\pi} \int_{\mathbb{D}} \frac{(1 - |w|^2)^{k-1}}{(1 - z\bar{w})^{k+1}} f(w) dm(w),$$

which is a holomorphic function in $z$ and, therefore, $\bar{\partial} f_k = \bar{\partial} f = \phi$.

**Corollary 10.35.** *If $f$ is holomorphic and integrable on $\mathbb{D}$, then*

$$f(z) = \frac{1}{\pi} \int_{\mathbb{D}} \frac{f(w)}{(1 - z\bar{w})^2} dm(w), \quad z \in \mathbb{D}.$$

Like the Cauchy integral formula, this is a reproducing formula for holomorphic functions, but there is an essential difference with respect to the Cauchy integral: here *all* the values of $f$ on $\mathbb{D}$ are involved and not just the values of $f$ on $\partial\mathbb{D}$. The kernel $1/\pi(1 - z\bar{w})^2$ is called the *Bergman kernel* of the unit disc (see Exercise 21 in Section 4.7).

Hence, for the unit disc $\mathbb{D}$, two kinds of solutions of the equation $\bar{\partial} f = \phi$ have been obtained. The first is the one given by the Cauchy integral,

$$C(\phi)(z) = \frac{1}{\pi} \int_{\mathbb{D}} \frac{\phi(w)}{z - w} dm(w),$$

defined when $\phi \in C(\mathbb{D}) \cap L^1(\mathbb{D})$. The second one is given by Proposition 10.33 with $k = 1$, that is,

$$f_1(z) = C_1(\phi)(z) = \frac{1}{\pi} \int_{\mathbb{D}} \frac{1 - |w|^2}{1 - z\bar{w}} \frac{\phi(w)}{z - w} dm(w)$$

whenever $\phi \in C(\mathbb{D})$ and

$$\int_{\mathbb{D}} |\phi(w)|(1 - |w|^2)dm(w) < +\infty.$$

We proceed to study them in more detail.

Clearly, $C(\phi) \in C(\mathbb{D})$; moreover if $\phi \in L^\infty(\mathbb{D}) \cap C(\mathbb{D})$, then $C(\phi)$ is continuous on $\bar{\mathbb{D}}$, by Theorem 10.28 a). This happens, in particular, if $\phi \in C(\bar{\mathbb{D}})$.

In the following statement the notations are those of Proposition 7.28.

**Theorem 10.36.** *If $\phi \in L^\infty(\mathbb{D}) \cap C(\mathbb{D})$, among all the solutions of the equation $\bar{\partial} f = \phi$, $f = C(\phi)$ is the only continuous solution on $\bar{\mathbb{D}}$ that satisfies $f \in \overline{A_0(\mathbb{T})}$, that is, which is perpendicular to the space $A(\mathbb{T})$ with respect to the scalar product of $L^2(\mathbb{T})$.*

*Proof.* According to Proposition 7.28 we need to check that $C(\phi)_{|\mathbb{T}}$ has vanishing Fourier coefficients at every non-negative integer:

$$\int_0^{2\pi} C(\phi)(e^{it})e^{-int} dt = 0, \quad n \geq 0.$$

Using Fubini's theorem, it is enough to show the same equalities for the function $(z - w)^{-1}$, $w \in \mathbb{D}$:

$$\int_0^{2\pi} \frac{e^{-int}}{e^{it} - w} = \int_0^{2\pi} e^{-int} \sum_{m=0}^{\infty} w^m e^{-i(m+1)t} dt = 0, \quad n \geq 0,$$

and clearly this is so.

Conversely, assume that $f \in C(\bar{\mathbb{D}})$ satisfies the conditions $\bar{\partial} f = \phi$ and $f_{|\mathbb{T}} \in \overline{A_0(\mathbb{T})}$. Then $\bar{\partial}(f - C(\phi)) = 0$, that is, $\varphi = f - C(\phi)$ is holomorphic. Hence $\varphi$ is a function of $A(\mathbb{T})$ and also belongs to $\overline{A_0(\mathbb{T})}$ because $f$ and $C(\phi)$ are in this space. Therefore, $\langle \varphi, \varphi \rangle = 0$ in $L^2(\mathbb{T})$, and so

$$\int_0^{2\pi} |\varphi(e^{it})|^2 dt = 0.$$

As a consequence $\varphi = 0$, that is, $f - C(\phi) = 0$ on $\mathbb{T}$, and being holomorphic, this yields $f - C(\phi) = 0$. $\qquad\square$

**Example 10.37.** Take $\phi(z) = \bar{z}^n, n \in \mathbb{N}$. Clearly, $f = \frac{1}{n+1}\bar{z}^{n+1}$ satisfies $\bar{\partial} f = \phi$ and restricted to $\mathbb{T}$ belongs to $\overline{A_0(\mathbb{T})}$. Then, $C(\phi) = f$ and

$$\frac{1}{\pi} \int_{\mathbb{D}} \frac{\bar{w}^n}{z - w} \, dm(w) = \frac{1}{n + 1} \bar{z}^{n+1}, \quad |z| < 1.$$

More generally, let $g$ be holomorphic on $\mathbb{D}$, continuous on $\bar{\mathbb{D}}$, $\phi = \bar{g}$ and let $h$ be a holomorphic antiderivative of $g$, $h' = g$, with $h(0) = 0$. Then the same argument proves that $C(\phi) = \bar{h}$ because $\bar{\partial}\bar{h} = \overline{h'} = \bar{g} = \phi$ and $\overline{C(\phi)} = h$ is holomorphic and vanishing at the origin. $\qquad\square$

The orthogonality of $C(\phi)$ with respect to $A(\mathbb{T})$ just described amounts to saying that among all solutions $f \in C(\bar{\mathbb{D}})$ of the equation $\bar{\partial} f = \phi$, $C(\phi)$ is the one minimizing the quantity

$$\int_{\mathbb{T}} |f(z)|^2 |dz|.$$

Regarding the solution $f_1(z)$, the corresponding statement is the following.

**Theorem 10.38.** *If $\phi \in L^\infty(\mathbb{D}) \cap C(\mathbb{D})$, among all the solutions of the equation $\bar{\partial} f = \phi$, the function $f = f_1 = C_1(\phi)$ is the only one being perpendicular to $A(\mathbb{D})$ with respect to the scalar product of $L^2(\mathbb{D})$: $\langle f, g \rangle = \int_{\mathbb{D}} f(z)\overline{g(z)}dm$.*

*Proof.* We need to prove that $\int_{\mathbb{D}} f_1(z)\overline{g(z)}dm(z) = 0$ for $g \in A(\mathbb{D})$. By Fubini's theorem it is enough to show that

$$\int_{\mathbb{D}} \frac{1}{(1 - z\bar{w})(z - w)} \overline{g(z)}dm(z) = 0, \quad w \in \mathbb{D}.$$

Consider the form $\eta = \mathrm{Log}\left|\frac{z-w}{1-z\bar{w}}\right| \overline{g(z)}d\bar{z}$, which is continuous on $\bar{\mathbb{D}} \setminus D(w, \varepsilon)$ with $\varepsilon > 0$ small enough and satisfies

$$d\eta = \partial\left(\overline{g(z)}\,\mathrm{Log}\left|\frac{z - w}{1 - z\bar{w}}\right|\right) dz \wedge d\bar{z} = \overline{g(z)}\partial\,\mathrm{Log}\left|\frac{z - w}{1 - z\bar{w}}\right| dz \wedge d\bar{z}.$$

Now, $\partial_z \operatorname{Log}\left|\frac{z-w}{1-z\bar{w}}\right| = \frac{1}{2}\partial_z \log\left(\frac{z-w}{1-z\bar{w}}\right) = \frac{1}{2}\frac{1-|w|^2}{(z-w)(1-z\bar{w})}$. Since $\eta = 0$ when $|z| = 1$ (because $|z - w| = |1 - z\bar{w}|$ if $|z| = 1$), Stokes' theorem yields

$$\frac{1}{2}\int_{\mathbb{D}\setminus D(w,\varepsilon)} \frac{(1-|w|^2)}{(z-w)(1-z\bar{w})}\overline{g(z)}dz \wedge d\bar{z} = -\int_{C(w,\varepsilon)} \operatorname{Log}\left|\frac{z-w}{1-z\bar{w}}\right|\overline{g(z)}d\bar{z}$$

and letting $\varepsilon \to 0$ we obtain the desired result, because the integral on the right-hand side is bounded by $\varepsilon \operatorname{Log}\frac{1}{\varepsilon}$.

Conversely, if $f$ is another solution of $\bar{\partial}f = \phi$ with the same properties, arguing as in Theorem 10.36 one proves

$$\int_{\mathbb{D}} |f - f_1|^2 dm(z) = 0,$$

and so $f = f_1$. $\qquad\square$

**Example 10.39.** If $g \in A(\mathbb{D})$, $\phi = \bar{g}$ and $h' = g$ with $h(0) = 0$, once again $f = \bar{h}$ satisfies $\bar{\partial}f = \phi$. So if $F \in A(\mathbb{D})$,

$$\int_{\mathbb{D}} F(z)\overline{f(z)}dm = \int_{\mathbb{D}} F(z)h(z)dm(z) = \pi F(0)h(0) = 0,$$

because $h(0) = 0$. Hence, $C_1(\phi) = \bar{h}$ as well. $\qquad\square$

Another way to express the orthogonality property is to say that, among all the solutions $f$ of $\bar{\partial}f = \phi$, the function $f_1 = C_1(\phi)$ is the one minimizing the quantity

$$\int_{\mathbb{D}} |f(z)|^2 dm(z).$$

Let us apply now Theorem 10.31 with $U = \mathbb{C}$ and

$$H(z, w) = \exp(z\bar{w} - |w|^2).$$

First, one needs to find a quite general condition on $\phi \in C(\mathbb{C})$ such that

$$A(z) = \int_{\mathbb{C}} |\exp(z\bar{w} - |w|^2)|\,|\phi(w)|dm(w)$$

is a locally bounded function. A suitable condition is the following one:

$$\|\phi\|_*^2 \overset{\text{def}}{=} \int_{\mathbb{C}} |\phi(w)|^2 \exp(-|w|^2)dm(w) < +\infty.$$

Indeed, by Schwarz's inequality, one has

$$|A(z)| \le \int_{\mathbb{C}} \exp(\operatorname{Re} z\bar{w} - |w|^2)|\phi(w)|dm(w)$$

$$\le \|\phi\|_* \left\{ \int_{\mathbb{C}} \exp(2\operatorname{Re} z\bar{w} - |w|^2)dm(w) \right\}^{1/2}$$

$$= \|\phi\|_* \exp |z|^2 \left\{ \int_{\mathbb{C}} \exp(-|z-w|^2)dm(w) \right\}^{1/2}$$

$$= C\|\phi\|_* \exp |z|^2 \le M,$$

where $M$ is a constant, if $z$ belongs to a compact set.

Hence, one obtains the following result.

**Proposition 10.40.** *If $\phi \in C(\mathbb{C})$ and $\|\phi\|_* < +\infty$, then*

$$f(z) = \frac{1}{\pi} \int_{\mathbb{C}} \exp(z\bar{w} - |w|^2) \frac{\phi(w)}{z-w} dm(w)$$

*is a continuous function on $\mathbb{C}$ that satisfies $\bar{\partial} f = \phi$ in the weak sense; if $\phi$ is locally Lipschitz, then $f \in C^1(\mathbb{C})$ and $\bar{\partial} f = \phi$ in the classical sense.*

The methods developed so far allow us to solve the equation $\bar{\partial} f = \phi$, on a domain $U$, for functions $\phi \in C(U)$ that are integrable on $U$ with respect to certain weights. However they do not work for an arbitrary function $\phi \in C(U)$, without any integrability condition. Next this question will be discussed in its more general formulation, using Runge's theorem.

**Theorem 10.41.** *Let $U$ be a domain of $\mathbb{C}$, bounded or not, and let $\mu$ be a measure of locally finite mass on $U$, that is, for every compact set $K \subset U$, one has $|\mu|(K) < +\infty$. Then there exists a function $f \in L^1_{\mathrm{loc}}(U)$ satisfying $\bar{\partial} f = \mu$ in the weak sense in $U$. Moreover, on every open set $V \subset U$ where $d\mu = \phi dw$ with $\phi$ locally Lipschitz, the function $f$ belongs to $C^1(V)$, and if $\phi \in C^k(V)$, then $f \in C^k(V)$, $1 \le k \le \infty$, as well.*

*Proof.* Consider the compact sets $K_n \subset U$ of Lemma 1.15 and let $\psi_n \in C_c^\infty(U)$ be functions with $\psi_n = 1$ on a neighborhood of $K_n$: take $\varphi_1 = \psi_1$ and $\varphi_n = \psi_n - \psi_{n-1}$ if $n > 1$, so that $\varphi_n = 0$ on a neighborhood of $K_{n-1}$ and $\sum_1^\infty \varphi_n = 1$ on $U$. Consider now the measure $\varphi_n d\mu$ with compact support and finite mass, and its Cauchy integral,

$$C(\varphi_n d\mu)(z) = \frac{1}{\pi} \int \frac{\varphi_n(w)}{w-z} d\mu(w),$$

which, as it is known, satisfies $\bar{\partial}(C(\varphi_n d\mu)) = \varphi_n d\mu$ in the weak sense, on $\mathbb{C}$.

Of course, $C(\varphi_n d\mu)$ is a holomorphic function on a neighborhood of $K_{n-1}$ and, by Theorem 10.7, there are functions $f_n \in H(U)$ such that

$$|C(\varphi_n d\mu)(z) - f_n(z)| \leq 2^{-n} \quad \text{if } z \in K_{n-1}.$$

Now let us show that the series

$$f(z) = \sum_{n=1}^{\infty} (C(\varphi_n d\mu)(z) - f_n(z))$$

defines the desired function $f$. A compact set $K \subset U$ being fixed, all the terms of the series, for $n$ big enough, are holomorphic functions on a neighborhood $W$ of $K$ and the series is uniformly convergent on compact sets of $W$. That is, for fixed $K$ there is a natural number $N$ and a neighborhood $W$ of $K$ such that

$$f = \sum_{n=1}^{N} C(\varphi_n d\mu) - f_n(z)) + \text{holomorphic function on } W$$

$$= \sum_{n=1}^{N} C(\varphi_n d\mu) + \text{holomorphic function on } W.$$

Hence, $f \in L^1_{\text{loc}}(U)$ and the same decomposition above shows that $\bar{\partial} f = d\mu$ in the weak sense. If $d\mu = \phi dm$ on an open set $V \subset U$, then $f - C(\phi)$ is holomorphic on $V$; therefore, any solution of $\bar{\partial} f = \phi$ has the same local regularity properties as $C(\phi)$. $\qquad\square$

An important difference with respect to solutions found in Theorems 10.36 and 10.38 is that here there is no linear operator giving the solution. That is, one cannot assure that $f$ depends linearly on $\mu$.

**Corollary 10.42.** *The Poisson equation* $\Delta u = \mu$ *has a solution* $u \in L^1_{\text{loc}}(U)$ *in the weak sense, for any measure* $\mu$ *of locally finite mass on* $U$.

*Proof.* Use the equality $\Delta = 4\partial\bar{\partial}$. First, with the help of Theorem 10.41 choose a function $f \in L^1_{\text{loc}}(U)$ such that $\bar{\partial} f = \mu$. Now one must solve the equation $4\partial u = f$, which is equivalent to $4\bar{\partial}\bar{u} = \bar{f}$, and this is done with the same theorem. $\qquad\square$

## 10.6 The Dirichlet problem for the $\bar{\partial}$ operator

As shown above, the Cauchy–Riemann equation $\bar{\partial} f = \phi$ in a domain $U$ always has a solution $f \in L^1_{\text{loc}}(U)$, for any data $\phi \in C(U)$. The general solution is then, $f + h$ with $h \in H(U)$. If we want to set a problem with a unique solution, it is necessary to impose some additional condition to $f$. For example, it was shown in Proposition 10.19 that, in the case $U = \mathbb{C}$, the Cauchy potential $C(\phi)$ is the only solution vanishing at infinity, and in the case of the unit disc two results in this sense are Theorem 10.36 and Theorem 10.38.

One possibility is to prescribe the values of $f$ on the boundary of the domain and to set a problem of Dirichlet kind: given $\varphi \in C(\partial U)$ and $\phi \in C(U)$, look for a function $f \in C(\bar{U})$ such that $f = \varphi$ on $\partial U$ and $\bar{\partial} f = \phi$ on $U$. The solution of this problem, if one exists, is unique, because if $f_1$, $f_2$ arc two solutions, then $f_1 - f_2$ is holomorphic on $U$ and vanishes on $\partial U$ and so $f_1 = f_2$. However, it is easy to convince oneself that this is a non-well posed problem that, in general, has no solution. The reason is that the data $\varphi$, $\phi$ must satisfy a compatibility condition.

From now on it will be assumed that $U$ is a bounded domain with piecewise positively oriented regular boundary and also that there are solutions on $U$, in the classical sense, of the equation $\bar{\partial} f = \phi$. Given $\varphi \in C(\partial U)$ and $\phi \in C(\bar{U})$, if $f \in C^1(\bar{U})$ satisfies $f = \varphi$ on $\partial U$ and $\bar{\partial} f = \phi$ on $U$, then for every function $h$ holomorphic on a neighborhood of $\bar{U}$ we will have

$$\int_{\partial U} \varphi h \, dz = \int_{\partial U} f h \, dz = \int_U d(f h \, dz) = \int_U h \phi \, d\bar{z} \wedge dz.$$

Consequently, in order for the Dirichlet type problem to have a solution we need the data $\varphi$, $\phi$ to satisfy the following compatibility equation:

$$\int_{\partial U} \varphi(z) h(z) dz = \int_U h(z) \phi(z) d\bar{z} \wedge dz, \tag{10.8}$$

for every function $h$ holomorphic on a neighborhood of $\bar{U}$.

In the homogeneous case, $\phi = 0$, the discussion is about which functions $\varphi \in C(\partial U)$ are the boundary values of continuous functions on $\bar{U}$ and holomorphic on $U$. The necessary condition is

$$\int_{\partial U} \varphi(z) h(z) dz = 0, \tag{10.9}$$

for every function $h$ holomorphic on a neighborhood of $\bar{U}$.

Now, by Runge's theorem, these functions $h$ are uniform limits on $\bar{U}$ of rational functions with (simple) poles in $\mathbb{C} \setminus \bar{U}$, so that condition (10.9) is equivalent to

$$\int_{\partial U} \frac{\varphi(z)}{z - w} dz = 0, \quad w \notin \bar{U}. \tag{10.10}$$

It will be shown now that this condition is a sufficient one for the existence of a solution of this holomorphic Dirichlet problem.

**Theorem 10.43.** *A function $\varphi \in C(\partial U)$ has a continuous extension to $\bar{U}$, holomorphic on $U$, if and only if the equivalent conditions* (10.9) *or* (10.10) *are satisfied.*

*Proof.* We just need to prove that the conditions are sufficient. The desired function, $F \in C(\bar{U}) \cap H(U)$ with $F_{|\partial U} = \varphi$, must be, by the Cauchy integral formula,

$$F(z) = \frac{1}{2\pi i} \int_{\partial U} \frac{\varphi(w)}{w - z} dw.$$

Using (10.10) we have to show that $\lim_{\substack{z \to a \\ z \in U}} F(z) = \varphi(a)$, for $a \in \partial U$. Every point $z \in U$, close enough to $\partial U$, has a well-defined projection $\pi(z) \in \partial U$, in the sense that $\pi(z)$ is the point of $\partial U$ nearest to $z$. That is,

$$d(z, \partial U) = \min\{|z - w| : w \in \partial U\} = |z - \pi(z)|$$

and this point is unique by the regularity of $\partial U$. Denote by $z^*$ a point located on the exterior normal to $\partial U$ at the point $\pi(z)$, with a distance of $d(z, \partial U)$ from $\pi(z)$ (Figure 10.4). Then $z^* \notin \bar{U}$ and one has

$$\int_{\partial U} \frac{\varphi(w)}{w - z^*} dw = 0.$$

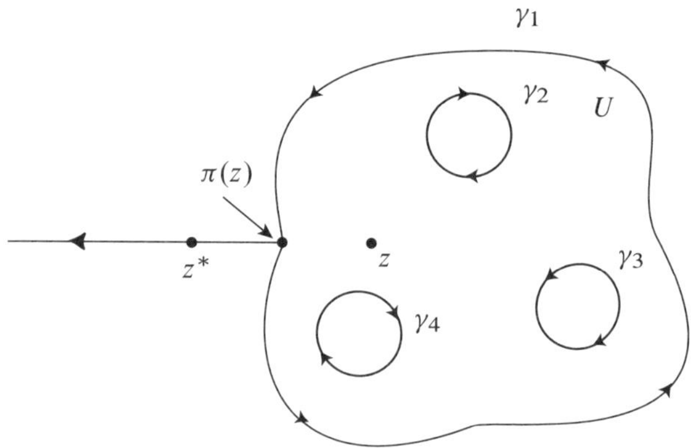


Figure 10.4

Consequently,

$$F(z) = \frac{1}{2\pi i} \int_{\partial U} \varphi(w) \left( \frac{1}{w - z} - \frac{1}{w - z^*} \right) dw$$

$$= \frac{1}{2\pi i} \int_{\partial U} \varphi(w) \frac{z - z^*}{(w - z)(w - z^*)} dw.$$

Consider now the kernel

$$P(z, w) = \frac{z - z^*}{(w - z)(w - z^*)}$$

and notice the following properties:

a) $\frac{1}{2\pi i} \int_{\partial U} P(z, w)dw = 1$ if $z \in U$,

b) $|P(z, w)| \leq c \frac{d(z, \partial U)}{|z - w|^2}$ if $w \in \partial U, z \in U$,

c) $\int_{\partial U} |P(z, w)||dw| \leq c$ if $z \in U$,

where $c$ is a constant.

The first one is a consequence of equalities

$$\frac{1}{2\pi i} \int_{\partial U} \frac{dw}{w - z} = 1, \quad \frac{1}{2\pi i} \int_{\partial U} \frac{dw}{w - z^*} = 0.$$

For the second one, observe that

$$|z - w| \leq d(z, \partial U) + |\pi(z) - w| \leq 3|z - w| \tag{10.11}$$

(because $|\pi(z) - w| \leq d(z, \partial U) + |z - w| \leq 2|z - w|$), and the same holds for $z^*$. Then,

$$|z^* - w| \geq \frac{1}{3} (d(z, \partial U) + |\pi(z) - w|) \geq \frac{1}{3}|z - w|,$$

proving b). Property c) is a consequence of b) (Exercise 21 of Section 10.7).

These properties are analogous to the ones of the Poisson kernel of the disc (Subsection 7.6.1). Now the proof follows the pattern of the solution of Dirichlet's problem in the disc (indeed, $F$ is also the solution of Dirichlet's problem). Write

$$F(z) - \varphi(a) = \frac{1}{2\pi i} \int_{\partial U} (\varphi(w) - \varphi(a)) P(z, w)dw$$

and note that, due to the continuity of $\varphi$, given $\varepsilon > 0$ there is a $\delta > 0$ such that $|\varphi(w) - \varphi(a)| < \varepsilon$, if $|w - a| < \delta$, $w \in \partial U$. The contribution of $\{w : |w - a| < \delta\}$ to the integral above is then, in absolute value, bounded by

$$\frac{\varepsilon}{2\pi} \int_{\partial U} |P(z, w)|dw \leq \frac{c\varepsilon}{2\pi}.$$

The contribution of $\{w : |w - a| \geq \delta\}$ is dominated by

$$\frac{2\|\varphi\|_\infty}{2\pi} \int_{|w-a|\geq\delta} |P(z, w)||dw| \leq \frac{\|\varphi\|_\infty d(z, \partial U)}{\pi} \int_{|w-a|\geq\delta} \frac{|dw|}{|z - w|^2}. \tag{10.12}$$

If $|z - a| < \delta/2$ and $|w - a| \geq \delta$, then $|w - z| \geq \delta/2$ and the right-hand side term of (10.12) is less than or equal to $C\delta^{-2}d(z, \partial U)$ with $C$ constant. Thus it is smaller than $\varepsilon$ if $z$ is close enough to the point $a$. $\qquad\square$

If $\bar{U}$ is simply connected (that is, $\mathbb{C} \setminus \bar{U}$ is connected), we know that every holomorphic function on a neighborhood of $\bar{U}$ is the uniform limit on $\bar{U}$ (therefore, on $\partial U$) of a sequence of polynomials. Then condition (10.9) becomes

$$\int_{\partial U} \varphi(z) z^n \, dz = 0, \quad n = 0, 1, 2, \ldots$$

as in the case of the disc (Subsection 7.6.1)

To finish, the non-homogeneous Dirichlet's type problem will be dealt with for the operator $\bar{\partial}$. If the data $\varphi \in C(\partial U)$ and $\phi \in C(\bar{U})$ satisfy the compatibility condition (10.8), choose first $f = C(\phi)$, which is a continuous function on $\bar{U}$ and satisfies $\bar{\partial} f = \phi$ in the weak sense. If $h$ is holomorphic on a neighborhood of $\bar{U}$, one has

$$\int_{\partial U} (\varphi(z) - f(z)) h(z) \, dz = \int_U h \phi \, d\bar{z} \wedge dz - \int_{\partial U} f(z) h(z) \, dz.$$

Now,

$$\begin{aligned}
\int_{\partial U} f(z) h(z) \, dz &= \int_{\partial U} \left( \frac{1}{\pi} \int_U \frac{\phi(w)}{z - w} \, dm(w) \right) h(z) \, dz \\
&= \frac{1}{\pi} \int_U \phi(w) \int_{\partial U} \frac{h(z)}{z - w} \, dz \, dm(w) \\
&= 2i \int_U \phi(w) h(w) \, dm(w) \\
&= \int_U h \phi \, d\bar{w} \wedge dw.
\end{aligned}$$

As a consequence, the function $\varphi - f$ satisfies condition (10.9), and by Theorem 10.43 it is the value on $\partial U$ of some $F \in C(\bar{U})$, $F$ holomorphic on $U$. With all this, $f + F$ is the solution of the non-homogeneous problem, because $f + F = \varphi$ on $\partial U$ and $\bar{\partial}(f + F) = \bar{\partial} f = \phi$ on $U$.

Hence, the following result has been proved:

**Theorem 10.44.** *Given the functions $\phi \in C(\bar{U})$ and $\varphi \in C(\partial U)$, the problem $\bar{\partial} f = \phi$ on $U$ and $f = \varphi$ on $\partial U$ has a solution $f \in C(\bar{U})$ if and only if $\phi$ and $\varphi$ satisfy the compatibility condition (10.8).*

Once again, if $\bar{U}$ is simply connected, it is enough to impose (10.8) to polynomials $h(z) = z^n$, $n \in \mathbb{N}$, so that (10.8) is equivalent to

$$\int_{\partial U} \varphi(z) z^n \, dz = \int_U \phi(z) z^n \, d\bar{z} \wedge dz, \quad n = 0, 1, 2, \ldots.$$

**Example 10.45.** If $\phi(z) = \bar{z}$ and $U$ is the unit disc $\mathbb{D}$, the condition on $\varphi$ is

$$\int_{\partial\mathbb{D}} \varphi(z) z^n \, dz = \int_{\mathbb{D}} \bar{z} z^n \, d\bar{z} \wedge dz = 2i \int_0^1 \int_0^{2\pi} r^{n+2} e^{i(n-1)\theta} \, dr \, d\theta,$$

for $n = 0, 1, 2, \ldots$. This last integral is $0$ for $n \neq 1$, and $\pi i$ for $n = 1$. If $\varphi(e^{i\theta}) = \sum_{k=-\infty}^{+\infty} \hat{\varphi}(k) e^{ik\theta}$ is the Fourier series expansion of $\varphi$, the integral in the left-hand side is $2\pi i \hat{\varphi}(-n-1)$. Hence, $\hat{\varphi}(-1) = 0$, $\hat{\varphi}(k) = 0$ for $k \leq -3$ and $\hat{\varphi}(-2) = \frac{1}{2}$, that is,

$$\varphi(e^{i\theta}) = \frac{1}{2} e^{-2i\theta} + \sum_{k=0}^{\infty} \hat{\varphi}(k) e^{ik\theta}.$$

The extension $f$ of $\varphi$ with $\bar{\partial} f = \phi$ is

$$f(z) = \frac{1}{2} \bar{z}^2 + \sum_{k=0}^{\infty} \hat{\varphi}(k) z^k. \qquad \square$$

Unlike Dirichlet's problem for the Laplacian, it makes no sense in general, to set the problem

$$\bar{\partial} f = \phi \text{ on } U, \quad f = 0 \text{ on } \partial U.$$

This problem just has a (unique) solution if $\phi$ satisfies the condition

$$\int_U \phi h \, dm = 0, \quad \text{for } h \text{ holomorphic on a neighborhood of } \bar{U}.$$

Instead of imposing the vanishing of the solution $f$ on $\partial U$, it is most natural to look for "the smallest possible one" in the sense, for example, of minimizing the value of the integral

$$\int_{\partial U} |f(z)|^2 |dz|$$

among all the solutions of $\bar{\partial} f = \phi$. An example of a solution of this problem in the case of the unit disc is Theorem 10.36.

## 10.7 Exercises

**1.** Decide whether there can exist a sequence of polynomials $(P_n)$ such that

$$P_n(z) \xrightarrow[n\to\infty]{} 1 \quad \text{if } \operatorname{Im} z > 0, \quad P_n(x) \xrightarrow[n\to\infty]{} 0 \quad \text{if } x \in \mathbb{R},$$

$$P_n(z) \xrightarrow[n\to\infty]{} -1 \quad \text{if } \operatorname{Im} z < 0.$$

**2.** Let $0 < a < b$ be fixed. For each $n \in \mathbb{N}$ construct a polynomial $P_n$ such that, for $|z| < n$,

$$|P_n(z)| \leq \frac{1}{n} \quad \text{if } \operatorname{Im} z \leq 0 \text{ or } \operatorname{Im} \geq b,$$

$$|P_n(z)| \geq n \quad \text{if } \operatorname{Im} z = a$$

holds. Use $(P_n)$ to find a sequence of polynomials converging pointwise to zero on $\mathbb{C}$ and uniformly on every compact set of $\mathbb{C} \setminus \mathbb{R}$, but not converging uniformly on any neighborhood of any point of the real axis.

**3.** Let $f$ be a continuous function on the disc $\bar{D}(a,r)$, $a \in \mathbb{C}$, $r > 0$ and holomorphic on $D(a,r)$. Show that $f(z)$ is the uniform limit on $\bar{D}(a,r)$ of a sequence of polynomials in $z - a$. Prove also that if $f$ is continuous on $\mathbb{C} \setminus D(a,r)$ and holomorphic on $\mathbb{C} \setminus \bar{D}(a,r)$, then $f(z)$ is the uniform limit on $\mathbb{C} \setminus D(a,r)$ of a sequence of polynomials in $1/(z-a)$.

**4.** Let $U$ be a bounded domain of the plane such that $\partial U$ is a finite union of circles. Show that every continuous function on $\bar{U}$ and holomorphic on $U$ can be uniformly approximated on $\bar{U}$ by rational functions with poles outside $\bar{U}$. What are the continuous functions on $\bar{U}$ that can be uniformly approximated on $\bar{U}$ by polynomials?

*Hint:* If $C(z_0, r_0)$ is the exterior boundary of $U$ and $C(z_i, r_i)$, $i = 1, \ldots, m$, the other components of $\partial U$, prove that a function $f$ under the hypotheses of the problem may be decomposed as $f = f_0 + f_1 + \cdots + f_m$ in such a way that each $f_i$ satisfies some of the hypotheses of Exercise 3 of this section with respect to $D(z_i, r_i)$.

**5.** Let $U = \{z : |z| < 1 \text{ and } |2z - 1| \geq 1\}$ and $f \in H(U)$. Answer the following questions:

   i) Are there polynomials $(P_n)_{n \in \mathbb{N}}$ such that $P_n \to f$ uniformly on compact sets of $U$?

   ii) Are there polynomials $(P_n)_{n \in \mathbb{N}}$ such that $P_n \to f$ uniformly on $U$?

   iii) What can be said if $f$ is holomorphic on a neighborhood of $\bar{U}$?

**6.** Let $K \subset U$, $K$ a compact set, $U$ an open set of $\mathbb{C}$. Define

$$\hat{K} = \hat{K}_U = \{z \in U : |f(z)| \leq \sup_{w \in K} |f(w)| \text{ for any function } f \in H(U)\}.$$

Prove the following facts:

   i) $\hat{K}$ is a compact set, $\hat{\hat{K}} = \hat{K}$ and $\hat{K}$ is contained in the convex hull of $K$.

ii) $\hat{K}$ is the union of $K$ and the connected components of $U \setminus K$ that are relatively compact in $U$.

**7.** Let $U_1 \subset U_2$ be two open sets of $\mathbb{C}$. Show the following statements are equivalent:

i) Every function $f \in H(U_1)$ can be approximated by functions in $H(U_2)$, uniformly on compact sets of $U_1$.

ii) If $U_2 \setminus U_1 = L \cup F$ with $L$ a compact set, $F$ closed in $U_2$ and $L \cap F = \emptyset$, then one must have $L = \emptyset$.

iii) For each $K \subset U_1$, $K$ compact, one has $\hat{K}_{U_2} = \hat{K}_{U_1}$.

iv) For each $K \subset U_1$, $K$ compact, one has $\hat{K}_{U_2} \cap U_1 = \hat{K}_{U_1}$.

v) For each $K \subset U_1$, $K$ compact, then $\hat{K}_{U_2} \cap U_1$ is compact.

**8.** Construct a meromorphic function on the unit disc with simple poles and residue 1 at the points $z_{nk} = (1 - 2^{-n})e^{\frac{2\pi i}{n}k}$, $k = 0, 1, \ldots, n - 1$ and $n = 0, 1, \ldots$.

**9.** Show that $f(z) = \sum_{n=-\infty}^{\infty} \frac{1}{z^3 - n^3}$ defines a meromorphic function on the whole plane. Find the poles of $f$ and principal parts at each pole. Write $f$ in terms of trigonometric functions.

**10.** Prove the formula

$$\pi \, \mathrm{tg}(\pi z) = 2z \sum_{n=1}^{\infty} \frac{1}{\left(n - \frac{1}{2}\right)^2 - z^2} \quad \text{for } z \neq k + \frac{1}{2}, \, n \in \mathbb{Z}.$$

**11.** For $a, b \in \mathbb{R}$, $a \notin \mathbb{Z}$, $b \neq 0$, compute

$$\sum_{-\infty}^{+\infty} \frac{(-1)^n}{(n + a)^2} \quad \text{and} \quad \sum_{-\infty}^{+\infty} \frac{(-1)^n}{n^2 + b^2}.$$

**12.** Prove the equality

$$\sum_{n=0}^{\infty} \frac{(-1)^n}{(2n + 1)^3} = \frac{\pi^3}{32}.$$

**13.** Prove the formulae

a) $\dfrac{\sin \alpha z}{\sin \pi z} = \dfrac{2}{\pi} \sum_{n=1}^{\infty} (-1)^n \dfrac{n \sin n\alpha}{z^2 - n^2}$,

b) $\dfrac{\cos \alpha z}{\cos \pi z} = \dfrac{1}{\pi z} + \dfrac{2z}{\pi} \sum_{n=1}^{\infty} (-1)^n \dfrac{\cos n\alpha}{z^2 - n^2}$

with $\alpha \in \mathbb{R}$ and $\alpha \neq k\pi$, $k$ integer, in item a).

**14.** Let $U = \bigcup_{j=1}^{\infty} U_j$ with $U$, $U_j$ open sets and $f_j$ a meromorphic function on $U_j$ for each $j = 1, 2, \ldots$ such that $f_j - f_k \in H(U_j \cap U_k)$ for every couple $j, k$. Show that there exists a function $f$, meromorphic on $U$, such that $f - f_j \in H(U_j)$ for $j = 1, 2, \ldots$.

**15.** Let $U = \bigcup_{j=1}^{\infty} U_j$ with $U$, $U_j$ open sets and $g_{jk} \in H(U_j \cap U_k)$ for $j, k = 1, 2, \ldots$ such that the following conditions hold:

$$g_{jk} = -g_{kj}, \quad g_{jk} + g_{kl} + g_{lj} = 0 \quad \text{on } U_j \cap U_k \cap U_l \text{ for all } j, k, l \in \mathbb{N}.$$

Prove there exist functions $g_j \in H(U_j)$ such that

$$g_{jk} = g_k - g_j \quad \text{on } U_j \cap U_k, \ j, k = 1, 2, \ldots.$$

**16.** For $\phi \in C^1(\mathbb{D})$ bounded on $\mathbb{D}$ show directly that the Cauchy integral of $\phi$, $C(\phi)(z) = \frac{1}{\pi} \int_{\mathbb{D}} \frac{\phi(w)}{w-z}\, dw$, is continuous on $\mathbb{C}$, is of class $C^2$ on $\mathbb{D}$ and satisfies $\bar{\partial} C(\phi) = 0$ on $\mathbb{D}$.

**17.** Generalize Propositions 10.33 and 10.34 taking

$$H(z, w) = \varphi\left(\frac{1 - |w|^2}{1 - z\bar{w}}\right),$$

where $\varphi$ is a holomorphic function on the half-plane $\Pi = \{z : \mathrm{Re}\, z > 0\}$ and continuous on $\bar{\Pi}$ satisfying $\varphi(1) = 1$. Show that under appropriate hypotheses on $\phi \in C(\mathbb{D})$, the function

$$u(z) = \frac{1}{\pi} \int_{\mathbb{D}} \varphi\left(\frac{1 - |w|^2}{1 - z\bar{w}}\right) \frac{\phi(w)}{z - w}\, dm(w)$$

is a solution of the equation $\bar{\partial} u = \phi$ on $\mathbb{D}$.

**18.** Generalize Proposition 10.40 taking $H(z, w) = \varphi(z\bar{w} - |w|^2)$ with $\varphi$ an entire function such that $\varphi(0) = 1$.

**19.** Prove the following decomposition formula, accompanying the formula of Proposition 10.40: if $f \in C^1(\mathbb{C})$ and $f$ and $\bar{\partial} f$ satisfy appropriate conditions (set which ones), then one has

$$f(z) = c \int_{\mathbb{C}} e^{z\bar{w} - |w|^2} f(w)\, dm(w) + \frac{1}{\pi} \int_{\mathbb{C}} e^{z\bar{w} - |w|^2} \frac{\bar{\partial} f(w)}{z - w}\, dm(w),$$

$c$ a constant. What is the value of the constant $c$?

**20.** If $f \in L^1(\mathbb{D})$ and $k \in \mathbb{N}$ define

$$T_k f(z) = \frac{k}{\pi} \int_{\mathbb{D}} \frac{(1 - |w|^2)^{k-1}}{(1 - z\bar{w})^{k+1}} f(w) \, dm(w) \quad \text{for } z \in \mathbb{D}.$$

a) Show that for $f \in L^1(\mathbb{D}) \cap H(\mathbb{D})$ one has

$$T_k f(z) = f(z), \quad T_k \bar{f}(z) = \overline{f(0)}, \quad z \in \mathbb{D}.$$

b) Show there is a constant $c > 0$ such that

$$\|T_k f\|_2 \le c \, \|f\|_2$$

for every function $f \in L^2(\mathbb{D})$.

*Hint:* Compute $T_k f(z)$ with $f(w) = w^n, n = 0, 1, 2, \ldots$.

**21.** Prove that property c) of the kernel $P(z, w)$ is a consequence of b), as stated in the proof of Theorem 10.43.

# Chapter 11
# Zeros of holomorphic functions

The first part of this chapter is devoted to the construction of holomorphic functions with prescribed zeros, either on the whole plane or on any domain. The solution of this problem is given by the Weierstrass factorization theorem, expressing the desired functions as an infinite product of elementary factors. For this reason properties of infinite products are studied in detail, both of complex numbers and of functions. Also, Weierstrass' theorem is used to interpolate a sequence of numbers by an entire function. This problem can also be solved using the existence of solutions of a non-homogeneous Cauchy–Riemann equation, proved in Chapter 10.

Next, the relationship between the zeros of holomorphic functions and the Poisson equation is analyzed. It turns out that having a method to solve the Poisson equation gives a procedure to construct holomorphic functions with prescribed zeros. In the case of the unit disc, the explicit formula to solve the Poisson equation leads to the so-called Jensen formula. This provides a relationship between the distribution of the zeros of a holomorphic function on the disc and its behavior on the unit circle. Jensen's formula is used to study the distribution of the zeros of an entire function of finite order.

At the end of the chapter we deal with the closed ideals of the algebra of holomorphic functions on a domain of the plane. The ideals of this algebra are closely related to the zero sets of holomorphic functions, and so their description depends on the results previously obtained.

## 11.1 Infinite products

The definition of a convergent infinite product given below, suggested by the one of convergent series, seems quite natural.

**Definition 11.1.** Given a sequence $(p_n)_{n=1}^{\infty}$ of complex numbers, the infinite product $\prod_{n=1}^{\infty} p_n$ is said to converge to $P$ and it is written $\prod_{n=1}^{\infty} p_n = P$, if the sequence of partial products $P_n = \prod_{k=1}^{n} p_k$ converges to $P$, when $n \to \infty$.

**Example 11.2.** $\prod_{n=2}^{\infty} \left(1 - \frac{1}{n^2}\right) = \frac{1}{2}$ because it is easy to check that $P_n = \frac{1}{2}\frac{n+1}{n}$.
$\qquad\square$

Observe that with the definition above, if there is a term $p_n = 0$, then the infinite product converges to zero. It is clear as well that if the product $\prod_{n=1}^{\infty} p_n$ converges to $P$, then $\prod_{n=1}^{\infty} |p_n|$ converges to $|P|$. The convergence of $\prod_{n=1}^{\infty} r_n$ with $r_n > 0$ is clearly equivalent to the condition

$$-\infty \le \sum_{n=1}^{\infty} \operatorname{Log} r_n < \infty$$

and, in this case, one has

$$\prod_{n=1}^{\infty} r_n = \exp \sum_{n=1}^{\infty} \operatorname{Log} r_n,$$

interpreting that $\exp(-\infty)$ is $0$. Note that we can have an infinite product $\prod_{n=1}^{\infty} p_n$ with $p_n \ne 0$ for all $n$ such that it converges and the product vanishes: this is so if $\sum_{n=1}^{\infty} \operatorname{Log} |p_n| = -\infty$. For example, the product

$$\prod_{n=2}^{\infty} \left(1 - \frac{1}{n}\right)$$

has this property because $\operatorname{Log}\left(1 - \frac{1}{n}\right) \sim -\frac{1}{n}$ and, hence, $\sum_{n=1}^{\infty} \operatorname{Log}\left(1 - \frac{1}{n}\right) = -\infty$. In this case, one may consider the partial products and it turns out that

$$P_n = \frac{1}{2} \cdot \frac{2}{3} \cdot \frac{3}{4} \cdot \ldots \cdot \frac{n-1}{n} = \frac{1}{n} \to 0, \quad n \to \infty.$$

Since $p_n = \frac{P_n}{P_{n-1}}$, it follows that if $\prod_{n=1}^{\infty} p_n$ converges to $P \ne 0$, then $p_n \to 1$, when $n \to \infty$. But a product can converge to zero without having $p_n \to 1$; for instance, this happens if $p_n = p$, for every $n \ge 1$ with $0 < p < 1$.

A finite product $\prod_{n=1}^{N} p_n$ is zero if and only if some of its factors are zero. For a finite product of polynomials, $P(z) = \prod_{n=1}^{N} p_n(z)$, a number is a zero of $P$ if and only if it is a zero of some of the factors $p_n$ and the multiplicity it has as a zero of $P$ is the sum of its multiplicities as a zero of each $p_n$. When studying zeros and multiplicities of an infinite product of functions, one would like this property to hold as well; for an infinite product of numbers we would like that

$$\prod_{n=1}^{\infty} p_n = 0 \iff \text{a finite number of factors are zero.} \tag{11.1}$$

In particular, in an infinite product $\prod_{n=1}^{\infty} p_n = P$ we would like that $p_n \ne 0$ for every $n \ge 1$ entails $P \ne 0$.

As usual, $\operatorname{Log}$ denotes the principal branch of the logarithm, corresponding to the principal branch of the argument, $-\pi < \operatorname{Arg} z \le \pi$.

**Proposition 11.3.** *If $(p_n)_{n=1}^{\infty}$ is a sequence of complex numbers with $p_n \neq 0$ for all $n$, then the infinite product $\prod_{n=1}^{\infty} p_n$ converges to $P \neq 0$ if and only if the series $\sum_{n=1}^{\infty} \operatorname{Log} p_n = S$ is convergent and, in this case, one has $P = e^S$.*

*Proof.* It is clear that if the series $\sum_{n=1}^{\infty} \operatorname{Log} p_n$ converges to $S$, then $\prod_{n=1}^{\infty} p_n$ converges to $P = e^S$. The converse is obvious if $p_n > 0$, as said above, but for complex $p_n$ a proof is needed. Let $S_n = \sum_{k=1}^{n} \operatorname{Log} p_k$ and assume $e^{S_n} = P_n = \prod_{k=1}^{n} p_k$ converges to $P \neq 0$; we must show that the sequence $S_n$ is convergent. This is not true for any sequence $S_n$ (for example, if $S_n = 2\pi n i$) and the additional information $S_{n+1} - S_n = \operatorname{Log} p_{n+1} \to 0$ when $n \to \infty$ (because $p_n \to 1$) must be used.

With the previous notation one has $\operatorname{Log} P_n = S_n + 2\pi i k_n$ for $k_n \in \mathbb{Z}$ and

$$
\begin{aligned}
2\pi i (k_{n+1} - k_n) &= \operatorname{Log} P_{n+1} - \operatorname{Log} P_n + S_n - S_{n+1} \\
&= \operatorname{Log} P_{n+1} - \operatorname{Log} P_n - \operatorname{Log} p_{n+1}.
\end{aligned}
\tag{11.2}
$$

If $P$ is not a negative real number, then $\operatorname{Log} P_n \to \operatorname{Log} P$ and, since $p_n \to 1$, the right-hand side term of (11.2) approaches zero when $n \to \infty$. Therefore, $k_n$ is constant for $n$ big enough and then $S_n = \operatorname{Log} P_n - 2\pi i k_n$ is convergent. If $P$ is real and negative, one can argue similarly with the branch of the logarithm given by $0 < \arg z < 2\pi$ and one gets $\log P_n = S_n + 2\pi i l_n$ with $l_n \in \mathbb{Z}$. This yields

$$
\begin{aligned}
2\pi i (l_{n+1} - l_n) &= \log P_{n+1} - \log P_n + S_n - S_{n+1} \\
&= \log P_{n+1} - \log P_n - \operatorname{Log} p_{n+1}
\end{aligned}
$$

and one concludes as before that $S_n$ converges. $\qquad\square$

If $\prod_{n=1}^{\infty} p_n$ is an infinite product with $\lim_{n\to\infty} p_n = 1$, then $p_n \neq 0$ for $n \geq n_0$, and if $\sum_{n \geq n_0} \operatorname{Log} p_n$ converges to $S$, we have

$$
\prod_{n=1}^{\infty} p_n = e^S \prod_{n=1}^{n_0-1} p_n
$$

and property (11.1) holds. In addition to this property, one also wants that infinite products are *unconditionally convergent*, that is, their character and their product are independent of the order of the factors. The following proposition gives a class of infinite products enjoying both desired properties, large enough for applications.

**Proposition 11.4.** a) *Let $(p_n)_{n=1}^{\infty}$ be a sequence of complex numbers such that $\lim_{n\to\infty} p_n = 1$ and $\sum_{n \geq n_0} |\operatorname{Log} p_n| < +\infty$ for some $n_0 \in \mathbb{N}$. Then $\prod_{n=1}^{\infty} p_n$ is an unconditionally convergent infinite product satisfying property (11.1).*

b) *Let $p_n \colon A \to \mathbb{C}$, $n \in \mathbb{N}$, be bounded functions of a variable $\lambda \in A$, where $A$ is an arbitrary set, such that $p_n(\lambda) \to 1$ uniformly on $A$ when $n \to \infty$, and assume* *the series $\sum_{n \geq n_0} |\operatorname{Log} p_n(\lambda)|$ is uniformly convergent on $A$ for some $n_0 \in \mathbb{N}$. Then the infinite product*

$$P(\lambda) = \prod_{n=1}^{\infty} p_n(\lambda)$$

*is uniformly convergent on $A$, that is, writing $P_n(\lambda) = \prod_{n=1}^{n} p_n(\lambda)$, the limit*

$$P(\lambda) = \lim_{n \to \infty} P_n(\lambda)$$

*exists uniformly on $A$. Moreover, property (11.1) holds, that is, $P(\lambda) = 0$ if and only if $p_n(\lambda) = 0$ for a finite number of values of $n$. Finally, equality $P(\lambda) = \prod_{k=1}^{\infty} p_{n_k}(\lambda)$ also holds, uniformly on $A$, for any permutation $k \to n_k$ of $\mathbb{N}$.*

*Proof.* The series $\sum_{n \geq n_0} \operatorname{Log} p_n$ converges, since it converges absolutely; therefore, $\prod_{n=1}^{\infty} p_n$ is convergent to $\exp\left(\sum_{n \geq n_0} \operatorname{Log} p_n\right) \prod_{n=1}^{n_0-1} p_n$. The same is true changing the order of the terms: $\prod_{k=1}^{\infty} p_{n_k}$ is convergent for any permutation $k \to n_k$. If some factor $p_n$ is zero, both infinite products above are zero; else they are also equal because $\sum_n \operatorname{Log} p_n = \sum_k \operatorname{Log} p_{n_k}$, since absolutely convergent series are unconditionally convergent. Now it remains to show that under the hypothesis of uniform convergence of the series $\sum_{n \geq n_0} |\operatorname{Log} p_n(\lambda)|$ on $A$ in b), the equality

$$P(\lambda) = \exp \sum_{n \geq n_0} \operatorname{Log} p_n(\lambda) \prod_{n=1}^{n_0-1} p_n(\lambda),$$

holds uniformly on $A$. Writing $S_n(\lambda) = \sum_{k=n_0}^{n} \operatorname{Log} p_k(\lambda)$, the hypothesis implies

$$|S_n(\lambda)| \leq \sum_{k=n_0}^{n} |\operatorname{Log} p_k(\lambda)| \leq M,$$

for a constant $M < +\infty$. Now the uniform convergence of $S_n$, say to $S$, yields the uniform convergence of $\exp(S_n)$ to $\exp(S)$ because

$$|\exp(S_n(\lambda)) - \exp(S(\lambda))| \leq c\,|S_n(\lambda) - S(\lambda)|,$$

$c$ being a constant such that $|\exp(z_1) - \exp(z_2)| \leq c\,|z_1 - z_2|$ if $|z_1|, |z_2| \leq M$. $\qquad \square$

The way Proposition 11.4 is used is explained in the following theorem.

**Theorem 11.5.** *Let $U$ be a domain of $\mathbb{C}$ and $f_n \in H(U)$, $n = 1, 2, \ldots$. Assume $f_n(z) \to 1$, when $n \to \infty$, uniformly on compact sets of $U$ and the series $\sum_{n \geq n_0(K)} |\operatorname{Log} f_n(z)|$ is uniformly convergent on each compact set $K \subset U$, for*

*$n_o(K)$ that may depend on $K$. Then the infinite product*

$$F(z) = \prod_{n=1}^{\infty} f_n(z)$$

*converges uniformly on compact sets of $U$, $F \in H(U)$, $Z(F) = \bigcup_n Z(f_n)$ and*

$$m(F, z) = \sum_{n=1}^{\infty} m(f_n, z), \quad \text{for all } z \in U.$$

*Proof.* On each compact set $K \subset U$ we can apply Proposition 11.4 to get that the infinite product converges uniformly on compact sets of $U$. Each partial product $F_n = \prod_{k=1}^{n} f_k$ is a holomorphic function, and Theorem 9.3 gives $F \in H(U)$. Since the infinite product satisfies property (11.1), $Z(F) = \bigcup_{n=1}^{\infty} Z(f_n)$, that is, a point $a \in U$ is a zero of $F$ if and only if $a$ is a zero of some function $f_n$ (for a finite number of values of $n$, because $f_n(z) \to 1$). Fix now a disc $\bar{D}(a.r) \subset U$; we will have $f_n(z) \neq 0$, if $n \geq n_0$ and $z \in D(a, r)$ and, thus, the function $\prod_{k=n_0+1}^{\infty} f_k(z) = G(z)$ is holomorphic without zeros on $D(a, r)$. Then, $F(z) = \prod_{n=1}^{n_0} f_n(z) G(z)$ and, clearly,

$$m(F, a) = \sum_{n=1}^{n_0-1} m(f_n, a) = \sum_{n=1}^{\infty} m(f_n, a). \qquad \square$$

The function $\mathrm{Log}(1 + w)$ has a power series expansion around the origin: $w - \frac{1}{2}w^2 + \cdots$; so, $|\mathrm{Log}(1 - w)| \sim |w|$ for $|w| \to 0$. Using this estimate the following statement equivalent to Proposition 11.4 and Theorem 11.5 is obtained.

**Theorem 11.6.** a) *If $\sum_{n=1}^{\infty} |1 - p_n| < +\infty$, then $\prod_{n=1}^{\infty} p_n$ is an unconditionally convergent infinite product for which property (11.1) holds.*

b) *If $U$ is a domain of $\mathbb{C}$ and $f_n \in H(U)$, $n = 1, 2, \ldots$, are functions such that the series $\sum_n |1 - f_n(z)|$ is uniformly convergent on compact sets of $U$, then the infinite product $F(z) = \prod_{n=1}^{\infty} f_n(z)$ has the properties of Theorem 11.5.*

In the previous chapters we have often used the fact that the logarithmic derivative $f'/f$ of $f \in H(U)$ is a meromorphic function on $U$ with simple poles at the zero set of $f$ and residues given by the multiplicity of these zeros. For a finite product $F = f_1 \cdots f_n$, one has

$$\frac{F'}{F} = \sum_{i=1}^{n} \frac{f_i'}{f_i},$$

that is, the logarithmic derivative is the sum of logarithmic derivatives. Now let us show that this result also holds for infinite products.

**Theorem 11.7.** *Let $U$ be a domain of $\mathbb{C}$, $f_n \in H(U)$, $n = 1, 2, \ldots$ such that $\sum_{n=1}^{\infty} |1 - f_n(z)|$ is uniformly convergent on compact sets of $U$, and define $F(z) = \prod_{n=1}^{\infty} f_n(z)$. Then the equality*

$$\frac{F'(z)}{F(z)} = \sum_{n=1}^{\infty} \frac{f_n'(z)}{f_n(z)}$$

*holds, uniformly on compact sets of $U$.*

*Proof.* Fix a compact set $K \subset U$; the hypothesis implies that $f_n(z) \to 1$ uniformly on $K$, and so

$$|1 - f_n(z)| \le \frac{1}{2}, \quad \text{for each } z \in K, \; n \ge n_0(K).$$

Therefore, $|f_n(z)| \ge \frac{1}{2}$, for $z \in K$, if $n \ge n_0 = n_0(K)$. Hence, the terms of the series $\sum \frac{f_n'}{f_n}$, for $n$ big enough, are holomorphic functions on a neighborhood of $K$. Let, as before,

$$F(z) = \prod_{n=1}^{n_0} f_n(z) G(z)$$

with $G(z) = \prod_{n_0+1}^{\infty} f_n(z)$. We then have

$$\frac{F'}{F} = \sum_{n=1}^{n_0} \frac{f_n'}{f_n} + \frac{G'}{G}$$

and, thus, it is enough to prove that the equality $\frac{G'}{G} = \sum_{n_0+1}^{\infty} \frac{f_n'}{f_n}$ holds uniformly on $K$.

Consider the partial products

$$G_l(z) = \prod_{n_0+1}^{n_0+l} f_n(z)$$

converging uniformly to $G$ on $K$, when $l \to \infty$. Since $G(z) \ne 0$ if $z \in K$, there is a $\delta > 0$ such that $1/\delta > |G(z)| \ge \delta$ if $z \in K$, and, by the uniform convergence, one has $\frac{2}{\delta} > |G_l(z)| \ge \frac{\delta}{2}$ if $z \in K$, for $l$ big enough. On the other hand, the fact that $G_l' \to G'$ uniformly on $K$ and the inequalities

$$\left| \frac{G'(z)}{G(z)} - \frac{G_l'(z)}{G_l(z)} \right| = \frac{|G_l(z)G'(z) - G(z)G_l'(z)|}{|G(z)||G_l(z)|}$$

$$\le \frac{|G_l(z)||G'(z) - G_l'(z)| + |G_l'(z)||G_l(z) - G(z)|}{(\delta/2)^2}$$

$$\le c \cdot \left( |G'(z) - G_l'(z)| + |G(z) - G_l(z)| \right), \quad c \text{ constant,}$$

show that $\frac{G'_l}{G_l} \to \frac{G'}{G}$ uniformly on $K$, when $l \to \infty$. Finally observing that $\frac{G'_l}{G_l} = \sum_{n_0+1}^{n_0+l} \frac{f'_n}{f_n}$, the theorem is proved. $\qquad\square$

## 11.2 The Weierstrass factorization theorem

We want to construct an entire function with zero set some prescribed set of points $z_n \in \mathbb{C}$ and given multiplicities, $m_n \in \mathbb{N}$. The zero set $A = \{z_n : n \in \mathbb{N}\}$ is any closed set of isolated points (therefore, $|z_n| \to \infty$ when $n \to \infty$, if $A$ is not finite) and the sequence of multiplicities $(m_n)$ is arbitrary. Without loss of generality, suppose $z_n \neq 0$.

Of course, if $A = \{z_1, \ldots, z_N\}$ is a finite set, just take the polynomial

$$P(z) = \prod_{n=1}^{N} \left(1 - \frac{z}{z_n}\right)^{m_n}.$$

In general it is quite natural to consider an infinite product of the type

$$F(z) = \prod_{n=1}^{\infty} \left(1 - \frac{z}{z_n}\right)^{m_n}$$

whenever it satisfies the hypotheses of Theorem 11.6. Considering that the factor $\left(1 - \frac{z}{z_n}\right)$ is repeated $m_n$ times, we find that the suitable condition is the convergence of the series

$$\sum_{n=1}^{\infty} m_n \left|\frac{z}{z_n}\right|$$

uniformly on every disc of the plane, or equivalently

$$\sum_{n=1}^{\infty} \frac{m_n}{|z_n|} < +\infty.$$

For instance, the product $\prod_{n=1}^{\infty} \left(1 - \frac{z}{n^2}\right)$ provides an entire function vanishing once at each point $z_n = n^2$, $n = 1, 2, \ldots$. In general, the previous condition does not hold (for example, if $z_n = n$) and we need to introduce some factors $A_n(z)$, without zeros, correcting the infinite product so that

$$F(z) = \prod_{n=1}^{\infty} \left(1 - \frac{z}{z_n}\right)^{m_n} A_n(z) \tag{11.3}$$

is uniformly convergent on compact sets. Since the factors $A_n(z)$ cannot have zeros, we look for them in the form $A_n(z) = \exp(-B_n(z))$. To search for the functions

$B_n$, observe that by Theorem 11.7 the logarithmic derivative of $F$ will be

$$\frac{F'(z)}{F(z)} = \sum_{n=1}^{\infty} \left( \frac{m_n}{z - z_n} - Q_n(z) \right) \quad \text{with } Q_n = B_n'.$$

Accordingly, the functions $Q_n = B_n'$ should be correction terms, so that the previous series represents a meromorphic function with simple poles at the points $z_n$ and residues $m_n$.

This observation establishes a link with Mittag-Leffler's theorem. In the proof of Theorem 10.9, it has been shown that the most natural choice of correction terms $Q_n$ is to take a partial sum of the expansion of $\frac{m_n}{z-z_n}$ in power series on the disc $D(0, |z_n|)$. So they will be written as

$$Q_n(z) = -m_n \sum_{k=0}^{\lambda_n} \frac{z^k}{z_n^{k+1}}.$$

Notice now that equality $Q_n = B_n'$ holds taking

$$B_n = -m_n \sum_{k=1}^{\lambda_n+1} \frac{1}{k} \left( \frac{z}{z_n} \right)^k.$$

Therefore, define

$$A_n(z) = \exp \left\{ m_n \sum_{k=1}^{\lambda_n+1} \frac{1}{k} \left( \frac{z}{z_n} \right)^k \right\}$$

and look for conditions that numbers $\lambda_n$ should satisfy so that the hypotheses of Theorem 11.5 hold. Write $p_n = 1 + \lambda_n$ and define

$$E_p(z) = (1 - z) \exp \sum_{k=1}^{p} \frac{z^k}{k}, \quad p \in \mathbb{N},$$

which are the so-called *Weierstrass elementary factors*. So

$$F(z) = \prod_{n=1}^{\infty} \left( 1 - \frac{z}{z_n} \right)^{m_n} A_n(z) = \prod_{n=1}^{\infty} E_{p_n}^{m_n} \left( \frac{z}{z_n} \right).$$

Now we need an estimate of $|\operatorname{Log} E_p(z)|$. Except for an integer multiple of $2\pi i$, one has

$$\operatorname{Log} E_p(z) = \operatorname{Log}(1 - z) + \sum_{k=1}^{p} \frac{z^k}{k} = - \sum_{k=p+1}^{\infty} \frac{z^k}{k}, \quad \text{if } |z| < 1.$$

If $|z| < \frac{1}{2}$ the quantity above is bounded in absolute value by

$$\frac{1}{p+1} \sum_{k=p+1}^{\infty} |z|^k = \frac{1}{p+1} \frac{|z|^{p+1}}{1-|z|} < \frac{2}{p+1} |z|^{p+1} < 2^{-p}.$$

Given $r > 0$, assuming $|z| \le r$ and considering only points $z_n$ with $|z_n| \ge 2r$, we have

$$\sum_{n \ge n_0(r)} \left| \operatorname{Log} E_{p_n}^{m_n} \left( \frac{z}{z_n} \right) \right| \le 2 \sum_{|z_n| \ge 2r} \frac{m_n}{p_n + 1} \left( \frac{r}{|z_n|} \right)^{p_n+1},$$

for some $n_0(r)$ depending on $r$. Thus we have proved the following result in which the series

$$\sum_{|z_n| \ge 2r} \frac{m_n}{p_n + 1} \left( \frac{r}{|z_n|} \right)^{p_n+1} \tag{11.4}$$

has a main role.

**Theorem 11.8** (Weierstrass). *If the series* (11.4) *converges for all* $r > 0$, *then the infinite product* $\prod_{n=1}^{\infty} E_{p_n}^{m_n} \left( \frac{z}{z_n} \right)$ *defines an entire function vanishing exactly at the points* $z_n \in \mathbb{C}$ *with multiplicities* $m_n \in \mathbb{N}$.

*There is always a suitable choice of* $p_n \in \mathbb{N}$, *for example* $p_n = n m_n$. *One can take* $p_n = p$, *independently of n, if*

$$\sum_{n=1}^{\infty} \frac{m_n}{|z_n|^{p+1}} < +\infty.$$

**Example 11.9.** Take $z_n = n$ and $m_n = 1$ for $n \in \mathbb{Z}$. The simplest entire function vanishing at integer points with multiplicity 1 is

$$F(z) = z \prod_{n \ne 0} \left( 1 - \frac{z}{n} \right) e^{z/n}$$

which will be computed explicitly next. At the same points, if one wants now $m_n = |n|$, one needs to take $p = 2$ to obtain

$$F(z) = \prod_{n \ne 0} \left[ \left( 1 - \frac{z}{n} \right) \exp \left( \frac{z}{n} + \frac{1}{2} \left( \frac{z}{n} \right)^2 \right) \right]^{|n|}$$

$$= \prod_{n=1}^{\infty} \left( 1 - \frac{z^2}{n^2} \right)^n \exp \left( \frac{z^2}{n} \right),$$

an entire function with $F(0) = 1$ and a zero of order $|n|$ at each point $n \in \mathbb{Z}$, $n \ne 0$. $\qquad\square$

In the same way that Mittag-Leffler's theorem implies a result about the decomposition of a meromorphic function into simple functions, Weierstrass' theorem leads to the factorization of an entire function.

**Theorem 11.10.** *Every entire function with a zero of multiplicity $k$ at the origin and zeros of multiplicity $m_n > 0$ at the points $z_n \in \mathbb{C}$ can be factorized as*

$$F(z) = z^k \prod_{n=1}^{\infty} E_{p_n}^{m_n}\left(\frac{z}{z_n}\right) \exp g(z),$$

*where the numbers $p_n \in \mathbb{N}$ make the series (11.4) convergent for each $r > 0$, and $g$ is an entire function.*

By means of the logarithmic derivative one can obtain factorization of several entire functions from the examples of Subsection 10.3.1.

**Example 11.11.** Consider the function $F(z) = \sin \pi z$ that has simple zeros at the integer points. Since $\sum \frac{1}{n^2} < +\infty$, one can take $p_n = 1$ in Theorem 11.10 to obtain

$$\sin \pi z = z \prod_{n \neq 0} \left(1 - \frac{z}{n}\right) e^{z/n} \exp g(z).$$

Taking the logarithmic derivative and using the expansion of $\pi \cot \pi z$ of Example 10.13, we get $g'(z) = 0$ and so $g$ is constant. Since $\lim_{z \to 0} \frac{\sin \pi z}{z} = \pi$, we must have $e^{g(z)} = \pi$, and so

$$\sin \pi z = \pi z \prod_{n \neq 0} \left(1 - \frac{z}{n}\right) e^{z/n}.$$

Also gathering the terms in $n$ and $-n$,

$$\sin \pi z = \pi z \prod_{n=1}^{\infty} \left(1 - \frac{z^2}{n^2}\right). \qquad \square$$

**Example 11.12.** The simplest entire function vanishing at the negative integers is

$$F(z) = \prod_{n=1}^{\infty} \left(1 + \frac{z}{n}\right) e^{-z/n}.$$

Comparing with the example above yields

$$z F(z) F(-z) = \frac{\sin \pi z}{\pi}.$$

The function $F(z - 1)$ vanishes at the negative integers and also at zero, so that

$$F(z - 1) = z F(z) \exp g(z),$$

for some entire function $g$. Taking logarithmic derivative, one has

$$\sum_{n=1}^{\infty}\left(\frac{1}{z-1+n}-\frac{1}{n}\right)=\frac{1}{z}+g'(z)+\sum_{n=1}^{\infty}\left(\frac{1}{z+n}-\frac{1}{n}\right),$$

which implies $g'(z)=0$ and, therefore, $g$ is a constant that will be called $\gamma$. Thus, $F(z-1)=zF(z)e^{\gamma}$ and the function $H(z)=F(z)e^{\gamma z}$ satisfies $H(z-1)=zH(z)$. Setting $z=1$ in $F(z-1)=zF(z)e^{\gamma}$ one obtains

$$1=F(0)=e^{\gamma}F(1),\quad e^{-\gamma}=\prod_{n=1}^{\infty}\left(1+\frac{1}{n}\right)e^{-\frac{1}{n}},$$

which means

$$\gamma=\lim_{n\to\infty}\left(1+\frac{1}{2}+\cdots+\frac{1}{n}-\operatorname{Log}n\right).$$

This constant $\gamma$ is called *Euler's constant*. The function $\Gamma(z)=\frac{1}{zH(z)}$ is meromorphic on $\mathbb{C}$ with simple poles at points $0,-1,-2,\ldots$, satisfying the equation $\Gamma(z+1)=z\Gamma(z)$ and $\Gamma(1)=1$. In particular, one has $\Gamma(n)=(n-1)!$ if $n\in\mathbb{N}$. From the equality

$$\Gamma(z)=\frac{e^{-\gamma z}}{z}\prod_{n=1}^{\infty}\left(1+\frac{z}{n}\right)^{-1}e^{\frac{z}{n}}$$

we get

$$\Gamma(z)\Gamma(1-z)=\frac{\pi}{\sin\pi z},$$

and, taking $z=\frac{1}{2}$, $\Gamma(\frac{1}{2})=\sqrt{\pi}$. Hence, one has

$$\Gamma(z)=\frac{e^{-\gamma z}}{z}\lim_{n\to\infty}\prod_{k=1}^{n}\left(1+\frac{z}{k}\right)^{-1}e^{\frac{z}{k}}$$

$$=e^{-\gamma z}\lim_{n\to\infty}\frac{n!}{z(z+1)\cdots(z+n)}e^{z\sum_{k=1}^{n}\frac{1}{k}}=\lim_{n\to\infty}\frac{n^{z}n!}{z(z+1)\cdots(z+n)}.$$

The function $\Gamma$ is called *Euler's $\Gamma$ function*. At points $z\in\mathbb{C}$ with $\operatorname{Re}z>0$ we claim that

$$\Gamma(z)=\int_{0}^{\infty}e^{-t}t^{z-1}dt.$$

To see this, let $G(z)$ denote the function defined by the integral above. Integrating by parts one may easily check that $G$ also satisfies the equation $G(z+1)=zG(z)$. To show $G(z)=\Gamma(z)$ if $\operatorname{Re}z>0$, it is enough to prove, for $z=s, 0<s<1$, the equality

$$\lim_{n\to\infty}\frac{G(s)s(s+1)\cdots(s+n)}{n^{s}n!}=\lim_{n\to\infty}\frac{G(s+n+1)}{n^{s}n!}=1,\quad 0<s<1.$$

From the definition of $G$ and breaking the integral into two pieces, from 0 to $n$ and from $n$ to $+\infty$, it turns out that

$$G(s+n+1) = \int_0^\infty e^{-t}t^{s+n}dt \leq n^s \int_0^n t^n e^{-t}dt + n^{s-1}\int_n^\infty t^{n+1}e^{-t}dt,$$

$$G(s+n+1) \geq n^{s-1}\int_0^n t^{n+1}e^{-t}dt + n^s \int_n^{+\infty} t^n e^{-t}dt.$$

Integrating by parts, one has $\int_n^\infty t^{n+1}e^{-t}dt = n^{n+1}e^{-n} + \int_n^\infty (n+1)t^n e^{-t}dt$ and $\int_0^n t^{n+1}e^{-t}dt = -n^{n+1}e^{-n} + \int_0^n (n+1)t^n e^{-t}$, which lead to

$$G(s+n+1) \leq n^s \int_0^\infty t^n e^{-t}dt + e^{-n}n^{n+s} + n^{s-1}\int_n^\infty t^n e^{-t}dt,$$

$$G(s+n+1) \geq n^s \int_0^\infty t^n e^{-t}dt - e^{-n}n^{n+s} + n^{s-1}\int_0^n t^n e^{-t}dt.$$

These inequalities show the equivalence $G(s+n+1) \sim n^s\, n!$ is true for $n \to \infty$, as claimed. $\qquad\square$

**Example 11.13.** Now an entire function will be constructed with simple zeros at points $z_{kl} \in \mathbb{C}$ of integer coordinates, $z_{kl} = k + li$, $k,l \in \mathbb{Z}$. We are interested in finding a number $p \in \mathbb{N}$ such that $\sum_{k,l} \frac{1}{|z_{kl}|^{p+1}} < +\infty$; according to Example 10.16, we may choose $p = 2$. Therefore, let us consider

$$\sigma(z) = z \prod_{k,l} \left(1 - \frac{z}{k+li}\right) \exp\left\{\frac{z}{k+li} + \frac{1}{2}\left(\frac{z}{k+li}\right)^2\right\},$$

called *Weierstrass' $\sigma$ function*. The logarithmic derivative of $\sigma$ is

$$\frac{\sigma'(z)}{\sigma(z)} = \frac{1}{z} + \sum_{k,l}\left(\frac{1}{z-(k+li)} + \frac{1}{k+li} + \frac{z}{k+li}\right),$$

that has a simple pole at each point $z_{kl}$, $k,l \in \mathbb{Z}$. $\qquad\square$

Next Weierstrass' theorem for an arbitrary domain will be proved; in fact, the statement below prescribes poles as well as zeros.

**Theorem 11.14.** *Let $U$ be a domain of $\mathbb{C}^*$, $U \neq \mathbb{C}^*$, let $A = \{z_n : n \in \mathbb{N}\}$ be a discrete and closed subset of $U$, and let $(m_n)$ a sequence of integers, $m_n \neq 0$. Then there exists a function $f$ meromorphic on $U$ and a neighborhood $V_n$ of $z_n$, so that*

$$f(z) = (z - z_n)^{m_n} g_n(z), \quad z \in V_n$$

*holds with $g_n$ holomorphic on $V_n$ and $g_n(z_n) \neq 0$ (hence, if $m_n > 0$, $f$ has a zero of order $m_n$ at the point $z_n$ and if $m_n < 0$, $f$ has a pole of order $|m_n|$ at the point $z_n$).*

*Proof.* Writing $A = A_1 \cup A_2$ so that points $z_n \in A_1$ correspond to $m_n > 0$, and points $z_n \in A_2$, to $m_n < 0$, it is enough to consider the case $m_n > 0$ for all $n$ (because if $f_1$ has zeros of multiplicity $m_n$ at points of $A_1$ and $f_2$ has zeros of multiplicity $-m_n$ at points of $A_2$ then $f = \frac{f_1}{f_2}$ works). Applying, if necessary, a linear transformation we can assume $\infty \in U$ but $\infty \notin A$, so that $\mathbb{C}^* \setminus U$ is a compact set of $\mathbb{C}$; finally, we can also assume that $A$ is infinite. For each point $z_n$, let $w_n \in \mathbb{C}^* \setminus U$ be such that $d(z_n, \mathbb{C}^* \setminus U) = |z_n - w_n|$. Since $A$ cannot have accumulation points in $U$, we must have $\lim_n |z_n - w_n| = 0$. Now we mimic the proof of Theorem 11.8 with $p_n = n m_n$ and write

$$f(z) = \prod_{n=1}^{\infty} E_{p_n}\left(\frac{z_n - w_n}{z - w_n}\right)^{m_n} = \prod_{n=1}^{\infty} f_n(z)^{m_n} \quad \text{with } f_n(z) = E_{p_n}\left(\frac{z_n - w_n}{z - w_n}\right).$$

Each factor $f_n$ is holomorphic on $U$, so we need to check that hypotheses of Theorem 11.5 are satisfied. If $K$ is a compact set of $U$, there is an integer $n_0(K)$ such that $|z - w_n| > 2|z_n - w_n|$ or, $\left|\frac{z_n - w_n}{z - w_n}\right| < \frac{1}{2}$, if $z \in K$ and $n \geq n_0(K)$. Just before the statement of theorem 11.8 it has been shown that $|\operatorname{Log} E_p(z)|$ is dominated by $\frac{2}{p+1}|z|^{p+1}$ when $|z| \leq \frac{1}{2}$, and so we obtain

$$\sum_{n \geq n_0(K)} m_n |\operatorname{Log} f_n(z)| \leq \sum_{n \geq n_0(K)} \frac{1}{n}\left(\frac{1}{2}\right)^n < +\infty. \qquad \square$$

**Corollary 11.15.** *Every meromorphic function on a domain $U$ of the plane is the quotient of two holomorphic functions on $U$.*

*Proof.* Let $f$ be meromorphic on $U$ with poles $z_n$ of order $m_n$ and let $F$ be holomorphic with zeros of multiplicities $m_n$ at the same points $z_n$; then $g = fF$ has only removable singularities; therefore, $g \in H(U)$ and $f = \frac{g}{F}$. $\qquad \square$

**Corollary 11.16.** *For each domain $U$ there is a function $f$ holomorphic on $U$ that cannot be analytically continued to any point of the boundary of $U$, that is to say $f$ cannot extend to a holomorphic function on $U'$ if $U'$ is any domain containing $U$ strictly.*

*Proof.* Consider a discrete and closed set $A \subset U$ such that every point of $\partial U$ is an accumulation point of $A$. For example, $A$ may be constructed in the following way: let $(z_n)_{n=1}^{\infty}$ be a sequence containing the points of $U$ with rational coordinates and put $r_n = d(z_n, \mathbb{C} \setminus U)$. Let $K_n$ be an increasing sequence of compact sets with $\bigcup_n K_n = U$ as in Lemma 1.15. For each $n$ let $w_n \notin K_n$, $w_n \in U$, such that $|z_n - w_n| < r_n$ and $w_n \neq w_m$ if $n \neq m$. Since $w_n \notin K_n$, $A = \{w_n : n \in \mathbb{N}\}$ is discrete. If $a \in \partial U$ and $\varepsilon > 0$ there is a point $z_n$ with $|a - z_n| < \frac{\varepsilon}{2}$; therefore,

$r_n \leq \frac{\varepsilon}{2}$ and $|a - w_n| \leq |a - z_n| + |z_n - w_n| < \frac{\varepsilon}{2} + r_n \leq \varepsilon$. Thus, every point $a \in \partial U$ is an accumulation point of $A$.

Now take as $f$ a function vanishing exactly at the points of $A$. We claim that there is no disc $D(a, \varepsilon)$ centered at a point $a \in \partial U$ such that $f$ extends analytically to a function $g \in H(D(a, \varepsilon))$. Indeed, in this case, the zeros of $g$ would have an accumulation point in $D(a, \varepsilon)$, hence $g \equiv 0$ and, therefore, $f \equiv 0$. $\qquad\square$

## 11.3 Interpolation by entire functions

If $z_1, \ldots, z_n$ are different points of $\mathbb{C}$ and $a_1, \ldots, a_n$ are arbitrary complex values, there is a unique polynomial $P$ of degree less than or equal to $n - 1$ such that $P(z_j) = a_j$, $1 \leq j \leq n$. Observe that this fact imposes $n$ conditions on the polynomial, while the space of polynomials of degree less than or equal to $n - 1$ has dimension precisely $n$. This statement may be proved writing

$$P(z) = c_1 + c_2 z + \cdots + c_n z^{n-1}$$

and imposing the conditions

$$P(z_j) = c_1 + c_2 z_j + \cdots + c_n z_j^{n-1} = a_j, \quad 1 \leq j \leq n.$$

The matrix of this linear system in the unknowns $c_1, c_2, \ldots, c_n$ has Vandermonde's determinant

$$\det \begin{pmatrix} 1 & z_1 & z_1^2 & \cdots & z_1^{n-1} \\ 1 & z_2 & z_2^2 & \cdots & z_2^{n-1} \\ \vdots & & & & \\ 1 & z_n & z_n^2 & \cdots & z_n^{n-1} \end{pmatrix} = \prod_{i<j} (z_i - z_j) \neq 0.$$

Therefore, there is a unique solution for all data. The polynomial $P$ is called the *Lagrange interpolating polynomial*. To exhibit $P$ it is not necessary to solve the above system. Setting $Q(z) = (z - z_1)(z - z_2) \ldots (z - z_n)$, one has

$$P(z) = \sum_{i=1}^{n} a_i \frac{Q(z)}{Q'(z_i)(z - z_i)}.$$

Indeed the polynomials $Q_i(z) = \frac{Q(z)}{Q'(z_i)(z-z_i)} = \frac{1}{Q'(z_i)} \prod_{j \neq i} (z - z_j)$ satisfy $Q_i(z_j) = \delta_{ij}$, their degree is less than or equal to $n - 1$ and, therefore,

$$P(z_j) = \sum_{i=1}^{n} a_i Q_i(z_j) = \sum_{i=1}^{n} a_i \delta_{ij} = a_j.$$

Now we want to solve the same problem but with infinitely many points $(z_n)_{n=1}^{\infty}$, with no finite accumulation point. The data are arbitrary complex values $(a_n)_{n=1}^{\infty}$ and we look for an entire function $f$ ("polynomial of infinite degree") such that $f(z_n) = a_n$, for $n \geq 1$. Let $F$ be an entire function with a simple zero at each one of the points $z_n$: $F(z_n) = 0$, $F'(z_n) \neq 0$. Then the functions

$$F_n(z) = \frac{1}{F'(z_n)} \frac{F(z)}{(z - z_n)}$$

are entire and $F_n(z_m) = \delta_{n,m}$. If the series

$$f(z) = \sum_{n=1}^{\infty} a_n F_n(z) = \sum_{n=1}^{\infty} \frac{a_n}{F'(z_n)} \frac{F(z)}{(z - z_n)}$$

converges uniformly on compact sets, as it happens when

$$\sum_{n=1}^{\infty} \frac{|a_n|}{|F'(z_n)||z_n|} < +\infty,$$

then $f$ will be a solution of the problem.

**Example 11.17.** Consider the interpolation problem $f(n) = a_n$, $n \in \mathbb{Z}$. The most simple entire function $F$ vanishing at integer points is $F(z) = \sin \pi z$, with $F'(n) = \pi \cos \pi n = \pi(-1)^n$. Hence, if the series

$$h(z) = \sum_{n \in \mathbb{Z}} (-1)^n \frac{a_n}{z - n}$$

converges uniformly on compact sets (in the usual sense, after canceling a finite number of terms for $z$ in a fixed compact set), then the function $\frac{1}{\pi} F(z)h(z)$ solves the interpolation problem. The uniform convergence on compact sets of the series defining the function $h$ holds if

$$\sum_{n \neq 0}^{\infty} \frac{|a_n|}{|n|} < +\infty.$$

By Hölder's inequality, if $\frac{1}{p} + \frac{1}{q} = 1$, one has

$$\sum_{n \neq 0}^{\infty} \frac{|a_n|}{|n|} \leq \left( \sum \frac{1}{|n|^p} \right)^{1/p} \left( \sum |a_n|^q \right)^{1/q}.$$

Since $\sum_{n=1}^{\infty} n^{-p}$ converges if $p > 1$, it is enough to impose $\sum |a_n|^q < +\infty$ for some value of $q$, $1 \leq q < \infty$. Observe that the series that defines $h$ may be uniformly convergent on compact sets without being absolutely convergent. In Example 10.13 it was shown that the series $\sum_{n\in\mathbb{Z}}\frac{(-1)^n}{z-n}$ is uniformly convergent on compact sets, with sum $\frac{\pi}{\sin\pi z}$:

$$\frac{\pi}{\sin\pi z} = \frac{1}{z} + \sum_{n=1}^{\infty}(-1)^n\frac{2z}{z^2-n^2}.$$

Thus another way to ensure the uniform convergence on compact sets of the series defining $h$ is imposing the condition $\sum_{n\in\mathbb{Z}}|a_n-a_{n+1}| < +\infty$, according to Abel's criterion (Theorem 2.14).

Ultimately, then, in the vector space of all sequences $(a_n)_{n\in\mathbb{Z}}$ that satisfy $\sum_{n\in\mathbb{Z}}|a_n|^q < +\infty$, $1 \le q < +\infty$, as well as in the one of all sequences with $\sum_n|a_n - a_{n+1}| < +\infty$, the operator

$$(a_n)_{n\in\mathbb{Z}} \longmapsto \frac{1}{\pi}\sin\pi z\sum_{n\in\mathbb{Z}}(-1)^n\frac{a_n}{z-n}$$

is an interpolating linear operator.     $\square$

The convergence condition stated just before Example 11.17 does not hold in general, and therefore we need to modify the definition of the function $f$. Suppose, without loss of generality, that $z_n \neq 0$ for all $n$ and $F$ is like in Weierstrass' theorem:

$$F(z) = \prod_{n=1}^{\infty} E_{p_n}\left(\frac{z}{z_n}\right)$$

with

$$\sum_{n=1}^{\infty}\frac{r^{p_n}}{p_n|z_n|^{p_n}} < +\infty \quad \text{if } r > 0.$$

Then writing

$$f(z) = \sum_{n=1}^{\infty}\frac{a_n}{F'(z_n)}\frac{F(z)}{z-z_n}\left(\frac{z}{z_n}\right)^{q_n},$$

with $q_n$ suitable natural numbers, we will have a solution of the interpolation problem. We need to choose $q_n$ such that

$$\sum_{n=1}^{\infty}\frac{|a_n|}{|F'(z_n)|}\frac{r^{q_n}}{|z_n|^{q_n+1}} < +\infty \quad \text{for all } r > 0 \tag{11.5}$$

and, to this end, it suffices to take $q_n \nearrow +\infty$ satisfying

$$\left(\frac{|a_n|}{|F'(z_n)|}\frac{1}{|z_n|}\right)^{\frac{1}{q_n}} \le C, \quad C \text{ constant.}$$

Summarizing, one has the following result.

**Theorem 11.18.** *If $\{z_n : n \in \mathbb{N}\}$ is a closed and discrete set of $\mathbb{C}$, that is, $\lim_n |z_n| = +\infty$, and $(a_n)_{n=1}^{\infty}$ is any sequence of complex numbers, then there is an entire function $f$ such that*

$$f(z_n) = a_n, \quad \text{for } n = 1, 2, \ldots.$$

**Example 11.19.** In the interpolation problem $f(n) = a_n, n \in \mathbb{Z}$, of Example 11.17 where $F(z) = \sin \pi z$, the condition that exponents $q_n$ must satisfy so that (11.5) holds is

$$\frac{|a_n|}{|n|} \leq C^{q_n}, \quad C \text{ constant.}$$

If the sequence $(a_n)_{n \in \mathbb{Z}}$ is bounded or, more generally, $|a_n| = O(|n|)$ when $|n| \to +\infty$, we may take any sequence $q_n \nearrow +\infty$, for instance, $q_n = |n|$. Hence,

$$(a_n)_{n \in \mathbb{Z}} \longmapsto \frac{1}{\pi} \sin \pi z \sum_{n \in \mathbb{Z}} \frac{(-1)^n a_n}{z - n} \left(\frac{z}{n}\right)^{|n|}$$

is an interpolating linear operator on the space of bounded sequences. In general, the choice of the sequence $(q_n)$ will depend on $(a_n)$ and one cannot exhibit an interpolating linear operator acting on *all* sequences. $\square$

We will show now that Theorem 11.18 may be improved prescribing also, at each point $z_n$, a finite number of derivatives. That is to say, suppose for each $n \geq 1$, a number $k_n \in \mathbb{N}$ and arbitrary values $a_n^0, a_n^1, \ldots, a_n^{k_n} \in \mathbb{C}$ are given and one wants to find an entire function $f$ such that

$$f^{(i)}(z_n) = a_n^i, \quad \text{for } i = 0, \ldots, k_n, \ n \geq 1.$$

This problem can be solved combining Weierstrass' theorem with Mittag-Leffler's theorem. However, a different method based on the existence of solutions of the inhomogeneous Cauchy–Riemann equation will be used. This is a useful method to solve interpolation problems in other contexts. The idea is to solve first the problem with a function $\phi \in C^{\infty}(\mathbb{C})$ having small $\bar{\partial}\phi$ on the neighborhood of $z_n$ and after that to correct the solution to get a holomorphic function. Consider the polynomials

$$\phi_n(z) = \sum_{i=0}^{k_n} \frac{a_n^i}{i!}(z - z_n)^i$$

with $\phi_n^{(i)}(z_n) = a_n^i, i = 0, 1, \ldots, k_n$. If $\varepsilon_n > 0$ are small enough numbers so that the discs $D(z_n, \varepsilon_n)$ are pairwise disjoint, consider

$$\phi(z) = \sum_{n=1}^{\infty} \phi_n(z) \, \chi\left(\frac{|z - z_n|}{\varepsilon_n}\right)$$

where $\chi$ is an even $C^\infty$ function on $\mathbb{R}$ with compact support contained on $[-1, 1]$, with $\chi(t) = 1$ if $|t| \leq \frac{1}{2}$. Then $\phi \in C^\infty(\mathbb{C})$ and $\phi(z) = \phi_n(z)$ on the disc $D(z_n, \varepsilon_n/2)$; therefore, $\phi^{(i)}(z_n) = a_n^i$. Let $F$ be an entire function with zeros of multiplicity $k_n$ at the points $z_n$. Let us look now for the solution $f$ of the interpolation problem, under the form

$$f = \phi + uF$$

with $u$ to be determined; since we want $f$ to be entire, that is, $\bar{\partial} f = 0$, we must have

$$0 = \bar{\partial} f = \bar{\partial}\phi + F\bar{\partial}u \quad \text{or} \quad \bar{\partial}u = -\frac{\bar{\partial}\phi}{F}.$$

Since $\phi = \phi_n$ on $D(z_n, \varepsilon/2)$, one has $\bar{\partial}\phi = 0$ on this disc and, therefore, $-\frac{\bar{\partial}\phi}{F}$ is a well-defined $C^\infty$ function on $\mathbb{C}$. Theorem 10.41 asserts that, indeed, there is a function $u \in C^\infty(\mathbb{C})$ satisfying $\bar{\partial}u = -\frac{\bar{\partial}\phi}{F}$, and consequently $f$ is a solution of the problem.

## 11.4 Zeros of holomorphic functions and the Poisson equation

This section is devoted to establishing a link between two apparently independent topics. On the one hand, it is known (Corollary 10.42) that on each domain $U \subset \mathbb{C}$ the Poisson equation

$$\Delta u = \mu,$$

where $\mu$ is a measure of locally finite mass in $U$, has a solution $u \in L^1_{\text{loc}}(U)$ in the weak sense; that is, for every function $\varphi \in C_c^\infty(U)$, the equality

$$\int_U u\Delta\varphi \, dA = \int_U \varphi \, d\mu$$

holds. On the other hand, for every closed and discrete set $\{z_n : n \in \mathbb{N}\}$ of $U$ and every sequence of multiplicities $(m_n)$ there is a function $f \in H(U)$ with a zero of multiplicity $m_n$ at the point $z_n$, for each $n$.

**Proposition 11.20.** *Let $f$ be a holomorphic function on the domain $U$, $f \not\equiv 0$ and let $Z(f) = \{z_n : n \in \mathbb{N}\}$ be the zero set of $f$ ( finite or countable) and $(m_n)$ the sequence of the respective multiplicities. Then, the function $u = \mathrm{Log}\,|f|$ is locally integrable on $U$ and one has, in the weak sense,*

$$\Delta u = 2\pi \sum_n m_n \delta_{z_n},$$

*where $\delta_{z_n}$ is the measure with mass 1 at the point $z_n$ (Dirac's delta at the point $z_n$).*

Observe that $\mu = \sum_n m_n \delta_{z_n}$ is a measure of locally finite mass because each compact set of $U$ contains only a finite number of points $z_n$.

*Proof.* Fix a compact set $K \subset U$. Then $z_n \notin K$ for $n \geq n_0(K)$. So in a neighborhood $V$ of $K$, we may write

$$f(z) = \prod_{k=1}^{n_0} (z - z_k)^{m_k} g(z)$$

with $g$ holomorphic on $U$ and $g(z) \neq 0$ if $z \in V$, and so

$$\mathrm{Log}\,|f(z)| = \sum_{k=1}^{n_0} m_k \,\mathrm{Log}\,|z - z_k| + \mathrm{Log}\,|g(z)|, \quad z \in V.$$

The function $\mathrm{Log}\,|g(z)|$ is harmonic on $V$ (Corollary 4.31) and, thus, it is locally integrable. By Lemma 7.11 each function $\mathrm{Log}\,|z - z_k|$ is also integrable on $K$ and, consequently, $\int_K |\mathrm{Log}(f(z)|\,dm(z) < +\infty$.

Theorem 7.42 states that $\Delta \,\mathrm{Log}\,|z - z_k| = 2\pi \delta_{z_k}$ in the weak sense, whence

$$\Delta \,\mathrm{Log}\,|f(z)| = \sum_{k=1}^{n_0} m_k 2\pi \delta_{z_k} + \Delta \,\mathrm{Log}\,|g| = 2\pi \sum_{k=1}^{n_0} m_k \delta_{z_k}$$

on $V$. $\qquad\qquad\square$

The interesting point in the proposition above is that the converse is also true, whenever the domain $U$ is simply connected.

**Theorem 11.21.** *Let $U$ be a simply connected domain, $\{z_n : n \in \mathbb{N}\}$ a closed and discrete set in $U$, $(m_n)$ a sequence of multiplicities, and consider the measure $\mu$ of locally finite mass, $\mu = \sum_n m_n \delta_{z_n}$. Then every solution $u \in L^1_{\mathrm{loc}}(U)$ of the Poisson equation $\Delta u = 2\pi \mu$ may be written as $u = \mathrm{Log}\,|f|$, with $f$ holomorphic on $U$, vanishing exactly at the points $z_n$ with multiplicity $m_n$.*

*Proof.* Let $u$ be a solution of $\Delta u = 2\pi \mu$. By Weierstrass' theorem, there is a function $g \in H(U)$ vanishing at points $z_n$ with multiplicities $m_n$ and, thus, $\Delta \,\mathrm{Log}\,|g| = 2\pi \mu$, by Proposition 11.20. Hence,

$$\Delta(u - \mathrm{Log}\,|g|) = 0.$$

By Weyl's lemma, $u - \mathrm{Log}\,|g|$ is a harmonic function on $U$. If $U$ is simply connected, according to Theorem 6.22 there is a function $h \in H(U)$, without zeros, such that $u - \mathrm{Log}\,|g| = \mathrm{Log}\,|h|$. Then one has $u = \mathrm{Log}\,|f|$ with $f = gh$ and $f$ has the same zeros and multiplicities as $g$. $\qquad\square$

Hence, in a simply connected domain, there is a one-to-one correspondence between solutions of the Poisson equation $\Delta u = 2\pi\mu$ and holomorphic functions with "encoded zeros" by the measure $\mu$. So every method for constructing holomorphic functions with prescribed zeros and multiplicities leads to a way of solving the Poisson equation, and conversely.

To justify what has been said above, apply Weierstrass' theorem, according to which the function

$$f = \prod_{n=1}^{\infty} E_p^{m_n}\left(\frac{z}{z_n}\right) = \prod_{n=1}^{\infty}\left(1 - \frac{z}{z_n}\right)^{m_n} \exp\left[m_n \sum_{k=1}^{p} \frac{1}{k}\left(\frac{z}{z_n}\right)^k\right]$$

vanishes at points $(z_n)$ with multiplicities $(m_n)$, whenever

$$\sum_n \frac{m_n}{|z_n|^{p+1}} < +\infty.$$

Rewrite everything – conclusion and hypothesis – in terms of $\mathrm{Log}\,|f|$ and the measure $\mu = \sum_n m_n \delta_{z_n}$. We have

$$\mathrm{Log}\,|f(z)| = \sum_{n=1}^{\infty} m_n \left\{ \mathrm{Log}\left|1 - \frac{z}{z_n}\right| + \sum_{k=1}^{p} \frac{1}{k} \mathrm{Re}\left(\frac{z}{z_n}\right)^k \right\} = \int_{\mathbb{C}} K_p(z, w) d\mu(w)$$

with

$$K_p(z, w) = \mathrm{Log}\left|1 - \frac{z}{w}\right| + \sum_{k=1}^{p} \frac{1}{k}\mathrm{Re}\left(\frac{z}{w}\right)^k, \quad p \in \mathbb{N}. \tag{11.6}$$

Moreover, we can also write

$$\sum_n \frac{m_n}{|z_n|^{p+1}} = \int_{\mathbb{C}} \frac{1}{|w|^{p+1}} d\mu(w).$$

The following result states the above conclusion for an arbitrary measure.

**Theorem 11.22.** *Let $\mu$ be a measure on $\mathbb{C}$ of locally finite mass, vanishing on a neighborhood of the origin, with $\int_{\mathbb{C}} |w|^{-p-1} d|\mu|(w) < +\infty$, for some $p \in \mathbb{N}$. Then the integral*

$$u(z) = \int_{\mathbb{C}} K_p(z, w) d\mu(w),$$

*with the kernel $K_p(z, w)$ given by (11.6), defines a function $u \in L^1_{\mathrm{loc}}(\mathbb{C})$ such that $\Delta u = 2\pi\mu$ in the weak sense.*

*Proof.* The identity

$$\mathrm{Log}\left(1 - \frac{z}{w}\right) + \sum_{k=1}^{p} \frac{1}{k}\left(\frac{z}{w}\right)^k = -\sum_{p+1}^{\infty} \frac{1}{k}\left(\frac{z}{w}\right)^k, \quad \text{if } |w| > |z|,$$

implies, taking real parts,

$$|K_p(z, w)| \leq \frac{1}{p+1} \sum_{p+1}^{\infty} \frac{|z|^k}{|w|^k} = \frac{1}{p+1} \left( \frac{|z|}{|w|} \right)^{p+1} \frac{1}{1 - \frac{|z|}{|w|}}.$$

Consequently, one has

$$|K_p(z, w)| \leq \frac{2}{p+1} \left( \frac{|z|}{|w|} \right)^{p+1} \quad \text{if } |w| > 2|z|.$$

For $r > 0$ fixed, decompose the function $u$, on the disc $D(0, r)$, into three parts:

$$\begin{aligned}
u(z) &= \int_{|w| \leq 2r} K_p(z, w) d\mu(w) + \int_{|w| > 2r} K_p(z, w) d\mu(w) \\
&= \int_{|w| \leq 2r} \text{Log } |z - w| d\mu(w) \\
&\quad + \int_{|w| \leq 2r} \left( -\text{Log } |w| + \sum_{k=1}^{p} \frac{1}{k} \text{Re} \left( \frac{z}{w} \right)^k \right) d\mu(w) \\
&\quad + \int_{|w| > 2r} K_p(z, w) d\mu(w) \\
&= u_1 + u_2 + u_3.
\end{aligned}$$

The integral defining $u_2$ is, in fact, extended to $\{w : \varepsilon < |w| < 2r\}$, $\varepsilon > 0$, by the hypothesis on $\mu$; the integrand is a harmonic function in $z$ and bounded in $w$ and, therefore, $u_2$ is a harmonic function ($\mu$ has finite mass by hypothesis). On the domain $\{w : |w| > 2r > 2|z|\}$ it has been shown that $|K_p(z, w)| = O(|w|^{-p-1})$ and $K_p(z, w)$ is harmonic for $w \in \mathbb{C} \setminus \{z\}$; as a consequence, $u_3$ is also well defined and harmonic. Finally, $u_1$ is $2\pi$ times the logarithmic potential of $\mu_{|D(0,2r)}$ and, by Theorem 7.42, $\Delta u = 2\pi \mu$ on $D(0, 2r)$. Since $r > 0$ is arbitrary, we conclude that $\Delta u = 2\pi \mu$. $\qquad \square$

In the opposite direction, a method for solving the Poisson equation leads to a procedure to construct holomorphic functions with prescribed zeros. This will be shown, for a bounded simply connected domain, in Section 11.6. Since these domains are conformally equivalent to the unit disc, it will suffice to consider this particular domain.

## 11.5 Jensen's formula

Formula (7.32) in Section 7.8 allows to solve the non-homogeneous Dirichlet problem on a disc $D(0, R)$. For $u \in C^2(\bar{D}(0, R))$ one has

$$u(z) = \frac{1}{2\pi R} \int_{C(0,R)} \frac{R^2 - |z|^2}{|w - z|^2} u(w)ds(w)$$
$$+ \frac{1}{2\pi} \int_{D(0,R)} \mathrm{Log}\, \frac{|w - z|R}{|R^2 - \bar{w}z|} \Delta u(w)dm(w). \tag{11.7}$$

If $f$ is holomorphic on a neighborhood of $\bar{D}(0, R)$, we will apply this formula to the function $u = \mathrm{Log}\,|f(z)|$. Especially interesting is the case in which $f$ has zeros; then $u = \mathrm{Log}\,|f(z)|$ does not belong to $C^2(\bar{D}(0, R))$, however, it will be shown that the formula obtained by formally replacing $u(z)$ with $\mathrm{Log}\,|f(z)|$ is correct. Since $\Delta\,\mathrm{Log}\,|f(z)| = 2\pi \sum_{n=1}^{N} m_n \delta_{z_n}$ on $D(0, R)$, where $z_1, \ldots, z_N$ are the zeros of $f$ in this disc with multiplicities $m_1, \ldots, m_N$ (the number of zeros of $f$ in $D(0, R)$ is finite because $f$ is holomorphic on a neighborhood of $\bar{D}(0, R)$), formula (11.7) gives

$$\mathrm{Log}\,|f(z)| = \frac{1}{2\pi R} \int_{C(0,R)} \frac{R^2 - |z|^2}{|z - w|^2} \mathrm{Log}\,|f(w)|ds(w)$$
$$+ \sum_{i=1}^{N} m_i \,\mathrm{Log}\, \frac{|z_i - z|R}{|R^2 - \bar{z}_i z|}, \quad |z| < R. \tag{11.8}$$

In order to justify this formula, we first need to show that the function $\mathrm{Log}\,|f|$ is integrable on the circle $C(0, R)$. If $f$ has no zeros there, it is clear; if $f(a) = 0$ with $|a| = R$ and $m$ is the multiplicity of $a$ as a zero of $f$, then $|f(w)| \sim |w - a|^m$ and, therefore, $\mathrm{Log}\,|f(w)| \sim m\,\mathrm{Log}\,|w - a|$. Now,

$$\int_{-\varepsilon}^{\varepsilon} |\mathrm{Log}\,|t||dt < +\infty, \quad \text{if } \varepsilon > 0,$$

so that $\mathrm{Log}\,|f|$ is integrable on $C(0, R)$.

To prove (11.8), observe that if $z$ is one of the zeros $z_1, z_2, \ldots, z_N$ of $f$, then both terms in (11.8) are $-\infty$. The interesting case is, then, when $z \neq z_1, \ldots, z_N$. Whether $f$ has zeros in $C(0, R)$ or not, there is always a $\delta > 0$ such that $f$ has no zeros in $\{w \colon R - \delta < |w| < R\}$. Consider the domain $U = D(0, R - \varepsilon) \setminus [\bigcup_{i=1}^{N} \bar{D}(z_i, \varepsilon) \cup \bar{D}(z, \varepsilon)]$, with $\varepsilon > 0$ small enough, and apply the second Green's identity (7.6) to $U$, taking $u = \mathrm{Log}\,|f(w)|$ (which is of class $C^2$ on $\bar{U}$) and for $v$, Green's function of $D(0, R)$ with pole at $z$, that is,

$$v(w) = \frac{1}{2\pi} \mathrm{Log}\, \frac{|w - z|R}{|R^2 - \bar{w}z|}.$$

Recall that $\frac{\partial v}{\partial \vec{N}} = \frac{1}{2\pi R}\frac{R^2 - |z|^2}{|w - z|^2}$ if $|w| = R$, which is the Poisson kernel of $D(0, R)$. Both $u$ and $v$ are harmonic on $U$ and of class $C^2$ on $\bar{U}$. One has, therefore,

$$\int_{C(0,R-\varepsilon)} \left( u\frac{\partial v}{\partial \vec{N}} - v\frac{\partial u}{\partial \vec{N}} \right) ds(w) = \int_{C(z,\varepsilon)} \left( u\frac{\partial v}{\partial \vec{N}} - v\frac{\partial u}{\partial \vec{N}} \right) ds(w)$$
$$+ \sum_{i=1}^{N} \int_{C(z_i,\varepsilon)} \left( u\frac{\partial v}{\partial \vec{N}} - v\frac{\partial u}{\partial \vec{N}} \right) ds(w).$$

Since $v(w) = 0$ if $|w| = R$, the limit of the left-hand side term, when $\varepsilon \to 0$, is

$$\int_{C(0,R)} u\frac{\partial v}{\partial \vec{N}} ds(w) = \frac{1}{2\pi R} \int_{C(0,R)} \text{Log}\,|f(w)|\frac{R^2 - |z|^2}{|w - z|^2} ds(w).$$

In order to deal with the limit of the first term in the right when $\varepsilon \to 0$, observe that $v$ is the difference of $\frac{1}{2}\text{Log}\,|w - z|$ and a function without a pole at the point $z$, and $u$ is regular. Then arguing as in the proof of Theorem 7.13 we get

$$\lim_{\varepsilon \to 0} \int_{C(z,\varepsilon)} \left( u\frac{\partial v}{\partial \vec{N}} - v\frac{\partial u}{\partial \vec{N}} \right) ds(w) = u(z) = \text{Log}\,|f(z)|.$$

Finally, for the last term we proceed in the same way, but interchanging the roles of $u$ and $v$. On the neighborhood of $z_i$, $v$ is regular and $u$ is the difference of $m_i \text{Log}\,|z - z_i|$ and a regular function.

This yields

$$\lim_{\varepsilon \to 0} \int_{C(z_i,\varepsilon)} \left( u\frac{\partial v}{\partial \vec{N}} - v\frac{\partial u}{\partial \vec{N}} \right) ds(w)$$
$$= -\lim_{\varepsilon \to 0} \int_{C(z_i,\varepsilon)} v\frac{\partial u}{\partial \vec{N}} ds(w)$$
$$= -\lim_{\varepsilon \to 0} \int_{C(z_i,\varepsilon)} v\frac{\partial}{\partial \vec{N}} (m_i \text{Log}\,|z - z_i|)\,ds(w)$$
$$= -2\pi m_i v(z_i) = -m_i \text{Log}\,\frac{|z_i - z|R}{|R^2 - \bar{z}_i z|}.$$

Setting $z = 0$ in formula (11.8) we obtain the following result.

**Theorem 11.23** (Jensen's formula). *Let $f$ be a holomorphic function on a neighborhood of the disc $\bar{D}(0, R)$, $f(0) \neq 0$, and $z_1, \ldots, z_N$ the zeros of $f$ in $D(0, R)$ with multiplicities $m_1, \ldots, m_N$. Then one has*

$$\text{Log}\,|f(0)| = \frac{1}{2\pi} \int_0^{2\pi} \text{Log}\,|f(Re^{it})|dt + \sum_{i=1}^{N} m_i \text{Log}\,\frac{|z_i|}{R}.$$

If $f$ has a zero of multiplicity $k$ at the origin, we can apply Theorem 11.23 to the function $g(z) = \frac{f(z)}{z^k}$, which satisfies $g(0) = \frac{f^{(k)}(0)}{k!}$, to obtain

$$\mathrm{Log}\left|\frac{f^{(k)}(0)}{k!}\right| = \frac{1}{2\pi}\int_0^{2\pi}\mathrm{Log}\,|f(Re^{it})|dt - k\,\mathrm{Log}\,R + \sum_{i=1}^{N}m_i\,\mathrm{Log}\,\frac{|z_i|}{R}.$$

**Example 11.24.** Jensen's formula with $f(z) = z - \alpha$, $|\alpha| < 1$, $\alpha \neq 0$, gives

$$\frac{1}{2\pi}\int_0^{2\pi}\mathrm{Log}\,|e^{it} - \alpha|dt = \mathrm{Log}\,|f(0)| - \mathrm{Log}\,|\alpha| = 0.$$

If $\alpha = 0$, trivially the integral is zero. Hence, if $P$ is a polynomial with all its zeros inside the unit disc, the following holds:

$$\int_0^{2\pi}\mathrm{Log}\,|P(e^{it})|dt = 0. \qquad \square$$

Another interesting consequence of formula (11.8) comes from the fact that all the terms of the sum are negative (every Green's function is negative).

**Corollary 11.25.** *If $f$ is a holomorphic function on a neighborhood of the disc $\bar{D}(0, R)$, then the function $\mathrm{Log}\,|f(z)|$ is subharmonic on $D(0, R)$, in the sense that*

$$\mathrm{Log}\,|f(z)| \leq \frac{1}{2\pi}\int_0^{2\pi}\frac{R^2 - |z|^2}{|z - Re^{it}|^2}\mathrm{Log}\,|f(Re^{it})|dt, \quad |z| < R.$$

Assume $f$ has no zeros on the circle $C(0, R)$; then formula (11.8) shows that the values of $|f|$ on this circle and the zeros of $f$ determine $|f(z)|$ for $z \in D(0, R)$. Now, it is known that every holomorphic function $f$ is determined by $|f|$, up to a constant with modulus 1; therefore, it seems natural to try to obtain the values of $f(z)$ in a straightforward manner. To this end consider the function

$$g(z) = f(z)\prod_{i=1}^{N}\frac{(R^2 - \bar{z}_i z)^{m_i}}{(z - z_i)^{m_i}R}$$

which does not vanish and satisfies $|g(z)| = |f(z)|$ if $|z| = R$. The function $\mathrm{Log}\,|g|$ is harmonic on a neighborhood of $\bar{D}(0, R)$ and (11.8) is written as

$$\mathrm{Log}\,|g(z)| = \frac{1}{2\pi R}\int_{C(0,R)}\frac{R^2 - |z|^2}{|z - w|^2}\mathrm{Log}\,|g(w)|ds(w),$$

expressing $\text{Log}\,|g|$ as the Poisson transform of its boundary values. Consider now the holomorphic function

$$
\begin{aligned}
h(z) &= \frac{1}{2\pi R} \int_{C(0,R)} \text{Log}\,|g(w)| \frac{w+z}{w-z}\,ds(w) \\
&= \frac{1}{2\pi} \int_0^{2\pi} \frac{Re^{it}+z}{Re^{it}-z} \text{Log}\,|g(Re^{it})|\,dt.
\end{aligned}
$$

If $|w| = R$, one has

$$
\text{Re}\,\frac{w+z}{w-z} = \frac{\text{Re}(w+z)(\bar w - \bar z)}{|w-z|^2} = \frac{R^2 - |z|^2}{|w-z|^2}.
$$

This means that the real part of $h$ is $\text{Log}\,|g|$ and, consequently, $h(z)$ differs from a branch of $\log g(z)$ by an imaginary constant:

$$
\log g(z) = \frac{1}{2\pi} \int_0^{2\pi} \frac{Re^{it}+z}{Re^{it}-z} \text{Log}\,|f(Re^{it})|\,dt + i\alpha, \quad \alpha \in \mathbb{R}.
$$

Hence, the following result has been proved.

**Theorem 11.26.** *Let $f$ be a holomorphic function on a neighborhood of the disc $\bar D(0,R)$, without zeros in $C(0,R)$. Then*

$$
f(z) = C \prod_{i=1}^{N} \frac{(z-z_i)^{m_i} R}{(R^2 - \bar z_i z)^{m_i}} \exp\left\{ \frac{1}{2\pi} \int_0^{2\pi} \frac{Re^{it}+z}{Re^{it}-z} \text{Log}\,|f(Re^{it})|\,dt \right\},
$$

*where $z_i$, $i = 1,\ldots,N$, are the zeros of $f$ in $D(0,R)$, $m_i$, $i = 1,\ldots,N$, their multiplicities and $C$ a constant with $|C| = 1$.*

## 11.6  Growth of a holomorphic function and distribution of the zeros. The Blaschke condition

As shown in Section 11.5, Jensen's formula is a consequence of the equation $\Delta \text{Log}\,|f| = 2\pi\mu$, with $\mu = \sum_n m_n \delta_{z_n}$, satisfied by a holomorphic function $f$ with zeros $z_n$ of multiplicities $m_n$ and, in fact, it establishes a quantitative link between $|f|$ and $\mu$.

An interesting example in which this link is explicit appears when dealing with bounded holomorphic functions on the unit disc $\mathbb{D}$. If $f$ is bounded and holomorphic on $\mathbb{D}$, with zeros $z_n$ of multiplicities $m_n$, Jensen's formula yields

$$
\sum_{0 < |z_n| < R} m_n \text{Log}\,\frac{R}{|z_n|} \le C, \quad \text{for } R < 1,
$$

with $C$ a constant independent from $R$. This means

$$\sum_{z_n \neq 0} m_n \operatorname{Log} \frac{1}{|z_n|} \leq C,$$

which is equivalent to

$$\sum_n m_n (1 - |z_n|) < +\infty.$$

This condition is called *Blaschke's condition*. Hence, if $f$ is holomorphic and bounded on the unit disc or, more generally, if $f \in H(\mathbb{D})$, $f \not\equiv 0$ and there is a constant $C \geq 0$ such that

$$\int_0^{2\pi} \operatorname{Log} |f(Re^{it})| dt \leq C, \quad 0 \leq R < 1,$$

then the zeros of $f$ satisfy the Blaschke condition.

As seen above, Jensen's formula is (11.7) for $z = 0$,

$$u(0) = \frac{1}{2\pi} \int_0^{2\pi} u(Re^{it}) dt + \frac{1}{2\pi} \int_{D(0,R)} \operatorname{Log} \frac{|w|}{R} \Delta u(w) dm(w),$$

applied to $u = \operatorname{Log}|f|$. For a function $u \in C^2(\mathbb{D})$ with $\Delta u \geq 0$ (or, more generally, with $\Delta u = \mu$, in the weak sense, where $\mu$ is a positive measure), the condition

$$\sup_R \int_0^{2\pi} u(Re^{it}) dt < +\infty,$$

that holds in particular if $u$ is bounded, implies the Blaschke condition, written now as

$$\int_{\mathbb{D}} \operatorname{Log} \frac{1}{|w|} \Delta u(w) dm(w) < +\infty.$$

So this is a condition related to the Laplace equation, which can be formulated in any dimension. For example, on the unit ball $\mathbb{B}$ of $\mathbb{R}^n$, if $u \in C^2(\mathbb{B})$ with $\Delta u \geq 0$ satisfies

$$\sup_{0 < R < 1} \int_{\mathbb{S}} u(Ry) d\sigma(y) < +\infty,$$

the Blaschke condition is

$$\int_{\mathbb{B}} (1 - |x|) \Delta u(x) dV(x) < +\infty.$$

Proposition 7.52 shows that the potential

$$u(z) = \frac{1}{2\pi} \int_{\mathbb{D}} \operatorname{Log} \left| \frac{z - w}{1 - \bar{w}z} \right| d\mu(w)$$

is well defined when $\mu$ is a positive measure satisfying the Blaschke condition

$$\int_{\mathbb{D}} (1 - |w|^2) d\mu(w) < +\infty.$$

Then $u$ is integrable on $\mathbb{D}$ and $\Delta u = \mu$ in the weak sense. Taking $\mu = 2\pi \sum m_n \delta_{z_n}$, the condition $\int_{\mathbb{D}} (1 - |w|^2) d\mu(w) = 2\pi \sum_n m_n (1 - |z_n|^2) < +\infty$ is the Blaschke condition for the points $z_n$ and the integers $m_n$, and then $u$ must be of the form $\text{Log} |f(z)|$ with $f \in H(\mathbb{D})$ vanishing at points $z_n$ with multiplicities $m_n$. Moreover since $u \le 0$, $f$ will be bounded by 1. Since now

$$u(z) = \sum_n m_n \text{Log} \left| \frac{z - z_n}{1 - \bar{z}_n z} \right|,$$

it is natural to consider the infinite product

$$\prod_n \left( \lambda_n \frac{z - z_n}{1 - \bar{z}_n z} \right)^{m_n}$$

with $|\lambda_n| = 1$ as a candidate for $f(z)$. The choice of the $\lambda_n$ will be such that the series $\sum_{n \ge n_0(r)} m_n \left| \text{Log} \left( \lambda_n \frac{z - z_n}{1 - \bar{z}_n z} \right) \right|$ converges uniformly on each disc $\bar{D}(0, r)$, $0 < r < 1$, or, equivalently, that

$$\sum_n m_n \left| 1 - \lambda_n \frac{z - z_n}{1 - \bar{z}_n z} \right|$$

converges uniformly on each compact disc $\bar{D}(0, r)$. For $|1 - \bar{z}_n z - \lambda_n z + \lambda_n z_n|$ to be dominated by $1 - |z_n|$, if $z_n = r_n w_n$ with $|w_n| = 1$, $r_n = |z_n| \ne 0$, the quantity

$$1 - r_n \bar{w}_n z - \lambda_n z + \lambda_n r_n w_n$$

must vanish when $r_n = 1$, that is, one must have $1 - \bar{w}_n z - \lambda_n z + \lambda_n w_n = 0$. Therefore, let us take $\lambda_n = -\bar{w}_n = -\frac{\bar{z}_n}{|z_n|}$ if $z_n \ne 0$. With this choice, one has that $1 - \bar{z}_n z - \lambda_n z + \lambda_n z_n = 1 - \bar{z}_n z + \frac{\bar{z}_n z}{|z_n|} - |z_n| = (1 - |z_n|) + z\bar{z}_n \left( \frac{1}{|z_n|} - 1 \right) = (1 - |z_n|)\left(1 + \frac{z\bar{z}_n}{|z_n|}\right)$ has absolute value less than or equal to $(1 - |z_n|)(1 + |z|)$. Then the series

$$\sum_{z_n \ne 0} m_n \left| 1 - \lambda_n \frac{z - z_n}{1 - \bar{z}_n z} \right|, \quad \text{dominated by} \quad \sum_{z_n \ne 0} m_n \frac{(1 - |z_n|)(1 + |z|)}{1 - |z|},$$

is, indeed, uniformly convergent on each disc $\bar{D}(0, r)$.

Summarizing, the following result has been proved.

**Theorem 11.27.** *Let $f$ be a bounded holomorphic function on the unit disc, $f \not\equiv 0$. Then the zero set $\{z_n : n \in \mathbb{N}\}$ with multiplicities $(m_n)$ of $f$ satisfies the Blaschke condition*

$$\sum_n m_n (1 - |z_n|) < +\infty.$$

*Conversely, given the points $z_n \in \mathbb{D}$ and the integers $m_n \geq 1$, $n \in \mathbb{N}$, satisfying this condition, the infinite product*

$$B(z) = z^k \prod_{z_n \neq 0} \left( \frac{\bar{z}_n}{z_n} \frac{z_n - z}{1 - \bar{z}_n z} \right)^{m_n}$$

*converges uniformly on compact sets of $\mathbb{D}$. The function $B$ is holomorphic on $\mathbb{D}$, satisfies $|B(z)| \leq 1$, $z \in \mathbb{D}$ and vanishes exactly at the points $z_n$ with multiplicities $m_n$.*

The infinite product defining $B$ is called a *Blaschke product* (see Example 8.29).

**Example 11.28.** The sequence $(z_n)_{n \in \mathbb{N}}$ with $z_n = 1 - 1/n$ cannot be the zero set of a bounded function $f \in H(\mathbb{D})$ because

$$\sum_n (1 - |z_n|) = \sum_n \frac{1}{n} = +\infty.$$

Instead, if $z_n = 1 - 1/n^\alpha$ with $\alpha \in \mathbb{R}$ and $\alpha > 1$, then $(z_n)_{n \in \mathbb{N}}$ is the sequence of zeros of a bounded function $f$ holomorphic on $\mathbb{D}$; more precisely,

$$f(z) = \prod_{n=1}^\infty \frac{n^{-\alpha} - z}{1 - n^{-\alpha} z} = \prod_{n=1}^\infty \frac{1 - n^\alpha z}{n^\alpha - z}. \qquad \square$$

## 11.7 Entire functions of finite order

Starting from an entire function $f$, Jensen's formula can be applied to $f$ on any disc centered at the origin. One may take advantage of this fact in order to analyze the relationship between the growth of $f$ and the distribution of its zeros.

Let $f$ be an entire function with $f(0) = 1$. The growth of $f$ is measured by the function

$$M(r) = M_f(r) = \max\{|f(z)| : |z| = r\}, \quad r \geq 0.$$

By the maximum modulus principle one has $M(r) = \max\{|f(z)| : |z| \leq r\}$ and, therefore, $M(r)$ is an increasing function of $r$. Indeed, it is strictly increasing if $f$ is not constant. To prove this let $r_1 < r_2$ and assume $M(r_1) = M(r_2)$; then $|f|$ would have a local maximum at a point of the disc $D(0, r_2)$ and, by the maximum modulus principle, $f$ would be constant on $D(0, r_2)$ and, therefore, on the whole of $\mathbb{C}$.

The growth of $M(r)$ can be of different kinds. By Cauchy's inequalities, if $M(r)$ has a polynomial growth, that is, $M(r) = O(r^N)$ for some natural $N$, then $f$ is a polynomial of degree less than or equal to $N$ (Section 4.7, Exercise 11). Other kinds of growth may be quantified by means of the *order of an entire function*.

In general, if $\varphi(r) \geq 0$ is an increasing function of $r$, for $r > 0$, the order of $\varphi$ is the number $\rho \geq 0$ defined by

$$\rho = \limsup_{r \to +\infty} \frac{\operatorname{Log} \varphi(r)}{\operatorname{Log} r}.$$

This means that $\rho$ is the infimum of the set of numbers $\lambda > 0$ such that

$$\varphi(r) = O(r^\lambda), \quad r \to +\infty.$$

**Definition 11.29.** The order $\rho = \rho(f)$ of an entire function $f$ is the order of the function $\operatorname{Log} M(r)$, that is to say

$$\rho = \limsup_{r \to +\infty} \frac{\operatorname{Log} \operatorname{Log} M(r)}{\operatorname{Log} r}.$$

Hence $\rho$ is the infimum of the set of numbers $\lambda > 0$ such that

$$|f(z)| = O(\exp(|z|^\lambda)), \quad |z| \to \infty.$$

**Example 11.30.** Polynomials have order zero and the exponential function, $f(z) = e^z$, has order 1. The order may be infinite; for example, $f(z) = \exp(\exp(z))$ has infinite order. $\quad\square$

**Example 11.31.** The function $f(z) = \sin z$ has order 1, because $|\sin z|^2 = \sin^2 x + \operatorname{sh}^2 y$ if $z = x + iy$, and so

$$\operatorname{sh}^2 r \leq M(r)^2 \leq 1 + \operatorname{sh}^2 r,$$

giving that $M(r)$ behaves like $e^r$. $\quad\square$

Let $(z_n)_{n=1}^\infty$ be the sequence of zeros of an entire function $f$ with $|z_n| \leq |z_{n+1}|$ and $(m_n)_{n=1}^\infty$ the respective multiplicities. The radial distribution of zeros of $f$ is encoded by means of the *counting function* $n_f(r)$, which computes (with multiplicities) the number of zeros in each disc $D(0, r)$:

$$n(r) = n_f(r) = \sum_{|z_n| \leq r} m_n, \quad \text{if } r > 0.$$

Hence, $n(r)$ is an increasing step function, constant between $|z_n|$ and $|z_{n+1}|$ (if $|z_n| < |z_{n+1}|$), jumping $m_n$ at the point $|z_n|$.

The behavior of $n(r)$ when $r$ increases controls the convergence or divergence of series of type $\sum_n m_n \psi(|z_n|)$ with $\psi : (0, \infty) \to (0, \infty)$ differentiable and decreasing. Actually, applying Fubini's theorem, for $r > 0$ fixed one has

$$
\sum_{|z_n| \le r} m_n \psi(|z_n|) = \sum_{|z_n| \le r} \int_0^\infty m_n \mathbb{1}_{[0, \psi(|z_n|)]}(t) dt
$$

$$
= \int_0^\infty \#\{n : |z_n| \le r, \ \psi(|z_n|) \ge t\} dt
$$

$$
= \int_0^\infty \#\{n : |z_n| \le \min(r, \psi^{-1}(t)\} dt \qquad (11.9)
$$

$$
= \int_{\psi(r)}^\infty n(\psi^{-1}(t)) dt + n(r)\psi(r)
$$

$$
= \int_0^r n(t)|\psi'(t)| dt + n(r)\psi(r)
$$

(the notation $\#A$ is used to denote the number of elements of a finite set $A$).

Taking $\psi(x) = \operatorname{Log} \frac{r}{x}$ one gets the equality $\sum_{|z_n| \le r} m_n \operatorname{Log} \frac{r}{|z_n|} = \int_0^r \frac{n(t)}{t} dt$, and Jensen's formula can be written as

$$
\operatorname{Log}|f(0)| + \int_0^r \frac{n(t)}{t} dt = \frac{1}{2\pi} \int_0^{2\pi} \operatorname{Log}|f(re^{i\theta})| d\theta. \qquad (11.10)
$$

**Lemma 11.32.** *Let $\psi : (0, \infty) \to (0, \infty)$ be decreasing and differentiable. Then the series $\sum_n m_n \psi(|z_n|)$ is convergent if and only if the integral $\int_0^\infty n(t)|\psi'(t)| dt$ is convergent and, in this case, $n(r)\psi(r) \xrightarrow[r \to +\infty]{} 0$ and $\sum_n m_n \psi(|z_n|) = \int_0^\infty n(t)|\psi'(t)| dt$.*

*Proof.* The equality between the first and the last term of (11.9) proves that the integral is convergent if the series is, because $n(r)\psi(r) \ge 0$. If the integral is convergent, then

$$
n(r)\psi(r) \le \int_r^\infty n(t)|\psi'(t)| dt \xrightarrow[r \to +\infty]{} 0,
$$

and the same equality proves the convergence of the series. $\qquad \square$

The *exponent of convergence* $\mu$ of a sequence of points $(z_n)_{n=1}^\infty$, $z_n \ne 0$, with multiplicities $(m_n)$ is defined as

$$
\mu = \inf \{\lambda > 0 : \sum m_n |z_n|^{-\lambda} < +\infty\}.
$$

By Lemma 11.32, one has

$$
\mu = \inf \{\lambda > 0 : \int_0^\infty \frac{n(t)}{t^{\lambda+1}} dt < +\infty\}.
$$

**Lemma 11.33.** *The exponent of convergence $\mu$ of a sequence of points $(z_n)$, $z_n \neq 0$, with multiplicities $(m_n)$, coincides with the order of the counting function $n(r)$,*
$$\tilde{\rho} = \limsup_{r \to +\infty} \frac{\operatorname{Log} n(r)}{\operatorname{Log} r}.$$

*Proof.* Let us show first that $\tilde{\rho} \leq \mu$. If $\mu = +\infty$ there is nothing to prove; if $\mu < +\infty$ and $\lambda > \mu$, then $\sum m_n |z_n|^{-\lambda} < +\infty$ implying $n(r) = O(r^{\lambda})$ and, therefore, $\tilde{\rho} \leq \lambda$. Since $\lambda$ is arbitrary, it turns out that $\tilde{\rho} \leq \mu$. For the reverse inequality, if $\tilde{\rho} < +\infty$ and $\lambda > \tilde{\rho}$, the fact that $n(r) = O(r^{\lambda})$ shows that the integral
$$\int_0^\infty \frac{n(t)}{t^{\lambda+1+\varepsilon}} dt$$
converges for $\varepsilon > 0$; thus one must have $\mu \leq \lambda + \varepsilon$, and since $\varepsilon > 0$ is arbitrary, it yields $\mu \leq \lambda$. This holds for all $\lambda > \tilde{\rho}$ and, therefore, $\mu \leq \tilde{\rho}$.  $\square$

**Example 11.34.** Let $k \in \mathbb{N}$ be fixed and take $z_n = n^k$, $n \in \mathbb{Z}$: the series $\sum |z_n|^{-\lambda} = \sum n^{-k\lambda}$ converges exactly when $k\lambda > 1$; therefore, the exponent of convergence of $(z_n)$ is $1/k$. Taking now as $(z_n)$ the sequence of prime numbers, the prime number distribution theorem implies that the corresponding function $n(r)$ behaves like $r/\operatorname{Log} r$, so that the order and the exponent of convergence are both equal to 1. For a geometric sequence $z_n = \rho^n$ with $\rho > 1$, trivially $\sum |z_n|^{-\lambda} = \sum \rho^{-n\lambda}$ converges if and only if $\rho^\lambda > 1$ and the exponent of convergence is 0.  $\square$

**Theorem 11.35** (Hadamard). *If $f$ is an entire function of order $\rho$ and $\mu$ is the exponent of convergence of the zeros of $f$, then it is $\mu \leq \rho$.*

*Proof.* Assuming $f(0) = 1$, Jensen's formula (11.10) gives
$$\int_0^r \frac{n(t)}{t} dt = \frac{1}{2\pi} \int_0^{2\pi} \operatorname{Log} |f(re^{i\theta})| d\theta \leq \operatorname{Log} M(r).$$

Since $n(t)$ is increasing, it turns out that
$$n(r) \leq \int_r^{er} \frac{n(t)}{t} dt \leq \int_0^{er} \frac{n(t)}{t} dt \leq \operatorname{Log} M(er).$$

From this we obtain
$$\mu = \limsup_{r \to +\infty} \frac{\operatorname{Log} n(r)}{\operatorname{Log} r} \leq \limsup_{r \to +\infty} \frac{\operatorname{Log} \operatorname{Log} M(er)}{\operatorname{Log} r}$$
$$= \limsup_{r \to +\infty} \frac{\operatorname{Log} \operatorname{Log} M(r)}{\operatorname{Log} r} = \rho.  \qquad \square$$

Thus, if an entire function $f$ has order $\rho$ and zeros $(z_n)$ with multiplicities $(m_n)$, the following inequality holds:

$$\sum_n m_n |z_n|^{-\rho-\varepsilon} < +\infty \quad \text{for any } \varepsilon > 0.$$

Let $p$ be the entire part of $\rho$, $p = [\rho]$; since $p+1 > \rho$, the series $\sum_n m_n |z_n|^{-p-1}$ converges and one can consider the *canonical product*

$$F(z) = \prod_{n=1}^{\infty} E_p^{m_n}\left(\frac{z}{z_n}\right). \tag{11.11}$$

**Theorem 11.36** (Hadamard). *Every entire function $f$ of finite order $\rho$ can be factorized as*

$$f(z) = z^m \exp(P(z)) F(z),$$

*where $m$ is the multiplicity of the origin as a zero of $f$, $P$ is a polynomial of degree less than or equal to $\rho$ and $F$ is the canonical product* (11.11).

*Proof.* Replacing $f$ by $f/c_m z^m$, where $c_m = \frac{f^{(m)}(0)}{m!}$, we may assume that $f(0) = 1$. Fix $R > 0$, suppose $f(z) \neq 0$ if $|z| = R$ and let $z_1, \ldots, z_N$ be the zeros of $f$ with $|z_n| < R$, $n = 1, 2, \ldots, N$. By Theorem 11.26 one has

$$f(z) = C \prod_{n=1}^{N} \frac{(z - z_n)^{m_n} R}{(R^2 - \bar{z}_n z)^{m_n}} \exp\left\{\frac{1}{2\pi} \int_0^{2\pi} \frac{Re^{it} + z}{Re^{it} - z} \operatorname{Log} |f(Re^{it})| dt\right\}.$$

Consider now, with $p = [\rho]$, the partial product

$$F_R(z) = \prod_{n=1}^{N} E_p^{m_n}\left(\frac{z}{z_n}\right) = C_R \prod_{n=1}^{N} (z - z_n)^{m_n} \exp\left[m_n \sum_{n=1}^{N} \sum_{k=1}^{p} \frac{1}{k}\left(\frac{z}{z_n}\right)^k\right],$$

where $C_R$ is constant. Consider as well the function $h_R(z) = \frac{f(z)}{F_R(z)}$, without zeros in $D(0, R)$, and its logarithm,

$$\log h_R(z) = d_R + \sum_{n=1}^{N} m_n \operatorname{Log} \frac{1}{R^2 - \bar{z}_n z} - m_n \sum_{n=1}^{N} \sum_{k=1}^{p} \frac{1}{k}\left(\frac{z}{z_n}\right)^k$$

$$+ \frac{1}{2\pi} \int_0^{2\pi} \frac{Re^{it} + z}{Re^{it} - z} \operatorname{Log} |f(Re^{it})| dt, \quad |z| < R,$$

for some constant $d_R$.

Differentiating $p + 1$ times the previous equality we get

$$(\log h_R(z))^{(p+1)} = \sum_{n=1}^{N} m_n \frac{p!\, \bar{z}_n^{p+1}}{(R^2 - \bar{z}_n z)^{p+1}}$$

$$+ \frac{(p+1)!}{2\pi} \int_0^{2\pi} \mathrm{Log}\,|f(Re^{it})|\,dt\, \frac{2Re^{it}}{(Re^{it} - z)^{p+2}}\,dt.$$

In terms of $M(R)$ and $n(R) = n_f(R)$ it turns out that, for $|z| = r < R$,

$$|(\log h_R(z))^{(p+1)}| \le \frac{p!\,n(R)}{(R-r)^{p+1}} + \frac{(p+1)!}{2\pi}\,\mathrm{Log}\,M(R)\frac{4\pi R}{(R-r)^{p+2}}.$$

When $R \to +\infty$, the functions $\log h_R$ converge uniformly on compact sets of $\mathbb{C}$ to $\log h$, where $h = \frac{f}{F}$ and, therefore, the derivatives converge as well. Since $\mathrm{Log}\,M(R) = O(R^{\rho+\varepsilon})$ and $n(R) = O(R^{\rho+\varepsilon})$ for all $\varepsilon > 0$, letting $R \to +\infty$ we get that $\log h$ has derivative of order $p + 1$ identically zero and so it must be a polynomial of degree $p$. $\qquad\square$

**Example 11.37.** If $\rho$ is an integer and $P$ a polynomial of degree smaller than or equal to $\rho$, the function $z^m \exp P(z)$ with $m \in \mathbb{N}$, has order $\rho$. If $\rho$ is not an integer, the factor $F$ in Theorem 11.36 with infinitely many terms must necessarily appear; otherwise the order of $f$ would be the degree of $P$. Thus, if an entire function has non-integer order, it has necessarily infinitely many zeros. $\qquad\square$

Remark that Hadamard's theorem implies that the canonical product $F$ has also finite order. The aim now is to find what is exactly the order of a canonical product. Recall $F$ is given by

$$F(z) = \prod_{n=1}^{\infty} E_p^{m_n}\left(\frac{z}{z_n}\right),$$

where $p$ is the smallest integer such that

$$\sum_n \frac{m_n}{|z_n|^{p+1}} < +\infty.$$

We need some estimates of functions $\mathrm{Log}\,|E_p(z)|$ more precise than the ones used in Section 11.2.

**Lemma 11.38.** *The Weierstrass elementary factors $E_p(z)$ satisfy the following inequalities:*

a) $\mathrm{Log}\,|E_p(z)| \le C_p \frac{|z|^{p+1}}{1+|z|}$, $p > 0$, $z \in \mathbb{C}$ *and* $C_p$ *constant.*

b) $\mathrm{Log}\,|E_0(z)| \le \mathrm{Log}(1 + |z|)$, $z \in \mathbb{C}$.

*Proof.* Part b) is evident. Recall now that for $|z| < 1$, one has

$$\operatorname{Log} E_p(z) = - \sum_{k=p+1}^{\infty} \frac{z^k}{k}.$$

It follows that

$$\operatorname{Log} |E_p(z)| \le \sum_{k=p+1}^{\infty} \frac{|z|^k}{k} \le \frac{1}{p+1} \frac{|z|^{p+1}}{1-|z|} \le \frac{2}{p+1} |z|^{p+1},$$

for $|z| < \frac{1}{2}$. If $|z| > 1$, one has, trivially,

$$|E_p(z)| \le (1+|z|) \exp \sum_{k=1}^{p} \frac{|z|^k}{k} \le (1+|z|) \exp\left(p|z|^p\right) \le \exp\left[(p+1)|z|^p\right].$$

Since $\frac{|z|^{p+1}}{1+|z|}$ is of the order of $|z|^{p+1}$ for $|z| < \frac{1}{2}$, of the order of $|z|^p$ for $|z| > 1$ and it is bounded below and above for $\frac{1}{2} \le |z| \le 1$, these estimates prove a) for a constant $C_p$. $\qquad\square$

**Proposition 11.39.** *If $\sum_n \frac{m_n}{|z_n|^{p+1}} < +\infty$, then the canonical product $F(z) = \prod_n E_p^{m_n}\left(\frac{z}{z_n}\right)$ satisfies the estimate*

$$\operatorname{Log} |F(z)| \le B_p r^p \left\{ \int_0^r \frac{n(t)}{t^{p+1}} dt + r \int_r^{\infty} \frac{n(t)}{t^{p+2}} dt \right\},$$

*where $r = |z|$, $n(t)$ is the counting function of $\{z_n; m_n\}$ and $B_p$ is a constant.*

*Proof.* If $p = 0$, using item (b) of Lemma 11.38 and Lemma 11.32, it turns out that

$$\operatorname{Log} |F(z)| \le \sum_{n=1}^{\infty} m_n \operatorname{Log}\left(1 + \frac{r}{|z_n|}\right) = r \int_0^{\infty} \frac{n(t)}{t(t+r)} dt$$

which in turn is bounded above by

$$\int_0^r \frac{n(t)}{t} dt + r \int_r^{\infty} \frac{n(t)}{t^2} dt.$$

If $p > 0$, item a) of Lemma 11.38 and Lemma 11.32, give

$$\operatorname{Log} |F(z)| \le C_p \sum_n m_n \frac{r^{p+1}}{|z_n|^p (r + |z_n|)}$$

$$= C_p r^{p+1} \int_0^{\infty} \left[ \frac{p}{t^{p+1}(t+r)} + \frac{1}{t^p(t+r)^2} \right] n(t) dt.$$

Separating the contributions of $0 \le t \le r$ and of $t > r$, the proof is finished. $\qquad\square$

**Theorem 11.40** (Borel). *The order of a canonical product is equal to the exponent of convergence of its zeros.*

*Proof.* Let $\mu$ be the exponent of convergence of the zeros of the product $F$ defined by (11.11); if $\mu$ is not an integer, $p = [\mu]$ works, and for $\mu$ an integer, take $p = \mu - 1$ if the series $\sum_n \frac{m_n}{|z_n|^\mu}$ converges, and $p = \mu$ if it diverges. Suppose first $\mu < p + 1$; if $\mu + \varepsilon < p + 1$, one has $n(t) = O(t^{\mu+\varepsilon})$ and Proposition 11.39 implies

$$\mathrm{Log}\,|F(z)| \le B_p\, r^p \left\{ O(1) + \int_0^r t^{\mu+\varepsilon-p-1}\,dt + r \int_r^\infty t^{\mu+\varepsilon-p-2}\,dt \right\}$$
$$= B_p r^p O(r^{\mu+\varepsilon-p}) = O(r^{\mu+\varepsilon}).$$

This shows $\rho \le \mu$ if $\rho$ is the order of the product $F$. Suppose now $\mu = p + 1$, that is, the series $\sum_n \frac{m_n}{|z_n|^\mu}$ converges. By Lemma 11.32, the convergence of this series is equivalent to the convergence of the integral

$$\int_0^\infty \frac{n(t)}{t^{\mu+1}}\,dt = \int_0^\infty \frac{n(t)}{t^{p+2}}\,dt.$$

This gives

$$\int_0^r \frac{n(t)}{t^{p+1}}\,dt + r \int_r^\infty \frac{n(t)}{t^{p+2}}\,dt = O(r)$$

and, by Proposition 11.39, we get

$$\mathrm{Log}\,|F(z)| = O(r^{p+1}) = O(r^\mu),$$

from where $\rho \le \mu$ follows. Theorem 11.35 gives the opposite inequality, $\mu \le \rho$. $\qquad\square$

Combining Hadamard's and Borel's theorems the following result is obtained.

**Theorem 11.41.** *A sequence of points of the plane $(z_n)$, $z_n \ne 0$ with multiplicities $(m_n)$ is the sequence of zeros of an entire function of order less than or equal to $\rho$ if and only if $\sum_n m_n |z_n|^{-\rho-\varepsilon} < +\infty$ for every $\varepsilon > 0$.*

## 11.8 Ideals of the algebra of holomorphic functions

One of the most interesting topics when dealing with an algebra is the knowledge of its ideals. A very natural and easy example is the algebra of the polynomials in one variable on an arbitrary field. Let us consider $\mathbb{C}[z]$ and let $I$ be an ideal of this algebra. If we choose a polynomial $P \in I$ of minimal degree among all the non-zero polynomials of $I$, it is easy to see that every other polynomial $H \in I$ is a multiple of $P$, that is, $H = P \cdot Q$, with $Q \in \mathbb{C}[z]$. Actually, we need only divide $H$ by $P$, which gives $H = P \cdot Q + R$ with $\deg(R) < \deg(P)$ and $R \in I$. So $R = 0$. Hence, the ideal $I$ is generated by $P$, that is $I = \langle P \rangle = \{H \in \mathbb{C}[z]\colon H = P \cdot Q,\ Q \in \mathbb{C}[z]\}$, and considering now the zeros of $P$, namely $z_1, z_2, \ldots, z_N$ counted with their multiplicities $m_1, m_2, \ldots, m_N$, it turns out that a polynomial $H$ is in $I$ if and only if $H$ vanishes at each of the points $z_i$, $i = 1, \ldots, N$, with multiplicity greater than or equal to $m_i$.

A natural generalization of this example is the algebra of power series with complex coefficient or algebra of formal power series $\mathbb{C}[[z]]$. It is easy to show that in this algebra, as well as in $\mathbb{C}[z]$, every ideal has just one generator. If now we consider power series with an infinite radius of convergence, we are dealing with the algebra of entire functions. The aim of this section is to study the ideals of the algebra of holomorphic functions on any domain of the complex plane. It will be shown that, in this case, we can only characterize the ideals that satisfy some restrictions.

Fix, then, a domain $U$ of the complex plane and consider the space of the holomorphic functions on $U$, $H(U)$. It is clear that $H(U)$ is a $\mathbb{C}$-algebra, that is to say, a vector space on complex numbers that, in addition, has ring structure. Recall also that $H(U)$ as a subspace of $C(U)$ has a topology coming from a complete metric. The convergence corresponding to this topology is the uniform convergence on compact sets of $U$ (see Subsection 9.1.3), and the operations of the algebra are continuous with respect to this topology.

The easiest ideals in any algebra are *principal ideals*, that is, ideals with only one generator. Choose a function $f \in H(U)$, $f \not\equiv 0$, and consider the principal ideal

$$I = \langle f \rangle = \{g = fh \colon h \in H(U)\}.$$

Let $(z_n)$ be the sequence of the zeros of $f$ and denote by $m_n \geq 1$ the multiplicity of $z_n$. Clearly each function $g \in I$ vanishes at each point $z_n$ with a multiplicity $m'_n \geq m_n$. Conversely, if a function $g \in H(U)$ vanishes at each $z_n$ with multiplicity $m'_n \geq m_n$, it follows that $g \in I$. Actually, the function $g/f$ is holomorphic on $U \setminus \{z_n\}$, and around each point $z_n$ one has $f(z) = (z - z_n)^{m_n} f_1(z)$ and $g(z) = (z - z_n)^{m'_n} g_1(z)$ with $f_1, g_1$ holomorphic and non-vanishing. Therefore, $\frac{g(z)}{f(z)} = (z - z_n)^{m'_n - m_n} \frac{g_1}{f_1}$ is holomorphic at the point $z_n$. Writing $h = \frac{g}{f} \in H(U)$ we get the desired conclusion. It has been shown, then, that the principal ideal $I = \langle f \rangle$ is formed, exactly, by the functions of $H(U)$ vanishing at each zero $z_n$ of $f$ with multiplicity greater than or equal to $m_n$. In particular, applying Theorem 9.3, it follows:

**Proposition 11.42.** *If $I$ is a principal ideal of $H(U)$, then $I$ is a closed set of $H(U)$.*

Start now from a set $Z = \{z_n; m_n\}_{n \in \mathbb{N}}$ composed by a closed and discrete set of points $\{z_n\}$ of $U$ and a sequence of integers $m_n \geq 1$, and consider the set of holomorphic functions

$$I(Z) = \{g \in H(U): g(z_n) = g'(z_n) = \cdots = g^{(m_n-1)}(z_n) = 0, \ n = 1, 2, \ldots\},$$
$$(11.12)$$

which vanish at each point $z_n$ with multiplicity greater than or equal to $m_n$. It is clear that $I(Z)$ is an ideal of $H(U)$. Is it principal? The answer is yes and follows from Weierstrass' theorem (Theorem 11.8). Actually, according to this theorem, there is a function $f \in H(U)$ vanishing at each point $z_n$ with multiplicity $m_n$. This means that $f \in I$ and, as said above, $I = \langle f \rangle$.

From now on, a set of type $Z = \{z_n; m_n\}_{n \in \mathbb{N}}$ with $\{z_n\}$ discrete and closed in $U$ and $m_n \geq 1$ integers, will be called a *zero set in $U$*. Then, the following result has been proved.

**Proposition 11.43.** *There is a one-to-one correspondence between principal ideals of $H(U)$ and zero sets in $U$. The correspondence is obtained by associating to each zero set $Z$ the ideal $I(Z)$ defined by (11.12).*

**Remark 11.1.** In order for this proposition to be true, the empty set $Z = \{\emptyset; \emptyset\}$ must be accepted as a zero set as corresponding to the ideal generated by a function non-vanishing on $U$. It is clear this ideal is the whole algebra $H(U)$. As usual, an ideal $I$ of $H(U)$ such that $\{0\} \neq I \neq H(U)$ will be called a *proper ideal*; then principal proper ideals correspond to non-empty zero sets.

As a next step in studying the ideals of $H(U)$ one considers ideals with a finite number of generators. They can be written as $I = \langle f_1, f_2, \ldots, f_n \rangle = \{g = \sum_{i=1}^{n} h_i f_i: h_i \in H(U)\}$. One may ask: is the ideal $I$ closed? Or even more: is the ideal $I$ principal? The positive answer to these questions will be given below.

Start by noting that $I$ being principal means that there exists a function $f_0 \in H(U)$ such that $f_i \in \langle f_0 \rangle, i = 1, \ldots, n$. If $f_0$ has some zero in $U$, then all functions $f_1, f_2, \ldots, f_n$ should vanish at the zeros of $f_0$. Therefore, if $f_1, \ldots, f_n$ have no common zeros, $f_0$ cannot have zeros and one should have $I = \langle f_0 \rangle = H(U)$. The following result tells that, indeed, this is the case.

**Theorem 11.44.** *Let $f_1, f_2, \ldots, f_n \in H(U)$ be holomorphic functions without common zeros in $U$. Then there exist $g_1, g_2, \ldots, g_n \in H(U)$ such that $\sum_{i=1}^{n} f_i g_i = 1$; that is, $\langle f_1, f_2, \ldots, f_n \rangle = H(U)$.*

*Proof.* Let us proceed by induction on the number $n$ of functions. The case $n = 1$ is trivial, but the case $n = 2$ is needed in order to carry out the induction.

Suppose $f_1, f_2 \in H(U)$ such that $f_1$ and $f_2$ do not vanish simultaneously at any point of $U$. Define the functions

$$\varphi_1 = \frac{\bar{f_1}}{|f_1|^2 + |f_2|^2}, \quad \varphi_2 = \frac{\bar{f_2}}{|f_1|^2 + |f_2|^2},$$

that are $C^\infty$ on $U$ because $|f_1|^2 + |f_2|^2 > 0$ on $U$ and in addition, satisfy the condition $f_1\varphi_1 + f_2\varphi_2 = 1$. The problem consists in replacing $\varphi_1$, $\varphi_2$ by two holomorphic functions $g_1$, $g_2$ so that this equality remains true. To this end look for a function $\psi$, $C^\infty$ on $U$ as well, such that defining

$$g_1 = \varphi_1 + \psi f_2, \quad g_2 = \varphi_2 - \psi f_1,$$

the functions $g_1$, $g_2$ work. The equation $f_1 g_1 + f_2 g_2 = 1$ is clearly satisfied and the question is to choose $\psi$ so that $g_1$, $g_2 \in H(U)$. Imposing $\bar\partial g_1 = \bar\partial g_2 = 0$ yields the equations

$$\bar\partial\varphi_1 + \bar\partial\psi f_2 = 0, \quad \bar\partial\varphi_2 - \bar\partial\psi f_1 = 0,$$

or,

$$\bar\partial\psi = -\frac{\bar\partial\varphi_1}{f_2}; \quad \bar\partial\psi = \frac{\bar\partial\varphi_2}{f_1}. \tag{11.13}$$

An easy computation, beginning with the definition of $\varphi_1$ and $\varphi_2$, gives

$$\frac{\bar\partial\varphi_1}{f_2} = \frac{\bar{f}_1'\bar{f}_2 - \bar{f}_1 \bar{f}_2'}{(|f_1|^2 + |f_2|^2)^2}; \quad \frac{\bar\partial\varphi_2}{f_1} = \frac{\bar{f}_2'\bar{f}_1 - \bar{f}_2 \bar{f}_1'}{(|f_1|^2 + |f_2|^2)^2}$$

so that these two functions are of class $C^\infty$ on $U$. Furthermore, the relation $f_1\varphi_1 + f_2\varphi_2 = 1$ makes both equations of (11.13) to be the same and, finally, one needs to find $\psi$ such that

$$\bar\partial\psi = -\frac{\bar\partial\varphi_1}{f_2} = \frac{\bar\partial\varphi_2}{f_1} \quad \text{on } U.$$

Theorem 10.41 ensures the existence of this function $\psi$, $C^\infty$ on $U$, and this finishes the proof for the case $n = 2$.

Let us suppose now that the theorem holds for $n - 1$ functions and let $f_1, f_2, \ldots$, $f_n \in H(U)$ without common zeros. If $f_1, \ldots, f_{n-1}$ do not have any common zeros, applying the induction hypothesis we are done. If not, let $Z = \{z_k; m_k\}$ be the set of common zeros to $f_1, \ldots, f_{n-1}$, where $m_k$ is the minimum of the multiplicities of $z_k$ as a zero of each $f_i$, $i = 1, \ldots, n - 1$. By Weierstrass' theorem, there is a function $f$ vanishing exactly at each point $z_k$ with multiplicity $m_k$. Then $f_1/f, \ldots, f_{n-1}/f$ are holomorphic on $U$ without common zeros. Therefore, by the induction hypothesis, one can find $h_1, \ldots, h_{n-1} \in H(U)$ such that

$$\sum_{i=1}^{n-1} f_i h_i = f.$$

But $f_n$ does not have common zeros with $f$ (because zeros of $f$ are also zeros of $f_1, \ldots, f_{n-1}$); therefore, by the case $n = 2$ there are functions $g_n, g \in H(U)$ such that $fg + f_n g_n = 1$. That is,

$$\sum_{i=1}^{n-1} f_i(h_i g) + f_n g_n = 1,$$

and this ends the proof. $\qquad\qquad\square$

**Corollary 11.45.** *Every ideal of $H(U)$ with a finite number of generators is principal and, therefore, closed.*

*Proof.* Let $I = \langle f_1, f_2, \ldots, f_n \rangle$ with $f_i \in H(U)$. If the functions $f_1, \ldots, f_n$ do not have common zeros, Theorem 11.44 gives $I = H(U) = \langle 1 \rangle$. If the generators of $I$ have common zeros, take a divisor $f$ of $f_1, \ldots, f_n$ so that $f_1/f, \ldots, f_n/f$ do not have common zeros, as in the last part of the proof of Theorem 11.44. Then there are functions, $g_1, \ldots, g_n \in H(U)$ such that $\sum_{i=1}^n f_i g_i = f$; that is, $f \in I$ and it is clear that every function $g = \sum_{i=1}^n f_i h_i \in I$ must be divisible by $f$ because $g$ must vanish at the common zeros of $f_1, \ldots, f_n$, which are the ones of $f$. $\qquad\square$

So far it has been shown that if $I$ is an ideal with a finite number of generators, then $I = I(Z)$ where $Z$ is a zero set in $U$. Now it is natural to ask if this result holds in general, that is, if every ideal $I$ of $H(U)$ is associated to a zero set $Z$ in $U$. In order for this to be possible, some restriction to $I$ must be imposed. Actually, the ideal $I$ must be closed because if the terms of a sequence of holomorphic functions on $U$, $(f_n)$, vanish on $Z$ and $f_n \to f$, uniformly on compact sets of $U$, then $f$ vanishes on $Z$ too. We will show that this necessary condition is also sufficient in order that $I$ be associated to a zero set in $U$. This implies, in particular, that if $I$ is a closed proper ideal, then the set of common zeros of all the functions of $I$ is not empty. In other words, the ideals without common zeros are to be excluded. What are these ideals? Before answering this, let us introduce a useful notation.

If $I$ is an ideal of $H(U)$, define the set $Z(I)$ of zeros of $I$ in the following way: $Z(I) = \{z_n; m_n\}_{n \in \mathbb{N}}$, where $\{z_n\} = \bigcap_{f \in I} Z(f)$ is the discrete and closed set of $U$ formed by points at which all functions of $I$ vanish; moreover for each $z_n$, take $m_n = \inf\{m(f, z_n) : f \in I\}$. We admit the possibility $Z(I) = \emptyset$, that is, the functions of $I$ do not vanish, simultaneously, at any point of $U$.

**Proposition 11.46.** *If $I$ is an ideal of $H(U)$ such that $Z(I) = \emptyset$, then $I$ is dense in $H(U)$.*

*Proof.* In order to see that $I$ is dense we have to show that given $f \in H(U)$, a compact set $K \subset U$ and $\varepsilon > 0$, there is a function $h \in I$ such that $\sup\{|f(z) - h(z)| : z \in K\} < \varepsilon$ (see Subsection 9.1.3). We may assume that no connected component of $U \setminus K$ is relatively compact in $U$ (if there were components with this property we would join them to $K$ passing to a bigger compact set). Since $Z(I) = \emptyset$, we have

$$Z(I) \cap K = \bigcap_{f \in I} (Z(f) \cap K) = \emptyset.$$

Now $K$ being compact this is so for $Z(f) \cap K$ and we get that there must be a finite number of functions $f_1, f_2, \ldots, f_n \in I$ such that $\bigcap_{i=1}^{n} Z(f_i) \cap K = \emptyset$. That is, we can find an open set $V \subset U$, with $K \subseteq V$ and $f_1, f_2, \ldots, f_n$ without common zeros on $V$. Now we may apply Theorem 11.44 to conclude that there exist functions $g_1, g_2, \ldots, g_n \in H(V)$ with $f_1 g_1 + \cdots + f_n g_n = 1$, on $V$ (if $V$ was not connected, we would apply the theorem to each connected component of $V$). Finally, a version of Runge's theorem (Theorem 10.7) allows us to approximate each $g_i \in H(V)$ by a function $h_i \in H(U)$ uniformly on $K$. Hence, $h = \sum_{i=1}^{n} f \cdot f_i h_i$ satisfies the required condition if the $h_i$ are quite close to the functions $g_i$. $\qquad \square$

**Corollary 11.47.** *If $I$ is a closed proper ideal of $H(U)$, then $Z(I) \neq \emptyset$.*

**Example 11.48.** Now one may show how a dense ideal in $H(U)$ can be. Take a sequence of points $(z_n)$ of $U$ converging to a point in the boundary of $U$. Defining $I$ by

$$I = \{f \in H(U): \text{ there is } n_0(f) \text{ such that } f(z_n) = 0 \text{ if } n \geq n_0(f)\},$$

then $I$ is an ideal with $Z(I) = \emptyset$ and, therefore, it is dense. $\qquad \square$

Finally, one can prove that closed ideals of $H(U)$ are associated to their zero sets.

**Theorem 11.49.** *Let $I$ be a closed ideal of $H(U)$. Then one has $I = I(Z(I))$. In particular, $I$ is a principal ideal.*

*Proof.* It is clear that $I \subseteq I(Z(I))$. For the converse, let $f$ be a generator of the principal ideal $I(Z(I))$ and write $I_1 = \{g \in H(U): fg \in I\}$. Clearly $I_1$ is an ideal and $I \subseteq I_1$. Therefore, $Z(I) \supseteq Z(I_1)$, but now let us show that $Z(I_1) = \emptyset$. Actually, if $z_0 \in Z(I)$, there is a function $f_0 \in I$ such that $z_0$ has the same multiplicity as a zero of $f_0$ as well as a zero of $f$. Therefore, $z_0$ is not a zero of the holomorphic function $g = \frac{f_0}{f}$ and $g \in I_1$; hence, $z_0 \notin Z(I_1)$. By Proposition 11.46, $I_1$ is a dense ideal in $H(U)$. If $T_f : H(U) \to H(U)$ is the operator given by $T_f(h) = fh$, $h \in H(U)$, one has

$$I(Z(I)) = T_f(H(U)) = T_f(\bar{I}_1) \subseteq \overline{T_f(I_1)} \subset \bar{I} = I,$$

so that $I = I(Z(I)) = \langle f \rangle$. $\qquad \square$

Hence, Proposition 11.43 may be reformulated with the same statement but replacing principal ideals by closed ideals.

In every ring it is interesting to find maximal ideals as well. In the case of $H(U)$ it is easy now to determine the maximal ideals that are also closed. Observe that, since $H(U)$ has dense ideals, it must also have dense maximal ideals. Now, if $M \subset H(U)$ is a closed maximal ideal, one must have $Z(M) \neq \emptyset$, by Proposition 11.46, and $Z(M)$ must consist of a unique point with multiplicity 1, that is, $Z(M) = \{z_0; 1\}$; otherwise $M$ would not be maximal. Hence, $M$ is generated by the function $z - z_0$ and it yields

$$M = \{h(z)(z - z_0) : h \in H(U)\}, \quad z_0 \in U.$$

It is clear that every ideal of this kind is maximal and closed and, therefore, these ideals are in a one-to-one correspondence with the points of $U$.

## 11.9 Exercises

1. Let $f$ be a holomorphic function on a neighborhood of the disc $\bar{D}(0, R)$ that has the zeros $z_1, z_2, \ldots, z_N$ in $D(0, R)$. Show the inequality

$$|f(z)| \leq R^N \left| \frac{z - z_1}{R^2 - \bar{z}_1 z} \right| \cdots \left| \frac{z - z_N}{R^2 - \bar{z}_N z} \right| \cdot \sup\{|f(z)| : |z| = R\} \text{ if } |z| < R,$$

   which improves Corollary 4.35 in the case $U = D(0, R)$.

2. Write $P_n(z) = \prod_{k=1}^{n} \left(1 - \frac{z}{k^2}\right)$, $n \in \mathbb{N}$, and let $a_n \in \mathbb{C}$, $n = 1, 2, \ldots$, with $\sum_{1}^{\infty} |a_n| < +\infty$.

   a) Prove that the infinite product $\prod_{k=1}^{\infty} \left(1 - \frac{z}{k^2}\right)$ and the series $f(z) = \sum_{k=1}^{\infty} a_k P_k(z)$ converge uniformly on compact sets of $\mathbb{C}$.

   b) Compute

   $$\int_{\gamma} \frac{z f(z^2)}{P_n(z^2)} \, dz \quad \text{if } \gamma = C(0, n + 1/2) \text{ for } n = 1, 2, \ldots.$$

3. Define on the unit disc $\mathbb{D}$ the sequence of holomorphic functions

   $$f_1(z) = \frac{z(1 - 2z)}{2 - z}, \quad f_{n+1} = f_1 \circ f_n, \quad n = 1, 2, \ldots.$$

   Show that the infinite product $\prod_{n=2}^{\infty} \frac{2 f_n(z)}{f_{n-1}(z)}$ and the sequence $(2^n f_n(z))$, converge uniformly on compact sets of $\mathbb{D}$ and calculate $f(0)$ and $f'(0)$ if $f(z) = \lim_n 2^n f_n(z)$.

   *Hint:* Use the inequalities $|f_n(z)| \leq |z| \left(\frac{1 + 2|z|}{2 + |z|}\right)^n$, $z \in \mathbb{D}$, $n \geq 1$.

**4.** Let $a \in \mathbb{C}$, $|a| < 1$. Prove that the function

$$\theta(z) = \prod_{n=1}^{\infty} (1 + a^{2n-1}e^z)(1 + a^{2n-1}e^{-z})$$

is entire and find its zeros. Show also that the equation

$$\theta(z) = ae^z \theta(z + 2\operatorname{Log} a), \quad z \in \mathbb{C}$$

holds.

*Hint:* Show that the function $\dfrac{e^{-z}\theta(z)}{\theta(z+2\operatorname{Log} a)}$ has two periods that are linearly independent.

**5.** Let $F$ be an entire function written as in Theorem 11.10 with $p_n = p \leq 1$, for all $n$, and $g$ a polynomial of degree less than or equal to 1. Suppose that $F(z)$ is real when $z$ is real and the zeros of $F$ are real. Prove that the zeros of $F'$ are also real and between each two consecutive zeros of $F$ there is exactly one zero of $F'$.

*Hint:* Show that the logarithmic derivative of $F$ has imaginary part different from zero outside the real axis.

**6.** Prove the following formulae for the derivative of the logarithmic derivative of function $\Gamma$:

$$\frac{d}{dz}\left(\frac{\Gamma'(z)}{\Gamma(z)}\right) = \sum_{n=0}^{\infty} \frac{1}{(z+n)^2} = \frac{1}{2z^2} + \int_0^{\infty} \cot \pi t \frac{2zt}{(z^2+t^2)^2}\, dt.$$

*Hint:* Integrate $w \to \dfrac{\pi \cot \pi w}{(z+w)^2}$ on the boundary of the rectangle with vertical sides on $\operatorname{Re} w = 0$ and $\operatorname{Re} w = n + 1/2$ and the horizontal ones on $\operatorname{Im} w = \pm M$; then let $M, n \to +\infty$.

**7.** Prove Legendre's formula

$$\Gamma(2z) = \frac{2^{2z-1}}{\sqrt{\pi}} \Gamma(z)\Gamma(z + 1/2).$$

**8.** Show the result about interpolation by an entire function and its derivatives, on page 476, using Weierstrass' theorem (Theorem 11.8) and Mittag-Leffler's theorem (Theorem 10.9).

**9.** Let $U$ be a bounded domain of $\mathbb{C}$ and $A \subset \partial U$ a countable set of points dense in $\partial U$. Let $(w_n)_{n \in \mathbb{N}}$ be a sequence containing each point of $A$ infinitely many times. For each $n \in \mathbb{N}$ choose a point $z_n \in U$ such that $\sum_n |z_n - w_n| < +\infty$. Prove that the infinite product

$$F(z) = \prod_{n=1}^{\infty} \frac{z - z_n}{z - w_n}$$

converges uniformly on compact sets of $U$ and represents a function $F$ holomorphic on $U$ which cannot be analytically continued to any neighborhood of any point of $\partial U$.

10. Let $f$ be an entire function, $\{z_n; m_n\}_{n \in \mathbb{N}}$ its zero set and $M(r) = \sup\{|f(z)| : |z| = r\}$, $r > 0$. Assume for $\lambda > 0$ that $\int_1^{\infty} \frac{\log M(r)}{r^{\lambda+1}} \, dr < +\infty$ and show that the series $\sum_n m_n |z_n|^{-\lambda}$ is convergent.

11. Let $f$ be holomorphic on the unit disc $\mathbb{D}$ satisfying

$$\sup_{0<r<1} \frac{1}{2\pi} \int_0^{2\pi} \mathrm{Log}^+ |f(re^{i\theta})| \, d\theta = M < +\infty.$$

Prove the inequality

$$|f(z)| \leq \exp\left(M \frac{1 + |z|}{1 - |z|}\right), \quad z \in \mathbb{D}$$

is satisfied. Let now $f$ be holomorphic on $\mathbb{D}$ with $|f(z)| \geq 1$ for $z \in \mathbb{D}$; show that

$$|f(z)| \leq |f(0)|^{2/1-|z|}, \quad z \in \mathbb{D}.$$

12. Deduce Weierstrass' theorem from Mittag-Leffler's theorem (Theorem 10.9):

If $z_n \in \mathbb{C}$, $(z_n) \to \infty$ and $m_n \in \mathbb{N}$, $n = 1, 2, \ldots$, then there exists an entire function vanishing exactly at the points $z_n$ with multiplicities $m_n$, for $n = 1, 2, \ldots$.

*Hint:* Consider a meromorphic function $h$ having a simple pole at each $z_n$ with residue $m_n$ and after that take $f$ with $f'/f = h$.

13. If $B$ is a Blaschke product, show that

$$\lim_{r \to 1} \int_{-\pi}^{\pi} \mathrm{Log} \, |B(re^{i\theta})| \, d\theta = 0.$$

14. Find all the entire functions $f$ satisfying $|f(z)| = 1$ when $|z| = 1$.

15. Prove that the order $\rho(f)$ of an entire function $f$ has the following properties:

i) If $\rho(f_1) < \rho(f_2)$, then $\rho(f_1 + f_2) = \rho(f_2)$.

ii) $\rho(f_1 f_2) \le \max\{\rho(f_1), \rho(f_2)\}$.

iii) If $\rho(f) = \rho$ and $P$ is a polynomial, then $\rho(fP) = \rho$. If $f/P$ is entire, then $\rho(f/P) = \rho$.

iv) $\rho(f) = \rho(f')$.

**16.** If $f_1$, $f_2$ are entire functions so that $f_1/f_2$ is also entire, show that $\rho(f_1/f_2) \le \max\{\rho(f_1), \rho(f_2)\}$.

**17.** Let $f(z) = \sum_0^\infty c_n z^n$ be an entire function of order $\rho$. Prove the formula

$$\frac{1}{\rho} = \liminf_{n \to \infty} \frac{\operatorname{Log} 1/|c_n|}{n \operatorname{Log} n}.$$

*Hint:*

i) The inequality $\sqrt[n]{|c_n|} \le n^{-\alpha}$ for $n \ge n_0$ implies $\rho \le 1/\alpha$.

ii) The inequality $M(r) \le e^{r^\alpha}$ for $r \ge r_0$ implies $\sqrt[n]{|c_n|} \le c\, n^{-1/\alpha}$ if $n \ge n_0$, $c$ constant.

**18.** Let $f$ be an entire function of order $\rho$, $\rho$ being a fractional number. Prove:

a) The exponent of convergence of the zeros of $f$ is equal to $\rho$.

b) The function $f$ takes all the complex values infinitely many times.

**19.** Let $U$ be a simply connected domain, $f \in H(U)$ and $n \in \mathbb{N}$. Show that $f$ has an $n$-th root holomorphic on $U$ if and only if the multiplicity of each zero of $f$ in $U$ is divisible by $n$.

**20.** Let $U$ be a domain of the plane, $f_1, \ldots, f_n \in H(U)$ and $\varphi = \gcd(f_1, \ldots, f_n)$. Show that there exists a unit $u$ in the ring $H(U)$ and functions $g_2, \ldots, g_n \in H(U)$ such that
$$u\varphi = f_1 + g_2 f_2 + \cdots + g_n f_n.$$

# Chapter 12
# The complex Fourier transform

The Fourier representation in the real field exhibits, under suitable conditions, any function as an infinite linear combination of sines and cosines of all frequencies. In this chapter the extension of the Fourier transform to the complex field will be studied; it is obtained by replacing the real frequencies by a complex parameter.

The complex Fourier transform identifies some subspaces of $L^2(\mathbb{R})$ with spaces of entire functions defined by growth conditions. The precise statement is given by the Paley–Wiener theorems. An important case is the one of bandlimited functions, which, according to the Shannon–Whittaker theorem, are determined by their samples on a discrete set of points. This fact has a great interest in signal processing theory.

The Laplace transform, which is a complex Fourier transform defined on the so-called causal functions, will be also studied in detail. The properties of this transformation make it very useful in applications. One of the more typical applications is outlined, namely the one to the solution of linear differential equations with constant coefficients.

The chapter ends by considering some discrete analogs of the Laplace transform, such as Dirichlet's series, that are important in number theory, or the $Z$-transform, which has applications in finite difference equations.

## 12.1 The complex extension of the Fourier transform. First Paley–Wiener theorem

Let $E$ be a measurable subset of $\mathbb{R}$ and $f$ a measurable function on $E$ with real or complex values. As usual, we will write

$$\|f\|_1 = \int_E |f(x)|dx; \quad \|f\|_2 = \left[\int_E |f(x)|^2 dx\right]^{1/2}$$

and

$$L^1(E) = \{f : E \to \mathbb{R} \text{ or } \mathbb{C} : f \text{ is measurable and } \|f\|_1 < +\infty\},$$
$$L^2(E) = \{f : E \to \mathbb{R} \text{ or } \mathbb{C} : f \text{ is measurable and } \|f\|_2 < +\infty\}.$$

The space $L^1(E)$ with $\|\cdot\|_1$ and the space $L^2(E)$ with $\|\cdot\|_2$ are Banach spaces. We will be usually interested in the case that $E$ is the whole line, or a half-line or an interval.

The Fourier transform, $\hat{f}$, of a function $f$ of the space $L^1(\mathbb{R})$ is defined by means of the formula

$$\hat{f}(\xi) = \int_{-\infty}^{+\infty} f(t)\, e^{-2\pi i \xi t}\, dt, \quad \xi \in \mathbb{R}.$$

The function $\hat{f}$ is continuous on $\mathbb{R}$ and vanishes at infinity, but is not necessarily integrable. If $\hat{f}$ is integrable on $\mathbb{R}$, the *inversion formula*

$$f(t) = \int_{-\infty}^{+\infty} \hat{f}(\xi)\, e^{2\pi i \xi t}\, dt \quad \text{for almost all } t \in \mathbb{R} \tag{12.1}$$

holds. The inversion formula holds also under other hypotheses.

If $f \in L^1(\mathbb{R}) \cap L^2(\mathbb{R})$, then $\hat{f} \in L^2(\mathbb{R})$ and *Parseval's identity*

$$\int_{-\infty}^{+\infty} |\hat{f}(\xi)|^2 d\xi = \int_{-\infty}^{+\infty} |f(t)|^2 dt \tag{12.2}$$

is satisfied. Then the Fourier transform of a function $f \in L^2(\mathbb{R})$ can be defined in the following way:

$$\hat{f}(\xi) = \lim_{h \to +\infty} \int_{-h}^{+h} f(t)\, e^{-2\pi i \xi t}\, dt,$$

where the limit is taken in $L^2(\mathbb{R})$. Actually, the last integral is $\hat{f}_h(\xi)$ with $f_h(t) = f(t)\mathbb{1}_{(-h,+h)}(t)$ and we have

$$\|\hat{f}_h - \hat{f}_k\|^2 = \|f_h - f_k\|_2^2 = \int_{h<|t|<k} |f(t)|^2 dt \xrightarrow[h,k \to +\infty]{} 0.$$

The pointwise equality

$$\hat{f}(\xi) = \lim_{h \to +\infty} \int_{-h}^{+h} f(t)\, e^{-2\pi i \xi t}\, dt \quad \text{for almost every } \xi \in \mathbb{R},$$

also holds, though this result is much deeper and more difficult to prove (*Carleson's theorem*).

With the above definition of $\hat{f}$ for $f \in L^2(\mathbb{R})$, equality (12.2) is true for every function $f \in L^2(\mathbb{R})$. The same equality also holds replacing $\hat{f}$ by the function $\check{f}$ defined by

$$\check{f}(t) = \lim_{h \to +\infty} \int_{-h}^{+h} f(\xi)\, e^{2\pi i \xi t}\, d\xi, \quad f \in L^2(\mathbb{R}),$$

with the limit taken in $L^2(\mathbb{R})$ as well.

The transformations $f \mapsto \hat{f}$ and $f \mapsto \check{f}$ are inverse to each other on a dense subspace of $L^2(\mathbb{R})$ (by the inversion formula) and we conclude that the Fourier transform is an isometry from $L^2(\mathbb{R})$ onto $L^2(\mathbb{R})$ (*Parseval's theorem*).

We will be interested in considering the *complex Fourier transform*, which consists in the formal substitution of the real frequency parameter $\xi$ by a *complex frequency $z$*. Hence we will write

$$\hat{f}(z) = \int_{-\infty}^{+\infty} f(t)\, e^{-2\pi itz}\, dt \tag{12.3}$$

whenever this integral makes sense. In this case, the function $\hat{f}(\xi)$ has a *complex extension*.

**Definition 12.1.** For $a > 0$, we denote by $H^2(B_a)$ the space of holomorphic functions $F$ on the strip $B_a = \{z \in \mathbb{C} : |\operatorname{Im} z| < a\}$, satisfying

$$\|F\|_2^2 = \sup_{|y|<a} \int_{-\infty}^{+\infty} |F(x+iy)|^2 dx < +\infty.$$

Further, by $L_a^2$ we denote the subspace of $L^2(\mathbb{R})$ consisting of functions $f \in L^2(\mathbb{R})$ such that

$$\|f\|_{2,a}^2 = \int_{-\infty}^{+\infty} |f(t)|^2\, e^{4\pi a|t|} dt < +\infty.$$

**Theorem 12.2** (First Paley–Wiener theorem). *The complex Fourier transformation given by* (12.3) *establishes, for each $a > 0$, a topological isomorphism between the spaces $L_a^2$ and $H^2(B_a)$; that is, $f \mapsto \hat{f} = F$ is a linear isomorphism satisfying $m\,\|f\|_{2,a} \le \|F\|_2 \le M\,\|f\|_{2,a}$, for some constants $m$, $M$ not depending on $a$.*

*Proof.* Writing $z = x + iy$, one has for $|y| < a$,

$$
\begin{aligned}
|\hat{f}(z)| &\le \int_{-\infty}^{+\infty} |f(t)|\, e^{2\pi ty} \\
&\le \left( \int_{-\infty}^{+\infty} |f(t)|^2\, e^{4\pi a|t|} dt \right)^{1/2} \left( \int_{-\infty}^{+\infty} e^{4\pi(ty-a|t|)} dt \right)^{1/2} \\
&< +\infty,
\end{aligned}
\tag{12.4}
$$

if $f \in L_a^2$.

So $F(z) = \hat{f}(z)$ is well defined if $|y| < a$. Furthermore, if $|y| \le b < a$, then

$$\sup_{x\in\mathbb{R},\, |y|\le b} |f(t)\, e^{-2\pi itz}| \le |f(t)|\, e^{2\pi tb}$$

and the right-hand side function is integrable by (12.4). The same inequality (12.4) guarantees that the integral defining $f(z)$ is continuous with respect to $z$, by the continuity of an integral with respect to a parameter. Therefore, the function $F$ is continuous on $B_a$ and using Morera's theorem and Fubini's theorem, it turns out that $F$ is holomorphic on $B_a$. Notice that $f \in L_a^2$ implies $f \in L^1(\mathbb{R})$, so that $\hat{f}$ is not only continuous, but also has a holomorphic extension to $B_a$. Writing

$$F(z) = \int_{-\infty}^{+\infty} f(t)\, e^{-2\pi i t z}\, dt = \int_{-\infty}^{+\infty} f(t)\, e^{2\pi t y}\, e^{-2\pi i t x}\, dt,$$

one gets that for fixed $y$, with $|y| < a$, the function $x \mapsto F(x + iy)$ is the Fourier transform of $f(t)\, e^{2\pi t y} \overset{\text{def}}{=} f_y(t)$ and $f_y \in L^2(\mathbb{R})$ because $f \in L_a^2$. By Parseval's theorem, one has

$$\int_{-\infty}^{+\infty} |F(x + iy)|^2\, dx = \int_{-\infty}^{+\infty} |f(t)|^2\, e^{4\pi t y}\, dt \le \int_{-\infty}^{+\infty} |f(t)|^2\, e^{4\pi a |t|}\, dt,$$

an inequality that gives $\|F\|_2 \le \|f\|_{2,a}$ and so $F \in H^2(B_a)$.

Conversely, let $F \in H^2(B_a)$ and write $F_y(x) = F(x + iy)$. It is known that $F(x) = \hat{f}(x)$ with

$$f(t) = \lim_h \int_{-h}^{+h} F(x)\, e^{2\pi i t x}\, dx \quad \text{on } L^2(\mathbb{R}).$$

Fix $y$, with $|y| < a$, and let $Q$ be the rectangle with vertices $\pm h$, $\pm h + iy$. By Cauchy's theorem applied to the function $F(z)\, e^{2\pi i t z}$ and to $\partial Q$, it turns out that

$$\int_{-h}^{+h} F(x)\, e^{2\pi i t x}\, dx - \int_{-h}^{+h} F(x + iy)\, e^{2\pi i t (x + iy)}\, dx$$

$$= \int_0^y F(h + is)\, e^{2\pi i t (h + is)} i\, ds - \int_0^y F(-h + is)\, e^{2\pi i t (-h + is)} i\, ds = 0.$$

$$\tag{12.5}$$

Now it will be shown that there is a sequence of numbers $h_j \to +\infty$ such that the last two integrals of (12.5) have limit zero for every value $t \in \mathbb{R}$, when $h = h_j$ and $j \to \infty$. For example, the first one is bounded by

$$\int_0^y |F(h + is)|\, e^{-2\pi t s}\, ds \le \left( \int_0^y |F(h + is)|^2\, ds \right)^{1/2} \left( \int_0^y e^{-4\pi t s}\, ds \right)^{1/2}$$

and it is enough to check that there are values $h_j \to +\infty$ such that

$$\int_0^y |F(\pm h_j + is)|^2\, ds \longrightarrow 0, \quad \text{if } j \to \infty.$$

Writing $\varphi(h) = \int_0^y |F(h+is)|^2 ds$, the hypothesis $F \in H^2(B_a)$ and Fubini's theorem give

$$\int_{-\infty}^{+\infty} \varphi(h)\,dh = \int_0^y \int_{-\infty}^{+\infty} |F(h+is)|^2 ds\,dh \leq y\|F\|_2^2.$$

Therefore, the function $\varphi(h) + \varphi(-h)$ is integrable and, consequently,

$$\int_j^{j+1} (\varphi(h) + \varphi(-h))\,dh \xrightarrow[j \to +\infty]{} 0.$$

Since $\varphi$ is continuous, this last integral coincides with $\varphi(h_j) + \varphi(-h_j)$ for some values $h_j$, $j \leq h_j \leq j+1$. Since

$$f(t) = \lim_j \int_{-h_j}^{+h_j} F(x)\,e^{2\pi itx}\,dx \quad \text{on } L^2(\mathbb{R}),$$

there exists a partial sequence $(h_{j_k})$ of $(h_j)$ providing pointwise convergence a. e. in the above equality. Therefore, by (12.1) and (12.5)

$$f(t) = \lim_k \int_{-h_{j_k}}^{+h_{j_k}} F(x+iy)\,e^{2\pi it(x+iy)}\,dx \quad \text{for a. e. } t \in \mathbb{R}.$$

This means that for the function $f$, inverse Fourier transform of $F$, the equality

$$f(t) = e^{-2\pi ty} \lim_k \int_{-h_{j_k}}^{+h_{j_k}} F(x+iy)\,e^{2\pi itx}\,dx$$

holds. Hence the function $e^{2\pi ty} f(t)$ is the inverse transform of $x \mapsto F(x+iy)$, and Parseval's theorem implies

$$\int_{-\infty}^{+\infty} e^{4\pi ty}|f(t)|^2\,dt = \int_{-\infty}^{+\infty} |F(x+iy)|^2\,dx \leq \|F\|_2^2.$$

Letting now $y \to a$, we obtain

$$\int_0^{+\infty} |f(t)|^2\,e^{4\pi ta}\,dt \leq \|F\|_2^2,$$

and letting $y \to -a$,

$$\int_{-\infty}^0 |f(t)|^2\,e^{4\pi a|t|}\,dt \leq \|F\|_2^2. \qquad \square$$

The easiest example of a pair of functions $f$, $F$ that satisfy the conditions of Theorem 12.2 is $f(t) = e^{-2\pi|t|}$ and

$$F(z) = \frac{1}{\pi}\frac{1}{1+z^2},$$

which restricted to $\mathbb{R}$ is the Poisson kernel.

## 12.2 The Poisson formula

One of the most important results in Harmonic Analysis is the Poisson formula stating the equality

$$\sum_{n\in\mathbb{Z}} f(n) = \sum_{n\in\mathbb{Z}} \hat{f}(n), \qquad (12.6)$$

under certain restrictions on $f$ and its Fourier transform $\hat{f}$, which must guarantee, in particular, that both members of the equation are well defined. In this section this formula will be proved, with methods of complex variables, when $f$ and $\hat{f} = F$ satisfy some additional hypothesis. Concretely assume that $F \in H^2(B_a)$ and

$$\sup_{|y|<a} |F(x+iy)| \le \varphi(x) \qquad (12.7)$$

with $\varphi \in L^1(\mathbb{R})$ and $\varphi(x) \to 0$ when $|x| \to +\infty$. Then $F$ is integrable on $\mathbb{R}$ and

$$f(t) = \int_{-\infty}^{+\infty} F(x)\, e^{2\pi itx}\, dx, \qquad (12.8)$$

is a continuous function.

**Theorem 12.3.** *If $F \in H^2(B_a)$ satisfies (12.7) with $\varphi \in L^1(\mathbb{R})$ and $\varphi(x) \to 0$ when $|x| \to +\infty$, then the function $f$ defined by (12.8) satisfies*

$$|f(t)| = O\!\left(e^{-2\pi b|t|}\right) \quad \text{for every } b < a.$$

*Furthermore, one has*

$$F(z) = \int_{-\infty}^{+\infty} f(t) e^{-2\pi itz}\, dt, \quad \text{for } |\operatorname{Im} z| < a$$

*and*

$$\sum_{n\in\mathbb{Z}} f(n) = \lim_{N\to\infty} \sum_{|n|\le N} F(n) \quad \text{(Poisson formula)}.$$

*Proof.* First, as in Theorem 12.2 we will show that the function $f$ is given by

$$f(t) = e^{-2\pi ty} \int_{-\infty}^{+\infty} F(x+iy)\, e^{2\pi itx}\, dx, \quad |y| < a \qquad (12.9)$$

(hypothesis (12.7) implies that $x \mapsto F(x+iy)$ is integrable for each $y$). With the same notations of Theorem 12.2, it is enough to prove that the integral

$$\int_I |F(h+is)|\, e^{-2\pi ts}\, ds$$

has limit zero when $h \to +\infty$, for each value of $t$, where $I$ is either the interval $[0, y]$ or $[y, 0]$. This is clear because it is dominated by $\varphi(h) \int_I e^{-2\pi t s} ds$. If $b < a$ and we take $y = b$ when $t > 0$ or $y = -b$ when $t < 0$ in (12.9) we obtain $|f(t)| = O(e^{-2\pi b|t|})$. So $f \in L^1(\mathbb{R})$ and the inversion formula shows that $F(x) = \hat{f}(x)$; alternatively we give now a proof of this fact using arguments of complex variable.

For $t \geq 0$ we use (12.9) with $y = b > 0$, $b < a$, obtaining

$$\int_0^\infty f(t) e^{-2\pi i t x} dt = \int_0^\infty e^{-2\pi t b} \left\{ \int_{-\infty}^{+\infty} F(s + ib) e^{2\pi i t s} ds \right\} e^{-2\pi i t x} dt,$$

which, by Fubini's theorem, equals

$$\int_{-\infty}^{+\infty} F(s + ib) \left\{ \int_0^\infty e^{-2\pi i t (x - s - ib)} dt \right\} = -\frac{1}{2\pi i} \int_{-\infty}^{+\infty} \frac{F(s + ib)}{s + ib - x} ds.$$

Similarly, for $t \leq 0$ (12.9) is used with $y = -b$ to find

$$\int_{-\infty}^0 f(t) e^{-2\pi i t x} dt = \frac{1}{2\pi i} \int_{-\infty}^{+\infty} \frac{F(s - ib)}{s - ib - x} ds.$$

Consider the rectangle $Q_h$ with vertices $\pm h \pm ib$. By the Cauchy integral formula, one has

$$F(x) = \frac{1}{2\pi i} \int_{\partial Q_h} \frac{F(z)}{z - x} dz, \quad \text{if } |x| < h.$$

Using (12.7) it is easy to see that the integrals on both vertical sides of $Q_h$ converge to zero when $h \to +\infty$, so that

$$F(x) = \frac{1}{2\pi i} \int_{-\infty}^{+\infty} \frac{F(s - ib)}{s - ib - x} ds - \frac{1}{2\pi i} \int_{-\infty}^{+\infty} \frac{F(s + ib)}{s + ib - x} ds$$

$$= \int_{-\infty}^{+\infty} f(t) e^{-2\pi i t x} dt.$$

Thus, $\hat{f}(z)$ and $F(z)$ are two holomorphic functions on $B_a$ that are equal for $z = x \in \mathbb{R}$. By the analytic continuation principle, it follows that $\hat{f}(z) = F(z)$, $z \in B_a$.

Next the Poisson formula will be proved. Consider $G(z) = F(z)/(e^{2\pi i z} - 1)$, which is meromorphic on $B_a$ with simple poles at the integer points $n \in \mathbb{Z}$ and residue $(2\pi i)^{-1} F(n)$. Apply the residue theorem to the function $G$ and to the rectangle $Q_N$ with vertices $\pm (N + 1/2) \pm ib$, with $N$ integer and $b > 0$. It yields

$$\sum_{|n| \leq N} F(n) = \int_{\partial Q_N} \frac{F(z)}{e^{2\pi i z} - 1}.$$

On a vertical segment $z = \pm(N + 1/2) + is$, $-b < s < b$, the quantity

$$e^{2\pi i z} - 1 = e^{-2\pi s} e^{\pm 2\pi i(N+1/2)} - 1 = -(1 + e^{-2\pi s})$$

has modulus greater than 1; the contribution of this segment to the integral is, therefore, controlled by $\varphi(\pm(N + 1/2))2b$ and goes to zero when $N \to +\infty$. On the other hand, if $z = s \pm ib$, one has $e^{2\pi i z} = e^{2\pi i s} e^{\pm 2\pi b}$, so that $|e^{2\pi i z} - 1|$ is bounded below and the integrals on the horizontal sides of $Q_N$ are convergent when $N \to +\infty$, $F(s \pm ib)$ being integrable. The conclusion is that the limit

$$\lim_{N \to +\infty} \sum_{|n| \le N} F(n) = \int_{L_1} \frac{F(z)}{e^{2\pi i z} - 1} \, dz - \int_{L_2} \frac{F(z)}{e^{2\pi i z} - 1} \, dz$$

exists, where $L_1$, $L_2$ are the lines $L_1 = \{z : \operatorname{Im} z = -b\}$, $L_2 = \{z : \operatorname{Im} z = b\}$ travelled from the left to the right. For $z \in L_1$ one has $z = s - ib$ and $e^{2\pi i z} = e^{2\pi b} e^{2\pi i s}$ has modulus $e^{2\pi b} \ge 1$ so that

$$\frac{1}{e^{2\pi i z} - 1} = e^{-2\pi i z} \sum_{n=0}^{\infty} e^{-2\pi i n z}$$

with uniform convergence for $z \in L_1$. For $z \in L_2$, $z = s + ib$ and $e^{2\pi i z} = e^{-2\pi b} e^{2\pi i s}$ has modulus $e^{-2\pi b} < 1$, and so

$$\frac{1}{e^{2\pi i z} - 1} = -\sum_{n=0}^{\infty} e^{2\pi i n z} \quad \text{uniformly for } z \in L_2.$$

The uniform convergence and the fact that $F(s \pm ib)$ is integrable in the variable $s$ allows us to integrate term by term both series yielding

$$\lim_{N \to +\infty} \sum_{|n| \le N} F(n) = \sum_{n=-\infty}^{-1} \int_{L_1} F(z) \, e^{2\pi i n z} \, dz + \sum_{n=0}^{\infty} \int_{L_2} F(z) \, e^{2\pi i n z} \, dz,$$

which equals $\sum_{n \in \mathbb{Z}} f(n)$, by (12.9). $\qquad\qquad\qquad\qquad\qquad\qquad\square$

It is worth noticing that the Poisson formula holds under more general hypotheses than the ones presented here (see Exercise 11 of Section 12.8).

Observe that if $F$ satisfies condition (12.7), then the same condition holds for each function $F(x + h)$, $h$ being any fixed number. Writing, as before, $F = \hat{f}$, then $F(x + h)$ is the Fourier transform of $f(t) = e^{-2\pi i t h}$. Hence, one has

$$\sum_n F(n + h) = \sum_n f(n) e^{-2\pi i n h},$$

showing that $f(-n)$ is the $n$-th Fourier coefficient of the 1-periodic function $\sum_n F(n + h)$.

**Example 12.4.** When $F$ is the Poisson kernel, $F(z) = \frac{1}{\pi} \frac{1}{1+z^2}$, then $f(t) = e^{-2\pi|t|}$ and

$$\frac{1}{\pi} \sum_{n \in \mathbb{Z}} \frac{1}{1 + (n+h)^2} = \sum_{n \in \mathbb{Z}} e^{-2\pi|n|} e^{-2\pi inh},$$

which is equivalent to

$$\frac{1}{\pi} \sum_{n \in \mathbb{Z}} \frac{1}{1 + (n+h)^2} = 2 \sum_{n=0}^{\infty} e^{-2\pi n} \cos 2\pi nh. \qquad \square$$

**Example 12.5.** If $F(z) = e^{-\pi z^2}$ then $f(t) = e^{-\pi t^2}$ and, therefore, taking $F(z) = a^{-1/2} e^{-\pi \frac{z^2}{a}} e^{2\pi ibz}$ with $a, b \in \mathbb{R}$, $a > 0$, one gets $f(t) = e^{-\pi a(t+b)^2}$. So the equality

$$\sum_{n \in \mathbb{Z}} e^{-\pi a(n+b)^2} = \sum_{n \in \mathbb{Z}} a^{-1/2} e^{-\pi \frac{n^2}{a}} e^{-2\pi inb}$$

follows. $\qquad \square$

The *theta function* is

$$\vartheta(t) = \sum_{n \in \mathbb{Z}} e^{-\pi n^2 t}.$$

It is important in analytic number theory. When $b = 0$, the equality of Example 12.5 gives the functional relation

$$\vartheta(t) = t^{-1/2} \vartheta(1/t), \quad t > 0.$$

## 12.3 Bandlimited functions. Second Paley–Wiener theorem

If the function $f \in L^2(\mathbb{R})$ has compact support, it is clear that the complex Fourier transform

$$\hat{f}(z) = \int_{-\infty}^{+\infty} f(t) e^{-2\pi itz} dt$$

is defined for every $z \in \mathbb{C}$ and is an entire function. More generally, $\hat{f}(z)$ is an entire function whenever

$$\int_{-\infty}^{+\infty} |f(t)| e^{2\pi ty} dt < +\infty, \quad \text{for all } y \in \mathbb{R}.$$

In this section we will characterize the functions $\hat{f}$ for $f \in L^2(\mathbb{R})$ with compact support. If $F(x) = \hat{f}(x)$, then $\hat{F}(x) = f(-x)$ and so the following definition is appropriate.

**Definition 12.6.** A function $F \in L^2(\mathbb{R})$ is said to be bandlimited if $\widehat{F}$ has compact support.

As said above every bandlimited function $F \in L^2(\mathbb{R})$ has an entire extension. Let $I$ be the smallest interval containing the support of $\widehat{F}$; if $2\tau$ denotes the length of $I$, some translation of $\widehat{F}$ has support contained in $(-\tau, \tau)$, so that for some $a \in \mathbb{R}$, the function $e^{2\pi i a x} F(x)$ can be written

$$e^{2\pi i a x} F(x) = \int_{-\tau}^{\tau} f(t)\, e^{-2\pi i x t} dt,$$

with $f \in L^2(-\tau, \tau)$. Without loss of generality, assume $a = 0$. First we will characterize the entire functions $F$ of type

$$F(z) = \int_{-\tau}^{+\tau} f(t)\, e^{-2\pi i t z} dt, \quad f \in L^2(-\tau, \tau),\ \tau > 0. \tag{12.10}$$

**Theorem 12.7.** *An entire function $F$ can be written as in (12.10) if and only if it satisfies*

$$\int_{-\infty}^{+\infty} |F(x+iy)|^2 dx \le C\, e^{4\pi\tau|y|}, \quad y \in \mathbb{R} \tag{12.11}$$

*with $C$ a constant.*

*Proof.* If $F$ is given by (12.10), $x \mapsto F(x+iy)$ is the Fourier transform of $f(t)\, e^{2\pi t y}$ and Parseval's theorem gives

$$\int_{-\infty}^{+\infty} |F(x+iy)|^2 dx = \int_{-\tau}^{+\tau} |f(t)|^2\, e^{4\pi t y} dt \le e^{4\pi\tau|y|} \int_{-\tau}^{+\tau} |f(t)|^2 dt.$$

Conversely, if $F$ is entire and satisfies (12.11), then $F \in H^2(B_a)$ for all $a > 0$, and according to Theorem 12.2 one gets $F = \widehat{f}$ with

$$\int_{-\infty}^{+\infty} |f(t)|^2\, e^{4\pi a|t|} dt \le \frac{1}{m} \sup_{|y|<a} \int_{-\infty}^{+\infty} |F(x+iy)|^2 dx \le C_1\, e^{4\pi a\tau},$$

$C_1$ a constant. Hence,

$$\int_{-\infty}^{+\infty} |f(t)|^2\, e^{4\pi a(|t|-\tau)} dt \le C_1$$

for all $a > 0$. Letting $a \to +\infty$, in this inequality we get $f(t) = 0$ a. e. if $|t| > \tau$. So, $F = \widehat{f}$ with $f \in L^2(-\tau, \tau)$. $\qquad\square$

The aim now is to describe the space of entire functions for which condition (12.11) holds in an alternative and more useful way.

**Definition 12.8.** For $\tau > 0$, the Paley–Wiener space $PW_\tau$ is the set of entire functions $F$ satisfying

$$|F(z)| = O(e^{2\pi\tau|z|}) \quad \text{and} \quad \|F\|_2^2 = \int_{-\infty}^{+\infty} |F(x)|^2 dx < +\infty.$$

It will be shown that this space is the same as the one defined by condition (12.11). A result of Phragmen–Lindelöf, related to the maximum modulus principle, will be needed.

**Theorem 12.9** (Phragmen–Lindelöf). *Let $F$ be a holomorphic function on a circular sector $S_\alpha$ of angle $\pi/\alpha$, with $\alpha > 0$, and continuous on $\overline{S}_\alpha$. Suppose that*

$$|F(z)| = O(\exp|z|^\beta), \quad z \in S_\alpha \text{ with } \beta < \alpha,$$

*and $|F(z)| \le M$ for a constant $M > 0$, if $z \in \partial S_\alpha$. Then one has $|F(z)| \le M$ for all $z \in \overline{S}_\alpha$.*

*Proof.* Without loss of generality, suppose that $S_\alpha$ is the sector $\{z : |\operatorname{Arg} z| < \pi/2\alpha\}$. Take $\beta < \gamma < \alpha$ and using the principal branch of $z^\gamma$ consider the holomorphic function $g(z) = F(z)\exp(-\varepsilon z^\gamma)$ with $\varepsilon > 0$. If $z = r\,e^{i\theta}$, then $|g(z)| = |F(z)|\exp(-\varepsilon r^\gamma \cos\gamma\theta)$; moreover, if $|\theta| \le \pi/2\alpha$, one has $|\gamma\theta| < \pi/2$ and there exists $\delta > 0$ such that $\cos\gamma\theta > \delta$; therefore,

$$|g(z)| \le C \exp(r^\beta - \delta\varepsilon r^\gamma), \quad C \text{ constant},$$

so that $|g(z)|$ tends to zero when $|z| \to +\infty$. This yields that $|g(z)|$ has a global maximum that, by the maximum principle, must be reached on the boundary of $S_\alpha$. But for $z \in \partial S_\alpha$ one has

$$|g(z)| = |F(z)|\exp\left(-\varepsilon r^\gamma \cos\gamma\frac{\pi}{2\alpha}\right) \le |F(z)| \le M.$$

In conclusion, $|g(z)| \le M$ if $z \in S_\alpha$, and letting $\varepsilon \to 0$ we get $|F(z)| \le M$. $\square$

**Corollary 12.10.** a) *Let $F$ be an entire function such that there are constants $C$, $M$ with $|F(z)| \le C\,e^{\tau|z|}$ for $z \in \mathbb{C}$, $\tau > 0$ and $|F(x)| \le M$ for $x \in \mathbb{R}$. Then one has $|F(x+iy)| \le M\,e^{\tau|y|}$, for $x, y \in \mathbb{R}$.*

   b) *Let $F$ be an entire function such that there are constants $C$, $M$ with $|F(z)| \le C\,e^{\tau|z|}$ for $z \in \mathbb{C}$, $\tau > 0$ and*

$$\int_{-\infty}^{+\infty} |F(x)|^p dx \le M^p, \quad 1 \le p < +\infty.$$

*Then*

$$\int_{-\infty}^{+\infty} |F(x+iy)|^p dx \le M^p\,e^{\tau p|y|} \quad \text{for } x, y \in \mathbb{R}.$$

*Proof.* In order to prove item a) assume $y > 0$; let $g(z) = F(z)\,e^{i(\tau+\varepsilon)z}$, $\varepsilon > 0$, so that $|g(x)| = |F(x)| \le M$ and $|g(iy)| = |F(iy)|\exp(-(\tau+\varepsilon)y) \le C\exp(-\varepsilon y) \to 0$ when $y \to +\infty$. Applying the Phragmen–Lindelöf theorem separately to the first and the second quadrants, one obtains $|g(z)| \le M$ if $y = \operatorname{Im} z > 0$, that is,

$$|F(z)| \le M\exp((\tau+\varepsilon)y).$$

Letting $\varepsilon \to 0$ ends the proof of a) for $y > 0$. For $y < 0$, we argue analogously.

For b) we must prove that $\varphi$ being a function of $C_c(\mathbb{R})$ with

$$\int_{-\infty}^{+\infty} |\varphi(x)|^q dx = 1, \quad \frac{1}{p} + \frac{1}{q} = 1,$$

one has

$$\left| \int_{-\infty}^{+\infty} |F(x+iy)|\varphi(x)dx \right| \le M\,e^{\tau|y|}.$$

Considering the entire function

$$G(z) = \int_{-\infty}^{+\infty} F(z+t)\varphi(t)dt, \quad z \in \mathbb{C},$$

we have to check the inequality $|G(iy)| \le M\,e^{\tau|y|}$. Now,

$$|G(z)| \le \int_{-\infty}^{+\infty} |F(z+t)|\,|\varphi(t)|dt \le C \int e^{\tau(|z|+|t|)}|\varphi(t)|dt \le C'\,e^{\tau|z|},$$

$C'$ constant, and, by Hölder's inequality,

$$|G(x)| \le \left( \int_{-\infty}^{+\infty} |F(x+t)|^p dt \right)^{1/p} \left( \int_{-\infty}^{+\infty} |\varphi(t)|^q dt \right)^{1/q}$$

$$\le \left( \int_{-\infty}^{+\infty} |F(x+t)|^p dt \right)^{1/p} = \|F\|_p^p \le M.$$

By item a), already proved, we obtain $|G(z)| \le M\,e^{\tau|y|}$ as claimed. $\qquad\square$

**Theorem 12.11** (Second Paley-Wiener theorem). *Let $F$ be an entire function and $\tau > 0$. Then the following conditions are equivalent:*

a) *$F$ satisfies condition (12.11).*

b) *$F \in PW_\tau$.*

c) *$F_{|\mathbb{R}} \in L^2(\mathbb{R})$ and $|F(z)| = O(e^{2\pi\tau|\operatorname{Im} z|})$.*

*Consequently, the complex Fourier transform establishes an isometry between the spaces $L^2(-\tau, \tau)$ and $PW_\tau$.*

*Proof.* If $F$ satisfies condition (12.11), $F$ can be written as in (12.10) and by Parseval's theorem one has

$$|F(z)| \leq \int_{-\tau}^{+\tau} |f(t)\, e^{2\pi t \operatorname{Im} z}| dt \leq e^{2\pi\tau |\operatorname{Im} z|} \int_{-\tau}^{+\tau} |f(t)| dt \tag{12.12}$$

$$\leq (2\tau)^{1/2} \|f\|_2\, e^{2\pi\tau |\operatorname{Im} z|}.$$

Hence, a) $\Rightarrow$ c). The implication c) $\Rightarrow$ b) is trivial. Finally, if $F \in PW_\tau$, $F$ satisfies the hypotheses of part b) of Corollary 12.10 with $2\pi\tau$ instead of $\tau$ and $p = 2$, so that (12.11) holds.

Now the last statement is a consequence of Theorem 12.7.     $\square$

The space $PW_\tau$ is a relevant space in signal processing. The reason is that bandlimited functions correspond to several signals that are important in the applications of Fourier analysis; for example acoustic signals that fall inside the range of perception of the ear. The Fourier representation

$$F(x) = \int_{-\infty}^{+\infty} \widehat{F}(\xi)\, e^{2\pi i x \xi} d\xi$$

exhibits that every signal of finite energy, described by a function $F \in L^2(\mathbb{R})$ is the superposition of the oscillations $e^{2\pi i x \xi} = \cos 2\pi x \xi + i \sin 2\pi x \xi$ with amplitude $\widehat{F}(\xi)$. Bandlimited functions are those in which only frequencies $\xi$ with $|\xi| \leq \tau$ for a $\tau > 0$ take part. This is the case of acoustic signals, where only frequencies under a certain threshold are perceptible.

So far it has been proved that every bandlimited function has an entire extension and when multiplied by $e^{2\pi i a z}$ for some $a \in \mathbb{R}$ it belongs to the space $PW_\tau$. Now the properties of these functions will be analyzed in more detail. The aim is the *Shannon–Whittaker theorem*, which is the theoretical basis of digitalization processes.

A fundamental result of Fourier will be used. It will be formulated in the context of Hilbert space. The space $L^2(-\tau, \tau)$, $\tau > 0$, is a Hilbert space with the (correlation) scalar product

$$\langle f, g \rangle = \langle f, g \rangle_{L^2(-\tau,\tau)} = \int_{-\tau}^{+\tau} f(x)\overline{g(x)}dx$$

and norm $\|f\|_2^2 = \langle f, f \rangle$. Naturally, $L^2(-\tau, \tau)$ is identified with the space of functions defined on $\mathbb{R}$, $2\tau$-periodic and that are square integrable on one period. The most characteristic of these functions are the $2\tau$-periodic cosine and sine, $\cos \frac{n\pi}{\tau} x$, $\sin \frac{n\pi}{\tau} x$, $n \in \mathbb{Z}$. The result of Fourier states that "every $2\tau$-periodic function is the superposition of the $2\tau$-periodic sine and cosine". We use complex exponentials $e^{\frac{n\pi}{\tau}xi} = \cos\frac{n\pi}{\tau}x + i\sin\frac{n\pi}{\tau}x$ and the notion of orthonormal basis to formulate Fourier's result. The complex exponentials are normalized:

$$e_n(x) = \frac{1}{\sqrt{2\tau}}\, e^{\frac{n\pi}{\tau}xi}, \quad n \in \mathbb{Z},$$

so that $\|e_n\|_2 = 1$. Then Fourier's result reads as follows.

**Theorem 12.12** (Fourier). *The family of functions $\{e_n : n \in \mathbb{Z}\}$ is an orthonormal basis of the space $L^2(-\tau, \tau)$.*

Recall this means that the functions $e_n$ satisfy $\langle e_n, e_m\rangle = \delta_{n,m}$, $n, m \in \mathbb{Z}$ and for each $f \in L^2(-\tau, \tau)$, one has

$$f = \lim_{N\to+\infty} \sum_{|n|\le N} \langle f, e_n\rangle e_n \quad \text{in } L^2(-\tau, \tau).$$

In particular,

$$\|f\|_2^2 = \lim_{N\to+\infty} \Big\| \sum_{|n|\le N} \langle f, e_n\rangle e_n \Big\|_2^2 = \lim_{N\to+\infty} \sum_{|n|\le N} |\langle f, e_n\rangle|^2 = \sum_{-\infty}^{+\infty} |\langle f, e_n\rangle|^2,$$

which is Parseval's theorem in the periodic case. When $\{e_n : n \in \mathbb{Z}\}$ is an orthonormal basis, we write symbolically

$$f \sim \sum_{-\infty}^{+\infty} \langle f, e_n\rangle e_n$$

and the series on the right-hand side is called *Fourier series of $f$*. The proof of Theorem 12.12 can be found in [6], p. 140.

Under hypotheses of regularity of the function $f$, the convergence of the Fourier series of $f$ to $f$ is with respect to more precise norms, e.g. to the uniform one. In some cases it is pointwise for every value of $x$:

$$f(x) = \lim_{N\to+\infty} \sum_{|n|\le N} \langle f, e_n\rangle e_n(x).$$

Carleson's deep theorem states that the above holds at almost every point of $(-\tau, \tau)$ if $f \in L^2(-\tau, \tau)$. Here only Fourier's statement in the form of Theorem 12.12 will be needed. Observe that this result can be also formulated in terms of the functions sine and cosine: the normalized $2\tau$-periodic functions

$$\frac{1}{\sqrt{2\tau}}, \quad \frac{1}{\sqrt{\tau}}\cos\frac{n\pi}{\tau}x, \quad \frac{1}{\sqrt{\tau}}\sin\frac{n\pi}{\tau}x, \quad n = 1, 2, 3, \ldots$$

are an orthonormal basis of $L^2(-\tau, \tau)$.

Being aware of the equalities

$$\langle f, e_n \rangle e_n(x) = \frac{1}{\sqrt{2\tau}} \left( \int_{-\tau}^{\tau} f(x)\, e^{-\frac{n\pi}{\tau}xi}\, dx \right) \frac{1}{\sqrt{2\tau}}\, e^{\frac{n\pi}{\tau}xi}, \quad n \in \mathbb{Z}$$

it is usual to write the Fourier series as

$$f \sim \sum_{-\infty}^{+\infty} \hat{f}(n)\, e^{\frac{n\pi}{\tau}xi}.$$

Now

$$\hat{f}(n) = \frac{1}{2\tau} \int_{-\tau}^{+\tau} f(x)\, e^{-\frac{n\pi}{\tau}xi}\, dx,$$

is the so-called *n-th Fourier coefficient* of $f$. Parseval's theorem is written

$$\int_{-\tau}^{+\tau} |f(x)|^2 dx = \int_{-\tau}^{+\tau} \left| \sum \hat{f}(n)\, e^{\frac{n\pi}{\tau}xi} \right|^2 dx = 2\tau \sum_n |\hat{f}(n)|^2.$$

According to Theorem 12.11, the complex Fourier transform

$$f \mapsto F(z) = \int_{-\tau}^{+\tau} f(x)\, e^{-2\pi i x z}\, dx$$

is an isometry between $L^2(-\tau, \tau)$ and $PW_\tau$, that is, writing $F = \hat{f}$ and $G = \hat{g}$ for $f, g \in L^2(-\tau, \tau)$, one has

$$\int_{-\tau}^{+\tau} f(x)\overline{g(x)}\, dx = \langle f, g \rangle_{L^2(-\tau,\tau)} = \langle F, G \rangle_{L^2(\mathbb{R})} = \int_{-\infty}^{+\infty} F(x)\overline{G(x)}\, dx.$$

Since $\{e_n,\ n \in \mathbb{Z}\}$ is an orthonormal basis of $L^2(-\tau, \tau)$, the entire functions $\{\hat{e}_n,\ n \in \mathbb{Z}\}$ will be an orthonormal basis of $PW_\tau$. We may write down these functions explicitly:

$$\hat{e}_n(z) = \frac{1}{\sqrt{2\tau}} \int_{-\tau}^{+\tau} e^{\frac{n\pi}{\tau}xi}\, e^{-2\pi i x z}\, dx = \frac{1}{\sqrt{2\tau}} \int_{-\tau}^{+\tau} e^{2\pi i x \left(\frac{n}{2\tau}-z\right)}\, dx$$

$$= \frac{1}{2\pi i \left(\frac{n}{2\tau} - z\right)} \frac{1}{\sqrt{2\tau}}\, e^{2\pi i x \left(\frac{n}{2\tau}-z\right)} \bigg|_{x=-\tau}^{x=+\tau} = \sqrt{2\tau}\, \frac{\sin \pi (n - 2\tau z)}{\pi (n - 2\tau z)}.$$

**Definition 12.13.** The sine cardinal function is the entire function

$$\operatorname{sinc} z = \frac{\sin \pi z}{\pi z}.$$

Thus, one has $\widehat{e_n}(z) = \sqrt{2\tau}\,\mathrm{sinc}(n - 2\tau z)$. Moreover if $\hat{f} = F$, the correlation $\langle f, e_n \rangle$ is written

$$\langle f, e_n \rangle = \frac{1}{\sqrt{2\tau}} \int_{-\tau}^{+\tau} f(x)\, e^{-\frac{n\pi}{\tau} xi}\, dx = \frac{1}{\sqrt{2\tau}} F\left(\frac{n}{2\tau}\right) = \langle F, \widehat{e_n} \rangle.$$

Hence, for every function $F \in PW_\tau$ one has

$$F(z) = \lim_{N \to +\infty} \sum_{|n| \leq N} F\left(\frac{n}{2\tau}\right) \mathrm{sinc}(n - 2\tau z) \quad \text{on } L^2(\mathbb{R}).$$

Now, inequality (12.12) has as a consequence that the convergence in $L^2(\mathbb{R})$ implies the uniform convergence on each horizontal strip and so for $F \in PW_\tau$, the equality

$$F(z) = \sum_{-\infty}^{+\infty} F\left(\frac{n}{2\tau}\right) \mathrm{sinc}(n - 2\tau z)$$

holds uniformly on each horizontal strip $\{z : |\,\mathrm{Im}\, z| < a\}$, $a > 0$. Parseval's theorem reads now

$$\int_{-\infty}^{+\infty} |F(x)|^2 dx = \sum_n |\langle F, \widehat{e_n} \rangle|^2 = \frac{1}{2\tau} \sum_{-\infty}^{+\infty} \left| F\left(\frac{n}{2\tau}\right) \right|^2.$$

The fact that $\{e_n,\ n \in \mathbb{Z}\}$ is an orthonormal basis of $L^2(-\tau, \tau)$ also means that $f = \sum_n \lambda_n e_n$ with $\sum_n |\lambda_n|^2 < +\infty$ is the *general expression* of a function $f \in L^2(-\tau, \tau)$. In other words,

$$F(z) = \sum_{n=-\infty}^{+\infty} \lambda_n \,\mathrm{sinc}(n - 2\tau z), \quad \sum_{-\infty}^{+\infty} |\lambda_n|^2 < +\infty$$

is the general expression of a function $F \in PW_\tau$. Summarizing, we get the following result, which in communication theory is known as Shannon–Whittaker's theorem.

**Theorem 12.14** (Shannon–Whittaker). *Every function $F$ bandlimited to $(-\tau, \tau)$ (that is, $F \in L^2(\mathbb{R})$ and $\mathrm{spt}(\hat{F}) \subset (-\tau, \tau)$) is completely determined by its values at the points $\frac{n}{2\tau}$, $n \in \mathbb{Z}$, by means of* cardinal series

$$F(z) = \sum_{-\infty}^{+\infty} F\left(\frac{n}{2\tau}\right) \mathrm{sinc}(n - 2\tau z),$$

*which is uniformly convergent on each horizontal strip $\{z : |\,\mathrm{Im}\, z| < a\}$, $a > 0$. The values $F\left(\frac{n}{2\tau}\right)$ determine $F$ in a stable way, that is,*

$$\int_{-\infty}^{+\infty} |F(x)|^2 = \sum_{-\infty}^{+\infty} \left| F\left(\frac{n}{2\tau}\right) \right|^2 \quad \text{if } F \in PW_\tau.$$

*Conversely, given a sequence of numbers $(\lambda_n)_{n\in\mathbb{Z}}$ with $\sum_n |\lambda_n|^2 < +\infty$, the series*

$$F(z) = \sum_{-\infty}^{+\infty} \lambda_n \operatorname{sinc}(n - 2\tau z)$$

*defines a function $F \in PW_\tau$, satisfying $F(n) = \lambda_n$ for each $n \in \mathbb{Z}$.*

A sequence $(a_n)_{n\in\mathbb{Z}}$ of real numbers is called a *sampling sequence* for the space $PW_\tau$ if there are constants $m, M > 0$ such that

$$m\,\|F\|_2^2 \le \sum_{-\infty}^{+\infty} |F(a_n)|^2 \le M\,\|F\|_2^2, \quad \text{for each } F \in PW_\tau.$$

This means, as before, that the values $\{F(a_n): n \in \mathbb{Z}\}$ determine $F$ in a stable way. A sequence $(a_n)_{n\in\mathbb{Z}}$ is called an *interpolating sequence* for the space $PW_\tau$ if for every sequence $(\lambda_n)_{n\in\mathbb{Z}}$ with $\sum_n |\lambda_n|^2 < +\infty$ there is a function $F \in PW_\tau$ such that $F(a_n) = \lambda_n, n \in \mathbb{Z}$.

Hence, Shannon–Whittaker's theorem asserts that the regularly spaced sequence $\left(a_n = \frac{n}{2\tau}\right)_{n\in\mathbb{Z}}$ is simultaneously sampling and interpolating for the space $PW_\tau$. The quantity $\frac{1}{2\tau}$ is called *Nyquist frequency*. It can be shown that if $\mu < \frac{1}{2\tau}$, the sequence $(\mu_n)_{n\in\mathbb{Z}}$ with $\mu_n = n\mu$ is sampling but not interpolating, and for $\mu > \frac{1}{2\tau}$, it is interpolating but not sampling.

This section ends with a reproducing formula for functions in the space $PW_\tau$.

**Theorem 12.15.** *If $F \in PW_\tau$, one has*

$$F(z) = 2\tau \int_{-\infty}^{+\infty} F(x) \operatorname{sinc} 2\tau(z - x)\,dt.$$

*Proof.* If $F = \hat{f}$ one has

$$F(z) = \int_{-\tau}^{+\tau} f(x)\,e^{-2\pi i x z}\,dz = \langle f, e_z \rangle_{L^2(-\tau,\tau)},$$

where $e_z(x) = e^{2\pi i x \bar{z}}$; therefore, $F(z) = \langle F, \widehat{e_z} \rangle_{L^2(\mathbb{R})}$. Computing $\widehat{e_z}$ it turns out that

$$\widehat{e_z}(w) = \int_{-\tau}^{+\tau} e^{2\pi i x(\bar{z}-w)}\,dx = \frac{1}{2\pi i(\bar{z} - w)}\,e^{2\pi i x(\bar{z}-w)}\Bigg|_{x=-\tau}^{x=+\tau}$$

$$= s\,\frac{\sin 2\pi\tau(\bar{z} - w)}{\pi(\bar{z} - w)}$$

and so

$$F(z) = \int_{-\infty}^{+\infty} F(x)\overline{\widehat{e_z}(x)}\,dx = \int_{-\infty}^{+\infty} F(x)\,\frac{\sin 2\pi\tau(z - x)}{\pi(z - x)}\,dx. \qquad \square$$

It is interesting to remark that the previous properties of functions in the space $PW_\tau$ can be obtained directly without using the representation (12.10). For example (assuming $\tau = \frac{1}{2}$) to obtain the convergence of the series $\sum_n |F(n)|^2$ when $F \in PW_\tau$ we may use the mean value property on the discs with center $n$ and radius $\frac{1}{2}$ to bound this sum by the integral of $|F|^2$ on the union of these discs. Since this union is inside the strip $\{z: |\operatorname{Im} z| < 1\}$, inequality (12.11) implies that this last integral is finite. A similar argument gives that $F(x) \to 0$ when $x \to +\infty$. Let us prove now directly the Shannon–Whittaker formula (Theorem 12.14); since $\sin \pi(n - z) = (-1)^{n+1} \sin \pi z$, we have to set the equality

$$\frac{F(z)}{\sin \pi z} = \sum_{-\infty}^{+\infty} (-1)^n \frac{F(n)}{\pi(z - n)}, \tag{12.13}$$

a formula that will be proved with the same methods used in Section 10.3. Observe first that the convergence of the series $\sum_n |F(n)|^2$ and Schwarz's inequality imply that the series of the right-hand side term of (12.13) converges uniformly on compact sets and defines, therefore, a meromorphic function on the whole plane, with poles at the points $n \in \mathbb{Z}$ and residues $(-1)^n F(n)/\pi$, the same as those of $F(z)/\sin \pi z$. Hence the difference between both terms of (12.13) is an entire function $h$. Let us show that $h$ is constant in a similar way as in Example 10.12, checking first that both terms of (12.13), and so $h$, are bounded on the boundary of the square $Q_N$ centered at the origin and with side $2N + 1$ by a constant independent from $N$. For the right-hand side term this fact has been already proved in Example 10.12, and for the function $F(z)/\sin \pi z$ it comes from $|F(z)| = O(e^{\pi|y|})$ and the fact that $|\sin \pi z|^2 = \sin^2 \pi x + \sinh^2 \pi y$ is bounded below by $ce^{2\pi|y|}$ on the boundary of $Q_N$, with $c$ a constant independent from $N$. So $h$ is constant. To see that this constant is zero take $z = m + \frac{1}{2}$ in both terms of (12.13) and let us check that they have limit zero when $m \to +\infty$. This has been shown before for $F(m + 1/2)$ while for the series

$$\sum_{n=-\infty}^{+\infty} (-1)^n \frac{F(n)}{\pi(m + \frac{1}{2} - n)},$$

the convergence to zero when $m \to +\infty$ is a consequence of $\sum_n |F(n)|^2 < +\infty$.

## 12.4 The Laplace transform

### 12.4.1 The Laplace transform of locally integrable functions

In this section the *Laplace transform* will be analyzed. It is an integral transformation defined on functions vanishing at almost all points of the half-line $(-\infty, 0)$. These are called *causal* functions and correspond to signals depending only on time (positive values of the parameter). Throughout this subsection as well as in Section 12.5 it will be assumed that

$$f : [0, +\infty) \longrightarrow \mathbb{C}$$

is a measurable function such that

$$\int_0^R |f(t)|dt < +\infty, \quad \text{for every } R > 0.$$

**Definition 12.16.** The Laplace transform of $f$, $\mathcal{L}f$, is defined by

$$\mathcal{L}f(z) = \int_0^\infty e^{-tz} f(t)dt = \lim_{R \to +\infty} \int_0^R e^{-zt} f(t)dt$$

for complex values of $z$ for which the limit above exists.

So formally, the Laplace transform of $f$ is the function $\hat{f}\left(\frac{z}{2\pi i}\right)$ whenever $f$ vanishes on $(-\infty, 0)$ and $\hat{f}$ denotes the complex Fourier transform. We can also write

$$\hat{f}(\xi) = \mathcal{L}f(2\pi i \xi).$$

Note that $\mathcal{L}$ is applied only to causal functions and that it is defined for complex arguments of the variable. This makes $\mathcal{L}$ having some additional properties and some advantages with respect to the Fourier transform. For example, the transform $\hat{f}$ may not exist, but $\mathcal{L}f$ does exist.

**Example 12.17.** Consider $f(t) = 1, t \geq 0$. Then

$$\int_0^R e^{-zt} f(t)dt = -\frac{1}{z} e^{-zt} \Big|_{t=0}^{t=R} = \frac{1}{z}(1 - e^{-Rz}), \quad z \neq 0.$$

If $\operatorname{Re} z > 0$, one has $|e^{-Rz}| = e^{-R\operatorname{Re} z} \to 0$ when $R \to +\infty$. Hence, $\mathcal{L}f(z)$ is defined if $\operatorname{Re} z > 0$ and has value $1/z$. Instead, $\hat{f}(\xi)$ is not defined. $\square$

**Example 12.18.** Consider $f(t) = e^{\alpha t}$ for $t \geq 0$, with $\alpha \in \mathbb{C}$. Then

$$\mathcal{L}(f)(z) = \int_0^\infty e^{-tz} f(t)dt = \int_0^\infty e^{t(\alpha - z)} dt = \frac{1}{z - \alpha},$$

for $\operatorname{Re} z > \operatorname{Re} \alpha$. If $f(t) = \cos \alpha t = \frac{1}{2}(e^{i\alpha t} + e^{-i\alpha t})$, then

$$\mathcal{L}(f)(z) = \frac{1}{2}\left(\frac{1}{z - i\alpha} + \frac{1}{z + i\alpha}\right) = \frac{z}{z^2 + \alpha^2}, \quad \text{for } \operatorname{Re} z > |\operatorname{Im} \alpha|$$

and if $f(t) = \sin \alpha t = \frac{1}{2i}(e^{i\alpha t} - e^{-i\alpha t})$, one obtains

$$\mathcal{L}(f)(z) = \frac{1}{2}\left(\frac{1}{z - i\alpha} + \frac{1}{z + i\alpha}\right) = \frac{\alpha}{z^2 + \alpha^2}, \quad \text{for } \operatorname{Re} z > |\operatorname{Im} \alpha|. \quad \square$$

**Example 12.19.** If $f = \mathbb{1}_{[0,T]}$, $T > 0$, it turns out that

$$\mathcal{L}(f)(z) = \int_0^T e^{-tz}\,dt = -\frac{1}{z}\Big[e^{-tz}\Big]_{t=0}^{t=T} = \frac{1}{z}(1 - e^{Tz}). \qquad \square$$

Examples 12.17 and 12.18 show an interesting feature of the Laplace transform. They deal with functions that are neither in $L^1(0,\infty)$ nor in $L^2(0,\infty)$ so that in principle they do not have a well-defined Fourier transform. However, the corresponding Laplace transforms are defined on a half plane and even make sense on the boundary of this domain. It has been already observed that formally $\hat{f}(\xi) = \mathcal{L}f(2\pi i\xi)$, and so $\mathcal{L}f(2\pi i\xi)$ should be a Fourier transform. For example, when $f = 1$ on $[0, +\infty)$, one has $\mathcal{L}f(z) = \frac{1}{z}$ and formally

$$\hat{f}(\xi) = \mathcal{L}f(2\pi i\xi) = \frac{1}{2\pi i\xi}.$$

If $f(t) = \sin t$ on $[0, +\infty)$, then $\mathcal{L}f(z) = \frac{1}{z^2+1}$ and

$$\hat{f}(\xi) = \frac{1}{(2\pi i\xi)^2 + 1} = \frac{1}{1 - 4\pi^2\xi^2}.$$

This extension of the Fourier transform by means of the Laplace transform can be rigorously done using the theory of distributions.

Another advantage of the Laplace transform is that it includes the values of $f$ and of the derivatives of $f$ at the origin, in the formulation of the differentiation and integration rules for $\mathcal{L}f$. This will be shown later on, when the applications of the Laplace transform are considered.

Of course, it may happen that $\mathcal{L}f(z)$ is not defined for any value of $z$, for example, if $f(t) = \exp(t^2)$. We say that the function $f$ has *exponential growth* if there is a real constant $a$ such that

$$|f(t)| = O(e^{at}), \quad t > 0.$$

Then one has

$$\int_0^{+\infty} |f(t)|\,|e^{-tz}|\,dt = \int_0^{+\infty} |f(t)|\,e^{-t\,\mathrm{Re}\,z}\,dt \leq C\int_0^{+\infty} e^{t(a-\mathrm{Re}\,z)}\,dt,$$

so that the left-hand side integral is absolutely convergent, so convergent, whence $\mathcal{L}f(z)$ is defined for $\mathrm{Re}\,z > a$.

The first question to be considered is to find the *domain of convergence* of the Laplace transform of $f$, that is, the region

$$C(f) = \{z \in \mathbb{C} : \mathcal{L}(f)(z) \text{ is convergent}\}.$$

As it will be seen, there is a parallelism with power series for which, as it is known, the domain of convergence is given by the radius of convergence of the series.

**Lemma 12.20.** *Suppose that $\mathcal{L}(f)(z_0)$ converges. Then $\mathcal{L}(f)(z)$ converges uniformly on the whole sector $S_\beta = \{z: |\mathrm{Arg}(z - z_0)| \leq \beta\}$ if $\beta < \pi/2$. That is, for every $\varepsilon > 0$ there is a number $R_0 = R_0(\varepsilon, \beta)$ such that for $R > R_0$, one has*

$$\left| \mathcal{L}f(z) - \int_0^R e^{-tz} f(t)dt \right| < \varepsilon, \quad z \in S_\beta.$$

*Proof.* Write, for $R_1 < R_2$,

$$\int_{R_1}^{R_2} e^{-tz} f(t)dt = \int_{R_1}^{R_2} e^{-t(z-z_0)} e^{-tz_0} f(t)dt$$

and consider

$$g(t) = \int_t^{+\infty} e^{-sz_0} f(s)ds = \mathcal{L}(f)(z_0) - \int_0^t e^{-sz_0} f(s)ds.$$

Then integration by parts gives,

$$\int_{R_1}^{R_2} e^{-tz} f(t)dt = -e^{-t(z-z_0)} g(t)\Big|_{t=R_1}^{t=R_2} - (z - z_0) \int_{R_1}^{R_2} e^{-t(z-z_0)} g(t)dt$$

(this equality can be directly proved by Fubini's theorem). Given $\varepsilon > 0$, let $R_0$ be such that $|g(t)| < \varepsilon$ if $t > R_0$. If $z \in S_\beta$ and $R_1 > R_0$, then

$$\left| \int_{R_1}^{R_2} e^{-tz} f(t)dt \right| \leq 2\varepsilon + |z - z_0| \frac{2\varepsilon}{\mathrm{Re}(z - z_0)} \leq 2\varepsilon\left(1 + \frac{1}{\cos \beta}\right)$$

and, by Cauchy's criterion, the result is obtained. $\qquad\square$

**Definition 12.21.** The abscissa of convergence of the Laplace transform $\mathcal{L}(f)$ is the number $\alpha_f$ defined as

$$\alpha_f = \inf\{\alpha \in \mathbb{R}: \text{ there exists } z \in \mathbb{C} \text{ with } \mathrm{Re}\, z = \alpha \text{ and } \mathcal{L}(f)(z) \text{ converges}\}.$$

Lemma 12.20 implies that $C(f)$ contains the half plane $\{z: \mathrm{Re}\, z > \alpha_f\}$ and that $C(f)$ is contained in $\{z: \mathrm{Re}\, z \geq \alpha_f\}$ if $\alpha_f$ is finite. If $C(f) = \emptyset$ one has $\alpha_f = +\infty$ and $\alpha_f = -\infty$ implies $C(f) = \mathbb{C}$.

As for power series, in general nothing can be said about the behavior of $\mathcal{L}f(z)$ for $\mathrm{Re}\, z = \alpha_f$.

**Example 12.22.** When $f(t) = 1$, $t \geq 0$ or, more generally, $f$ coincides on $(0, +\infty)$ with a polynomial, one has $\alpha_f = 0$ and $\mathcal{L}(f)(z)$ does not converge if $\mathrm{Re}\, z = 0$. If $f$ is integrable on $(0, +\infty)$,

$$\int_0^{+\infty} |f(t)|dt < +\infty,$$

then

$$\int_0^{+\infty} |f(t)|\,|e^{-tz}|\,dt = \int_0^{+\infty} |f(t)|\,e^{-t\,\mathrm{Re}\,z}\,dt \le \int_0^{+\infty} |f(t)|\,dt, \quad \mathrm{Re}\,z \ge 0$$

so that $C(f)$ contains $\{z : \mathrm{Re}\,z \ge 0\}$. For example, the Laplace transform of $f(t) = \frac{1}{1+t^2}$ has abscissa of convergence equal to $0$ and $\mathcal{L}(f)(z)$ converges if and only if $\mathrm{Re}\,z \ge 0$. $\qquad\square$

Remark that $\alpha_f \le 0$ whenever $f \in L^p(0,\infty)$ with $1 \le p \le +\infty$, because for $\mathrm{Re}\,z > 0$,

$$\int_0^{+\infty} |f(t)|\,e^{-t\,\mathrm{Re}\,z}\,dt \le \left(\int_0^{+\infty} |f(t)|^p\,dt\right)^{1/p}\left(\int_0^{+\infty} e^{-qt\,\mathrm{Re}\,z}\,dt\right)^{1/q} < +\infty$$

if $1 < p < +\infty$ and $\frac{1}{p} + \frac{1}{q} = 1$, and

$$\int_0^{+\infty} |f(t)|\,e^{-t\,\mathrm{Re}\,z}\,dt \le \int_0^{+\infty} |f(t)|\,dt \quad \text{if } p = 1,$$

$$\int_0^{+\infty} |f(t)|\,e^{-t\,\mathrm{Re}\,z}\,dt \le \|f\|_\infty \int_0^{+\infty} e^{-t\,\mathrm{Re}\,z}\,dt \quad \text{if } p = +\infty.$$

One can also introduce the notion of *abscissa of absolute convergence*:

$$\beta_f = \inf\left\{\beta \in \mathbb{R} : \int_0^{+\infty} |f(t)|\,e^{-\beta t}\,dt < +\infty\right\}.$$

It is easy to prove that $\mathcal{L}(f)(z)$ converges absolutely if $\mathrm{Re}\,z > \beta_f$ and that it does not if $\mathrm{Re}\,z < \beta_f$. Clearly, $\alpha_f \le \beta_f$, but as well as an integral may be convergent without being absolutely convergent, it may happen that $\alpha_f < \beta_f$. This fact establishes a difference to power series, for which both notions coincide. When $f \in L^p(0,+\infty)$, $1 \le p \le +\infty$, it has already been noticed that $\beta_f \le 0$.

Hence, if $-\infty \le \alpha_f < +\infty$, the set $\{z : \mathrm{Re}\,z > \alpha_f\}$ is the biggest open set on which $\mathcal{L}(f)$ is defined. It is the analog of the disc of convergence of a power series.

**Proposition 12.23.** *If $\alpha_f < +\infty$, then $\mathcal{L}(f)$ is a holomorphic function on the half plane $\{z : \mathrm{Re}\,z > \alpha_f\}$.*

*Proof.* Let

$$F_n(z) = \int_0^n e^{-tz} f(t)\,dt, \quad z \in \mathbb{C},\ n \in \mathbb{N}.$$

By Lemma 12.20, the sequence $(F_n)$ converges uniformly on every compact set of $\{z : \mathrm{Re}\,z > \alpha_f\}$. Since each $F_n$ is an entire function, it follows from Theorem 9.3 that $\mathcal{L}(f)$ is holomorphic on $\{z : \mathrm{Re}\,z > \alpha_f\}$. $\qquad\square$

The analogous result to Theorem 2.31 is the following one.

**Proposition 12.24.** *Suppose $\alpha_f < +\infty$ and consider the functions $f_k(t) = t^k f(t)$ for $k \in \mathbb{N}$. Then the abscissa of convergence of $\mathcal{L}(f_k)$ is less than or equal to $\alpha_f$ and one has*

$$\mathcal{L}(f_k)(z) = (-1)^k \mathcal{L}(f)^{(k)}(z), \quad \text{if } \operatorname{Re} z > \alpha_f.$$

*Proof.* This equality is obtained by differentiating formally $k$ times the equality defining $\mathcal{L}(f)(z)$. In order to prove it rigorously observe that, with the notations of Proposition 12.23, Theorem 9.3 implies

$$\mathcal{L}(f)^{(k)}(z) = \lim_{n \to \infty} F_n^{(k)}(z) = \lim_{n \to \infty} \int_0^n (-t)^k e^{-tz} f(t)\,dt$$

uniformly on compact sets of $\{z : \operatorname{Re} z > \alpha_f\}$. This already shows that the abscissa of convergence of $\mathcal{L}(f_k)$ is less than or equal to $\alpha_f$ and finishes the proof.     $\square$

**Example 12.25.** Taking $f(t) = 1$, $t \geq 0$, in Proposition 12.24 and using Example 12.17 which gives $\mathcal{L}(1) = \frac{1}{z}$, for $\operatorname{Re} z > 0$, one obtains that the Laplace transform of a polynomial $P(t) = \sum_{k=0}^n a_k t^k$ is

$$\mathcal{L}(P)(z) = \sum_{k=0}^n \frac{k!\, a_k}{z^{k+1}}.$$     $\square$

The analog of Abel's theorem (Theorem 2.20) for the Laplace transform is the following result:

**Proposition 12.26.** *Suppose $\alpha_f < +\infty$ and $\mathcal{L}(f)(z_0)$ converges at the point $z_0 \in \mathbb{C}$ with $\operatorname{Re} z_0 = \alpha_f$. Then one has*

$$\lim_{z \to z_0} \mathcal{L}(f)(z) = \mathcal{L}(f)(z_0)$$

*for $z$ approaching $z_0$ inside a sector $\{z : |\operatorname{Arg}(z - z_0)| \leq \beta\}$, $\beta < \frac{\pi}{2}$.*

*Proof.* It is an immediate consequence of Lemma 12.20, because $\int_0^R e^{-tz} f(t)\,dt$ is an entire function for each $R < +\infty$.     $\square$

As a consequence of Lemma 12.20 we get that $\mathcal{L}f(z)$ vanishes at infinity.

**Proposition 12.27.** *If $\alpha_f < +\infty$, one has $\lim_{|z| \to +\infty} \mathcal{L}f(z) = 0$ for $z$ converging to $\infty$ inside a sector $\{z : |\operatorname{Arg}(z - z_0)| \leq \beta\}$ with $\beta < \frac{\pi}{2}$ and $\operatorname{Re} z_0 = \alpha_f$.*

*Proof.* It has been proved that the equality

$$\mathcal{L}f(z) = \lim_{R \to +\infty} \int_0^R e^{-tz} f(t)dt$$

holds uniformly on the circular sector. Hence, we just need to show that for fixed $R$ this integral converges to zero when $\mathrm{Re}\, z \to +\infty$ and this is a consequence of the estimate

$$\left| \int_0^R e^{-tz} f(t)\,dt \right| \leq \int_0^R e^{-t\,\mathrm{Re}\,z} |f(t)|\,dt$$

and the dominated convergence theorem.     $\square$

**Example 12.28.** Consider $f(t) = \frac{\sin \alpha t}{t}$. Since $tf(t) = \sin \alpha t$ has as Laplace transform $\frac{\alpha}{z^2+\alpha^2}$, (Example 12.18), Proposition 12.24 shows that $\mathcal{L}f(z)$ has as derivative the function $-\frac{\alpha}{z^2+\alpha^2}$. Hence $\mathcal{L}f(x) = -\arctan \frac{x}{\alpha} + k$ and since it must converge to zero when $x \to +\infty$, Proposition 12.27 yields $k = \frac{\pi}{2}$. It is well known that the integral

$$\int_0^\infty \frac{\sin \alpha t}{t}\, dt$$

is convergent but not absolutely convergent, that is, $0 \in C(f)$. By Proposition 12.26, it turns out that

$$\int_0^\infty \frac{\sin \alpha t}{t}\, dt = \mathcal{L}f(0) = \lim_{x \to 0^+} \mathcal{L}f(x) = \lim_{x \to 0} \left( \frac{\pi}{2} - \arctan \frac{x}{\alpha} \right) = \frac{\pi}{2}.     \square$$

Let $\sum c_n z^n$ be a power series with $\sum |c_n| < +\infty$. Then the radius of convergence $R$ satisfies $R \geq 1$ and $f(z) = \sum_n c_n z^n$ is continuous on $\overline{D}(0, R)$. The analog of this fact for the Laplace transform is the following proposition.

**Proposition 12.29.** *If* $f \in L^1(0, \infty)$, *then* $\beta_f \leq 0$ *and* $\mathcal{L}f(z)$ *is continuous on* $\{z : \mathrm{Re}\, z \geq 0\}$.

*Proof.* It has already been observed that $\beta_f \leq 0$. Since

$$|f(t)|\,|e^{-tz}| = |f(t)|\,e^{-t\,\mathrm{Re}\,z} \leq |f(t)|,$$

the statement is a consequence of the dominated convergence theorem.     $\square$

Next the uniqueness theorem for the Laplace transform will be proved.

**Theorem 12.30.** *The Laplace transformation is one-to-one; more generally, if* $\alpha_f < +\infty, \alpha_g < +\infty$ *and* $\mathcal{L}f(z) = \mathcal{L}g(z)$ *for* $\mathrm{Re}\, z$ *big enough, then* $f(t) = g(t)$ *for a. e.* $t > 0$.

*Proof.* Working with $h = f - g$, we need to show that $\mathcal{L}h(z) = 0$ for $\operatorname{Re} z$ large enough entails $h(t) = 0$ for a. e. $t > 0$. Obviously, by the principle of analytic continuation, we have $\mathcal{L}h(z) = 0$ for $\operatorname{Re} z > \alpha_h$, that is,

$$\int_0^\infty e^{-tz} h(t)\,dt = 0, \quad \operatorname{Re} z > \alpha_h.$$

The idea is simply to specify $z$ at the points of the kind $\alpha_h + n, n = 1, 2, \ldots$, which implies

$$\int_0^\infty e^{-t\alpha_h}\, e^{-tn} h(t)\,dt = 0, \quad n = 1, 2, \ldots,$$

next to make the change of variable $x = e^{-t}, dx = -e^{-t}\,dt$, which yields

$$\int_0^1 x^{n-1} h\,(-\operatorname{Log} x)\, x^{\alpha_h}\,dx = 0, \quad n = 1, 2, \ldots,$$

and then to use the density of the polynomials in the space $L^1(0, 1)$. However, since the integrals that appear need not be absolutely convergent, we have to integrate by parts before. So, introduce, for $z_0$ fixed with $\operatorname{Re} z_0 > \alpha_h$, the function

$$g(t) = \int_0^t e^{-sz_0} h(s)\,ds$$

and integrate by parts to obtain

$$\int_0^R e^{-sz} h(s)\,ds = \int_0^R e^{-s(z-z_0)}\, e^{-sz_0} h(s)\,ds$$
$$= e^{-R(z-z_0)} g(R) + (z - z_0) \int_0^R e^{-s(z-z_0)} g(s)\,ds. \tag{12.14}$$

The function $g$ is continuous on $[0, +\infty)$; furthermore, $\lim_{t \to +\infty} g(t)$ exists and equals $\mathcal{L}h(z_0) = 0$ and equality (12.14) implies

$$\mathcal{L}h(z) = (z - z_0) \int_0^\infty e^{-s(z-z_0)} g(s)\,ds.$$

Therefore, this last integral is zero if $\operatorname{Re} z > \operatorname{Re} z_0$. Specifying at $z = z_0 + n$, $n = 1, 2, \ldots$, we find

$$\int_0^\infty e^{-sn} g(s)\,ds = 0$$

and writing $x = e^{-s}$,

$$\int_0^1 x^n g\,(-\operatorname{Log} x)\,dx = 0, \quad n = 0, 1, 2, \ldots.$$

Now $G(x) = g(-\mathrm{Log}\, x)$ is a continuous function on $[0, 1]$ and so we can conclude that $G \equiv 0$. Therefore, $g(t) = 0$ for all $t > 0$ and so $h(t) = 0$ for a. e. $t > 0$, because $g(t)$ is absolutely continuous with derivative $e^{-tz_0}h(t)$ a. e. $\qquad\square$

Next we analyze which explicit inversion formulae can be found to express $f$ in terms of $\mathcal{L}f$. It is clear this is a question deeply related with the inversion of the Fourier transform. For $x > \alpha_f$,

$$\mathcal{L}f(x+iy) = \int_0^{+\infty} e^{-(x+iy)t} f(t)dt = \int_0^{+\infty} e^{-tx} f(t)\, e^{-iyt}\, dt$$

formally equals $\hat{g}_x\left(\frac{y}{2\pi}\right)$, where $g_x(t) = e^{-tx} f(t)$. Therefore one has

$$e^{-tx} f(t) = g_x(t) = \int_{-\infty}^{+\infty} \hat{g}_x(\xi)\, e^{2\pi it\xi} d\xi$$

$$= \frac{1}{2\pi} \int_{-\infty}^{+\infty} \hat{g}_x\left(\frac{y}{2\pi}\right) e^{ity}\, dy$$

$$= \frac{1}{2\pi} \int_{-\infty}^{+\infty} \mathcal{L}f(x+iy)\, e^{ity}\, dy.$$

So, independently of the value of $x$, $x > \alpha_f$, we get

$$f(t) = \frac{1}{2\pi i} \int_{\mathrm{Re}\, z = x} \mathcal{L}f(z)e^{tz}\, dz = \lim_{T \to +\infty} \frac{1}{2\pi i} \int_{x-iT}^{x+iT} \mathcal{L}f(z)\, e^{tz}\, dz.$$

This formal inversion formula holds if $\mathcal{L}f$ satisfies some additional hypotheses, as in the case of the Fourier transform. A result of this kind is the following one, analogous to Theorem 12.3.

**Theorem 12.31.** *Assume $\beta_f < +\infty$ and for $\beta > \beta_f$,*

$$\sup_{x > \beta} |\mathcal{L}f(x+iy)| \le \varphi(y)$$

*with $\varphi \in L^1(\mathbb{R})$ and $\varphi(y) \to 0$ when $|y| \to +\infty$. Then $f(t)$ equals for almost every $t > 0$ the continuous function*

$$e^{tx}\frac{1}{2\pi} \int_{-\infty}^{+\infty} \mathcal{L}f(x+iy)\, e^{ity}\, dy = \frac{1}{2\pi i} \int_{\mathrm{Re}\, z = x} \mathcal{L}f(z)\, e^{tz}\, dz$$

*independently of the value of $x$, $x > \beta$.*
*For $t < 0$ one has $\int_{\mathrm{Re}\, z = x} \mathcal{L}f(z)\, e^{tz}\, dz = 0$.*

*Proof.*  As in the proof of Theorem 12.3, Cauchy's theorem and the hypothesis imply that the function

$$g(t) = \frac{1}{2\pi i} \int_{\mathrm{Re}\, z = x} \mathscr{L} f(z)\, e^{tz}\, dz$$

is independent from $x$, $x > \beta$ and continuous on $\mathbb{R}$.  Letting $x \to +\infty$ in the estimate

$$|g(t)| \leq \frac{e^{tx}}{2\pi} \int_{-\infty}^{+\infty} |\mathscr{L} f(x+iy)|\, dy \leq \frac{e^{tx}}{2\pi} \int_{-\infty}^{+\infty} \varphi(y)\, dy$$

we get $g(t) = 0$ if $t < 0$.  Now the equality $\mathscr{L} g(z) = \mathscr{L} f(z)$ if $\mathrm{Re}\, z > \beta$ will be proved; then the uniqueness theorem gives $f = g$ a. e.  Fixing $z$ with $\mathrm{Re}\, z > \beta$, we choose $x$ with $\mathrm{Re}\, z > x > \beta$ such that

$$\mathscr{L} g(z) = \int_0^{+\infty} g(t)\, e^{-tz}\, dt = \int_0^{+\infty} \left( \frac{1}{2\pi i} \int_{\mathrm{Re}\, w = x} \mathscr{L} f(w)\, e^{tw}\, dw \right) e^{-tz}\, dt.$$

The double integral is absolutely convergent and we may interchange the order of integration to obtain

$$\mathscr{L} g(z) = \frac{1}{2\pi i} \int_{\mathrm{Re}\, w = x} \mathscr{L} f(w) \left( \int_0^{+\infty} e^{t(w-z)}\, dt \right) dw$$

$$= -\frac{1}{2\pi i} \int_{\mathrm{Re}\, w = x} \frac{\mathscr{L} f(w)}{w - z}\, dw.$$

Now, applying Cauchy's formula to the square $Q_R$ with vertices $x \pm iR$, $R \pm iR$ with $R > x$ big enough such that $z$ is in its interior, one has

$$\mathscr{L} f(z) = -\frac{1}{2\pi i} \int_{x-iR}^{x+iR} \frac{\mathscr{L} f(w)}{w - z}\, dw + I_1 + I_2 + I_3,$$

where $I_1$ and $I_2$ correspond to the horizontal sides and $I_3$ to the vertical side $\{w \colon \mathrm{Re}\, w = R\}$ of $Q_R$.  The following estimates hold:

$$|I_1|, |I_2| \leq \varphi(R)\frac{R - x}{d(z, \partial Q_R)} \leq C\varphi(R) \xrightarrow[R \to +\infty]{} 0, \quad C \text{ constant},$$

$$|I_3| \leq \frac{1}{d(z, \partial Q_R)} \int_{-\infty}^{+\infty} \varphi(y)\, dy \xrightarrow[R \to +\infty]{} 0$$

and, letting $R \to +\infty$, we find

$$\mathscr{L} f(z) = -\frac{1}{2\pi i} \int_{\mathrm{Re}\, w = x} \frac{\mathscr{L} f(w)}{w - z}\, dw,$$

which implies $\mathscr{L} f = \mathscr{L} g$.  $\qquad\square$

The hypothesis on the growth of $\mathcal{L}f$ in Theorem 12.31 holds, for example, if $|\mathcal{L}f(z)| = O(|z|^{-2})$, taking $\varphi(y) = (\beta^2 + y^2)^{-1}$. Note that it has been proved that every function $F$ holomorphic on a half plane $\{z: \operatorname{Re} z > \beta\}$ and satisfying the conditions of Theorem 12.31 is the Laplace transform of a function $f$ with $\beta_f < +\infty$. For example, if $F$ is a rational function vanishing at infinity, $F(z)$ is the sum of terms of type $C(z - \alpha)^{-1}$ with $C, \alpha \in \mathbb{C}$ and other terms of order $O(|z|^{-2})$. Consequently $F$ is the Laplace transform of some function.

**Example 12.32.** Let us compute $\mathcal{L}^{-1}F$ for $F(z) = \frac{1}{z^2-3z+2}$. We use the inversion formula of Theorem 12.31, that is, we will compute the integral

$$f(t) = \frac{1}{2\pi i} \int_{x-i\infty}^{x+i\infty} \frac{e^{tz}}{z^2 - 3z + 2} \, dz =$$
$$= \lim_{T\to+\infty} \frac{1}{2\pi i} \int_{x-iT}^{x+iT} \frac{e^{tz}}{z^2 - 3z + 2} \, dz$$

by residues, with $x$ big enough. The poles are $z = 1$ and $z = 2$, $F(z)$ is holomorphic for $\operatorname{Re} z > 2$, and so we take, $x > 2$. For big $T$ consider the closed path $C_T$ of Figure 12.1. On the circular part of $C_T$, $z = x + R e^{i\theta}$ with $\frac{\pi}{2} \le \theta \le \frac{3\pi}{2}$ and

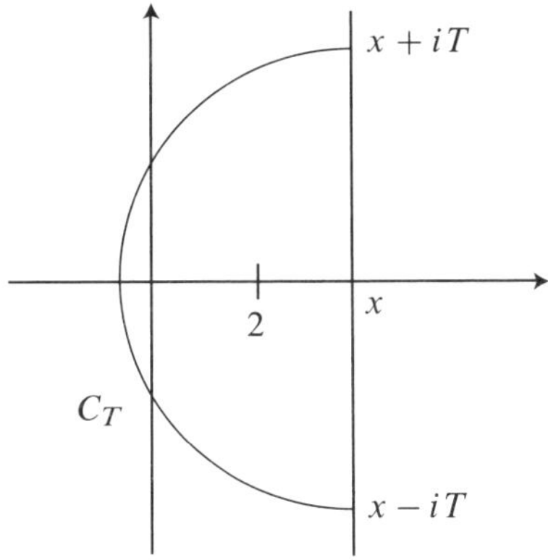


Figure 12.1

$$|F(z) e^{tz}| = e^{tx} \frac{e^{tT\cos\theta}}{|z^2 - 3z + 2|} \le C \frac{e^{tx}}{T^2}, \quad \text{for } T \text{ big enough and } C \text{ constant,}$$

so that in the limit, when $T \to +\infty$, the contribution of this part is zero. Moreover,

$$\operatorname{Res}\left(\frac{e^{tz}}{z^2 - 3z + 2}, z = 1\right) = -e^t, \quad \operatorname{Res}\left(\frac{e^{tz}}{z^2 - 3z + 2}, z = 2\right) = e^{2t}.$$

Consequently, $f(t) = e^{2t} - e^t$. We can reach the same conclusion by decomposing $F$ into simple fractions

$$F(z) = \frac{1}{z-2} - \frac{1}{z-1}$$

and recalling that $e^{\alpha t}$ has $1/z - \alpha$ as its Laplace transform. $\qquad\square$

It has been observed that every rational function vanishing at infinity is the Laplace transform of a function $f$. To find $f$ we can decompose the rational function as a sum of simple fractions and remark that by Proposition 12.24, the function $\frac{t^k}{k!} e^{\alpha t}$ has as Laplace transform the function $\frac{1}{(z-\alpha)^{k+1}}$. The functions $f$ having as Laplace transforms rational functions vanishing at infinity are, in consequence, the linear combinations of functions $t^k e^{\alpha t}$ with $k \in \mathbb{N}$, $\alpha \in \mathbb{C}$, that is,

$$f(t) = \sum_{i=1}^{N} P_i(t) e^{\alpha_i t}$$

with $\alpha_i \in \mathbb{C}$ and $P_i$ polynomial functions. These functions are called *exponential polynomials*.

## 12.4.2  The Laplace transform of square integrable functions

The behavior of the Laplace transform on the space $L^2(0, +\infty)$ is particularly simple, as it is the case with the Fourier transform on $L^2(\mathbb{R})$. In this subsection the corresponding version of Plancherel's theorem will be stated.

**Definition 12.33.** The Hardy space $H^2(\Pi)$ on the half plane $\Pi = \{z : \operatorname{Re} z > 0\}$ is the set of functions $F \in H(\Pi)$ satisfying

$$\|F\|_2^2 = \sup_{x>0} \int_{-\infty}^{+\infty} |F(x + iy)|^2 dy < +\infty.$$

**Theorem 12.34.** *The Laplace transform,* $F = \mathcal{L}f$, *is a bijection between* $L^2(0, +\infty)$ *and* $H^2(\Pi)$ *that satisfies, in addition,*

$$2\pi \int_0^{+\infty} |f(t)|^2 dt = \|F\|_2^2.$$

*Proof.* It has been observed before that $\alpha_f \leq 0$ if $f \in L^2(0, \infty)$, so that $\mathcal{L}f$ is defined on $\Pi$. We also know that $\mathcal{L}f(x + iy) = \hat{g}_x\left(\frac{y}{2\pi}\right)$ with $g_x(t) = e^{-tx} f(t)$. Therefore, by Parseval's identity, one has

$$\int_{-\infty}^{+\infty} |\mathcal{L}f(x + iy)|^2 dy = 2\pi \int_{-\infty}^{+\infty} |\hat{g}_x(\xi)|^2 d\xi$$

$$= 2\pi \int_{-\infty}^{+\infty} e^{-2tx} |f(t)|^2 dt \leq 2\pi \int_{-\infty}^{+\infty} |f(t)|^2 dt,$$

implying $F = \mathscr{L}f \in H^2(\Pi)$ and $\|F\|_2^2 \le 2\pi \|f\|_2^2$. Conversely, let now $F \in H^2(\Pi)$ be given. Using Cauchy's theorem and an argument analogous to the one in the proof of Theorem 12.2, we get that the limit

$$\lim_{T \to +\infty} \frac{1}{2\pi i} \int_{x-iT}^{x+iT} F(z)\, e^{tz}\, dz = e^{tx} \lim_{T \to +\infty} \frac{1}{2\pi} \int_{-T}^{T} F(x+iy)\, e^{ity}\, dy$$
$$= e^{tx} h_x(t)$$

is independent from $x$, where $2\pi h_x(2\pi t)$ is the inverse Fourier transform of the function $y \to F(x+iy)$. By Plancherel's theorem, then

$$2\pi \int_{-\infty}^{+\infty} |h_x(t)|^2 dt = \int_{-\infty}^{+\infty} |F(x+iy)|^2 dy \le \|F\|_2^2.$$

Writing $f(t) = e^t h_1(t) = e^{tx} h_x(t)$, we get

$$\int_{-\infty}^{+\infty} e^{-2tx} |f(t)|^2 dt = \int_{-\infty}^{+\infty} |h_x(t)|^2 dt \le \frac{1}{2\pi} \|F\|_2^2.$$

Letting $x \to +\infty$ we obtain $f(t) = 0$ a. e. on $(-\infty, 0)$, and letting $x \to 0^+$ it follows that $f \in L^2(0, +\infty)$ and

$$2\pi \int_0^\infty |f(t)|^2 \le \|F\|_2^2.$$

Since $f \in L^2(0, +\infty)$, we have $h_x \in L^1(\mathbb{R})$. Now, taking the Fourier transform of the function $2\pi h_x(2\pi t)$ yields

$$F(x+iy) = \int_{-\infty}^{+\infty} 2\pi h_x(2\pi t)\, e^{-2\pi ty}\, dt$$
$$= \int_0^\infty h_x(t)\, e^{-ity}\, dt = \int_0^\infty e^{-tz} f(t)\, d,$$

that is, $F$ is the Laplace transform of $f$.                                $\square$

## 12.5 Applications of the Laplace transform

The applications of the transformation $f \mapsto \mathscr{L}f$ are based on its properties with respect to operations between causal functions. The most important are given here.

Recall that $f \colon [0, +\infty) \to \mathbb{C}$ is assumed to be measurable satisfying $\int_0^R |f(t)|\, dt < +\infty$, for all $R > 0$.

P1. $\mathscr{L}(t^n f(t)) = (-1)^n (\mathscr{L}f)^{(n)}(z)$ if $n \in \mathbb{N}$.

This is Proposition 12.24. For example, the function $\frac{1}{z-a}$ corresponds by the Laplace transformation to the function $e^{\alpha t}$, and $\frac{n!}{(z-a)^{n+1}}$ corresponds to $t^n e^{\alpha t}$.

P2. $\mathcal{L}(f(\lambda t))(z) = \frac{1}{\lambda} \mathcal{L}f\left(\frac{z}{\lambda}\right), for \ \lambda > 0.$

P3. *Assume that $f$ is of class $C^1$ on $(0, +\infty)$ and right continuous at zero. If $\alpha_{f'} < +\infty$, then $\alpha_f < +\infty$ too and for $\mathrm{Re}\, z$ big enough one has*

$$\mathcal{L}(f')(z) = z\mathcal{L}(f)(z) - f(0).$$

*Proof.* Integrating by parts yields, for $R > 0$,

$$\int_0^R e^{-tz} f'(t)dt = e^{-Rz} f(R) - f(0) + z \int_0^R e^{-tz} f(t)dt$$

and thus it suffices to prove that $\alpha_{f'} < +\infty$ implies $e^{-Rz} f(R) \to 0$ when $R \to +\infty$, for $\mathrm{Re}\, z > 0$ big enough. By hypothesis, $\lim_R \int_0^R e^{-tx} f'(t)dt$ exists for some value of $x$, for which we write

$$g(t) = \int_0^t e^{-sx} f'(s)ds.$$

Then

$$e^{-Rz} f(R) = e^{-Rz} f(0) + e^{-Rz} \int_0^R f'(t)dt$$

$$= e^{-Rz} f(0) + e^{-Rz} \int_0^R e^{xt} g'(t)dt$$

$$= e^{-Rz} f(0) + e^{-(z-x)R} g(R) - e^{-Rz} x \int_0^R e^{xt} g(t)dt.$$

Since $g$ is bounded, the equality above implies

$$|e^{-Rz} f(R)| \le e^{-R\,\mathrm{Re}\, z}|f(0)| + C\, e^{-R(\mathrm{Re}\, z - x)} + C\, e^{-R(\mathrm{Re}\, z - x)}, \quad C \text{ constant,}$$

and the right-hand term converges to zero when $R \to +\infty$, if $\mathrm{Re}\, z > x$.   $\square$

P4. *If $f$ is of class $C^n$ on $(0, +\infty)$ and the right derivatives $f^{(i)}(0^+)$ for $i = 0, \ldots, n-1$ exist, and moreover $\alpha_{f^{(n)}} < +\infty$, then*

$$\mathcal{L}(f^{(n)})(z) = z^n \mathcal{L}(f)(z) - z^{n-1} f(0) - z^{n-2} f'(0) - \cdots - f^{(n-1)}(0).$$

P5. *Under the hypotheses of P3, one has*

$$f(0) = \lim_{x \to +\infty} x\mathcal{L}(f)(x).$$

Actually, apply Proposition 12.27.

Analogously, under the hypotheses of P4, one has

$$f'(0) = \lim_{x \to +\infty} x^2 \mathcal{L}(f)(x) - x f(0) = \lim_{x \to +\infty} x(x \mathcal{L}(f)(x) - f(0)),$$

$$\vdots$$

$$f^{(k)}(0) = \lim_{x \to +\infty} x^{k+1} \mathcal{L}(f)(x) - x^k f(0) - x^{k-1} f'(0) - \cdots - x f^{(k-1)}(0).$$

The meaning of these equalities is that the asymptotic expansion of $\mathcal{L} f(x)$ when $x \to +\infty$ is

$$\mathcal{L} f(x) = \frac{f(0)}{x} + \frac{f'(0)}{x^2} + \frac{f''(0)}{x^3} + \cdots + \frac{f^{(k)}(0)}{x^{k+1}} + \cdots .$$

If $\mathcal{L} f(z)$ is holomorphic at infinity, then this series, in the variable $z$, is the expansion around the point at infinity of $\mathcal{L} f(z)$.

P6. *Let $f$ be continuous and define $g(t) = \int_0^t f(s)ds$. Then*

$$\mathcal{L} g(z) = \frac{1}{z} \mathcal{L} f(z).$$

This is a consequence of P3.

P7. *Assume the function $g(t) = \frac{f(t)}{t}$ is also integrable on each interval $[0, R]$ with $R > 0$ and, in addition, $\alpha_g < +\infty$. Then $(\mathcal{L} g)'(z) = -\mathcal{L} f(z)$.*

This assertion is a consequence of P1.

For example, consider the function $g(t) = \frac{\cos \alpha t - \cos \beta t}{t}$, with $\alpha$, $\beta$ real parameters. Then $\mathcal{L} g(z)$ has as derivative

$$\frac{z}{z^2 + \beta^2} - \frac{z}{z^2 + \alpha^2}$$

and it vanishes at infinity. Hence, for $x \in \mathbb{R}$, one has

$$\mathcal{L} g(x) = \int_x^\infty \left( \frac{t}{t^2 + \alpha^2} - \frac{t}{\alpha^2 + \beta^2} \right) dt = \frac{1}{2} \text{Log} \frac{x^2 + \alpha^2}{x^2 + \beta^2}$$

and, by analytic continuation, it turns out that $\mathcal{L} g(z) = \frac{1}{2} \text{Log} \frac{z^2 + \alpha^2}{z^2 + \beta^2}$. Also $0 \in C(g)$, and, by Proposition 12.29,

$$\int_0^\infty \frac{\cos \alpha t - \cos \beta t}{t} dt = \mathcal{L} g(0) = \lim_{x \to 0^+} \mathcal{L} g(x) = \text{Log} \frac{\alpha}{\beta}.$$

P8. *If $\lambda > 0$ and interpreting that the translated function $\tau_\lambda f(t) = f(t - \lambda)$ has value $0$ for $t < \lambda$, then $\mathcal{L}(f(t - \lambda)(z) = e^{-\lambda z} \mathcal{L} f(z)$.*

**P9.** *If $z_0 \in C(f)$ and $z - z_0 \in C(f)$, then*

$$\mathcal{L} f(z - z_0) = \mathcal{L}(e^{z_0 t} f(t))(z).$$

**P10.** *If $f$ is $\lambda$-periodic, then $\alpha_f \le 0$ and for $\operatorname{Re} z > 0$ one has*

$$\mathcal{L}(f)(z) = \frac{1}{1 - e^{-\lambda z}} \int_0^\lambda e^{-tz} f(t)dt.$$

Actually, if $g$ denotes the function that equals $f$ on $(0, \lambda)$ and vanishes outside this interval, we can write $f(t) = \sum_{n=0}^\infty g(t - n\lambda)$ and applying P8, it turns out that

$$\mathcal{L} f(z) = \sum_{n=0}^\infty e^{-n\lambda z} \mathcal{L} g(z) = (1 - e^{-\lambda z})^{-1} \mathcal{L} g(z).$$

The *convolution* of two causal functions $f$, $g$ is defined as

$$(f * g)(t) = \int_0^t f(s)g(t - s)ds.$$

Observe that it is the usual convolution of $f$ and $g$ if these functions are considered to be extended by zero to $(-\infty, 0)$.

**P11.** *Assume $\beta_f < +\infty$ and $\beta_g < +\infty$. Then $\beta_{f*g} \le \max(\beta_f, \beta_g)$ and, if $\operatorname{Re} z > \max(\beta_f, \beta_g)$, one has*

$$\mathcal{L}(f * g)(z) = \mathcal{L}(f)(z) \cdot \mathcal{L}(g)(z).$$

Indeed, let $x > \beta_f, \beta_g$ such that

$$I = \int_0^{+\infty} e^{-tx}(|f(t)| + |g(t)|)dt < +\infty.$$

Then

$$\int_0^{+\infty} e^{-tx}|(f * g)(t)|dt$$

$$\le \int_0^{+\infty} e^{-tx}\left(\int_0^t |f(s)|\,|g(t - s)|\,ds\right)dt$$

$$= \int_0^{+\infty} |f(s)|\left(\int_s^{+\infty} |g(t - s)|\,e^{-tx}dt\right)ds$$

$$= \int_0^{+\infty} e^{-sx}|f(s)|\left(\int_s^{+\infty} |g(t - s)|\,e^{-(t-s)x}dt\right)ds$$

$$= \int_0^{+\infty} e^{-sx}|f(s)|\left(\int_0^{+\infty} |g(t)|\,e^{-tx}dt\right)ds \le I^2.$$

Therefore, $\beta_{f*g} \leq \max(\beta_f, \beta_g)$, and repeating the computation above without absolute values, the equality of P.11 is found.

Once the list of properties of the Laplace transform is stated, we proceed to its main applications.

A typical application of the Laplace transform is the solution of ordinary linear differential equations with constant coefficients. As known, the solution of the equation

$$P(D)f = g \tag{12.15}$$

with $g(t)$ a given function and $P(D) = D^n + a_{n-1}D^{n-1} + \cdots + a_1 D + a_0$, $D = \frac{d}{dt}$, $a_0, a_1, \ldots, a_{n-1} \in \mathbb{R}$, is completely determined by the initial conditions at a point $t_0$:

$$f^{(k)}(t_0) = b_k, \quad k = 0, \ldots, n-1, \tag{12.16}$$

with $b_0, b_1, \ldots, b_{n-1}$ constants.

Since the problem is invariant under translations, we may assume that $t_0 = 0$. In order to find the solution $f(t)$ for $t > 0$ the Laplace transform will be used. For $t < 0$ just remark that $\tilde{f}(t) = f(-t)$ satisfies the equation

$$\tilde{P}(D)\tilde{f}(t) = \tilde{g}(t) = g(-t)$$

with $\tilde{P} = \sum_{i=0}^{n} a_i(-1)^i D^i$. It is convenient to extend a little bit the concept of *solution*. So, $g$ will be assumed to be continuous on $(0, +\infty)$, except for a set of isolated points at each of which $g$ has a jump. A *generalized solution $f$* is then a function of class $C^{n-1}$ on $(0, +\infty)$, being of class $C^n$ on the set of points where $g$ is continuous and satisfying $P(D)f = g$ on this set. The initial conditions are interpreted as

$$f^{(k)}(0+) = b_k, \quad k = 0, \ldots, n-1.$$

As it will be shown, the use of the transformation $\mathcal{L}$ has two advantages: first the data $g$ and $b_0, \ldots, b_{n-1}$ are simultaneously handled, and afterwards the problem is reduced to an algebraic equation. The idea is to first find $\mathcal{L}f$ and then look for $f$ with the inversion theorem, so that the equation becomes a problem about functions of the complex variable $z$.

Suppose $f$ is a solution of the problem (12.15) and (12.16). Write $F = \mathcal{L}f$ and $G = \mathcal{L}g$, which is a data. Due to rule P4,

$$\mathcal{L}(P(D)f) = P(z)F(z) - b_0 p_{n-1}(z) - \cdots - b_{n-1}p_0(z),$$

where $P(z) = z^n + \sum_{i=0}^{n-1} a_i z^i$ and $p_k(z)$ are the polynomials

$$p_0(z) = 1; \quad p_k(z) = z^k + a_{n-1}z^{k-1} + \cdots + a_{n-k}, \quad k = 1, \ldots, n-1.$$

Therefore,

$$F(z) = \frac{G(z) + b_0 p_{n-1}(z) + \cdots + b_{n-1} p_0(z)}{P(z)}.$$

Now we must invert the Laplace transform. Observe, first, that we have the decomposition

$$F = F_1 + F_2, \quad \text{with } F_1 = \frac{G}{P} \text{ and } F_2 = \frac{b_0 p_{n-1} + \cdots + b_{n-1} p_0}{P}.$$

This decomposition corresponds to $f = f_1 + f_2$, where $f_1$ is the solution of $P(D)f_1 = g$ with vanishing initial values and $f_2$ is the solution of $P(D)f_2 = 0$ with initial values $f_2^{(k)} = b_k$, $k = 0, \ldots, n-1$. The solution $F_2$ is rational and vanishes at infinity, and so $f_2$ is an exponential polynomial.

The function $\frac{1}{P}$ is rational and vanishes at infinity and so there is an exponential polynomial $h$ with $\mathcal{L}h = \frac{1}{P}$. From $\mathcal{L}f_1 = F_1 = \frac{G}{P} = \mathcal{L}(h)\mathcal{L}(g)$, using P11 and the uniqueness theorem, we obtain $f_1 = g * h$, that is,

$$f_1(t) = \int_0^t h(t-s)f(s)ds, \quad t > 0.$$

When $g$ is continuous, the next result follows from the existence and uniqueness theorem of solutions for ordinary differential equations. Here an independent proof is given, without assuming the continuity of $g$.

**Theorem 12.35.** *The problem*

$$P(D)f = (D^n + a_{n-1}D^{n-1} + \cdots + a_1 D + a_0)f = g,$$

$f^{(k)}(0) = b_k$, $k = 0, \ldots, n-1$, *has a unique generalized solution given by*

$$f(t) = f_1(t) + f_2(t) = \int_0^t h(t-s)g(s)ds + f_2.$$

*Here $f_2$ is the exponential polynomial satisfying*

$$\mathcal{L}f_2(z) = \frac{1}{P(z)} \sum_{k=0}^{n-1} b_k \, p_{n-1-k}(z)$$

*with $p_k(z) = z^k + a_{n-1}z^{k-1} + \cdots + a_{n-k}$ ($p_0(z) = 1$) and $h$ is the exponential polynomial such that $\mathcal{L}h = \frac{1}{P}$.*

*Proof.* The previous considerations prove that if there is a solution, then it is the one given in the statement. To prove that it is indeed a solution, it is enough to check the initial conditions $f^{(k)}(0) = b_k$, $k = 0,\ldots,n-1$, because then $\mathcal{L}(P(D)f) = \mathcal{L}(g)$ by construction and $P(D)f = g$. Check first that

$$f_1^{(k)}(0) = 0, \quad k = 0,\ldots,n-1.$$

By property P5 and observing that $P(z) = z^n + \cdots$, we deduce

$$h(0) = \cdots = h^{n-1}(0) = 0, \quad h^{(n)}(0) = 1.$$

It is clear that $f_1(0) = 0$ and, differentiating successively, it follows that

$$f_1^{(k)}(t) = \int_0^t h^{(k)}(t-s)g(s)ds, \quad f_1^{(k)}(0) = 0, \quad k = 0,\ldots,n-1$$

(since $h^{(k)}(0) = 0, k = 0,\ldots,n-1$, these equalities hold even at the points where $g$ has a jump).

Next we need to prove that $f_2^{(k)}(0) = b_k$ for $k = 0, 1,\ldots,n-1$. The function $\frac{P_{n-1-k}(z)}{P(z)}$ has order $z^{-1-k}$ when $z \to \infty$ and it is therefore clear that $\mathcal{L} f_2(z) \sim \frac{b_0}{z}$. Now, in the equality

$$z\mathcal{L} f_2(z) - b_0 = \frac{z\sum b_k\, p_{n-1-k}(z) - b_0 P(z)}{P(z)}$$

the numerator starts with $b_1 z^{n-1}$ and in consequence it is of the order of $\frac{b_1}{z}$, and so on. $\qquad\square$

In many practical cases, to find the solution $f$ there is no need to compute $h$ and $h * g$ separately. For example, if $g$ is an exponential polynomial, then $G = \mathcal{L}g$ is a rational function vanishing at infinity, so this holds for $F_1 = \frac{G}{P}$ as well, and we may straightforwardly compute $f_1$ by decomposing into simple fractions. This procedure can also be applied if $g$ is a linear combination of translations of exponential polynomials in the sense of P8. This is shown in the first of the following examples.

**Example 12.36.** Let us solve the equation

$$f'(t) + 2f(t) = \begin{cases} 1 & \text{if } 0 < t < 1, \\ 0 & \text{if } t > 1 \end{cases}$$

with the initial value $f(0) = 1$.

In this case one has

$$\mathcal{L}(f' + 2f) = \mathcal{L}\left(\mathbb{1}_{[0,1]}\right) = \frac{1}{z}\left(1 - e^{-z}\right).$$

But $\mathcal{L}(f')(z) = -f(0) + z\mathcal{L}(f)(z)$ and, hence,

$$-1 + (z+2)\mathcal{L}(f)(z) = \frac{1}{z}\left(1 - e^{-z}\right),$$

and so

$$\mathcal{L}(f)(z) = \frac{1}{z+2} + \frac{1}{z(z+2)} - \frac{-e^{-z}}{z(z+2)}.$$

Decomposing into simple fractions

$$\frac{1}{z(z+2)} = \frac{1}{2}\left(\frac{1}{z} - \frac{1}{z+2}\right)$$

we get

$$\mathcal{L}(f)(z) = \frac{1}{2}\left(\frac{1}{z} + \frac{1}{z+2}\right) - \frac{1}{2}e^{-z}\left(\frac{1}{z} - \frac{1}{z+2}\right).$$

Recall that $\frac{1}{z-\alpha}$ is the transform of $e^{\alpha t}$. Applying P8 one finds

$$f(t) = \frac{1}{2}\left(1 + e^{-2t}\right) - \frac{1}{2}\tau_1\left(1 - e^{-2t}\right),$$

that is, $f(t) = \frac{1}{2}(1 + e^{-2t})$ if $0 < t < 1$ and

$$f(t) = \frac{1}{2}\left(1 + e^{-2t}\right) - \frac{1}{2}\left(1 - e^{-2(t-1)}\right) = \frac{1}{2}\left(e^{-2t} + e^{-2(t-1)}\right) \quad \text{if } t > 1.$$

Observe that $f$ is continuous but not differentiable at the point 1, because $g = \mathbb{1}_{[0,1]}$ is not continuous at this point. $\qquad\square$

**Example 12.37.** Let us solve the equation $f''(t) - 2f'(t) + 5f(t) = g(t)$ with $f(0) = 1$, $f'(0) = 0$, where $g$ is the $2L$-periodic function with value 1 on $[0, L]$ and 0 on $[L, 2L]$.

By P10, we have

$$\mathcal{L}g(z) = \frac{\int_0^L e^{-tz}\,dt}{1 - e^{-2Lz}} = \frac{1}{z}\frac{1 - e^{-Lz}}{1 - e^{-2Lz}} = \frac{1}{z(1 + e^{-Lz})}.$$

Hence, we must have

$$z^2\mathcal{L}(f) - z - 2(z\mathcal{L}(f) - 1) + 5\mathcal{L}f(z) = \frac{1}{z(1 + e^{-Lz})},$$

$$\mathcal{L}f(z)(z^2 - 2z + 5) = z - 2 + \frac{1}{z(1 + e^{-Lz})},$$

$$\mathcal{L}f(z) = \frac{z - 2}{z^2 - 2z + 5} + \frac{1}{(z^2 - 2z + 5)z(1 + e^{-Lz})} = F_2 + F_1.$$

Decompose into simple fractions:

$$\frac{1}{z^2 - 2z + 5} = \frac{1}{4i}\left(\frac{1}{z - \alpha} - \frac{1}{z - \bar{\alpha}}\right), \quad \alpha = 1 + 2i,$$

$$\frac{z - 2}{z^2 - 2z + 5} = \frac{\frac{1}{2} + \frac{i}{4}}{z - \alpha} + \frac{\frac{1}{2} - \frac{i}{4}}{z - \bar{\alpha}}.$$

The rational function $\frac{z-2}{z^2-2z+5}$ is the transform of

$$f_2(t) = \left(\frac{1}{2} + \frac{i}{4}\right) e^{\alpha t} + \left(\frac{1}{2} - \frac{i}{4}\right) e^{\bar{\alpha}t} = \mathrm{Re}\left[\left(1 + \frac{i}{2}\right) e^{(1+2i)t}\right]$$

$$= e^t \,\mathrm{Re}[(1 + \frac{i}{2})\, e^{2it}] = e^t \cos 2t - \frac{1}{2} e^t \sin 2t.$$

The rational function $\frac{1}{z^2-2z+5}$ is the transform of

$$h(t) = \frac{1}{4i}\left(e^{\alpha t} - e^{\bar{\alpha}t}\right) = \frac{1}{2} e^t \sin 2t$$

and, as shown, $\frac{1}{z(1+e^{-Lz})}$ is the transform of $g$. Hence, we obtain

$$f_1(t) = (g * h)(t) = \frac{1}{2}\int_0^t g(t - s)\, e^s \sin 2s\, ds.$$

The solution is then $f = f_1 + f_2$. □

Next we focus on the problem

$$P(D)f = g, \quad f^{(k)}(0) = 0, \quad k = 0, \ldots, n - 1. \tag{12.17}$$

As we know, it has a unique solution given by

$$f(t) = \int_0^t h(t - s)g(s)ds,$$

where $h$ is an exponential polynomial with $h^{(k)}(0) = 0, k = 0, \ldots, n - 1$, so that $f$ is $C^{n-1}$ and

$$f^{(k)}(t) = \int_0^t h^{(k)}(t - s)g(s)ds, \quad k = 0, \ldots, n - 1.$$

On the set of points where $g$ is continuous, $f$ is of class $C^n$ since $h^{(n)}(0) = 1$ and we have

$$f^{(n)}(t) = g(t) + \int_0^t h^{(n)}(t - s)g(s)ds.$$

The function $h$ is the *Green's function* of the problem. It is also called an *impulse response* because, when formally replacing $g$ by the impulse $\delta$ (Dirac delta at the origin), we get $f = h$. The function $\frac{1}{P}$, which is the Laplace transform of $h$, $\mathcal{L}h = \frac{1}{P}$, is called the *transfer function*.

These notions apply to all linear operators $T$, transforming causal functions $g$ into causal functions $f = Tg$ that, moreover, are invariant under translations in the sense that

$$f = Tg \text{ and } \lambda > 0 \text{ imply } \tau_\lambda f = T(\tau_\lambda g).$$

These operators are usually called *filters*.

The problem $P(D)f = g$, $f^{(k)}(0) = 0$, $k = 0, \ldots, n-1$, is invariant under translations. It may be proved under general hypotheses that to every operator invariant under translations $T$ corresponds an impulse response $h$ such that

$$Tg = g * h, \quad \text{that is,} \quad Tg(t) = \int_0^t h(t-s)g(s)ds.$$

As said, $h$ may be interpreted as $h = T(\delta)$. The intuitive idea is that every function $g$ is an (infinite) linear combination of translations of the Dirac delta at the origin $\delta$:

$$g = \int g(s)\delta_s = \int g(s)\tau_s\delta.$$

So by linearity and invariancy by translations, we get

$$Tg = \int g(s)\tau_s T\delta = \int g(s)\tau_s h\, ds,$$

which means

$$Tg(t) = \int_0^t g(s)h(t-s)ds.$$

A filter $T$ is called *stable* if $Tg$ is bounded whenever $g$ is bounded. For the problem (12.17) it is easy to decide when the corresponding operator $T = P(D)$ is stable.

**Theorem 12.38.** *If the transfer function of a filter is a function $R$, rational and vanishing at infinity, then the filter is stable if and only if all the poles of $R$ have negative real part.*

*Proof.* Let us prove first that the condition is necessary. Taking $g(t) = e^{itw}$ for a fixed real number $w$, which is a bounded function, then $Tg = f$ must be bounded. Now, since $Tg = g * h$, $h$ being the impulse response and $\mathcal{L}h = R$ the transfer function, we obtain $\mathcal{L}f(z) = \mathcal{L}g(z)R(z) = \frac{R(z)}{z-iw}$. Let $\alpha_1, \ldots, \alpha_n$ be the poles of $R$ with multiplicities $m_1, \ldots, m_n$. If $\alpha_j \neq iw$, $j = 1, \ldots, n$, we have the decomposition in simple fractions:

$$\mathcal{L} f(z) = \frac{R(iw)}{z - iw} + \sum_{j=1}^{n} R_j(z),$$

where $R_j$ is the principal part of $R$ at the point $\alpha_j$. Then $\mathcal{L}^{-1}(R_j)$ can be written as $P_j(t) e^{\alpha_j t}$, where $P_j$ is a polynomial of degree $m_j - 1$, and we find

$$f(t) = R(iw) e^{itw} + \sum_{j=1}^{n} P_j(t) e^{\alpha_j t}.$$

This expression is bounded if and only if $\operatorname{Re} \alpha_j \leq 0$ and if the poles $\alpha_j$ with $\operatorname{Re} \alpha_j = 0$ have multiplicity $m_j = 1$. Now, if $\alpha_j = iw_0$ and we take $w = w_0$, then $iw_0$ is a pole of order greater than or equal to 2 of $\mathcal{L} f(z)$ and $f(t)$ would contain an unbounded term. Therefore, $\operatorname{Re} \alpha_j < 0$, $j = 1, 2, \ldots, n$, is a necessary condition.

To prove that the condition is sufficient, we use the equality

$$Tg(t) = f(t) = \int_0^t h(t - s) g(s) d(s)$$

with $\mathcal{L} h = R$, assuming that $g$ is bounded. The function $h$ will be written as

$$h(t) = \sum_{j=1}^{n} Q_j(t) e^{\alpha_j t}$$

with $Q_j$ polynomials and $\alpha_1, \ldots, \alpha_n$ the poles of $R$. If $\operatorname{Re} \alpha_j \leq -\varepsilon$ with $\varepsilon > 0$ for $j = 1, \ldots, n$, and $N$ is the maximum of the degrees of the polynomials $Q_j$, we have

$$|f(t)| \leq \|g\|_\infty \int_0^t |h(t)| ds \leq C \|g\|_\infty \int_0^\infty |t|^N e^{-\varepsilon t} dt < +\infty,$$

with $C$ constant. $\qquad\square$

Note that the methods based on the Laplace transform to solve ordinary differential equations do not require the function $\mathcal{L} f(z)$ to be defined for complex values. One could define $\mathcal{L} f(x)$ only for real $x$ and the uniqueness theorem, stating that $\mathcal{L} f(x) = \mathcal{L} g(x)$ implies $f \equiv g$, would have the same proof as Theorem 12.30. However, if we want to have inversion formulae, it is necessary to consider complex arguments.

## 12.6 Dirichlet series

Dirichlet series are a special type of series of holomorphic functions very important in analytic number theory. The most well-known example is the series defining the *Riemann $\zeta$ function,*

$$\zeta(z) = \sum_{n=1}^{\infty} \frac{1}{n^z}.$$

**Definition 12.39.** A Dirichlet series is a function series of type

$$\sum_{n=1}^{\infty} c_n\, e^{-\lambda_n z}$$

with $c_n \in \mathbb{C}$ and $\lambda_n \in \mathbb{R}$ satisfying $\lambda_n \le \lambda_{n+1}$, $n \in \mathbb{N}$, and $\lim_{n \to +\infty} \lambda_n = +\infty$.

When $\lambda_n = n$, the previous series is a power series in $w = e^{-z}$. When $\lambda_n = \operatorname{Log} n$, the series is written as

$$\sum_{n=1}^{\infty} \frac{c_n}{n^z}$$

and it is called a *special Dirichlet series.*

The Dirichlet series are a discrete analog of the Laplace transform, and there is a parallelism between both theories.

**Proposition 12.40.** *If the Dirichlet series $\sum_{n=1}^{\infty} c_n\, e^{-\lambda_n z}$ converges at the point $z_0 \in \mathbb{C}$, then it converges at every point $z$ with $\operatorname{Re} z > \operatorname{Re} z_0$. Moreover, the convergence is uniform on the whole circular sector $S_\beta = \{z : |\operatorname{Arg}(z - z_0)| \le \beta\}$, $\beta < \pi/2$.*

*Proof.* Without loss of generality, we may suppose $z_0 = 0$ and $\lambda_n \ge 0$. Hence $\sum c_n$ is convergent and we have to show that $\sum c_n\, e^{-\lambda_n z}$ is convergent for $\operatorname{Re} z > 0$, uniformly on the sector $\{z : |\operatorname{Arg} z| \le \beta\}$, $\beta < \pi/2$. Applying Abel's criterion (Theorem 2.14), we must check that

$$\sum_n \left| e^{-\lambda_n z} - e^{-\lambda_{n-1} z} \right| \le C_\beta, \quad \text{if } |\operatorname{Arg} z| \le \beta < \pi/2, \text{ for a constant } C_\beta.$$

Now, if $z = x + iy$, one has

$$\left| e^{-\lambda_n z} - e^{-\lambda_{n-1} z} \right| = \left| z \int_{\lambda_{n-1}}^{\lambda_n} e^{-\lambda z}\, d\lambda \right|$$

$$\le |z| \int_{\lambda_{n-1}}^{\lambda_n} e^{-\lambda \operatorname{Re} z}\, d\lambda = \frac{|z|}{x} \left( e^{-\lambda_{n-1} x} - e^{-\lambda_n x} \right)$$

and then the series above is dominated by $\frac{|z|}{x} \le \frac{1}{\cos \beta}$. $\qquad \square$

Proposition 12.40 implies:

**Theorem 12.41.** *For every Dirichlet series $\sum_{n=1}^{\infty} c_n e^{-\lambda_n z}$ there is a real number $\alpha$ such that the series converges if $\mathrm{Re}\, z > \alpha$ and diverges if $\mathrm{Re}\, z < \alpha$. The function $D(z) = \sum c_n e^{-\lambda_n z}$ is holomorphic on the half plane $\{z \in \mathbb{C} : \mathrm{Re}\, z > \alpha\}$.*

The number $\alpha$ is called the *abscissa of convergence* of the series.

There is also an *abscissa of absolute convergence* $\alpha_a$ defined by the property that the series is absolutely convergent for $\mathrm{Re}\, z > \alpha_a$ and not absolutely convergent for $\mathrm{Re}\, z < \alpha_a$. Trivially one has $\alpha_a \geq \alpha$, but it could happen that $\alpha < \alpha_a$; then the set $\{z : \alpha < \mathrm{Re}\, z < \alpha_a\}$ is called *strip of conditional convergence*.

For example, for the series

$$\sum_{n=1}^{\infty} \frac{(-1)^{n+1}}{n^z} = 1 - \frac{1}{2^z} + \frac{1}{3^z} - \frac{1}{4^z} + \cdots$$

one has $\alpha_a = 1$ and $\alpha = 0$.

Of course, the derivative of the function $D(z)$ of Theorem 12.41 can be also represented by a Dirichlet series,

$$D'(z) = -\sum c_n \lambda_n e^{-\lambda_n z}, \quad \text{if } \mathrm{Re}\, z > \alpha.$$

**Theorem 12.42** (Uniqueness theorem). *If $D(z) = \sum c_n e^{-\lambda_n z}$ and $D(z) = 0$ for $\mathrm{Re}\, z > \alpha$, where $\alpha$ is the abscissa of convergence of the Dirichlet series, then $c_n = 0$ for $n = 0, 1, 2, \ldots$. More generally, if $D(z)$ has infinitely many zeros in a sector $S_\beta = \{z : |\mathrm{Arg}\, z| \leq \beta\}$ with $\beta < \pi/2$, then $c_n = 0$ for every $n \in \mathbb{N}$ and $D(z)$ is identically zero.*

*Proof.* We can write

$$D(z)\, e^{\lambda_1 z} = c_1 + \sum_{n=2}^{\infty} c_n\, e^{-(\lambda_n - \lambda_1)z}.$$

The series of the right-hand side term is uniformly convergent on the sector $S_\beta = \{z : |\mathrm{Arg}\, z| < \beta\}$ and each term has limit zero when $z \to \infty$ in this sector. This means that

$$c_1 = \lim_{\substack{z \to \infty \\ z \in S_\beta}} D(z)\, e^{\lambda_1 z}.$$

Similarly $c_2 = \lim_{z \to \infty} \left(D(z) - c_1 e^{-\lambda_1 z}\right) e^{\lambda_2 z}$ and so on. Now if $D(w_n) = 0$ for infinitely many $w_n$ in the sector $S_\beta$, we have $c_n = 0$ for each $n$. $\qquad \square$

Many of the properties of the Laplace transforms can be transferred to Dirichlet series by means of the following observation. Start writing

$$\frac{D(z)}{z} = \sum_n c_n \frac{e^{-\lambda_n z}}{z}.$$

Now the function $\frac{1}{z}$ equals $\mathcal{L}(1)(z)$ and the function $\frac{e^{-\lambda_n z}}{z}$ equals $\mathcal{L}(\tau_{\lambda_n} 1)(z)$. So writing $H = \mathbb{1}_{(0,+\infty)}$ (called *Heaviside function*), one has

$$\frac{D(z)}{z} = \sum_n c_n \mathcal{L}(H(t - \lambda_n)) = \mathcal{L}(f)(z)$$

with

$$f(t) = \sum_{\lambda_n \le t} c_n.$$

Hence, the equation $D(z) = z\mathcal{L}(f)(z)$ holds.

## 12.7 The $Z$-transform

The $Z$-transform is another discrete version of the Laplace transform which is suitable for the solution of finite difference equations. These equations correspond to ordinary differential equations in a discrete context. The idea behind the $Z$-transform is the same as for power series or Laurent series, although traditionally a slightly different language is used.

**Definition 12.43.** The $Z$-transform of a sequence $c = (c_n)_{n=0}^{\infty}$ of complex numbers is the function

$$C(z) = \sum_{n=0}^{\infty} c_n z^{-n},$$

defined for the complex values of $z$ for which the series converges.

So it is a power series in $w = \frac{1}{z}$. More generally, one can consider the $Z$-transform of a sequence $(c_n)_{n\in\mathbb{Z}}$ indexed by $n \in \mathbb{Z}$, which will be a Laurent series in $w = \frac{1}{z}$.

If $\rho = \rho(c) = \limsup_{n\to\infty} |c_n|^{1/n}$, the series $\sum_{n=0}^{\infty} c_n w^n$ has radius of convergence $R = 1/\rho$. This series converges absolutely if $|w| < R$ and diverges if $|w| > R$. This means that $C(z)$ converges absolutely if $|z| > \rho$ and diverges if $|z| < \rho$.

**Example 12.44.** If $c_n = \lambda$, constant, then $C(z) = \lambda \sum_{n=0}^{\infty} z^{-n} = \frac{\lambda z}{z-1}$ for $|z| > 1$. If $c_n = \frac{1}{n!}$, then $C(z) = \sum \frac{z^{-n}}{n!} = \exp\left(\frac{1}{z}\right)$ for $|z| > 0$.     $\square$

A way for interpreting the $Z$-transform is to consider it as a discrete version of the Laplace transform. The integral

$$\mathcal{L}f(z) = \int_0^\infty f(t)\,e^{-tz}\,dt$$

is approximated by the Riemann sums

$$\sum_{n=0}^\infty f(n\tau)\tau\,e^{-n\tau z} = \sum_{n=0}^\infty c_n\,(e^{\tau z})^{-n} = C(e^{\tau z}) \quad \text{with } c_n = \tau f(n\tau),\ \tau > 0.$$

The function $z \mapsto e^{\tau z}$ transforms a half plane $\{z:\ \mathrm{Re}\,z > \alpha\}$ into the exterior of a disc.

The properties of the $Z$-transform are analogous to the properties of the Laplace transform given in Section 12.5 and will not be specified here. Only one, corresponding to P.11, will be stressed. Accordingly, let $a = (a_n)$, $b = (b_n)$ be two sequences with $\rho(a) < +\infty$, $\rho(b) < +\infty$, and let $(c_n)$ be the *convolution* of those two sequences, given by

$$c_n = \sum_{k=0}^n b_k c_{n-k}, \quad n = 0, 1, 2, \ldots.$$

Then $\rho(c) \le \max(\rho(a), \rho(b))$ and the $Z$-transform $C(z)$ satisfies $C(z) = A(z)B(z)$, where $A(z)$ and $B(z)$ are, respectively, the $Z$-transforms of $a$ and $b$.

The $Z$-transform is a very useful tool to solve *difference equations*. They are equations of the kind

$$\sum_{j=0}^p \alpha_j y_{n+j} = \sum_{j=0}^q \beta_j x_{n+j}, \quad n = 0, 1, 2, \ldots$$

with initial values $y_j = \gamma_j$, $j = 0, 1, \ldots, p-1$, where $\gamma_j$ are given constants. The numbers $\alpha_j$ for $j = 0, 1, \ldots, p$, $\beta_j$ for $j = 0, 1, \ldots, q$ and the sequence $x = (x_n)_{n=0}^\infty$ are the data, and the sequence $y = (y_n)_{n=0}^\infty$ is the unknown.

Adopting the notation $\tau_j(c) = (c_{n+j})_{n=0}^\infty$ when $c = (c_n)_{n=0}^\infty$ the equation above is written:

$$\sum_{j=0}^p \alpha_j \tau_j(y) = \sum_{j=0}^q \beta_j \tau_j(x).$$

In order to find $y$, we apply the $Z$-transform and use that the $Z$-transform of $\tau_j(c)$ is the function

$$z^j\left(C(z) - \sum_{l=0}^{j-1} \frac{c_l}{z^l}\right),$$

if $C(z)$ is the $Z$-transform of $c = (c_n)_{n=0}^{\infty}$. We obtain, then, for the $Z$-transforms $X(z)$, $Y(z)$ of $x = (x_n)$, $y = (y_n)$ the equation

$$\sum_{j=0}^{p} \alpha_j z^j \left( Y(z) - \sum_{l=0}^{j-1} \frac{\gamma_l}{z^l} \right) = \sum_{j=0}^{q} \beta_j z^j \left( X(z) - \sum_{l=0}^{j-1} \frac{x_l}{z^l} \right)$$

from which $Y(z)$ can be isolated. The solution is finally obtained expanding $Y(z)$ at the point at infinity.

**Example 12.45.** Consider the equation $3y_n + 2y_{n+1} = x_n - x_{n+2}$, $n \geq 0$ with $y_0 = 1$. We have

$$3Y(z) + 2z(Y(z) - 1) = X(z) - z^2 \left( X(z) - \left( x_0 + \frac{x_1}{z} \right) \right)$$

and from here,

$$Y(z) = \frac{\left( x_0 + \frac{x_1}{z} + \frac{x_2}{z^2} + \cdots \right) - \left( x_2 + \frac{x_3}{z} + \frac{x_4}{z^2} + \cdots \right) + 2z}{2z + 3}.$$

The expansion of $\frac{1}{2z+3}$ at infinity, which holds if $|z| > \frac{3}{2}$, is

$$\frac{1}{2z + 3} = \frac{1}{2z \left( 1 + \frac{3}{2z} \right)} = \frac{1}{2z} \sum_{n=0}^{\infty} \frac{(-1)^n 3^n}{(2z)^n} = \sum_{n=0}^{\infty} \frac{(-1)^n 3^n}{(2z)^{n+1}}.$$

This gives

$$y_n = \frac{(-1)^n 3^n}{2^n} + \sum_{j=1}^{n} \frac{(-1)^{j-1} 3^{j-1}}{2^j} \left( x_{n-j} - x_{n-j+2} \right), \quad n \geq 1. \qquad \square$$

As done in Section 12.5 for the equation $P(D)f = g$, consider now the case of all initial values being zero, that is, $\gamma_j = 0$, $j = 0, \ldots, p-1$. Suppose also that $q = 0$ and $\beta_0 = 1$. Then $y$ depends linearly on $x$ and the transform $Y(z)$ is written in terms of $X(z)$ as

$$Y(z) = \frac{1}{\sum_{j=0}^{p} \alpha_j z^j} X(z).$$

The function $\frac{1}{P(z)}$, with $P(z) = \sum_{j=0}^{p} \alpha_j z^j$, is called the *transfer function*. The sequence $h = (h_n)_{n=0}^{\infty}$ whose $Z$-transform is $\frac{1}{P(z)}$ allows us to find the solution of the finite difference equation by means of the convolution

$$y_n = \sum_{k=0}^{n} h_{n-k} x_k.$$

When the impulse is $x = (x_n) = (1, 0, 0, \ldots, 0, \ldots)$, then $y = h$, and for this reason the sequence $h$ is called *impulse response*.

As in the continuous case, a convolution $y = h * x$ is the general expression of a *discrete filter*, that is, a linear transformation taking causal sequences $(x_n)_{n=0}^{\infty}$ into causal sequences $(y_n)_{n=0}^{\infty}$ and invariant under translations. The filter is called *stable* if it transforms bounded sequences into bounded sequences.

**Theorem 12.46.** *A filter with rational transfer function $R$ is stable if and only if all the poles of $R$ have absolute value less than 1.*

*Proof.* If all the poles of $R$ have modulus less than 1, the expansion at infinity of $R$,

$$R(z) = \sum_{n \geq 0} \frac{h_n}{z^n},$$

is absolutely convergent for $|z| = 1$ and, therefore, $\sum_{n \geq 0} |h_n| = C < +\infty$. So if there is a constant $D > 0$ such that $|x_n| \leq D$, $n = 0, 1, 2, \ldots$, then

$$|y_n| \leq \sum_{k=0}^{n} |h_{n-k}| \, |x_k| \leq CD$$

and the filter is stable. We show now that the condition is necessary. Suppose that whenever $|x_n| \leq D$, $n = 0, 1, 2, \ldots$, the sequence $(y_n)$ is bounded. Note that for $|y_n| \leq D'$, $n = 0, 1, 2, \ldots$, then $Y(z)$, the $Z$-transform of $(y_n)$, is defined on $\{z : |z| > 1\}$ and satisfies

$$|Y(z)| \leq \sum_{n=0}^{\infty} \frac{|y_n|}{|z|^n} \leq D' \sum_{n=0}^{\infty} \frac{1}{|z|^n} = D' \frac{|z|}{|z| - 1}, \quad \text{for } |z| > 1. \tag{12.18}$$

Since $Y(z) = R(z)X(z)$, it is clear that $R$ cannot have poles with modulus greater than 1. Suppose now that $\lambda$ is a pole of $R$ with $|\lambda| = 1$. Write

$$R(z) = \frac{F(z)}{(z - \lambda)^m}, \quad m \geq 1$$

with $F$ analytic on a neighborhood of $\lambda$ and $F(\lambda) \neq 0$. Taking as $x$ the bounded sequence given by $x_n = \lambda^n$, one has $X(z) = \sum_{n=0}^{\infty} \frac{\lambda^n}{z^n} = \frac{z}{z - \lambda}$ and

$$Y(z) = R(z)X(z) = \frac{z F(z)}{(z - \lambda)^{m+1}}.$$

Since $m \geq 1$, this equality implies that $Y(z)$ does not satisfy (12.18) for any constant $D'$ and, then, it cannot be bounded. Hence, all the poles of $R$ have modulus less than 1. $\qquad \square$

## 12.8  Exercises

1. Let $F$ be a bounded holomorphic function on a strip $U = \{z : a < |\operatorname{Im} z| < b\}$, $a, b \in \mathbb{R}$. For $a < y < b$, write

$$M(y) = \sup\{|F(x + iy)| : x \in \mathbb{R}\}.$$

Show that $\operatorname{Log} M(y)$ is a convex function of $\operatorname{Log} y$, that is, for $a < c < y < d < b$, one has

$$M(y)^{d-c} \le M(c)^{d-y} M(d)^{y-c}.$$

*Hint:* First consider the case $M(c) = M(d) = 1$ using the function $F(z)/(1 + \varepsilon(z - ia))$. For the general case, take

$$g(z) = M(c)^{\frac{d-z}{d-c}} M(d)^{\frac{z-c}{d-c}}$$

and apply the particular case to $F/g$.

2. Let $F$ be a holomorphic function on the annulus $C(0, R_2, R_1)$ and let

$$M(r) = \sup\{|F(z)| : |z| = r\}, \qquad R_2 < r < R_1.$$

Prove that $\operatorname{Log} M(r)$ is a convex function of $\operatorname{Log} r$.

3. Prove that Phragmen–Lindelöf's theorem (Theorem 12.9) also holds if $F$ satisfies the estimate $|F(z)| = O(e^{\varepsilon|z|^{\alpha}})$ for every $\varepsilon > 0$, $z \in S_\alpha$ and $|F(z)| \le M$, if $z \in \partial S_\alpha$.

4. Let $F$ be an entire function satisfying $|F(z)| = O(e^{a|z|^2})$, $z \in \mathbb{C}$ and $|F(x)| = O(e^{-b|x|^2})$, $x \in \mathbb{R}$, for $a, b > 0$, constants. Show that

$$|F(x + iy)| = O(e^{-cx^2 + dy^2})$$

with some constants $c, d$ depending on $a, b$.

5. Let $F$ be an entire function satisfying the estimates of the hypothesis in Exercise 4 of this section. Prove that the Fourier transform of $F(x)$ is an entire function satisfying the same estimates.

6. Let $F$ be a holomorphic function on a neighborhood of a regular polygon with $n$ sides, $L_1, L_2, \ldots, L_n$, centered at the origin. Let $M_i = \sup\{|F(z)| : z \in L_i\}$, $i = 1, 2, \ldots, n$. Show the inequality

$$|F(0)| \le (M_1 M_2 \cdots M_n)^{\frac{1}{n}}.$$

7. Let $F$ be an entire function of exponential type $\tau \geq 0$, that is, satisfying $|F(z)| \leq C\, e^{\tau|z|}$, $z \in \mathbb{C}$, for $C$ constant. Assume that $F$ belongs to the spaces $L^p(\mathbb{R})$, corresponding to two non-parallel lines, for $p$ fixed with $1 < p < +\infty$. Prove that $F$ is identically zero. Show also that every entire function of exponential type that is bounded on two non-parallel lines is constant.

8. Let $F$ be an entire function of exponential type $\tau$ (Exercise 7 of this section) that, in addition, satisfies the condition

$$\int_{-\infty}^{+\infty} |F(x)|^p \, dx < +\infty, \quad 1 \leq p < \infty.$$

Show there is a constant $A(\tau, p)$ such that

$$\int_{-\infty}^{+\infty} |F'(x)|^p \, dx \leq A(\tau, p) \int_{-\infty}^{+\infty} |F(x)|^p \, dx.$$

Find the best value of the constant $A(\tau, 2)$ using the second Paley–Wiener theorem.

9. The Bernstein space $B_\tau$ is the space of entire functions of exponential type $\tau$ (Exercise 7 of this section) that are bounded on the real axis $\mathbb{R}$. Show Bernstein's inequality

$$\sup\{|F'(x)| : x \in \mathbb{R}\} \leq \tau \sup\{|F(x)| : x \in \mathbb{R}\}, \quad \text{if } F \in B_\tau.$$

*Hint:* First prove it for $F$ given by

$$F(z) = \int_{-\rho}^{+\rho} e^{2\pi t z} f(t) \, dt$$

with $f$ integrable and $2\pi\rho = \tau$, using the equality

$$2\pi i t e^{-\pi i t} = \frac{4}{\pi} \sum_{n=-\infty}^{\infty} \frac{(-1)^n}{2n+1} e^{2\pi i n t}, \quad |t| \leq \frac{1}{2}.$$

In the general case, consider $F_\varepsilon(z) = F(z)\frac{\sin \varepsilon z}{\varepsilon z}$.

10. Let $F$ be a function bandlimited to $(-\frac{1}{2}, \frac{1}{2})$. Instead of the corresponding Nyquist frequency (equal to 1, see page 520), consider a frequency $\rho < 1$. Find a formula expressing $F(z)$ in terms of the values $F(n\rho)$, $n \in \mathbb{Z}$, of type

$$F(z) = \sum_{n \in \mathbb{Z}} F(n\rho)\varphi(z - n\rho),$$

with $\varphi$ rapidly decreasing to infinity ($\lim_{|x|\to\infty} |x|^n \, |\varphi^{(k)}(x)| = 0$, for $n, k = 0, 1, 2, \ldots$).

*Hint:* With $\tau = \frac{1}{2\rho}$, consider the $2\tau$-periodic Fourier expansion of $\widehat{F}$, multiply it by a function in $C_c^\infty$ with value 1 on $[-\frac{1}{2}, \frac{1}{2}]$ and support inside $[-\tau, \tau]$ and argue as in the proof of Shannon–Whittaker's theorem.

**11.** Let $f \in L^1(\mathbb{R})$ and consider the 1-periodic function

$$F(x) = \sum_{k \in \mathbb{Z}} f(x + k).$$

Show that the $n$-th Fourier coefficient of $F$,

$$\widehat{F}(n) = \int_0^1 F(x) e^{-2\pi i n x} \, dx,$$

is $\hat{f}(n)$. Therefore, the Poisson formula

$$\sum_{k \in \mathbb{Z}} f(x + k) = \sum_{n \in \mathbb{Z}} \hat{f}(n) e^{2\pi i n x}$$

holds if the Fourier series of $F$ is pointwise convergent; for example when $F$ is continuous with bounded variation. Show that this is the case when $f$ is, in addition, absolutely continuous, that is,

$$f(x) = \int_{-\infty}^x g(t) \, dt$$

with $g$ integrable.

**12.** Find conditions on $f$ in order that the equality

$$\sum_{n \in \mathbb{Z}} f(n) = \int_{-\infty}^{+\infty} f(x) \, dx$$

holds. Show that this equality is true if $f$ is the restriction to $\mathbb{R}$ of an entire function of exponential type $\tau$ with $\tau \leq 1$ (Exercise 7 of this section) and integrable on $\mathbb{R}$.

**13.** Let us consider the Dirichlet series

$$\sum_{n \geq 2} \frac{(-1)^n}{\sqrt{n}(\mathrm{Log}\, n)^z}.$$

Prove that $\alpha = -\infty$ and $\beta = +\infty$, where $\alpha$ is the abscissa of convergence and $\beta$ the abscissa of absolute convergence.

**14.** Let $\sum_{n=1}^{\infty} c_n e^{-\lambda_n z}$ be a Dirichlet series with coefficients $\lambda_n$ satisfying the condition $\frac{\operatorname{Log} n}{\lambda_n} \to 0$, for $n \to \infty$. Show that

$$\alpha = \beta = \limsup_{n \to \infty} \frac{\operatorname{Log} |c_n|}{\lambda_n},$$

with $\alpha$ and $\beta$ equal, respectively, to the abscissas of convergence and of absolute convergence of the series.

This formula is analogous to the one giving the radius of convergence of a power series.

**15.** Prove Parseval's formula for the Dirichlet series $\sum_{n=1}^{\infty} c_n e^{-\lambda_n z}$: if $x > \alpha$,

$$\sum_{n=1}^{\infty} |c_n|^2 e^{-2\lambda_n x} = \lim_{N \to \infty} \frac{1}{2N} \int_{-N}^{N} |D(x + it)|^2 \, dt,$$

where $D(z) = \sum c_n e^{-\lambda_n z}$ and $\alpha$ is the abscissa of convergence of the series.

**16.** The aim of this exercise is to introduce the *Mellin transform*, corresponding to the Fourier transform in the multiplicative group $\mathbb{R}^+$, equipped with the measure $d\mu = dt/t$. If $f$ is integrable with respect to $d\mu$, the Mellin transform is defined as

$$f^*(z) = \int_0^{\infty} t^{-iz-1} f(t) \, dt.$$

With the change of variables $t = e^{2\pi s}$ it becomes essentially the complex Fourier transform.

a) State Parseval's formula for the Mellin transform.

b) Show that $f^* g^* = (f \& g)^*$, where $f \& g$ denotes the convolution associated to the multiplicative group $\mathbb{R}^+$, that is,

$$(f \& g)(x) = \int_0^{\infty} f\left(\frac{x}{t}\right) g(t) \, \frac{dt}{t}.$$

c) State the inversion formula for the Mellin transform.

**17.** Use the Laplace transform to solve the following problems:

a) $f''(t) - 2f'(t) + 2f(t) = e^{-t}$; $f(0) = 0$, $f'(0) = 2$.

b) $f''(t) + tf'(t) + f(t) = 0$; $f(0) = 1$, $f'(0) = 0$.

c) $f''(t) + 2f'(t) + 2f(t) = \delta(t - 1)$; $f(0) = 1$, $f'(0) = 1$ ($\delta$ denotes the Dirac delta at the origin).

18. Let $f$ be a continuous function on $(0, +\infty)$ and $k$ locally integrable on $(0, +\infty)$. Solve the Volterra equation in the unknown function $\Phi$:

$$\Phi(t) + \int_0^t k(t-s)\Phi(s)\,ds = f(t), \quad t > 0,$$

using the Laplace transform.

19. Solve the following finite difference equations:

a) $y_n - y_{n+1} = x_n + x_{n+2}$, $y_0 = 1$;

b) $y_n + 2y_{n+1} + y_{n+2} = x_n - x_{n+1}$, $y_0 = 1$, $y_1 = 0$;

c) $y_n - y_{n+1} - 4y_{n+2} + 4y_{n+3} = x_n + x_{n+1}$, $y_0 = y_1 = 0$, $y_2 = 1$.

20. Find which of the following linear systems are stable:

a) $y_n - y_{n+1} - 4y_{n+2} + 4y_{n+3} = x_n - x_{n+1}$;

b) $y_n - y_{n+1} + 3y_{n+2} + 5y_{n+3} = x_n - x_{n+1}$;

c) $y_n - y_{n+1} - 4y_{n+2} - 4y_{n+3} = x_n$;

d) $2y_n + y_{n+1} = x_n + 2x_{n+1}$;

e) $6y_n - 20y_{n+1} + 6y_{n+2} + 10y_{n+3} = x_n + x_{n+1}$.

# Bibliography


[1] Ahlfors, L. V., *Complex analysis*, 3rd edition, Mc Graw-Hill Int. Book Co., New York, 1978.

[2] Blair, D. E., *Inversion theory and conformal mapping*, AMS Publications, Providence, RI, 2000.

[3] Burckel, R. B., *An introduction to classical complex analysis*, Vol. 1, Academic Press, New York - London, 1979.

[4] Carmona, J. J., and J. Cufí, The index of a plane curve and Green's formula, *Rend. Circolo Matem. Palermo* 53 (2004), 103–128.

[5] Christenson, C. O. and W. L. Voxman, *Aspects of topology*, Marcel Dekker, Inc., New York–Basel, 1977.

[6] Dieudonné, J., *Éléments d'analyse 1*, Gauthier-Villars, Paris, 1969.

[7] Gamelin, T. W., *Complex analysis*, Springer-Verlag, New York, 2001.

[8] Muñoz, J., *Curso de teoría de funciones 1*, Tecnos, Madrid, 1978.

[9] Nevanlinna, R., and V. Paatero, *Introduction to complex analysis*, 2nd edition, Chelsea Publishing Co., New York, 1982.

[10] Rudin, W., *Principles of mathematical analysis*, 3rd edition, Mc Graw-Hill Book Co., New York–Auckland–Düsseldorf, 1976.

[11] Saks, S., and A. Zygmund, *Fonctions analitiques*, Masson et Cie. Éditeurs, Paris, 1970.

[12] Schwartz, L., *Cours d'analyse 1*, 2nd edition, Hermann, Paris, 1981.

[13] Warner, F. W., *Foundations of differentiable manifolds and Lie groups*, Scott, Foresman and Co., Glenview, Ill.–London, 1971.

# Symbols